Advanced Lectures in Mathematics
Volume 39

Hodge Theory and L^2-analysis

edited by

Lizhen Ji

International Press
www.intlpress.com

高等教育出版社
HIGHER EDUCATION PRESS

Advanced Lectures in Mathematics, Volume 39
Hodge Theory and L^2-analysis

Editor:
Lizhen Ji (Department of Mathematics, University of Michigan)

ISBN: 978-1-57146-351-7

Printed in the United States of America.

ADVANCED LECTURES IN MATHEMATICS

Dedicated to Steven Zucker for his 65th Birthday

Conference on Hodge Theory and L^2-cohomology, Johns Hopkins University, November 21—23, 2014

Contents

Preface

Steven (Steve) Zucker has made major contributions to the following three areas of mathematics and the interactions among them:

1. Hodge theory in algebraic geometry (e.g., normal functions, variation of mixed Hodge structure, degeneration of Hodge structure);
2. L^2-cohomology, and also L^p-cohomology for $p \neq 2$;
3. Compactification of locally symmetric spaces (Satake, reductive Borel-Serre, toroidal, etc.).

The famous Zucker conjecture from 1980 relates items (2) and (3) above. It was resolved independently in 1987 by Saper-Stern and by Looijenga, using very different methods. For a broader, detailed description of Steve's work, see his own narrative which follows.

In November 2014, a conference titled "Hodge Theory and L^2-cohomology" was held in Steve's honor at Johns Hopkins University. This book is based on the conference. Besides contributions from most of the speakers, several other people were also invited to contribute. Since Steve's work and some of the contributed papers involve L^2-analysis, which includes L^2-cohomology, the title of the book was changed to "Hodge Theory and L^2-analysis". By consensus, the conference was of high quality. We believe that this book reflects the conference well and is also of high quality.

It perhaps helpful to emphasize the intimate connection between Hodge theory and L^2-analysis. In his thesis in 1851, Riemann tried to use the Dirichlet form and the Dirichlet principle to prove the Riemann mapping theorem. In his very influential paper on abelian functions in 1857, he also used harmonic functions and the Dirichlet principle to prove the Riemann inequality in the Riemann-Roch theorem. These results were made rigorous and systematically presented in Weyl's classical book *The Concept of a Riemann Surface* which was published in 1913. Some of the results for Riemann surfaces were later generalized to higher dimension Riemannian and Kähler manifolds by Hodge, Weyl and Kodaira, whose works gave birth to the Hodge theory.

The contributors to this book had been asked to make their papers expository to the extent possible. We hope that this book will serve as a valuable introduction to Hodge theory, L^2-analysis and the interaction between the two. All papers have been carefully refereed, and we would like to thank the referees for their help.

The conference was held at approximately the 65th birthday of Steve. We wish him many happy and productive years to come.

Lizhen Ji

September, 2016

The Research Career of Steven Zucker: An Autobiographical Account

Acknowledgments. I thank Rafe Mazzeo, Arvind Nair, and Morihiko Saito for their constructive comments on earlier versions of this narrative. I must thank in particular Benjamin Diamond, who served as the copy editor of its penultimate version.

For the reader's convenience, I divide the references to my work to date among the three areas mentioned in the Preface.

(1) Hodge theory in algebraic geometry (normal functions, variation of mixed Hodge structure, degeneration of Hodge structure, etc.): [Z01]–[Z06], [Z09], [Z12]–[Z23];

(2) L^2-cohomology, and also L^p-cohomology for $p \neq 2$: [Z03], [Z07], [Z10], [Z24], [Z28];

(3) Compactification of locally symmetric spaces (Satake, reductive Borel-Serre, toroidal, etc.): [Z08], [Z20]–[Z22], [Z24]–[Z28].

In what follows, there will be no false modesty, and I will avoid caustic remarks. In addition to a description of my work, I will acknowledge the major influences of other mathematicians, as well as including the impact of my own work on that of others. I wish to put in the records that I gave general (moral) support to two first-rate mathematicians who were getting a bit discouraged while in graduate school: Kari Vilonen and Teruyoshi Yoshida. I know now that Yoshida remembers this well and with appreciation; Vilonen is less clear, but he believes it is accurate. I mention this to remind the reader that there are ways of helping a novice beyond the imparting of mathematical ideas.

I was born in New York City on September 12, 1949. A year later, my parents (and I) moved to Queens Village, located in the Borough of Queens, NYC, where my father had set up his optometric practice. I went to public elementary, junior high school (a.k.a. middle school) and high school. All were in walking distance from our home, but we lived even closer to the city line. That precluded the idea of going to a selective high school, for the commute would have been way too long; besides, the local high school was rather good. I have one sibling, a brother, born in 1952.

In 1966, I headed off for Brown University (that was *before* the "new curriculum" was enacted) and I graduated in 1970 with a degree in Mathematics. From there, I went to Princeton University, receiving a Ph.D. in Mathematics in 1974.

Working towards my dissertation was over a rocky road. Phillip Griffiths had given me the problem of fixing [BlGr], which was unfortunately dependent on strong ampleness conditions, so that it would apply to the example they really had in mind. Towards the end of my second year, when I heard that Griffiths was moving to Harvard, he recommended that I continue to work under Spencer Bloch, who was junior faculty and also of an algebraic bent (I was more analytic in nature). The problem was, given the mapping $f : Y \to P^1$ of a smooth variety Y of dimension $2n$ onto P^1 associated to a Lefschetz pencil of hyperplane sections, to show that every primitive integral Hodge class in $H^{2n}(Y)$ comes from a normal function. I became aware that the obstruction to a complete solution to the assigned problem was the inability to see the Hodge structure on $H^1(P^1, R^{2n-1}f_*\mathbb{C}) \simeq H^1(P^1, j_*\mathbb{V})$, where $j : S \hookrightarrow P^1$ is the inclusion of the Zariski-open subset over which f is smooth, and $\mathbb{V}$ is the restriction of to S of $R^{2n-1}f_*\mathbb{C}$ (underlying its variation of Hodge structure) from a construction $\underline{\text{on } P^1}$. Not only was there an essential difficulty, but it was only resolved much later by [Z03]. Fortunately, I could produce a theorem that was almost as good by technical "magic". For reason of that, I was hired as assistant professor at Rutgers University; my thesis, trimmed down a little, was published as [Z01].

I am grateful that my next publication [Z02] was remembered by Pierre Deligne when he published [D2]. One of the appendices in that article got nearly buried in the literature. That appendix came about because Fred Almgren (of geometric measure theory) had heard about a possible counterexample to the Hodge conjecture, and he wanted to hear more about it, for the use of integral currents might have been relevant. It concerned a special class of abelian varieties with a three-dimensional space of Hodge classes. I recognized that if one had a compact complex torus (not an abelian variety) with only the corresponding two-dimensional subspace of Hodge classes, it was impossible that these classes were representable by analytic subvarieties, for there were no *positive* classes. Thus, there is no formulation of the Hodge conjecture for general Kähler manifolds.

I have already presented [Z03] with due fanfare. The article was strongly influenced by my interaction with Deligne. I had gone to IHES during the summer of 1975, where both Deligne and Joseph Steenbrink were in residence. I knew of Steenbrink because we had finished our dissertations at about the same time, and we exchanged preprints (his was published as [St]). It was determined that Deligne had an unpublished manuscript that provided a proof of the theorem on normal functions when f is smooth. After I finished looking at it—it seemed so complicated to me at the time—Deligne sketched an approach to proving it when f had singular fibers. I was awestruck! He suggested using some sort of L^2-cohomology; the features of the problem led to the Poincaré metric. Various technical issues had to get resolved, and I finished work on it after about two years. The main result is roughly:

Let $\overline{f} : \overline{X} \to \overline{S}$ be a proper morphism of varieties, where $\overline{X}$ is smooth, $\overline{S}$ is a smooth complete curve, f is smooth over $S \subset \overline{S}$ and $j : S \hookrightarrow \overline{S}$ is the inclusion.

Then for any variation of Hodge structure $\mathbb{V}$ *over* S — *in particular* $\mathbb{V} = R^q f_* \mathbb{C}$ — $H^1(\overline{S}, j_*\mathbb{V})$ *has a natural Hodge structure that is compatible with the Leray spectral sequence for* $\overline{f}$.

As pointed out by Donu Arapura in [Ar], his work supersedes in a big way the argument in [Z03] for the compatibility with the Leray spectral sequence; keep in mind, though, that as much as I like [Ar], it was published more than 35 years later.

As asserted in [CGM], Deligne was wondering about the generalization of this main theorem to higher dimensional $\overline{S}$. This led to the famous iterated truncation formula for intersection cohomology in [GM], as well as the L^2-cohomology of a variation of Hodge structure for a product of punctured discs (in [CKS] and [KK]).

It made me both angry and anguished that [Z03] (available as a preprint) was judged insufficient to warrant my promotion to tenure at Rutgers in the Spring Semester of 1979. My supporters in the department were able to arrange, however, that I be granted *two* terminal years and that proved to be very useful. Before getting into that, let me say that I had told David DeGeorge (an assistant professor working in representation theory) what I had been working on, and he pointed out that [MM] seemed similar. The difference was that in [MM] Hodge type was taken only from the base, whereas Deligne's construction mixed in the Hodge filtration of the variation of Hodge structure (i.e., of the fibers). It didn't take long to see a 4-fold decomposition of the cohomology emerging (in [Z06]). Robert Langlands expressed an early interest in [Z03] that I didn't quite understand at the time; in retrospect he was looking towards [Z07] and [Z10] and their generalization. Little did I know that I was venturing close to the terrain of Shimura varieties [D1].

Virtually immediately after we started graduate school together in Princeton, it became apparent that David Cox and I *had* to write a paper together, for reason of the juxtaposition of names. When he was appointed assistant professor at Rutgers the year after I was, we were together at the same place again. Our article [Z04] is, however, a serious piece of mathematics that was put together to address a question posed by our senior colleague William Hoyt. An application of [Z03] appears in the last section.

The article [Z05] was written almost immediately after [Ft] appeared. It was viewed by me to be an exercise in limit Hodge theory. At first it didn't occur to me that I should publish it, but once I did, it had a significant impact on subsequent work in algebraic geometry (see [Fn:§3]).

I want to describe the benefits of having had that second terminal year. Princeton is only a half-hour's drive from Rutgers, so it is relatively easy to attend talks and seminars at Princeton University and the Institute for Advanced Study. Given my emerging interests (in the direction of [Z07]), both DeGeorge and our senior colleague Nolan Wallach urged me to attend Armand Borel's seminar at IAS. It was extremely useful for getting oriented in the subject, and after [Z07] was available as a preprint, for unearthing a notorious error in [BS].[1] It also highlighted two techni-

[1] There was a gap in an important proof in [Z07:§4]. Borel was helpful in suggesting how it might get fixed. Though his manner was often quite gruff, he had a robust sense of humor and was at bottom warm-hearted.

cal points: to use sheaf theory (local to global methods) to study L^2-cohomology on a Riemannian manifold, one had to invoke a compactification so as to "store" the global L^2 condition at the boundary; moreover, the fineness of sheaves had to be specially verified at the boundary, where cut-off functions needed to have bounded differentials. The former article is best-known for the so-called Zucker conjecture, which asserts a natural isomorphism between the L^2-cohomology of a locally symmetric variety X and the (middle) intersection cohomology of its Baily-Borel Satake compactification X^{BB} (from [BB]). Because X is understood to have its canonical class of quasi-isometric complete, locally-invariant Kähler metrics, this imparts a Hodge structure to the intersection cohomology $IH(X^{BB})$. It had been reported that Langlands was excited by the conjecture, for (if true) he thought it might allow for one to manage without the complicated toroidal resolution(s) of singularities $X^{tor} \to X^{BB}$ from [AMRT] (no such luck!)[2].

Moreover, I could apply selectively in 1980 for my next position, as there was always the following year. One of those applications became an offer, an *untenured* associate professorship at Indiana University, and that had to begin in January of 1981. With an advance offer of membership in the IAS 1981–1982 special year in Algebraic Geometry, and a promise of a vote on tenure while I would be away on leave, Daniel Gorenstein, then Chair at Rutgers, encouraged me to view it as a very good deal. Besides, by the end of the Fall Semester of 1980, I had finished my business with Borel; I was ready to move onward.

I was on the faculty of Indiana University for a total of three and a half years, and in residence for one and a half. During the Spring Semester of 1981, I wrote almost all of [Z08], a work so technical that I find it hard to believe I was the author! I was later told by C. Skinner that it was regarded as the definitive treatment of Satake compactifications. I had entered the arena in order to show the presumably useful fact that the Borel-Serre compactification X^{BS} (the manifold with corners constructed in [BS]) mapped onto X^{BB} as a morphism of compactifications of the locally symmetric variety; indeed, it does so for every Satake compactification, even in the absence of complex structure, and these are always finite in number.

Much later, it was reported to me by R. Mazzeo that [Z07:§2] had applications beyond arithmetic groups. It showed how sheaf theory could be used on a manifold with corners. Also, I learned directly from [ChSh] of its relevance on product domains. Indeed, that Section of [Z07] was written with the hope of general application.

Two very different proofs of the Zucker conjecture were offered at about the same time, in 1987; they appeared in print as [Lo] and [SS]. I wrote an expository article [Z11] that presented both proofs. For each, the goal consisted of showing that truncations by degree coincided with truncations by weight, and that led to the notion of weighted cohomology in [GHM]. Both proofs take off from [Z10] and diverge from there. In the case of [SS], the authors did hard analytic calculations; [Lo] used softer methods. With the conjecture proved, it followed that the spaces of L^2 harmonic forms were finite-dimensional in all degrees, and that d had closed

[2] Recent work points to ways around the use of the toroidal compactifications after all. See, e.g., [V].

range. These facts[3] could also be deduced by $(\mathfrak{g}, K)$-cohomological methods from representation theory (see [BG] and [BC]), but the geometric interpretation was missing. The isomorphism allows one to compare structures on the two sides, e.g., the nature of the action of the Hecke algebra.

I went on leave at IAS as planned for 1981–1982. I remember that when the Fall Semester began, there was a superabundance of talks, something that my constitution could not quite bear. I recall a case where I could not decide whether to attend a certain lecture. In vacillating over that, I passed the beginning time of the talk, and my decision was made de facto; I was happy! I must concede, though, that I had acquired a good sense of the state of Algebraic Geometry at the time. In the Spring, I heard rumors that Johns Hopkins University wanted to hire me as an associate professor. This came rather late, and I returned to Indiana University for AY 1982–1983. The situation was uncomfortable because tenure had been granted at IU as expected, whereas the position of associate professor at JHU never carried tenure in those days; moreover, the Mathematics Department was not given the authority to hire at the Full Professor level. To remain at all competitive, it was a departmental decision that hiring at, or promoting to, the level of Associate Professor was a virtual promise to promote to Full Professor. I spent AY 1983–1984 on leave at Johns Hopkins.

Writing up the article [Z09] with El Zein was a bit tricky because of our different styles (American versus French), so we patiently worked it out line by line. One of the noteworthy features of this article was the use of Deligne-Beilinson cohomology to give a simple criterion under which the normal function of an algebraic cycle (over a curve) extends across singularities.

The collaboration with Steenbrink was a model of efficiency. The content of [Z11] was carried out in his office in the mathematics institute of the University of Leiden, over a pair of four-week periods in the late spring of 1981 and 1982. He had a long table in his office. We would sit at opposite ends, scribbling out our thoughts. Whenever one of us had something useful to say, he would go up to the blackboard. I recall that, as it happened, one of us would lead the progress on one day, the other the next day, and so on. I kept a strict set of notes, which made for a quick answer to questions like "Do you remember how we ...?" The thrust of our discussions in 1981 was the mixed analogue of [St] (where $\overline{f}$ was no longer proper), and it seemed as though by 1982 "everyone under the sun" had studied this problem. In 1982 we shifted to finding the naive linear algebra behind the existence of relative weights (from [D3]). Since they do not exist in general, it was saying that local monodromy in algebraic geometry satisfied some non-trivial equations, and we proceeded to determine them. After that, we could prove the mixed Hodge version of Hodge theory with degenerating coefficients. There was one item that Steenbrink was supposed to finish, but he conceded defeat; the collaboration was finished. Somewhat later on, I found I could take care of it myself, and that became the seed for [Z13].

The source of the collaboration in [Z16] is a bit humorous. Morihiko Saito had sent me a preprint under the same title and told me it had been submitted for publication. I pointed out to him that the first non-trivial case was contained

[3] as well as a formula for the L^2-cohomology

in [Z13], and he proceeded to offer me coauthorship! I protested that it was unnecessary, but if he insisted I would want to rewrite the paper. That became our understanding, but I was soon contacted by the editor about signing the copyright agreement. I told the editor about the deal with Saito, and he contacted Saito to confirm it. Well, how long would it take? I was asked. A couple of months, and I set out to perform my end of the bargain. After two false attempts I got it right, and that is what appears in [Z16].

When [S1] and [S2] appeared, I felt that I had been dethroned as "King of Hodge Theory". True, Y. Kawamata's work in higher dimensional geometry had already given a vast generalization of [Z05] (see [Fn]), but I meant in that title the development of Hodge theory, not its applications. Saito's work gave a non-L^2 Hodge structure for intersection cohomology (with coefficients in a variation of Hodge structure); that is the "jewel" in the crown. The method uses a complicated, recursive construction using $\mathcal{D}$-modules. However, it was [Z03] that defined the ground level, viz., over a curve (see [Sb]), and that provided some consolation. The outcome is realized algebro-geometrically in [dCM] and [dC] for geometric variations, using perverse cohomology sheaves, because of the role of the decomposition theorem in both formulations. This I consider to be state of the art.[4]

I had a nice collaboration with another Saito, namely Masa-Hiko. The objective was to produce [Z19], an interesting result about deformations and the infinitesimal Torelli problem for fiber spaces. While working on it, he asked me to explain Faltings' classification of non-rigid families of principally polarized abelian varieties, with its finiteness theorem of Arakelov type. It became a model for [Z18]. The question arose as to whether [Z18] should be considered joint work. It was decided that it was, despite the fact that I did not involve myself with K3 surfaces. (This should be contrasted with the situation concerning [Z13].)

The way by which [Z14] came about is also somewhat interesting. Both Richard Hain and I wanted to try to produce a solo proof of the theorem that classified unipotent variations of mixed Hodge structure, and we set out on our own. Eventually, each of us realized that he was stuck on one main point. But these main points were different! We settled on publishing it as a joint article.

A recurring theme will be deciding that a candidate for the Hodge structure for the intersection cohomology (middle perversity throughout) of a variety Y is "the right one". When Y has only isolated singularities, the answer is clear, because this intersection cohomology can be expressed in terms of the (ordinary) cohomology of Y and its regular locus. By this method, I could verify the Hodge theoretic Zucker conjecture, viz., that the isomorphism in the Zucker conjecture (from [Z7]) respected the Hodge structures in such cases [Z15]. In general, this problem is wide open. The people working in representation theory and automorphic forms would like the L^2 version to be the right one.

The three publications [Z20] through [Z22] were the output of a major collaboration. It began when Michael Harris had started working on [Z20]. He then sought a collaborator, and I became a natural choice. In contrast to [Z12], the

[4] See, however, [S3].

communication between the coauthors was almost entirely by e-mail. That was probably the best way to proceed, for our styles of doing mathematics are so different. On the other hand, our outputs were combined without difficulty. The main issue was to understand the boundary of the smooth toroidal compactifications X^{tor}. These are constructed over X^{BB}, stratum by stratum, so as to form a resolution of singularities of the latter (they also depend on a combinatorial parameter, but that does not affect the nature of what follows). We also had to compare this with the boundary of the Borel-Serre compactification X^{BS}, despite the non-existence of a morphism of compactifications between X^{tor} and X^{BS}. This took a lot of effort, but it ultimately led to the Hodge-theoretic isomorphism of the spectral sequences for the deleted neighborhood cohomology of the two. Of course, on a manifold with corners, the deleted neighborhood cohomology coincides, parabolic by parabolic subgroup, with that of the boundary itself. This led to a workable formula for the E_1-term. We had hoped that this would allow us to rule out the existence of so-called ghost classes (see [Z22: (4.6)] for the definition) for Hodge-theoretic reasons, but alas they exist in abundance [KR].

[Z23] was my only publication with more than one coauthor. Indeed, I knew of Johan Dupont only through Hain as a hub. My contributions may or may not be obvious, but I will confess that I forced myself to write a clear exposition of Dupont's proof of the second main theorem, which expresses the Borel regulator as two times the reduction of the universal Cheeger-Simons class modulo $\mathbb{R}$.

Articles [Z24] through [Z26] comprised an in-depth examination of "my favorite space", the reductive Borel-Serre compactification X^{red} of an arithmetic quotient of a symmetric space of non-compact type (these are the locally symmetric varieties when the symmetric space is Hermitian). As with the Borel-Serre manifold with corners, the construction makes sense for the non-Hermitian case. The (open) boundary faces of X^{BS} are nilpotent fibrations over a lower-rank symmetric space, and one simply collapses each fiber to a point. With this operation effected, one gets a stratified space X^{red}, whose closed boundary strata are the reductive Borel-Serre compactifications of the aforementioned lower-rank symmetric spaces. Making this construction was a useful technical trick in [Z07], where it was done without ado. The name "reductive Borel-Serre" was attached a few years later. In [Al], Albin considered X^{BS} from the point of view of manifolds with corners having iterated fibration structure à la R. Melrose. If one first modifies X^{BS} as a manifold with corners, one gets a Melrose manifold with corners, which blows down to yield a resolution of the Thom-Mather stratified space X^{red}.

Item [Z24] was my first solo research article to be written in ten years, and it reestablished my status as a research mathematician. The main theorem was something that was waiting to be plucked some twenty years after the Zucker conjecture was made, though it was much easier to prove. For contrast, the Zucker conjecture was about L^2-cohomology, and it was a difficult matter to prove it, even after it was posed (see [Z11]); the main theorem in [Z24] admitted a direct proof nearly immediately after it was asserted: $H_{(p)}(X)$ is isomorphic, when p is finite and sufficiently large, to the (ordinary) cohomology of the compactification X^{red} of X. Though determining the L^∞-cohomology would have been more natural, that had to be pushed down into $H_{(p)}(X)$ to gain a topological interpretation (one can do this because the locally invariant volume of X is finite).

For [Z25], I introduced the notions of an excentric quotient of the Borel-Serre compactification and of the toroidal compactification(s) (constructed in [AMRT] in the Hermitian case). These involve taking the quotient of their boundary strata by a real torus (product of circles), yielding the compactifications $X^{BS,exc} \to X^{red}$ and $X^{tor,exc}$ of X. Though they are not homeomorphic (when X is Hermitian), they have a striking resemblance, and one finds that the two are homotopy equivalent. On the other hand, the greatest common quotient of X^{red} and X^{tor} is X^{BB} [Ji]. This was carelessly conjectured for X^{BS} instead of X^{red} in [Z21]; there was a counterexample known to the authors that passed our notice when the conjecture was made. In that instance, X^{tor} happens to be homeomorphic to $X^{BS,exc}$.

Article [Z26] started a venture into "outer space". In that article, a mixed Hodge structure is constructed on the cohomology of X^{red} (when X is Hermitian), satisfying all obvious compatibilities, e.g., with the mixed Hodge structure on the cohomology of X^{BB}. The construction uses an inelegant recursion that is based on the description of its closed boundary strata given three paragraphs back. It made sense to attempt to find an underlying motive. With the suggestion of Marc Levine, Joseph Ayoub and I met in 2007. I presented Ayoub with the task of motivizing [Z26], in the sense of V. Voevodsky. He preferred and needed a more canonical construction, one that involved motives over a diagram of schemes, and the like. Despite the length of the resulting article [Z27], he wanted to submit it to a top journal. I thought that would be unreasonable, but he insisted. I had no strong objection, though, because he was the junior author, interested in promoting his contributions to mathematics. Besides, the result was *that* remarkable, showing that something that was not an algebraic variety could have motivic cohomology groups (over $\mathbb{Q}$). As I like to say, it justifies the occurrence of X^{red} in the literature; as M. Goresky put it, X^{red} thinks it is an algebraic variety. Because of the heavy use of motives, most of the writing was done by Ayoub, and it was important that the article be written in English — *sorry, Joseph; he was more comfortable in French* — and he really did an excellent job.

Article [Z28] was the only collaboration involving the organizers of the conference. Lizhen has told me that he considers himself, to a certain extent, to be a student of mine. What I make of this is that his involvement in the vast ocean [BJ] of compactifications of symmetric and locally symmetric spaces, and the relations among them, stems in part from his work for [Ji].

Last but not least, I mention the collaboration with Jean-Luc Brylinski [Z17], which provided an account of advances in Hodge theory that occurred roughly in the years 1975–1990, the fifteen years after the expository article [GS] came out.

References

I am listing as [Z**] *those publications of which I was an author. Coauthors (if any) are noted thereafter as:* "(with +++)".

[Al] P. Albin, On the Hodge theory of stratified spaces. In this volume.

[AMRT] A. Ash, D. Mumford, M. Rapoport and Y.-S. Tai, Smooth Compactifications of Locally symmetric varieties, 2nd ed. Cambridge University Press, 2010.

[Ar] D. Arapura, The Leray spectral sequence is motivic. Invent. Math. **160** (2005) 567–589.

[BB] W. Baily and A. Borel, Compactification of arithmetic quotients of bounded symmetric domains. Ann. of Math. **84** (1966) 442–528.

[BlGr] S. Bloch and P. Griffiths, A theorem about normal functions associated to Lefschetz pencils on algebraic varieties, 1971.

[BC] A. Borel and W. Casselman, L^2-cohomology of locally symmetric manifolds of finite volume. Duke Math. J. **50** (1983) 625–647.

[BG] A. Borel and H. Garland, Laplacian and the discrete spectrum of an arithmetic group. Amer. J. Math. **105** (1983) 309–335.

[BJ] A. Borel and L. Ji, Compactifications of Symmetric and Locally Symmetric Spaces. Birkhäuser, 2006.

[BS] A. Borel and J.-P. Serre, Corners and arithmetic groups. Comm. Math. Helv. **48** (1973) 436–491.

[CKS] E. Cattani, A. Kaplan and W. Schmid, L^2 and intersection cohomologies for a polarizable variation of Hodge structure. Invent. Math. **87** (1987) 217–252.

[ChSh] D. Chakrabarti and M.C. Shaw, The Cauchy-Riemann equations on product domains. Math. Annalen **349** (2011) 977–998.

[CGM] J. Cheeger, M. Goresky and R. MacPherson, L^2 cohomology and intersection homology of singular algebraic varieties. In: S.-T. Yau (ed.), Seminar on Differential Geometry, Ann. of Math. Study **102**, Princeton University Press, 1982, 303–339.

[dC] M.A. de Cataldo, The perverse filtration and the Lefschetz hyperplane theorem, II. J. Alg. Geom. **21** (2012) 305–345.

[dCM] M.A. de Cataldo and L. Migliorini, The perverse filtration and the Lefschetz hyperplane theorem. Annals of Math. **171** (2010) 2089–2113.

[D1] P. Deligne, Variétés de Shimura: interprétation modulaire, et techniques de construction de modèles canoniques. Proc. of Symp. in Pure Math. **33**, part 2, AMS, 1979, 247–290.

[D2] P. Deligne, The Hodge conjecture. In: J. Carlson, A. Jaffe and A. Wiles, eds., The Millennium Prize Problems, Clay Math. Institute and Amer. Math. Soc., 2006, 45–53.

[D3] P. Deligne, La conjecture de Weil, II. Publ. Math. IHES **52** (1980) 137–252.

[Fn] O. Fujino, On semipositivity, injectivity and vanishing theorems. In this volume.

[Ft] T. Fujita, On Kähler fiber spaces over curves. J. Math. Soc. Japan **30** (1978), no. 4, 779–794.

[GHM] M. Goresky, G. Harder and R. MacPherson, Weighted cohomology. Invent. Math. **116** (1994) 139–213.

[GM] M. Goresky and R. MacPherson, Intersection homology, II. Invent. Math. **71** (1983) 77–129.

[GS] P. Griffiths and W. Schmid, Recent developments in Hodge theory: A discussion of techniques and results. In: Proc. Bombay Colloq. on Discrete Subgroups of Lie Groups, 1975, 31–127.

[Ji] L. Ji, The greatest common quotient of Borel-Serre and the toroidal compactifications. Geom. and Functional Anal. **8** (1998) 978–1015.

[KK] M. Kashiwara and T. Kawai, The Poincaré lemma for a variation of Hodge structure. Publ. RIMS **23** (1987) 345–407.

[KR] A. Kewenig and T. Rieband, Geisterklassen im Bild der Borelabbildung für symplektische und orthogonale Gruppen. Diplomarbeit, Uni. Bonn, 1997.

[Lo] E. Looijenga, L^2-cohomology of locally symmetric varieties. Compositio Math. **67** (1988) 3–20.

[MM] Y. Matsushima and S. Murakami, On vector bundle valued harmonic forms and automorphic forms on symmetric Riemannian manifolds. Ann. of Math. **78** (1963) 365–416.

[Sb] C. Sabbah, Hodge theory, singularities and $\mathcal{D}$-modules. Lecture notes, CIRM Luminy, March 2007.

[S1] Mo. Saito, Modules de Hodge polarisable. Publ. RIMS **24** (1988) 849–995.

[S2] Mo. Saito, Mixed Hodge modules. Publ. RIMS **26** (1990) 221–333.

[S3] Mo. Saito, Decomposition theorem for proper Kähler morphisms. Tohoku Math. J. **42** (1990) 127–147.

[SS] L. Saper and M. Stern, L^2-cohomology of arithmetic varieties. Ann. of Math. **132** (1990) 1–69.

[St] J. Steenbrink, Limits of Hodge structures. Invent. Math. **31** (1976) 229–257.

[V] V. Vaish, Motivic weightless complex and the relative Artin motive, Adv. in Math. **292** (2016) 316–373.

[Z01] Generalized intermediate Jacobians and the theorem on normal functions. Invent. Math. **33** (1976) 185–222.

[Z02] The Hodge conjecture for cubic fourfolds. Compos. Math. **34** (1977) 199–209.

[Z03] Hodge theory with degenerating coefficients: L^2-cohomology in the Poincaré metric. Annals of Math. **109** (1979) 415–476.

[Z04] (with D. Cox) Intersection numbers of sections of elliptic surfaces. Invent. Math. **53** (1979) 1–44.

[Z05] Remarks on a theorem of Fujita. J. Math. Soc. Japan **34** (1982) 47–54.

[Z06] Locally homogeneous variations of Hodge structure. L'Enseignement Math. **27** (1981) 243–276.

[Z07] L^2-cohomology of warped products and arithmetic groups. Invent. Math. **70** (1982) 169–218.

[Z08] Satake compactifications. Commentarii Math. Helv. **58** (1983) 312–343.

[Z09] (with F. El Zein) Extendability of normal functions associated to algebraic cycles. In Griffiths, ed., Topics in Transcendental Algebraic Geometry, Princeton Univ. Press, 1984, 269–288.

[Z10] L^2-cohomology and intersection homology of locally symmetric varieties, II. Comp. Math. **59** (1986) 339–398.

[Z11] L^2-cohomology and intersection homology of locally symmetric varieties, III. In: Hodge Theory: Proceedings Luminy, 1987, Astérisque **179–180** (1989) 245–278.

[Z12] (with J. Steenbrink) Variation of mixed Hodge structure, I. Invent. Math. **80** (1985) 489–542.

[Z13] Variation of mixed Hodge structure, II. Invent Math. **80** (1985) 543–565.

[Z14] (with R. Hain) Unipotent variations of mixed Hodge structure. Invent. Math. **88** (1987) 83–124.

[Z15] The Hodge structures on the intersection homology of varieties with isolated singularities. Duke Math. J. **55** (1987) 603–616.

[Z16] (with Mo. Saito) The kernel spectral sequence of vanishing cycles. Duke Math. J. **61** (1990) 329–339.

[Z17] (with J.-L. Brylinski) An overview of recent advances in Hodge theory. In: W. Barth and R. Narasimhan, eds., Several Complex Variables VI, Encyclopaedia of Math. Sci., vol. **69**, Springer-Verlag, 1990, 39–142.

[Z18] (with M.-H. Saito) Classification of non-rigid families of K3 surfaces and a finiteness theorem of Arakelov type. Math. Annalen. **289** (1991) 1–31.

[Z19] (with M.-H. Saito) On the Torelli problem for fiber spaces. In: Complex Geometry and Lie Theory, Sundance, Utah, 1989, Proc. of Symposia in Pure Math. **53** (1991) 269–282.

[Z20] (with M. Harris) Boundary cohomology of Shimura varieties, I: Coherent cohomology on toroidal compactifications. Ann. Sci. ENS **27** (1994) 249–344.

[Z21] (with M. Harris) Boundary cohomology of Shimura varieties, II: Hodge theory at the boundary. Invent. Math. **116** (1994; dedicated to A. Borel) 243–308. Erratum, Invent. Math. **121** (1995) 437.

[Z22] (with M. Harris) Boundary cohomology of Shimura varieties, III: Coherent cohomology on higher-rank boundary strata and applications to Hodge theory. Mem. Soc. Math. France **85**, 2001 (116 pp.).

[Z23] (with J. Dupont and R. Hain) Regulators and characteristic classes of flat bundles. In: The Arithmetic and Geometry of Algebraic Cycles, CRM Proceedings & Lecture Notes, vol. **24**, 2000, 47–92.

[Z24] On the reductive Borel-Serre compactification: L^p-cohomology of arithmetic groups (for large p). Amer. J. Math. **123** (2001) 951–984.

[Z25] On the reductive Borel-Serre compactification II: Excentric quotients and least common modifications. Amer. J. Math. **130** (2008) 859–912.

[Z26] On the reductive Borel-Serre compactification III: Mixed Hodge structures. Asian J. Math. **8** (2004) 881–912 (Volume in memory of Armand Borel).

[Z27] (with J. Ayoub) Relative Artin motives and the reductive Borel-Serre compactification of a locally symmetric variety. Invent. Math. **188** (2012) 277–427.

[Z28] (with L. Ji) On the L^p-cohomology and the geometry of metrics on moduli spaces of curves. In: A. Papadopoulos, ed., Handbook of Teichmueller Theory, Vol. III, European Math. Soc., 2012, 747–775.

Hodge Theory and L^2-analysis
ALM 39, pp. 1–78

On the Hodge Theory of Stratified Spaces*

Pierre Albin†

Abstract

This article is a survey of recent work of the author, together with Markus Banagl, Eric Leichtnam, Rafe Mazzeo, and Paolo Piazza, on the Hodge theory of stratified spaces. We discuss how to resolve a Thom-Mather stratified space to a manifold with corners with an iterated fibration structure and the generalization of a perversity in the sense of Goresky-MacPherson to a mezzoperversity. We define Cheeger spaces and their signatures and describe how to carry out the analytic proof of the Novikov conjecture on these spaces. Finally we review the reductive Borel-Serre compactification of a locally symmetric space to a stratified space and describe its resolution to a manifold with corners.

1 Introduction

The purpose of this paper is to describe some recent advances in the Hodge theory of pseudomanifolds.

Algebraic varieties and L^2-metrics have been integral aspects of Hodge theory from the very beginning. Hodge appealed to the Dirichlet principle [60] and the Levi parametrix method [61, 62] to find harmonic forms in each de Rham cohomology class and used these forms to study the cohomology of complex projective algebraic varieties. Weyl [116] referred to [62] as "one of the great landmarks in the history of our science in the present century" (and corrected Hodge's existence proof in [115]). An existence proof using purely Hilbert space methods was carried out by Gaffney [47, 48] who, for instance, considered the exterior derivative d on a smooth manifold X as an unbounded operator on square integrable differential forms with domain

$$\mathcal{D}_{\max}(d) = \{\omega \in L^2(X; \Lambda^\bullet T^*X) : d\omega \in L^2(X; \Lambda^\bullet T^*X)\},$$

*Dedicated to Steven Zucker on the occasion of his 65th birthday.

†University of Illinois at Urbana-Champaign. E-mail: palbin@illinois.edu

where $d\omega$ is computed distributionally. This domain forms a complex whose cohomology, known as the L^2-cohomology of X, was introduced independently in [118] on a complete manifold and in [36] on an incomplete manifold.

Algebraic varieties are often singular but Whitney [117, 67] showed that they are stratified spaces. This means that they can be decomposed into smooth manifold pieces of various dimensions, called *strata*, in such a way that different points on the same stratum have similar neighborhoods in the variety. In particular one of these pieces is an open dense smooth manifold. Endowing this 'regular part' with a conic-type Riemannian metric, Cheeger studied the de Rham complex of differential forms in L^2. He singled out a topological condition that guaranteed that the exterior derivative had an unambiguous definition as an unbounded operator, every cohomology class had a harmonic representative, and the resulting cohomology groups satisfy Poincaré duality. In the case of conic singularities he explained how, if this topological condition failed, one can choose ideal boundary conditions and still establish these results. In this note I will review recent work [3, 1, 4, 5] extending Cheeger's results to general stratified *'pseudomanifolds'*.

Specifically, let $\widehat{X}$ denote a pseudomanifold (a stratified space 'without boundary') and endow its regular part $\widehat{X}^{\mathrm{reg}}$ with a 'wedge metric' g. The exterior derivative d and its formal adjoint δ make up the de Rham operator

$$\eth_{\mathrm{dR}} = d + \delta : \mathcal{C}^\infty_c(\widehat{X}^{\mathrm{reg}}; \Lambda^\bullet T^*\widehat{X}^{\mathrm{reg}}) \longrightarrow \mathcal{C}^\infty_c(\widehat{X}^{\mathrm{reg}}; \Lambda^\bullet T^*\widehat{X}^{\mathrm{reg}}).$$

There are two canonical extensions of $\eth_{\mathrm{dR}}$ to a closed operator on $L^2(\widehat{X}^{\mathrm{reg}}; \Lambda^\bullet T^*\widehat{X}^{\mathrm{reg}})$. The first has domain

$$\mathcal{D}_{\max}(\eth_{\mathrm{dR}}) = \{\omega \in L^2(\widehat{X}^{\mathrm{reg}}; \Lambda^\bullet T^*\widehat{X}^{\mathrm{reg}}) : \eth_{\mathrm{dR}}\omega \in L^2(\widehat{X}^{\mathrm{reg}}; \Lambda^\bullet T^*\widehat{X}^{\mathrm{reg}})\}$$

where $\eth_{\mathrm{dR}}\omega$ is computed distributionally, and the second has domain

$$\begin{aligned}\mathcal{D}_{\min}(\eth_{\mathrm{dR}}) = \{\omega \in L^2(\widehat{X}^{\mathrm{reg}}; \Lambda^\bullet T^*\widehat{X}^{\mathrm{reg}}) : \exists(\omega_n) \subseteq \mathcal{C}^\infty_c(\widehat{X}^{\mathrm{reg}}; \Lambda^\bullet T^*\widehat{X}^{\mathrm{reg}}) \\ \text{s.t. } \omega_n \to \omega \text{ in } L^2 \text{ and } (\eth_{\mathrm{dR}}\omega_n) \text{ is } L^2\text{-Cauchy}\}\end{aligned}$$

and $\eth_{\mathrm{dR}}\omega$ can be computed both distributionally and as $\lim \eth_{\mathrm{dR}}\omega_n$.

Cheeger [36] singled out a class of spaces, now known as *Witt spaces* [106], where for a suitably chosen metric we have

$$\mathcal{D}_{\min}(\eth_{\mathrm{dR}}) = \mathcal{D}_{\max}(\eth_{\mathrm{dR}}).$$

This implies that many of the classical Hodge theory results will hold on $\widehat{X}$. For example, the L^2-cohomology is finite dimensional, represented by harmonic forms, and satisfies Poincaré duality. Moreover, Cheeger showed that this cohomology coincides with the (middle perversity) intersection homology of Goresky-MacPherson and so in particular one can study the signature of intersection homology using analytic tools.

The signature is an important invariant of smooth manifolds, e.g., in [56], Gromov says that it is "*the* invariant, which can be matched in the beauty and power only by the Euler characteristic". The bordism invariance of the signature implies that it is a polynomial in the rational Pontryagin classes, and Hirzebruch's signature theorem identified the polynomial with the 'L-class'. The signature is also invariant under (oriented) homotopies and Novikov proved that, for a simply connected oriented manifold, the only rational Pontryagin characteristic class that is (oriented) homotopy invariant is the 'top order part' of the L-class. He defined 'higher signatures' on non-simply connected manifolds and the celebrated, still open, Novikov conjecture claims that these are homotopy invariant and they are the only ones.

Due to the importance of the conjecture and the ubiquity of singular spaces, it is natural to study the Novikov conjecture on pseudomanifolds. A fruitful approach due to Lusztig [82], Mishchenko [93], and Kasparov [69] involves the Atiyah-Singer index theorem for families of operators but with C^*-algebra bundles (see [100] for a recent survey of the Novikov conjecture). In [3] we carried out this approach to the Novikov conjecture on Witt spaces. In the process we established Cheeger's results on Witt spaces by an approach that lent itself to allowing C^*-algebra bundles.

In [4, 5] we carried out the analytic approach to the Novikov conjecture on a larger class of stratified spaces. In [36], Cheeger considered non-Witt spaces with isolated conic singularities and explained that one could still define an L^2 cohomology theory satisfying Poincaré duality by choosing 'ideal boundary conditions' at the cone points. In [4] we explain how to use ideal boundary conditions on a general pseudomanifold to define a self-adjoint Fredholm domain for the de Rham operator, and an associated L^2-de Rham cohomology complex. We call the spaces that admit ideal boundary conditions for the signature operator *Cheeger spaces* in recognition of Cheeger's contributions. In [5] we carry out the analytic approach to the Novikov conjecture on Cheeger spaces.

Our first step, described in §6, is to resolve the geometry in the sense of Melrose (e.g., [91]) by replacing a stratified space $\widehat{X}$ with a smooth manifold with corners $\widetilde{X}$, so that the regular part of $\widehat{X}$ becomes the interior of $\widetilde{X}$. The idea of resolving a stratified space goes back at least as far as Thom [111] and continues to be an object of study [9]. In our formulation the stratification is replaced by a *boundary fibration structure*: each boundary hypersurface of $\widetilde{X}$ is the total space of a fiber bundle and these fibrations are compatible over corners. Collapsing the fibers of these fibrations in the appropriate order recovers the singular space from the smooth space, while 'blowing-up' the singular strata constructs the smooth space from the singular one. Having replaced one compactification $\widehat{X}$ of the regular part by another one $\widetilde{X}$, we next replace the cotangent bundle $T^*\widetilde{X}$ by a 'rescaled bundle', the wedge cotangent bundle ${}^wT^*\widetilde{X}$, that reflects the original degeneration of $\widehat{X}$. This replacement brings out the iterative structure of the de Rham operator as we describe fully in §3.1. A key fact is that these replacements do not change the regular part of the stratified space nor the cotangent bundle over the regular part; since analysis on a stratified space means analysis on the regular part, these convenient replacements have not changed the problem we wish to solve.

Once these changes are made, $\eth_{\mathrm{dR}}$ is amenable to study using the microanalytic tools developed in, e.g., [92, 90, 85, 87, 76, 105]. We explain how to use these tools first in the case of a single singular stratum in §3 and then for an arbitrary pseudomanifold in §4. One of the features of our approach is to take advantage of the iterative structure of the space and the operator which means that most of the analytic difficulties are already present for a single singular stratum.

If $\widehat{X}$ has a single singular stratum Y, then $\widetilde{X}$ is a smooth manifold with boundary and its boundary participates in a fiber bundle of smooth closed manifolds

$$Z - \partial\widetilde{X} \xrightarrow{\phi} Y$$

with base Y and typical fiber Z. Using the microlocal tools mentioned above, we see in §3.3 that elements in $\mathcal{D}_{\max}(\eth_{\mathrm{dR}})$ have (distributional) asymptotic expansions as they approach the boundary of $\widetilde{X}$,

$$\omega \in \mathcal{D}_{\max}(\eth_{\mathrm{dR}}) \implies \omega \sim r^{-v/2}(\alpha(\omega) + dr \wedge \beta(\omega)) + \widetilde{\omega} \text{ as } r \to 0$$

where $\widetilde{\omega}$ is a 'remainder' and $\alpha(\omega)$, $\beta(\omega)$ are distributional forms on Y,

$$\alpha(\omega), \beta(\omega) \in \mathcal{C}^{-\infty}(Y; \Lambda^\bullet T^*Y \otimes \mathcal{H}^{\dim Z/2}(\partial\widetilde{X}/Y)),$$

with coefficients in $\mathcal{H}^{\dim Z/2}(\partial\widetilde{X}/Y)$ the vertical Hodge cohomology of the boundary fibration. These forms serve as 'Cauchy data' for ω. For example we show that the minimal domain is determined by the vanishing of the Cauchy data. Thus if v is odd or $\mathrm{H}^{\dim Z/2}_{\mathrm{dR}}(Z) = 0$ (which is the 'Witt condition' in this context) then the maximal domain is the same as the minimal domain. Otherwise $\mathcal{H}^{\dim Z/2}(\partial\widetilde{X}/Y)$ is a flat bundle and for every flat subbundle W,

$$\begin{array}{ccc} W & \longrightarrow & \mathcal{H}^{\dim Z/2}(\partial\widetilde{X}/Y) \\ & \searrow \quad \swarrow & \\ & Y & \end{array},$$

we define a domain for $\eth_{\mathrm{dR}}$ by

$$\begin{aligned} \mathcal{D}_W(\eth_{\mathrm{dR}}) = \{&\omega \in \mathcal{D}_{\max}(\eth_{\mathrm{dR}}) : \\ &\alpha(\omega) \in \mathcal{C}^{-\infty}(Y; \Lambda^\bullet T^*Y \otimes W), \quad \beta(\omega) \in \mathcal{C}^{-\infty}(Y; \Lambda^\bullet T^*Y \otimes W^\perp)\}. \end{aligned}$$

We refer to W as a *mezzoperversity* and to $\mathcal{D}_W(\eth_{\mathrm{dR}})$ as imposing *Cheeger ideal boundary conditions*. With this domain, we see in §3.4 that $\eth_{\mathrm{dR}}$ is a self-adjoint Fredholm operator with compact resolvent and hence the Hodge cohomology groups $\mathcal{H}^*_W(\widehat{X})$ are finite dimensional. In §3.5 we explain how W determines a domain for the exterior derivative, $\mathcal{D}_W(d)$, and a Fredholm Hilbert complex in the sense of [31]. The corresponding de Rham cohomology is denoted $\mathrm{H}^*_{W,\mathrm{dR}}(\widehat{X})$ and it satisfies the Hodge theorem

$$\mathrm{H}^*_{W,\mathrm{dR}}(\widehat{X}) \cong \mathcal{H}^*_W(\widehat{X}).$$

In §4 we explain how these constructions can be carried out on a general pseudomanifold. This involves an inductive procedure: assuming we have carried out this analysis of Hodge and de Rham cohomology on pseudomanifolds of depth less than k, we can carry out it out on spaces of depth k. A mezzoperversity in this context consists of system of flat bundles, $\mathcal{W} = (W_1, \ldots, W_\ell)$, one over each non-Witt stratum, satisfying a compatibility condition at the corners of $\widetilde{X}$. To each mezzoperversity we assign finite dimensional de Rham and Hodge cohomology groups and we prove their equality,

$$\mathrm{H}^*_{\mathcal{W},\mathrm{dR}}(\widehat{X}) \cong \mathcal{H}^*_{\mathcal{W}}(\widehat{X}).$$

If $\widetilde{X}$ is oriented we associate to each mezzoperversity $\mathcal{W}$ a dual mezzoperversity $\mathbb{D}\mathcal{W}$ and we show that the intersection pairing on differential forms descends to a non-degenerate pairing

$$\mathrm{H}^*_{\mathcal{W},\mathrm{dR}}(\widehat{X}) \times \mathrm{H}^*_{\mathbb{D}\mathcal{W},\mathrm{dR}}(\widehat{X}) \longrightarrow \mathbb{R} \tag{1.1}$$

yielding a generalized Poincaré duality in the sense of [53]. This will be a 'true' Poincaré duality if $\mathcal{W} = \mathbb{D}\mathcal{W}$ but there are topological obstructions to the existence of such self-dual mezzoperversities, notably the signatures of the fibers of the boundary fibrations. We refer to a pseudomanifold with a self-dual mezzoperversity as a *Cheeger space.*

Section 5 is devoted to explaining that these cohomology theories share many interesting properties with the usual cohomology theories of smooth manifolds. For example we show that the de Rham cohomology is independent of the metric used to define it. The key fact is that the boundary conditions can be defined using de Rham cohomology instead of Hodge cohomology, allowing us (inductively) to define the boundary conditions without reference to a metric.

We also show that the cohomology is invariant under stratified homotopy equivalences. This is a bit tricky since pull-back by a homotopy equivalence need not define a bounded operator on L^2. We make use of a clever construction of Hilsum and Skandalis [59] to define a bounded pull-back map, and then explain how it acts on mezzoperversities and on cohomology.

On Cheeger spaces self-dual mezzoperversities define, through the quadratic form (1.1), a signature. In §5.3 we explain how a self-dual mezzoperversity determines a domain for the signature operator on which it is Fredholm with index the signature of (1.1). If two Cheeger spaces, endowed with self-dual mezzoperversities, are bordant (with a Cheeger space bordism and an extension of the self-dual mezzoperversities) then the signature does not change. Together with a theorem of Banagl [11], this implies that the signature of (1.1) depends only on $\widehat{X}$ and not on the self-dual mezzoperversity! Thus every Cheeger space has a signature that is invariant under Cheeger space bordism and stratified homotopy equivalences.

A beautiful approach to characteristic classes of Thom [110] allows us to use this signature to define an L-class in homology. (This same approach was used by Goresky-MacPherson [53] and Banagl [11]; Cheeger [38] used η-invariants of the links to give a combinatorial formula for the L-class of a Witt space.) In §5.4 we define the L-class of a Cheeger space and use it to define the higher signatures. In [5] we carry out the analytic approach to the Novikov conjecture and prove that these higher signatures are stratified homotopy invariant for a large class of Cheeger spaces.

Cheeger was able to relate L^2-cohomology to the intersection cohomology of Goresky-MacPherson [53, 54]. Specifically, the complex $\mathcal{D}_{\min}(d)$ has cohomology isomorphic to $\mathrm{IH}^*_{\overline{m}}(\widehat{X})$, the 'lower middle perversity' intersection cohomology of $\widehat{X}$, and the complex $\mathcal{D}_{\max}(d)$ has cohomology isomorphic to $\mathrm{IH}^*_{\overline{n}}(\widehat{X})$, the 'upper middle perversity' intersection cohomology of $\widehat{X}$. These can be described as the hypercohomology groups of two sheaf complexes

$$\mathbf{IC}^\bullet_{\overline{m}}, \quad \mathbf{IC}^\bullet_{\overline{n}}$$

respectively. Banagl [10] took the sheaf-theoretic description of intersection cohomology from [54] and studied the category of all self-dual sheaf complexes consistent with $\mathbf{IC}^\bullet_{\overline{m}}$ and $\mathbf{IC}^\bullet_{\overline{n}}$. In [1] we considered the category of sheaf complexes consistent with $\mathbf{IC}^\bullet_{\overline{m}}$ and $\mathbf{IC}^\bullet_{\overline{n}}$ but not necessarily self-dual. As explained in §5.5, we showed that the cohomology groups $\mathrm{H}^\bullet_{\mathcal{W},\mathrm{dR}}(\widehat{X})$ corresponding to a mezzoperversity are the hypercohomology groups of one of these sheaf complexes. Also self-dual mezzoperversities precisely correspond to the self-dual sheaves studied by Banagl. This yields the 'de Rham theorem' to complement our 'Hodge theorem' and completes the analogy between our new cohomology theories and the classical cohomology of closed manifolds.

The final two sections discuss the geometry of stratified spaces. Most of §6 is devoted to showing the equivalence of Thom-Mather stratified spaces and manifolds with corners and boundary fibration structures. To exemplify the usefulness of this equivalence we include a description of a smooth de Rham complex that computes the intersection cohomology of any (Goresky-MacPherson) perversity function. One of the most interesting sources of stratified spaces is the reductive Borel-Serre compactification of a locally symmetric space, introduced by Zucker in [122], which we discuss in the final section, §7. By resolving this stratified space we obtain a new compactification, the resolved Borel-Serre compactification, of a locally symmetric space to a manifold with corners and a boundary fibration structure.

Acknowledgements. I thank Lizhen Ji for the invitation to write this article. I am grateful to my collaborators on the work being reported on, Markus Banagl, Eric Leichtnam, Rafe Mazzeo, Richard Melrose, and Paolo Piazza. I am also grateful for conversations on these topics with Francesco Bei and Jeff Cheeger. Melinda Lanius and Greg Friedman made suggestions that improved the exposition and Steven Zucker was kind enough to read the manuscript closely and suggest many improvements. Part of this project was completed while the author was in residence at the Mathematical Sciences Research Institute in Berkeley, CA, made possible by the N.S.F. grant 0932078000. This work was partially supported by Simons Foundation grant #317883.

2 Background and context

As described in [41, §1.10], in the mid 70's three independent sources gave rise to a new (co)homology theory for singular spaces:

- Goresky and MacPherson while working to develop a theory of characteristic classes of algebraic varieties developed intersection homology.
- Cheeger's proof of the Ray-Singer conjecture on the equality of analytic torsion and Reidemeister torsion inspired his study of the L^2-cohomology of differential forms on the regular part of a variety.
- A generalization by Deligne of work of Zucker on extending a variation of Hodge structure from the complement of a divisor to the entirety of a non-singular complex algebraic variety.

The very fruitful history of this development is discussed in detail through 1989 in [72] and has continued apace. Our focus in this expository paper is on the L^2 cohomology of singular spaces, continuing the work of Cheeger [36, 37, 41, 38].

2.1 Stratified spaces

Let us start by considering a particular example of a singular space. *Whitney's umbrella* is given by

$$\mathscr{W} = \{(x, y, z) \in \mathbb{R}^3 : x^2 - zy^2 = 0\}.$$

The gradient of the defining polynomial is $(2x, -2yz, -y^2)$, so the critical points make up the $\{z\text{-axis}\}$. At every point $p \in \mathscr{W} \setminus \{z\text{-axis}\}$, the inverse function theorem guarantees that there is a neighborhood of p in $\mathscr{W}$ that is diffeomorphic to an open ball in $\mathbb{R}^2$. Thus, this '*regular part*' of $\mathscr{W}$, $\mathscr{W}^{\text{reg}}$, forms a two dimensional smooth manifold and we have a decomposition of $\mathscr{W}$ into two smooth manifolds

$$\mathscr{W} = \{z\text{-axis}\} \cup \mathscr{W}^{\text{reg}}.$$

This is the first feature we will ask of a stratified space: it is a union of smooth manifolds, often of different dimensions. Each of these manifolds is called a *stratum* and we talk about the collection of *strata*.

Manifolds have a local homogeneity: every point has a neighborhood diffeomorphic to the same Euclidean ball. Local homogeneity is a second feature we will require of each stratum of a stratified space. Of course every point on the z-axis has a neighborhood on the z-axis diffeomorphic to an interval, but what we want is to look at neighborhoods of points on the z-axis as subsets of $\mathscr{W}$. The intersections of $\mathscr{W}$ with the planes $z = c$ consist of

$$\begin{cases} \text{two lines: } x = cy, x = -cy & \text{if } c > 0, \\ \text{one line: } x = 0 & \text{if } c = 0, \\ \text{one point: } x = y = 0 & \text{if } c < 0, \end{cases}$$

so we see that a point on the negative z-axis has a neighborhood in $\mathscr{W}$ homeomorphic to an interval, while a point on the positive z-axis has a neighborhood in $\mathscr{W}$ homeomorphic to an interval (on the z-axis) times a small cross, which we think of as a truncated cone on four points, $(0,1) \times C(\{\text{ four points }\})$. Thus to have local homogeneity we have to refine the decomposition of $\mathscr{W}$ to

$$\mathscr{W} = \{\text{ origin }\} \cup \{\text{ negative } z\text{-axis }\} \cup \{\text{ positive } z\text{-axis }\} \cup (\mathscr{W} \setminus \{z\text{-axis}\}).$$

This exhibits $\mathscr{W}$ as a disjoint union of four strata: one 0-dimensional, two 1-dimensional, and one 2-dimensional, and each stratum is locally homogeneous.

We defer the formal definition of a stratified space to §6 and meanwhile content ourselves with an informal description and some examples. For us a stratified space will always mean a "Thom-Mather stratification"; for other notions of stratification we refer the interested reader to the excellent surveys [64, 73]. Roughly speaking this means that we have a topological space $\widehat{X}$ with a decomposition into smooth manifold pieces called strata. The stratum of largest dimension, $\widehat{X}^{\text{reg}}$, is called the regular part and is assumed to be dense; the other strata are called singular. Each stratum Y is 'locally homogeneous' in that each point has a neighborhood in $\widehat{X}$ homeomorphic to $\mathbb{B}^h \times C(Z)$, for some stratified space Z called the *link* of Y in $\widehat{X}$. Moreover each stratum Y has a 'tubular neighborhood' $\mathcal{T}$ in $\widehat{X}$ that is the total space of a fiber bundle

$$C(Z) \,\text{---}\, \mathcal{T} \longrightarrow Y.$$

A stratified space is called a *pseudomanifold* if all of the singular strata have codimension at least two. (The Whitney umbrella is not a pseudomanifold in this sense, both because it has a stratum of codimension one and because the regular part is not dense.)

A key feature of stratified spaces is that they show up even when our main object of interest is smooth.

Example 1 (Algebraic varieties). *Suppose $\widehat{X}$ is the zero set of a finite set of polynomials $\{f_1, \ldots, f_k\}$. The Jacobian matrix of $F = (f_1, \ldots, f_k)$ has rank in $\{0, \ldots, k\}$ at each point in $\widehat{X}$. The sets*

$$Y_k = \{\zeta \in \widehat{X} : \operatorname{rank}(DF)(\zeta) = k\}$$

partition $\widehat{X}$ into smooth manifolds, but do not always form a stratification. Nevertheless, Whitney [117] showed that this partition can always be refined to yield a stratification (see [67]).

Example 2 (Group Actions). *Another common example of a stratified set is the orbit space of a smooth manifold under the action of a compact Lie group. Suppose that M is a smooth manifold with an action of the compact Lie group G, meaning a homomorphism $G \longrightarrow \operatorname{Diff}(M)$. Recall that to each point $\zeta \in M$ we associate an orbit $G \cdot \zeta$ and a stabilizer G_ζ,*

$$G \cdot \zeta = \{g(\zeta) \in M : g \in G\} \subseteq M, \quad G_\zeta = \{g \in G : g(\zeta) = \zeta\} \subseteq G.$$

Note that the stabilizer group changes along an orbit since

$$G_{g(\zeta)} = g G_\zeta g^{-1},$$

suggesting that a better invariant is the conjugacy class of the stabilizer. This is borne out by a classical result of Borel: the orbit space $M/G = \{G \cdot \zeta : \zeta \in M\}$ *inherits a smooth manifold structure along the projection* $M \longrightarrow M/G$ *if all of the stabilizer groups* G_ζ *are conjugate to each other in* G.

In general the space M/G *is a stratified space. Given a subgroup* H *of* G, *let* $[H]$ *denote its conjugacy class in* G, *and let*

$$M_{[H]} = \{\zeta \in M : G_\zeta \in [H]\}.$$

It's clear from the above that $M_{[H]}$ *is closed under the action of* G *and Borel's theorem implies that* $M_{[H]}/G$ *is a smooth manifold. This exhibits* M/G *as a union of smooth manifolds. The fact that this decomposition is a stratification follows from the fact that the* $M_{[H]}$ *have equivariant tubular neighborhoods in* M.

(In [6] we discuss the resolution of a group action. The minimal isotropy type X *can be identified with the interior of a smooth manifold with corners* $Y(M)$ *endowed with a* G*-action and an equivariant map*

$$\beta : Y(M) \longrightarrow M.$$

The manifold $Y(M)$ *has an iterated fibration structure in the sense of §6.4 where all spaces have* G*-actions and all maps are equivariant. Moreover* $Y(M)$ *and the bases of its boundary fibrations have unique isotropy types and hence manifold orbit spaces. The orbit space of* $Y(M)$ *thus inherits an iterated fibration structure and can be identified with the resolution of the stratified space* M/G.*)*

As explained in [64], If M *and* G *are not smooth, or the action of* G *on* M *is not smooth, then the orbit space* M/G *need not admit a Thom-Mather stratification. This is the case even for a finite group and was one of the main motivations for introducing more general notions of stratification.*

Example 3 (Moduli spaces). *A moduli space is a space, usually an open manifold, whose points represent (isomorphism classes of) geometric objects of some fixed kind. Moduli spaces often involve singular spaces in two ways. On the one hand natural compactifications of moduli spaces are often singular. On the other hand, points in the boundary of the compactification can often be interpreted as moduli of singular versions of the original objects.*

For example, let Σ_g *be a smooth oriented surface of genus* g, *say* $g > 1$, *and let* C_g *denote the space of complex structures on* Σ_g. *The orientation preserving diffeomorphisms of* Σ_g, $\mathrm{Diff}^+(\Sigma_g)$, *act naturally on* C_g *and Teichmüller space is the quotient by the subgroup* $\mathrm{Diff}^+_0(\Sigma_g)$ *of diffeomorphisms isotopic to the identity,*

$$\mathcal{T}_g = C_g / \mathrm{Diff}^+_0(\Sigma_g).$$

This has a natural manifold structure, for which $\mathcal{T}_g \cong \mathbb{R}^{6g-6}$ *as real manifolds. The* augmented Teichmüller space, $\widehat{\mathcal{T}}_g$, *is a* partial *compactification of* $\mathcal{T}_g$ *obtained by attaching a stratified boundary to* $\mathcal{T}_g$, *with boundary hypersurfaces corresponding to allowing geodesic lengths to converge to zero. Thus* $\widehat{\mathcal{T}}_g$ *is singular and points on the singular strata correspond to Riemann surfaces with geodesics pinched to points. The action of the 'mapping class group',* $\mathrm{Diff}^+(\Sigma_G)/\mathrm{Diff}^+_0(\Sigma_g)$, *extends from* $\mathcal{T}_g$

to $\widehat{\mathcal{T}}_g$ and the quotient yields the Deligne-Mumford (stratified) compactification of $\mathcal{M}_g$, see e.g., [63, 112].

Example 4 (Unstable manifolds). *Following [96], suppose M is a smooth manifold of dimension m, $f : M \longrightarrow \mathbb{R}$ a Morse function, and $p \in M$ is a critical point of f. Let V be a gradient-like function on M (e.g., the gradient of f with respect to a Riemannian metric) and let Φ_t denote the flow generated by $-V$. The stable/unstable manifolds associated to this data are respectively*

$$W_p^{\pm} = \{\zeta \in M : \lim_{t\to\pm\infty} \Phi_t(\zeta) = p\}.$$

These are both smooth manifolds and, up to homeomorphism,

$$W_p^- \cong \mathbb{R}^\lambda, \quad W_p^+ \cong \mathbb{R}^{m-\lambda},$$

where $\lambda = \lambda(f,p)$ is the Morse index of f at p. The flow satisfies the Morse-Smale transversality condition if the stable and unstable manifolds associated to all of the critical points of f intersect each other transversely.

Let

$$Y_k = \bigcup \{W_p^- : \lambda(f,p) = k\}.$$

Nicolaescu shows in [96, Chapter 4] that the decomposition of M into $\{Y_k\}$ is a stratification if and only if the flow satisfies the Morse-Smale transversality condition.

Example 5 (Homology classes). *If X is a topological space, we say that a homology class $h \in H_*(X)$ is realizable by a smooth manifold if there is a closed orientable manifold M and a continuous map $f : M \longrightarrow X$ such that*

$$h = f_*[M].$$

Steenrod's problem consists of describing the realizable homology classes when X is a smooth manifold. Thom [109] proved that all classes are realizable with $\mathbb{Z}_2$ or $\mathbb{Q}$ coefficients. He showed that all elements of $H_j(X;\mathbb{Z})$ are realizable for $j \leq 6$ and for $\dim X - 2 \leq j < \dim X$ and provided an explicit example of a non-realizable class. Goresky showed that you can represent every homology class of a smooth manifold by a stratified subset [52, 95].

(Thom's results do not hold if we impose conditions on f beyond being continuous, e.g., [55] shows that there are homology classes with $\mathbb{Z}_2$ coefficients that can not be realized by immersed submanifolds. See also [101, 29, 99, 22].)

Example 6 (Locally symmetric spaces). *Let $\mathbf{G}$ be a connected semisimple algebraic group defined over $\mathbb{Q}$, $G = \mathbf{G}(\mathbb{R})$ its real points, regarded as a Lie group, K a maximal compact subgroup of G, Γ a torsion-free arithmetic subgroup, and*

$$X = \Gamma \backslash G / K$$

the corresponding locally symmetric space. This is generally non-compact but there are many natural compactifications of X, see [25]. The reductive Borel-Serre compactification is a stratified space obtained by adding one stratum per Γ-conjugacy class of rational parabolic subgroups P of G. We will review this and related compactifications in §7.

2.2 Intersection homology and Witt spaces

One of the primary motivations for [53] was a question of Sullivan: what is the largest class of spaces with singularities to which the signature of manifolds extends as a cobordism invariant?

Recall that on a smooth oriented manifold Poincaré duality defines a non-degenerate pairing on (co)homology whose signature is the signature of the manifold. The singular homology of a stratified space does not satisfy Poincaré duality.

Example 7. *If M is a smooth manifold, its (unreduced) suspension is the quotient space*

$$SM = (M \times [0,1])/\{(m_1,0) \sim (m_2,0) \text{ and } (m_1,1) \sim (m_2,1) \text{ for all } m_i \in M\}$$

which is naturally a pseudomanifold with singular stratum the two cone points. A Mayer-Vietoris argument shows that

$$\widetilde{H}_k(SM) \cong \widetilde{H}_{k-1}(M), k \geq 0.$$

In particular if M is a two dimensional torus,

$$H_0(SM) = \mathbb{Z}, \quad H_1(SM) = 0, \quad H_2(SM) = \mathbb{Z} \oplus \mathbb{Z}, \quad H_3(SM) = \mathbb{Z}$$

and we see that Poincaré duality can not hold.

Goresky-MacPherson came up with the brave idea of redefining homology to recover Poincaré duality. The key is to keep track of the failure of transversality of the chains with the singular set. We keep track of this with a perversity function $\overline{p}$, a function from $\{2,3,\ldots\}$ to $\mathbb{N}_0$ such that $\overline{p}(c)$ and $\overline{q}(c) = 2 - c - \overline{p}(c)$ are both non-negative and non-decreasing. The perversity $\overline{q}$ is known as the dual perversity to $\overline{p}$. (For more general perversity functions, see [44].)

Intersection homology can be defined analogously to singular homology but maintaining control of the dimensions of the intersections of chains and singularities [70]. Let $\widehat{X}$ be a stratified space of dimension n and let Y_k be the union of the strata of dimension k. Given a perversity function $\overline{p}$ let the admissible chains, $AC_i^{\overline{p}}(\widehat{X})$, be the free group with generators the singular i-simplices

$$f : \Delta^i \longrightarrow \widehat{X} \text{ s.t. } f^{-1}(Y_{n-j}) \subseteq \; i - j + p(j)\text{-skeleton of } \Delta^i \text{ for all } j,$$

and let

$$SC_i^{\overline{p}}(\widehat{X}) = \{f \in AC_i^{\overline{p}}(\widehat{X}) : \partial f \in AC_{i-1}^{\overline{p}}(\widehat{X})\}.$$

The intersection homology groups of perversity $\overline{p}$ are the homology groups of the complex $SC_*^{\overline{p}}(\widehat{X})$,

$$\mathrm{IH}_i^{\overline{p}}(\widehat{X}) = H_i(SC_*^{\overline{p}}(\widehat{X})).$$

(For the definition of intersection homology through complexes of sheaves see our §5.5 below.)

Example 8. *Consider again the suspension of a smooth m-dimensional manifold M, SM. Here Y_0 consists of two points and the regular part is*

$$X = SM \setminus Y_0 = M \times (0,1).$$

A perversity consists of a single integer $p(m+1) \in [0, m-1]$, and the intersection homology is given by

$$\mathrm{IH}_k^{\overline{p}}(SM) = \begin{cases} H_k(M) & k < m - p(m+1), \\ 0 & k = m - p(m+1), \\ H_{k-1}(M) & k > m - p(m+1). \end{cases}$$

The intuition is that allowable chains avoid the suspension points in low dimensions, and so really map into X, but not in high dimensions (cf. [71, proof of Proposition 4.7.2]).

Goresky-MacPherson proved that the intersection pairing induces a pairing

$$\mathrm{IH}_i^{\overline{p}}(\widehat{X}) \times \mathrm{IH}_i^{\overline{q}}(\widehat{X}) \longrightarrow \mathbb{Z}$$

that is non-degenerate when tensored with the rational numbers. In general different perversities give rise to different groups. The closest dual perversities,

$$\overline{m} = (0,0,1,1,2,2,\ldots), \quad \overline{n} = (0,1,1,2,2,\ldots),$$

are known respectively as the upper and lower middle perversities and differ only with respect to strata of odd codimension. Thus if $\widehat{X}$ has only even codimensional strata (e.g., a complex variety) then $\mathrm{IH}_*^{\overline{m}}(\widehat{X})$ is dually paired with itself.

More generally, Siegel [106, Theorem I.3.4] showed that if every even dimensional link Z satisfies $\mathrm{IH}^{\overline{m}}_{\dim Z/2}(Z) = 0$, then $\mathrm{IH}_*^{\overline{m}}(\widehat{X})$ is dually paired with itself. He called these spaces *Witt spaces* and proved that the signature is a Witt-bordism invariant, i.e., if two Witt spaces are cobordant via a Witt space (with boundary) then their signatures coincide. Banagl [10] managed to extend Poincaré duality to stratified spaces that are not necessarily Witt by studying the category of sheaves that are self-dual and compatible with the sheaf-theoretic approach to intersection homology (see §5.5). Spaces that carry a self-dual sheaf of this type (not every space does) are known as *L-spaces.* Banagl mentions in [10] that the 'idea of employing Lagrangian subspaces in order to obtain self-duality' is invoked in unpublished work of Morgan and is also present in Cheeger's ideal boundary conditions. Indeed in [1] we show that a Thom-Mather stratified space is an L-space if and only if it is a Cheeger space.

2.3 L^2-cohomology

If (M, g) is a Riemannian manifold, the metric defines a volume form and a bundle metric on the exterior powers of the cotangent bundle, $\Lambda^\bullet T^*M$. The space of L^2 differential forms can be defined to be the closure of $\mathcal{C}_c^\infty(M; \Lambda^\bullet T^*M)$ with respect to the norm

$$\|\omega\|_{L^2}^2 = \int_M |\omega(\zeta)|_g^2 \, \mathrm{dvol}(\zeta)$$

and is then a Hilbert space. The exterior derivative is an unbounded operator

$$d : \mathcal{C}_c^\infty(M; \Lambda^\bullet T^*M) \longrightarrow \mathcal{C}_c^\infty(M; \Lambda^\bullet T^*M)$$

and its formal adjoint is the operator

$$\delta : \mathcal{C}_c^\infty(M; \Lambda^\bullet T^*M) \longrightarrow \mathcal{C}_c^\infty(M; \Lambda^\bullet T^*M)$$

characterized by

$$\langle d\omega, \eta \rangle_{L^2} = \langle \omega, \delta\eta \rangle_{L^2}.$$

If ω and η are L^2 differential forms we say that the $d\omega = \eta$ holds weakly (or distributionally) if

$$\langle \omega, \delta\alpha \rangle_{L^2} = \langle \eta, \alpha \rangle_{L^2} \text{ for all } \alpha \in \mathcal{C}_c^\infty(M; \Lambda^\bullet T^*M).$$

The maximal domain of d is

$$\mathcal{D}_{\max}(d) = \{\omega \in L^2(M; \Lambda^\bullet T^*M) : d\omega \in L^2(M; \Lambda^\bullet T^*M)\}$$

where $d\omega$ is computed distributionally. If we denote the k-forms in the maximal domain of d by

$$\mathcal{D}_{\max}(d)_k = \mathcal{D}_{\max}(d) \cap L^2(M; \Lambda^k T^*M)$$

then we obtain a complex

$$0 \longrightarrow \mathcal{D}_{\max}(d)_0 \xrightarrow{d} \mathcal{D}_{\max}(d)_1 \xrightarrow{d} \cdots \xrightarrow{d} \mathcal{D}_{\max}(d)_{\dim M} \longrightarrow 0.$$

The cohomology of this complex is the L^2-cohomology of M (see, e.g., [34] for an introduction to L^2-cohomology).

If (M, g) is not complete then typically there are many closed domains for d other than $\mathcal{D}_{\max}(d)$. A *Hilbert complex for* d consists of a choice of domains $\mathcal{D}$ such that $(d, \mathcal{D})$ is a closed operator and

$$0 \longrightarrow \mathcal{D}(d)_0 \xrightarrow{d} \mathcal{D}(d)_1 \xrightarrow{d} \cdots \xrightarrow{d} \mathcal{D}(d)_{\dim M} \longrightarrow 0$$

forms a complex. An excellent study of Hilbert complexes is carried out in [31].

One of the motivations behind [4, 5, 1] was to find an analytic version of Banagl's cohomology groups by defining appropriate Hilbert complexes. As mentioned above, Cheeger showed in [37, §6] that the L^2-cohomology groups coincide with intersection cohomology with perversity $\overline{m}$. He also came across the Witt condition independently in [36] and described ideal boundary conditions for isolated conic singularities as a way of recovering Poincaré duality. We explain below (§3.4) how to use the analytic properties of the de Rham operator $\eth_{\mathrm{dR}}$ to impose ideal boundary conditions, and then how to use the Kodaira decomposition and these ideal boundary conditions to define Hilbert complexes (§3.5).

Let us mention some related but different approaches. Signatures for general pseudomanifolds are defined in [65, 45, 15]. These can not be bordism invariant (since the cone over a smooth manifold would exhibit every smooth manifold as a boundary) and it would be interesting to understand their relation with Cheeger spaces. There are also analytic studies of other perversity functions in, e.g., [65, 14].

We mention a few instances of the vast literature on the L^2 cohomology of complete metrics. Already in [8], Atiyah, Patodi, and Singer explain that the 'correct analytical context' for their signature theorem is the signature of the L^2 harmonic forms on a manifold with a cylindrical end. Zucker [119], in order to analyze the variation of Hodge structure on a degenerating family of curves, worked out the L^2 cohomology on a manifold with ends modeled on the Poincaré metric of a punctured disk; this *is* the intersection cohomology with coefficients. Still in the setting of complete metrics, Zucker [120] conjectured that the L^2 cohomology of a Hermitian locally symmetric space arising from the real points of a semisimple connected algebraic group defined over $\mathbb{Q}$ would equal the intersection homology of its Baily-Borel-Satake compactification (see [25]), which is a singular space. This celebrated conjecture was verified independently by Looijenga and Saper-Stern [81, 103]. Among many others, let us list some of the results which are closest in spirit to the subject and methods here

- the L^2 cohomology of spaces with asymptotically cylindrical ends was determined by Melrose [90],
- cusp ends by Müller [94],
- asymptotically geometrically finite hyperbolic quotients in [84, 86] by Mazzeo and Phillips,
- flat ends by Carron [35],
- asymptotically fibered cusp ends and asymptotically fibered boundary ends by Hausel, Hunsicker, and Mazzeo [58],
- foliated analogues of the latter by Gell-Redman and Rochon [49], and
- iterated fibered cusp metrics on stratified spaces by Hunsicker and Rochon [66].

Finally we point out that closely related to the L^2-cohomology is the study of index theorems. Cheeger's study of L^2-cohomology was part of his wider study of the spectral geometry of singular spaces in [38] and found application to index theory problems even on smooth spaces cf. [39, 17, 18, 19, 20]. There has been a lot of work on index theory on spaces with isolated conic singularities e.g., [42, 43, 32, 79] and recently also on index theory on spaces with a single singular stratum [7, 30, 40, 2].

3 Depth one stratifications

A pseudomanifold $\widehat{X}$ is said to have *depth one* if there is a single singular stratum. Thus there is one regular stratum, $\widehat{X}^{\text{reg}}$, and one singular stratum, Y. We denote the link of $\widehat{X}$ at a point in Y by Z, and we will denote dimensions by

$$n = \dim \widehat{X}^{\text{reg}}, \quad h = \dim Y, \quad v = \dim Z$$

related by $n = h + v + 1$. We assume that $\widehat{X}$ is compact and, for notational convenience, also that $\widehat{X}^{\text{reg}}$, Y, and Z are connected. This situation presents most of the analytic complications of the general case but has the advantage that the geometry is much simpler than the general case. There are many treatments of analysis in this setting, for example [87, 50, 78, 74, 77, 75, 104].

Our main object of study is the operator $d + \delta$ on differential forms on $\widehat{X}^{\text{reg}}$. We will see that the analysis of this operator is facilitated by first resolving the geometry.

3.1 Resolving the geometry

Let us describe our geometric setup in detail. We have a smooth open manifold $\widehat{X}^{\text{reg}}$ and a compactification of it to a singular space

$$\widehat{X} = \widehat{X}^{\text{reg}} \cup Y.$$

The smooth manifold Y has a tubular neighborhood $\mathcal{T} \subseteq \widehat{X}$ that fibers over Y with fiber the cone over a smooth manifold Z,

$$C(Z) - \mathcal{T} \longrightarrow Y.$$

Let r be a radial variable for the cone and let

$$\widetilde{X} = \widehat{X} \setminus \{r \leq \varepsilon\}.$$

For sufficiently small $\varepsilon > 0$ this is a smooth manifold with boundary $\partial\widetilde{X} = \{r = \varepsilon\}$ and, up to diffeomorphism, is independent of the choice of ε. The boundary of $\widetilde{X}$ inherits a smooth fiber bundle structure

$$Z - \partial\widetilde{X} \xrightarrow{\phi} Y.$$

The interior of $\widetilde{X}$ is naturally identified with $\widehat{X}^{\text{reg}}$ and so $\widetilde{X}$ is a compactification of $\widehat{X}^{\text{reg}}$ to a smooth manifold with boundary. A key fact for us is that analysis on $\widehat{X}$ refers to analysis on $\widehat{X}^{\text{reg}}$ and hence can equally well be carried out on $\widetilde{X}$, with the advantage that this is a smooth manifold.

A *rigid wedge metric* is a Riemannian metric on $\widehat{X}^{\text{reg}} = \widetilde{X}^{\circ}$ that, in a collar neighborhood of $\partial\widetilde{X}$, takes the form

$$g = dr^2 + r^2 g_Z + \phi^* g_Y$$

with g_Z and g_Y independent of r. (Precisely, it has this form after we choose a diffeomorphism of a collar neighborhood with $[0, \varepsilon)_r \times \partial\widetilde{X}$ and then we extend ϕ to this product in the obvious way.) This metric is ill-behaved as we approach the boundary of $\widetilde{X}$, but there is a natural way to resolve this. Let us denote the induced bundle metric on the cotangent bundle still by g, and define

$$\mathcal{E} = \{\omega \in \mathcal{C}^\infty(\widetilde{X}; T^*\widetilde{X}) : |\omega|_g \text{ is bounded } \}.$$

It turns out that the space $\mathcal{E}$ is the space of sections of a new bundle and g defines a non-degenerate bundle metric on this bundle.

Precisely, there is a bundle which we call the *wedge cotangent bundle* and denote ${}^wT^*\widetilde{X}$, together with a map

$$j : {}^wT^*\widetilde{X} \longrightarrow T^*\widetilde{X},$$

such that

$$j \circ \mathcal{C}^\infty(\widetilde{X}; {}^wT^*\widetilde{X}) = \mathcal{E} \subseteq \mathcal{C}^\infty(\widetilde{X}; T^*\widetilde{X}).$$

We can show this either abstractly, through the Serre-Swan theorem, or concretely as in [90, Proposition 8.1] by showing that the vector spaces

$${}^wT^*_\zeta\widetilde{X} = \mathcal{E}/(\mathcal{J}_\zeta \cdot \mathcal{E}),$$

where $\mathcal{J}_\zeta = \{f \in \mathcal{C}^\infty(\widetilde{X}) : f(\zeta) = 0\}$, inherit a vector bundle structure from $T^*\widetilde{X}$ and a map j from the inclusion of $\mathcal{E}$ into $\mathcal{C}^\infty(\widetilde{X}; T^*\widetilde{X})$. The map j is an isomorphism over $\widetilde{X}^\circ$, but over $\partial\widetilde{X}$ fits into a short exact sequence

$$0 \longrightarrow T^*(\partial\widetilde{X}/Y) \longrightarrow {}^wT^*\widetilde{X}\big|_{\partial\widetilde{X}} \xrightarrow{\ j\ } {}^wN^*\partial\widetilde{X} \cong N^*\partial\widetilde{X} \oplus \phi^*T^*Y \longrightarrow 0$$

of bundles over $\partial\widetilde{X}$ (here $T^*\partial\widetilde{X}/Y\big|_{\phi^{-1}(y)} \cong T^*Z_y$, see §3.2.1).

In local coordinates trivializing the fiber bundle ϕ in a neighborhood of a point $q \in \partial\widetilde{X}$,

$$r, y_1, \ldots, y_h, z_1, \ldots, z_v, \tag{3.1}$$

a section of $T^*\widetilde{X}$ is locally spanned by dr, dy_i, dz_j, while a section of ${}^wT^*\widetilde{X}$ is locally spanned by

$$dr, dy_1, \ldots, dy_h, r\, dz_1, \ldots, r\, dz_v.$$

Crucially, $r\, dz_j$ does not vanish at $\partial\widetilde{X}$ as a section of ${}^wT^*\widetilde{X}$ since it is not equal to r times a section of ${}^wT^*\widetilde{X}$.

The replacement of $T^*\widetilde{X}$ by ${}^wT^*\widetilde{X}$ is very convenient. On the one hand, as anticipated above, the definition of $\mathcal{E}$ implies that the metric g extends from $\widetilde{X}^\circ$ to a non-degenerate metric over all of $\widetilde{X}$. On the other hand, let us consider the effect of this change on the behavior of $d + \delta$ near $\widetilde{X}$. For simplicity of comparing $d + \delta$ in these two settings, let us briefly assume that Y is a single point, i.e., that $\widehat{X}$ has an isolated conic singularity. After a choice of collar neighborhood of the boundary $c(\partial\widetilde{X})$, we have

$$\mathcal{C}^\infty(c(\partial\widetilde{X}); \Lambda^i(T^*\widetilde{X})) = \mathcal{C}^\infty(\partial\widetilde{X}; \Lambda^iT^*Z) \oplus dr \wedge \mathcal{C}^\infty(\partial\widetilde{X}; \Lambda^{i-1}T^*Z),$$
$$\mathcal{C}^\infty(c(\partial\widetilde{X}); \Lambda^i({}^wT^*\widetilde{X})) = \mathcal{C}^\infty(\partial\widetilde{X}; r^i\Lambda^iT^*Z) \oplus dr \wedge \mathcal{C}^\infty(\partial\widetilde{X}; r^{i-1}\Lambda^{i-1}T^*Z)$$

and correspondingly

$$\text{on } \mathcal{C}^\infty(c(\partial\widetilde{X}); \Lambda^i(T^*\widetilde{X})), \quad d+\delta = \begin{pmatrix} d^Z + \frac{1}{r^2}\delta^Z & -\partial_r + \frac{2i-v-1}{r} \\ \partial_r & -d^Z - \frac{1}{r^2}\delta^Z \end{pmatrix},$$

$$\text{on } \mathcal{C}^\infty(c(\partial\widetilde{X}); \Lambda^i({}^wT^*\widetilde{X})), \quad d+\delta = \begin{pmatrix} \frac{1}{r}(d^Z + \delta^Z) & -\partial_r + \frac{i-v}{r} \\ \partial_r + \frac{i}{r} & -\frac{1}{r}(d^Z + \delta^Z) \end{pmatrix}.$$

There are two things to note about these expressions. First the form of the metric is going to lead to operators with coefficients that blow-up at the boundary no matter how we look at it. Secondly, when we work with ${}^wT^*\widetilde{X}$, the link Z enters the description of the de Rham operator $d+\delta$ only through its de Rham operator $d^Z + \delta^Z$. This is very convenient for inductive arguments!

(A related approach, espoused in, e.g., [32, 30] is to conjugate the de Rham operator on exterior powers of $T^*\widetilde{X}$ with a suitable scaling operator; we realize this rescaling at the level of bundles.)

3.2 Model operators of the de Rham operator

Having resolved the geometry by replacing the singular space $\widehat{X}$ with the smooth manifold with boundary $\widetilde{X}$, and resolved the degeneracy of the metric by replacing the cotangent bundle $T^*\widetilde{X}$ with the wedge cotangent bundle ${}^wT^*\widetilde{X}$, we now proceed to study the de Rham operator $d+\delta$.

3.2.1 Cohomology of fiber bundles

Since the boundary of $\widetilde{X}$ is the total space of a fiber bundle, let us briefly review the behavior of d and δ on fiber bundles (cf. [21], [16, §10.1] [58, §5.3]). Let

$$F - H \xrightarrow{\varphi} B$$

be a fiber bundle of smooth closed manifolds. The functions $\varphi^*\mathcal{C}^\infty(B) \subseteq \mathcal{C}^\infty(H)$ are called horizontal, and a vector field on H is called vertical if it annihilates horizontal functions. Vertical vector fields are the sections of a vector bundle $TH/B \longrightarrow H$, which participates in a short exact sequence of bundles

$$0 \longrightarrow TH/B \longrightarrow TH \longrightarrow \varphi^*TB \longrightarrow 0.$$

The Lie bracket of two vertical vector fields is again a vertical vector field and hence the Cartan formula defines a vertical exterior derivative on the exterior powers of the bundle T^*H/B dual to TH/B,

$$d^Z : \mathcal{C}^\infty(H; \Lambda^k T^*H/B) \longrightarrow \mathcal{C}^\infty(H; \Lambda^{k+1}T^*H/B).$$

The cohomology of the resulting complex, $\mathrm{H}^*_{\mathrm{dR}}(H/B)$, is called the vertical de Rham cohomology and forms a vector bundle over B with typical fiber $\mathrm{H}^*_{\mathrm{dR}}(F)$. Indeed, given a transition function $G : \mathcal{U} \longrightarrow \mathrm{Diff}(F)$ for φ, defined on a neighborhood $\mathcal{U} \subseteq B$, there is an induced map on cohomology

$$G^* : \mathcal{U} \longrightarrow \mathrm{End}(\mathrm{H}^*_{\mathrm{dR}}(F))$$

giving a transition function for $\mathrm{H}^*_{\mathrm{dR}}(H/B)$. Moreover if $\mathcal{U}$ is connected then homotopy invariance of de Rham cohomology implies that this transition function is constant, which exhibits $\mathrm{H}^*_{\mathrm{dR}}(H/B)$ as a flat vector bundle.

Now, let $g_H = g_{H/B} + \varphi^* g_B$ be a Riemannian submersion metric on H with corresponding splittings

$$TH = TH/B \oplus \varphi^* TB, \quad T^*H = T^*H/B \oplus \varphi^* T^*B,$$
$$\Lambda^k T^*H = \bigoplus_{i+j=k} \Lambda^{i,j} T^*H, \text{ where } \Lambda^{i,j}T^*H = \Lambda^i T^*H/B \otimes \Lambda^j \varphi^* T^*B.$$

We refer to i in $\Lambda^{i,j}T^*H$ as the *vertical form degree.* The exterior derivative interacts with this splitting via

$$\begin{aligned} d : \mathcal{C}^\infty(H; \Lambda^{i,j}T^*H) &\longrightarrow \mathcal{C}^\infty(H; \Lambda^{i+1,j}T^*H) \oplus \mathcal{C}^\infty(H; \Lambda^{i,j+1}T^*H) \\ &\qquad \oplus \mathcal{C}^\infty(H; \Lambda^{i-1,j+2}T^*H), \\ &\qquad d = d^Z + \widetilde{d}^B + R \end{aligned}$$

where the first summand is the vertical exterior derivative, the second summand extends the exterior derivative of B from sections of $\Lambda^{0,*}T^*H$, and the third is a tensorial map built up from the curvature of φ. In the special case where $H = F \times B$ and φ is the projection onto B, this decomposition is just $d = d^Z + d^B$. There is a corresponding splitting of the adjoint of d,

$$\begin{aligned} \delta : \mathcal{C}^\infty(H; \Lambda^{i,j}T^*H) &\longrightarrow \mathcal{C}^\infty(H; \Lambda^{i-1,j}T^*H) \oplus \mathcal{C}^\infty(H; \Lambda^{i,j-1}T^*H) \\ &\qquad \oplus \mathcal{C}^\infty(H; \Lambda^{i+1,j-2}T^*H), \\ &\qquad \delta = \delta^Z + \widetilde{\delta}^B + R^* \end{aligned}$$

which reduces to $\delta^Z + \delta^B$ when $H = F \times B$.

Next let us point out that $\widetilde{d}^B$ induces a flat connection, ∇^{H}, on $\mathrm{H}^\bullet_{\mathrm{dR}}(H/B) \longrightarrow B$ compatible with the flat structure described above. Indeed, we have various relationships among d^Z, $\widetilde{d}^B$, and R stemming from the fact that their sum squares to zero, including

$$d^Z \widetilde{d}^B + \widetilde{d}^B d^Z = 0, \quad (\widetilde{d}^B)^2 + d^Z R + R d^Z = 0.$$

If $\omega \in \mathcal{C}^\infty(H; \Lambda^k T^*H/Z)$ then

$$d^Z \widetilde{d}^B \omega = -\widetilde{d}^B d^Z \omega$$

shows that $\widetilde{d}^B$ preserves both the image and the null space of d^Z, and hence induces a map on $\mathrm{H}^\bullet_{\mathrm{dR}}(H/B)$. The Leibnitz rule of d implies that

$$\widetilde{d}^B((\varphi^* f)\omega) = (\varphi^* df) \wedge \omega + \varphi^* f \widetilde{d}^B \omega$$

which shows that the induced map on $\mathrm{H}^\bullet_{\mathrm{dR}}(H/B)$ is a connection. Finally, if $d^Z \omega = 0$, then

$$(\widetilde{d}^B)^2 \omega = d^Z(-R\omega)$$

which shows that the induced connection on $\mathrm{H}^\bullet_{\mathrm{dR}}(H/B)$ is flat. Since this connection coincides with d^B over trivializations of the fiber bundle φ, it follows that this flat structure is the same as the one above.

Finally note that the Hodge isomorphism on the fibers of φ, $\mathrm{H}^\bullet_{\mathrm{dR}}(F) \cong \mathcal{H}^\bullet(F)$, can be used to import the flat bundle structure to the vertical Hodge cohomology

$$\mathcal{H}^\bullet(H/B) \longrightarrow B.$$

We denote the flat connection on this bundle by $\nabla^{\mathcal{H}}$.

3.2.2 Model operators

Now let us return to $\widetilde{X}$, with boundary fibration

$$Z - \partial\widetilde{X} \xrightarrow{\phi} Y.$$

Let

$$c : [0,\varepsilon)_r \times \partial\widetilde{X} \longrightarrow \widetilde{X}$$

be a collar neighborhood of $\partial\widetilde{X}$, extend ϕ trivially to $[0,\varepsilon)_r \times \partial\widetilde{X}$, and assume that g is a wedge metric that is rigid on $c(\widetilde{\partial X})$ so that $c^*g = dr^2 + r^2 g_Z + \phi^* g_Y$. We abbreviate $c([0,\varepsilon)_r \times \partial\widetilde{X})$ by $c(\partial\widetilde{X})$.

Near $\partial\widetilde{X}$ we have a splitting

$${}^wT^*\widetilde{X}\big|_{c(\partial\widetilde{X})} = N^*\partial\widetilde{X} \oplus rT^*\partial\widetilde{X}/Y \oplus T^*Y$$

and we will make frequent use of the decomposition

$$\begin{aligned}&\mathcal{C}^\infty(c(\partial\widetilde{X}); \Lambda^\bullet({}^wT^*\widetilde{X}))\\ &= \mathcal{C}^\infty(c(\partial\widetilde{X}); \Lambda^\bullet(rT^*\partial\widetilde{X}/Y \oplus T^*Y)) \oplus dr \wedge \mathcal{C}^\infty(c(\partial\widetilde{X}); \Lambda^\bullet(rT^*\partial\widetilde{X}/Y \oplus T^*Y)).\end{aligned}$$

With respect to this decomposition, the de Rham operator $\eth_{\mathrm{dR}} = d + \delta$ of g takes the form

$$\eth_{\mathrm{dR}} = \begin{pmatrix} \frac{1}{r}\eth^Z_{\mathrm{dR}} + \widetilde{\eth}^Y_{\mathrm{dR}} + r\mathbf{R} & -\partial_r + \frac{1}{r}(\mathbf{N} - v) \\ \partial_r + \frac{1}{r}\mathbf{N} & -\frac{1}{r}\eth^Z_{\mathrm{dR}} - \widetilde{\eth}^Y_{\mathrm{dR}} - r\mathbf{R} \end{pmatrix} \tag{3.2}$$

where, in the notation of §3.2.1, $\widetilde{\eth}^Y_{\mathrm{dR}} = \widetilde{d}^Y + \widetilde{\delta}^Y$, $\mathbf{R} = R + R^*$, and $\mathbf{N}$ is the 'vertical number operator', i.e., the integer valued map with the property that

$$\mathbf{N}\big|_{\Lambda^i N^*\partial\widetilde{X} \wedge \Lambda^j(rT^*\partial\widetilde{X}/Y) \wedge \Lambda^k(T^*Y)} = j.$$

The local coordinate expression exhibits $\eth_{\mathrm{dR}}$ as an example of a *wedge differential operator* (see, e.g., [3, 76]). A vector field on $\widetilde{X}$ is said to be an *edge vector field* if its restriction to $\partial\widetilde{X}$ is tangent to the fibers of ϕ. The space of edge vector fields is denoted $\mathcal{V}_e$ and its enveloping algebra is the space of *edge differential operators*

[85]. Put another way, a differential operator of order m is an edge differential operator if in every choice of local coordinates like (3.1) it has the form

$$\sum_{j+|\alpha|+|\beta|\leq m} a_{j,\alpha,\beta}(r,y,z)(r\partial_r)^j(r\partial_y)^\alpha\partial_z^\beta$$

where α and β are multi-indices and the coefficients $a_{j,\alpha,\beta}$ are smooth on all of $\widetilde{X}$. A differential operator L of order m is a *wedge differential operator* if $r^m L$ is an edge differential operator. Thus, since the above expression shows that $r\eth_{\mathrm{dR}}$ can be written in terms of the vector fields $r\partial_r$, $r\partial_y$, ∂_z, we can conclude that $\eth_{\mathrm{dR}}$ is first order wedge differential operator. As such, its analysis proceeds by understanding three model operators known as the (wedge) principal symbol, the normal operator, and the indicial operator.

Just as the principal symbol of a differential operator on a closed manifold is a section defined on the cotangent bundle, the principal symbol of $\eth_{\mathrm{dR}}$ as a wedge differential operator will be a section of $\mathrm{End}(\Lambda^\bullet({}^wT^*\widetilde{X}))$ over ${}^wT^*\widetilde{X}$. We can find this symbol in one of the standard ways, given $\xi \in {}^wT_p^*\widetilde{X}$ choose a function $f \in \mathcal{C}^\infty(\widetilde{X})$ such that $df(p) = \xi$ and define

$$\sigma(\eth_{\mathrm{dR}})(\xi) = [\eth_{\mathrm{dR}}, f](p) \in \mathrm{End}(\Lambda^\bullet({}^wT^*\widetilde{X})).$$

It is easy to see that

$$(\sigma(\eth_{\mathrm{dR}})(\xi))^2 = |\xi|_g^2 \,\mathrm{Id} \in \mathrm{End}(\Lambda^\bullet({}^wT^*\widetilde{X}))$$

so we see that $\eth_{\mathrm{dR}}$ is *uniformly elliptic* on $\widetilde{X}$ as a wedge differential operator.

The normal operator of $\eth_{\mathrm{dR}}$ is a family of operators parametrized by $q \in Y$. At each point $q \in Y$ we look at then model wedge $\mathbb{R}_r^+ \times Z_q \times T_qY$ with the metric $dr^2 + r^2 g_{Z_q} + g_{T_qY}$ obtained from g by freezing coefficients at q. The de Rham operator of this metric is given by

$$N_q(\eth_{\mathrm{dR}}) = \begin{pmatrix} \frac{1}{r}\eth_{\mathrm{dR}}^{Z_q} + \eth_{\mathrm{dR}}^{\mathbb{R}^h} & -\partial_r + \frac{1}{r}(\mathbf{N} - v) \\ \partial_r + \frac{1}{r}\mathbf{N} & -\frac{1}{r}\eth_{\mathrm{dR}}^{Z_q} - \eth_{\mathrm{dR}}^{\mathbb{R}^h} \end{pmatrix}, \tag{3.3}$$

known as the *normal operator of* $\eth_{\mathrm{dR}}$ *at* $q \in Y$, and it is these operators that model the behavior of $\eth_{\mathrm{dR}}$ near $\partial\widetilde{X}$. A key fact, already mentioned above in the case of conic singularities, is that *the link* Z_q *only enters through its de Rham operator,* $\eth_{\mathrm{dR}}^{Z_q}$.

The final model operator, the indicial family, determines the leading order behavior of elements in the null space of $\eth_{\mathrm{dR}}$. Assume that we have an element of the null space that in local coordinates has the form $r^\zeta a(y,z) + \mathcal{O}(r^{\zeta+1})$ for some $\zeta \in \mathbb{C}$. Applying $\eth_{\mathrm{dR}}$ yields

$$\eth_{\mathrm{dR}}(r^\zeta a(y,z) + \mathcal{O}(r^{\zeta+1})) = r^{\zeta-1}\begin{pmatrix} \eth_{\mathrm{dR}}^Z & -\zeta + \mathbf{N} - v \\ \zeta + \mathbf{N} & -\eth_{\mathrm{dR}}^Z \end{pmatrix} a(y,z) + \mathcal{O}(r^\zeta)$$

and in order for this to vanish we need to have $a(q,\cdot)$ in the null space of the *indicial family of* $\eth_{\mathrm{dR}}$ *at* $q \in Y$, $\zeta \in \mathbb{C}$,

$$I_q(\eth_{\mathrm{dR}};\zeta) = \begin{pmatrix} \eth_{\mathrm{dR}}^Z & -\zeta + \mathbf{N} - v \\ \zeta + \mathbf{N} & -\eth_{\mathrm{dR}}^Z \end{pmatrix}. \tag{3.4}$$

For each q the set of ζ for which $I_q(\eth_{\mathrm{dR}};\zeta)$ has a non-trivial null space are called the *indicial roots* or *boundary spectrum* of $\eth_{\mathrm{dR}}$ and denoted

$$\operatorname{spec}_b(\eth_{\mathrm{dR}};q).$$

Ellipticity implies that, for a fixed q, this is a discrete subset of $\mathbb{C}$.

3.3 The maximal and minimal domain

The model operators described above can be used to understand the behavior of $\eth_{\mathrm{dR}}$ on various different function spaces. For our purposes the most important will be (a) the L^2 space of differential forms determined by the Riemannian metric and (b) the edge Sobolev spaces. The latter are defined by

$$\begin{aligned} &H_e^m(\widetilde{X};\Lambda^\bullet({}^wT^*\widetilde{X})) = \{\omega \in L^2(\widetilde{X};\Lambda^\bullet({}^wT^*\widetilde{X})) : \\ &\quad P\omega \in L^2(\widetilde{X};\Lambda^\bullet({}^wT^*\widetilde{X})) \text{ for every edge differential operator } P \text{ of order } m\}. \end{aligned}$$

In fact an elliptic edge differential operator has a unique (closed) domain as an operator on L^2, namely the appropriate edge Sobolev space. As we will now explore, this is usually not true for wedge differential operators such as $\eth_{\mathrm{dR}}$.

The de Rham operator unambiguously defines an operator

$$\eth_{\mathrm{dR}} : \mathcal{C}_c^\infty(\widetilde{X};\Lambda^\bullet({}^wT^*\widetilde{X})) \longrightarrow \mathcal{C}_c^\infty(\widetilde{X};\Lambda^\bullet({}^wT^*\widetilde{X})),$$

but if we wish to study $\eth_{\mathrm{dR}}$ as an operator on L^2-differential forms we need to be careful about its domain. Recall that an unbounded operator is said to be *closed* if its graph is a closed set, and that a well-behaved operator (e.g., Fredholm or even non-empty resolvent set) is necessarily closed. There are two canonical closed extensions of $\eth_{\mathrm{dR}}$ from smooth compactly supported forms. First there is the *minimal domain*,

$$\begin{aligned} &\mathcal{D}_{\min}(\eth_{\mathrm{dR}}) = \{\omega \in L^2(\widetilde{X};\Lambda^\bullet({}^wT^*\widetilde{X})) : \\ &\quad \exists(\omega_n) \subseteq \mathcal{C}_c^\infty(\widetilde{X};\Lambda^\bullet({}^wT^*\widetilde{X})) \text{ s.t. } \omega_n \to \omega \text{ in } L^2 \text{ and } \eth_{\mathrm{dR}}\omega_n \text{ is } L^2\text{-Cauchy}\}, \end{aligned}$$

on which we define $\eth_{\mathrm{dR}}\omega = \lim \eth_{\mathrm{dR}}\omega_n$. This domain is characterized by the fact that the graph of $(\eth_{\mathrm{dR}}, \mathcal{D}_{\min}(\eth_{\mathrm{dR}}))$ is the closure of the graph of $(\eth_{\mathrm{dR}}, \mathcal{C}_c^\infty(\widetilde{X};\Lambda^\bullet({}^wT^*\widetilde{X})))$. Secondly, there is the *maximal domain*,

$$\mathcal{D}_{\max}(\eth_{\mathrm{dR}}) = \{\omega \in L^2(\widetilde{X};\Lambda^\bullet({}^wT^*\widetilde{X})) : \eth_{\mathrm{dR}}\omega \in L^2(\widetilde{X};\Lambda^\bullet({}^wT^*\widetilde{X}))\}$$

where $\eth_{\mathrm{dR}}\omega$ is computed distributionally. Clearly, if $\mathcal{D} \subseteq L^2(\widetilde{X};\Lambda^\bullet({}^wT^*\widetilde{X}))$ is a closed domain for $\eth_{\mathrm{dR}}$ containing $\mathcal{C}_c^\infty(\widetilde{X};\Lambda^\bullet({}^wT^*\widetilde{X}))$ then it must satisfy

$$\mathcal{D}_{\min}(\eth_{\mathrm{dR}}) \subseteq \mathcal{D} \subseteq \mathcal{D}_{\max}(\eth_{\mathrm{dR}})$$

justifying the labels 'minimal' and 'maximal'.

These domain are related to the edge Sobolev space of order one by the fact that $r\eth_{\mathrm{dR}}$ is an elliptic edge operator. In fact this yields that

$$rH^1_e(\widetilde{X};\Lambda^\bullet({}^wT^*\widetilde{X})) \subseteq \mathcal{D}_{\min}(\eth_{\mathrm{dR}}) \subseteq \mathcal{D}_{\max}(\eth_{\mathrm{dR}}) \subseteq H^1_e(\widetilde{X};\Lambda^\bullet({}^wT^*\widetilde{X})).$$

It turns out that the key to distinguishing between closed extensions of $\eth_{\mathrm{dR}}$ lies in the indicial operator. To see intuitively why this happens, suppose we have a differential form supported near the boundary of the form $r^\zeta a(y,z)$. The volume form of a wedge metric in local coordinates is essentially of the form $r^v\, drdydz$ so that

$$r^\zeta a(y,z) \in L^2 \iff \operatorname{Re}\zeta + \tfrac{1}{2}(v+1) > 0.$$

Now if we also want to have

$$\eth_{\mathrm{dR}}(r^\zeta a(y,z)) = r^{\zeta-1} I_y(\eth_{\mathrm{dR}};\zeta)a(y,z) + \mathcal{O}(r^\zeta)$$

in L^2 we will need to have

$$\operatorname{Re}\zeta + \tfrac{1}{2}(v+1) > 1 \quad \text{or} \quad \operatorname{Re}\zeta + \tfrac{1}{2}(v+1) > 0 \text{ and } I_y(\eth_{\mathrm{dR}};\zeta)a(y,z) = 0.$$

This suggests that a key rôle will be played by

$$\operatorname{spec}_b(\eth_{\mathrm{dR}};q) \cap \{\zeta : 0 < \operatorname{Re}\zeta + \tfrac{1}{2}(v+1) \le 1\},$$

and we now make an extra assumption on the metric in order to make this set as simple as possible. Directly from (3.4) it is easy to see that the set $\operatorname{spec}_b(\eth_{\mathrm{dR}};q)$ can be determined from the spectrum of $\eth^{Z_q}_{\mathrm{dR}}$, the de Rham operator on Z_q with respect to the metric g_{Z_q}. By rescaling this metric, we can assume that there is an interval around zero with no non-zero eigenvalues of $\eth^{Z_q}_{\mathrm{dR}}$ for any $q \in Y$. We will say that a wedge metric is *suitably scaled* if

$$\operatorname{spec}_b(\eth_{\mathrm{dR}};q) \cap \{\zeta : 0 \le \operatorname{Re}\zeta + \tfrac{1}{2}(v+1) \le 1\} \subseteq \begin{cases} \{-\frac{1}{2}(v)\} & \text{if } v \text{ is even,} \\ \{-\frac{1}{2}(v \pm 1)\} & \text{if } v \text{ is odd.} \end{cases}$$

These are the indicial roots that come from the null space of $\eth^{Z}_{\mathrm{dR}}$ and can not be removed by scaling the metric. Note that the null space of the indicial operator at these roots is cohomological, e.g.,

$$\ker(I_q(\eth_{\mathrm{dR}};-\tfrac{v}{2})) = \mathcal{H}^{v/2}(Z_q) \oplus \mathcal{H}^{v/2}(Z_q) = \{\text{harmonic forms on } Z_q \text{ of degree } v/2\}^2,$$

and so these null spaces fit together to form a bundle over Y, isomorphic to two copies of the flat bundle described in §3.2.1.

The intuitive analysis about the leading terms of elements in the maximal domain from the last paragraph can be made precise, but it requires distributional asymptotic expansions [4, Lemma 2.1], cf. [85, Theorem 7.3], [74, §5], [92]:

Lemma 3.1. *If g is a suitably scaled wedge metric on $\widetilde{X}$, then any $\omega \in \mathcal{D}_{\max}(\eth_{\mathrm{dR}})$ has a partial distributional asymptotic expansion*

$$\omega \sim r^{-v/2}(\alpha(\omega) + dr \wedge \beta(\omega)) + \widetilde{\omega}$$

where $\alpha,\beta \in H^{-1/2}(Y;\Lambda^\bullet T^*Y \otimes \mathcal{H}^{v/2}(\partial\widetilde{X}/Y))$, $\widetilde{\omega} \in \bigcap_{\varepsilon>0} r^{1-\varepsilon} H^{-1}_e(\widetilde{X};\Lambda^\bullet({}^wT^*\widetilde{X}))$

in the sense that the L^2-pairing of ω with a sufficiently regular form is equal to the distributional pairing with this expansion. If v is odd or $\mathrm{H}^{v/2}_{\mathrm{dR}}(Z) = 0$ (i.e., if $\widehat{X}$ is Witt) then this expansion is just $\omega \sim \widetilde{\omega}$ and our convention is that $\alpha(\omega) = \beta(\omega) = 0$.

As anticipated, we can use this lemma to distinguish between the domains of closed extensions of $\eth_{\mathrm{dR}}$ [4, Proposition 2.2].

Proposition 3.2. *If g is a suitably scaled wedge metric on $\widetilde{X}$ and $\omega \in \mathcal{D}_{\max}(\eth_{\mathrm{dR}})$ then*

$$\omega \in \mathcal{D}_{\min}(\eth_{\mathrm{dR}}) \iff \alpha(\omega) = \beta(\omega) = 0.$$

In particular, if $\widehat{X}$ is a Witt space then the minimal and maximal domains of $\eth_{\mathrm{dR}}$ coincide so we say that $\eth_{\mathrm{dR}}$ is *essentially closed* (in fact *essentially self-adjoint* since this closed extension will necessarily be self-adjoint). On a non-Witt space, this proposition shows that domains between the minimal and the maximal are determined by the coefficients α and β. We can think of these as being Cauchy data on which we can impose ideal boundary conditions.

3.4 Cheeger ideal boundary conditions

Given an element $\omega \in \mathcal{D}_{\max}(\eth_{\mathrm{dR}})$ of the maximal domain, its 'Cauchy data' consists of two distributional sections $\alpha(\omega)$, $\beta(\omega)$ of the vector bundle

$$\mathcal{H}^{v/2}(\partial\widetilde{X}/Y) \longrightarrow Y.$$

This vector bundle has a canonical flat structure reviewed in §3.2.1. We say that its subbundle W,

$$\begin{array}{ccc} W & \longrightarrow & \mathcal{H}^{v/2}(\partial\widetilde{X}/Y) \\ & \searrow \quad \swarrow & \\ & Y & \end{array}$$

is a *mezzoperversity* if it is parallel with respect to the flat connection of $\mathcal{H}^{v/2}(\partial\widetilde{X}/Y)$. Note that the bundle $\mathcal{H}^{v/2}(\partial\widetilde{X}/Y)$ inherits a bundle metric (from the L^2 metric on harmonic forms); we denote the orthogonal complement of W by $W^{\perp}$. To each mezzoperversity we associate *Cheeger ideal boundary conditions* to define the domain

$$\begin{aligned} \mathcal{D}_W(\eth_{\mathrm{dR}}) = \{&\omega \in \mathcal{D}_{\max}(\eth_{\mathrm{dR}}) : \\ &\alpha(\omega) \in H^{-1/2}(Y; \Lambda^{\bullet}T^*Y \otimes W),\ \beta(\omega) \in H^{-1/2}(Y; \Lambda^{\bullet}T^*Y \otimes W^{\perp})\}. \end{aligned}$$

Note that mezzoperversities always exist: two canonical choices are $W = \{0\}$ and $W = \mathcal{H}^{v/2}(\partial\widetilde{X}/Y)$.

To avoid having to continually distinguish between the Witt and non-Witt cases, let us adopt the convention of writing the unique closed domain of $d + \delta$ on a Witt space by

$$\mathcal{D}_W(\eth_{\mathrm{dR}})$$

with $W = \emptyset$, and refer to this as the empty mezzoperversity.

Theorem 3.3. *Let $\widehat{X}$ be a stratified space of depth one endowed with a suitably scaled wedge metric g and let $\mathcal{D}_W(\eth_{\mathrm{dR}}) \subseteq \mathcal{D}_{\max}(\eth_{\mathrm{dR}})$ be the domain corresponding to a (possibly empty) mezzoperversity. The operator $(\eth_{\mathrm{dR}}, \mathcal{D}_W(\eth_{\mathrm{dR}}))$ is closed, self-adjoint, and Fredholm with compact resolvent, and*

$$\mathcal{D}_W(\eth_{\mathrm{dR}}) \subseteq \bigcap_{\varepsilon>0} r^{1/2-\varepsilon} H^1_e(\widetilde{X}; \Lambda^\bullet({}^wT^*\widetilde{X})).$$

This theorem is proved in [4, §4]. The fact that the domain is closed follows easily from Lemma 3.1 and continuity of the Cauchy data map. The proof that the operator $(\eth_{\mathrm{dR}}, \mathcal{D}_W)$ is Fredholm and has compact resolvent uses the normal operators $N_q(\eth_{\mathrm{dR}})$ defined in §3.2.2. The domain $\mathcal{D}_W$ determines a domain for $N_q(\eth_{\mathrm{dR}})$ as an unbounded operator on the appropriate L^2 space. With this domain $N_q(\eth_{\mathrm{dR}})$ is invertible and we can put these inverses together to construct a parametrix for $(\eth_{\mathrm{dR}}, \mathcal{D}_W)$. This parametrix shows that $(\eth_{\mathrm{dR}}, \mathcal{D}_W)$ is Fredholm and also that $\mathcal{D}_W$ is compactly contained in $L^2(\widetilde{X}; \Lambda^\bullet({}^wT^*\widetilde{X}))$ which implies that the resolvent will be compact. Finally the proof of that this operator is self-adjoint in the non-Witt case follows from facts about domains of d that we will discuss in the next subsection (Theorem 3.4(iii)).

An important consequence of the theorem is that the null space of $(\eth_{\mathrm{dR}}, \mathcal{D}_W(\eth_{\mathrm{dR}}))$ is finite dimensional. We refer to this null space as the *Hodge cohomology with Cheeger ideal boundary conditions* and we denote it by

$$\ker(\eth_{\mathrm{dR}}, \mathcal{D}_W(\eth_{\mathrm{dR}})) = \mathcal{H}^*_W(\widehat{X}).$$

With this theorem we have established the most important analytic properties of $\eth_{\mathrm{dR}}$ and Hodge cohomology. We next look for a corresponding de Rham theory.

3.5 De Rham cohomology

The operators d and δ individually define unbounded operators on $L^2(\widetilde{X}; \Lambda^\bullet({}^wT^*\widetilde{X}))$ and have minimal and maximal domains,

$$\mathcal{D}_{\min}(d) \subseteq \mathcal{D}_{\max}(d), \quad \mathcal{D}_{\min}(\delta) \subseteq \mathcal{D}_{\max}(\delta).$$

Since d and δ are formally adjoint, these domains are related by

$$(d, \mathcal{D}_{\min}(d))^* = (\delta, \mathcal{D}_{\max}(\delta)), \quad (d, \mathcal{D}_{\max}(d))^* = (\delta, \mathcal{D}_{\min}(\delta))$$

as follows directly from the definition of adjoint and the distributional action of a differential operator.

Given a domain $\mathcal{D}$ for d, let us denote

$$\mathcal{D}_p = \mathcal{D} \cap L^2(\widetilde{X}; \Lambda^p({}^wT^*\widetilde{X})).$$

It is easy to see that

$$0 \longrightarrow \mathcal{D}_{\max}(d)_0 \xrightarrow{d} \mathcal{D}_{\max}(d)_1 \xrightarrow{d} \cdots \xrightarrow{d} \mathcal{D}_{\max}(d)_n \longrightarrow 0 \tag{3.5}$$

forms a complex, indeed a *Hilbert complex* in the sense of [31], and similarly for $\mathcal{D}_{\min}(d)$.

If $\widehat{X}$ is a Witt space, then $\eth_{\mathrm{dR}}$ is essentially self-adjoint and [31, Lemma 3.8] shows that

$$\mathcal{D}_{\max}(d) = \mathcal{D}_{\min}(d).$$

If we denote the cohomology of the complex (3.5) by $\mathrm{H}^*_{\mathrm{dR}}(\widehat{X})$ then Theorem 3.3 and [31, Corollary 2.5] show that

$$\mathrm{H}^p_{\mathrm{dR}}(\widehat{X}) \cong \mathcal{H}^p(\widehat{X}).$$

Consider next the situation when $\widehat{X}$ is not a Witt space. Since d is not elliptic, we can not expect arbitrary elements in the maximal domain of d to have a distributional asymptotic expansion at $\widetilde{X}$. However there is a way around this using the Kodaira decomposition (see, e.g., [31, Lemma 2.1]) corresponding to $(d, \mathcal{D}_{\min}(d))$,

$$L^2(\widetilde{X}; \Lambda^\bullet({}^wT^*\widetilde{X})) = \overline{d(\mathcal{D}_{\min}(d))} \oplus \overline{\delta(\mathcal{D}_{\max}(\delta))} \oplus \ker(d, \mathcal{D}_{\min}(d)) \cap \ker(\delta, \mathcal{D}_{\max}(\delta)).$$

Given $\omega \in \mathcal{D}_{\max}(d)$, let ω_δ be the projection onto $\overline{\delta(\mathcal{D}_{\max}(\delta))}$, and notice that

$$d\omega_\delta = d\omega, \quad \delta\omega_\delta = 0,$$

so $(d+\delta)\omega_\delta \in L^2(\widetilde{X}; \Lambda^\bullet({}^wT^*\widetilde{X}))$ and hence

$$\omega_\delta \in \mathcal{D}_{\max}(\eth_{\mathrm{dR}}).$$

In particular Lemma 3.1 determines Cauchy data $\alpha(\omega_\delta)$ and $\beta(\omega_\delta)$, and given a mezzoperversity W we define

$$\mathcal{D}_W(d) = \{\omega \in \mathcal{D}_{\max}(d) : \alpha(\omega_\delta) \in H^{-1/2}(Y; \Lambda^\bullet T^*Y \otimes W)\}.$$

It turns out that [4, Lemma 5.3]

$$\alpha((d\omega)_\delta) = \nabla^{\mathcal{H}}\alpha(\omega_\delta)$$

where $\nabla^{\mathcal{H}}$ is the flat connection on $\mathcal{H}^{v/2}(\partial\widetilde{X}/Y)$, and hence the fact that W is a flat bundle implies

$$d(\mathcal{D}_{W(d)}) \subseteq \mathcal{D}_W(d).$$

Thus each mezzoperversity yields a complex

$$0 \longrightarrow \mathcal{D}_W(d)_0 \xrightarrow{d} \mathcal{D}_W(d)_1 \xrightarrow{d} \cdots \xrightarrow{d} \mathcal{D}_W(d)_n \longrightarrow 0$$

whose cohomology we denote

$$\mathrm{H}^*_{W,\mathrm{dR}}(\widehat{X}).$$

This is the de Rham cohomology associated to a mezzoperversity, and we next establish a Hodge theorem showing that it is equal to the Hodge cohomology.

Theorem 3.4.

i) $\mathcal{D}_W(d)$ *is the closure of* $\mathcal{D}_W(\eth_{\mathrm{dR}})$ *with respect to the graph norm of* d.

ii) If $\omega \in \mathcal{D}_{\max}(\delta)$ *and* ω_d *is its projection onto* $\overline{d(\mathcal{D}_{\max}(d))}$ *then* ω_d *is an element of* $\mathcal{D}_{\max}(\eth_{\mathrm{dR}})$ *and hence has Cauchy data. The adjoint domain of* $\mathcal{D}_W(d)$ *is*

$$\mathcal{D}_W(\delta) = \{\omega \in \mathcal{D}_{\max}(\delta) : \beta(\omega_d) \in H^{-1/2}(Y; \Lambda^\bullet T^*Y \otimes W^\perp)\}.$$

iii) $\mathcal{D}_W(\eth_{\mathrm{dR}}) = \mathcal{D}_W(d) \cap \mathcal{D}_W(\delta)$ *and hence*

$$\mathrm{H}^\bullet_{W,\mathrm{dR}}(\widehat{X}) \cong \mathcal{H}^\bullet_W(\widehat{X}).$$

3.6 Generalized Poincaré duality

Recall that on any smooth oriented closed manifold there is a natural intersection pairing on differential forms

$$(\eta, \omega) \mapsto \int \eta \wedge \omega$$

that descends to a non-degenerate pairing on cohomology. The signature of this pairing is the signature of the manifold.

Assume that $\widehat{X}$ is an oriented pseudomanifold of depth one, and let

$$\begin{aligned} L^2(\widetilde{X}; \Lambda^\bullet({}^wT^*\widetilde{X})) \times L^2(\widetilde{X}; \Lambda^\bullet({}^wT^*\widetilde{X})) &\xrightarrow{\;B\;} \mathbb{R}, \\ (\eta, \omega) &\longmapsto \int_{\widetilde{X}} \eta \wedge \omega. \end{aligned}$$

Note that the Hodge star of a wedge metric is an isometric involution of $L^2(\widetilde{X}; \Lambda^\bullet({}^wT^*\widetilde{X}))$ and hence defines an involution

$$* : \mathcal{D}_{\max}(\eth_{\mathrm{dR}}) \longrightarrow \mathcal{D}_{\max}(\eth_{\mathrm{dR}}).$$

We have $B(\eta, *\eta) = \|\eta\|^2_{L^2}$, and hence B is non-degenerate on $\mathcal{D}_{\max}(\eth_{\mathrm{dR}})$.

For an oriented Witt space, B descends to a non-degenerate pairing on cohomology

$$B : \mathrm{H}^\bullet_{\mathrm{dR}}(\widehat{X}) \times \mathrm{H}^\bullet_{\mathrm{dR}}(\widehat{X}) \longrightarrow \mathbb{R},$$

and the signature of this form is the *signature of the Witt space* $\widehat{X}$.

If $\widehat{X}$ is oriented but not a Witt space, then $*$ will generally not preserve $\mathcal{D}_W(\eth_{\mathrm{dR}})$ and we need to bring in a different mezzoperversity. Define

$$\begin{gathered} Q : \mathcal{C}^\infty(Y; \mathcal{H}^{v/2}(\partial\widetilde{X}/Y)) \times \mathcal{C}^\infty(Y; \mathcal{H}^{v/2}(\partial\widetilde{X}/Y)) \longrightarrow \mathcal{C}^\infty(Y), \\ Q(\eta, \omega) = \int_{\partial\widetilde{X}/Y} \eta \wedge \omega. \end{gathered}$$

Directly from the definition of the flat connection $\nabla^{\mathcal{H}}$ we see that

$$dQ(\eta,\omega) = Q(\nabla^{\mathcal{H}}\eta,\omega) \pm Q(\eta,\nabla^{\mathcal{H}}\omega).$$

This shows that if $W \longrightarrow Y$ is a flat subbundle of $\mathcal{H}^{v/2}(\partial\widetilde{X}/Y)$ and $\mathbb{D}W \longrightarrow Y$ is the Q-orthogonal complement of W, then $\mathbb{D}W$ is also a flat subbundle. We call $\mathbb{D}W$ the *dual mezzoperversity* to W. We can rewrite Q using the Hodge star (on Z) and see that

$$\mathbb{D}W = *W^{\perp}.$$

In turn this makes it easy to see that the Hodge star defines a map

$$* : \mathcal{D}_W(\eth_{\mathrm{dR}}) \longrightarrow \mathcal{D}_{\mathbb{D}W}(\eth_{\mathrm{dR}})$$

and that B descends to non-degenerate pairing on cohomology,

$$B : \mathrm{H}^{\bullet}_{W,\mathrm{dR}}(\widehat{X}) \times \mathrm{H}^{\bullet}_{\mathbb{D}W,\mathrm{dR}}(\widehat{X}) \longrightarrow \mathbb{R},$$

which is a generalization of Poincaré duality to this setting.

If $W = \mathbb{D}W$ we say that W is a *self-dual mezzoperversity.* There are topological obstructions to the existence of a self-dual mezzoperversity, for example the existence of a self-dual mezzoperversity implies that the signature of the link Z vanishes. If a space admits a self-dual mezzoperversity (or is a Witt space) then we say that it is a *Cheeger space.* If W is a self-dual mezzoperversity then B descends to a non degenerate form on $\mathrm{H}^{\bullet}_{W,\mathrm{dR}}(\widehat{X})$ and so these groups satisfy Poincaré duality and define a signature.

In §5.3 we will explain why this signature is independent of the choice of self-dual mezzoperversity and hence only depends on the Cheeger space. Also in this section we describe this signature as the Fredholm index of the signature operator with an appropriate domain.

4 General stratified spaces

Now we consider the general case where $\widehat{X}$ is a union of several smooth manifolds,

$$\widehat{X} = Y_0 \cup Y_1 \cup \cdots \cup Y_\ell.$$

For simplicity let us *assume that,* for each k, the closure $\overline{Y}_k$ of Y_k in $\widehat{X}$ is $\widehat{Y}_k = Y_k \cup Y_{k+1} \cup \cdots \cup Y_\ell$. It is always true that the closure of a stratum is a union of strata, so our assumption is that there is a single chain of strata. The points of depth zero, Y_0, form $\widehat{X}^{\mathrm{reg}}$, the regular part of $\widehat{X}$, while the other strata form the singular part of $\widehat{X}$. We denote the link of $\widehat{X}$ at points in Y_k by $\widehat{Z}_k$, these are themselves stratified spaces (of depth $k-1$). We denote the dimensions of these spaces by

$$n = \dim \widehat{X}^{\mathrm{reg}}, \quad h_k = \dim Y_k, \quad v_k = \dim \widehat{Z}_k^{\mathrm{reg}}.$$

Note that h_k is a decreasing function of k. We assume throughout that all strata and links are connected. (Neither of these assumptions is significant and they mainly serve to simplify the notation.)

In §6 we will recall the precise definition of a stratified space and, more importantly for our purposes, explain how to resolve $\widehat{X}$ to a smooth manifold with corners $\widetilde{X}$. For now we content ourselves with describing the structure of the resolved space where our analysis is carried out. The resolved structure is called a *boundary fibration structure.* The boundary hypersurfaces of $\widetilde{X}$, which we denote $M_1, \ldots, M_\ell$, are embedded and in one to one correspondence with the singular strata. Indeed, each M_k is the total space of a fiber bundle,

$$\widetilde{Z}_k — M_k \xrightarrow{\phi_k} \widetilde{Y}_k,$$

with base $\widetilde{Y}_k$ the resolution of $\overline{Y}_k$, and typical fiber $\widetilde{Z}_k$, the resolution of $\widehat{Z}_k$. If $M_j \cap M_k \neq \emptyset$, and $j < k$, the fiber bundles are compatible in that there is a boundary hypersurface K_{jk} of Y_j and a fiber bundle map $\phi_{jk} : K_{jk} \longrightarrow Y_j$ such that the diagram

$$\begin{array}{ccc} M_j \cap M_k & \xrightarrow{\quad \phi_j \quad} & K_{jk} \subseteq Y_j \\ & {\scriptstyle \phi_k} \searrow \quad \swarrow {\scriptstyle \phi_{jk}} & \\ & Y_k & \end{array} \tag{4.1}$$

commutes.

A rigid wedge metric g on $\widehat{X}^{\text{reg}} = \widetilde{X}^\circ$ is a metric that in some collar neighborhood of M_k (for each k) has the form

$$dr_k^2 + r_k^2 g_{M_k/Y_k} + \phi_k^* g_{Y_k}$$

where r_k is a smooth function vanishing linearly on M_k and positive elsewhere, g_{Y_k} is a rigid wedge metric on $\widetilde{Y}_k$ and g_{M_k/Y_k} is a family of rigid wedge metrics on the fibers $\widetilde{Z}_k$. The covectors of bounded pointwise length with respect to g are the sections of a vector bundle, the *(iterated) wedge cotangent bundle,*

$${}^wT^*\widetilde{X} \longrightarrow \widetilde{X},$$

that is canonically isomorphic to $T^*\widetilde{X}$ over $\widetilde{X}^\circ$. A wedge metric extends to a non-degenerate bundle metric on the iterated wedge cotangent bundle. We denote the de Rham operator of a rigid wedge metric by $\eth_{\text{dR}} = d + \delta$. Its minimal and maximal domains $\mathcal{D}_{\min}(\eth_{\text{dR}})$, $\mathcal{D}_{\max}(\eth_{\text{dR}})$ are defined just as in the case of a single singular stratum.

Our plan will be to study $\eth_{\text{dR}}$ one stratum at a time. For this purpose, a useful class of functions on $\widetilde{X}$ is given by

$$\mathcal{C}^\infty_\Phi(\widetilde{X}) = \{f \in \mathcal{C}^\infty(\widetilde{X}) : f\big|_{M_k} \in \phi_k^* \mathcal{C}^\infty(\widetilde{Y}_k)\}.$$

Note that any such function descends to a continuous function on $\widehat{X}$, so we can think of $\mathcal{C}^\infty_\Phi(\widetilde{X})$ as providing a notion of smooth function on $\widehat{X}$. If $f \in \mathcal{C}^\infty_\Phi(\widetilde{X})$ then df extends from $\widetilde{X}^\circ$ to a smooth section of ${}^wT^*\widetilde{X}$ and this can be used to show

$$\omega \in \mathcal{D}_{\max}(\eth_{\text{dR}}), f \in \mathcal{C}^\infty_\Phi(\widetilde{X}) \implies f\omega \in \mathcal{D}_{\max}(\eth_{\text{dR}}).$$

It is easy to see that there is a partition of unity of $\widetilde{X}$ consisting entirely of functions in $\mathcal{C}^\infty_\Phi(\widetilde{X})$ ([1, Lemma 5.2]). This allows us to focus on elements of $\mathcal{D}_{\max}(\eth_{\mathrm{dR}})$ supported on a *distinguished neighborhood* of a point in a singular stratum $q \in Y_k = \widetilde{Y}_k^\circ$, by which we mean a subset $\mathcal{U}_q$ of a collar neighborhood $[0,1)_x \times M_k$ on which the fiber bundle is trivial and diffeomorphic to $[0,1)_x \times \widetilde{Z}_k \times \mathbb{B}^h$ where $\mathbb{B}^h$ is a ball in Y_k centered around q. Let

$$A(Y_k) = \{\chi \in \mathcal{C}^\infty_\Phi(\widetilde{X}) : \\ \chi \text{ is supported in a distinguished neighborhood of a point in } Y_k\}. \quad (4.2)$$

4.1 Cheeger ideal boundary conditions

The stratum Y_1 has depth one, meaning that the link at each point is a smooth manifold. If $q \in Y_1$ and $\mathcal{U}_q \subseteq \widetilde{X}$ is a distinguished neighborhood, then any element $u \in \mathcal{D}_{\max}(\eth_{\mathrm{dR}})$ supported in $\mathcal{U}_q$ is subject to the analysis carried out in §3. Thus if g is suitably scaled then u has a partial distributional asymptotic expansion as in Lemma 3.1. If Y_1 is not a Witt stratum then we choose W_1, a flat subbundle of $\mathcal{H}^{v/2}(M_1/Y_1)$, and we define

$$\mathcal{D}_{\max,W_1}(\eth_{\mathrm{dR}}) = \{u \in \mathcal{D}_{\max}(\eth_{\mathrm{dR}}) : \text{ for all } \chi \in A(Y_1), \\ \alpha(\chi u) \in H^{-1/2}(Y_1; \Lambda^\bullet({}^wT^*Y_1) \otimes W_1),\ \beta(\chi u) \in H^{-1/2}(Y_1; \Lambda^\bullet({}^wT^*Y_1) \otimes W_1^\perp)\}.$$

If Y_1 is a Witt stratum, then there is no need to choose boundary conditions at Y_1. For notational consistency, when Y_1 is a Witt stratum we set $W_1 = \emptyset$ and define $\mathcal{D}_{\max,W_1}(\eth_{\mathrm{dR}}) = \mathcal{D}_{\max}(\eth_{\mathrm{dR}})$.

Having chosen W_1, we can apply Theorem 3.3 (or rather its proof) and see that

$$u \in \mathcal{D}_{\max,W_1}(\eth_{\mathrm{dR}}) \implies \chi u \in r_1^a H^1_e(\widetilde{X}; \Lambda^\bullet({}^wT^*\widetilde{X})) \text{ for all } a < 1/2, \chi \in A(Y_1).$$

Note that $\mathcal{D}_{\max,W_1}(\eth_{\mathrm{dR}})$ is closed under multiplication by functions in $\mathcal{C}^\infty_\Phi(\widetilde{X})$, so we may continue to localize sections to consider one stratum at a time.

Next consider Y_2. This is a stratum of depth two, meaning that its link $\widehat{Z}_2$ is a stratified space of depth one. We can see, e.g., from the compatibility condition (4.1), that the link of a singular point in $\widehat{Z}_2$ must be Z_1. Thus in our choice of boundary conditions at Y_1, we have determined a domain for the de Rham operator on $\widehat{Z}_2$. Now recall the fundamental observation from (3.3) that the link only enters into the normal operator through its de Rham operator. This allows us to carry out the analysis from §3 keeping in mind that $\eth^{Z_2}_{\mathrm{dR}}$ comes with its domain $\mathcal{D}_{W_1}(\eth^{Z_2}_{\mathrm{dR}})$. Thus (the analogue of) Lemma 3.1 applies to show that

$$u \in \mathcal{D}_{\max,W_1}(\eth_{\mathrm{dR}}), \quad \chi \in A(Y_2) \\ \implies \chi u \sim r_2^{-v_2/2}(\alpha_2(\chi u) + dr_2 \wedge \beta_2(\chi u)) + \chi\widetilde{u}, \text{ where}$$

$$\alpha_2(\chi u), \beta_2(\chi u) \in H^{-1/2}(Y_2; \Lambda^\bullet({}^wT^*Y_2) \otimes \mathcal{H}^{v_2/2}_{W_1}(M_2/Y_2)),$$
$$\widetilde{u} \in \bigcap_{\varepsilon>0} r_2^{1-\varepsilon} H_e^{-1}(\widetilde{X}; \Lambda^\bullet({}^wT^*\widetilde{X}))$$

and $\mathcal{H}^{v_2/2}_{W_1}(M_2/Y_2)$ denotes the bundle over Y_2 whose typical fiber is $\ker(\eth_{\mathrm{dR}}^{Z_2}, \mathcal{D}_{W_1}(\eth_{\mathrm{dR}}^{Z_2}))$, [4, Lemma 2.1]. It is not obvious that these null spaces fit together to form a bundle but this follows from Corollary 5.2 below, and this bundle comes with a natural flat connection. If v_2 is odd or this bundle has rank zero, let us set $W_2 = \emptyset$. Otherwise, let W_2 be a flat subbundle of $\mathcal{H}^{v_2/2}_{W_1}(M_2/Y_2)$ and set

$$\mathcal{D}_{\max,(W_1,W_2)}(\eth_{\mathrm{dR}}) = \{u \in \mathcal{D}_{\max,W_1}(\eth_{\mathrm{dR}}) : \text{ for all } \chi \in A(Y_2),$$
$$\alpha_2(\chi u) \in H^{-1/2}(Y_2; \Lambda^\bullet({}^wT^*Y_2) \otimes W_2),\ \beta_2(\chi u) \in H^{-1/2}(Y_2; \Lambda^\bullet({}^wT^*Y_2) \otimes W_2^\perp)\}.$$

Another application of Theorem 3.3 shows that

$$u \in \mathcal{D}_{\max,(W_1,W_2)}(\eth_{\mathrm{dR}}) \implies \chi u \in r_2^a H_e^1(\widetilde{X}; \Lambda^\bullet({}^wT^*\widetilde{X})) \text{ for all } a < 1/2, \chi \in A(Y_2).$$

We proceed in this way inductively stratum by stratum. At each step, imposing boundary conditions at the first $k-1$ strata allows us to establish a distributional asymptotic expansion at the kth stratum. Using this expansion we can impose boundary conditions at the kth stratum, invert the normal operators, construct a local parametrix, and show that elements satisfying the boundary conditions vanish to order almost $1/2$ at the first k-strata. Once we reach the ℓth stratum we can conclude with the analogue of Theorem 3.3, [4, Theorem 4.3].

Theorem 4.1. *Let $\widehat{X}$ be a stratified space endowed with a suitably scaled rigid iterated wedge metric. Let $\mathbf{W} = (W_1, W_2, \dots, W_\ell)$ be Cheeger ideal boundary conditions at the singular strata of $\widehat{X}$, and let*

$$\mathcal{D}_{\mathbf{W}}(\eth_{\mathrm{dR}})$$

be the corresponding domain of the de Rham operator. Then $(\eth_{\mathrm{dR}}, \mathcal{D}_{\mathbf{W}}(\eth_{\mathrm{dR}}))$ is a closed, self-adjoint, Fredholm operator with compact resolvent and

$$\mathcal{D}_{\mathbf{W}}(\eth_{\mathrm{dR}}) \subseteq \bigcap_{\varepsilon>0} (r_1 \cdots r_\ell)^{1/2-\varepsilon} H_e^1(\widetilde{X}; \Lambda^\bullet({}^wT^*\widetilde{X})).$$

We denote the resulting Hodge cohomology groups by

$$\mathcal{H}^\bullet_{\mathbf{W}}(\widehat{X}).$$

4.2 De Rham cohomology

Given Cheeger ideal boundary conditions $\mathbf{W}$, we can define domains for d and δ by

$$\mathcal{D}_{\mathbf{W}}(d) = d\text{-graph closure of } \mathcal{D}_{\mathbf{W}}(\eth_{\mathrm{dR}}), \quad \mathcal{D}_{\mathbf{W}}(\delta) = \delta\text{-graph closure of } \mathcal{D}_{\mathbf{W}}(\eth_{\mathrm{dR}}).$$

To see that these form complexes and that their intersection is $\mathcal{D}_{\mathbf{W}}(\eth_{\mathrm{dR}})$, it is important to characterize these domains in terms of distributional asymptotic expansions.

Let Y be a stratum of $\widehat{X}$ of depth k, and assume inductively that we have carried out the analysis of Hodge and de Rham cohomology on all space of depth less than k. The domain

$$\mathcal{D}_{\max,(W_1,\ldots,W_{k-1})}(d)$$

forms a Hilbert complex and so we have a Kodaira decomposition

$$\begin{aligned} L^2(\widetilde{X};\Lambda^\bullet({}^wT^*\widetilde{X})) = \overline{d(\mathcal{D}_{\max,(W_1,\ldots,W_{k-1})}(d))} \oplus \overline{\delta(\mathcal{D}_{\max,(W_1,\ldots,W_{k-1})}(d)^*)} \\ \oplus \ker(d, \mathcal{D}_{\max,(W_1,\ldots,W_{k-1})}(d)) \cap \ker(\delta, \mathcal{D}_{\max,(W_1,\ldots,W_{k-1})}(d)^*). \end{aligned}$$

Given $\omega \in \mathcal{D}_{\max,(W_1,\ldots,W_{k-1})}(d)$, let ω_δ denote the orthogonal projection onto the second summand, and note that

$$\omega_\delta \in \mathcal{D}_{\max,(W_1,\ldots,W_{k-1})}(\eth_{\mathrm{dR}}).$$

This allows us to define

$$\begin{aligned} \mathcal{D}_{\max,(W_1,\ldots,W_k)}(d) = \{\omega \in \mathcal{D}_{\max,(W_1,\ldots,W_{k-1})}(d) : \\ \alpha_k((\chi\omega)_\delta) \in H^{-1/2}(Y_k;\Lambda^\bullet({}^wT^*Y_k)\otimes W_k) \text{ for all } \chi \in A(Y_k)\}. \end{aligned}$$

The fact that W_k is a flat subbundle of $\mathcal{H}^{v_k/2}_{(W_1,\ldots,W_{k-1})}(M_k/Y_k)$ implies that $\mathcal{D}_{\max,(W_1,\ldots,W_k)}(d)$ forms a complex.

Similar reasoning allows us to define

$$\begin{aligned} \mathcal{D}_{\max,(W_1,\ldots,W_k)}(\delta) = \{\omega \in \mathcal{D}_{\max,(W_1,\ldots,W_{k-1})}(\delta) : \\ \beta_k((\chi\omega)_d) \in H^{-1/2}(Y_k;\Lambda^\bullet({}^wT^*Y_k)\otimes W_k^\perp) \text{ for all } \chi \in A(Y_k)\} \end{aligned}$$

where $(\chi\omega)_d$ denotes the orthogonal projection of $\chi\omega$ onto $\overline{d(\mathcal{D}_{\max,(W_1,\ldots,W_{k-1})}(\delta)^*)}$. Together these expressions show that

$$\mathcal{D}_{\max,(W_1,\ldots,W_k)}(\eth_{\mathrm{dR}}) = \mathcal{D}_{\max,(W_1,\ldots,W_k)}(d) \cap \mathcal{D}_{\max,(W_1,\ldots,W_k)}(\delta)$$

and establish the analogue of Theorem 3.4, [4, Theorem 5.1].

Theorem 4.2. *Let $\widehat{X}$ be a stratified space endowed with a suitably scaled rigid iterated wedge metric. For any choice of Cheeger ideal boundary conditions, $\mathbf{W} = (W_1, W_2, \ldots, W_\ell)$, the complex*

$$0 \longrightarrow \mathcal{D}_{\mathbf{W}}(d)_0 \xrightarrow{d} \mathcal{D}_{\mathbf{W}}(d)_1 \xrightarrow{d} \cdots \xrightarrow{d} \mathcal{D}_{\mathbf{W}}(d)_n \longrightarrow 0$$

is a Fredholm Hilbert complex whose cohomology $\mathrm{H}^\bullet_{\mathbf{W},\mathrm{dR}}(\widehat{X})$ is canonically isomorphic to the Hodge cohomology of $\mathbf{W}$,

$$\mathrm{H}^\bullet_{\mathbf{W},\mathrm{dR}}(\widehat{X}) \cong \mathcal{H}^\bullet_{\mathbf{W}}(\widehat{X}).$$

4.3 Cheeger spaces

If $\widehat{X}$ is oriented then to each mezzoperversity we can associate a dual mezzoperversity. Inductively, assume that we have chosen compatible flat bundles $W_1, \ldots, W_k$ over $Y_1, \ldots, Y_k$ respectively, and their dual bundles $\mathbb{D}W_1, \ldots, \mathbb{D}W_k$. Then, over Y_k, we let

$$\begin{gathered} Q_k : \mathcal{C}^\infty(Y_k; \mathcal{H}^{v_k/2}_{(W_1,\ldots,W_{k-1})}(M_k/Y_k)) \times \mathcal{C}^\infty(Y_k; \mathcal{H}^{v_k/2}_{(\mathbb{D}W_1,\ldots,\mathbb{D}W_{k-1})}(M_k/Y_k)) \\ \longrightarrow \mathcal{C}^\infty(Y_k), \\ Q_k(\eta, \omega) = \int_{M_k/Y_k} \eta \wedge \omega \end{gathered} \tag{4.3}$$

and we define the dual bundle of a flat subbundle W_k of $\mathcal{H}^{v_k/2}_{(W_1,\ldots,W_k)}(M_k/Y_k)$ to be its Q-orthogonal complement. As before, this is again a flat bundle and satisfies

$$*W_k^\perp = \mathbb{D}W_k.$$

In this way, given a mezzoperversity $\mathbf{W} = (W_1, \ldots, W_\ell)$ we have defined a dual mezzoperversity

$$\mathbb{D}\mathbf{W} = (\mathbb{D}W_1, \ldots, \mathbb{D}W_\ell)$$

and shown that the Hodge start intertwines the respective domains,

$$* : \mathcal{D}_{\mathbf{W}}(\eth_{\mathrm{dR}}) \longrightarrow \mathcal{D}_{\mathbb{D}\mathbf{W}}(\eth_{\mathrm{dR}}).$$

We say that a stratified space is a *Cheeger space* if it admits a mezzoperversity that coincides with its dual mezzoperversity (by convention this includes Witt spaces).

Theorem 4.3 (Generalized Poincaré duality). *Let $\widehat{X}$ be an oriented pseudomanifold endowed with a rigid suitably scaled iterated wedge metric. The pairing*

$$B : L^2(\widetilde{X}; \Lambda^\bullet({}^wT^*\widetilde{X})) \times L^2(\widetilde{X}; \Lambda^\bullet({}^wT^*\widetilde{X})) \longrightarrow \mathbb{R}, \quad B(\eta, \omega) = \int \eta \wedge \omega$$

descends, for any choice of mezzoperversity $\mathbf{W}$, to a non-degenerate pairing

$$\mathrm{H}^\bullet_{\mathbf{W},\mathrm{dR}}(\widehat{X}) \times \mathrm{H}^\bullet_{\mathbb{D}\mathbf{W},\mathrm{dR}}(\widehat{X}) \longrightarrow \mathbb{R}.$$

Thus on a Cheeger space the groups $\mathrm{H}^\bullet_{\mathbf{W},\mathrm{dR}}(\widehat{X})$ satisfy Poincaré duality and define a signature. We will discuss the properties of this signature in §5.3.

5 Properties of cohomology

5.1 Metric independence of de Rham cohomology

Our definition of a mezzoperversity uses Hodge cohomology groups and hence is metric dependent. In this section we will explain how to define mezzoperversities in a metric-independent way, and why this leads to metric independent de

Rham cohomology groups. For this section we will incorporate the metric into the notation of the domain, e.g.,

$$\mathcal{D}_{\max}(d; g).$$

To start with, note that if g and g' are two iterated wedge metrics on the same pseudomanifold $\widehat{X}$ then they both induce bundle metrics on ${}^wT^*\widetilde{X}$ over $\widetilde{X}$ and, since $\widetilde{X}$ is compact, these must be quasi-isometric. That is, there is a positive constant C such that

$$Cg(\cdot,\cdot) \leq g'(\cdot,\cdot) \leq \frac{1}{C}g(\cdot,\cdot).$$

It follows that L^2 norms defined by g and g' on $\mathcal{C}^\infty_c(\widetilde{X}; \Lambda^\bullet({}^wT^*\widetilde{X}))$ are equivalent and hence

$$L^2(\widetilde{X}; \Lambda^\bullet({}^wT^*\widetilde{X}))$$

as a set is independent of the choice of metric and inherits equivalent norms from g and g'. Then directly from the definitions of the minimal and maximal domains we see that

$$\mathcal{D}_{\min}(d; g) = \mathcal{D}_{\min}(d; g'), \quad \mathcal{D}_{\max}(d; g) = \mathcal{D}_{\max}(d; g')$$

so we can denote these unambiguously as $\mathcal{D}_{\min}(d)$ and $\mathcal{D}_{\max}(d)$. If $\widehat{X}$ is a Witt space, then this shows that its de Rham cohomology is independent of the choice of metric.

For domains defined by Cheeger ideal boundary conditions, we proceed inductively. So first *assume that $\widehat{X}$ is a pseudomanifold of depth one* and adopt the notation of §3. A mezzoperversity is a flat subbundle W of $\mathcal{H}^{v/2}(M/Y) \longrightarrow Y$ and hence depends on the metric. Let us emphasize this by referring to W as a *Hodge mezzoperversity* or even a *g-Hodge mezzoperversity*. Analogously a *de Rham mezzoperversity* will refer to a flat subbundle of $\mathrm{H}^{v/2}_{\mathrm{dR}}(M/Y) \longrightarrow Y$. Using the Hodge isomorphism on the smooth manifold Z, we can associate a de Rham mezzoperversity to each Hodge mezzoperversity. We denote the de Rham mezzoperversity corresponding to W by $[W]$. The advantage is that the bundle $\mathrm{H}^{v/2}_{\mathrm{dR}}(M/Y) \longrightarrow Y$ and, from §3.2.1 its flat connection, are defined independently of the metric. Thus it makes sense to compare the two domains,

$$\mathcal{D}_{[W]}(d; g) \text{ and } \mathcal{D}_{[W]}(d; g').$$

Consider the subspace $\mathcal{D}^{\mathrm{reg}}_{\max}(d) \subseteq \mathcal{D}_{\max}(d)$ consisting of wedge differential forms with an asymptotic expansion near Y,

$$\chi\omega = r^{-v/2}(\alpha(\omega) + dr \wedge \beta(\omega)) + \widetilde{\omega}, \text{ for all } \chi \in \mathcal{A}(Y),$$

where $\widetilde{\omega} \in \mathcal{D}_{\min}(d)$ and $\alpha(\omega)$, $\beta(\omega)$ are smooth differential forms defined near $\partial\widetilde{X}$, of vertical degree $v/2$, in the null space of d^Z. This decomposition when it exists is not unique, since elements of $\mathcal{D}_{\min}(d)$ may themselves have asymptotic expansions, but one can show that the de Rham cohomology class of $\alpha(\omega)$ is well-defined. It turns out [4, §5] that $\mathcal{D}^{\mathrm{reg}}_{\max}(d)$ is dense in $\mathcal{D}_{\max}(d)$ with respect to the graph norm of d, and

$$\mathcal{D}^{\mathrm{reg}}_{[W]}(d; g) = \mathcal{D}_{[W]}(d, g) \cap \mathcal{D}^{\mathrm{reg}}_{\max}(d)$$

is dense in $\mathcal{D}_{[W]}(d;g)$ with respect to the graph norm of d. Moreover, if ω is in $\mathcal{D}^{\mathrm{reg}}_{\max}(d)$ then $\omega \in \mathcal{D}_{[W]}(d)$ precisely when the vertical de Rham cohomology class of $\alpha(\omega)$ as a section over Y is in $[W]$. Thus $\mathcal{D}^{\mathrm{reg}}_{[W]}(d;g) = \mathcal{D}^{\mathrm{reg}}_{[W]}(d;g')$ and, since taking the closure in L^2 with respect to g is the same as taking the closure with respect to g', we have

$$\mathcal{D}_{[W]}(d;g) = \mathcal{D}_{[W]}(d;g').$$

It follows that on a pseudomanifold of depth one, the domain of the exterior derivative corresponding to a de Rham mezzoperversity is independent of the choice of rigid suitably scaled wedge metric and hence

$$\mathrm{H}^{\bullet}_{[W],\mathrm{dR}}(\widehat{X}) \text{ is independent of the choice of } g.$$

The same thing can be shown on a general pseudomanifold proceeding inductively one stratum at a time.

Theorem 5.1 (Metric independence). *Let $\widehat{X}$ be a pseudomanifold and let $[\mathbf{W}]$ be a de Rham mezzoperversity, i.e., a sequential choice of flat subbundles of the vertical de Rham cohomology with Cheeger ideal boundary conditions at each non-Witt stratum. If g and g' are two rigid, suitably scaled, iterated wedge metrics on $\widehat{X}$ then*

$$\mathcal{D}_{[\mathbf{W}]}(d;g) = \mathcal{D}_{[\mathbf{W}]}(d;g')$$

and, in particular, the de Rham cohomology groups $\mathrm{H}^{\bullet}_{[\mathbf{W}],\mathrm{dR}}(\widehat{X})$ are independent of the choice of metric.

An easy corollary is that de Rham cohomology is invariant under pull-back by a diffeomorphism of manifolds with corners $\widetilde{X} \longrightarrow \widetilde{N}$ that intertwines iterated boundary fibration structures on $\widetilde{X}$ and $\widetilde{N}$ (i.e., an isomorphism in the sense of §6.4).

Corollary 5.2. *Let $\widehat{X}$ and $\widehat{N}$ be pseudomanifolds and $F : \widetilde{X} \longrightarrow \widetilde{N}$ a diffeomorphism intertwining the boundary fibration structures. Given a de Rham mezzoperversity $\mathbf{W}_{\widehat{N}}$ on $\widehat{N}$, pull-back by F determines a de Rham mezzoperversity $F^*\mathbf{W}_{\widehat{N}}$ on $\widehat{X}$, and an isomorphism*

$$F^* : \mathrm{H}^{\bullet}_{[\mathbf{W}_{\widehat{N}}],\mathrm{dR}}(\widehat{N}) \longrightarrow \mathrm{H}^{\bullet}_{[F^*\mathbf{W}_{\widehat{N}}],\mathrm{dR}}(\widehat{X}).$$

This is proved in [4, Theorem 5.4], we do not go into the details as in the next section we discuss the more general case of stratified homotopy equivalences.

5.2 Stratified homotopy invariance of de Rham cohomology

If X and N are smooth closed manifolds and $F : X \longrightarrow N$ is a smooth homotopy equivalence between them, then F defines a pull-back map of differential forms that commutes with the exterior derivative. Thus F^* descends to a map on de Rham cohomology which is necessarily an isomorphism,

$$F^* : \mathrm{H}^{\bullet}_{\mathrm{dR}}(N) \xrightarrow{\cong} \mathrm{H}^{\bullet}_{\mathrm{dR}}(X).$$

We want to prove an analogous statement for the L^2-de Rham cohomology of a pseudomanifold corresponding to a mezzoperversity.

One problem is that, even on smooth Riemannian manifolds, the pull-back by a smooth homotopy equivalence need not induce a bounded map on L^2-differential forms. For example, if F is an embedding and $u_\varepsilon \in \mathcal{C}^\infty(N)$ is the indicator function of an ε-ball around $F(X)$, then the L^2-norm of u_ε goes to zero with ε, while $F^* u_\varepsilon = 1$ has constant L^2 norm. Thus

$$F^* : \{ \text{ step functions } \} \subseteq L^2(N) \longrightarrow L^2(X)$$

is not even closable. There is a clever way around this, developed by Hilsum-Skandalis [59] for smooth manifolds, and extended in [3, 5] to pseudomanifolds. We replace F^* by a map on L^2-differential forms called the *Hilsum-Skandalis replacement* and denoted $HS(F)$.

First let us explain our assumptions on maps between stratified spaces. By a *smooth map between stratified spaces* we mean a continuous map $\widehat{F} : \widehat{X} \longrightarrow \widehat{N}$ that lifts to a smooth map between the resolutions of the stratified spaces, $F : \widetilde{X} \longrightarrow \widetilde{N}$. Equivalently, these are the smooth maps $\widetilde{X} \longrightarrow \widetilde{N}$ such that

$$F^* \mathcal{C}^\infty_\Phi(\widetilde{N}) \subseteq \mathcal{C}^\infty_\Phi(\widetilde{X}).$$

We denote these maps by

$$\mathcal{C}^\infty(\widehat{X}, \widehat{N}) \equiv \mathcal{C}^\infty_\Phi(\widetilde{X}, \widetilde{N}).$$

We say that $\widehat{F} \in \mathcal{C}^\infty(\widehat{X}, \widehat{N})$ is *stratum-preserving* if the inverse image of a stratum is a union of strata, and *strongly stratum preserving* if the codimension of a stratum of $\widehat{N}$ coincides with the codimension of its inverse image in $\widehat{X}$. We say that $\widehat{F} \in \mathcal{C}^\infty(\widehat{X}, \widehat{N})$ is a *stratified homotopy equivalence* if there is a smooth strongly stratum preserving map $\widehat{G} \in \mathcal{C}^\infty(\widehat{N}, \widehat{X})$ such that $\widehat{F} \circ \widehat{G}$ and $\widehat{G} \circ \widehat{F}$ are smoothly connected to the identity through smooth strongly stratum preserving maps.

We will assign a Hilsum-Skandalis replacement $HS(F)$ to a smooth strongly stratum preserving map. These replacements will descend to cohomology and only depend on the homotopy class of F within smooth strongly stratum preserving maps. The construction of $HS(F)$ is slightly complicated, but the main idea is that we want to 'spread out' the image of F to avoid precisely the sort of counterexample above. In [59] this 'spreading out' is done by embedding N into a Euclidean space and averaging among nearby points. We will do something similar using the geodesic flow of a wedge metric.

A smooth stratum-preserving map induces a map of the iterated wedge tangent bundles

$$DF : {}^wT\widetilde{X} \longrightarrow {}^wT\widetilde{N}$$

and hence a pull-back map

$$F^* : \mathcal{C}^\infty(\widetilde{L}; \Lambda^\bullet({}^wT^*\widetilde{N})) \longrightarrow \mathcal{C}^\infty(\widetilde{X}; \Lambda^\bullet({}^wT^*\widetilde{X})),$$

but as mentioned above, this map need not be bounded in L^2. To construct a replacement, let us start with ${}^w\mathbb{B}\widetilde{N} \xrightarrow{\pi_N} \widetilde{N}$, the ball subbundle of the dual bundle to ${}^wT^*\widetilde{N}$, ${}^wT\widetilde{N} \longrightarrow \widetilde{N}$, and the pull-back diagram

$$\begin{array}{ccc} F^*({}^w\mathbb{B}\widetilde{N}) & \xrightarrow{\mathbb{B}(F)} & {}^w\mathbb{B}\widetilde{N} \\ \downarrow{\scriptstyle \pi_X} & & \downarrow{\scriptstyle \pi_N} \\ \widetilde{X} & \xrightarrow{F} & \widetilde{N} \end{array}$$

Next, choose a Thom form $\mathcal{T}_N$ for ${}^w\mathbb{B}\widetilde{N} \xrightarrow{\pi_N} \widetilde{N}$, and let $\overline{\exp}_N : {}^w\mathbb{B}\widetilde{N} \longrightarrow \widetilde{N}$ denote the continuous extension of the exponential map of a (complete metric conformal to an) iterated wedge metric from $\widetilde{N}^\circ$ to $\widetilde{N}$. Then we set

$$\begin{gathered} HS(F) : \mathcal{C}^\infty(\widetilde{N}; \Lambda^\bullet({}^wT^*\widetilde{N})) \longrightarrow \mathcal{C}^\infty(\widetilde{X}; \Lambda^\bullet({}^wT^*\widetilde{X})), \\ HS(F)(\omega) = (\pi_X)_*(\mathbb{B}(F)^*\mathcal{T}_N \wedge (\overline{\exp}_N \circ \mathbb{B}(F))^*\omega). \end{gathered}$$

This map commutes with the exterior derivative

$$HS(F)d_N = d_X HS(F)$$

directly from its definition. If we endow ${}^w\mathbb{B}\widetilde{N}$ with a π_N-submersion metric compatible with g_N and $F^*({}^w\mathbb{B}\widetilde{N})$ with a π_X-submersion metric compatible with g_X, which is always possible, then it is easy to see that $HS(F)$ is a bounded map with respect to the L^2-norms and so extends to

$$HS(F) : L^2(\widetilde{N}; \Lambda^\bullet({}^wT^*\widetilde{N})) \longrightarrow L^2(\widetilde{X}; \Lambda^\bullet({}^wT^*\widetilde{X})).$$

Moreover, since $HS(F)$ sends forms with compact support in $\widetilde{N}^\circ$ to forms with compact support in $\widetilde{X}^\circ$, it induces maps

$$HS(F) : \mathcal{D}_{\max}(d_N) \longrightarrow \mathcal{D}_{\max}(d_X), \quad HS(F) : \mathcal{D}_{\min}(d_N) \longrightarrow \mathcal{D}_{\min}(d_X).$$

It turns out that $HS(F)$ defines a map on mezzoperversities. The key fact is that, when $\widehat{F}$ sends the stratum $Y \subseteq \widehat{X}$ to the stratum $Q \subseteq \widehat{N}$, the map F restricts to the corresponding boundary hypersurfaces to a fiber bundle map

$$\begin{array}{ccccccc} \widetilde{X} & \supseteq & M & \xrightarrow{F} & P & \subseteq & \widetilde{N} \\ & & \downarrow & & \downarrow & & \\ & & Y & \xrightarrow{\widehat{F}} & Q & & \end{array}$$

and the induced map between the link of a point in Y and the link of the corresponding point in Q will be a smooth strongly stratum preserving map. The upshot is that, compatibly with $HS(F)$, there is a family of maps $HS(F\big|_{Z_q})$, $q \in Y$, between vertical differential forms over Y and vertical differential forms over Q.

If $\widehat{X}$ and $\widehat{N}$ are depth one non-Witt spaces, then $HS(F\big|_{Z_q})$ induces a map between $\mathrm{H}^{v/2}_{\mathrm{dR}}(M/Y)$ and $\mathrm{H}^{v/2}_{\mathrm{dR}}(P/Q)$ (note that v is the same in both spaces because F is strongly stratum preserving). The origin of the flat connections in the exterior derivatives described in §3.2.1 can be used to show that they are intertwined by $HS(F\big|_{Z_q})$. Thus the boundary behavior of $HS(F)$ defines a 'pull-back' map between mezzoperversities on $\widehat{N}$ and mezzoperversities on $\widehat{X}$, which we denote

$$W_{\widehat{N}} \mapsto F^{\sharp}W_{\widehat{N}}.$$

By working with elements in the regular domain from §5.1, $\mathcal{D}^{\mathrm{reg}}_{[W_{\widehat{N}}]}(d_N)$, we can show that

$$HS(F)(\mathcal{D}_{[W_{\widehat{N}}]}(d_N)) \subseteq \mathcal{D}_{[F^{\sharp}W_{\widehat{N}}]}(d_X)$$

and hence there is a map on cohomology,

$$HS(F) : \mathrm{H}^{\bullet}_{[W_{\widehat{N}}],\mathrm{dR}}(\widehat{N}) \longrightarrow \mathrm{H}^{\bullet}_{[F^{\sharp}W_{\widehat{N}}],\mathrm{dR}}(\widehat{X}).$$

An argument in [59], extended in [3, Lemma 9.1], shows that this map is unchanged if F is allowed to vary among smooth strongly stratum preserving maps, and that $HS(F \circ G) = HS(G) \circ HS(F)$ as maps on cohomology, for all smooth strongly stratum preserving maps $G : \widehat{N} \longrightarrow \widehat{R}$. Iterating these constructions we obtain [5, Theorem 3.7].

Theorem 5.3. *Let $\widehat{X}$ and $\widehat{N}$ be pseudomanifolds, $F : \widehat{X} \longrightarrow \widehat{N}$ a smooth strongly stratum preserving map, and*

$$HS(F) : L^2(\widetilde{N}; \Lambda^{\bullet}({}^{w}T^*\widetilde{N})) \longrightarrow L^2(\widetilde{X}; \Lambda^{\bullet}({}^{w}T^*\widetilde{X}))$$

the Hilsum-Skandalis replacement of F. Given a mezzoperversity $\mathbf{W}_{\widehat{N}}$ on $\widehat{N}$, $HS(F)$ induces a mezzoperversity $F^{\sharp}\mathbf{W}_{\widehat{N}}$ on $\widehat{X}$ such that

$$HS(F)(\mathcal{D}_{[\mathbf{W}_{\widehat{N}}]}(d_N)) \subseteq \mathcal{D}_{[F^{\sharp}\mathbf{W}_{\widehat{N}}]}(d_X).$$

The induced map in de Rham cohomology

$$HS(F) : \mathrm{H}^{\bullet}_{[\mathbf{W}_{\widehat{N}}],\mathrm{dR}}(\widehat{N}) \longrightarrow \mathrm{H}^{\bullet}_{[F^{\sharp}\mathbf{W}_{\widehat{N}}],\mathrm{dR}}(\widehat{X})$$

is unchanged by varying F smoothly through smooth strongly stratum preserving maps and is an isomorphism if F is a stratified homotopy equivalence.

5.3 Properties of the signature

Let $\widehat{X}$ be an oriented Cheeger space and $[\mathbf{W}]$ a self-dual de Rham mezzoperversity. As mentioned above, the intersection pairing

$$B : \mathrm{H}^{\bullet}_{[\mathbf{W}],\mathrm{dR}}(\widehat{X}) \times \mathrm{H}^{\bullet}_{[\mathbf{W}],\mathrm{dR}}(\widehat{X}) \longrightarrow \mathbb{R}$$

is a non-degenerate quadratic form, and we denote its signature by

$$\mathrm{sign}(\widehat{X}, \mathbf{W}).$$

A careful inspection of the Hilsum-Skandalis replacement shows that the pull-back of a self-dual mezzoperversity is a self-dual mezzoperversity (indeed the forms Q_k from (4.3) are compatible with $HS(F)$) and in fact:

Proposition 5.4. *If $F : \widehat{X} \longrightarrow \widehat{N}$ is a stratified homotopy equivalence and $\mathbf{W}_N$ is a self-dual mezzoperversity on $\widehat{N}$, then*

$$\operatorname{sign}(\widehat{X}, F^\sharp \mathbf{W}_{\widehat{N}}) = \operatorname{sign}(\widehat{N}, \mathbf{W}_{\widehat{N}})$$

If we endow $\widehat{X}$ with a rigid suitably scaled iterated wedge metric g, then we can identify the signature with the Fredholm index of an elliptic operator. Assume $n = \dim \widehat{X}$ is even. Let

$$L^2(\widetilde{X}; \Lambda^\bullet({}^wT^*_{\mathbb{C}}\widetilde{X}))$$

denote the L^2-differential forms with complex coefficients, let

$$\tau : L^2(\widetilde{X}; \Lambda^\bullet({}^wT^*_{\mathbb{C}}\widetilde{X})) \longrightarrow L^2(\widetilde{X}; \Lambda^\bullet({}^wT^*_{\mathbb{C}}\widetilde{X})),$$
$$\tau\big|_{L^2(\widetilde{X}; \Lambda^p({}^wT^*_{\mathbb{C}}\widetilde{X}))} = i^{p(p-1)+n/2}*$$

and note that $\tau^2 = \operatorname{Id}$. Denote the ± 1 eigenspaces of τ by

$$L^2(\widetilde{X}; \Lambda^\bullet({}^wT^*_{\mathbb{C}}\widetilde{X}))_\pm.$$

Since $\eth_{\mathrm{dR}}$ anticommutes with τ, it interchanges these two eigenspaces, and the restriction to either of these eigenspaces is known as the signature operator, $\eth_{\mathrm{sign}}$. If $\mathbf{W}$ is a self-dual mezzoperversity then

$$\tau : \mathcal{D}_{\mathbf{W}}(\eth_{\mathrm{dR}}) \longrightarrow \mathcal{D}_{\mathbf{W}}(\eth_{\mathrm{dR}})$$

(with $\mathcal{D}_{\mathbf{W}}(\eth_{\mathrm{dR}}) \subseteq L^2(\widetilde{X}; \Lambda^\bullet({}^wT^*_{\mathbb{C}}\widetilde{X}))$ after complexification) and we define

$$\mathcal{D}_{\mathbf{W}}(\eth_{\mathrm{sign}}) = \mathcal{D}_{\mathbf{W}}(\eth_{\mathrm{dR}}) \cap L^2(\widetilde{X}; \Lambda^\bullet({}^wT^*_{\mathbb{C}}\widetilde{X}))_+.$$

Proposition 5.5. *With notation as above, the operator*

$$\eth_{\mathrm{sign}} : \mathcal{D}_{\mathbf{W}}(\eth_{\mathrm{sign}}) \subseteq L^2(\widetilde{X}; \Lambda^\bullet({}^wT^*_{\mathbb{C}}\widetilde{X}))_+ \longrightarrow L^2(\widetilde{X}; \Lambda^\bullet({}^wT^*_{\mathbb{C}}\widetilde{X}))_-$$

is Fredholm and its index is equal to $\operatorname{sign}(\widehat{X}, \mathbf{W})$.

One of the most important properties of the signature of a smooth manifold is that it is unchanged under oriented bordism. That is, if X and N are oriented smooth manifolds, and there is an oriented smooth manifold with boundary $\mathscr{M}$ such that

$$\partial \mathscr{M} = X \sqcup (-N),$$

then $\operatorname{sign}(X) = \operatorname{sign}(N)$.

Now suppose that $\mathbf{W}_{\widehat{X}}$, $\mathbf{W}_{\widehat{N}}$ are self-dual mezzoperversities on the oriented Cheeger spaces $\widehat{X}$ and $\widehat{N}$, respectively. A bordism between $(\widehat{X}, \mathbf{W}_{\widehat{X}})$ and $(\widehat{N}, \mathbf{W}_{\widehat{N}})$ is a Cheeger space $\widehat{\mathscr{M}}$ with collared boundary,

$$\partial \widehat{\mathscr{M}} = \widehat{X} \sqcup (-\widehat{N})$$

together with a mezzoperversity $\mathbf{W}_{\widehat{\mathscr{M}}}$ that restricts to $\mathbf{W}_{\widehat{X}}$ in a collar of $\widehat{X}$ and restricts to $\mathbf{W}_{\widehat{N}}$ in a collar of $\widehat{N}$. By showing that the signature operator of a Cheeger space defines a K-homology class and then applying K-homology theory, we proved in [5, Theorem 4.7] the bordism invariance of the signature.

Proposition 5.6. *If there is a bordism between $(\widehat{X}, \mathbf{W}_{\widehat{X}})$ and $(\widehat{N}, \mathbf{W}_{\widehat{N}})$ then*

$$\operatorname{sign}(\widehat{X}, \mathbf{W}_{\widehat{X}}) = \operatorname{sign}(\widehat{N}, \mathbf{W}_{\widehat{N}}).$$

A clever observation of Banagl [11] (in the context of L-spaces and self-dual sheaves, cf. [5, Theorem 4.8]) is that any two mezzoperversities on a single Cheeger space are bordant. Thus this proposition shows that the signature of a Cheeger space is independent of the choice of self-dual mezzoperversity! We can restate the results so far in a more satisfying fashion:

Theorem 5.7. *The signature of an oriented Cheeger space is unchanged under stratified homotopy equivalences or Cheeger space bordism.*

5.4 L-class and higher signatures

Let $\widehat{X}$ be a Cheeger space, $\mathbf{W}$ a self-dual mezzoperversity, and let $\mathbb{S}^\ell$ denote the ℓ-dimensional sphere. Goresky-MacPherson [53] showed that $\widehat{X}$ can be embedded into $\mathbb{R}^m$ for $m \gg 1$ and any

$$F : \widehat{X} \longrightarrow \mathbb{S}^{2q}$$

can be arranged via homotopy to be the restriction of a map defined on all of $\mathbb{R}^m$ and to be transverse to a given point $\mathrm{pt} \in \mathbb{S}^{2q}$.

Lemma 5.8. *The inverse image of* pt *under F is a Cheeger space and hence has a signature. Homotopic maps produce cobordant Cheeger spaces and hence the same signature.*

A trick due to Thom [108, 110] uses this construction to define an L-class in homology. Namely, if $4q > n+1$, then we have

$$\mathrm{H}^{2q}(\widehat{X};\mathbb{Q}) \cong [\widehat{X}, \mathbb{S}^{2q}] \otimes \mathbb{Q}$$

and so we have a map

$$\begin{aligned} \mathrm{H}^{2q}(\widehat{X};\mathbb{Q}) &\longrightarrow \mathbb{Q}, \\ [F] &\longmapsto \operatorname{sign}(F^{-1}(\mathrm{pt})) \end{aligned}$$

which defines an element in $H_{2q}(\widehat{X};\mathbb{Q})$. If $4q \leq n+1$, we choose any $\ell > n+1$ such that $4(q+\ell) > n+\ell+1$, and then $\widehat{X} \times \mathbb{S}^{2\ell}$ is a Cheeger space and

$$\mathrm{H}^{2q+2\ell}(\widehat{X} \times \mathbb{S}^{2\ell};\mathbb{Q}) \cong \mathrm{H}^{2q}(\widehat{X};\mathbb{Q})$$

so we can proceed as before.

The homology class defined in this way is called the *L-class* and denoted

$$\mathcal{L}(\widehat{X}) \in \mathrm{H}_{\mathrm{even}}(\widehat{X};\mathbb{Q}).$$

Now let $r : \widehat{X} \longrightarrow B\pi_1\widehat{X}$ be the classifying map for the universal cover of $\widehat{X}$. The higher signatures of $\widehat{X}$ are defined to be the rational numbers

$$\{\langle \gamma, r_*\mathcal{L}(\widehat{X})\rangle : \gamma \in \mathrm{H}^\bullet(B\pi_1\widehat{X};\mathbb{Q})\}$$

where $\langle\cdot,\cdot\rangle$ denotes the pairing between cohomology and homology of $B\pi_1\widehat{X}$.

The Novikov conjecture for $\widehat{X}$ is the statement that the higher signatures are stratified homotopy invariant (see, e.g., [100]). The strong Novikov conjecture for $\pi_1\widehat{X}$ states that an 'assembly map' between two K-homology groups is rationally injective. The latter conjecture is known for many classes of groups. The analytic approach to the Novikov conjecture is to use index theory to 'reduce' it to the strong Novikov conjecture. This reduces stratified homotopy invariance to the homotopy invariance of the index of a 'higher' signature operator and, by using the Hilsum-Skandalis replacement construction, we were able to prove this analytically on both the Witt and Cheeger space settings. (For Witt spaces the index of this higher signature can also be computed topologically, at least after changing coefficients to $\mathbb{Q}$, as a 'symmetric signature'. There are now two constructions of the symmetric signature of a Witt space, [13, 46], and the latter is stratified homotopy invariant.) In [5, Theorem 5.8] we carry out the analytic approach to the Novikov conjecture for Cheeger spaces.

Theorem 5.9. *Let $\widehat{X}$ be a Cheeger space. If the strong Novikov conjecture is true for $\pi_1 X$, then the higher signatures of $\widehat{X}$ are stratified homotopy invariants.*

5.5 Refined intersection (co)homology

In this subsection we explain a 'de Rham theorem' showing that the de Rham cohomology groups defined analytically above have sheaf-theoretic descriptions. Thus we will shift gears and work more topologically. We will show that the cohomology corresponding to a mezzoperversity lies 'in between' the two middle perversities. To make this precise we will first recall the sheaf-theoretic approach to intersection homology.

It is convenient to use a notation that emphasizes dimension as opposed to depth. So given $\widehat{X}$, an n-dimensional pseudomanifold, let

$$\emptyset = X_{-1} \subseteq X_0 \subseteq \cdots \subseteq X_{n-2} \subseteq X_n = \widehat{X}$$

be an increasing sequence of closed subsets such that the stratum

$$S_k = X_k \setminus X_{k-1}$$

is either empty or a smooth manifold of dimension k. The regular part of $\widehat{X}$ is equal to $S_n = \widehat{X} \setminus X_{n-2}$. Let

$$\mathcal{V}_k = \widehat{X} \setminus X_{n-k}$$

be the increasing sequence of open subsets of $\widehat{X}$, starting with $\mathcal{V}_2 = S_n = \widehat{X}^{\mathrm{reg}}$ and satisfying

$$\mathcal{V}_{k+1} = \mathcal{V}_k \cup S_{n-k}.$$

Denote the inclusions by

$$\mathcal{V}_k \xhookrightarrow{i_k} \mathcal{V}_{k+1} \xhookleftarrow{j_k} S_{n-k}.$$

Let $\mathrm{Sh}(\widehat{X})$ denote the category of sheaves of $\mathbb{R}$-vector spaces on $\widehat{X}$. A constructible differential graded sheaf is a complex $\mathbf{A}^\bullet$ such that $\mathbf{A}^\bullet|_{S_k}$ has locally constant cohomology sheaves and has finitely generated stalk cohomology. We denote the category of constructible differential graded sheaves by $\mathrm{Sh}^\bullet(\widehat{X})$. The derived category of bounded constructible complexes, $D(\widehat{X})$, has as objects the cochain complexes of bounded constructible sheaves but quasi-isomorphic chain complexes are now considered isomorphic. Every object $\mathbf{A}^\bullet$ in $D(\widehat{X})$ has associated to it a derived sheaf: the sheafification of the presheaf

$$\mathcal{U} \mapsto \mathrm{H}^\bullet(\mathbf{A}^\bullet(\mathcal{U})),$$

which we denote $\mathbf{H}^\bullet(\mathbf{A}^\bullet) \in \mathrm{Sh}^\bullet(\widehat{X})$. The global hypercohomology of the derived sheaf is the cohomology of the global sections of an injective (more generally, any acyclic) resolution and is denoted $\mathbb{H}^\bullet(\widehat{X}; \mathbf{A}^\bullet)$. For a review of these concepts, see [54, 26, 12, 71].

A perversity $\overline{p}$ is a function from $\{2, 3, \ldots\}$ to $\mathbb{N}_0$ such that $\overline{p}(c)$ and also $\overline{q}(c) = c - 2 - \overline{p}(c)$ are positive and nondecreasing functions of c. We refer to $\overline{q}(c)$ as the dual perversity to $\overline{p}(c)$. The sheaf theoretic formulation of intersection cohomology in [54] assigns to each perversity $\overline{p}$ an object in the derived category

$$\mathbf{IC}^\bullet_{\overline{p}} \in D(\widehat{X})$$

with the property that its hypercohomology is the intersection cohomology of $\widehat{X}$ with perversity $\overline{p}$,

$$\mathrm{IH}^i_{\overline{p}}(\widehat{X}) = \mathbb{H}^i(\widehat{X}; \mathbf{IC}^\bullet_{\overline{p}}).$$

There are axioms characterizing $\mathbf{IC}^\bullet_{\overline{p}}$. For example: any constructible bounded complex of sheaves of real vector spaces, $\mathbf{S}^\bullet$, that satisfies

(a) Normalization: There is an isomorphism of the restriction of $\mathbf{S}^\bullet$ to the regular part $\mathcal{V}_2$ of $\widehat{X}$ and the constant rank 1 sheaf over $\mathcal{V}_2$,

$$\nu^{\mathbf{S}} : \mathbb{R}_{\mathcal{V}_2} \xrightarrow{\cong} \mathbf{S}^\bullet\big|_{\mathcal{V}_2},$$

(b) Lower bound: $\mathrm{H}^\ell(i_x^* \mathbf{S}^\bullet) = 0$ for any $x \in \widehat{X}$ and $\ell < 0$,

(c) $\overline{p}$-stalk vanishing: $\mathrm{H}^\ell(i_x^* \mathbf{S}^\bullet) = 0$ for any $x \in \mathcal{V}_{k+1} \setminus \mathcal{V}_2$ and $\ell > \overline{p}(k)$,

(d) $\overline{p}$-costalk vanishing: $\mathrm{H}^\ell(i_x^! \mathbf{S}^\bullet) = 0$ for any $x \in S_{n-k}$ and $\ell \leq \overline{p}(k) + n - k + 1$, defines $\mathbf{IC}^\bullet_{\overline{p}}$ in $D(\widehat{X})$.

There is also a beautiful construction of $\mathbf{IC}^\bullet_{\overline{p}}$ in $D(\widehat{X})$ due to Deligne using only standard constructions of sheaf theory, namely

$$\mathbf{IC}^\bullet_{\overline{p}} = \tau_{\leq \overline{p}(n)} Ri_{n*} \cdots \tau_{\leq \overline{p}(2)} Ri_{2*} \mathbb{R}_{\mathcal{V}_2}.$$

We can think of this as efficiently patching together the local computation:

$$x \in S_{n-k} \implies H^i(\mathbf{IC}^\bullet_{\overline{p}})_x \cong \begin{cases} \mathbb{H}^i(Z; \mathbf{IC}^\bullet_{\overline{p}}) & i \leq \overline{p}(k), \\ 0 & i > \overline{p}(k). \end{cases}$$

Now let us modify this theory to admit mezzoperversities. Say that a constructible bounded complex of sheaves $\mathbf{S}^\bullet$ is a *refined middle perversity complex of sheaves* if it satisfies the axioms:

(a) Normalization, (b) Lower bound,
(c) $\overline{n}$-stalk vanishing, (d) $\overline{m}$-costalk vanishing,

and denote by $RP(\widehat{X})$ the full subcategory of $D(\widehat{X})$ with these objects.

The category $RP(\widehat{X})$ contains $\mathbf{IC}^\bullet_{\overline{n}}$ and $\mathbf{IC}^\bullet_{\overline{m}}$. Any complex $\mathbf{S}^\bullet \in RP(\widehat{X})$ has unique maps [1, Proposition 1.6]

$$\mathbf{IC}^\bullet_{\overline{m}} \longrightarrow \mathbf{S}^\bullet \longrightarrow \mathbf{IC}^\bullet_{\overline{n}}$$

compatible with the normalizations on $\mathcal{V}_2$, and factoring the canonical map $\mathbf{IC}^\bullet_{\overline{m}} \longrightarrow \mathbf{IC}^\bullet_{\overline{n}}$. In this sense $\mathbf{S}^\bullet$ is 'in between' $\mathbf{IC}^\bullet_{\overline{m}}$ and $\mathbf{IC}^\bullet_{\overline{n}}$. If $\widehat{X}$ is Witt then $\mathbf{IC}^\bullet_{\overline{m}} \cong \mathbf{IC}^\bullet_{\overline{n}}$ and this is the only object in $RP(\widehat{X})$.

If $\widehat{X}$ is not Witt then every object $\mathbf{S}^\bullet \in RP(\widehat{X})$ is obtained from $\mathbf{IC}^\bullet_{\overline{m}}$ by the 'addition' of a suitably compatible collection, $\mathcal{L}$, of locally constant sheaves, one on each stratum where the Witt condition does not hold. We refer to this collection of locally constant sheaves as a *topological mezzoperversity* [1, Definition 2.1]. Given a topological mezzoperversity $\mathcal{L} = (W(S_{n-3}), W(S_{n-5}), \ldots)$, Banagl [10] has constructed a modified truncation function $\tau_{\leq \mathcal{L}(\cdot)}$ and we can define

$$\mathbf{IC}^\bullet_{\mathcal{L}} = \tau_{\leq \mathcal{L}(n)} Ri_{n*} \cdots \tau_{\leq \mathcal{L}(2)} Ri_{2*} \mathbb{R}_{\mathcal{U}_2}.$$

This satisfies the local computation

$$x \in S_{n-k},\ k\,\text{even} \implies H^i(\mathbf{IC}^\bullet_{\mathcal{L}})_x \cong \begin{cases} \mathbb{H}^i(Z; \mathbf{IC}^\bullet_{\mathcal{L}}) & i \leq \overline{m}(k), \\ 0 & i > \overline{m}(k), \end{cases}$$

$$x \in S_{n-k},\ k\,\text{odd} \implies H^i(\mathbf{IC}^\bullet_{\mathcal{L}})_x \cong \begin{cases} \mathbb{H}^i(Z; \mathbf{IC}^\bullet_{\mathcal{L}}) & i < \overline{m}(k), \\ W(S_{n-k}) & i = \frac{k-1}{2}, \\ 0 & i > \overline{n}(k) \end{cases}$$

and so is a refinement of Deligne's construction in which we truncate the cohomology not just at a given degree but at a given locally constant sheaf. We prove

[1, Theorem 2.2] that every refined middle perversity complex of sheaves is given by $\mathbf{IC}^\bullet_{\mathcal{L}}$ for some topological mezzoperversity $\mathcal{L}$.

Now endow $\widehat{X}$ with a rigid suitably scaled iterated wedge metric g, and let $\mathcal{W}$ be a de Rham mezzoperversity. Define a presheaf on $\widehat{X}$ by assigning to each open set $\mathcal{U} \subseteq \widehat{X}$ the vector space

$$\mathcal{D}_{\mathcal{W}}(\mathcal{U}) = \{\omega \in \mathcal{D}_{\mathcal{W}}(d) : \operatorname{supp}\omega \subseteq (\mathcal{U} \cap \widehat{X}^{\mathrm{reg}})\}$$

and assigning to each inclusion $j : \mathcal{U} \subseteq \mathcal{V}$ the restriction map

$$j^* : \mathcal{D}_{\mathcal{W}}(\mathcal{V}) \longrightarrow \mathcal{D}_{\mathcal{W}}(\mathcal{U}).$$

The exterior derivative makes this into a complex of presheaves. Denote the sheafification by

$$\mathbf{L}^2_{\mathcal{W}}\mathbf{\Omega}^\bullet \in \operatorname{Sh}^\bullet(\widehat{X}).$$

(Note that, for $i \hookrightarrow \widehat{X}^{reg} \to \widehat{X}$, we are using $i_!(D_{\mathcal{W}}(d))$.)

Theorem 5.10. *Let $\widehat{X}$ be a pseudomanifold endowed with a rigid suitably scaled iterated wedge metric. Given any de Rham mezzoperversity $\mathcal{W}$, the sheaf $\mathbf{L}^2_{\mathcal{W}}\mathbf{\Omega}^\bullet$ defines an object in $RP(\widehat{X})$, and hence we have an equality in $D(\widehat{X})$,*

$$\mathbf{L}^2_{\mathcal{W}}\mathbf{\Omega}^\bullet \cong \mathbf{IC}^\bullet_{\mathcal{L}(\mathcal{W})}$$

for some topological mezzoperversity $\mathcal{L}(\mathcal{W})$. This establishes a one-to-one correspondence between de Rham mezzoperversities and topological mezzoperversities

$$\mathcal{W} \leftrightarrow \mathcal{L}(\mathcal{W}).$$

Thus we have an equivalence between the hypercohomology of a refined middle perversity complex of sheaves and the L^2-de Rham cohomology with Cheeger ideal boundary conditions,

$$\mathrm{H}^*_{\mathcal{W},\mathrm{dR}}(\widehat{X}) = \mathbb{H}^\bullet(\widehat{X}; \mathbf{IC}^\bullet_{\mathcal{L}(\mathcal{W})}).$$

On any pseudomanifold there are two canonical choices of mezzoperversity. Indeed, given a flat bundle $\mathcal{H} \longrightarrow Y$, two canonical flat subbundles are the zero bundle and the entire bundle $\mathcal{H}$ itself. Let $\mathcal{W}_0$ be the mezzoperversity corresponding to always choosing the zero bundle, and $\mathcal{W}_{\max}$ the mezzoperversity corresponding to always choosing the entire bundle of vertical cohomology. This theorem shows that

$$\mathrm{H}^*_{\mathcal{W}_0,\mathrm{dR}}(\widehat{X}) = \mathrm{IH}^*_{\overline{m}}(\widehat{X}), \quad \mathrm{H}^*_{\mathcal{W}_{\max},\mathrm{dR}}(\widehat{X}) = \mathrm{IH}^*_{\overline{n}}(\widehat{X}).$$

(The latter can be identified with the cohomology of $\mathcal{D}_{\max}(d)$ and the former with the cohomology of $\mathcal{D}_{\min}(d)$, so in these cases the isomorphisms follow from the work of Cheeger described in §2.)

The definition of $RP(\widehat{X})$ only requires a *topological stratification.* In this generality, in [10], Banagl studied a category $SD(\widehat{X})$ defined just like $RP(\widehat{X})$ but with the extra requirement of self-duality. Spaces $\widehat{X}$ with $SD(\widehat{X})$ non-empty are called L-spaces. We show in [1] that a Thom-Mather stratified space is an L-space if and only if it is a Cheeger space.

6 Stratified spaces and their resolutions

In this section we will review the relation between stratified spaces and manifolds with corners.

6.1 Thom-Mather stratified spaces

We follow [83, 114], see also [97]. Our aim in this section is to discuss the definition and some basic concepts culminating in the 'local cone structure' of a Thom-Mather stratified space.

All topological spaces will be 'nice' meaning Hausdorff, locally compact topological space with a countable basis for its topology. A subset A of a topological space V is locally closed if every point $a \in A$ has a neighborhood $\mathcal{U}$ in V such that $A \cap \mathcal{U}$ is closed in $\mathcal{U}$. A collection $\mathcal{S}$ of subsets of V is locally finite if every point $v \in V$ has a neighborhood that intersects only finitely many sets in $\mathcal{S}$.

Definition 1. *A (Thom-Mather)* stratified space *is a triple* $(\widehat{X}, \mathscr{S}, \mathscr{T})$ *in which* $\widehat{X}$ *is the space,* $\mathscr{S}$ *is a decomposition of* $\widehat{X}$ *into 'strata', and* $\mathscr{T}$ *is a collection of 'tubes', one per stratum. Specifically:*

- $\widehat{X}$ *is a 'nice' topological space.*
- $\mathscr{S}$ *is a locally finite collection of locally closed subsets of* $\widehat{X}$*. Each* $Y \in \mathscr{S}$ *is a topological manifold in the induced topology and is provided with a smooth structure. The family* $\mathscr{S}$ *satisfies the frontier condition*

$$Y, Y' \in \mathscr{S} \text{ and } Y \cap \overline{Y'} \neq \emptyset \implies Y \subseteq \overline{Y'}. \tag{6.1}$$

- $\mathscr{T}$ *contains, for each* $Y \in \mathscr{S}$*, a triple* $(\mathcal{T}_Y, \pi_Y, \rho_Y)$ *in which* $\mathcal{T}_Y$ *is an open neighborhood of* Y *in* $\widehat{X}$*, and*

$$\pi_Y : \mathcal{T}_Y \longrightarrow Y, \quad \rho_Y : \mathcal{T}_Y \longrightarrow \mathbb{R}^+$$

are continuous functions such that $\pi_Y\big|_Y = \mathrm{id}_Y$ *and* $\rho_Y^{-1}(0) = Y$*. The neighborhood* $\mathcal{T}_Y$ *is known as a tubular neighborhood of* Y *in* $\widehat{X}$ *and the triple* $(\mathcal{T}_Y, \pi_Y, \rho_Y)$ *as the control data of* Y *in* $\widehat{X}$*.*

The control data are compatible in that if $Y, Y' \in \mathscr{S}$ *are distinct strata such that* $\mathcal{T}_Y \cap Y' \neq \emptyset$ *then*

$$(\pi_Y, \rho_Y)\big|_{\mathcal{T}_Y \cap Y'} : \mathcal{T}_Y \cap Y' \longrightarrow Y \times (0, \infty) \tag{6.2}$$

is a smooth submersion, and

$$\pi_Y \circ \pi_{Y'} = \pi_Y, \quad \rho_Y \circ \pi_{Y'} = \rho_Y \tag{6.3}$$

when both sides are defined (i.e., on $\mathcal{T}_Y \cap \pi_{Y'}^{-1}(\mathcal{T}_Y \cap Y')$*).*

A stratified space is a *pseudomanifold* if the codimension of each singular stratum is at least two.

Example 9. *The* trivial stratification of a smooth manifold L *has a single stratum equal to* L *with control data* $\mathcal{T}_L = L$, $\pi_L = \mathrm{id}$, $\rho_L \equiv 0$.

If $(\widehat{X}, \mathscr{S}, \mathscr{T})$ *is a stratified space and* L *is a smooth manifold then* $\widehat{X} \times L$ *is naturally a stratified space with strata* $Y \times L$, $Y \in \mathscr{S}$.

If $(\widehat{X}, \mathscr{S}, \mathscr{T})$ *is a stratified space and* $A \subseteq \widehat{X}$ *is either an open subset or a closed subset that is a union of strata, then* A *is naturally a stratified space with strata* $Y \cap A$, $Y \in \mathscr{S}$.

Example 10 (Whitney stratified spaces). *Let* L *be a smooth manifold and let* R, S *be smooth submanifolds of* L. *We say that* (R, S) *satisfies Whitney's condition* (**B**) *if for every*

- $x \in R$,
- *sequence* $(y_k) \subseteq S$ *such that* $y_k \to x$ *and* $T_{y_k} S \to \tau \subseteq T_x L$ *(as sections of a Grassmanian bundle over* L*),*
- *sequence* $(x_k) \subseteq R$ *such that* $x_k \to x$, $x_k \neq y_k$ *and the secant lines* $\overline{x_k y_k}$ *(defined in a coordinate chart) converge to* $\ell \subseteq T_x L$

we have $\ell \subseteq \tau$.

A Whitney stratification of a subset $A \subseteq L$ *is a locally finite collection* $\mathscr{S}$ *of pairwise disjoint smooth submanifolds of* L *covering* A, *such that any* $Y, Y' \in \mathscr{S}$ *satisfy the frontier condition* (6.1) *and Whitney's condition* (**B**). *Spaces admitting a Whitney stratification include algebraic subvarieties of* $\mathbb{R}^n$ *or* $\mathbb{C}^n$ *[117], semianalytic and sub analytic sets [80], and 'o-minimal structures' [113].*

Whitney stratified spaces are intimately connected to Thom-Mather stratified spaces. Mather proved [83] that every Whitney stratified space admits a Thom-Mather stratification. Moreover, Teufel [107] showed that every Thom-Mather stratified space can be embedded into a Euclidean space so that its image is Whitney stratified.

Given $Y, Y' \in \mathscr{S}$ we write

$$Y < Y' \iff Y \cap \overline{Y'} \neq \emptyset \iff Y \subseteq \overline{Y'}.$$

This is an order on $\mathscr{S}$ and (6.2) shows that $Y < Y' \implies \dim Y < \dim Y'$. Another useful consequence of (6.1) is that the closure of a stratum is a union of strata. Indeed, this is equivalent to the frontier condition (6.1).

The *depth of a stratum* is

$$\mathrm{depth}(Y) = \sup\{n \in \mathbb{N}_0 : \exists Y_i \in \mathscr{S} \text{ s.t. } Y = Y_0 < Y_1 < \cdots < Y_n\}$$

and the *depth of a stratification* is

$$\mathrm{depth}(\mathscr{S}) = \sup\{\mathrm{depth}(Y) : Y \in \mathscr{S}\}.$$

The regular part of $\widehat{X}$ consists of those strata of depth zero,

$$\widehat{X}^{\mathrm{reg}} = \bigcup\{Y \in \mathscr{S} : \mathrm{depth}(Y) = 0\}.$$

Two stratified spaces $(\widehat{X}, \mathscr{S}, \mathscr{T})$ and $(\widehat{X}', \mathscr{S}', \mathscr{T}')$ are *equivalent* if $\widehat{X} = \widehat{X}'$ with the same strata (as smooth manifolds) and the control data at a stratum Y,

$$(\mathcal{T}_Y, \pi_Y, \rho_Y), \quad (\mathcal{T}'_Y, \pi'_Y, \rho'_Y),$$

coincide in a neighborhood of Y.

A *morphism between stratified spaces* $(\widehat{X}, \mathscr{S}_X, \mathscr{T}_X)$ and $(\widehat{N}, \mathscr{S}_N, \mathscr{T}_N)$ is a continuous map

$$f : \widehat{X} \longrightarrow \widehat{N}$$

that maps strata into strata such that, whenever $Y \in \mathscr{S}_X$, and $Q \in \mathscr{S}_N$ are such that $f(Y) \subseteq Q$, we have

$$f\big|_Y : Y \longrightarrow Q \text{ is smooth}$$

and, for some $(\widehat{X}, \mathscr{S}_X, \mathscr{T}'_X)$ equivalent to $(\widehat{X}, \mathscr{S}_X, \mathscr{T}_X)$,

$$f(\mathcal{T}'_Y) \subseteq \mathcal{T}_Q, \quad f \circ \pi'_Y = \pi_Q \circ f, \quad \rho'_Y = \rho_Q \circ f.$$

An *isomorphism of stratified spaces* is defined to be a bijective morphism whose inverse is also a morphism and hence is a homeomorphism that restricts to a diffeomorphism on each stratum.

A *submersive morphism* $f : \widehat{X} \longrightarrow \widehat{N}$ is a morphism such that, whenever $Y \in \mathscr{S}_X$, and $Q \in \mathscr{S}_N$, are such that $f(Y) \subseteq Q$, we have

$$f\big|_Y : Y \longrightarrow Q \text{ is a submersion.}$$

If $\widehat{X}_0$ is a stratified space and L is a smooth manifold, then $\widehat{X}_0 \times L$ has a natural stratification with strata $\{Y \times L : Y \in \mathscr{S}\}$, and the projection onto the second factor

$$\widehat{X}_0 \times L \xrightarrow{\pi_L} L$$

is a submersive morphism. We say that a submersive morphism $f : \widehat{X} \longrightarrow L$ is a *trivial fibration* if it participates in a commutative diagram

$$\begin{array}{ccc} \widehat{X} & \xrightarrow{F} & \widehat{X}_0 \times L \\ & {\scriptstyle f}\searrow \quad \swarrow{\scriptstyle \pi_L} & \\ & L & \end{array}$$

with F an isomorphism of stratified spaces F.

A submersive morphism $f : \widehat{X} \longrightarrow L$ onto a smooth manifold (with its trivial stratification) is a *locally trivial fibration* if every point $x \in L$ has a neighborhood $\mathcal{U}$ such that $f\big|_{f^{-1}(\mathcal{U})} : f^{-1}(\mathcal{U}) \longrightarrow \mathcal{U}$ is a trivial fibration. If L is connected then all of the fibers are isomorphic stratified spaces and we write

$$\widehat{X}_0 - \widehat{X} \xrightarrow{f} L$$

where $\widehat{X}_0$ is a fiber $f^{-1}(\mathrm{pt})$. (Note that the 'structure group' consists of isomorphisms of stratified spaces $\widehat{X}_0 \longrightarrow \widehat{X}_0$.)

Theorem 6.1 (Thom's first isotopy lemma for stratified spaces). *If $f : \widehat{X} \longrightarrow L$ is a proper submersive morphism between a stratified space $\widehat{X}$ and a smooth manifold L (with its trivial stratification), then f is a locally trivial fibration.*

This can be thought of as the version of Ehresmann's theorem for stratified spaces, and it indeed reduces to that theorem if $\widehat{X}$ is smooth. A proof can be found in [114, Theorem 2.6] (see also [83, Corollary 10.2, Proposition 11.1], [97, 3.9.2]).

The main tool in the proof of Theorem 6.1 is the notion of a controlled vector field on a stratified space (see [83, §9], [114, §2]). A collection of smooth vector fields, one per stratum,

$$\xi = \{\xi_Y \in \mathcal{C}^\infty(Y, TY) : Y \in \mathscr{S}\}$$

is a *controlled vector field* if whenever $Y, Y' \in \mathscr{S}$, $Y < Y'$, and $\zeta \in \mathcal{T}_Y \cap Y'$ is sufficiently close to Y,

$$D\pi_Y(\xi_{Y'}(\zeta)) = \xi_Y(\pi_Y(\zeta)), \quad \xi_{Y'}\rho_Y = 0.$$

These are analogous to smooth vector fields on a smooth manifold in at least two ways: First, the flow of a controlled vector field generates a (local) one parameter group of stratified space morphisms from an open subset $D_\xi \subseteq \widehat{X} \times \mathbb{R}$ to $\widehat{X}$. Secondly, if L is a smooth manifold and $f : \widehat{X} \longrightarrow L$ is a submersive morphism, then every vector field on L has a lift to a controlled vector field on $\widehat{X}$.

Let us reexamine the control data of a stratified space in light of Theorem 6.1. If Y is a stratum of $\widehat{X}$ with control data $(\mathcal{T}_Y, \pi_Y, \rho_Y)$ then $\mathcal{T}_Y$, as an open subset of $\widehat{X}$, inherits a stratification with strata

$$\mathscr{S}_{\mathcal{T}_Y} = \{\mathcal{T}_Y \cap Y' : Y' \in \mathscr{S}, Y \leq Y'\}.$$

The maps

$$(\pi_Y, \rho_Y)\big|_{\mathcal{T}_Y \setminus Y} : \mathcal{T}_Y \setminus Y \longrightarrow Y \times (0, \varepsilon), \quad \pi_Y : \mathcal{T}_Y \longrightarrow Y$$

are (for ε sufficiently small) proper submersive morphisms and hence locally trivial fibrations.

Let $a \in (0, \varepsilon)$ and $K_Y = \rho_Y^{-1}(a)$. The canonical vector field on $(0, \varepsilon)$ lifts to a controlled vector field on $\mathcal{T}_Y \setminus Y$ and we can use its flow to identify, as stratified spaces, $\mathcal{T}_Y \setminus Y$ with $K_Y \times (0, \varepsilon)$. The restriction $\psi_Y = \pi_Y\big|_{\rho_Y^{-1}(a)}$ is a locally trivial fibration

$$\psi_Y : K_Y \longrightarrow Y$$

and the fiber over a point $q \in Y$ is a stratified space Z_q called the *link of $\widehat{X}$ at Y*. Local triviality implies that the links over points in the same connected component of Y are isomorphic as stratified spaces (this explains the name 'isotopy lemma'). It follows that the fiber over q of the locally trivial fibration π_Y is the cone over Z_q,

$$C_\varepsilon(Z_q) = Z_q \times [0, \varepsilon)/Z_q \times \{0\}$$

with its natural stratification. Thus Thom's first isotopy lemma shows that a Thom-Mather stratified space is, locally near each stratum, a bundle of cones over simpler stratified spaces.

6.2 Resolving a single stratum

Given $(\widehat{X}, \mathscr{S}, \mathscr{T})$ a stratified space let

$$\mathscr{S}^* = \{Y \in \mathscr{S} : \mathrm{depth}_{\mathscr{S}}(Y) = \mathrm{depth}(\mathscr{S})\},$$
$$\widehat{X}^* = \bigcup\{Y \in \mathscr{S}^*\}, \quad \mathcal{T}_{\widehat{X}^*} = \bigcup\{\mathcal{T}_Y : Y \in \mathscr{S}^*\}.$$

We assume, replacing $(\widehat{X}, \mathscr{S}, \mathscr{T})$ with an equivalent stratified space if necessary, that

$$Y, Y' \in \mathcal{S}^* \implies \mathcal{T}_Y \cap \mathcal{T}_{Y'} = \emptyset$$

and we let $\pi_{\widehat{X}^*} : \mathscr{T}_{\widehat{X}^*} \longrightarrow \widehat{X}^*$, $\rho_{\widehat{X}^*} : \mathscr{T}_{\widehat{X}^*} \longrightarrow \mathbb{R}^+$ be the obvious maps. Note that by the frontier condition $\widehat{X}^*$ is a closed subset of $\widehat{X}$.

As explained above, Thom's first isotopy lemma implies that $\mathscr{T}_{\widehat{X}^*} \setminus \widehat{X}^*$ is isomorphic, as a stratified space, to $K_{\widehat{X}^*} \times (0, \varepsilon)$ for a stratified space participating in a locally trivial fibration

$$\psi_{\widehat{X}^*} : K_{\widehat{X}^*} \longrightarrow \widehat{X}^*.$$

For $a \in (0, \varepsilon)$, let

$$\widehat{X}^a = \widehat{X} \setminus \{\zeta \in \mathscr{T}_{Y^*} : \rho_{\widehat{X}^*}(\zeta) < a\}.$$

This is an example of a stratified space with boundary (formalized below in Definition 2). Note that the particular value of a is immaterial since different values yield isomorphic stratified spaces $K_{\widehat{X}^*} \times (a, \varepsilon)$. We will not distinguish between different values of a, but denote this space by

$$[\widehat{X}; \widehat{X}^*]$$

and refer to it as the *radial blow-up of* $\widehat{X}$ *along* $\widehat{X}^*$.

Note that the boundary $K_{\widehat{X}^*}$ inherits from $\widehat{X}$ the locally trivial fibration $\psi_{\widehat{X}^*}$ and also a *collar neighborhood.* By which we will mean an open neighborhood of $K_{\widehat{X}^*}$, $\mathcal{C}_{K_{\widehat{X}^*}} \subseteq [\widehat{X}; \widehat{X}^*]$, and a map

$$\Theta : \mathcal{C}_{K_{\widehat{X}^*}} \longrightarrow K_{\widehat{X}^*} \times \mathbb{R}^+$$

which is a homeomorphism onto its image and such that

$$\Theta(\zeta) = (\zeta, 0) \text{ for all } \zeta \in K_{\widehat{X}^*} \text{ and}$$
$$\Theta(Y \cap \mathcal{C}_{K_{\widehat{X}^*}}) \subseteq (Y \cap K_{\widehat{X}^*}) \times \mathbb{R}^+ \text{ for all } Y \in \mathscr{S} \setminus \mathscr{S}^*.$$

Finally note that

$$\mathrm{depth}([\widehat{X}; \widehat{X}^*]) = \mathrm{depth}(\widehat{X}) - 1,$$

so that we have simplified the stratification. On the other hand, we can recover $\widehat{X}$ by collapsing the fibers of the boundary fibration

$$\widehat{X} = [\widehat{X}; \widehat{X}^*] / \{K_{\widehat{X}} \ni \zeta \sim \psi_{\widehat{X}^*}(\zeta) \in \widehat{X}^*\}.$$

We denote the projection map by

$$\beta : [\widehat{X}; \widehat{X}^*] \longrightarrow \widehat{X}$$

and refer to it as the 'blow-down' map.

Hence for many purposes working on $\widehat{X}$ is equivalent to working on $[\widehat{X}; \widehat{X}^*]$ together with its boundary fibration $\psi_{\widehat{X}^*}$. The resolution of $\widehat{X}$ is the result of iterating this process to replace strata by boundaries.

6.3 Stratifications with collared corners

In this section we summarize part of §5 of [114] (where the terminology 'abstract stratifications with faces' is used). By a manifold with corners we mean a smooth manifold locally modeled on $(\mathbb{R}^+)^\ell$ with embedded boundary hypersurfaces, e.g., see [89]. (For notational convenience, we allow the empty set as a boundary hypersurface.) A manifold with corners always has (compatible) collar neighborhoods of its boundary hypersurfaces (see e.g., [6, Proposition 1.2]). By a manifold with collared corners we mean that we have fixed a choice of collars.

Definition 2. A stratified space with collared corners *is a quintuple* $(\widehat{X}, \mathscr{S}, \mathscr{T}, \mathscr{F}, \mathscr{C})$ *in which* $\widehat{X}$ *is the space,* $\mathscr{S}$ *is a decomposition of* $\widehat{X}$ *into 'strata',* $\mathscr{T}$ *is a collection of 'tubes', one per stratum,* $\mathscr{F}$ *are the 'boundary hypersurfaces' of* $\widehat{X}$*, and* $\mathscr{C}$ *is a collection of 'collars', one per boundary hypersurface. Specifically:*

- $\widehat{X}$ *is a 'nice' topological space.*
- $\mathscr{F}$ *is a collection of closed subsets of* $\widehat{X}$ *and* $\mathscr{S}$ *is a locally finite collection of locally closed subsets of* $\widehat{X}$*.*
- $\mathscr{C}$ *consists, for each* $K \in \mathscr{F}$*, of an open neighborhood* $\mathcal{U}_K \subseteq \widehat{X}$ *of* K *and a homeomorphism*

$$\Theta_K : \mathcal{U}_K \longrightarrow K \times \mathbb{R}^+$$

such that $\Theta_K\big|_K = \mathrm{id}_K$*, and*

$$\Theta_K(Y \cap \mathcal{U}_K) \subseteq (Y \cap K) \times \mathbb{R}^+ \text{ for every } Y \in \mathscr{S}.$$

The pair $(\mathcal{U}_K, \Theta_K)$ *is known as a collar of* K*. We denote the composition of* Θ_K *with the projection onto* K *by* θ_K *and the composition with the projection onto* $\mathbb{R}^+$ *by* x_K*, so that*

$$\Theta_K(\zeta) = (\theta_K(\zeta), x_k(\zeta)).$$

- *Each* $Y \in \mathscr{S}$ *is a topological manifold with boundary in the induced topology and is provided with the structure of a smooth manifold with corners whose boundary hypersurfaces are* $Y \cap K$ *for* $K \in \mathscr{F}$*. These are collared by* $\Theta_{K \cap Y} = \Theta_K\big|_{Y \cap \mathcal{U}_K}$*. The family* $\mathscr{S}$ *satisfies the frontier condition*

$$Y, Y' \in \mathscr{S} \text{ and } Y \cap \overline{Y'} \neq \emptyset \implies Y \subseteq \overline{Y'}.$$

- *$\mathscr{T}$ contains, for each $Y \in \mathscr{S}$, a triple $(\mathcal{T}_Y, \pi_Y, \rho_Y)$ in which $\mathcal{T}_Y$ is an open neighborhood of Y in $\widehat{X}$, and*

$$\pi_Y : \mathcal{T}_Y \longrightarrow Y, \quad \rho_Y : \mathcal{T}_Y \longrightarrow \mathbb{R}^+$$

are continuous functions such that $\pi_Y\big|_Y = \mathrm{id}_Y$ and $\rho_Y^{-1}(0) = Y$. The neighborhood $\mathcal{T}_Y$ is known as a tubular neighborhood of Y in $\widehat{X}$ and the triple $(\mathcal{T}_Y, \pi_Y, \rho_Y)$ as the control data of Y in $\widehat{X}$.

The control data are compatible in that if $Y, Y' \in \mathscr{S}$ are distinct strata such that $\mathcal{T}_Y \cap Y' \neq \emptyset$ then

$$(\pi_Y, \rho_Y)\big|_{\mathcal{T}_Y \cap Y'} : \mathcal{T}_Y \cap Y' \longrightarrow Y \times (0, \infty)$$

is a smooth submersion, and

$$\pi_Y \circ \pi_{Y'} = \pi_Y, \quad \rho_Y \circ \pi_{Y'} = \rho_Y$$

when both sides are defined (i.e., on $\mathcal{T}_Y \cap \pi_{Y'}^{-1}(\mathcal{T}_Y \cap Y')$). Finally the control data are compatible with the collar data in that for all $Y \in \mathscr{S}$, and $K \in \mathscr{F}$

$$\pi_Y^{-1}(Y \cap K) = \mathcal{T}_Y \cap K, \quad \Theta_{Y \cap K} \circ \pi_X = (\pi_X \times \mathrm{id}_{\mathbb{R}^+}) \circ \Theta_K, \quad \rho_Y = \rho_Y \circ \theta_K$$

whenever both sides make sense.

The depth, dimension, and regular part of a stratified space with collared corners are defined as before. The interior of a stratified space with collared corners is the complement of the boundary hypersurfaces and is itself a stratified space.

Example 11. *Any stratified space $(\widehat{X}, \mathscr{S}, \mathscr{T})$ is naturally a stratified space with collared corners.*

If $(\widehat{X}, \mathscr{S}, \mathscr{T}, \mathscr{F}, \mathscr{C})$ is a stratified space with collared corners and L is a smooth manifold with collared corners then $\widehat{X} \times L$ is naturally a stratified space with collared corners. Its strata are $Y \times L$, $Y \in \mathscr{S}$, and its boundary hypersurfaces are $K \times L^\circ$, $K \in \mathscr{F}$, and $\widehat{X}^\circ \times H$, with H a boundary hypersurface of L.

In particular $\widehat{X} \times \mathbb{R}^+$ is a stratified space with collared corners where $\mathbb{R}^+$ has boundary $\{0\}$ with collar the obvious map $\mathbb{R}^+ \longrightarrow \{0\} \times \mathbb{R}^+$.

If $(\widehat{X}, \mathscr{S}, \mathscr{T}, \mathscr{F}, \mathscr{C})$ is a stratified space with collared corners and $A \subseteq \widehat{X}$ is either an open set, a closed subset that is a union of strata, or a boundary hypersurface, then A is naturally a stratified space with strata $Y \cap A$, $Y \in \mathscr{S}$, and boundary hypersurfaces $K \cap A$, $K \in \mathscr{F}$.

Two stratified spaces with collared corners $(\widehat{X}, \mathscr{S}, \mathscr{T}, \mathscr{F}, \mathscr{C})$ and $(\widehat{X}', \mathscr{S}', \mathscr{T}', \mathscr{F}', \mathscr{C}')$ are *equivalent* if

$$\widehat{X} = \widehat{X}', \quad \mathscr{S} = \mathscr{S}', \quad \mathscr{F} = \mathscr{F}',$$

the control data coincide in a neighborhood of each stratum, and the collar data coincide in a neighborhood of each boundary hypersurface. Up to equivalence we can assume that

$$\mathcal{T}_Y \cap \mathcal{T}_{Y'} \neq \emptyset \implies Y \leq Y' \text{ or } Y' \leq Y.$$

If $(\widehat{X}, \mathscr{S}_X, \mathscr{T}_X, \mathscr{F}_X, \mathscr{C}_X)$ and $(\widehat{N}, \mathscr{S}_N, \mathscr{T}_N, \mathscr{F}_N, \mathscr{C}_N)$ are stratified spaces with collared corners and $f : \widehat{X} \longrightarrow \widehat{N}$ is a continuous map, we say that it is *compatible with faces* if we can decompose the boundary hypersurfaces into 'horizontal' and 'vertical',

$$\mathscr{F}_X = \mathscr{F}_X^h(f) \sqcup \mathscr{F}_X^v(f), \quad \mathscr{F}_N = \mathscr{F}_N^h(f) \sqcup \mathscr{F}_N^v(f)$$

so that

a) the inverse image of a boundary hypersurface in $\mathscr{F}_N^h(f)$ is either empty or all of $\widehat{X}$,

b) f induces a bijection between $\mathscr{F}_X^v(f)$ and $\mathscr{F}_N^v(f)$ and if $K \in \mathscr{F}_N^v(f)$ then, near $f^{-1}(K)$,
$$\Theta_K \circ f = (f\big|_{f^{-1}(K)} \times \mathrm{id}_{\mathbb{R}^+}) \circ \Theta_{f^{-1}(K)},$$

c) near a boundary hypersurface $H \in \mathscr{F}_X^h(f)$,
$$f = (f\big|_H) \circ \theta_H.$$

A continuous map $f : \widehat{X} \longrightarrow \widehat{N}$ is a *morphism between stratified spaces with collared corners,* $(\widehat{X}, \mathscr{S}_X, \mathscr{T}_X, \mathscr{F}_X, \mathscr{C}_X)$ and $(\widehat{N}, \mathscr{S}_N, \mathscr{T}_N, \mathscr{F}_N, \mathscr{C}_N)$, if it is compatible with faces, sends strata to strata smoothly and intertwines control data (as in §6.1). A morphism is *submersive* if its restriction to each stratum is a submersion. A morphism is an *isomorphism* if it is a bijection and a morphism whose inverse is a morphism.

For example, if Y, Y' are strata of a stratified space with collared corners with $Y < Y'$ then, after potentially shrinking $\mathcal{T}_Y$, the maps

$$\pi_Y\big|_{\mathcal{T}_Y \cap Y'} : \mathcal{T}_Y \cap Y' \longrightarrow Y, \quad \rho_Y\big|_{\mathcal{T}_Y \cap Y'} : \mathcal{T}_Y \cap Y' \longrightarrow \mathbb{R}$$

are compatible with faces, with all of the faces being vertical for π_Y and horizontal for ρ_Y.

If K is a boundary hypersurface of a stratified space with collared corners and $(\mathcal{U}_K, \Theta_K)$ is its collar then, after potentially shrinking $\mathcal{U}_K$, the collar map

$$\Theta_K : \mathcal{U}_K \longrightarrow K \times \mathbb{R}^+$$

is an isomorphism of stratified spaces with collared corners.

The notions of trivial fibration and locally trivial fibration are entirely analogous to §6.1.

Theorem 6.2 (Thom's first isotopy lemma for stratified spaces with collared corners [114, Corollary 5.8]). *If $f : \widehat{X} \longrightarrow L$ is a proper submersive morphism between a stratified space with collared corners $\widehat{X}$ and a smooth manifold with collared corners L, then f is a locally trivial fibration.*

Proceeding as in §6.1, Thom's first isotopy lemma applied to the control data of a stratified space with collared boundary implies that such a space is, locally near each stratum, a bundle of cones over simpler stratified space with collared boundary. As anticipated above, this structure allows us to iterate the radial blow-up construction from §6.2. Starting from a stratified space, blowing-up the strata of greatest depth produces a stratified space with collared boundary, and iterating produces a stratified space with collared corners. For spaces of finite depth this leads to what Verona called a 'total decomposition' [114, §6.3.1] and we call a resolution [3, 6].

6.4 The resolution of a stratified space

In [3], we used Verona's constructions to prove an equivalence between the class of Thom-Mather stratified spaces and the class of manifolds with corners and iterated fibration structures, as introduced by Melrose. This resolution is already claimed in Thom [111] but was not phrased in this way. As explained above, we have found the presentation of a stratified space as a manifold with corners and iterated fibration structures very useful for carrying out analysis.

By a *collective boundary hypersurface* we mean a finite union of non-intersecting boundary hypersurfaces.

Definition 3 (Melrose [6, 3]). *A manifold with corners with an iterated fibration structure if a triple* $(L, \mathscr{F}, \Phi)$ *where* L *is a smooth manifold with corners,* $\mathscr{F}$ *consists of disjoint collective boundary hypersurfaces with every boundary hypersurface of* L *occurring in some element of* $\mathscr{F}$*, and* Φ *is a collection of locally trivial fibrations, one per element* $H \in \mathscr{F}$,

$$Z_H - H \xrightarrow{\phi_H} Y_H$$

where Z_H *and* Y_H *are smooth manifolds with corners. These satisfy:*

- *If* $H, K \in \mathscr{F}$ *intersect then* $\dim Y_H \neq \dim Y_K$ *and we write*

$$H < K \text{ if } \dim Y_H < \dim Y_K.$$

- *If* $H < K$ *then* Y_K *has a collective boundary hypersurface* Y_{HK} *participating in a fiber bundle* $\phi_{HK} : Y_{HK} \longrightarrow Y_H$ *such that the diagram*

$$\begin{array}{ccccc} H \cap K & \xrightarrow{\phi_K} & Y_{HK} & \subseteq & Y_K \\ & {\scriptstyle\phi_H} \searrow & & \swarrow {\scriptstyle\phi_{HK}} & \\ & & Y_H & & \end{array}$$

commutes.

Two useful properties of these structures are: (a) a manifold with corners and an iterated fibration structure admits collar neighborhoods compatible with the boundary fibrations, see [6, Proposition 3.7], and (b) the bases of the boundary fibrations inherit iterated fibration structures [6, Lemma 3.4].

If L is a smooth manifold with corners and H is a boundary hypersurface then, since H is embedded, there exists a smooth function

$$x_H : L \longrightarrow [0,1]$$

with $H = x_H^{-1}(\{0\})$ such that dx_H has no zeroes on H. We call any such a function a *boundary defining function* for H.

Let L and M be smooth manifolds with corners and let $\{x_{H_i}\}$, $\{\widetilde{x}_{G_j}\}$ be complete sets of defining functions for the boundary hypersurfaces of L and M respectively. A smooth map $f : L \longrightarrow M$ is a *b-map* if for each boundary hypersurface $G_j \subseteq M$,

$$\text{either } f^*\widetilde{x}_{G_j} = 0 \text{ or}$$

$$f^*\widetilde{x}_{G_j} = a_{G_j} \prod x_{H_i}^{e(i,j)}, \qquad \text{with } a_{G_j} \in \mathcal{C}^\infty(L,(0,\infty)) \text{ and } e(i,j) \in \mathbb{N}_0.$$

The first alternative occurs only if $f(L) \subseteq \partial M$, in which case we say that f is a boundary b-map; otherwise we say that f is an interior b-map.

A *morphism between manifolds with corners and boundary fibration structures*, $(L, \mathscr{F}_L, \Phi_L)$ and $(M, \mathscr{F}_M, \Phi_M)$ is a b-map

$$f : L \longrightarrow M$$

that sends the collective boundary hypersurface in $\mathscr{F}_L$ to collective boundary hypersurfaces in $\mathscr{F}_M$ such that, whenever $H \in \mathscr{F}_L$ and $G \in \mathscr{F}_M$ are such that $f(H) \subseteq G$, we have

$$\begin{array}{ccc} H & \xrightarrow{\ f\ } & G \\ {\scriptstyle \phi_H}\downarrow & & \downarrow{\scriptstyle \phi_G} \\ Y_H & \xrightarrow{\ \overline{f}\ } & Y_G \end{array}$$

for some map $\overline{f}$. (Note that the maps $\overline{f}$ inherit the structure of morphisms between manifolds with corners and boundary fibration structures.)

A smooth action of a Lie group on a compact manifold gives rise to a manifolds with corners and boundary fibration structures, see [6], but for us the main class of examples comes from resolving a stratified space. Let us work out in detail how a stratified space gives rise to a manifold with corners and a boundary fibration structure in the two simplest non-trivial cases:

If $\widehat{X}$ has a single singular stratum Y, then blowing-up Y as described in §6.2 produces a smooth manifold with boundary $[\widehat{X}; Y]$ with a boundary fibration over Y. (Note that the boundary of $[\widehat{X}; Y]$ need not be connected, which is why the definition involves collective boundary hypersurfaces.) We denote the blown-up space by $\widetilde{X}$ and call it the resolution of $\widehat{X}$.

If $\widehat{X}$ has two singular strata $Y_1 < Y_2$ (assumed connected) then blowing-up Y_1 results in $[\widehat{X}; Y_1]$, a depth one stratified space with boundary K_1 and a locally trivial fibration

$$Z_1 - K_1 \xrightarrow{\psi_{Y_1}} Y_1.$$

The stratum Y_1 is a closed smooth manifold while K and Z_1 are stratified spaces (without boundary) of depth one. The stratum Y_2 of $\widehat{X}$ lifts (or restricts) to a stratum $\mathcal{Y}$ of $[\widehat{X}; Y_1]$. The compatibility conditions between the strata imply that $K_1 \cap \mathcal{Y}$ is the singular stratum of K and, for each $q \in Y_1$, $\psi_{Y_1}^{-1}(q) \cap \mathcal{Y}$ is the singular stratum of $\psi_{Y_1}^{-1}(q)$. The theory developed in §6.3 allows us to blow-up $\mathcal{Y}$ in $[\widehat{X}; Y_1]$ resulting in a manifold with corners $[[\widehat{X}; Y_1]; \mathcal{Y}]$.

We denote this space by $\widetilde{X}$ and call it the resolution of $\widehat{X}$. It has two boundary hypersurfaces. One of them, K_2, is created by the second blow-up and hence has a locally trivial fibration over $\mathcal{Y}$,

$$Z_2 - K_2 \xrightarrow{\psi_{\mathcal{Y}}} \mathcal{Y}.$$

Note that K_2 and $\mathcal{Y}$ are smooth manifolds with boundary and Z_2 is a smooth manifold (without boundary). In fact we can identify $\mathcal{Y}$ with $[\overline{Y}_2; Y_1]$ and, since Y_1 is the unique singular stratum in $\overline{Y}_2$, this is precisely $\widetilde{Y}_2$, the resolution of $\overline{Y}_2$. Thus this boundary fibration

$$Z_2 - K_2 \xrightarrow{\psi_{\mathcal{Y}}} \widetilde{Y}_2$$

has as base the resolution of Y_2 and as fiber the link of Y_2 in $\widehat{X}$.

The other boundary hypersurface of $\widetilde{X}$ is $[K_1; K_1 \cap \mathcal{Y}]$. The set $K_1 \cap \mathcal{Y}$ is transverse to the fibers of the fibration ψ_{Y_1} and so this fibration survives the blow-up to yield a fibration

$$[\psi_{Y_1}^{-1}(q); \psi_{Y_1}^{-1}(q) \cap \mathcal{Y}] - [K_1; K_1 \cap \mathcal{Y}] \xrightarrow{\phi_{Y_1}} Y_1.$$

As explained above, the subsets being blown-up in the fiber and the total space of this fibration are precisely the singular parts of the respective sets. Hence we can identify these sets with the corresponding resolutions,

$$\widetilde{Z}_1 - \widetilde{K}_1 \xrightarrow{\phi_{Y_1}} Y_1.$$

Finally note that the two boundary hypersurfaces $\widetilde{K}_1$, K_2 intersect

$$\widetilde{K}_1 \cap K_2 = \partial \widetilde{K}_1 = \partial K_2.$$

The boundary fibrations restricted to this intersection satisfy

$$\phi_{Y_1}(\widetilde{K}_1 \cap K_2) = Y_1, \quad \psi_{\mathcal{Y}}(\widetilde{K}_1 \cap K_2) = K_1 \cap \mathcal{Y} = \partial \widetilde{Y}_2$$

and hence fit into the commutative diagram

$$\begin{array}{ccc} \widetilde{K}_1 \cap K_2 & \xrightarrow{\psi_{\mathcal{Y}}} & K_1 \cap \mathcal{Y} \\ & {\scriptstyle \phi_{Y_1}} \searrow \quad \swarrow {\scriptstyle \psi_{Y_1}} & \\ & Y_1 & \end{array}$$

Thus $\widetilde{X}$ has an iterated fibration structure and we can recover $\widehat{X}$ by collapsing the fibers of ψ_Y and then collapsing the fibers of ψ_{Y_1}. We denote the resulting projection by

$$\beta : \widetilde{X} \longrightarrow \widehat{X}$$

and refer to it as the blow-down map.

This procedure can be carried out inductively on any compact space with a Thom-Mather stratification as in [3, Propositions 2.3 and 2.5], yielding the following result.

Theorem 6.3.

a) Let $\widehat{X}$ be a compact Thom-Mather stratified space. Iteratively blowing-up the deepest stratum produces a manifold with corners with an iterated fibration structure $\widetilde{X}$, known as the resolution of $\widehat{X}$, together with a continuous map

$$\beta : \widetilde{X} \longrightarrow \widehat{X}$$

such that:

i) β restricts to a diffeomorphism between the interior of $\widetilde{X}$ and the regular part of $\widehat{X}$,

ii) β determines a bijection between the collective boundary hypersurfaces $\mathscr{F}(\widetilde{X})$ and the singular strata of $\widehat{X}$,

$$\mathscr{S}(\widehat{X}) \ni Y \mapsto \overline{\beta^{-1}(Y)} \in \mathscr{F}(\widetilde{X}),$$

iii) the boundary fibration on $\overline{\beta^{-1}(Y)} \in \mathscr{F}(\widetilde{X})$ has base space the resolution of $\overline{Y}$, $\widetilde{Y}$, and fiber over a point $q \in Y$ equal to the resolution of the link of q in $\widehat{X}$.

b) Let $\widetilde{X}$ be a manifold with corners with an iterated fibration structure. Choosing compatible collar neighborhoods and then iteratively collapsing the fibers of the boundary fibration with largest dimensional base produces a Thom-Mather stratified space $\widehat{X}$, called the blow-down of $\widetilde{X}$.

c) These procedures are mutually inverse: the blow-down of the resolution of $\widehat{X}$ is isomorphic to $\widehat{X}$, and the resolution of the blow-down of L is isomorphic to L.

Resolution of stratified spaces is a functor (cf. [3, Propositions 2.6, 2.7]): If $\widehat{X}$ and $\widehat{N}$ are stratified spaces with resolutions $\widetilde{X}$ and $\widetilde{N}$, then any morphism $f : \widehat{X} \longrightarrow \widehat{N}$ lifts to a morphism $\widetilde{f} : \widetilde{X} \longrightarrow \widetilde{N}$ and compositions lift to compositions. Blow-down is also a functor but from manifolds with corners and iterated fibration structures to *isomorphism classes* of stratified spaces because we do not fix collar neighborhoods of the boundary hypersurfaces.

'Controlled' objects on a stratified space correspond to objects on the resolution that are compatible with the boundary fibration structure. For example given a controlled vector field on $\widehat{X}$,

$$\xi = \{\xi_Y \in \mathcal{C}^\infty(Y; TY) : Y \in \mathscr{S}(\widehat{X})\},$$

we get a vector field on $\widetilde{X}$ by lifting $\xi_{\widehat{X}^{\mathrm{reg}}}$ and taking the unique continuous extension from $\widetilde{X}^\circ$ (which exists by the control conditions on ξ). Conversely, given a vector field η on $\widetilde{X}$, the collection of vector fields

$$\xi = \{\beta_*(\eta\big|_{\widetilde{X}^\circ})\} \cup \{(\phi_H)_*(\eta\big|_H) : H \in \mathscr{F}(\widetilde{X})\}$$

is a controlled vector field on $\widehat{X}$.

Similarly the class of functions

$$\mathcal{C}^\infty_\Phi(\widetilde{X}) = \{f \in \mathcal{C}^\infty(\widetilde{X}) : f\big|_H \in \phi_H^\bullet \mathcal{C}^\infty(Y_H) \text{ for all } H \in \mathscr{F}(\widetilde{X})\}$$

corresponds to the 'controlled functions' on $\widehat{X}$ from [114, §1.2.8]. (See [122, §4] for a discussion of controlled vector bundles and connections and their corresponding Chern classes.)

Let us mention another approach to obtaining the resolution of a stratified space. After blowing-up the deepest stratum as in §6.2, one can make use of the collar of the boundary to define the *double of* $[\widehat{X}; \widehat{X}^*]$. This will be a Thom-Mather stratified space (without boundary) whose depth is now less than the depth of $\widehat{X}$. It comes equipped with an action of $\mathbb{Z}_2$ and a fundamental domain for this action recovers $[\widehat{X}; \widehat{X}^*]$. Inductively one can show that the double can be resolved consistently with the $\mathbb{Z}_2$ action and then restricting to a fundamental domain yields $\widetilde{X}$. This doubling technique is developed in [28] where it is called *déplissage*. This approach to resolution was carried out in [3].

Finally let us comment on the use of collective boundary hypersurfaces in Definition 3. A stratified space if called *normal* [53, §4] if all of its links are connected. In the resolution of a normal stratified space it is not necessary to use collective boundary hypersurfaces and each boundary hypersurface is endowed with a boundary fibration. On the other hand, given a boundary fibration structure with collective boundary hypersurfaces, one can restrict the boundary fibrations to the individual boundary hypersurfaces. This corresponds to switching from the resolution of $\widehat{X}$ to the resolution of its *normalization* (as in [53, §4.1]). For the purposes of carrying out analysis on the regular part of $\widehat{X}$, normalization makes only a notational difference and so for simplicity we often assume that the collective boundary hypersurfaces consist of a single boundary hypersurface. This is the approach taken in [3].

6.5 A smooth de Rham complex computing intersection cohomology

In [33], Brylinski introduced a de Rham complex due to Goresky-MacPherson on the regular part of a stratified space that computed the intersection cohomology (cf. [98, 28, 27]). This complex has a very natural description once we are working on the resolution of the stratified space. We give a brief description of this complex as an example of the usefulness of resolving a stratified space.

First, given a locally trivial fiber bundle of smooth manifolds with corners

$$F - H \xrightarrow{\varphi} B,$$

let us say that a differential form ω on H has *φ-vertical degree at most p* if

$$\text{for all } V_1, \ldots, V_{p+1} \in \mathcal{C}^\infty(H, TH/B), \quad \mathrm{i}_{V_1} \cdots \mathrm{i}_{V_{p+1}} \omega = 0.$$

Let $(\widehat{X}, \mathscr{S}, \mathscr{T})$ be a Thom-Mather stratified space, $(\widetilde{X}, \mathscr{F}, \Phi)$ its resolution and

$$\beta : \widetilde{X} \longrightarrow \widehat{X}$$

the blow-down function. Given $Y \in \mathscr{S}$ let us denote the collective boundary hypersurface associated to Y (i.e., the set $\overline{\beta^{-1}(Y)}$) by M_Y, and by

$$\phi_Y : M_Y \longrightarrow \widetilde{Y}, \quad j_Y : M_Y \hookrightarrow \widetilde{X}$$

the boundary fibration of M_Y and its inclusion into $\widetilde{X}$, respectively.

Given $\overline{p}$, a perversity function on $\widehat{X}$, let

$$\begin{aligned}\Omega^*_{\overline{p}}(\widetilde{X}) = \{\omega \in \mathcal{C}^\infty(\widetilde{X}; \Lambda^\bullet(T^*\widetilde{X})) :\ &\text{for each } Y \in \mathscr{S}, \\ &j_Y^*\omega \text{ and } j_Y^* d\omega \text{ have } \phi_Y\text{-vertical degree at most } \overline{p}(\operatorname{codim}_{\widehat{X}}(Y))\}.\end{aligned}$$

It is immediate that $\Omega^*_{\overline{p}}(\widetilde{X})$ defines a complex,

$$0 \longrightarrow \Omega^0_{\overline{p}}(\widetilde{X}) \xrightarrow{d} \Omega^1_{\overline{p}}(\widetilde{X}) \xrightarrow{d} \cdots \xrightarrow{d} \Omega^n_{\overline{p}}(\widetilde{X}) \longrightarrow 0.$$

Theorem 6.4 (Brylinski-Goresky-MacPherson). *Let $\widehat{X}$ be a Thom-Mather pseudomanifold and $\overline{p}$ a perversity. The cohomology of the complex $\Omega^*_{\overline{p}}(\widetilde{X})$ is isomorphic to the intersection cohomology of $\widehat{X}$ with perversity $\overline{p}$ and real coefficients.*

Proof. Define a sheaf complex $\mathbf{\Omega}^\bullet_{\overline{p}}$ over $\widehat{X}$ by

$$\widehat{X} \supseteq \mathcal{U} \mapsto \{\omega \in \Omega^*_{\overline{p}}(\widetilde{X}) : \operatorname{supp}\omega \subseteq \beta^{-1}(\mathcal{U})\}.$$

That the sheaf $\mathbf{\Omega}^\bullet_{\overline{p}}$ is quasi-isometric to $\mathbf{IC}^\bullet_{\overline{p}}$ as complexes of sheaves of $\mathbb{R}$-vector spaces will follow after we check the axioms (a)—(d) from §5.5.

Adopting the notation from §5.5, first note that the existence of an appropriate partition of unity in $\mathcal{C}^\infty_\Phi(\widetilde{X})$ implies that the sheaves are soft. The normalization and lower bound are automatic, so we need to check that, whenever $\mathcal{U} \subseteq \widehat{X}$ is a distinguished neighborhood of a point $x \in S_{n-k}$, we have

$$\begin{cases} \mathrm{H}^j(\boldsymbol{\Omega}^{\bullet}_{\overline{p}}(\mathcal{U})) = 0 & \text{if } j > \overline{p}(k), \\ \mathrm{H}^j(\boldsymbol{\Omega}^{\bullet}_{\overline{p}}(\mathcal{U})) \xrightarrow{\cong} \mathrm{H}^j(\boldsymbol{\Omega}^{\bullet}_{\overline{p}}(\mathcal{U} \setminus S_{n-k})) & \text{if } j \leq \overline{p}(k) \end{cases}$$

where the map in the latter condition is induced by restriction.

If $\mathcal{U}$ is as above and $\widetilde{\mathcal{U}} = \beta^{-1}(\mathcal{U})$ then

$$\widetilde{\mathcal{U}} \cong \mathbb{B}^{n-k} \times [0,1) \times \widetilde{Z}_x$$

where $\widetilde{Z}_q = \beta^{-1}(q)$ is the resolution of Z_q, hence a smooth manifold with corners, and $\boldsymbol{\Omega}^{\bullet}_{\overline{p}}(\mathcal{U})$ is, directly from its definition, given by

$$\begin{aligned} \boldsymbol{\Omega}^{\bullet}_{\overline{p}}(\mathcal{U}) = \{\omega \in \Omega^*(\mathbb{B}^{n-k} \times [0,1)_x \times \widetilde{Z}_x) : \\ \omega\big|_{x=0}, d\omega\big|_{x=0} \in \Omega^*(\mathbb{B}^{n-k} \times [0,1)) \wedge \boldsymbol{\Omega}^{\bullet}_{\overline{p}}(\widetilde{Z}_x)\big|_{\deg \leq \overline{p}(k)}\}. \end{aligned}$$

Since we are on a smooth manifold with corners, standard de Rham theory implies that the cohomology of this complex vanishes above degree $\overline{p}(k)$ and that, for other degrees, restriction to

$$\widetilde{\mathcal{U}} \setminus Y_{n-k} \cong \mathbb{B}^{n-k} \times (0,1) \times \widetilde{Z}_x$$

is an isomorphism in cohomology; indeed that the cohomology is equal to

$$\mathbf{H}^j(\boldsymbol{\Omega}^{\bullet}_{\overline{p}}(\mathcal{U})) = \begin{cases} \mathbf{H}^j(\boldsymbol{\Omega}^{\bullet}_{\overline{p}}(\widetilde{Z}_x)) & \text{if } j \leq \overline{p}(k), \\ 0 & \text{if } j > \overline{p}(k). \end{cases} \qquad \square$$

7 Borel-Serre compactifications and their resolutions

The Borel-Serre compactification and what is now known as the reductive Borel-Serre compactification of a locally symmetric space into a manifold with corners and a stratified space, respectively, were introduced in [23] and [120]. In this section we will define another compactification of a locally symmetric space into a manifold with corners, the *resolved Borel-Serre compactification*. We will show that while the Borel-Serre compactification does not have an iterated fibration structure, the resolved Borel-Serre compactification does and that it can be obtained by resolving the reductive Borel-Serre compactification in the sense of Theorem 6.3.

It may be interesting to compare the resolved Borel-Serre compactification to other compactifications of locally symmetric spaces.

7.1 Linear algebraic groups

We review enough of the theory of linear algebraic groups to describe compactifications of locally symmetric spaces. We follow [23, 25] to which we refer for proofs and longer explanations; see also, e.g., [51, 102, 121].

A linear algebraic group is a Zariski-closed subgroup of some $GL_n(\mathbb{C})$. If it can be defined by polynomials with $\mathbb{Q}$-coefficients we say that it is defined over $\mathbb{Q}$. We will denote such groups by bold letters such as $\mathbf{G}$. Their real points are then denoted by $\mathbf{G}(\mathbb{R})$ or the non-bold letter G and form a real Lie group. The connected component of the identity will be denoted $\mathbf{G}^\circ$.

A linear algebraic group $\mathbf{T}$ is called an *algebraic torus* if it is isomorphic to a product of copies of $\mathbb{C}^*$. If the isomorphism is defined over a field $\mathbb{F}$, we say that it is *split* over $\mathbb{F}$.

An element $r \in \mathbf{G}$ is *unipotent* if some positive power of $(r - \mathrm{Id})$ is equal to the identity, and a linear algebraic group is called unipotent if all of its elements are unipotent. Every unipotent group is solvable.

The *radical* $\mathbf{R}(\mathbf{G})$ of an algebraic group $\mathbf{G}$ is the maximal connected normal solvable subgroup of $\mathbf{G}$, and the *unipotent radical,* $\mathbf{R_U}(\mathbf{G})$, is the maximal connected unipotent normal subgroup of $\mathbf{G}$. If $\mathbf{G}$ is defined over $\mathbb{Q}$, then so are $\mathbf{R}(\mathbf{G})$ and $\mathbf{R_U}(\mathbf{G})$. We say that $\mathbf{G}$ is *semisimple* if $\mathbf{R}(\mathbf{G})$ is trivial and *reductive* if $\mathbf{R_U}(\mathbf{G})$ is trivial. We denote the real points in $\mathbf{R_U}(\mathbf{G})$ by $R_U G$.

A *Levi subgroup* $\mathbf{L}$ of $\mathbf{G}$ is a maximal reductive subgroup; any two such are conjugate under $\mathbf{R_U}(\mathbf{G})$ and $\mathbf{G} = \mathbf{L} \ltimes \mathbf{R_U}(\mathbf{G})$. By a Levi subgroup L of $G = \mathbf{G}(\mathbb{R})$ we will mean the real points of a Levi subgroup $\mathbf{L}$ of $\mathbf{G}$. Every compact subgroup of G is the group of real points of a reductive $\mathbb{R}$-group and is hence contained in a Levi subgroup.

The maximal tori of $\mathbf{G}$, split over a field $\mathbb{F}$, are all conjugate over $\mathbb{F}$, and their common dimension is called the $\mathbb{F}$-rank of $\mathbf{G}$, $\mathrm{rk}_{\mathbb{F}}(G)$. A group $\mathbf{G}$ is *split over* $\mathbb{Q}$ if its $\mathbb{Q}$-rank coincides with its $\mathbb{C}$-rank. A group $\mathbf{G}$ is *anisotropic over* $\mathbb{Q}$ if it is of $\mathbb{Q}$-rank zero.

A closed subgroup $\mathbf{P}$ of a connected linear algebraic group $\mathbf{G}$ is called a *parabolic subgroup* if $\mathbf{G}/\mathbf{P}$ is a projective variety or if $\mathbf{P}$ contains a maximal connected solvable subgroup (known as a *Borel subgroup*). By a parabolic subgroup P of G we will mean the real points of a parabolic subgroup $\mathbf{P}$ of $\mathbf{G}$, defined over $\mathbb{Q}$. A subgroup H of G is parabolic if and only if G/H is compact. We denote the set of *proper* parabolic subgroups of G by $\mathcal{P}(G)$.

If K is a maximal compact subgroup of G and P is a parabolic subgroup of G then $K_P = K \cap P$ is a maximal compact subgroup of P and $G = K \cdot P$.

If $\mathbf{G}$ is reductive and K is a maximal compact subgroup of G, there exists one and only one involutive automorphism θ_K of G whose fixed point set is K and which is algebraic (in a suitable sense [23, Proposition 1.6]). This automorphism is called the *Cartan involution* of G with respect to K.

If $\mathbf{N}$ is a normal subgroup of $\mathbf{G}$ defined over $\mathbb{R}$, then $\theta_K(\mathbf{N}(\mathbb{R})) = \mathbf{N}(\mathbb{R})$. If P is a parabolic subgroup of G and L a Levi subgroup of G containing K, then $L \cap P$ contains a unique Levi subgroup of P stable under θ_K.

We will assume from now on that $\mathbf{G}$ is a connected reductive linear algebraic group defined over $\mathbb{Q}$ whose center is (defined and) anisotropic over $\mathbb{Q}$.

For any parabolic subgroup $\mathbf{P}$ of $\mathbf{G}$ defined over $\mathbb{Q}$ both $\mathbf{U_P}$, the unipotent radical of $\mathbf{P}$, and $\mathbf{L_P} = \mathbf{P}/\mathbf{U_P}$, the *Levi quotient* of $\mathbf{P}$, are also defined over $\mathbb{Q}$. Let $\chi(\mathbf{L_P})$ denote the group of algebraic rationally defined characters of $\mathbf{L_P}$ and

$$\mathbf{M_P} = \bigcap_{\alpha \in \chi(\mathbf{L_P})} \ker \alpha^2.$$

Then $\mathbf{M_P}$ is a reductive algebraic group defined over $\mathbb{Q}$ whose center is anisotropic over $\mathbb{Q}$ (it may fail to be connected but see [24, Remark 2.10]). Let $\mathbf{S_P}$ be the maximal $\mathbb{Q}$-split torus in the center of $\mathbf{L_P}$, and denote

$$U_P = \mathbf{U_P}(\mathbb{R}), \quad L_P = \mathbf{L_P}(\mathbb{R}), \quad M_\mathbf{P} = \mathbf{M_P}(\mathbb{R}), \quad S_\mathbf{P} = \mathbf{S_P}(\mathbb{R}), \quad A_\mathbf{P} = S_P^\circ.$$

There is a canonical isomorphism

$$A_\mathbf{P} = (\mathbb{R}_+^*)^{\mathrm{park}_\mathbf{G}(\mathbf{P})}$$

where $\mathrm{park}_\mathbf{G}(\mathbf{P})$, the parabolic rank of $\mathbf{P}$ in $\mathbf{G}$, is the length of a maximal chain

$$\mathbf{P} = \mathbf{P}_{\mathrm{park}_\mathbf{G}(\mathbf{P})} \subset \mathbf{P}_{\mathrm{park}_\mathbf{G}(\mathbf{P})-1} \subset \cdots \subset \mathbf{P}_1 \subset G$$

of proper parabolic subgroups defined over $\mathbb{Q}$. The real points of the Levi quotient admit a decomposition

$$L_P = A_\mathbf{P} M_\mathbf{P} \cong A_\mathbf{P} \times M_\mathbf{P}.$$

For each choice of maximal compact subgroup K of G, there is a unique lift $L_{P,\theta_K} \subseteq P$ of L_P to a subgroup of P stable under θ_K. This lift splits the exact sequence

$$1 \longrightarrow U_P \longrightarrow P \longrightarrow L_P \longrightarrow 1.$$

It is possible to choose K so that L_{P,θ_K} is defined over $\mathbb{Q}$, and then so are the images $A_{\mathbf{P},\theta_K}$ and $M_{\mathbf{P},\theta_K}$ of its subgroups $A_\mathbf{P}$ and $M_\mathbf{P}$. This yields the *rational Langlands decomposition* of P :

$$P = U_P A_{\mathbf{P},\theta_K} M_{\mathbf{P},\theta_K} \cong U_P \times A_{\mathbf{P},\theta_K} \times M_{\mathbf{P},\theta_K}.$$

The intersection of K and P, $K_P = K \cap P$ is a maximal compact subgroup of P. Its image under the map $P \mapsto L_P$ is a maximal compact subset of M_P and the lift of this image coincides with K_P.

Next, let us consider the parabolic subgroups of $\mathbf{G}$, defined over $\mathbb{Q}$, contained in $\mathbf{P}$. These are the parabolic subgroups of $\mathbf{P}$ and [23, §4.1] are in one-to-one correspondence with the parabolic subgroups of $\mathbf{L_P}$ (in fact $\mathbf{M_P}$) by the projection $\mathbf{P} \longrightarrow \mathbf{P}/\mathbf{U_P}$. If $\mathbf{P}, \mathbf{Q} \in \mathcal{P}(\mathbf{G})$ and $\mathbf{Q} \subseteq \mathbf{P}$ then

$$Q_\mathbf{P} = Q \cap M_{\mathbf{P},\theta_K} \in \mathcal{P}(\mathbf{M_P}).$$

The Langlands decompositions of Q (with respect to θ_K) and $Q_{\mathbf{P}}$ (with respect to $\theta_{K\cap P}$),

$$Q = U_Q A_{\mathbf{Q},\theta_K} M_{\mathbf{Q},\theta_K}, \quad Q_{\mathbf{P}} = U_{Q_{\mathbf{P}}} A_{\mathbf{Q}_{\mathbf{P}},\theta_{K\cap P}} M_{\mathbf{Q}_{\mathbf{P}},\theta_{K\cap P}},$$

are related by

$$U_Q = U_P \rtimes U_{Q_P}, \quad A_{\mathbf{Q},\theta_K} = A_{\mathbf{P},\theta_K} \times A_{\mathbf{Q}_{\mathbf{P}},\theta_{K\cap P}}, \quad M_{\mathbf{Q},\theta_K} = M_{\mathbf{Q}_{\mathbf{P}},\theta_{K\cap P}}. \quad (7.1)$$

Note that $K_Q = K \cap M_{\mathbf{Q},\theta_K} = K_{Q_P}$. Below we will omit the Cartan involution θ in the notation for the Langlands decompositions.

The notion of an *arithmetic* subgroup of $\mathbf{G}(\mathbb{Q})$ is a sort of 'substitute' for $\mathbf{G}(\mathbb{Z})$. Given a representation $\rho : \mathbf{G} \longrightarrow \mathbf{GL}_n$ defined over $\mathbb{Q}$ with discrete null space (i.e., an 'almost faithful' representation), we can set

$$\mathbf{G}(\mathbb{Z})_\rho = \rho^{-1}(\mathbf{GL}_n(\mathbb{Z})).$$

If ρ' is any other almost faithful representation then $\mathbf{G}(\mathbb{Z})_{\rho'}$ is commensurable with $\mathbf{G}(\mathbb{Z})_\rho$, meaning that the intersection of these two groups is of finite index in either one. We call any group $\Gamma < \mathbf{G}(\mathbb{Q})$ commensurable with $\mathbf{G}(\mathbb{Z})_\rho$ an arithmetic subgroup.

An arithmetic subgroup Γ is *neat* if, whenever $\mathbf{H}_2 < \mathbf{H}_1$ is a pair of algebraic $\mathbb{Q}$-subgroups of $\mathbf{G}$,

$$\Gamma_{\mathbf{H}_1/\mathbf{H}_2} = (\Gamma \cap \mathbf{H}_1(\mathbb{Q}))/(\Gamma \cap \mathbf{H}_2(\mathbb{Q}))$$

is torsion-free (and arithmetic). Every arithmetic group contains a neat normal subgroup of finite index.

7.2 The partial compactification of a symmetric space to a manifold with corners

As above, we assume that $\mathbf{G}$ is a connected reductive linear algebraic group defined over $\mathbb{Q}$ whose center is (defined and) anisotropic over $\mathbb{Q}$. Let K be a maximal compact subgroup of G and let

$$X = G/K$$

be the corresponding symmetric space. In this section we describe the Borel-Serre partial compactification [23] of X to a non-compact manifold with corners, $\overline{X}$. We will also describe a 'resolution' of $\overline{X}$ to a non-compact manifold with corners $\widetilde{X}$ with a boundary fibration structure.

Given $P \in \mathcal{P}(G)$, the group $K_P = K \cap P$ is a maximal compact subgroup of P. It coincides with $K \cap M_{\mathbf{P}}$ and is also a maximal compact subgroup of $M_{\mathbf{P}}$. The quotient space

$$X_P = M_{\mathbf{P}}/K_P = P/K_P A_{\mathbf{P}} U_P$$

is a symmetric space of non-compact type for $M_{\mathbf{P}}$ called the *boundary symmetric space* associated to P.

For each $P \in \mathcal{P}(G)$ let

$$e_G(P) = U_P \times X_P.$$

The Borel-Serre partial compactification of X, as a set, is

$$\overline{X} = X \cup \bigsqcup_{P \in \mathcal{P}(G)} e_G(P), \tag{7.2}$$

where $e_G(P)$ is attached to X using the horospherical decomposition $X = U_P \times A_{\mathbf{P}} \times X_{\mathbf{P}}$ (see [23, 25]) and is a paracompact Hausdorff manifold with corners. Each $e_G(P)$ is the interior of a boundary face of $\overline{X}$, with codimension equal to the parabolic rank of P, and with closure in $\overline{X}$ given by

$$\overline{e_G(P)} = e_G(P) \cup \bigsqcup_{\substack{Q \in \mathcal{P}(G) \\ Q \subseteq P}} e_G(Q). \tag{7.3}$$

Note that $K_Q = K \cap M_Q = K_{Q_P}$, and hence the boundary symmetric space associated to Q as a subgroup of G is the same as the boundary symmetric space associated to Q_P as a subgroup of M_P,

$$X_Q = M_Q / K_Q = M_{Q_P} / K_{Q_P} = X_{Q_P}.$$

Proposition 7.1 (Boundary fibrations of the partial Borel-Serre compactification)**.** *Let X be the symmetric space of the real points of a connected reductive group $\mathbf{G}$ as above and let $\overline{X}$ be its Borel-Serre partial compactification to a manifold with corners. Each boundary face of $\overline{X}$ is of the form*

$$\overline{e_G(P)} = U_P \times \overline{X_P}$$

where $P \in \mathcal{P}(G)$, U_P is the unipotent radical of P, X_P is the boundary symmetric space corresponding to P and $\overline{X_P}$ is its Borel-Serre partial compactification. Let $\phi_P : \overline{e_G(P)} \longrightarrow \overline{X_P}$ denote the natural projection.

If $P, Q \in \mathcal{P}(G)$ then the boundary faces $\overline{e_G(P)}$, $\overline{e_G(Q)}$ intersect if and only if

$$R = P \cap Q \in \mathcal{P}(G)$$

in which case $\overline{e_G(P)} \cap \overline{e_G(Q)} = \overline{e_G(R)}$ and we have a commutative diagram,

$$\begin{array}{ccccccc}
\overline{e_G(Q)} & \hookleftarrow & & \overline{e_G(R)} & & \hookrightarrow & \overline{e_G(P)} \\
\downarrow{\scriptstyle \phi_Q} & & \swarrow{\scriptstyle \phi_Q|_{\overline{e_G(R)}}} & \downarrow{\scriptstyle \phi_R} & \searrow{\scriptstyle \phi_P|_{\overline{e_G(R)}}} & & \downarrow{\scriptstyle \phi_P} \\
\overline{X_Q} & \hookleftarrow & \overline{e_{M_Q}(R_Q)} & & \overline{e_{M_P}(R_P)} & \hookrightarrow & \overline{X_P} \\
 & & \searrow{\scriptstyle \phi_{R_Q}} & \downarrow & \swarrow{\scriptstyle \phi_{R_P}} & & \\
 & & & \overline{X_R} & & &
\end{array} \tag{7.4}$$

Proof. We have described the closure $\overline{e_G(P)}$ of $e_G(P)$ in $\overline{X}$ in (7.3). The boundary symmetric space X_P itself has a partial compactification as a symmetric space for M_P,

$$\overline{X_P} = X_P \cup \bigsqcup_{O\in\mathcal{P}(M_P)} e_{M_P}(O) = X_P \cup \bigsqcup_{\substack{R\in\mathcal{P}(G)\\ R\subseteq P}} e_{M_P}(R_P),$$

where $e_{M_P}(R_P) = U_{R_P} \times X_{R_P}$. Thus to each $R \in \mathcal{P}(G)$, $R \subseteq P$, we assign a boundary face $e_G(R)$ of $\overline{e(P)}$ and a boundary face $e_{M_P}(R_P)$ of $\overline{X_P}$, given respectively by

$$\begin{aligned} e_G(R) &= U_R \times X_R = U_P \times U_{R_P} \times X_R, \\ e_{M_P}(R_P) &= U_{R_P} \times X_{R_P} = U_{R_P} \times X_R. \end{aligned}$$

It follows that

$$\overline{e_G(P)} = e_G(P) \cup \bigsqcup_{\substack{R\in\mathcal{P}(G)\\ R\subseteq P}} e_G(R) = U_P \times X_P \cup \bigsqcup_{\substack{R\in\mathcal{P}(G)\\ R\subseteq P}} (U_P \times U_{R_P} \times X_R)$$

$$= U_P \times \Big(X_P \cup \bigsqcup_{\substack{R\in\mathcal{P}(G)\\ R\subseteq P}} (U_{R_P} \times X_R) \Big) = U_P \times \Big(X_P \cup \bigsqcup_{\substack{R\in\mathcal{P}(G)\\ R\subseteq P}} e_{M_P}(R_P) \Big) = U_P \times \overline{X_P}.$$

Next let $P, Q \in \mathcal{P}(G)$. Corollary 7.4 of [23], shows that

$$\overline{e_G(P)} \cap \overline{e_G(Q)} = \begin{cases} \overline{e_G(R)} & \text{if } R = P \cap Q \in \mathcal{P}(G), \\ \emptyset & \text{otherwise.} \end{cases}$$

From (7.1) we see that

$$\overline{e_G(R)} = U_R \times \overline{X_R} = U_P \times U_{R_P} \times \overline{X_R}$$

and, since $\phi_P : U_P \times \overline{X_P} \longrightarrow X_P$ is just the projection off of U_P,

$$\phi_P(\overline{e_G(R)}) = U_{R_P} \times \overline{X_R} = \overline{e_{M_P}(R_P)}.$$

Moreover

$$\overline{e_G(R)} \xrightarrow{\phi_P|_{\overline{e_G(R)}}} \overline{e_{M_P}(R_P)} \xrightarrow{\phi_{R_P}} \overline{X_R}$$

is the projection off of U_R and so coincides with ϕ_R. This gives that the right 'half' of (7.4) commutes, and by symmetry so does the left half. □

Proposition 7.1, particularly diagram (7.4), shows that the Borel-Serre partial compactification $\overline{X}$ does not have the structure of a manifold with corners with an iterated fibration structure (except when it is a manifold with boundary). It is always possible to modify $\overline{X}$ to obtain an iterated fibration structure.

Recall (e.g., [88, Chapter 5], [6]) that if M is a manifold with corners and $F \subseteq M$ is a boundary face, then

$$[M; F],$$

the radial blow-up of M along F is a manifold with corners in which F is replaced by its inward-pointing spherical normal bundle, forming a new boundary hypersurface (i.e., a boundary face of codimension one). There is a natural 'blow-down map' $\beta : [M; F] \longrightarrow M$ which collapses the new boundary hypersurface back to F.

Consider the simple case of two maximal proper parabolic subgroups P and Q such that $R = P \cap Q$ is itself parabolic. The space $[\overline{X}; \overline{e_G(R)}]$, replaces $\overline{e_G(R)}$ with a new boundary hypersurface

$$\overline{E_G(R)}.$$

The blow-down map $\beta : [\overline{X}; \overline{e_G(R)}] \longrightarrow \overline{X}$ restricts to $\overline{E_G(R)}$ to a (trivial) fiber bundle

$$\mathbb{S}_+^{\mathrm{park}(R)-1} - \overline{E_G(R)} \xrightarrow{\beta|_{\overline{E_G(R)}}} \overline{e_G(R)}$$

since $\overline{E_G(R)}$ is the total space of the inward pointing spherical normal bundle of $\overline{e_G(R)}$ in $\overline{X}$. We can relabel this in terms of the rational Langlands decomposition of P. Indeed, recall [23], that $e_G(R)$ has a neighborhood of the form

$$X(R) = X \times_{A_R} \overline{A_R}$$

where A_R acts on X by a 'geodesic action', and that we can identify $A_R \cong (\mathbb{R}_+^*)^{\mathrm{park}(R)}$. There is an action of $\mathbb{R}_+^*$ on A_R by scaling in all factors of $\mathbb{R}_+^*$ and

$$\mathbb{S}_+^{\mathrm{park}(R)-1} \cong A_R/\mathbb{R}_+^*.$$

Thus the new boundary hypersurface $E_G(R)$ participates in a (trivial) fiber bundle

$$A_R/\mathbb{R}_+^* - \overline{E_G(R)} \xrightarrow{\beta|_{\overline{E_G(R)}}} \overline{e_G(R)}$$

as well as a (trivial) fiber bundle over $\overline{X_R}$,

$$U_P \times (A_R/\mathbb{R}_+^*) - \overline{E_G(R)} \xrightarrow{\Phi_P} \overline{X_R}.$$

By blowing-up $\overline{e_G(R)}$, we have separated the boundary hypersurfaces $\overline{e_G(P)}$ and $\overline{e_G(Q)}$ of $\overline{X}$. Now at the corner of the intersection of $\beta^{-1}(\overline{e_G(P)})$ and $\overline{E_G(R)}$ we have a commutative diagram of fiber bundle maps

$$\begin{array}{ccc} \beta^{-1}(\overline{e_G(P)}) \cap \overline{E_G(R)} & \xrightarrow{\phi_P} & \overline{e_{M_P}(R_P)} \\ & \searrow^{\Phi_R} \quad \swarrow^{\phi_{R_P}} & \\ & \overline{X_R} & \end{array}$$

as required in the definition of a boundary fibration structure.

In order to meet the requirements of a boundary fibration structure at every corner, we need to blow-up every boundary face of $\overline{X}$ of codimension greater than one. This procedure is referred to as the *total boundary blow-up* in [88, §5.13], [57,

§2.6] and denoted by a subscript 'tb'. We define the *resolved partial Borel-Serre compactification* of X to be

$$\widetilde{X} = \overline{X}_{tb}.$$

It has a natural blow-down map $\beta_{tb} : \widetilde{X} \longrightarrow \overline{X}$, and one boundary hypersurface $\overline{E_G(P)}$ for each proper parabolic subgroup of G, where

$$E_G(P) = \beta_{tb}^{-1}\left(e_G(P)\right), \quad \overline{E_G(P)} = \overline{\beta_{tb}^{-1}\left(e_G(P)\right)}.$$

(In general $\overline{E_G(P)} \neq \beta_{tb}^{-1}\left(\overline{e_G(P)}\right)$ since the latter contains one boundary hypersurface for each parabolic subgroup of G contained in P.)

Proposition 7.2 (Boundary fibration structure of the resolved partial Borel-Serre compactification). *Let X be the symmetric space of the real points of a connected reductive group $\mathbf{G}$ as above and let $\widetilde{X}$ be its resolved Borel-Serre partial compactification to a manifold with corners. Each boundary hypersurface of $\widetilde{X}$ is of the form*

$$\overline{E_G(P)} = U_P \times (A_P/\mathbb{R}_+^*)_{tb} \times \widetilde{X_P}$$

where $P \in \mathcal{P}(G)$, U_P is the unipotent radical of P, A_P is the connected component of the identity in the real points of the split center of $U_P \backslash P$ over $\mathbb{Q}$, X_P is the boundary symmetric space corresponding to P and $\widetilde{X_P}$ is its resolved Borel-Serre partial compactification.

If $P, Q \in \mathcal{P}(G)$ then $\overline{E_G(P)} \cap \overline{E_G(Q)} \neq \emptyset$ if and only if one of the parabolic groups is a subgroup of the other.

The natural projections $\Phi_P : \overline{E_G(P)} \longrightarrow \widetilde{X_P}$ endow $\widetilde{X}$ with an iterated fibration structure. That is, whenever $\overline{E_G(P)} \cap \overline{E_G(Q)} \neq \emptyset$, we have (after relabeling if necessary) $\operatorname{park}(P) < \operatorname{park}(Q)$ *and a commutative diagram of fiber bundle maps*

$$\begin{array}{ccc}
\overline{E_G(P)} \cap \overline{E_G(Q)} & \xrightarrow{\quad \Phi_P \quad} & \overline{E_{M_P}(Q_P)} \\
& {\scriptstyle \Phi_Q} \searrow \qquad \swarrow {\scriptstyle \Phi_{Q_P}} & \\
& \widetilde{X_Q} &
\end{array}$$

Proof. The fact that $\overline{X}$ is a Hausdorff manifold with corners implies the same for $\widetilde{X}$.

For each $k \in \mathbb{N}$, let $\overline{X}_{tb>k}$ denote the space obtained from $\overline{X}$ by blowing-up all of the boundary faces of codimension greater than k and let

$$\beta_{tb>k} : \overline{X}_{tb>k} \longrightarrow \overline{X}$$

denote the blow-down map. Thus

$$\widetilde{X} = \overline{X}_{tb} = \overline{X}_{tb>1}.$$

Consider $P \in \mathcal{P}(G)$ of parabolic rank ℓ. The face $\overline{e_G(P)}$ has codimension ℓ in $\overline{X}$, so if

$$F_G(P) = \beta_{tb>\ell}^{-1}(e_G(P))$$

then $\overline{F_G(P)}$ is obtained from $\overline{e_G(P)}$ by blowing-up all of its boundary faces in order of increasing codimension. Thus we have

$$\overline{F_G(P)} = \overline{e_G(P)}_{tb} = U_P \times (\overline{X_P})_{tb} = U_P \times \widetilde{X_P}.$$

When we pass to $\overline{X}_{tb>\ell-1}$, the boundary face $\overline{F_G(P)}$ itself gets blown-up so it lifts to a boundary hypersurface

$$H_G(P) = \beta^{-1}_{tb>\ell-1}(e_G(P))$$

whose closure is

$$\overline{H_G(P)} = \overline{F_G(P)} \times (\mathbb{R}^*_+)^\ell/\mathbb{R}^*_+.$$

Finally when we pass to $\overline{X}_{tb}$ we blow-up the intersection of $\overline{H_G(P)}$ with (lifts of) boundary faces of $\overline{X}$ corresponding to parabolic subgroups of X that contain P, and these intersect $\overline{H_G(P)}$ in $\overline{F_G(P)}$ times a boundary face of $(\mathbb{R}^*_+)^\ell/\mathbb{R}^*_+$ so we end up with

$$\overline{E_G(P)} = \overline{F_G(P)} \times \left((\mathbb{R}^*_+)^\ell/\mathbb{R}^*_+\right)_{tb} = U_P \times (A_P/\mathbb{R}^*_+)_{tb} \times \widetilde{X_P}.$$

From this description we see that the boundary hypersurfaces that intersect $\overline{E_G(P)}$ come either from the blow-ups of faces $e_G(Q)$ of Q that are contained in P (accounting for the boundary faces of $\overline{F_G(P)}$) or from the blow-ups of faces $e_G(Q)$ of Q that contain P (accounting for the boundary faces of $\left((\mathbb{R}^*_+)^\ell/\mathbb{R}^*_+\right)_{tb}$).

The compatibility of the boundary fibrations over a corner now follows directly from (7.4), as explained above. □

7.3 The compactification of a locally symmetric space to a manifold with corners

By Theorem 9.3 of [23], any arithmetic subgroup Γ of G acts properly on $\overline{X}$ with compact orbit space $\Gamma\backslash\overline{X}$, known as the *Borel-Serre compactification of* $\Gamma\backslash X$. Similarly any arithmetic subgroup of M_P acts properly on $\overline{X_P}$ with compact orbit space.

If Γ is torsion-free it will act freely on X, but not necessarily on $\overline{X}$. However if Γ is also neat, then it will act freely on $\overline{X}$.

For $P \in \mathcal{P}$ let

$$\Gamma_P = P \cap \Gamma, \quad \Gamma_{U_P} = U_P \cap \Gamma.$$

These are arithmetic subgroups of P and U_P, respectively. From [23, Proposition 1.2], we know that the image of Γ_P under the projection $\pi_{U_P} : P \longrightarrow U_P\backslash P$ is contained in M_P and is an arithmetic subgroup of M_P which we will denote Γ_{M_P}. These groups fit into a short exact sequence

$$0 \longrightarrow \Gamma_{U_P} \longrightarrow \Gamma_P \xrightarrow{\pi_{U_P}} \Gamma_{M_P} \longrightarrow 0.$$

Since U_P is normal in P, the product decomposition $e_G(P) = U_P \times X_P$ descends to a fiber bundle

$$\Gamma_{U_P}\backslash U_P - \Gamma_P\backslash e_G(P) \xrightarrow{\psi_P} \Gamma_{M_P}\backslash X_P$$

with fiber a compact nilmanifold. We write the total space as

$$e'_G(P) = (\Gamma_{U_P} \backslash U_P) \times_{\Gamma_{M_P}} X_P$$

where Γ_{M_P} acts on the first factor by conjugation.

The stabilizer of $e_G(P)$ under the action of Γ on $\overline{X}$ is Γ_P, and hence $e'_G(P)$ is the image of $e_G(P)$ under the projection $\overline{X} \longrightarrow \Gamma\backslash\overline{X}$. Thus we have

$$\Gamma\backslash\overline{X} = \Gamma\backslash X \cup \bigsqcup_{P \in \mathcal{P}(G)/\Gamma} e'_G(P)$$

where $\mathcal{P}(G)/\Gamma$ denotes a set of representatives of the conjugacy classes of elements of $\mathcal{P}(G)$ under the action of Γ. The image of the closure of $e_G(P)$ is equal to the closure of the image of $e_G(P)$ and is given by

$$\overline{e'_G(P)} = \Gamma_P \backslash \overline{e_G(P)} = e'_G(P) \cup \bigsqcup_{Q \in \mathcal{P}(M_P)/\Gamma_{M_P}} e'_G(Q).$$

Proposition 7.3 (Boundary fibrations of the Borel-Serre compactification). *Let X be the symmetric space of the real point of a connected reductive group $\mathbf{G}$ as above, let Γ be a neat arithmetic subgroup of $\mathbf{G}(\mathbb{Q})$, and let $\Gamma\backslash\overline{X}$ be the Borel-Serre compactification of $\Gamma\backslash X$ to a manifold with corners. Each boundary face of $\Gamma\backslash\overline{X}$ is the total space of a fiber bundle*

$$\overline{e'_G(P)} = (\Gamma_{U_P} \backslash U_P) \times_{\Gamma_{M_P}} \overline{X_P} \xrightarrow{\psi_P} \Gamma_{M_P} \backslash \overline{X_P}$$

with fiber the nil manifold $\Gamma_{U_P} \backslash U_P$ and base the Borel-Serre compactification of $\Gamma_{M_P} \backslash X_P$.

If $P, Q \in \mathcal{P}(G)/\Gamma$ then the boundary faces $\overline{e'_G(P)}$, $\overline{e'_G(Q)}$ intersect if and only if, after replacing P with a Γ-conjugate subgroup,

$$R = P \cap Q \in \mathcal{P}(G)$$

in which case $\overline{e'_G(P)} \cap \overline{e'_G(Q)} = \overline{e'_G(R)}$ and we have a commutative diagram,

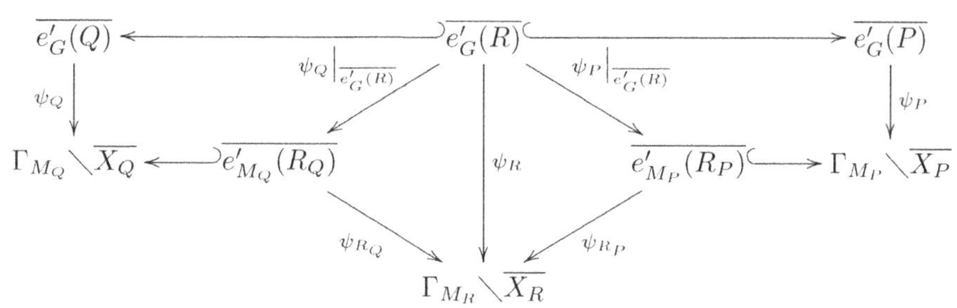

(7.5)

Proof. If $\mathbf{P}, \mathbf{Q} \in \mathcal{P}(G)$ and $\mathbf{Q} \subseteq \mathbf{P}$, then from (7.1) we have a short exact sequence

$$1 \longrightarrow U_P \longrightarrow U_Q \longrightarrow U_{Q_P} \longrightarrow 1,$$

and hence a fiber bundle

$$\Gamma_{U_P}\backslash U_P - \Gamma_{U_Q}\backslash U_Q \longrightarrow \Gamma_{U_{Q_P}}\backslash U_{Q_P}.$$

In particular, we can identify

$$\Gamma_{U_Q}\backslash U_Q = (\Gamma_{U_P}\backslash U_P) \times_{\Gamma_{U_{Q_P}}} U_{Q_P}$$

and hence

$$e'_G(Q) = (\Gamma_{U_Q}\backslash U_Q) \times_{\Gamma_{M_Q}} X_Q = \Big((\Gamma_{U_P}\backslash U_P) \times_{\Gamma_{U_{Q_P}}} U_{Q_P}\Big) \times_{\Gamma_{M_Q}} X_Q.$$

Then using the short exact sequence

$$1 \longrightarrow \Gamma_{U_{Q_P}} \longrightarrow \Gamma_{Q_P} \longrightarrow \Gamma_{M_Q} \longrightarrow 1$$

we can further identify this with

$$e'_G(Q) = (\Gamma_{U_P}\backslash U_P) \times_{\Gamma_{Q_P}} (U_{Q_P} \times X_Q) = (\Gamma_{U_P}\backslash U_P) \times_{\Gamma_{Q_P}} e_{M_Q}(Q_P).$$

The space $\overline{X_P}$ is a symmetric space for M_P and under the action of Γ_{M_P} on $\overline{X_P}$ the stabilizer of $e_{M_P}(R_P)$ is $\Gamma_{M_{R_P}}$, so we have

$$\Gamma_{M_P}\backslash\overline{X_P} = \Gamma_{M_P}\backslash X_P \cup \bigsqcup_{\substack{R\in\mathcal{P}(G)/\Gamma \\ R\subseteq P}} \Gamma_{M_{R_P}}\backslash e_{M_P}(R_P).$$

Thus

$$\begin{aligned}\overline{e'_G(P)} &= e'_G(P) \cup \bigsqcup_{\substack{R\in\mathcal{P}(G) \\ R\subseteq P}} e'_G(R) \\ &= (\Gamma_{U_P}\backslash U_P) \times_{\Gamma_{M_P}} X_P \cup \bigsqcup_{\substack{R\in\mathcal{P}(G) \\ R\subseteq P}} (\Gamma_{U_P}\backslash U_P) \times_{\Gamma_{R_P}} e_{M_R}(R_P) \\ &= (\Gamma_{U_P}\backslash U_P) \times_{\Gamma_{M_P}} \overline{X_P}.\end{aligned}$$

This proves the first part of the proposition, the rest follows from Proposition 7.1 and the equivariance of the projections ϕ_P. □

Just as the action of Γ on X extends to an action of Γ on $\overline{X}$, there is a unique action of Γ on $\widetilde{X}$ that extends the action of Γ on X. Indeed, for compact spaces and groups this was proved in [6, Proposition 6.2], and the extension to proper actions is straightforward (see, e.g., [68] for collar and tubular neighborhood theorems).

We define the *resolved Borel-Serre compactification of* $\Gamma\backslash X$ to be

$$\Gamma\backslash\widetilde{X}.$$

Let $\overline{E'_G(P)}$ denote the image of $\overline{E_G(P)}$ under the natural projection $\widetilde{X} \longrightarrow \Gamma\backslash\widetilde{X}$.

Theorem 7.4 (Boundary fibrations of the resolved Borel-Serre compactification). *Let X be the symmetric space of the real point of a connected reductive group $\mathbf{G}$ as above, let Γ be a neat arithmetic subgroup of $\mathbf{G}(\mathbb{Q})$, and let $\Gamma\backslash\widetilde{X}$ be the resolved Borel-Serre compactification of $\Gamma\backslash X$.*

There is one boundary hypersurface $\overline{E'_G(P)}$ for each Γ-conjugacy class of $\mathbb{Q}$-parabolic subgroups of $\mathbf{G}$. Each of these boundary hypersurfaces is the total space of a fiber bundle

$$\overline{E'_G(P)} = (A_P/\mathbb{R}^*_+)_{tb} \times \Big((\Gamma_{U_P}\backslash U_P) \times_{\Gamma_{M_P}} \widetilde{X_P}\Big) \xrightarrow{\Psi_P} \Gamma_{M_P}\backslash\widetilde{X_P}$$

with fiber the compact manifold with corners $(A_P/\mathbb{R}^*_+)_{tb} \times \Gamma_{U_P}\backslash U_P$ and base the resolved Borel-Serre compactification of $\Gamma_{M_P}\backslash X_P$. Here U_P denotes the unipotent radical of P, A_P is the connected component of the identity in the real points of the split center of $U_P\backslash P$ over $\mathbb{Q}$, and X_P is the boundary symmetric space corresponding to P.

If $P, Q \in \mathcal{P}(G)/\Gamma$ then $\overline{E'_G(P)} \cap \overline{E'_G(Q)} \neq \emptyset$ if and only if one of the parabolic groups is Γ-conjugate to a subgroup of the other.

The fiber bundle maps $\Psi_P : \overline{E'_G(P)} \longrightarrow \Gamma_{M_P}\backslash\widetilde{X_P}$ endow $\Gamma\backslash\widetilde{X}$ with an iterated fibration structure. That is, whenever $\overline{E'_G(P)} \cap \overline{E'_G(Q)} \neq \emptyset$, we have (after relabeling if necessary) $\operatorname{park}(P) < \operatorname{park}(Q)$ *and a commutative diagram of fiber bundle maps*

$$\begin{array}{ccc} \overline{E'_G(P)} \cap \overline{E'_G(Q)} & \xrightarrow{\quad\Psi_P\quad} & \overline{E'_{M_P}(Q_P)} \\ & {\scriptstyle\Psi_Q}\searrow \qquad \swarrow{\scriptstyle\Psi_{Q_P}} & \\ & \Gamma_Q\backslash\widetilde{X_Q} & \end{array}$$

Moreover, the resolved Borel-Serre compactification is the total boundary blow-up of the Borel-Serre compactification,

$$\Gamma\backslash\widetilde{X} = \left(\Gamma\backslash\overline{X}\right)_{tb}.$$

Proof. In terms of the rational Langlands decomposition of a parabolic subgroup

$$P = U_P A_{\mathbf{P}} M_{\mathbf{P}}$$

the face $E_G(P)$ of $\widetilde{X}$ is obtained by taking the quotient of A_P by a diagonal $\mathbb{R}^*_+$-action and the quotient of $M_{\mathbf{P}}$ by the action of a maximal compact subgroup. It is this decomposition of P that induces the decomposition

$$\overline{E_G(P)} = U_P \times (A_P/\mathbb{R}^*_+)_{tb} \times \widetilde{X_P}.$$

Thus if we take a quotient of this face by $\Gamma_{U_P} = \Gamma \cap U_P$ it will act entirely on the fist factor,

$$\Gamma_{U_P}\backslash\overline{E_G(P)} = (\Gamma_{U_P}\backslash U_P) \times (A_P/\mathbb{R}^*_+)_{tb} \times \widetilde{X_P}.$$

Subsequently taking the quotient by the action of Γ_{M_P} yields

$$\Gamma_P\backslash\overline{E_G(P)} = \big((\Gamma_{U_P}\backslash U_P) \times (A_P/\mathbb{R}^*_+)_{tb}\big) \times_{\Gamma_{M_P}} \widetilde{X_P}$$

with the group Γ_{M_P} acting on the first factor by conjugation. However, since A_P is in the center of L_P, this action does not affect it and we can pull-out the factor of A_P to obtain

$$\Gamma_P\backslash\overline{E_G(P)} = (A_P/\mathbb{R}^*_+)_{tb} \times \left((\Gamma_{U_P}\backslash U_P) \times_{\Gamma_{M_P}} \widetilde{X_P}\right)$$

as required.

Identifying this face with the lift of the face $\overline{e'_G(P)}$ after performing the total boundary blow-up of $\Gamma\backslash\overline{X}$ can be done as in the proof of Proposition 7.2.

The boundary fibration structure is inherited from that of $\widetilde{X}$ thanks to the equivariance of the maps Φ_P. □

7.4 The compactification of a locally symmetric space to a stratified space

In [120], Zucker introduced the *reductive Borel-Serre compactification* of a locally symmetric space. This is the stratified space $\widehat{M}$ obtained from $\Gamma\backslash\overline{X}$ by collapsing the fibers of the fiber bundle maps

$$\Gamma_{U_P}\backslash U_P - e'_G(P) \longrightarrow \Gamma_{M_P}\backslash X_P.$$

Note that the natural map $\Gamma\backslash\overline{X} \longrightarrow \widehat{M}$ restricts to a diffeomorphism of the interior, $\Gamma\backslash X$, and the regular part of $\widehat{M}$.

Alternatively, one obtains the same space by collapsing the fibers of the fiber bundle maps

$$\Gamma_{U_P}\backslash U_P - \overline{e'_G(P)} \longrightarrow \Gamma_{M_P}\backslash\overline{X_P}$$

in an order of non-decreasing codimension. Similarly, one could start with the space $\Gamma\backslash\widetilde{X}$ and collapse the fibers of the fiber bundle maps

$$(A_P/\mathbb{R}^*_+)_{tb} \times \Gamma_{U_P}\backslash U_P - \overline{E'_G(P)} \longrightarrow \Gamma_{M_P}\backslash\overline{X_P}$$

in an order of non-decreasing codimension and again end up with $\widehat{M}$. This is precisely the procedure outlined above for going from a manifold with corners and an iterated fibration structure to a stratified space.

Corollary 7.5. *Let X be the symmetric space of the real points of a connected reductive group $\mathbf{G}$ as above, let Γ be a neat arithmetic subgroup of $\mathbf{G}(\mathbb{Q})$, and let $\widehat{M}$ be the reductive Borel-Serre compactification of $\Gamma\backslash X$.*

The resolution of $\widehat{M}$ to a manifold with corners and an iterated fibration structure is $\Gamma\backslash\widetilde{X}$, the resolved Borel-Serre compactification of $\Gamma\backslash X$.

References

[1] Pierre Albin, Markus Banagl, Eric Leichtnam, Rafe Mazzeo, and Paolo Piazza, *Refined intersection homology on non-Witt spaces*, J. Topol. Anal. **7** (2015), no. 1, 105–133.

[2] Pierre Albin and Jesse Gell-Redman, *The index of Dirac operators on incomplete edge spaces*, available online at arXiv:1312.4241, 2013.

[3] Pierre Albin, Eric Leichtnam, Rafe Mazzeo, and Paolo Piazza, *The signature package on Witt spaces*, Annales Scientifiques de l'Ecole Normale Supèrieure **45** (2012), no. 2, 241–310.

[4] ——, *Hodge theory on Cheeger spaces*, to appear in Crelle's journal, 2013.

[5] ——, *The Novikov conjecture on Cheeger spaces*, to appear in Journal of Noncommutative Geometry, 2013.

[6] Pierre Albin and Richard B. Melrose, *Resolution of smooth group actions*, Spectral theory and geometric analysis, Contemp. Math., vol. 535, Amer. Math. Soc., Providence, RI, 2011, pp. 1–26.

[7] Michael F. Atiyah and Claude Lebrun, *Curvature, cones and characteristic numbers*, Math. Proc. Cambridge Philos. Soc. **155** (2013), no. 1, 13–37.

[8] Michael F. Atiyah, Vijay K. Patodi, and Isadore M. Singer, *Spectral asymmetry and Riemannian geometry*, Bull. London Math. Soc. **5** (1973), 229–234.

[9] David Ayala, John Francis, and Hiro Lee Tanaka, *Local structures on stratified spaces*, available online at arXiv:1409.0501, 2014.

[10] Markus Banagl, *Extending intersection homology type invariants to non-Witt spaces*, Mem. Amer. Math. Soc. **160** (2002), no. 760, x+83 pages.

[11] ——, *The L-class of non-Witt spaces*, Ann. of Math. (2) **163** (2006), no. 3, 743–766.

[12] ——, *Topological invariants of stratified spaces*, Springer Monographs in Mathematics, Springer, Berlin, 2007.

[13] ——, *The signature of singular spaces and its refinements to generalized homology theories*, Topology of Stratified Spaces, MSRI Publications, vol. 58, 2011.

[14] Francesco Bei, *General perversities and L^2 de Rham and Hodge theorems for stratified pseudomanifolds*, Bull. Sci. Math. **138** (2014), no. 1, 2–40.

[15] ——, *Poincaré duality, Hilbert complexes and geometric applications*, Adv. Math. **267** (2014), 121–175.

[16] Nicole Berline, Ezra Getzler, and Michèle Vergne, *Heat kernels and Dirac operators*, Grundlehren Text Editions, Springer-Verlag, Berlin, 2004, Corrected reprint of the 1992 original.

[17] Jean-Michel Bismut and Jeff Cheeger, *η-invariants and their adiabatic limits*, J. Amer. Math. Soc. **2** (1989), no. 1, 33–70.

[18] ——, *Families index for manifolds with boundary, superconnections, and cones. I. Families of manifolds with boundary and Dirac operators*, J. Funct. Anal. **89** (1990), no. 2, 313–363.

[19] ——, *Families index for manifolds with boundary, superconnections and cones. II. The Chern character*, J. Funct. Anal. **90** (1990), no. 2, 306–354.

[20] ——, *Remarks on the index theorem for families of Dirac operators on manifolds with boundary*, Differential geometry, Pitman Monogr. Surveys Pure Appl. Math., vol. 52, Longman Sci. Tech., Harlow, 1991, pp. 59–83.

[21] Jean-Michel Bismut and John Lott, *Flat vector bundles, direct images and higher real analytic torsion*, J. Amer. Math. Soc. **8** (1995), no. 2, 291–363.

[22] Christian Bohr, Bernhard Hanke, and Dieter Kotschick, *Cycles, submanifolds, and structures on normal bundles*, Manuscripta Math. **108** (2002), no. 4, 483–494.

[23] A. Borel and J.-P. Serre, *Corners and arithmetic groups*, Comment. Math. Helv. **48** (1973), 436–491, Avec un appendice: Arrondissement des variétés à coins, par A. Douady et L. Hérault.

[24] Armand Borel and Lizhen Ji, *Compactifications of locally symmetric spaces*, J. Differential Geom. **73** (2006), no. 2, 263–317.

[25] ——, *Compactifications of symmetric and locally symmetric spaces*, Mathematics: Theory & Applications, Birkhäuser Boston, Inc., Boston, MA, 2006.

[26] Armand Borel and Nicolas Spaltenstein, *Sheaf theoretic intersection cohomology*, Intersection cohomology (Bern, 1983), Progr. Math., vol. 50, Birkhäuser Boston, Boston, MA, 1984, pp. 47–182.

[27] Jean-Paul Brasselet, *de Rham theorems for singular varieties*, Differential topology, foliations, and group actions (Rio de Janeiro, 1992), Contemp. Math., vol. 161, Amer. Math. Soc., Providence, RI, 1994, pp. 95–112.

[28] Jean-Paul Brasselet, Gilbert Hector, and Martin Saralegi, *Théorème de de Rham pour les variétés stratifiées*, Ann. Global Anal. Geom. **9** (1991), no. 3, 211–243.

[29] Gregory W. Brumfiel, *Realizing homology classes by PL manifolds*, Algebraic and geometric topology (Proc. Sympos., Univ. California, Santa Barbara, Calif., 1977), Lecture Notes in Math., vol. 664, Springer, Berlin, 1978, pp. 43–60.

[30] Jochen Brüning, *The signature operator on manifolds with a conical singular stratum*, Astérisque (2009), no. 328, 1–44 (2010).

[31] Jochen Brüning and Matthias Lesch, *Hilbert complexes*, J. Funct. Anal. **108** (1992), no. 1, 88–132.

[32] Jochen Brüning and Robert Seeley, *An index theorem for first order regular singular operators*, Amer. J. Math. **110** (1988), no. 4, 659–714.

[33] Jean-Luc Brylinski, *Equivariant intersection cohomology*, Kazhdan-Lusztig theory and related topics (Chicago, IL, 1989), Contemp. Math., vol. 139, Amer. Math. Soc., Providence, RI, 1992, pp. 5–32.

[34] Gilles Carron, *L^2 harmonic forms on non-compact Riemannian manifolds*, Surveys in analysis and operator theory (Canberra, 2001), Proc. Centre Math. Appl. Austral. Nat. Univ., vol. 40, Austral. Nat. Univ., Canberra, 2002, pp. 49–59.

[35] ——, *L^2-cohomology of manifolds with flat ends*, Geom. Funct. Anal. **13** (2003), no. 2, 366–395.

[36] Jeff Cheeger, *On the spectral geometry of spaces with cone-like singularities*, Proc. Nat. Acad. Sci. U.S.A. **76** (1979), no. 5, 2103–2106.

[37] ——, *On the Hodge theory of Riemannian pseudomanifolds*, Geometry of the Laplace operator (Proc. Sympos. Pure Math., Univ. Hawaii, Honolulu, Hawaii, 1979), Proc. Sympos. Pure Math., XXXVI, Amer. Math. Soc., Providence, R.I., 1980, pp. 91–146.

[38] ——, *Spectral geometry of singular Riemannian spaces*, J. Differential Geom. **18** (1983), no. 4, 575–657 (1984).

[39] ——, *On the formulas of Atiyah-Patodi-Singer and Witten*, Proceedings of the International Congress of Mathematicians, Vol. 1, 2 (Berkeley, Calif., 1986), Amer. Math. Soc., Providence, RI, 1987, pp. 515–521.

[40] Jeff Cheeger and Xianzhe Dai, *L^2-cohomology of spaces with nonisolated conical singularities and nonmultiplicativity of the signature*, Riemannian topology and geometric structures on manifolds, Progr. Math., vol. 271, Birkhäuser Boston, Boston, MA, 2009, pp. 1–24.

[41] Jeff Cheeger, Mark Goresky, and Robert MacPherson, *L^2-cohomology and intersection homology of singular algebraic varieties*, Seminar on Differential Geometry, Ann. of Math. Stud., vol. 102, Princeton Univ. Press, Princeton, N.J., 1982, pp. 303–340.

[42] Arthur W. Chou, *The Dirac operator on spaces with conical singularities and positive scalar curvatures*, Trans. Amer. Math. Soc. **289** (1985), no. 1, 1–40.

[43] ——, *Criteria for selfadjointness of the Dirac operator on pseudomanifolds*, Proc. Amer. Math. Soc. **106** (1989), no. 4, 1107–1116.

[44] Greg Friedman, *An introduction to intersection homology with general perversity functions*, Topology of stratified spaces, Math. Sci. Res. Inst. Publ., vol. 58, Cambridge Univ. Press, Cambridge, 2011, pp. 177–222.

[45] Greg Friedman and Eugenie Hunsicker, *Additivity and non-additivity for perverse signatures*, J. Reine Angew. Math. **676** (2013), 51–95.

[46] Greg Friedman and James McClure, *The symmetric signature of a Witt space*, Journal of Topology and Analysis **5** (2013), 121–159.

[47] Matthew P. Gaffney, *The harmonic operator for exterior differential forms*, Proc. Nat. Acad. Sci. U. S. A. **37** (1951), 48–50.

[48] ——, *Hilbert space methods in the theory of harmonic integrals*, Trans. Amer. Math. Soc. **78** (1955), 426–444.

[49] Jesse Gell-Redman and Frédéric Rochon, *Hodge cohomology of some foliated boundary and foliated cusp metrics*, Math. Nachr. **288** (2015).

[50] Juan B. Gil, Thomas Krainer, and Gerardo A. Mendoza, *On the closure of elliptic wedge operators*, J. Geom. Anal. **23** (2013), no. 4, 2035–2062.

[51] M. Goresky, G. Harder, and R. MacPherson, *Weighted cohomology*, Invent. Math. **116** (1994), no. 1-3, 139–213.

[52] Mark Goresky, *Whitney stratified chains and cochains*, Trans. Amer. Math. Soc. **267** (1981), no. 1, 175–196.

[53] Mark Goresky and Robert MacPherson, *Intersection homology theory*, Topology **19** (1980), no. 1, 135–162.

[54] ——, *Intersection homology. II*, Invent. Math. **72** (1983), no. 1, 77–129.

[55] Mark Grant and András Szücs, *On realizing homology classes by maps of restricted complexity*, Bull. Lond. Math. Soc. **45** (2013), no. 2, 329–340.

[56] Misha Gromov, *Positive curvature, macroscopic dimension, spectral gaps and higher signatures*, Functional analysis on the eve of the 21st century, Vol. II (New Brunswick, NJ, 1993), Progr. Math., vol. 132, Birkhäuser Boston, Boston, MA, 1996, pp. 1–213.

[57] Andrew Hassell, Rafe Mazzeo, and Richard B. Melrose, *Analytic surgery and the accumulation of eigenvalues*, Comm. Anal. Geom. **3** (1995), no. 1-2, 115–222.

[58] Tamás Hausel, Eugenie Hunsicker, and Rafe Mazzeo, *Hodge cohomology of gravitational instantons*, Duke Math. J. **122** (2004), no. 3, 485–548.

[59] Michel Hilsum and George Skandalis, *Invariance de la signature à coefficients dans un fibré presque plat*, J. Reine Angew. math. **423** (1990), 73–99.

[60] William V. D. Hodge, *A Dirichlet problem for harmonic functionals, with applications to analytic varieties*, Proc. Lond. Math. Soc. **36** (1932), no. 2, 257–303.

[61] ——, *The existence theorem for harmonic integrals*, Proc. Lond. Math. Soc. **41** (1936), no. 2, 483–496.

[62] ——, *The theory and applications of harmonic integrals*, Cambridge Univ. Press, Cambridge, England; Macmillan Company, New York, 1941.

[63] John H. Hubbard and Sarah Koch, *An analytic construction of the Deligne-Mumford compactification of the moduli space of curves*, J. Differential Geom. **98** (2014), no. 2, 261–313.

[64] Bruce Hughes and Shmuel Weinberger, *Surgery and stratified spaces*, Surveys on surgery theory, Vol. 2, Ann. of Math. Stud., vol. 149, Princeton Univ. Press, Princeton, NJ, 2001, pp. 319–352.

[65] Eugenie Hunsicker, *Hodge and signature theorems for a family of manifolds with fibre bundle boundary*, Geom. Topol. **11** (2007), 1581–1622.

[66] Eugenie Hunsicker and Frédéric Rochon, *Weighted hodge cohomology of iterated fibred cusp metrics*, Annales mathématiques du Québec (2015), 1–8.

[67] Vadim Yu Kaloshin, *A geometric proof of the existence of Whitney stratifications*, Mosc. Math. J. **5** (2005), no. 1, 125–133.

[68] Marja Kankaanrinta, *Equivariant collaring, tubular neighbourhood and gluing theorems for proper Lie group actions*, Algebr. Geom. Topol. **7** (2007), 1–27.

[69] Gennadi Kasparov, *Novikov's conjecture on Higher Signatures: The Operator K-Theory Approach.*, Contem. Math., vol. 145, 1993.

[70] Henry C. King, *Topological invariance of intersection homology without sheaves*, Topology Appl. **20** (1985), no. 2, 149–160.

[71] Frances Kirwan and Jonathan Woolf, *An introduction to intersection homology theory*, second ed., Chapman & Hall/CRC, Boca Raton, FL, 2006.

[72] Steven L. Kleiman, *The development of intersection homology theory*, Pure Appl. Math. Q. **3** (2007), no. 1, Special Issue: In honor of Robert D. MacPherson. Part 3, 225–282.

[73] Benoît Kloeckner, *Quelques notions d'espaces stratifiés*, Actes du Séminaire de Théorie Spectrale et Géométrie. Vol. 26. Année 2007–2008, Sémin. Théor. Spectr. Géom., vol. 26, Univ. Grenoble I, Saint, 2009, pp. 13–28.

[74] Thomas Krainer and Gerardo A. Mendoza, *Boundary value problems for first order elliptic wedge operators*, available online at arXiv:1307.2398, to appear in Amer. J. Math., 2013.

[75] ——, *The kernel bundle of a holomorphic Fredholm family*, Comm. Partial Differential Equations **38** (2013), no. 12, 2107–2125.

[76] ——, *Boundary value problems for elliptic wedge operators: the first order case*, Elliptic and Parabolic Equations (J. Escher, E. Schrohe, J. Seiler, and C. Walker, eds.), vol. 119, Springer Proceedings in Mathematics & Statistics, 2014, pp. 209–232.

[77] ——, *Elliptic systems of variable order*, Rev. Mat. Iberoam. **31** (2015), no. 1, 127–160.

[78] ——, *The Friedrichs extension for elliptic wedge operators of second order*, available online at arXiv:1509.01842, 2015.

[79] Matthias Lesch, *Operators of Fuchs type, conical singularities, and asymptotic methods*, Teubner-Texte zur Mathematik [Teubner Texts in Mathematics], vol. 136, B. G. Teubner Verlagsgesellschaft mbH, Stuttgart, 1997.

[80] Stanisław Łojasiewicz, *On semi-analytic and subanalytic geometry*, Panoramas of mathematics (Warsaw, 1992/1994), Banach Center Publ., vol. 34, Polish Acad. Sci., Warsaw, 1995, pp. 89–104.

[81] Eduard Looijenga, *L^2-cohomology of locally symmetric varieties*, Compositio Math. **67** (1988), no. 1, 3–20.

[82] Gheorghe Lusztig, *Novikov's higher signature and families of elliptic operators*, J. Differential Geometry **7** (1972), 229–256.

[83] John Mather, *Notes on topological stability*, Bulletin of the American Mathematical Society **49** (2012), no. 4, 475–506.

[84] Rafe Mazzeo, *The Hodge cohomology of a conformally compact metric*, J. Differential Geom. **28** (1988), no. 2, 309–339.

[85] ——, *Elliptic theory of differential edge operators I*, Comm. in P.D.E. **16** (1991), 1615–1664.

[86] Rafe Mazzeo and Ralph S. Phillips, *Hodge theory on hyperbolic manifolds*, Duke Math. J. **60** (1990), no. 2, 509–559.

[87] Rafe Mazzeo and Boris Vertman, *Elliptic theory of differential edge operators, II: Boundary value problems*, Indiana Univ. Math. J. **63** (2014), no. 6, 1911–1955.

[88] Richard B. Melrose, *Differential analysis on manifolds with corners*, available online at `http://www-math.mit.edu/ rbm/book.html`.

[89] ——, *Calculus of conormal distributions on manifolds with corners*, Internat. Math. Res. Notices (1992), no. 3, 51–61.

[90] ——, *The Atiyah-Patodi-Singer index theorem*, Research Notes in Mathematics, vol. 4, A K Peters Ltd., Wellesley, MA, 1993.

[91] ——, *Fibrations, compactifications and algebras of pseudodifferential operators*, Partial differential equations and mathematical physics (Copenhagen, 1995; Lund, 1995), Progr. Nonlinear Differential Equations Appl., vol. 21, Birkhäuser Boston, Boston, MA, 1996, pp. 246–261.

[92] Richard B. Melrose and Gerardo Mendoza, *Elliptic operators of totally characteristic type*, MSRI preprint 047-83, 1983.

[93] A. S. Mishchenko, *Infinite-dimensional representations of discrete groups, and higher signatures*, Izv. Akad. Nauk SSSR Ser. Mat. **38** (1974), 81–106.

[94] Werner Müller, *Spectral theory for Riemannian manifolds with cusps and a related trace formula*, Report MATH 1982, vol. 3, Akademie der Wissenschaften der DDR, Institut für Mathematik, Berlin, 1982, With German and Russian summaries.

[95] Claudio Murolo, *Whitney homology, cohomology and Steenrod squares*, Ricerche Mat. **43** (1994), no. 2, 175–204.

[96] Liviu Nicolaescu, *An invitation to Morse theory*, second ed., Universitext, Springer, New York, 2011.

[97] Markus J. Pflaum, *Analytic and geometric study of stratified spaces*, Lecture Notes in Mathematics, vol. 1768, Springer-Verlag, Berlin, 2001.

[98] Guido Pollini, *Intersection differential forms*, Rend. Sem. Mat. Univ. Padova **113** (2005), 71–97.

[99] Elmer Rees and Emery Thomas, *Realizing homology classes by almost-complex submanifolds*, Math. Z. **172** (1980), no. 2, 195–201.

[100] Jonathan Rosenberg, *Novikov's Conjecture*, available online at arXiv:1506.05408, 2015.

[101] Yuli B. Rudyak, *The problem of realization of homology classes from Poincaré up to the present*, Global analysis—studies and applications, V, Lecture Notes in Math., vol. 1520, Springer, Berlin, 1992, pp. 93–110.

[102] Leslie Saper, *On the cohomology of locally symmetric spaces and of their compactifications*, Lie groups and automorphic forms, AMS/IP Stud. Adv. Math., vol. 37, Amer. Math. Soc., Providence, RI, 2006, pp. 169–239.

[103] Leslie Saper and Mark Stern, *L_2-cohomology of arithmetic varieties*, Ann. of Math. (2) **132** (1990), no. 1, 1–69.

[104] Bert-Wolfgang Schulze, *Pseudo-differential operators on manifolds with singularities*, Studies in Mathematics and its Applications, vol. 24, North-Holland Publishing Co., Amsterdam, 1991.

[105] ——, *Iterative structures on singular manifolds*, Geometric and spectral analysis, Contemp. Math., vol. 630, Amer. Math. Soc., Providence, RI, 2014, pp. 173–221.

[106] Paul Siegel, *Witt spaces: A geometric cycle theory for KO-homology theory at odd primes*, Amer. J. Math. **105** (1983), 1067–1105.

[107] Michael Teufel, *Abstract prestratified sets are (b)-regular*, J. Differential Geom. **16** (1981), no. 3, 529–536 (1982).

[108] René Thom, *Espaces fibrés en sphères et carrés de Steenrod*, Ann. Sci. Ecole Norm. Sup. **3** (1952), no. 69.

[109] ——, *Quelques propriétés globales des variétés différentiables*, Comment. Math. Helv. **28** (1954), 17–86.

[110] ——, *Les classes caractéristiques de Pontryagin des variétés triangulées*, Symp. Int. de Top. Alg. UNESCO (1959).

[111] ——, *Ensembles et morphismes stratifiés*, Bull. Amer. Math. Soc. **75** (1969), 240284.

[112] Ravi Vakil, *The moduli space of curves and its tautological ring*, Notices Amer. Math. Soc. **50** (2003), no. 6, 647–658.

[113] Lou van den Dries and Chris Miller, *Geometric categories and o-minimal structures*, Duke Math. J. **84** (1996), no. 2, 497–540.

[114] Andrei Verona, *Stratified mappings—structure and triangulability*, Lecture Notes in Mathematics, vol. 1102, Springer-Verlag, Berlin, 1984.

[115] Hermann Weyl, *On Hodge's theory of harmonic integrals*, Ann. of Math. (2) **44** (1943), 1–6.

[116] ——, *Address of the President of the Fields Medal Comittee 1954*, Proceedings of the International Congress of Mathematicians 1954, Groningen and North-Holland, Amsterdam, 1957, pp. 161–174.

[117] Hassler Whitney, *Tangents to an analytic variety*, Ann. of Math. (2) **81** (1965), 496–549.

[118] Steven Zucker, *Théorie de Hodge à coefficients dégénérescents*, C. R. Acad. Sci. Paris Sér. A-B **286** (1978), no. 23, A1137–A1140.

[119] ——, *Hodge theory with degenerating coefficients: L_2 cohomology in the Poincaré metric*, Ann. of Math. (2) **109** (1979), no. 3, 415–476.

[120] ——, *L_2 cohomology of warped products and arithmetic groups*, Invent. Math. **70** (1982/83), no. 2, 169–218.

[121] ——, *On the boundary cohomology of locally symmetric varieties*, Vietnam J. Math. **25** (1997), no. 4, 279–318.

[122] ——, *On the reductive Borel-Serre compactification: L^p-cohomology of arithmetic groups (for large p)*, Amer. J. Math. **123** (2001), no. 5, 951–984.

Hodge Theory and L^2-analysis
ALM 39, pp. 79–86

Simpson's Construction of Varieties with Many Local Systems*

Donu Arapura†

Abstract

The purpose of this largely expository note is to give some examples of smooth projective varieties with zero first Betti number but nontrivial families of local systems of higher ranks.

One of the goals of this note is to say something about the fundamental group of a smooth complex projective variety in terms of the quantity of local systems on it. Given a finitely generated group Γ, let $d_N(\Gamma)$ be the dimension of the space of irreducible rank N representations. The number $d_1(\Gamma)$ coincides with the first Betti number, so one may think of $d_N(\Gamma)$ as a nonabelian generalization. The basic problem is to see how these numbers behave when Γ is the fundamental group of a smooth projective variety X. In this case, these numbers are always even [A]. If X is a curve of genus at least two, or even if it maps onto such a curve, then $d_N(\Gamma) > 0$ for all N. If X is an abelian variety, then $d_1(\Gamma) > 0$ but $d_N(\Gamma) = 0$ for all $N > 0$. I want to consider examples which have the opposite behaviour, in that $d_1 = 0$ but some higher $d_N > 0$. Some cheap examples are given in the first section. However, they are not very interesting in the sense that they are very close to the examples we already know. In the second section I will turn to a beautiful construction due to Carlos Simpson [S], which also produces smooth projective varieties such that $d_N(\pi_1(X)) > 0$ for some $N > 1$. In fact, the real purpose of this article is to make Simpson's construction a bit more accessible and explicit, with the hope that these examples will be studied more thoroughly in the future. Some specific problems are suggested in the last section.

*To Steve Zucker. Partially supported by the NSF.

†Department of Mathematics, Purdue University, West Lafayette, IN 47907,U.S.A.

1 Representation varieties

For Γ a group with generators $g_1, \ldots, g_n$, an element of $Hom(\Gamma, GL_N(\mathbb{C}))$ is given by n matrices subject to the relations of the group. In this way, the set becomes an affine scheme of finite type, called the representation "variety". (For the present purposes, a scheme will be identified with the set of its closed points.) The algebraic group $GL_N(\mathbb{C})$ acts on the representation variety by conjugation, and the GIT quotient

$$\begin{aligned} M(\Gamma, N) &= Hom(\Gamma, GL_N(\mathbb{C}))//GL_N(\mathbb{C}) \\ &:= \operatorname{Spec} \mathcal{O}(Hom(\Gamma, GL_N(\mathbb{C})))^{GL_N(\mathbb{C})} \end{aligned}$$

can be identified with the set of isomorphism classes of *semisimple* representations of rank N [LM]. This is often called the character variety. Let

$$M(\Gamma, N)^{irred} \subset M(\Gamma, N)$$

denote the possibly empty open subset of irreducible representations. We have quasifinite (i.e. set theoretically finite to one) morphisms

$$M(\Gamma, N_1) \times M(\Gamma, N_2) \to M(\Gamma, N_1 + N_2)$$

given by direct sum. We can decompose

$$M(\Gamma, N) = \bigcup_{N_1 + \cdots + N_r = N} \operatorname{im} M(\Gamma, N_1)^{irred} \times \cdots \times M(\Gamma, N_r)^{irred}. \tag{1}$$

Let

$$d_N(\Gamma) = \dim M(\Gamma, N)^{irred}$$

where we take it to be zero if it is empty. From (1), we obtain:

Lemma 1.1. *We have*

$$\dim M(\Gamma, N) = \max_{N_1 + \cdots + N_r = N} d_{N_1}(\Gamma) + \cdots + d_{N_r}(\Gamma).$$

Therefore $\dim M(\Gamma, N) > 0$ *if and only if* $d_M(\Gamma) > 0$ *for some* $M \leq N$.

We have $M(\Gamma, 1) = \dim Hom(\Gamma, \mathbb{C}^*)$, therefore $d_1(\Gamma) = \operatorname{rank} \Gamma/[\Gamma, \Gamma]$. For higher N, these numbers are usually very difficult to calculate, although there are some easy cases. We have $d_N(\Gamma) = 0$ when $N > 1$ and Γ is abelian simply because in this case there are no irreducible representations of higher rank. If Γ surjects onto a nonabelian free group then a bit of thought shows that $d_N(\Gamma) > 0$ for all N. This remark applies to the fundamental group of a smooth projective curve of genus at least two.

When $\Gamma = \pi_1(X)$ is the fundamental group of a smooth projective variety X, Hodge theory tells us that $d_1(\Gamma) = \dim H^1(X)$ is even. More generally, nonabelian Hodge theory implies that $M(\Gamma, N)^{irred}$ carries a quaternionic or hyperkähler structure, therefore every $d_N(\Gamma)$ is even [A, thm 3.1]. Here is the example promised in the introduction.

Theorem 1.2. *There exists a smooth projective variety X with $d_1(\pi_1(X)) = 0$ and $d_N(\pi_1(X)) \geq 2d$ for any given $N > 1$ and $d > 0$.*

Proof. Let $C \to \mathbb{P}^1$ be a cyclic cover of the form $y^N = f(x)$, where f has distinct roots. Let x_0 denote one of the roots. By choosing $\deg f$ sufficiently large, we can assume that the genus g of C is greater than or equal to d. The group $G = \mathbb{Z}/N\mathbb{Z}$ will act on C with $C/G \cong \mathbb{P}^1$. If follows that $H^1(C, \mathbb{Q})^G = 0$. Consequently, if $\gamma \in G$ denotes a generator, it will act nontrivially on $H_1(C, \mathbb{Z})$. By Serre [Se, prop. 15], there exists a simply connected variety Y on which G acts freely. Let $X = (C \times Y)/G$, where G acts diagonally. The projection $X \to Y/G$ is a fibration with fibre C and section given by $y \mapsto (x_0, y)$. Therefore we have split exact sequence

$$1 \to \pi_1(C) \to \pi_1(X) \leftrightarrows G \to 1.$$

Using the Hochschild-Serre spectral sequence, we obtain an exact sequence

$$H^1(G, H^0(\pi_1(C), \mathbb{Q})) \to H^1(\pi_1(X), \mathbb{Q}) \to H^0(G, H^1(\pi_1(C), \mathbb{Q})).$$

As noted above, the group on the right vanishes. Since G is finite, the group on the left also vanishes. Therefore $H^1(\pi_1(X), \mathbb{Q}) = 0$, which means that $d_1(\pi_1(X)) = 0$.

Let $\rho \in Hom(\pi_1(C), \mathbb{C}^*) = (\mathbb{C}^*)^{2g}$ be a one dimensional character. For a generic choice of ρ, the characters $\rho, \rho \circ \gamma, \ldots, \rho \circ \gamma^{N-1}$ are all distinct. Let $\mathbb{C}_\rho$ denote the $\mathbb{C}[\pi_1(C)]$-module associated to ρ. The induced representation $V_\rho = \operatorname{Ind} \mathbb{C}_\rho$ gives a rank N $\mathbb{C}[\pi_1(X)]$-module. As a $\mathbb{C}[\pi_1(C)]$-module

$$V_\rho = \mathbb{C}_\rho \oplus \mathbb{C}_{\rho \circ \gamma} \oplus \cdots \tag{2}$$

and γ acts by cyclically permuting the factors. It follows easily that V_ρ is an irreducible $\pi_1(X)$-module for generic ρ. Also by computing characters, using (2), we see that $V_\rho \cong V_{\rho'}$ only if $\rho' \in \{\rho, \gamma\rho, \ldots\}$. Therefore the map $\rho \mapsto V_\rho$ is a quasifinite morphism from an open subset of $(\mathbb{C}^*)^{2g}$ to $M(\pi_1(X), N)^{irred}$. Thus $d_N(\pi_1(X)) \geq 2g \geq 2d$. □

The drawback of this method is that it does not produce any really new examples of fundamental groups of smooth projective varieties. I will describe a more subtle construct in the next section, but first I want to record the following useful fact which was used implicitly above.

Lemma 1.3. *Suppose that $\Gamma_1 \subset \Gamma$ is a subgroup of index $r < \infty$.*

(a) If W_ρ is a nontrivial (i.e. nonconstant) family of representations in $M(\Gamma, N)$, then the restrictions $\operatorname{Res} W_\rho$ *give a nontrivial family in $M(\Gamma_1, N)$.*

(b) Conversely if $\operatorname{Res} W_\rho$ *is a nontrivial family, then so is W_ρ.*

(c) If V_ρ is a nontrivial family of representations in $M(\Gamma_1, N)$, then $\operatorname{Ind} V_\rho$ *is a nontrivial family in $M(\Gamma, rN)$*

Proof. The first two items are the content of Lemma 1.5 of [S]. For (c), we have that $\operatorname{Res}(\operatorname{Ind} V_\rho) = V_\rho \oplus \cdots$ is nontrivial, so $\operatorname{Ind} V_\rho$ is nontrivial by (b). □

2 Simpson's construction

Let Z be a smooth projective variety with dimension $2n+1 \geq 3$ and positive first Betti number. Fix an embedding $Z \subset \mathbb{P}^K$ such that $\mathcal{O}_Z(1)$ is sufficiently ample in the sense that it is a high enough power of a given ample bundle. Sufficient ampleness is needed for the proofs of Proposition 2.1 and Theorem 2.4. Let $P \subset \check{\mathbb{P}}^K$ be a general linear subspace of the dual space of dimension $d \geq 2$. Then we can form the incidence variety

$$Y = \{(z, H) \in Z \times P \mid z \in H\}$$

with projections and inclusions labelled as follows

$$\begin{array}{ccc} Y & \xrightarrow{f} & Z \\ {\scriptstyle\pi}\downarrow & \searrow^{\iota} & \uparrow{\scriptstyle F} \\ P & \xleftarrow[\Pi]{} & Z \times P \end{array}$$

Denote the fibre of π over t by X_t. Let $D_1 = \{t \in P \mid X_t \text{ is singular}\}$ be the (reduced) discriminant. The following standard fact is stated in [DL] and various other places. A proof, assuming sufficient ampleness, can be found in [S, prop. 6.1].

Proposition 2.1. *The discriminant D_1 is an irreducible hypersurface and for a generic 2 dimensional plane $Q \subset P$, the singularities of $D_1 \cap Q$ are nodes and cusps.*

The next step is to form a double cover branched over D_1. If $g(x_1, \ldots, x_d) = 0$ is an affine equation of D_1, then the cover $y^2 = g$ may acquire additional ramification at infinity. It is better to control this in advance by defining

$$D = \begin{cases} D_1 & \text{if } \deg D_1 \text{ is even,} \\ D_1 + D_2 & \text{otherwise, where } D_2 \text{ is a hyperplane in general position.} \end{cases}$$

Let $U = P - D$. Let $p' : X' \to P$ be the double cover branched along D. As a scheme

$$X' = \mathbf{Spec}\left(\mathcal{O}_P \oplus \mathcal{O}_P\left(\frac{\deg D}{2}\right)\right)$$

where the sheaf in parantheses is made into an algebra in the standard way (cf. [EV, p. 22]). This will usually be singular but the singularities are normal local complete intersections. The singular set $X'_{sing} \subseteq \Sigma = p'^{-1} D_{sing}$. Let $q : X \to X'$ be a desingularization which is an isomorphism on the complement of Σ. This variety is what we are after. It is very similar to, although not identical to, Simpson's construction in [S, lemma 6.3]. The difference is that Simpson's variety is a branched cover of P of indeterminate degree, on which, by design, the local systems V_ρ constructed below extend. This makes it simpler for the purpose of constructing local systems. However, the lack of explicitness makes it harder to do precise computations.

Theorem 2.2. *The first Betti number of X is zero. For some $M > 1$, $d_M(\pi_1(X)) > 0$.*

The rest of this section will be devoted to the proof of this theorem.

Proposition 2.3. *The first Betti number of X is zero.*

Proof. By Hodge theory, the proposition is equivalent to $H^1(X, \mathcal{O}_X) = 0$. We prove the last equation by induction on d starting with $d = 2$. In this case, Σ consists of a finite set of singular points. The local analytic germ of X' at $p \in \Sigma$ is either of the form $y^2 = x_1x_2$ or $y^2 = x_1^2 - x_2^3$. These are the well known singularities of type A_n for $n = 1, 2$ [D]. These are rational singularities which implies that $H^1(X, \mathcal{O}_X) = H^1(X', \mathcal{O}_{X'})$. The last group

$$H^1(X', \mathcal{O}_{X'}) \cong H^1(\mathbb{P}^2, \mathcal{O}_{\mathbb{P}^2}) \oplus H^1(\mathbb{P}^2, \mathcal{O}_{\mathbb{P}^2}(\deg D/2)) = 0.$$

For $d > 2$, choose a general hyperplane $H \subset P$. By the Bertini, $G = p^{-1}H$ is smooth. By induction, we can assume that $H^1(G, \mathcal{O}_G) = 0$. We have an exact sequence

$$H^1(X, \mathcal{O}_X(-G)) \to H^1(X, \mathcal{O}_X) \to H^1(X, \mathcal{O}_G) = 0.$$

The first group $H^1(X, \mathcal{O}_X(-G)) = H^{d-1}(X, \omega_X(G))$ is zero by the Kawamata-Viehweg vanishing theorem [EV, p. 49]. □

We turn to the second part of theorem. By assumption Z carries a positive dimensional family of rank one local systems. Fix a generic such system $\mathbb{C}_\rho$, and consider the sheaf

$$V_\rho = \operatorname{coker}(R^n\Pi_*(F^*\mathbb{C}_\rho) \xrightarrow{\iota^*} R^n\pi_*(f^*\mathbb{C}_\rho))|_U.$$

This is a local system of some rank $N > 1$. The stalk of V_ρ over t is the primitive nth cohomology of X_t with coefficients in $\mathbb{C}_\rho$. The rank N is just the dimension of this space. Let $R_\rho : \pi_1(U) \to GL_N\mathbb{C}$ denote the representation corresponding to V_ρ.

Theorem 2.4 (Simpson [S, thm. 5.1])**.** *As ρ varies, V_ρ gives a nontrivial family in $M(\pi_1(U), N)$.*

The proof is rather involved, so we will be content to make a few brief comments about it. The key ingredient is nonabelian Hodge theory, which sets up a correspondence between semisimple local systems and certain Higgs bundles, which for our purposes can be viewed as sheaves on the cotangent bundle. Simpson then checks that as the ρ vary, the supports of the Higgs bundles corresponding to V_ρ, called spectral varieties, also vary nontrivially. When Z is an abelian variety, there is a more elementary argument which avoids Higgs bundles [S, p. 358], and this already suffices for constructing nontrivial examples.

Let γ_1 be a loop going once around a smooth point D_1. This involves a choice, but any two choices are conjugate because D_1 is irreducible. We have

$$R_\rho(\gamma_1)^2 = I \tag{3}$$

by the Picard-Lefschetz formula or see [S, lemma 6.5]. Let γ_2 be a loop around D_2 when it exists. Then

$$R_\rho(\gamma_2) = I \tag{4}$$

because V_ρ extends to a local system on $P - D_1$. Let $p = p' \circ q$ and $\tilde{U} = p^{-1}U$. We can identify $\tilde{U} = {p'}^{-1}U \subset X'$. This is an étale double cover of U corresponding to an index two subgroup $\pi_1(\tilde{U}) \subset \pi_1(U)$. This subgroup contains γ_i^2. We can identify $\pi_1(X' - \Sigma)$ with the quotient of $\pi_1(\tilde{U})$ by the normal subgroup generated by the γ_i^2. Combining this with (3) and (4) yields

Lemma 2.5. *The pullback of the local system* ${p'}^*V_\rho$ *extends to* $X' - \Sigma$.

Let $X'_Q = X' \cap {p'}^*Q$ where $Q \subset P$ is a general 2-plane.

Lemma 2.6. $\pi_1(X') \cong \pi_1(X'_Q)$.

Proof. Since X has local complete intersection singularities, we can apply the Lefschetz theorem of [FL, p. 28] to deduce the above isomorphism. □

To simplify notation, replace Σ by its restriction to X'_Q. Then Σ consists of a finite set of points. For each $p \in \Sigma$, let L_p denote the link which is the boundary of a small contractible neighbourhood of p. The group $\pi_1(X') = \pi_1(X'_Q)$ is the quotient of $\pi = \pi_1(X'_Q - \Sigma)$ by the normal subgroup N generated by $\bigcup_p \pi_1(L_p)$. For any group Γ, let

$$K(\Gamma) = \ker[\Gamma \to \widehat{\Gamma}]$$

where $\widehat{\Gamma}$ is the profinite completion. This can also be characterized as the intersection of all finite index subgroups, or as the smallest normal subgroup for which $\Gamma/K(\Gamma)$ is residually finite.

Lemma 2.7. *There exists a normal subgroup of finite index* $\Gamma \subseteq \pi$ *such that* $\pi_1(L_p) \cap \Gamma \subseteq K(\pi)$ *for each* $p \in \Sigma$.

Proof. As noted above, Σ consists of a finite set of singular points of type A_1 or A_2. These singularities can also be described as quotients of $(\mathbb{C}^2, 0)$ by an action of $\mathbb{Z}/2\mathbb{Z}$ or $\mathbb{Z}/3\mathbb{Z}$ [D]. Therefore $\pi_1(L_p)$ must either be $\mathbb{Z}/2\mathbb{Z}$ or $\mathbb{Z}/3\mathbb{Z}$ and in particular finite. Since $\pi/K(\pi)$ is residually finite, we can find a finite index subgroup $\bar{\Gamma}$ of it avoiding the nonzero elements $\mathrm{im}(\pi_1(L_p))$. Let Γ be the preimage. □

Let $\Psi \subseteq \pi_1(X')$ denote the image of Γ.

Lemma 2.8.

(a) *If* $\bar{\Gamma}$ *and* $\bar{N}$ *denote the images of* Γ *and* N *in* $\pi/K(\pi)$, *then* $\bar{\Gamma} \cap \bar{N} = 1$.

(b) $\Gamma/K(\pi) \cong \Psi/K(\pi_1(X'))$.

Proof. Item (a) follows immediately from Lemma 2.7. The canonical map $\Gamma/K(\pi) \to \Psi/K(\pi_1(X'))$ is clearly surjective. The kernel is $\bar{\Gamma} \cap \bar{N}$. So (b) follows from (a). □

Lemma 2.9. *The restriction of* R_ρ *to* Γ *is the pull back of a representation of* Ψ.

Proof. By a theorem of Malčev [M, p. 309], any finitely generated linear group is residually finite. Therefore the restriction $\operatorname{Res} V_\rho = R_\rho|_\Gamma$ factors through $\Gamma/K(\pi) \cong \Psi/K(\pi_1(X'))$. □

To finish the proof of Theorem 2.2, observe that by the above results, the restriction $\operatorname{Res} V_\rho$ comes from a Ψ-module W_ρ. We can form the induced $\pi_1(X')$-module $\operatorname{Ind} W_\rho$. This corresponds to a nontrivial family of semisimple local systems on X' by Lemma 1.3, which pulls back to a nontrivial family on X. Therefore by Lemma 1.1 $d_M(\pi_1(X)) > 0$ for some M. By Proposition 2.3, $M > 1$, and this concludes the proof.

3 Problems

I will end by discussing a few follow up problems.

Problem 3.1. *Determine (a presentation for) the fundamental group of X, constructed in section two, for some explicit choice of $Z \subset \mathbb{P}^K$, such as when it is an abelian variety.*

My hope is that this will give a genuinely new and interesting example of a group in the class of fundamental groups of smooth projective varieties. It is clear that it would differ from most of the standard known examples which either have positive first Betti number or are rigid in the sense that all $d_N = 0$. Furthermore $\pi_1(X)$ would be different from the examples constructed in section one. Simpson's arguments [S] show that in his terminology that X, with the local system $\mathcal{I} = q^* \operatorname{Ind} W_\rho$ above, has the nonfactorization property NF_1. This means that $\mathcal{I}$ is not the pull back of a local system on a curve even if we allow X to be replaced by another variety mapping surjectively to it. This will imply that $\pi_1(X)$ cannot contain the fundamental group of a curve as a subgroup of finite index.

Problem 3.2. *Find an example of a smooth projective variety with an infinite family of irreducible* unitary *representations which do not come from curves, i.e. that satisfy NF_1.*

This is equivalent to asking for a variety with an infinite family of stable vector bundles, with vanishing Chern classes, which do not come from curves. This can be rephrased as asking for an infinite family of stable Higgs bundles of the above type with zero Higgs fields. Simpson's construction described above yields Higgs bundles with nonzero Higgs fields. This is clear from his proof of Theorem 2.4.

For applications to the fundamental group, it suffices to stick with dimension $d = 2$. One reason for allowing $d > 2$ is that I feel that these varieties should be interesting from other points of view.

Problem 3.3. *Study the birational geometry of these varieties.*

For instance, although they have zero first Betti number, I suspect that they behave like varieties with large Albanese. One way to try to make this precise is by using the notion of Shafarevich maps in the sense of Campana and Kollár [K]. In most cases, I suspect that this map should be birational. This would be an analogue of the Albanese map being generically finite.

References

[A] D. Arapura, *Higgs bundles, integrability, and holomorphic forms*, Motives, polylogarithms and Hodge theory, Part II (Irvine, CA, 1998), 605–624, Int. Press Lect. Ser., 3, II, Int. Press, Somerville, MA, 2002.

[DL] I. Dolgachev, A. Libgober, *On the fundamental group of the complement to a discriminant variety*, Algebraic geometry (Chicago, Ill., 1980), 1–25, Lecture Notes in Math., 862, Springer, 1981.

[D] A. Durfee, *Fifteen characterizations of rational double points*, Enseign. Math. (2) 25 (1979), no. 1-2, 131–163.

[EV] H. Esnault, E. Viehweg, *Lectures on vanishing theorems*, DMV Seminar, 20. Birkhäuser Verlag, 1992.

[FL] W. Fulton, R. Lazarsfeld, *Connectivity and its applications in algebraic geometry*, Algebraic geometry (Chicago, Ill., 1980), 26–92, Lecture Notes in Math., 862, Springer, 1981.

[K] J. Kollár, *Shafarevich maps and automorphic forms*, Princeton University Press, 1995.

[LM] A. Lubotzky, A. Magid, *Varieties of representations of finitely generated groups*, Mem. Amer. Math. Soc. 58 (1985), no. 336.

[M] W. Magnus, *Residually finite groups*, Bull. Amer. Math. Soc. 75 (1969), 305–316.

[Se] J.P. Serre, *Sur la topologie des variiétés algébriques en caractéristique p*, Symp. Int. Top. Alg, Mexico (1958), 24–53.

[S] C. Simpson, *Some families of local systems over smooth projective varieties*, Annals of Math 138 (1993), no. 2, 337–425.

Hodge Theory and L^2-analysis
ALM 39, pp. 87–125

Motives and Algebraic Cycles: A Selection of Conjectures and Open Questions*

Joseph Ayoub[†]

Abstract

This article discusses a selection of conjectures and open questions in the study of motives and algebraic cycles. Special emphasis will be placed on the interplay between abstract properties of motives and concrete properties of Chow groups of algebraic cycles modulo rational equivalence.

2000 Mathematics Subject Classification: 14F42, 18E10, 19E15.

Keywords and Phrases: Algebraic cycles, Bloch conjecture, Chow groups, conservativity conjecture, homotopy t-structure, motives, motivic t-structure, slice filtration, vanishing conjectures.

Introduction

Since their invention by Grothendieck in the 1960's, motives have served as a guide and a source of inspiration in Arithmetic Algebraic Geometry. However, up to the 1990's, the theory of motives was lacking a solid foundation and the notion of motive was an elusive one. The situation changed drastically thanks to the work of Voevodsky and others, and we now have precise and natural mathematical entities that deserve to be called motives.[1] However, it is fair to say that the theory of

---

*Für Steven mit Zuckerzusatz.

[†]Institut für Mathematik Universität Zürich, Winterthurerstr. 190 CH-8057 Zürich Switzerland & CNRS LAGA Université Paris 13, 99 avenue J.B. Clément 93430 Villetaneuse France. E-mail: joseph.ayoub@math.uzh.ch. The author was supported in part by the Swiss National Science Foundation (NSF), Grant No. 2000201-124737/1.

[1]The reader might disagree with these statements. Indeed, it is possible to argue that Grothendieck's construction of the category of Chow motives yields a solid foundation for pure motives just as Voevodsky's construction of his triangulated category of motives yields a solid foundation for mixed motives. However, it is our personal opinion that the definition of Chow motives is ad hoc and not propitious to further structural developments in the theory of motives, whereas Voevodsky's construction is natural — I would even say inevitable — and its true potential is yet to be unlocked. Let the future tell if this opinion is naive or not!

motives is still consisting of more conjectures than theorems. Perhaps this is especially true only for "motives with rational coefficients" as our understanding of motivic cohomology with finite coefficients has been incredibly advanced with the solution of the Bloch–Kato conjecture by Voevodsky [52] with a crucial input of Rost [47].

We are mainly concerned with motives and motivic cohomology with rational coefficients. Unfortunately, in this context, breakthroughs are still to come, and, at present, we can only dream and fantasise about how beautiful and powerful the theory is ought to be. The goal of this article is to present a selection of conjectures and open problems about motives, and to explain their significance and impact on algebraic cycles. We hope to convey our enthusiasm and fascination about this beautiful subject. We have not tried to be exhaustive: there are many other fascinating conjectures that will not be mentioned at all. Also, we have refrained from extending too much the web of motivic conjectures: all the conjectures that we will discuss are folklore in one way or another.

In Section 1, we give a quick review of the theory of motives à la Voevodsky. The reader who is familiar with this theory can jump directly to Section 2 where we discuss the conservativity conjecture, one of the author's favorite conjectures. In Section 3, we discuss the vanishing conjecture for the motivic Hopf algebra; this is different from the Beilinson–Soulé vanishing conjecture in motivic cohomology, although the two conjectures are undeniably related. In Section 4, we discuss two more conjectures around the slice filtration and the homotopy t-structure. These conjectures are less established than the previous ones and should be rather considered as open questions.

Notations and conventions

We fix a base-field k and a choice of an algebraic closure $\bar{k}/k$ whose Galois group is denoted by $\mathrm{Gal}(\bar{k}/k)$. The symbol Λ always denotes the commutative ring of coefficients and $\mathbf{D}(\Lambda)$ denotes the derived category of the abelian category of Λ-modules. We are mainly interested in the case where k has characteristic zero and where $\Lambda = \mathbb{Q}$ (and, on one occasion, where $\Lambda = \overline{\mathbb{Q}}$). Therefore, even though some of what we will say remains valid in some greater generality, we assume from now on that k has characteristic zero and that Λ is a $\mathbb{Q}$-algebra unless otherwise stated.

By *k-variety* we mean a finite type separated k-scheme. We denote by Sch/k the category of k-varieties and by Sm/k its full subcategory consisting of smooth k-varieties. Given a k-variety X, its Chow groups with coefficients in Λ are denoted by $\mathrm{CH}^n(X;\Lambda)$. Similarly, we denote by $\mathrm{CH}^n(X, m;\Lambda)$ Bloch's higher Chow groups of X. Given a complex embedding $\sigma : k \hookrightarrow \mathbb{C}$, and a k-variety X, we denote by X^{an} the associated complex analytic space. The singular cohomology groups of X^{an} with coefficients in Λ are denoted by $\mathrm{H}^n(X^{\mathrm{an}};\Lambda)$.

By *mixed Hodge structure* we mean a rational graded-polarisable mixed Hodge structure, i.e., a triple (H, W, F) consisting of a finite dimensional $\mathbb{Q}$-vectorspace H, an increasing weight filtration W on H and a decreasing Hodge filtration F on $H \otimes \mathbb{C}$ satisfying the usual conditions (see [24, Définition 2.3.1]) and such that the graded pieces for the weight filtration are polarisable pure Hodge structures (see

[24, Définition 2.1.15]). The category of mixed Hodge structures is denoted by MHS. We will also consider the category of ind-objets in MHS which we denote by Ind-MHS.

An additive category is called *pseudo-abelian* if every projector has a kernel. In a monoidal category, the tensor product is denoted by $-\otimes-$ and the internal Hom, when it exists, is denoted by $\underline{\mathsf{Hom}}(-,-)$. The unit objet of a general monoidal category is denoted by $\mathbf{1}$. The dual of an object M is given by $M^{\vee} = \underline{\mathsf{Hom}}(M, \mathbf{1})$.

Acknowledgements

I thank Bruno Klingler for telling me about the Hodge-theoretic argument used in Remark 2.34. I thank Giuseppe Ancona for pointing a missing hypothesis in the original statement of Theorem 2.28. I also thank the referee and Giuseppe Ancona for improving my awareness of the literature by mentioning a few more references related to the subject of the paper. Finally, I thank Mikhail Bondarko for an interesting exchange of emails around Conjecture 4.22.

1 Review of motives

As usual, $\mathbf{DM}(k;\Lambda)$ denotes Voevodsky's category of motives with coefficients in a commutative ring Λ.[2] Almost everywhere in this article, the category of motives can be used as a black box. Therefore, we will not recall its construction. Instead, we devote this section to a list of useful facts that should help the novice reader to grasp quickly the part of the theory that is needed later on. All these facts, except the last one, are due to Voevodsky and al., and their proofs can be found in [53, 36].

1.1 Abstract categorical properties

The category $\mathbf{DM}(k;\Lambda)$ has, or is expected to have, formal properties analogous to those of the derived category of ind-mixed Hodge structures $\mathbf{D}(\text{Ind-MHS})$.[3] In this subsection, we list some of these properties. We will concentrate on what is known and leave the conjectures to later sections.

Fact A — $\mathbf{DM}(k;\Lambda)$ *has the structure of a closed monoidal triangulated category. Moreover, there is a natural functor* $\mathrm{M} : \mathrm{Sm}/k \longrightarrow \mathbf{DM}(k;\Lambda)$ *which takes a smooth k-variety X to its associated motive* $\mathrm{M}(X)$.[4]

Therefore, in $\mathbf{DM}(k;\Lambda)$, we can take the cone of a morphism, the tensor product of two objects and the internal Hom from an object with value in another

[2] The knowledgeable reader will soon guess which of the variants of Voevodsky's category of motives we have in mind. However, just in case, we spell it out: $\mathbf{DM}(k;\Lambda)$ is the homotopy category with respect to the stable $(\mathbb{A}^1, \text{Nis})$-local model structure of the category of Tate spectra of complexes of presheaves with transfers on the category of smooth k-varieties with values in the category of Λ-modules. In particular, we do not impose any boundedness conditions.

[3] This is not totally true because of the failure of conservativity; see Remark 2.3 and Lemma 2.4.

[4] The functor M is naturally extendable to Sch/k but we do not need this extension in this article.

one. Moreover, we have a supply of objets in $\mathbf{DM}(k;\Lambda)$ coming from smooth k-varieties. In this paper, objets in $\mathbf{DM}(k;\Lambda)$ will be referred to as *motives*. Elsewhere in the literature, the reader may also find the terms: mixed motives, triangulated motives, Voevodsky motives, etc.

The unit object of $\mathbf{DM}(k;\Lambda)$ is usually denoted by $\Lambda(0)$. The motive $\mathrm{M}(X)$ serves as the universal homology object associated to the smooth k-variety X. For this reason, it is sometimes called the *homological motive* of X by contrast to its dual, $\mathrm{M}(X)^{\vee}$, which is called the *cohomological motive* of X.

Notation 1.1 — We set

$$\Lambda(1) := \mathrm{Cone}\left\{\mathrm{M}(\mathrm{Spec}(k)) \xrightarrow{\infty} \mathrm{M}(\mathbb{P}^1_k)\right\}[-2].$$

We think about $\Lambda(1)$ as the reduced homology of the projective line.

The following property is imposed in the construction of $\mathbf{DM}(k;\Lambda)$.

Lemma 1.2 — *The objet $\Lambda(1)$ is invertible for the tensor product. Its inverse, which is also its dual $\underline{\mathsf{Hom}}(\Lambda(1),\Lambda(0))$, is denoted by $\Lambda(-1)$.*

Notation 1.3 — As usual, we set for $n \in \mathbb{Z}$:

$$\Lambda(n) = \left\{ \begin{array}{ll} \Lambda(1)^{\otimes n} & \text{if} \quad n \geqslant 0, \\ \Lambda(-1)^{\otimes -n} & \text{if} \quad n < 0. \end{array} \right.$$

These are the *pure Tate motives*. Given an objet $M \in \mathbf{DM}(k;\Lambda)$, we set

$$M(n) = M \otimes \Lambda(n).$$

These are the *Tate twists* of M.

Fact B — *The motive $\mathrm{M}(X)$ of a smooth k-variety X is a compact object of $\mathbf{DM}(k;\Lambda)$. Moreover, $\mathbf{DM}(k;\Lambda)$ is compactly generated by the Tate twists of motives of smooth k-varieties. In fact, it is enough to restrict to Tate twists of motives of smooth and projective k-varieties.*

We recall the meaning of compact objets and compactly generated triangulated categories.

Definition 1.4 — *Let $\mathcal{T}$ be a triangulated category with infinite sums.*

(a) *An object $M \in \mathcal{T}$ is said to be* compact *if for every (possibly infinite) family $(N_i)_{i\in I}$ of objects in $\mathcal{T}$, the obvious homomorphism*

$$\bigoplus_{i\in I} \hom_{\mathcal{T}}(M, N_i) \longrightarrow \hom_{\mathcal{T}}(M, \bigoplus_{i\in I} N_i)$$

is a bijection.

(b) *Let $\mathcal{B}$ be a set of objects in $\mathcal{T}$. We say that $\mathcal{T}$ is* compactly generated *by $\mathcal{B}$ if every objet in $\mathcal{B}$ is compact and if the following implication holds for every object $N \in \mathcal{T}$:*

$$\left[\hom_{\mathcal{T}}(M[i], N) = 0,\ \forall M \in \mathcal{B} \text{ and } \forall i \in \mathbb{Z}\right] \quad \Rightarrow \quad \left[N \simeq 0\right].$$

Remark 1.5 — If $\mathcal{T}$ is compactly generated by $\mathcal{B}$, then $\mathcal{T}$ coincides with its smallest triangulated subcategory closed under infinite sums and containing the elements of $\mathcal{B}$. In other words, every objet of $\mathcal{T}$ can be obtained by a (possibly infinite) iteration of cones and infinite sums from the objects belonging to $\mathcal{B}$. We leave it to the reader to contemplate the meaning of this in the case of $\mathbf{DM}(k;\Lambda)$.

The following lemma is contained in [41, Theorem 2.1].

Lemma 1.6 — *Let $\mathcal{T}$ be a triangulated category with infinite sums and assume that $\mathcal{T}$ is compactly generated by a set of objects $\mathcal{B}$. Let $\mathcal{T}^{\mathrm{co}} \subset \mathcal{T}$ be the full subcategory of compact objects of $\mathcal{T}$. Then $\mathcal{T}^{\mathrm{co}}$ is a triangulated subcategory of $\mathcal{T}$ which coincides with the smallest triangulated subcategory of $\mathcal{T}$ containing the elements of $\mathcal{B}$ and which is pseudo-abelian.*

Definition 1.7 — *We denote by $\mathbf{DM}_{\mathrm{gm}}(k;\Lambda)$ the full subcategory of compact objects in $\mathbf{DM}(k;\Lambda)$. The objets of $\mathbf{DM}_{\mathrm{gm}}(k;\Lambda)$ are called* geometric motives, *aka.,* constructible motives.

Remark 1.8 — By Fact B, the functor $\mathrm{M} : \mathrm{Sm}/k \longrightarrow \mathbf{DM}(k;\Lambda)$ factors through $\mathbf{DM}_{\mathrm{gm}}(k;\Lambda)$. Moreover, by Lemma 1.6, $\mathbf{DM}_{\mathrm{gm}}(k;\Lambda)$ is the smallest triangulated subcategory of $\mathbf{DM}(k;\Lambda)$ which is pseudo-abelian, is closed under Tate twists and contains the motives of smooth projective k-varieties.

Remark 1.9 — Under the analogy between $\mathbf{DM}(k;\Lambda)$ and $\mathbf{D}(\text{Ind-MHS})$, the subcategory $\mathbf{DM}_{\mathrm{gm}}(k;\Lambda)$ corresponds to $\mathbf{D}^{\mathrm{b}}(\mathrm{MHS})$ which we identify with the triangulated subcategory of $\mathbf{D}(\text{Ind-MHS})$ consisting of complexes of ind-mixed Hodge structures with finite dimensional total homology.

Fact C — *An object of $\mathbf{DM}(k;\Lambda)$ is compact if and only if it is strongly dualisable. In particular, the tensor category $\mathbf{DM}_{\mathrm{gm}}(k;\Lambda)$ is rigid. Moreover, if X is a smooth projective k-variety of pure dimension n, there is a motivic Poincaré duality isomorphism*

$$\mathrm{M}(X)^{\vee} \simeq \mathrm{M}(X)(-n)[-2n].$$

We recall the notion of strong duality.

Definition 1.10 — *Let $\mathcal{M}$ be a tensor category and M an object of $\mathcal{M}$. We say that M admits a* strong dual *if there exist an object $N \in \mathcal{M}$ and maps*

$$u : \mathbf{1} \longrightarrow M \otimes N \qquad \text{and} \qquad v : N \otimes M \longrightarrow \mathbf{1}$$

making the following triangles commutative

$$\begin{array}{ccc} N\otimes \mathbf{1} & \xrightarrow{N\otimes u} & N\otimes M\otimes N \\ & \searrow\!\!\!\!= & \downarrow{\scriptstyle v\otimes N} \\ & & \mathbf{1}\otimes N, \end{array} \qquad \begin{array}{ccc} \mathbf{1}\otimes M & \xrightarrow{u\otimes M} & M\otimes N\otimes M \\ & \searrow\!\!\!\!= & \downarrow{\scriptstyle M\otimes v} \\ & & M\otimes \mathbf{1}. \end{array}$$

Equivalently, N is a strong dual for M if there exists a natural isomorphism

$$\underline{\mathsf{Hom}}(M,-) \simeq N \otimes -.$$

We say that $\mathcal{M}$ is rigid *if every object of $\mathcal{M}$ is strongly dualisable.*

1.2 Relation with algebraic cycles

The category $\mathbf{DM}(k;\Lambda)$ bears a strong relation with algebraic cycles. Let $\mathbf{Chow}(k;\Lambda)$ be the category of Chow motives with coefficients in Λ.

Fact D — *There is a fully faithful embedding* $\mathbf{Chow}(k;\Lambda) \hookrightarrow \mathbf{DM}_{\mathrm{gm}}(k;\Lambda)$ *which is compatible with the tensor structures.*

In this paper, we use the homological convention for Chow motives. Objets in $\mathbf{Chow}(k;\Lambda)$ are given by triples (X,γ,m) where X is a smooth projective variety, $\gamma \in \mathrm{CH}_{\dim(X)}(X \times X;\Lambda)$ an idempotent and $m \in \mathbb{Z}$. Morphisms are given by[5]

$$\hom_{\mathbf{Chow}(k;\,\Lambda)}((X,\gamma,m),(Y,\delta,n)) = \delta \circ \mathrm{CH}_{\dim(X)+m-n}(X \times Y;\Lambda) \circ \gamma.$$

The embedding in Fact D takes a Chow motive (X,γ,m) to the image of the idempotent γ acting on the Voevodsky motive $\mathrm{M}(X)(m)[2m]$.

Remark 1.11 — By Fact B and Lemma 1.6, we know that the pseudo-abelian triangulated category $\mathbf{DM}_{\mathrm{gm}}(k;\Lambda)$ is generated by the image of the fully faithful embedding $\mathbf{Chow}(k;\Lambda) \hookrightarrow \mathbf{DM}_{\mathrm{gm}}(k;\Lambda)$. Therefore, we can think about Voevodsky's category $\mathbf{DM}_{\mathrm{gm}}(k;\Lambda)$ as being a "triangulated envelop" of Grothendieck's category of Chow motives.

Definition 1.12 — *Let X be a smooth k-variety. For $(p,q) \in \mathbb{Z}^2$, we set*

$$\mathrm{H}^p(X;\Lambda(q)) := \hom_{\mathbf{DM}(k;\,\Lambda)}(\mathrm{M}(X),\Lambda(q)[p]).$$

These are the motivic cohomology groups *of X. Similarly, we set*

$$\mathrm{H}_p(X;\Lambda(q)) := \hom_{\mathbf{DM}(k;\,\Lambda)}(\Lambda(-q)[p],\mathrm{M}(X)).$$

These are the motivic homology groups *of X. In the special case $q=0$, these are also known as the* Suslin homology groups *of X.*

Fact E — *There is a natural isomorphism between motivic cohomology and Bloch's higher Chow groups [17] with coefficients in Λ:*

$$\mathrm{H}^p(X;\Lambda(q)) \simeq \mathrm{CH}^q(X,2q-p;\Lambda).$$

In particular, $\mathrm{H}^{2n}(X;\Lambda(n))$ coincides with the usual Chow group $\mathrm{CH}^n(X;\Lambda)$ of codimension n cycles.

A proof of Fact E can be found in [50]. As a special case, and thanks to [42] and [49], one has the following relation between motivic cohomology and Milnor K-theory [37].

Corollary 1.13 — *For $n \geqslant 0$, we have a natural isomorphism*

$$\mathrm{H}^n(\mathrm{Spec}(k);\Lambda(n)) \simeq \mathrm{K}^M_n(k) \otimes \Lambda$$

where $\mathrm{K}^M_n(k)$ is the n-th Milnor K-theory group of the field k.

---

[5] This formula only makes sense when X has pure dimension. Otherwise, one has to interpret $\dim(X)$ as a locally constant function on $X \times Y$.

There is also a cycle-theoretic description for the motivic homology groups. To explain this, we first recall the notion of finite correspondence which is basic in the theory of motives à la Voevodsky.

DEFINITION 1.14 — *Let X and Y be smooth k-varieties. An* elementary finite correspondence *from X to Y is an integral closed subscheme $Z \subset X \times Y$ such that the projection from Z to X is finite and surjective onto a connected component of X. A* finite correspondence *is a Λ-linear combination of elementary finite correspondences. We denote by* $\mathrm{Cor}(X,Y;\Lambda)$ *the Λ-module of correspondences from X to Y. When $\Lambda = \mathbb{Z}$, we simply write* $\mathrm{Cor}(X,Y)$. *Clearly,*

$$\mathrm{Cor}(X,Y;\Lambda) \simeq \mathrm{Cor}(X,Y) \otimes \Lambda.$$

There is a natural composition law on finite correspondences. Moreover, smooth k-varieties and finite correspondences form an additive category denoted by

$$\mathrm{SmCor}(k;\Lambda).$$

There is a natural functor $\mathrm{Sm}/k \longrightarrow \mathrm{SmCor}(k;\Lambda)$ *which is the identity on objects and which takes a morphism of k-varieties to its graph.*

Notation 1.15 — Given a pair (X,D) consisting of a smooth k-variety X and a normal crossing divisor $D = \bigcup_{i\in I} D_i$, with D_i irreducible, we denote by

$$\mathrm{Cor}((X,D),V;\Lambda)$$

the kernel of the homomorphism

$$\mathrm{Cor}(X,V;\Lambda) \longrightarrow \bigoplus_{i\in I} \mathrm{Cor}(D_i,V;\Lambda).$$

Similarly, we denote by

$$\mathrm{Cor}(U,(X,D);\Lambda)$$

the cokernel of the homomorphism

$$\bigoplus_{i\in I} \mathrm{Cor}(U,D_i;\Lambda) \longrightarrow \mathrm{Cor}(U,X;\Lambda).$$

Notation 1.16 — We denote by $\mathbb{E}^1 = \mathrm{Spec}(k[t,t^{-1}])$ the complement of the zero section in $\mathbb{A}^1 = \mathrm{Spec}(k[t])$. We consider the pair $(\mathbb{E}^1,1)$ and we denote by $(\mathbb{E}^1,1)^n$ the pair consisting of $\mathbb{E}^n = \mathrm{Spec}(k[t_1,t_1^{-1},\ldots,t_n,t_n^{-1}])$ and its normal crossing divisor defined by the equation $(t_1-1)\cdots(t_n-1)=0$.

For $n \in \mathbb{N}$, we consider the algebraic simplex

$$\Delta^n = \mathrm{Spec}(k[t_0,\ldots,t_n]/(t_0+\cdots+t_n-1)).$$

Varying n, one gets a cosimplicial scheme $\Delta^\bullet$.

FACT F — *There are natural isomorphisms*

$$\mathrm{H}_p(X;\Lambda(q)) \simeq \begin{cases} \mathrm{H}_{p+q}(\mathrm{Cor}(\Delta^\bullet, X\times(\mathbb{E}^1,1)^q;\Lambda)) & \text{if } q \geqslant 0,\\ \mathrm{H}_{p+q}(\mathrm{Cor}((\mathbb{E}^1,1)^{-q}\times\Delta^\bullet, X;\Lambda)) & \text{if } q \leqslant 0.\end{cases}$$

1.3 Realisations

Motives are rather abstract mathematical entities. Fortunately, we have at our disposal *realisation functors* that convert a motive into more familiar objects such as mixed Hodge structures or Galois representations. We only quote the following result (see [29] for the compact contravariant version).

Fact G — *Given a complex embedding $\sigma : k \hookrightarrow \mathbb{C}$, there exists a triangulated monoidal functor*

$$\mathrm{R}_\sigma : \mathbf{DM}(k;\mathbb{Q}) \longrightarrow \mathbf{D}(\text{Ind-MHS})$$

which takes the motive $\mathrm{M}(X)$ of a smooth k-variety X to a complex of ind-mixed Hodge structures computing the singular homology groups of X^{an} together with their natural mixed Hodge structures. This functor is called the Hodge realisation.

Remark 1.17 — We will also consider a more basic realisation functor, namely the Betti realisation

$$\mathrm{B}_\sigma : \mathbf{DM}(k;\Lambda) \longrightarrow \mathbf{D}(\Lambda)$$

which takes the motive $\mathrm{M}(X)$ of a smooth k-variety X to the singular chain complex of X^{an} with coefficients in Λ. When $\Lambda = \mathbb{Q}$, the functor B_σ is, up to a natural isomorphism, the composition of R_σ with the obvious forgetful functor.

Remark 1.18 — Given a smooth k-variety X, we set $\mathrm{C}_{\mathrm{Hdg}}(X) = \mathrm{R}_\sigma(\mathrm{M}(X))$. It is known that

$$\hom_{\mathbf{D}(\text{Ind-MHS})}(\mathrm{C}_{\mathrm{Hdg}}(X), \mathbb{Q}(q)[p])$$

coincides with the Deligne–Beilinson cohomology group $\mathrm{H}^p(X;\mathbb{Q}_{\mathcal{D}}(q))$. Therefore, the functor R_σ induces a homomorphism

$$\mathrm{H}^p(X,\mathbb{Q}(q)) \longrightarrow \mathrm{H}^p(X;\mathbb{Q}_{\mathcal{D}}(q))$$

from motivic cohomology to Deligne–Beilinson cohomology. This is the so-called "regulator map" which extends the classical "cycle map" on Chow groups

$$\mathrm{CH}^n(X;\mathbb{Q}) \longrightarrow \mathrm{H}^{2n}(X;\mathbb{Q}_{\mathcal{D}}(n)).$$

This is to say that the functor R_σ encodes deep and complicated information which are not always easy to unravel.

2 The conservativity conjecture

There are many longstanding conjectures about motives but, arguably, the most basic one is the so-called *conservativity conjecture.*[6]

[6]The author is aware that "conservativity" is not an English word as much as "conservativité" is not a French word. However, its seems that there is no better alternative for naming this conjecture.

2.1 Statement and general remarks

Recall that a functor $f : \mathcal{C} \longrightarrow \mathcal{D}$ is *conservative* if it detects isomorphisms, i.e., a morphism $\alpha : A \longrightarrow B$ in $\mathcal{C}$ is an isomorphism if and only if $f(\alpha) : f(A) \longrightarrow f(B)$ is an isomorphism. In the triangulated setting, f is conservative if and only if it detects the zero objects.

Fix a complex embedding $\sigma : k \hookrightarrow \mathbb{C}$. The *conservativity conjecture* states the following.

CONJECTURE 2.1 — *The Betti realisation functor*

$$\mathrm{B}_\sigma : \mathbf{DM}_{\mathrm{gm}}(k;\Lambda) \longrightarrow \mathbf{D}(\Lambda)$$

is conservative.

Remark 2.2 — Our convention that Λ is a $\mathbb{Q}$-algebra is essential for Conjecture 2.1. Indeed, it is easy to construct nonzero geometric motives $M \in \mathbf{DM}_{\mathrm{gm}}(k;\mathbb{Z})$ with zero Betti realisation. The simplest example is obtained as follows. Assume that k contains the n-th roots of unity for some integer $n \geqslant 2$. Given that

$$\hom_{\mathbf{DM}(k;\,\mathbb{Z})}(\mathbb{Z}(0),\mathbb{Z}(1)[r]) = \begin{cases} 0 & \text{if} \quad r \neq 1, \\ k^\times & \text{if} \quad r = 1, \end{cases}$$

we deduce that

$$\hom_{\mathbf{DM}(k;\,\mathbb{Z})}(\mathbb{Z}/n\mathbb{Z}(0),\mathbb{Z}/n\mathbb{Z}(1)) = \mu_n(k).$$

Therefore, the choice of a generator $\xi \in \mu_n(k)$ induces a morphism in $\mathbf{DM}(k;\mathbb{Z})$

$$\xi : \mathbb{Z}/n\mathbb{Z}(0) \longrightarrow \mathbb{Z}/n\mathbb{Z}(1).$$

It is easy to see that $\mathrm{B}_\sigma(\xi)$ is an isomorphism whereas ξ is not.

Remark 2.3 — Conjecture 2.1 has an obvious analogue in Hodge theory. Indeed, the counterpart in Hodge theory of the restriction of the Betti realisation to geometric motives is given by the forgetful functor

$$\mathbf{D}^{\mathrm{b}}(\mathrm{MHS}) \longrightarrow \mathbf{D}(\mathbb{Q})$$

which is obviously conservative. In fact, more generally, the forgetful functor

$$\mathbf{D}(\text{Ind-MHS}) \longrightarrow \mathbf{D}(\mathbb{Q})$$

is also conservative; this is particularly puzzling when contrasted with the next result which shows that the restriction to geometric motives is essential in Conjecture 2.1.

LEMMA 2.4 — *If k has infinite transcendence degree over its prime field, then there are nonzero motives $F \in \mathbf{DM}(k;\mathbb{Q})$ such that $\mathrm{B}_\sigma(F) \simeq 0$.*

Proof. Recall by Corollary 1.13 that

$$\mathrm{H}^n(\mathrm{Spec}(k);\mathbb{Q}(n)) = \mathrm{K}_n^M(k) \otimes \mathbb{Q}$$

where $\mathrm{K}_n^M(k)$ is the n-th Milnor K-theory group of the field k. Let $(a_n)_{n\in\mathbb{N}}$ be a family of elements in $k^\times$ which are algebraically independent. For each $n \in \mathbb{N}$, we consider the map

$$a_n : \mathbb{Q}(n)[n] \longrightarrow \mathbb{Q}(n+1)[n+1]$$

which corresponds to $a_n \in k^\times$ modulo the chain of isomorphisms

$$\begin{aligned}\hom_{\mathbf{DM}(k;\,\mathbb{Q})}(\mathbb{Q}(n)[n], \mathbb{Q}(n+1)[n+1]) &\simeq \hom_{\mathbf{DM}(k;\,\mathbb{Q})}(\mathbb{Q}(0), \mathbb{Q}(1)[1]) \\ &= \mathrm{H}^1(\mathrm{Spec}(k), \mathbb{Q}(1)) \\ &\simeq k^\times \otimes \mathbb{Q}.\end{aligned}$$

This gives an $\mathbb{N}$-inductive system $\{\mathbb{Q}(n)[n]\}_{n\in\mathbb{N}}$ and we take F to be its homotopy colimit:

$$F = \mathrm{hocolim}_{n\in\mathbb{N}}\, \mathbb{Q}(n)[n].$$

As B_σ commutes with direct sums, it also commutes with homotopy colimits of $\mathbb{N}$-inductive systems. Thus, we have

$$\mathrm{B}_\sigma(F) \simeq \mathrm{hocolim}_{n\in\mathbb{N}}\, \mathrm{B}_\sigma(\mathbb{Q}(n)[n]) \simeq \mathrm{hocolim}_{n\in\mathbb{N}}\, \mathbb{Q}[n] \simeq 0.$$

To finish, it remains to show that F is nonzero.

More precisely, we will show that the natural map

$$\alpha_\infty : \mathbb{Q}(0) \longrightarrow F$$

is nonzero. As $\mathbb{Q}(0)$ is compact, we have an identification

$$\hom_{\mathbf{DM}(k;\,\mathbb{Q})}(\mathbb{Q}(0), F) \simeq \mathrm{colim}_{n\in\mathbb{N}}\, \hom_{\mathbf{DM}(k;\,\mathbb{Q})}(\mathbb{Q}(0), \mathbb{Q}(n)[n]).$$

Therefore, α_∞ is zero if and only if there exists $n \in \mathbb{N}$ such that the natural map

$$\alpha_n : \mathbb{Q}(0) \longrightarrow \mathbb{Q}(n)[n]$$

is zero. By the identification

$$\hom_{\mathbf{DM}(k;\,\mathbb{Q})}(\mathbb{Q}(0), \mathbb{Q}(n)[n]) \simeq \mathrm{K}_n^M(k) \otimes \mathbb{Q},$$

α_n corresponds to the symbol $\{a_0, \ldots, a_n\} \in \mathrm{K}_n^M(k) \otimes \mathbb{Q}$. This symbol is nonzero under the hypothesis that the a_i's are algebraically independent. □

Remark 2.5 — Lemma 2.4 indicates that the analogy between $\mathbf{DM}(k;\mathbb{Q})$ and $\mathbf{D}$(Ind-MHS) breaks down when dealing with "big" motives. Said differently, $\mathbf{DM}(k;\mathbb{Q})$ is not the correct candidate for the big derived category of mixed motives. Here is an obvious way to fix the problem.

Definition 2.6 — *We call* fantom motive *an objet of* $\mathbf{DM}(k;\Lambda)$ *whose Betti realisation is zero.*[7] *Fantom motives form a localising triangulated subcategory*

[7]The notion of fantom motive is independent of the choice of the complex embedding. This can be shown using the comparison theorem between the Betti and ℓ-adic realisations. (Here, we have in mind a covariant triangulated functor $\mathbf{DM}(k;\mathbb{Q}) \longrightarrow \mathbf{D}(\mathbb{Q}_\ell)$, commuting with arbitrary direct sums and sending the motive of a smooth k-variety X to a complex of $\mathbb{Q}_\ell$-vectorspaces computing the ℓ-adic homology groups $\mathrm{H}_*(X \otimes_k \bar{k}; \mathbb{Q}_\ell)$. The restriction of such a functor to geometric motives is constructed in [8] and there are standard ways to extend this to $\mathbf{DM}(k;\mathbb{Q})$ in a continuous manner.)

of $\mathbf{DM}(k;\Lambda)$ *that we denote by* $\mathcal{F}ant$. *We define* $\mathbf{DM}^\flat(k;\Lambda)$ *to be the Verdier quotient:*

$$\mathbf{DM}^\flat(k;\Lambda) := \mathbf{DM}(k;\Lambda)/\mathcal{F}ant.$$

We call $\mathbf{DM}^\flat(k;\Lambda)$ *the category of* true motives.

Remark 2.7 — The Betti realisation on $\mathbf{DM}(k;\Lambda)$ induces a Betti realisation functor

$$\mathrm{B}_\sigma : \mathbf{DM}^\flat(k;\Lambda) \longrightarrow \mathbf{D}(\Lambda)$$

which is conservative by construction.

In the present state of knowledge, calling $\mathbf{DM}^\flat(k;\Lambda)$ the category of true motives might sound a bit pretentious. However, here is a plausible conjecture that, if true, would justify such a terminology.

Conjecture 2.8 — *Let* $F \in \mathcal{F}ant$ *be a fantom motive and let* $M \in \mathbf{DM}_{\mathrm{gm}}(k;\Lambda)$ *be a geometric motive. Then*

$$\hom_{\mathbf{DM}(k;\,\Lambda)}(F, M) = 0.$$

Remark 2.9 — As a meagre evidence, we note that Conjecture 2.8 is satisfied for the motive F constructed in the proof of Lemma 2.4. This follows immediately from the following well-known property of geometric motives. Given $M \in \mathbf{DM}_{\mathrm{gm}}(k;\Lambda)$, there exists an integer n_0 (depending on M) such that

$$\hom_{\mathbf{DM}(k;\,\Lambda)}(\Lambda(n)[m], M) = 0$$

for all $n \geqslant n_0$ and $m \in \mathbb{Z}$.

Proposition 2.10 — *Assume Conjecture 2.8. Then the composite functor*

$$\mathbf{DM}_{\mathrm{gm}}(k;\Lambda) \hookrightarrow \mathbf{DM}(k;\Lambda) \twoheadrightarrow \mathbf{DM}^\flat(k;\Lambda)$$

is a fully faithful embedding.

Proof. This is a consequence of the construction of the Verdier localisation. □

Corollary 2.11 — *Conjecture 2.8 implies Conjecture 2.1.*

Lemma 2.4 indicates that the failure of conservativity might be a phenomenon related to big base fields. Indeed, as far as we know, there is no reason to disbelieve the following conjecture which, if true, lies probably much deeper than Conjecture 2.1.

Conjecture 2.12 — *Assume that* k *has finite transcendence degree over its prime field. Then the Betti realisation functor*

$$\mathrm{B}_\sigma : \mathbf{DM}(k;\Lambda) \longrightarrow \mathbf{D}(\Lambda)$$

is conservative.

Remark 2.13 — In a way, the state of affair that we are describing is to be expected. Indeed, the motivic cohomological dimension of a field k, whatever this means, is expected to be $\mathrm{trdeg}(k/\mathbb{Q}) + 1$ and, in particular, should be infinite when the transcendence degree of k over its prime field is infinite. Moreover, it is well-known that "infinite cohomological dimension" in a Grothendieck abelian category can prevent its (unbounded) derived category from being compactly generated. Therefore, Fact B indicates that $\mathbf{DM}(k;\Lambda)$ is not a reasonable candidate for the unbounded derived category of the abelian category of ind-mixed motives unless $\mathrm{degtr}(k/\mathbb{Q}) < \infty$. On the contrary, when $\mathrm{degtr}(k/\mathbb{Q}) = \infty$, the category $\mathbf{DM}^{\flat}(k;\Lambda)$ is presumably not compactly generated as it was obtained by a Verdier localisation with respect to the triangulated subcategory $\mathcal{F}ant$ which is not expected to be compactly generated (unless Conjecture 2.8 is totally wrong). Therefore, from this perspective, there is no obstacles for $\mathbf{DM}^{\flat}(k;\Lambda)$ being the derived category of the abelian category of ind-mixed motives.

Another way of extending the conservativity conjecture to non-necessary geometric objets is given by the following conjecture (which, perhaps, deserves better the status of an open question).

CONJECTURE 2.14 — *Denote by* $\mathbf{DM}^{\mathrm{eff}}_{\leqslant n}(k;\Lambda)$ [8] *the smallest triangulated subcategory of* $\mathbf{DM}(k;\Lambda)$ *closed under infinite sums and containing the motives* $\mathrm{M}(X)$ *for any smooth k-variety X of dimension $\leqslant n$. Then, the Betti realisation functor*

$$\mathrm{B}_\sigma : \mathbf{DM}^{\mathrm{eff}}_{\leqslant n}(k;\Lambda) \longrightarrow \mathbf{D}(\Lambda)$$

is conservative.

2.2 Some consequences

The conservativity conjecture has many concrete consequences on algebraic cycles. One of these consequences is a famous 40 years old conjecture of Bloch on 0-cycles on surfaces.[9]

PROPOSITION 2.15 — *Assume Conjecture 2.1. Let X be a smooth projective surface over an algebraically closed field k of characteristic zero. If $p_g(X) = 0$, then the Albanese homomorphism*

$$\mathrm{CH}_0(X) \longrightarrow \mathrm{Alb}(X)(k)$$

is injective (where $\mathrm{Alb}(X)$ *denotes the Albanese scheme of X).*[10]

Proof. By Rojtman's theorem [46], the kernel of the Albanese map is torsion free. Therefore, it is enough to prove that the map

$$\mathrm{CH}_0(X;\mathbb{Q}) \longrightarrow \mathrm{Alb}(X)(k) \otimes \mathbb{Q}$$

is injective. Recall that $p_g(X) = \dim(\Omega^2(X))$ is the dimension of the space of global (holomorphic) differential 2-forms. Thus, the condition $p_g(X) = 0$ implies

[8] The subcategories $\mathbf{DM}^{\mathrm{eff}}_{\leqslant n}(k;\Lambda)$ will be considered in more details in Subsection 4.2.

[9] See [22] for an interesting story about how Severi had stumbled upon the statement of the Bloch conjecture a few decades before the conjecture was stated.

[10] The converse of this statement is true (unconditionally) and is a well-known theorem of Mumford [38]: if the Albanese homomorphism is injective, then necessary $p_g(X) = 0$.

that the Hodge structure on $\mathrm{H}_2(X^{\mathrm{an}};\mathbb{Q})$ is of type $(-1,-1)$. By the (dual of the) Lefschetz $(1,1)$ Theorem, this is also equivalent to the condition that the cycle class map

$$\mathrm{NS}_1(X;\mathbb{Q}) \longrightarrow \mathrm{H}_2(X^{\mathrm{an}};\mathbb{Q})$$

is an isomorphism. (Here, $\mathrm{NS}_1(X;\mathbb{Q})$ is the Néron–Severi group of 1-dimensional cycles, or equivalently divisors, on X with rational coefficients.)

The motive of the surface X has a Chow–Kunneth decomposition

$$\mathrm{M}(X) = \bigoplus_{i=0}^{4} \mathrm{M}_i(X)[i]$$

such that $\mathrm{B}_\sigma(\mathrm{M}_i(X))$ is isomorphic to $\mathrm{H}_i(X^{\mathrm{an}};\mathbb{Q})$. Furthermore, according to [33], $\mathrm{M}_2(X)$ decomposes into an algebraic and a transcendental part

$$\mathrm{M}_2(X) = \mathrm{M}_2^{\mathrm{alg}}(X) \oplus \mathrm{M}_2^{\mathrm{tr}}(X).$$

The algebraic part $\mathrm{M}_2^{\mathrm{alg}}(X)$ is given by

$$\mathrm{M}_2^{\mathrm{alg}}(X) = \mathrm{NS}_1(X;\mathbb{Q})(1)[2]$$

where the finite dimensional $\mathbb{Q}$-vectorspace $\mathrm{NS}_1(X;\mathbb{Q})$ is considered as an Artin motive in the obvious way. The transcendental part $\mathrm{M}_2^{\mathrm{tr}}(X)$ determines the kernel of the Albanese map via the formula

$$\hom_{\mathbf{DM}(k;\,\mathbb{Q})}(\mathbb{Q}(0), \mathrm{M}_2^{\mathrm{tr}}(X)) \simeq \ker\left\{\mathrm{CH}_0(X;\mathbb{Q}) \to \mathrm{Alb}(X)(k) \otimes \mathbb{Q}\right\}.$$

Now, the isomorphisms

$$\mathrm{NS}_1(X;\mathbb{Q}) \simeq \mathrm{H}_2(X^{\mathrm{an}};\mathbb{Q}) \qquad \text{and} \qquad \mathrm{M}_2^{\mathrm{alg}}(X) \simeq \mathrm{NS}_1(X;\mathbb{Q})(1)[2]$$

show that

$$\mathrm{B}_\sigma(\mathrm{M}_2^{\mathrm{alg}}(X)) \simeq \mathrm{B}_\sigma(\mathrm{M}_2(X)) \simeq \mathrm{H}_2(X^{\mathrm{an}};\mathbb{Q}).$$

It follows that $\mathrm{B}_\sigma(\mathrm{M}_2^{\mathrm{tr}}(X)) = 0$. Applying Conjecture 2.1, one deduces that $\mathrm{M}_2^{\mathrm{tr}}(X) \simeq 0$. This finishes the proof thanks to the above description of the kernel of the Albanese map. □

Remark 2.16 — Bloch's conjecture for surfaces with $p_g = 0$ has been checked for all surfaces which are not of general type in [18]. It is also known for some surfaces of general type admitting nice cyclic étale covers, see for example [31], [54], [55], [15], [16], etc. Perhaps, the most intriguing surfaces with $p_g = 0$ for which Bloch's conjecture seems intractable are the fake projective planes (aka., Mumford surfaces). Recall that a fake projective plane is a smooth proper surface with the same Betti numbers as $\mathbb{P}^2$ but which is not isomorphic to $\mathbb{P}^2$. The first example of a fake projective plane was constructed by Mumford [39].

The conservativity conjecture applies much beyond the case of surfaces. For instance, one can prove the following statement using the same reasoning as in the proof of Proposition 2.15.

PROPOSITION 2.17 — *Assume Conjecture 2.1. Let X be a smooth projective k-variety with algebraic cohomology, i.e., satisfying the following two conditions:*

(a) *for any* $m \in \mathbb{N}$, $\mathrm{H}^{2m+1}(X^{\mathrm{an}};\mathbb{Q}) = 0$;

(b) *for any* $m \in \mathbb{N}$, *the cycle class map*

$$\mathrm{CH}^m(X;\mathbb{Q}) \longrightarrow \mathrm{H}^{2m}(X^{\mathrm{an}};\mathbb{Q})$$

is surjective.

Then, $\mathrm{M}(X)$ *is a Tate motive. In particular, the cycle class maps in* (b) *are also injective.*

Proof. For $m \in \mathbb{Z}$, choose a family of elements $(\alpha_{m,i})_{i\in I_m}$ in $\mathrm{CH}^m(X;\mathbb{Q})$ whose images in singular cohomology form a basis of $\mathrm{H}^{2m}(X^{\mathrm{an}};\mathbb{Q})$. The $\alpha_{m,i}$'s induce a morphism in $\mathbf{Chow}(k;\mathbb{Q})$:

$$\beta : \mathrm{M}(X) \longrightarrow \bigoplus_{m\geqslant 0} \bigoplus_{i\in I_m} \mathbb{Q}(m)[2m].$$

Moreover, the conditions (a) and (b) of the statement implies that $\mathrm{B}_\sigma(\beta)$ is an isomorphism. Applying Conjecture 2.1, we deduce that β was already an isomorphism. □

Remark 2.18 — The conclusion of Proposition 2.17 is reminiscent to the following (unconditional) theorem of Jannsen (see [26] where this theorem is discussed and generalised): if the cycle class map from the Chow ring of a smooth projective k-variety to its singular cohomology is injective, then this cycle class map is in fact an isomorphism and the Chow motive of X is Tate. (Here, one needs to assume that k is algebraically closed with infinite transcendence degree.)

Here is another amusing consequence of the conservativity conjecture.

Proposition 2.19 — *Assume Conjecture 2.1. Assume also that the* $\mathbb{Q}$-*algebra* Λ *is connected (i.e., has no nontrivial idempotents). Let* $M \in \mathbf{DM}_{\mathrm{gm}}(k;\Lambda)$ *be a geometric motive. Then the following two conditions are equivalent:*

(i) M *is invertible (for the tensor product),*

(ii) *there exists a unique integer* m *such that* $\mathrm{H}_i(\mathrm{B}_\sigma(M))$ *is zero unless* $i = m$ *in which case it is an invertible* Λ-*module.*

Proof. The implication (i) ⇒ (ii) is unconditional on Conjecture 2.1 and we leave it to the reader. (It is here where the connectedness of Λ is used.)

To prove that (ii) ⇒ (i), we remark that (ii) implies that the natural evaluation morphism in $\mathbf{D}(\Lambda)$

$$\mathrm{B}_\sigma(M) \otimes \mathrm{B}_\sigma(M)^\vee \longrightarrow \Lambda$$

is an isomorphism. Now, B_σ is a monoidal functor and M is a strongly dualisable motive (because it is assumed to be geometric). This implies that $\mathrm{B}_\sigma(M)^\vee \simeq \mathrm{B}_\sigma(M^\vee)$. This shows that the evaluation map

$$M \otimes M^\vee \longrightarrow \Lambda(0)$$

becomes an isomorphism after applying B_σ. By Conjecture 2.1, we are done. □

Remark 2.20 — In Proposition 2.19, it is useful to allow general coefficients rings. Indeed, it is expected that all invertible motives in $\mathbf{DM}(k;\mathbb{Q})$ are Artin–Tate, i.e., of the form $U(n)[m]$ where U is an Artin motive corresponding to a representation of $\mathrm{Gal}(\bar{k}/k)$ on a 1-dimensional $\mathbb{Q}$-vectorspace. (Of course, such a representation factors through a character $\mathrm{Gal}(\bar{k}/k) \longrightarrow \{\pm 1\}$.) On the contrary, it is easy to construct invertible motives in $\mathbf{DM}(k;\overline{\mathbb{Q}})$ which are not Artin–Tate; here is an example. Let E be an elliptic curve with complex multiplication by an order $\mathcal{O}$ in an imaginary quadratic extension $L/\mathbb{Q}$. The motive $\mathrm{M}_1(E)$ (corresponding to the first singular homology group of E^{an}) decomposes in $\mathbf{DM}(k;L)$ in two parts

$$\mathrm{M}_1(E) = \mathrm{M}_1^+(E) \oplus \mathrm{M}_1^-(E).$$

The part $\mathrm{M}_1^+(E)$ (resp. $\mathrm{M}_1^-(E)$) is characterized by the property that $\mathcal{O}$ acts on the L-vectorspace $\mathrm{B}_\sigma(\mathrm{M}_1^+(E))$ (resp. $\mathrm{B}_\sigma(\mathrm{M}_1^-(E))$) via the embedding $\mathcal{O} \hookrightarrow L$ (resp. the composite embedding $\mathcal{O} \hookrightarrow L \xrightarrow{c} L$ where c is the complex conjugation). It is easy to show that the motives $\mathrm{M}_1^\pm(E)$ are invertible.

To go further, we recall the following definitions from [34] (see also [3, §9] and [40, Chapter 4]).

Definition 2.21 — *Let $\mathcal{C}$ be a $\mathbb{Q}$-linear pseudo-abelian tensor category. An objet $A \in \mathcal{C}$ is said to be* even *(resp.* odd*) if $\bigwedge^m(A) = 0$ (resp. $\mathrm{S}^m(A) = 0$) for m large enough. An objet $A \in \mathcal{C}$ is said to be* Kimura finite *if it decomposes as the direct sum of an even object and an odd object.*

Proposition 2.22 — *Assume Conjecture 2.1. For simplicity, assume also that Λ is a field extension of $\mathbb{Q}$. Let $M \in \mathbf{DM}_{\mathrm{gm}}(k;\Lambda)$ be a geometric motive. Then the following conditions are equivalent:*

(a) *M is even (resp. odd);*

(b) *$\mathrm{H}_i(\mathrm{B}_\sigma(M)) = 0$ for i odd (resp. i even).*

In particular, a motive M is Kimura finite if and only if there exists a decomposition $M = M_{\mathrm{eve}} \oplus M_{\mathrm{odd}}$ such that the complexes $\mathrm{B}_\sigma(M_{\mathrm{eve}})$ and $\mathrm{B}_\sigma(M_{\mathrm{odd}})$ have their homology in even and odd degrees respectively.

Proof. This is clear. Indeed, a complex of Λ-vectorspaces with finite dimensional homology is even (resp. odd) if and only if its homology is in even (resp. odd) degrees. □

Remark 2.23 — A conjecture of Kimura–O'Sullivan states that every Chow motive is Kimura finite. This conjecture does not follow from Conjecture 2.1. However, Proposition 2.22 shows that the conjecture of Kimura–O'Sullivan follows from the combination of Conjecture 2.1 and a weak version of the existence of a Chow–Kunneth decomposition (see [40, Definition 6.1.1]). Using this principle, one can deduce the following corollary.

Corollary 2.24 — *Assume Conjecture 2.1.*

(i) *The Chow motive of a smooth projective surface is Kimura finite.*

(ii) *The Chow motive of a smooth hypersurface in a projective space is Kimura finite.*

The property of being Kimura finite is not closed under extensions in a triangulated category. As a consequence, it is easy to produce examples of geometric motives in $\mathbf{DM}_{\mathrm{gm}}(k;\Lambda)$ which are not Kimura finite. A weaker property which behaves better in a triangulated context is the following one.

Definition 2.25 — *Let $\mathcal{C}$ be a $\mathbb{Q}$-linear pseudo-abelian tensor category. An object $A \in \mathcal{C}$ is said to be* Schur finite *if there exists an integer $n \in \mathbb{N}$ and a nonzero projector $\gamma \in \mathbb{Q}[\Sigma_n]$ in the group algebra of the symmetric group Σ_n which acts by zero on $A^{\otimes n}$.*

Clearly, a Kimura finite object is also Schur finite. If $\mathcal{C}$ is a tensor triangulated category obtained as the homotopy category of a stable monoidal model category, we know by [28] that Schur finiteness has the "2 out 3" property in distinguished triangles. In particular, a perfect complex of Λ-modules is a Schur finite object of $\mathbf{D}(\Lambda)$. This immediately implies the following.

Proposition 2.26 — *Assume Conjecture 2.1. Any object of $\mathbf{DM}_{\mathrm{gm}}(k;\Lambda)$ is Schur finite.*

We close our list of consequences of the conservativity conjecture with the following statement.

Proposition 2.27 — *Assume Conjecture 2.1. Let K be a field endowed with a discrete valuation with residue field k of characteristic zero. Choose a uniformizer element $\pi \in K$. Then the "nearby motive" functor*

$$\Psi_\pi : \mathbf{DM}_{\mathrm{gm}}(K;\mathbb{Q}) \longrightarrow \mathbf{DM}_{\mathrm{gm}}(k;\mathbb{Q})$$

is conservative.

Proof. The functor Ψ_π is analogous to the formation of the limiting mixed Hodge structure in Hodge theory. We will not recall the construction of Ψ_π here. We direct the reader to [5, Chapitre 3], complemented by [11, Appendice 1.A], where the theory of nearby motivic sheaves is developed; see also [12, §2.3] for a quick recollection.

The idea of the proof of Proposition 2.27 is to use the compatibility of Ψ_π with the Betti realisation functors (see [6, §4]). This is possible when K is the fraction field of a curve endowed with the valuation associated to a close point. However, for more general K's, one runs into some technical problems that are presumably solvable with some effort. Alternatively, one can use the compatibility of Ψ_π with the ℓ-adic realisation functors [8, §10] which holds for general K's. Indeed, thanks to the comparison theorem between Betti and ℓ-adic cohomology, Conjecture 2.1 implies that the ℓ-adic realisation functor is also conservative. □

2.3 More about conservativity and Kimura finiteness

In this subsection, we discuss further the relation between the conservativity conjecture and the Kimura finiteness for Chow motives. Our goal is to explain the following theorem.

Theorem 2.28 — *Assume that the Standard Conjecture D is true, i.e., that homological equivalence and numerical equivalence coincide (see [40, Chapter 3]). Assume also that the Kimura–O'Sullivan conjecture is true, i.e., that every Chow motive is Kimura finite. Then Conjecture 2.1 is also true.*

The main ingredient we need is a beautiful construction of Bondarko [19], based on his notion of *weight structure*, and generalising the construction of weight complexes of Gillet–Soulé [27]. Given an additive category $\mathcal{C}$, we denote by $\mathbf{K}^{\mathrm{b}}(\mathcal{C})$ the homotopy category of bounded complexes in $\mathcal{C}$.

Proposition 2.29 — *There exists a triangulated functor*

$$\mathrm{W} : \mathbf{DM}_{\mathrm{gm}}(k;\mathbb{Q}) \longrightarrow \mathbf{K}^{\mathrm{b}}(\mathbf{Chow}(k;\mathbb{Q}))$$

which is conservative and which makes the following triangle commutative (up to a canonical natural isomorphism)

$$\begin{array}{ccc} \mathbf{Chow}(k;\mathbb{Q}) & \longrightarrow & \mathbf{DM}_{\mathrm{gm}}(k;\mathbb{Q}) \\ & \searrow & \downarrow{\scriptstyle \mathrm{W}} \\ & & \mathbf{K}^{\mathrm{b}}(\mathbf{Chow}(k;\mathbb{Q})). \end{array}$$

Proof. This is essentially [19, Proposition 6.3.1 and Remark 6.3.2]. □

There is a similar construction for mixed Hodge structures. Let CHS denotes the full additive subcategory of $\mathbf{D}^{\mathrm{b}}(\mathrm{MHS})$ whose objects are the direct sums $\bigoplus_{n\in\mathbb{Z}} M_n[n]$ with M_n a polarisable pure Hodge structure of weight $-n$. By the decomposition theorem, the mixed Hodge complex associated to a smooth projective variety lies in CHS. Therefore, one has a commutative square

$$\begin{array}{ccc} \mathbf{Chow}(k;\mathbb{Q}) & \xrightarrow{\mathrm{R}_\sigma} & \mathrm{CHS} \\ \downarrow & & \downarrow \\ \mathbf{DM}_{\mathrm{gm}}(k;\mathbb{Q}) & \xrightarrow{\mathrm{R}_\sigma} & \mathbf{D}^{\mathrm{b}}(\mathrm{MHS}). \end{array}$$

From [20, 21], and especially [21, Proposition 2.1.1], one gets the following.

Proposition 2.30 — *There exists a triangulated functor*

$$\mathrm{W} : \mathbf{D}^{\mathrm{b}}(\mathrm{MHS}) \longrightarrow \mathbf{K}^{\mathrm{b}}(\mathrm{CHS}),$$

which is conservative and which makes the following diagrams commutative (up to canonical natural isomorphisms)

$$\begin{array}{ccc} \mathrm{CHS} & \longrightarrow & \mathbf{D}^{\mathrm{b}}(\mathrm{MHS}) \\ & \searrow & \downarrow{\scriptstyle \mathrm{W}} \\ & & \mathbf{K}^{\mathrm{b}}(\mathrm{CHS}), \end{array} \qquad \begin{array}{ccc} \mathbf{DM}_{\mathrm{gm}}(k;\mathbb{Q}) & \xrightarrow{\mathrm{W}} & \mathbf{K}^{\mathrm{b}}(\mathbf{Chow}(k;\mathbb{Q})) \\ \downarrow{\scriptstyle \mathrm{R}_\sigma} & & \downarrow{\scriptstyle \mathrm{R}_\sigma} \\ \mathbf{D}^{\mathrm{b}}(\mathrm{MHS}) & \xrightarrow{\mathrm{W}} & \mathbf{K}^{\mathrm{b}}(\mathrm{CHS}). \end{array}$$

As a corollary, Theorem 2.28 follows from the following statement.

Lemma 2.31 — *Assume that the Standard Conjecture D is true. Assume also that the Kimura–O'Sullivan conjecture is true. Then the functor*

$$\mathrm{R}_\sigma : \mathbf{K}^{\mathrm{b}}(\mathbf{Chow}(k;\mathbb{Q})) \longrightarrow \mathbf{K}^{\mathrm{b}}(\mathrm{CHS})$$

is conservative.

Proof. Consider the category $\mathbf{Num}(k;\mathbb{Q})$ of numerical motives (with rational coefficients). By [32], $\mathbf{Num}(k;\mathbb{Q})$ is abelian semi-simple. The Kimura–O'Sullivan conjecture implies that the natural functor

$$\mathbf{K}^{\mathrm{b}}(\mathbf{Chow}(k;\mathbb{Q})) \longrightarrow \mathbf{K}^{\mathrm{b}}(\mathbf{Num}(k;\mathbb{Q})) = \mathbf{D}^{\mathrm{b}}(\mathbf{Num}(k;\mathbb{Q}))$$

is conservative. Indeed, let $M = M_\bullet$ be a bounded complex of Chow motives whose image in $\mathbf{K}^{\mathrm{b}}(\mathbf{Num}(k;\mathbb{Q}))$ is acyclic. We need to show that M is also acyclic. We argue by induction on the length of M. (If M is zero, there is nothing to prove.) Let $n \in \mathbb{Z}$ be the smallest integer such that $M_n \neq 0$. We know that the differential $\delta : M_{n+1} \longrightarrow M_n$ induces a split surjective morphism in $\mathbf{Num}(k;\mathbb{Q})$. Therefore, we may find a morphism of Chow motives $\eta : M_n \longrightarrow M_{n+1}$ such that $\delta\circ\eta - \mathrm{id}_{M_n}$ is numerically equivalent to zero. By [34, Proposition 7.5], this implies that $\delta\circ\eta - \mathrm{id}_{M_n}$ is a nilpotent endomorphism of M_n, which in turn implies that $\delta\circ\eta$ is an automorphism of M_n. Therefore, replacing η by $\eta\circ(\delta\circ\eta)^{-1}$, we may assume that $\delta\circ\eta = \mathrm{id}_{M_n}$. This gives a decomposition $M_{n+1} \simeq M'_{n+1} \oplus M_n$. Let $M' = M'_\bullet$ be the complex obtained from M by replacing the portion $\{M_{n+1} \to M_n\}$ by $\{M'_{n+1} \to 0\}$. By construction, there is a map $M' \longrightarrow M$ which is a homotopy equivalence (i.e., an isomorphism in $\mathbf{K}^{\mathrm{b}}(\mathbf{Chow}(k;\mathbb{Q}))$). We now use induction to conclude.

To go further, let PHS be the full subcategory of MHS consisting of pure Hodge structures. There is a functor CHS $\longrightarrow$ PHS which takes $\bigoplus_{n\in\mathbb{Z}} M_n[n]$ to $\bigoplus_{n\in\mathbb{Z}} M_n$. (Note that this is not an equivalence of categories!) By the Standard Conjecture D, there is a functor $\mathbf{Num}(k;\mathbb{Q}) \longrightarrow$ PHS and it is easy to see that the square

$$\begin{array}{ccc} \mathbf{K}^{\mathrm{b}}(\mathbf{Chow}(k;\mathbb{Q})) & \longrightarrow & \mathbf{D}^{\mathrm{b}}(\mathbf{Num}(k;\mathbb{Q})) \\ \downarrow & & \downarrow \\ \mathbf{K}^{\mathrm{b}}(\mathrm{CHS}) & \longrightarrow & \mathbf{D}^{\mathrm{b}}(\mathrm{PHS}) \end{array}$$

commutes up to a natural isomorphism. As $\mathbf{Num}(k;\mathbb{Q}) \longrightarrow$ PHS is clearly conservative, we are done. □

Remark 2.32 — It is very unlikely that Theorem 2.28 is of any use for proving the conservativity conjecture in general. Indeed, proving new cases of the Kimura–O'Sullivan conjecture and/or the Standard Conjecture D by direct means seems quite hopeless and we expect things to go in the other way round. Nevertheless, one can use the known supply of motives for which these conjectures are known to give some evidence for the conservativity conjecture. This gives the following result of Wildeshaus [56, Theorem 1.12].

Corollary 2.33 — *Let $\mathcal{A} \subset \mathbf{DM}_{\text{gm}}(k;\mathbb{Q})$ be the smallest triangulated subcategory closed under direct summands, tensor product and duality, and containing the motives of smooth projective curves. Then the restriction of the Betti realisation to $\mathcal{A}$ is conservative.*

Proof. Repeat the proof of Theorem 2.28 restricting to $\mathcal{A}$ and using that the Standard Conjecture D and the Kimura–O'Sullivan conjecture are known for Chow motives which are direct summands of tensor products of motives of smooth projective curves. □

Remark 2.34 — At the time of writing, the category $\mathcal{A}$ is the largest subcategory of $\mathbf{DM}_{\text{gm}}(k;\mathbb{Q})$ for which the conservativity conjecture is known to hold. Note that $\mathcal{A}$ contains a lot of objects. For example, the motive of a finite type group-scheme, such as a semi-abelian variety, belongs to $\mathcal{A}$. Nevertheless, $\mathcal{A}$ is still a very small portion of $\mathbf{DM}_{\text{gm}}(k;\mathbb{Q})$. To appreciate the gap to be filled, note that an effective polarisable pure Hodge structure of weight 2 with Hodge number $h^{2,0} \geqslant 2$ and with generic Mumford–Tate group does not belong to the tannakian subcategory of MHS generated by those Hodge structures of type $\{(1,0),(0,1)\}$; this follows from [25, Proposition 7.3]. Nevertheless, such weight 2 Hodge structures abound in algebraic geometry. Indeed, using [25, Proposition 7.5] (see also [2]), it can be shown that they appear as the middle primitive cohomology of general hypersurfaces $S \subset \mathbb{P}^3$ of degree high enough. In particular, the motive of a general hypersurface $S \subset \mathbb{P}^3$ of degree high enough is not in $\mathcal{A}$. For a more detailed analysis, we refer the reader to [25, §7].

3 The vanishing conjecture for the motivic period algebra

In this section we discuss a vanishing conjecture concerning the so-called motivic period algebra. This conjecture can be stated in a very elementary way, so we decided to first give its statement and explain later its motivic origin and how it fits in a broader context.

3.1 A concrete formulation

We start by introducing some notation. For $n \in \mathbb{N}$, we denote by $\overline{\mathbb{D}}^n = \overline{\mathbb{D}}^n(o,1)$ the closed unit polydisc in $\mathbb{C}^n$:

$$\overline{\mathbb{D}}^n = \{(t_1,\ldots,t_n) \in \mathbb{C}^n;\ |t_i| \leqslant 1,\ \forall 1 \leqslant i \leqslant n\}.$$

We will always consider $\overline{\mathbb{D}}^n$ as a pro-analytic variety in the obvious way, i.e., as being the pro-system $\{\mathbb{D}^n(o,r)\}_{r>1}$ of open polydiscs in $\mathbb{C}^n$ of constant polyradius $r > 1$. In particular, $\mathcal{O}(\overline{\mathbb{D}}^n)$ will denote the ring of overconvergent holomorphic functions on $\overline{\mathbb{D}}^n$, i.e., of those functions which are given by power series in n

variables converging on $\mathbb{D}^n(o, r)$ for some $r > 1$. Note that $\mathcal{O}(\overline{\mathbb{D}}^n)$ is a noetherian regular ring.

As before, we fix a complex embedding $\sigma : k \hookrightarrow \mathbb{C}$. We consider a "semi-algebraic" version of $\overline{\mathbb{D}}^n$ (see [9, §2.2.4]).

Definition 3.1 — *We denote by $\overline{\mathbb{D}}^n_{\text{ét}}$ the pro-k-variety of étale neighbourhoods of $\overline{\mathbb{D}}^n$ in $\mathbb{A}^n$. More precisely, consider the category $\mathcal{V}_{\text{ét}}(\overline{\mathbb{D}}^n/\mathbb{A}^n)$ whose objects are pairs (U, u) consisting of an étale $\mathbb{A}^n$-scheme U and a morphism of pro-analytic spaces $u : \overline{\mathbb{D}}^n \longrightarrow U^{\text{an}}$ making the following triangle commutative*

$$\begin{array}{ccc} \overline{\mathbb{D}}^n & \xrightarrow{u} & U^{\text{an}} \\ & \searrow & \downarrow \\ & & \mathbb{C}^n. \end{array}$$

Then $\mathcal{V}_{\text{ét}}(\overline{\mathbb{D}}^n/\mathbb{A}^n)$ is a cofiltered category and the pro-k-variety $\overline{\mathbb{D}}^n_{\text{ét}}$ is given by the forgetful functor

$$\begin{array}{rcl} \overline{\mathbb{D}}^n_{\text{ét}} : \mathcal{V}_{\text{ét}}(\overline{\mathbb{D}}^n/\mathbb{A}^n) & \longrightarrow & \mathrm{Sch}/k \\ (U, u) & \mapsto & U. \end{array}$$

By construction, we have a pro-étale morphism of pro-k-varieties $\overline{\mathbb{D}}^n_{\text{ét}} \longrightarrow \mathbb{A}^n$.

Note the following result.

Lemma 3.2 — *The pro-k-variety $\overline{\mathbb{D}}^n_{\text{ét}}$ is affine (i.e., isomorphic to a pro-object in the category of affine k-varieties) and $\mathcal{O}(\overline{\mathbb{D}}^n_{\text{ét}})$ is isomorphic to the sub-k-algebra of $\mathcal{O}(\overline{\mathbb{D}}^n)$ consisting of those overconvergent power series*

$$f = \sum_{\underline{e} \in \mathbb{N}^n} c_{\underline{e}} \cdot \underline{t}^{\underline{e}} \in \mathcal{O}(\overline{\mathbb{D}}^n)$$

which are algebraic over the field of rational functions $k(t_1, \ldots, t_n)$.

Proof. This is the combination of [9, Propositions 2.58 and 2.102]. □

We set

$$\mathcal{O}(\overline{\mathbb{D}}^\infty_{\text{ét}}) = \bigcup_{n \in \mathbb{N}} \mathcal{O}(\overline{\mathbb{D}}^n_{\text{ét}}).$$

(In the union above, we identify $\mathcal{O}(\overline{\mathbb{D}}^n_{\text{ét}})$ with a subset of $\mathcal{O}(\overline{\mathbb{D}}^{n+1}_{\text{ét}})$ by considering a function of the variables $(t_1, \ldots, t_n)$ as a function of the variables $(t_1, \ldots, t_{n+1})$ which is constant with respect to t_{n+1}.)

Next, we consider differential forms of finite co-degree on $\overline{\mathbb{D}}^\infty_{\text{ét}}$.

Definition 3.3 — *For $d \in \mathbb{N}$, we denote by $\Omega^{\infty-d}(\overline{\mathbb{D}}^\infty_{\text{ét}})$ the $\mathcal{O}(\overline{\mathbb{D}}^\infty_{\text{ét}})$-module freely generated by symbols $\widehat{\mathrm{d}t_I}$, one for each finite subset $I \subset \mathbb{N} \smallsetminus \{0\}$ of cardinality d:*

$$\Omega^{\infty-d}(\overline{\mathbb{D}}^\infty_{\text{ét}}) = \bigoplus_{I \subset \mathbb{N} \smallsetminus \{0\};\ |I| = d} \mathcal{O}(\overline{\mathbb{D}}^\infty_{\text{ét}}) \cdot \widehat{\mathrm{d}t_I}.$$

(If $I = \{i_1 < \cdots < i_d\}$, we think about $\widehat{\mathrm{d}t_I}$ as the differential form of infinite degree

$$\mathrm{d}t_1 \wedge \cdots \wedge \widehat{\mathrm{d}t_{i_1}} \wedge \cdots \wedge \widehat{\mathrm{d}t_{i_2}} \wedge \cdots \wedge \widehat{\mathrm{d}t_{i_d}} \wedge \cdots$$

where the $\mathrm{d}t_{i_s}$ have been removed for $1 \leqslant s \leqslant d$.) Elements of $\Omega^{\infty-d}(\overline{\mathbb{D}}^{\infty}_{\text{ét}})$ are called differential forms of co-degree d *on* $\overline{\mathbb{D}}^{\infty}_{\text{ét}}$.

Remark 3.4 — Perhaps, a more natural way to define $\Omega^{\infty-d}(\overline{\mathbb{D}}^{\infty}_{\text{ét}})$ is as follows. One has an inductive system $\{\Omega^{n-d}(\overline{\mathbb{D}}^{n}_{\text{ét}})\}_{n\geqslant d}$ with transition maps given by

$$- \wedge \mathrm{d}t_{n+1} : \Omega^{n-d}(\overline{\mathbb{D}}^{n}_{\text{ét}}) \longrightarrow \Omega^{n+1-d}(\overline{\mathbb{D}}^{n+1}_{\text{ét}}).$$

Then $\Omega^{\infty-d}(\overline{\mathbb{D}}^{\infty}_{\text{ét}})$ is the colimit of this inductive system.

Notation 3.5 — For $d \in \mathbb{N}$, we denote by $\widetilde{\Omega}^{\infty-d}(\overline{\mathbb{D}}^{\infty}_{\text{ét}})$ the subspace of $\Omega^{\infty-d}(\overline{\mathbb{D}}^{\infty}_{\text{ét}})$ consisting of those differential forms that vanish at all the faces of $\overline{\mathbb{D}}^{\infty}_{\text{ét}}$. More precisely, a differential form of co-degree d on $\overline{\mathbb{D}}^{\infty}_{\text{ét}}$

$$\omega = \sum_{I \subset \mathbb{N}\smallsetminus\{0\};\ |I|=d} f_I \cdot \widehat{\mathrm{d}t_I}$$

belongs to $\widetilde{\Omega}^{\infty-d}(\overline{\mathbb{D}}^{\infty}_{\text{ét}})$ if and only if $f_I|_{t_i=\epsilon} = 0$ for all $I \subset \mathbb{N} \smallsetminus \{0\}$ of cardinality d, $i \in I$ and $\epsilon \in \{0,1\}$. (For a function f depending on a complex variable t, the notation $f|_{t=c}$ stands for the function obtained from f by substituting the variable t by the constant $c \in \mathbb{C}$.)

Definition 3.6 — *There is an obvious de Rham differential*

$$\mathrm{d} : \Omega^{\infty-(d+1)}(\overline{\mathbb{D}}^{\infty}_{\text{ét}}) \longrightarrow \Omega^{\infty-d}(\overline{\mathbb{D}}^{\infty}_{\text{ét}})$$

which takes $\widetilde{\Omega}^{\infty-(d+1)}(\overline{\mathbb{D}}^{\infty}_{\text{ét}})$ inside $\widetilde{\Omega}^{\infty-d}(\overline{\mathbb{D}}^{\infty}_{\text{ét}})$. We thus obtain a complex

$$\mathcal{P}^{\text{eff}}_{\text{mot}}(k,\sigma) := \widetilde{\Omega}^{\infty-\bullet}(\overline{\mathbb{D}}^{\infty}_{\text{ét}}),$$

concentrated in positive homological degrees, which is called the (effective) motivic period algebra *(aka.,* algebra of motivic periods*).*

Remark 3.7 — Contrary to what our terminology indicates, the complex $\mathcal{P}^{\text{eff}}_{\text{mot}}(k,\sigma)$ is not an algebra on the nose. The algebra structure exists only in the derived category $\mathbf{D}(k)$. We refer the interested reader to [9, Proposition 2.108] for a description of the algebra product on $\mathrm{H}_0(\mathcal{P}^{\text{eff}}_{\text{mot}}(k,\sigma))$.

We can now formulate the main conjecture of this section.

Conjecture 3.8 — *The motivic period algebra $\mathcal{P}^{\text{eff}}_{\text{mot}}(k,\sigma)$ has no homology except in degree zero, i.e., $\mathrm{H}_i(\mathcal{P}^{\text{eff}}_{\text{mot}}(k,\sigma)) = 0$ for $i \in \mathbb{Z} \smallsetminus \{0\}$.*

Remark 3.9 — In degree zero, the homology of $\mathcal{P}^{\text{eff}}_{\text{mot}}(k,\sigma)$ is expected to be very large. Indeed, there exists an "evaluation map"

$$\int_{[0,1]^{\infty}} : \mathrm{H}_0(\mathcal{P}^{\text{eff}}_{\text{mot}}(k,\sigma)) \longrightarrow \mathbb{C}$$

which takes the class of a co-degree zero differential form

$$f \cdot \widehat{\mathrm{d}t_{\emptyset}} = f \cdot \mathrm{d}t_1 \wedge \cdots,$$

with $f \in \mathcal{O}(\overline{\mathbb{D}}^n_{\text{ét}})$, to the complex number $\int_{[0,1]^n} f$. It can be shown that the image of this evaluation map is the ring of effective periods obtained from the relative cohomology of pairs of k-varieties. When k is a number field (or if the complex embedding $\sigma : k \hookrightarrow \mathbb{C}$ is generic), the evaluation map is expected to be injective; this is the famous period conjecture of Kontsevich–Zagier (see [35] and [10]).

Remark 3.10 — There is a canonical lifting of $2\pi\boldsymbol{i} \in \mathbb{C}$ to a class $\underline{2\pi\boldsymbol{i}} \in \mathrm{H}_0(\mathcal{P}^{\text{eff}}_{\text{mot}}(k;\sigma))$. This class can be represented by the co-degree zero differential form

$$\mathrm{d}\log f \wedge \mathrm{d}t_2 \wedge \cdots$$

where $f \in \mathcal{O}(\overline{\mathbb{D}}^1_{\text{ét}})^{\times}$ is an invertible algebraic function in the variable t_1 such that $f(0) = f(1) = 1$ and $f|_{[0,1]} : [0,1] \longrightarrow \mathbb{C}^{\times}$ is counterclockwise simple loop around zero. (Note that the plane $\mathbb{C}$ is oriented by the choice of $\boldsymbol{i}$.) We set

$$\mathcal{P}_{\text{mot}}(k,\sigma) := \mathcal{P}^{\text{eff}}_{\text{mot}}(k,\sigma)[(\underline{2\pi\boldsymbol{i}})^{-1}];$$

this is the (non-effective) *motivic period algebra* (aka., *algebra of motivic periods*).

For later references, we state the non-effective version of Conjecture 3.8.

Conjecture 3.11 — *The non-effective motivic period algebra $\mathcal{P}_{\text{mot}}(k,\sigma)$ has no homology except in degree zero, i.e., $\mathrm{H}_i(\mathcal{P}_{\text{mot}}(k,\sigma)) = 0$ for $i \in \mathbb{Z} \smallsetminus \{0\}$.*

The following statement is obvious.

Lemma 3.12 — *Conjecture 3.8 implies Conjecture 3.11.*

3.2 The significance of Conjecture 3.8

In this subsection, we explain the origin and significance of Conjecture 3.8. We start by recalling the weak tannakian formalism as developed in [9, §1].

Theorem 3.13 — *Let $f : \mathcal{M} \longrightarrow \mathcal{E}$ be a monoidal functor between two monoidal categories. (Everything is symmetric and unitary.) Assume the following conditions.*

(i) *f has a monoidal section $e : \mathcal{E} \longrightarrow \mathcal{M}$, i.e., $f \circ e \simeq \mathrm{id}_{\mathcal{E}}$.*

(ii) *f has a right adjoint g and e has a right adjoint u.*

(iii) *The natural coprojection morphism*

$$g(A') \otimes B \longrightarrow g(A' \otimes f(B))$$

is an isomorphism for all $A' \in \mathcal{E}$ and $B \in \mathcal{M}$.

Then, we have the following conclusions.

(A) *$H = f(g(\mathbf{1}_{\mathcal{E}}))$ has a natural structure of a Hopf algebra in $\mathcal{E}$.*

(B) *For every object $M \in \mathcal{M}$, $f(M)$ is naturally a left H-comodule. This gives a monoidal functor*

$$\widetilde{f} : \mathcal{M} \longrightarrow \mathrm{coMod}(H)$$

making the following triangle commutative

$$\begin{array}{ccc} \mathcal{M} & \xrightarrow{\widetilde{f}} & \mathrm{coMod}(H) \\ & \searrow^{f} & \downarrow \\ & & \mathcal{E}. \end{array}$$

(C) *H is the universal Hopf algebra satisfying* (B).

As explained in [9, §2.1.3], Theorem 3.13 can be applied to the Betti realisation. Starting from now, we will denote by

$$\mathrm{B}^*_\sigma : \mathbf{DM}(k; \Lambda) \longrightarrow \mathbf{D}(\Lambda),$$

instead of B_σ, the Betti realisation functor. As $\mathbf{DM}(k;\Lambda)$ is compactly generated and B^*_σ commutes with infinite sums, the Brown representability theorem implies that B^*_σ has a right adjoint $\mathrm{B}_{\sigma,*}$ (see [41, Theorem 4.1]). On the other hand, the "Artin motive" functor

$$\mathbf{D}(\Lambda) \longrightarrow \mathbf{DM}(k; \Lambda)$$

is a monoidal section to B^*_σ and admits a right adjoint, again by the Brown representability theorem. Condition (iii) is also satisfied; it is an easy consequence of rigidity, in the monoidal sense (see [9, Lemme 2.8] and the proof of [9, Proposition 2.7]).

Definition 3.14 — *The Hopf algebra*

$$\mathcal{H}_{\mathrm{mot}}(k, \sigma; \Lambda) := \mathrm{B}^*_\sigma \mathrm{B}_{\sigma,*}\Lambda$$

is called the motivic Hopf algebra *of k (associated to $\sigma : k \hookrightarrow \mathbb{C}$). Note that this is a Hopf algebra in* $\mathbf{D}(\Lambda)$.

Remark 3.15 — We simply write $\mathcal{H}_{\mathrm{mot}}(k,\sigma)$ instead of $\mathcal{H}_{\mathrm{mot}}(k,\sigma;\mathbb{Q})$. In fact, one gains nothing by allowing more general rings of coefficients as

$$\mathcal{H}_{\mathrm{mot}}(k, \sigma; \Lambda) \simeq \mathcal{H}_{\mathrm{mot}}(k, \sigma) \otimes \Lambda.$$

Remark 3.16 — The motivic Hopf algebra co-acts on the singular homology of motives. More precisely, given a motive $M \in \mathbf{DM}(k;\mathbb{Q})$, $\mathrm{B}^*_\sigma(M)$ is naturally a comodule over $\mathcal{H}_{\mathrm{mot}}(k,\sigma)$. This co-action contains a lot of arithmetico-geometric information. Indeed, it determines the mixed Hodge structure on the Betti homology of M and the Galois action on the ℓ-adic homology of M.

Conjecture 3.17 — *The Hopf algebra $\mathcal{H}_{\mathrm{mot}}(k;\sigma)$ has no homology except in degree zero.*

Remark 3.18 — According to Conjecture 3.17, the motivic Hopf algebra is the coordinate ring of an honest affine group-scheme. This group-scheme is the so-called *motivic Galois group* of k (associated to $\sigma : k \hookrightarrow \mathbb{C}$).

There is a nice interpretation of Conjecture 3.17 in terms of operations on homology. We first make a definition.

Definition 3.19 — *Assume for simplicity that the* $\mathbb{Q}$*-algebra* Λ *is a field. An* operation *of degree* $d \in \mathbb{N}$ *is a* Λ*-linear transformation*

$$\mathrm{H}^0(\mathrm{B}^*_\sigma(M)) \longrightarrow \mathrm{H}^d(\mathrm{B}^*_\sigma(M))$$

which is natural in $M \in \mathbf{DM}(k;\Lambda)$.

There are plenty of operations of degree zero. For instance, when $\Lambda = \mathbb{Q}_\ell$, every element of $\mathrm{Gal}(\bar{k}/k)$ induces an operation of degree zero. (Here $\bar{k}$ is the algebraic closure of $\sigma(k)$ in $\mathbb{C}$.) On the contrary, for operations of nonzero degrees, we conjecture the following.

Conjecture 3.20 — *There are no nonzero operations of nonzero degree.*

Proposition 3.21 — *Conjecture 3.17 and Conjecture 3.20 are equivalent.*

Proof. Indeed, by [9, Proposition 1.32], an operation γ of degree d corresponds to a linear form on $\mathcal{H}_{\mathrm{mot}}(k,\sigma;\Lambda)$ of degree d, i.e., to a map

$$\ell : \mathrm{H}_d(\mathcal{H}_{\mathrm{mot}}(k,\sigma;\Lambda)) \longrightarrow \Lambda.$$

The correspondence works as follows. Given a linear form ℓ, the corresponding operation is given by the composition of

$$\begin{array}{ccc} \mathrm{H}^0(\mathrm{B}^*_\sigma(M)) \xrightarrow{\ \mathrm{ca}\ } & \mathrm{H}^0(\mathcal{H}_{\mathrm{mot}}(k,\sigma;\Lambda)\otimes \mathrm{B}^*_\sigma(M)) & \\ & \downarrow & \\ & \mathrm{H}_d(\mathcal{H}_{\mathrm{mot}}(k,\sigma;\Lambda))\otimes \mathrm{H}^d(\mathrm{B}^*_\sigma(M)) & \xrightarrow{\ \ell\otimes\mathrm{id}\ } \mathrm{H}^d(\mathrm{B}^*_\sigma(M)). \end{array}$$

(In the above diagram, ca denotes the co-action map of $\mathcal{H}_{\mathrm{mot}}(k,\sigma;\Lambda)$ on $\mathrm{B}^*_\sigma(M)$.) Conversely, given an operation λ, the corresponding linear form is given by the composition of

$$\mathrm{H}^{-d}(\mathrm{B}^*_\sigma \mathrm{B}_{\sigma,*}\Lambda) \xrightarrow{\ \lambda\ } \mathrm{H}^0(\mathrm{B}^*_\sigma \mathrm{B}_{\sigma,*}\Lambda) \xrightarrow{\ \mathrm{cu}\ } \Lambda.$$

(In the above diagram, cu denotes the co-unit map of $\mathcal{H}_{\mathrm{mot}}(k,\sigma;\Lambda)$.) The claimed result is now clear. □

The link with the vanishing conjectures of Subsection 3.1 follows from the following result.

Proposition 3.22 — *There is a natural quasi-isomorphism of complexes of* $\mathbb{C}$*-vectorspaces:*

$$\mathcal{H}_{\mathrm{mot}}(k,\sigma)\otimes\mathbb{C} \simeq \mathcal{P}_{\mathrm{mot}}(k,\sigma)\otimes_{k,\sigma}\mathbb{C}.$$

Proof. The Grothendieck comparison theorem between singular cohomology and algebraic de Rham cohomology can be stated in a compact form as an isomorphism in $\mathbf{DM}(k;\mathbb{Q})$:

$$(\mathrm{B}_{\sigma,*}\mathbb{Q})\otimes\mathbb{C}\simeq\boldsymbol{\Omega}_{/k}\otimes_{k,\sigma}\mathbb{C},$$

where $\boldsymbol{\Omega}_{/k}$ is an object representing algebraic de Rham cohomology and constructed using the usual algebraic de Rham complexes of smooth k-varieties. This isomorphism induces an isomorphism in $\mathbf{D}(\mathbb{Q})$

$$(\mathrm{B}_\sigma^*\mathrm{B}_{\sigma,*}\mathbb{Q})\otimes\mathbb{C}\simeq(\mathrm{B}_\sigma^*\boldsymbol{\Omega}_{/k})\otimes_{k,\sigma}\mathbb{C}.$$

The left-hand side is $\mathcal{H}_{\mathrm{mot}}(k,\sigma)\otimes\mathbb{C}$ by construction and it remains to identify $\mathrm{B}_\sigma^*\boldsymbol{\Omega}_{/k}$ with $\mathcal{P}_{\mathrm{mot}}(k,\sigma)$. This is a computation which can be found in [9, §2.3.1]; it relies on an approximation theorem for singular chains [9, Théorème 2.61] whose proof relies on Popescu's theorem [44, 45]. □

Corollary 3.23 — *Conjecture 3.11 and Conjecture 3.17 are equivalent.*

Remark 3.24 — As $\mathrm{H}_d(\mathcal{P}_{\mathrm{mot}}(k,\sigma))=0$ for $d<0$, we deduce from Proposition 3.22 that $\mathrm{H}_d(\mathcal{H}_{\mathrm{mot}}(k,\sigma))=0$ for $d<0$. In particular, there are no nonzero operations of strictly negative degrees.

Remark 3.25 — There exists an effective version of the motivic Hopf algebra which is denoted by $\mathcal{H}^{\mathrm{eff}}_{\mathrm{mot}}(k,\sigma)$. This is only a bialgebra, i.e., it does not have an antipode. It is defined by the same method while restricting to effective motives. More precisely, one considers the restriction $\mathrm{B}_\sigma^{\mathrm{eff},*}$ of the Betti realisation to the subcategory $\mathbf{DM}^{\mathrm{eff}}(k;\mathbb{Q})$ of effective motives (see Definition 4.1 below). This functor has a right adjoint that we denote by $\mathrm{B}^{\mathrm{eff}}_{\sigma,*}$. The effective motivic bialgebra is given by

$$\mathcal{H}^{\mathrm{eff}}_{\mathrm{mot}}(k,\sigma):=\mathrm{B}_\sigma^{\mathrm{eff},*}\mathrm{B}^{\mathrm{eff}}_{\sigma,*}\mathbb{Q}.$$

It can be shown that $\mathcal{H}_{\mathrm{mot}}(k,\sigma)\simeq\mathcal{H}^{\mathrm{eff}}_{\mathrm{mot}}(k,\sigma)[\varsigma^{-1}]$ for a canonical element $\varsigma\in\mathrm{H}_0(\mathcal{H}^{\mathrm{eff}}_{\mathrm{mot}}(k,\sigma))$. Moreover, as in Proposition 3.22, there exists a quasi-isomorphism of complexes of $\mathbb{C}$-vectorspaces

$$\mathcal{H}^{\mathrm{eff}}_{\mathrm{mot}}(k,\sigma)\otimes\mathbb{C}\simeq\mathcal{P}^{\mathrm{eff}}_{\mathrm{mot}}(k,\sigma)\otimes_{k,\sigma}\mathbb{C}.$$

(Under this isomorphism, the element $\varsigma\otimes 1$ corresponds to $\underline{2\pi\boldsymbol{i}}\otimes(2\pi\boldsymbol{i})^{-1}$.) In particular, this shows that Conjecture 3.8 is equivalent to the effective version of Conjecture 3.17, i.e., to the property that $\mathcal{H}^{\mathrm{eff}}_{\mathrm{mot}}(k,\sigma)$ has no homology except in degree zero.

3.3 Relation with the conjectural motivic t-structure

Conjecture 3.17 is intimately related to the problem of constructing the so-called motivic t-structure. The existence of the motivic t-structure is one of the most central open problems in the theory of motives.

Conjecture 3.26 — *Given a complex embedding $\sigma:k\hookrightarrow\mathbb{C}$, let $\mathcal{T}^{\mathcal{M}}_{\geqslant 0}$ be the full subcategory of $\mathbf{DM}(k;\Lambda)$ consisting of those motives M such that $\mathrm{B}_\sigma^*(M)\in$*

$\mathbf{D}_{\geqslant 0}(\Lambda)$. *Define* $\mathcal{T}^{\mathcal{M}}_{<0}$ *to be the right orthogonal to* $\mathcal{T}^{\mathcal{M}}_{\geqslant 0}$, *i.e., the full subcategory whose objects are the* $N \in \mathbf{DM}(k;\Lambda)$ *such that*

$$\hom_{\mathbf{DM}(k;\Lambda)}(M,N) = 0$$

for all $M \in \mathcal{T}^{\mathcal{M}}_{\geqslant 0}$. *Then the following properties should hold:*

(i) *The pair* $(\mathcal{T}^{\mathcal{M}}_{\geqslant 0}, \mathcal{T}^{\mathcal{M}}_{<0})$ *is a t-structure which is independent from the choice of the complex embedding* σ.

(ii) *The Betti realisation* $\mathrm{B}^*_\sigma : \mathbf{DM}(k;\Lambda) \longrightarrow \mathbf{D}(\Lambda)$ *is t-exact. This is equivalent to saying that* B^*_σ *takes a motive in* $\mathcal{T}^{\mathcal{M}}_{<0}$ *to a complex in* $\mathbf{D}_{<0}(\Lambda)$.

(iii) *Assuming that* Λ *is a regular ring,*[11] *the t-structure* $(\mathcal{T}^{\mathcal{M}}_{\geqslant 0}, \mathcal{T}^{\mathcal{M}}_{<0})$ *restricts to* $\mathbf{DM}_{\mathrm{gm}}(k;\Lambda)$, *i.e., the pair*

$$(\mathcal{T}^{\mathcal{M}}_{\geqslant 0} \cap \mathbf{DM}_{\mathrm{gm}}(k;\Lambda), \mathcal{T}^{\mathcal{M}}_{<0} \cap \mathbf{DM}_{\mathrm{gm}}(k;\Lambda))$$

is a t-structure on $\mathbf{DM}_{\mathrm{gm}}(k;\Lambda)$. *Said differently, the truncation functors associated to the t-structure* $(\mathcal{T}^{\mathcal{M}}_{\geqslant 0}, \mathcal{T}^{\mathcal{M}}_{<0})$ *preserve geometric motives.*

If it exists, the t-structure of Conjecture 3.26 is called the *motivic t-structure.*

Remark 3.27 — It is not difficult to construct a t-structure on $\mathbf{DM}(k;\Lambda)$ which, if Conjecture 3.26 was true, coincides with the motivic t-structure, at least after restricting to $\mathbf{DM}_{\mathrm{gm}}(k;\Lambda)$ (under the assumption that Λ is regular). Indeed, let $\mathcal{T}_{\geqslant 0}$ be the smallest subcategory of $\mathbf{DM}(k;\Lambda)$ closed under infinite sums, extensions and suspensions, and containing the geometric motives whose Betti realisations belong to $\mathbf{D}_{\geqslant 0}(\Lambda)$. Also, let $\mathcal{T}_{<0}$ be the right orthogonal to $\mathcal{T}_{\geqslant 0}$. Then, by a general argument (see for example [4, Proposition 2.1.70]), $(\mathcal{T}_{\geqslant 0}, \mathcal{T}_{<0})$ is a t-structure, which is the alluded candidate for the motivic t-structure. Unfortunately, with such a definition, it is totally unclear how to establish properties (ii) and (iii) of Conjecture 3.26.

Remark 3.28 — If k has infinite transcendence degree over its prime field, the motivic t-structure on $\mathbf{DM}(k;\Lambda)$ degenerates. Indeed, from the definition, $\mathcal{T}^{\mathcal{M}}_{\geqslant 0}$ contains the triangulated subcategory $\mathcal{F}ant$ of fantom motives (see Definition 2.6). However, under the assumption that Λ is regular, the conservativity conjecture (i.e., Conjecture 2.1) implies that the restriction of the motivic t-structure to $\mathbf{DM}_{\mathrm{gm}}(k;\Lambda)$ is non-degenerate. Indeed, if

$$M \in \mathbf{DM}_{\mathrm{gm}}(k;\Lambda) \cap \left(\bigcap_{n \in \mathbb{N}} \mathcal{T}^{\mathcal{M}}_{\geqslant n} \right),$$

then necessarily $\mathrm{B}^*_\sigma(M) = 0$, and thus $M = 0$.

Proposition 3.29 — *Conjecture 3.26 implies Conjecture 3.17.*

[11] This assumption is necessary. To see this, note that, at least when Λ is noetherian, the canonical t-structure on $\mathbf{D}(\Lambda)$ restricts to a t-structure on the subcategory of perfect complexes if and only if Λ is regular.

Proof. By Remark 3.24, we know that $\mathcal{H}_{\mathrm{mot}}(k,\sigma) \in \mathbf{D}_{\geqslant 0}(\mathbb{Q})$. Therefore, we only need to show that $\mathcal{H}_{\mathrm{mot}}(k,\sigma) \in \mathbf{D}_{\leqslant 0}(\mathbb{Q})$.

Now, assuming Conjecture 3.26, we have at our disposal a t-structure on $\mathbf{DM}(k;\mathbb{Q})$ for which B_σ^* is t-exact. It follows, by a general argument, that the right adjoint $\mathrm{B}_{\sigma,*}$ is left t-exact, i.e., takes $\mathbf{D}_{\leqslant 0}(\mathbb{Q})$ to $\mathcal{T}^{\mathcal{M}}_{\leqslant 0}$. In particular, we have

$$\mathrm{B}_{\sigma,*}\mathbb{Q} \in \mathcal{T}^{\mathcal{M}}_{\leqslant 0}.$$

Using again that B_σ^* is t-exact, we deduce that

$$\mathrm{B}_\sigma^*\mathrm{B}_{\sigma,*}\mathbb{Q} \in \mathbf{D}_{\leqslant 0}(\mathbb{Q})$$

as needed. □

Remark 3.30 — As far as we know, the vanishing conjecture for the motivic Hopf algebra (i.e., Conjecture 3.17) does not imply any longstanding conjecture on algebraic cycles, as was the case for the conservativity conjecture (i.e., Conjecture 2.1). Nonetheless, we believe that this is an important conjecture for at least two reasons. First, as explained above, the validity of Conjecture 3.17 is a necessary condition for the existence of the motivic t-structure. But more importantly, it is our strong believe that Conjecture 3.17 should constitute a crucial step in a future construction of the motivic t-structure. Indeed, in [9, §2.4], we stated a conjecture, [9, Conjecture B on page 127], which, if true, enables one to reconstruct a geometric motive from its Betti realisation using a descent procedure. We believe that Conjecture 3.26 follows, modulo some routine work, from the combination of [9, §2.4, Conjecture B] and Conjecture 3.17 (which is also [9, §2.4, Conjecture A]).

4 Homotopy t-structure, slice filtration and dimension

In this section, we will formulate some plausible conjectures around the homotopy t-structure and the slice filtration, two standard tools in the theory of motives à la Voevodsky. Contrary to the ones stated previously, these conjectures are not part of the classical motivic paradise and rather originate from unexpected features of Voevodsky's construction of his category of motives. We will start by recalling the basic notion of effectiveness for motives.

4.1 Effective motives, homotopy t-structure and slice filtration

Recall that a mixed Hodge structure is said to be *effective* if its Hodge numbers $h^{p,q}$ are zero unless both p and q are non-negative. Unfortunately, for motives, the notion of effectivity (see Definition 4.1 below) corresponds to the dual one: the dual of the Hodge realisation of a geometric effective motive belongs to the subcategory of $\mathbf{D}^{\mathrm{b}}(\mathrm{MHS})$ generated by effective mixed Hodge structures.

Definition 4.1 — *We denote by* $\mathbf{DM}^{\mathrm{eff}}(k;\Lambda)$ *the smallest triangulated subcategory of* $\mathbf{DM}(k;\Lambda)$ *closed under infinite sums and containing the homological motives* $\mathrm{M}(X)$ *for all smooth quasi-projective k-varieties X. (The crucial point here is that we do not allow negative Tate twists of the* $\mathrm{M}(X)$*'s.) A motive is said to be* effective *if it lies in* $\mathbf{DM}^{\mathrm{eff}}(k;\Lambda)$.

We denote by $\mathbf{DM}^{\mathrm{eff}}_{\mathrm{gm}}(k;\Lambda)$ *the subcategory of compact objets of* $\mathbf{DM}^{\mathrm{eff}}(k;\Lambda)$. *This is the category of effective geometric motives.*

Remark 4.2 — Almost by definition, every geometric motive $M \in \mathbf{DM}_{\mathrm{gm}}(k;\Lambda)$ can be made effective after applying a sufficiently positive Tate twist. This is analogous (or rather dual) to the fact that every mixed Hodge structure becomes effective after a sufficiently negative Tate twist.

Remark 4.3 — One interesting feature of the category $\mathbf{DM}^{\mathrm{eff}}(k;\Lambda)$ is that it admits a more concrete description than the larger $\mathbf{DM}(k;\Lambda)$. Indeed, there is a fully faithful embedding

$$\mathbf{DM}^{\mathrm{eff}}(k;\Lambda) \hookrightarrow \mathbf{D}(\mathbf{Str}_{\mathrm{Nis}}(\mathrm{Sm}/k;\Lambda))$$

into the derived category of the abelian category $\mathbf{Str}_{\mathrm{Nis}}(\mathrm{Sm}/k;\Lambda)$ of *Nisnevich sheaves with transfers* on Sm/k. (Recall that a Nisnevich sheaf with transfers is a Λ-linear functor from $\mathrm{SmCor}(k;\Lambda)$ (see Definition 1.14) to the category of Λ-modules whose restriction to Sm/k is a sheaf for the Nisnevich topology.) Moreover, a complex of Nisnevich sheaves with transfers is in the image of this embedding if and only if its homology sheaves are homotopy invariant. (Recall that a Nisnevich sheaf F on Sm/k is homotopy invariant if the map $F(X) \longrightarrow F(\mathbb{A}^1 \times X)$, induced by the natural projection to X, is an isomorphism for every $X \in \mathrm{Sm}/k$.) As a consequence, one obtains the following result.

Lemma 4.4 — *The canonical t-structure on* $\mathbf{D}(\mathbf{Str}_{\mathrm{Nis}}(\mathrm{Sm}/k;\Lambda))$ *restricts to a t-structure on* $\mathbf{DM}^{\mathrm{eff}}(k;\Lambda)$ *called the* homotopy t-structure. *The heart of the homotopy t-structure is equivalent to the category of* homotopy invariant sheaves with transfers *that we denote by* $\mathbf{HI}(k;\Lambda)$. *(Here and in the rest of the article, the word "sheaf" will always refer to the Nisnevich topology.)*

Remark 4.5 — It is possible to extend the homotopy t-structure to the full triangulated category of motives $\mathbf{DM}(k;\Lambda)$. The heart of this extended homotopy t-structure is more complicated to describe. Its objects are *Tate spectra*[12] with some properties. We direct the interested reader to [23] where the heart of the homotopy t-structure on $\mathbf{DM}(k;\mathbb{Z})$ is described and shown to be equivalent to the category of Rost modules [47].

Remark 4.6 — It should be noted that the homotopy t-structure is *not* the motivic t-structure of Conjecture 3.26. In fact, these two t-structures have nothing in common. The homotopy t-structure does not correspond under realisations to any classical construction. Although, the Betti realisation

$$\mathrm{B}^{\mathrm{eff},*}_{\sigma} : \mathbf{DM}^{\mathrm{eff}}(k;\Lambda) \longrightarrow \mathbf{D}(\Lambda)$$

[12] The notion of Tate spectra is analogous to the notion of spectra in stable homotopy theory, see for example [1].

is right t-exact when $\mathbf{DM}^{\text{eff}}(k;\Lambda)$ is endowed with the homotopy t-structure, it is very far from being left t-exact. Moreover, the homotopy t-structure does not restrict to $\mathbf{DM}^{\text{eff}}_{\text{gm}}(k;\Lambda)$. Indeed, the truncation functors with respect to the homotopy t-structure do not preserve geometric effective motives.

For later use, we introduce the following notation.

Notation 4.7 — Given an effective motive $M \in \mathbf{DM}^{\text{eff}}(k;\Lambda)$, we denote by $\mathrm{h}_n(M)$ the homology objects associated to M with respect to the homotopy t-structure. These are homotopy invariant sheaves with transfers. As a Nisnevich sheaf, $\mathrm{h}_n(M)$ coincides with the sheafification of the following presheaf

$$U \in \mathrm{Sm}/k \quad \mapsto \quad \hom_{\mathbf{DM}^{\text{eff}}(k;\,\Lambda)}(\mathrm{M}(U)[n], M).$$

Given a smooth k-variety X, we write $\mathrm{h}_n(X)$ instead of $\mathrm{h}_n(\mathrm{M}(X))$. Note that $\mathrm{h}_n(X) = 0$ for $n < 0$ and that the group of global sections of $\mathrm{h}_n(X)$ coincides with the Suslin homology group $\mathrm{H}_n(X;\Lambda(0))$ (see Definition 1.12).

We will also need the slice filtration on motives (compare with [51]).

Definition 4.8 — *By the Brown representability theorem, the inclusion of* $\mathbf{DM}^{\text{eff}}(k;\Lambda)$ *into* $\mathbf{DM}(k;\Lambda)$ *admits a right adjoint*

$$\nu^{\geqslant 0} : \mathbf{DM}(k;\Lambda) \longrightarrow \mathbf{DM}^{\text{eff}}(k;\Lambda).$$

For $n \in \mathbb{Z}$, we set

$$\nu^{\geqslant n}(M) := \nu^{\geqslant 0}(M(-n))(n).$$

(Note that $\nu^{\geqslant n}$ takes values in the subcategory $\mathbf{DM}^{\text{eff}}(k;\Lambda)(n)$ consisting of motives whose $(-n)$-th Tate twists are effective.) There are natural transformations

$$\cdots \longrightarrow \nu^{\geqslant n} \longrightarrow \nu^{\geqslant n-1} \longrightarrow \cdots .$$

The system $\{\nu^{\geqslant n}\}_{n\in\mathrm{N}}$ is called the slice filtration.

Remark 4.9 — The slice filtration does not correspond via realisations to any classical construction. In fact, the functors $\nu^{\geqslant n}$ are not well-behaved; see [30] for some conditional and unconditional negative properties. In particular, it is known that geometric motives are not preserved by these functors.

For later use, we record the following simple lemma.

Lemma 4.10 — *The triangulated category $\mathbf{DM}^{\text{eff}}(k;\Lambda)$ is monoidal and closed. Given two effective motives M and N, we denote by $\underline{\mathsf{Hom}}^{\text{eff}}(M,N)$ the internal hom from M to N computed in $\mathbf{DM}^{\text{eff}}(k;\Lambda)$. We then have the formula*

$$\underline{\mathsf{Hom}}^{\text{eff}}(M,N) = \nu^{\geqslant 0}\underline{\mathsf{Hom}}(M,N).$$

Proof. The fact that the tensor product of two effective motives is also effective follows form the formula

$$\mathrm{M}(X) \otimes \mathrm{M}(Y) \simeq \mathrm{M}(X \times Y)$$

for all $X, Y \in \mathrm{Sm}/k$. To prove the formula for $\underline{\mathsf{Hom}}^{\mathrm{eff}}(M, N)$, it is enough to show that $\nu^{\geqslant 0}\underline{\mathsf{Hom}}(M, -)$ is right adjoint to $M \otimes -$. Given an effective motive L, we have the identifications

$$\begin{aligned}\hom_{\mathbf{DM}^{\mathrm{eff}}(k;\,\Lambda)}(L, \nu^{\geqslant 0}\underline{\mathsf{Hom}}(M, N)) &\simeq \hom_{\mathbf{DM}(k;\,\Lambda)}(L, \underline{\mathsf{Hom}}(M, N)) \\ &\simeq \hom_{\mathbf{DM}(k;\,\Lambda)}(M \otimes L, N) \\ &= \hom_{\mathbf{DM}^{\mathrm{eff}}(k;\,\Lambda)}(M \otimes L, N).\end{aligned}$$

This proves what we want. □

Proposition 4.11 — *Let $n \geqslant 0$. For $M \in \mathbf{DM}^{\mathrm{eff}}(k; \Lambda)$, there is a natural isomorphism*

$$\nu^{\geqslant 0}(M(-n)) \simeq \underline{\mathsf{Hom}}^{\mathrm{eff}}(\Lambda(n), M).$$

Proof. Use that $M(-n) = \underline{\mathsf{Hom}}(\Lambda(n), M)$. □

Remark 4.12 — Using a similar reasoning, one sees that

$$\mathrm{B}^{\mathrm{eff}}_{\sigma,\,*}\Lambda \simeq \nu^{\geqslant 0}(\mathrm{B}_{\sigma,\,*}\Lambda).$$

The Betti realisation of the motive $\mathrm{B}^{\mathrm{eff}}_{\sigma,\,*}\Lambda$ is the motivic bialgebra $\mathcal{H}^{\mathrm{eff}}_{\mathrm{mot}}(k, \sigma; \Lambda)$; see Remark 3.25.

4.2 Filtration by dimension

In order to formulate our conjectures, we introduce the following subcategories.

Definition 4.13 — *For $n \in \mathbb{N}$, we denote by $\mathbf{DM}^{\mathrm{eff}}_{\leqslant n}(k; \Lambda)$ the smallest subcategory of $\mathbf{DM}^{\mathrm{eff}}(k; \Lambda)$ closed under infinite sums and containing the motives* $\mathrm{M}(X)$ *for all smooth quasi-projective k-varieties X of dimension at most n.*

Remark 4.14 — The categories $\mathbf{DM}^{\mathrm{eff}}_{\leqslant n}(k; \Lambda)$ are well understood for $n \in \{0, 1\}$. Indeed, $\mathbf{DM}^{\mathrm{eff}}_{\leqslant 0}(k; \Lambda)$ is equivalent to the derived category of continuous representations of $\mathrm{Gal}(\bar{k}/k)$ with coefficients in Λ (endowed with the discrete topology). Equivalently, there is an equivalence of categories

$$\mathbf{D}(\mathbf{Shv}_{\text{ét}}(\mathrm{Et}/k; \Lambda)) \simeq \mathbf{DM}^{\mathrm{eff}}_{\leqslant 0}(k; \Lambda)$$

where Et/k is the small étale site of $\mathrm{Spec}(k)$. The case $n = 1$ is also related to classical objects, namely Deligne 1-motives. Indeed, there is an equivalence of categories

$$\mathbf{D}(\text{Ind-}\mathcal{M}_1(k; \mathbb{Q})) \simeq \mathbf{DM}^{\mathrm{eff}}_{\leqslant 1}(k; \mathbb{Q})$$

where $\mathcal{M}_1(k; \mathbb{Q})$ is the abelian category of Deligne 1-motives up to isogeny. (In the geometric case, this equivalence is proven in [43].) It is possible to see in these statements some evidence, rather meagre admittedly, for the existence of the motivic t-structure (see Conjecture 3.26).

We note the following fact which shows that the situation is really very nice when $n \in \{0,1\}$.

PROPOSITION 4.15 — *For $n \in \{0,1\}$, the inclusion of $\mathbf{DM}^{\text{eff}}_{\leqslant n}(k;\Lambda)$ into $\mathbf{DM}^{\text{eff}}(k;\Lambda)$ admits a left adjoint. When $n = 0$, this left adjoint is denoted by*

$$\mathrm{L}\pi_0 : \mathbf{DM}^{\text{eff}}(k;\Lambda) \longrightarrow \mathbf{DM}^{\text{eff}}_{\leqslant 0}(k;\Lambda).$$

When $n = 1$, this left adjoint is denoted by

$$\mathrm{LAlb} : \mathbf{DM}^{\text{eff}}(k;\Lambda) \longrightarrow \mathbf{DM}^{\text{eff}}_{\leqslant 1}(k;\Lambda).$$

Proof. This is proven in [13]. See also [14] where the geometric case is treated using a different approach. □

Remark 4.16 — For a geometric effective motive $M \in \mathbf{DM}^{\text{eff}}_{\text{gm}}(k;\Lambda)$, one has the following formula

$$\mathrm{LAlb}(M) = \underline{\mathsf{Hom}}^{\text{eff}}\left(\underline{\mathsf{Hom}}^{\text{eff}}(M, \Lambda(1)), \Lambda(1)\right).$$

This formula was taken as a definition in [14, §5].

Remark 4.17 — Contrary to right adjoints, the existence of left adjoints is often a nontrivial property. In fact, the functors $\mathrm{L}\pi_0$ and LAlb are very nice and they do correspond to classical constructions in Hodge theory. For instance, as the notation indicates, LAlb is somehow a "derived" version of the Albanese variety.

Remark 4.18 — The situation becomes quickly extremely complicated starting form $n = 2$. Indeed, there is no known description of $\mathbf{DM}^{\text{eff}}_{\leqslant 2}(k;\Lambda)$ using classical objects. Conjecturally, $\mathbf{DM}^{\text{eff}}_{\leqslant 2}(k;\Lambda)$ should be the derived category of the abelian category of mixed 2-motives whose existence is out of reach (but see [7] for an attempt). There are also negative results. For instance, it is shown in [13, §2.5] that the inclusion of $\mathbf{DM}^{\text{eff}}_{\leqslant 2}(k;\mathbb{Q})$ into $\mathbf{DM}^{\text{eff}}(k;\mathbb{Q})$ cannot have a left adjoint (at least if k has infinite transcendence degree). Nevertheless, we will offer in Subsection 4.3 below two positive conjectures shedding light on the possibility of some beautiful structures in the theory of algebraic cycles.

We will also need the following related notion. (Recall Notation 4.7 for the significance of $\mathrm{h}_0(X)$.)

DEFINITION 4.19 — *Let F be an object of $\mathbf{HI}(k;\Lambda)$, i.e., a homotopy invariant sheaf with transfers. We say that F in n-*generated *if there exists a surjection in $\mathbf{HI}(k;\Lambda)$:*

$$\bigoplus_{i\in I} \mathrm{h}_0(X_i) \twoheadrightarrow F$$

*where the X_i's have dimensions at most n. We say that F is n-*presented *if there exists an exact sequence in $\mathbf{HI}(k;\Lambda)$:*

$$\bigoplus_{j\in J} \mathrm{h}_0(Y_j) \longrightarrow \bigoplus_{i\in I} \mathrm{h}_0(X_i) \longrightarrow F \longrightarrow 0$$

where the X_i's and Y_j's have dimensions at most n. (Above, we do not assume that I nor J is finite.) We denote by $\mathbf{HI}_{\leqslant n}(k;\Lambda)$ the full subcategory of $\mathbf{HI}(k;\Lambda)$ consisting of n-presented homotopy invariant sheaves with transfers.

Remark 4.20 — In [13], an n-presented homotopy invariant sheaf with transfers was called an *n-motivic sheaf*. It is also conjectured in *loc. cit.* that the notions of n-generated and n-presented are equivalent (see [13, Conjecture 1.4.1 and Lemma 1.4.3]).

Remark 4.21 — It is maybe useful to explain the idea of Definition 4.19 in a geometrically meaningful case. A basic example of an object in $\mathbf{HI}(k;\mathbb{Q})$ is the Nisnevich sheaf $\underline{\mathrm{CH}}^d(X;\mathbb{Q})$ associated to the presheaf

$$\mathrm{CH}^d(X\times -;\mathbb{Q})\ :\ U\in \mathrm{Sm}/k \quad\mapsto\quad \mathrm{CH}^d(X\times U;\mathbb{Q}).$$

This makes sense for any k-variety X but, for simplicity, we will assume that X is smooth and proper. Saying that $\underline{\mathrm{CH}}^d(X;\mathbb{Q})$ is n-generated amounts to saying the following. Every family of codimension d cycles on X parametrised by a quasi-projective smooth variety U is, up to rational equivalence and maybe after shrinking U, obtained from a family of codimension d cycles on X parametrised by a quasi-projective smooth variety V of dimension at most n by pulling-back along a finite correspondence from U to V (in the sense of Definition 1.14). In fact, knowing that $\underline{\mathrm{CH}}^d(X;\mathbb{Q})$ is n-generated implies that there exists a denumbrable family $(S_\alpha)_\alpha$ of smooth and projective varieties of dimension at most n, and codimension d-cycles $Z_\alpha\subset S_\alpha\times X$, such that the induced map

$$\bigoplus_\alpha \mathrm{CH}_0(S_\alpha;\mathbb{Q})\longrightarrow \mathrm{CH}^d(X;\mathbb{Q})$$

is universally surjective (i.e., remains surjective after extending the base field, say to $\mathbb{C}$). Such properties are very nice and this is perhaps the next best thing to hope for, knowing the failure of representability of Chow groups in general [38].

4.3 The conjectures

In this last subsection, we will formulated two conjectures describing the behaviour of the homotopy t-structure and the slice filtration with respect to the filtration by dimension on the category of motives.

Conjecture 4.22 — *The endofunctor $\underline{\mathsf{Hom}}^{\mathrm{eff}}(\Lambda(1),-)$ of $\mathbf{DM}^{\mathrm{eff}}(k;\Lambda)$ takes the subcategory $\mathbf{DM}^{\mathrm{eff}}_{\leqslant n}(k;\Lambda)$ to the subcategory $\mathbf{DM}^{\mathrm{eff}}_{\leqslant n-1}(k;\Lambda)$.*

Remark 4.23 — The endofunctor $\underline{\mathsf{Hom}}^{\mathrm{eff}}(\Lambda(1)[1],-)$ is also known as Voevodsky's *contraction* and is usually denoted by $(-)_{-1}$. Voevodsky's contraction is indeed a remarkable operation. A notable property is that $(-)_{-1}$ is exact with respect to the homotopy t-structure. The above conjecture is simply saying that Voevodsky's contraction sends $\mathbf{DM}^{\mathrm{eff}}_{\leqslant n}(k;\Lambda)$ to $\mathbf{DM}^{\mathrm{eff}}_{\leqslant n-1}(k;\Lambda)$.

Remark 4.24 — Using that

$$\underline{\mathsf{Hom}}^{\mathrm{eff}}(\Lambda(1),-)^{\circ m}\simeq \underline{\mathsf{Hom}}^{\mathrm{eff}}(\Lambda(m),-)$$

we see that, under Conjecture 4.22, $\underline{\mathsf{Hom}}^{\mathrm{eff}}(\Lambda(m),-)$ induces a functor

$$\underline{\mathsf{Hom}}^{\mathrm{eff}}(\Lambda(m),-):\mathbf{DM}^{\mathrm{eff}}_{\leqslant n}(k;\Lambda)\longrightarrow \mathbf{DM}^{\mathrm{eff}}_{\leqslant n-m}(k;\Lambda)$$

(with the convention that $\mathbf{DM}^{\mathrm{eff}}_{\leqslant r}(k;\Lambda)=0$ for $r<0$).

The following proposition gives some (meagre) evidence for Conjecture 4.22.

Proposition 4.25 — *Let $n\in\mathbb{N}$ and $M\in\mathbf{DM}^{\mathrm{eff}}_{\leqslant n}(k;\Lambda)$. Then*

$$\underline{\mathsf{Hom}}^{\mathrm{eff}}(\Lambda(m),M)\in\mathbf{DM}^{\mathrm{eff}}_{\leqslant n-m}(k;\Lambda)$$

for all $m\geqslant n-1$. (In particular, $\underline{\mathsf{Hom}}^{\mathrm{eff}}(\Lambda(m),M)=0$ if $m\geqslant n+1$.)

Proof. We will assume that $m\in\{n,n-1\}$. Indeed, the case $m\geqslant n+1$ follows from the case $m=n$ and the first equivalence of categories in Remark 4.14 using the easy fact that the contraction of an étale-locally constant sheaf is zero.

It is enough to treat the case where $M=\mathrm{M}(X)$ with X a smooth projective k-variety of pure dimension n. Recall that the strong dual of M is given by $M(-n)[-2n]$. It follows that (Poincaré motivic duality)

$$M\simeq\underline{\mathsf{Hom}}^{\mathrm{eff}}(M,\Lambda(n)[2n]).$$

But we have natural isomorphisms

$$\begin{aligned}\underline{\mathsf{Hom}}^{\mathrm{eff}}(\Lambda(m),\underline{\mathsf{Hom}}^{\mathrm{eff}}(M,\Lambda(n)))&\simeq\underline{\mathsf{Hom}}^{\mathrm{eff}}(M,\underline{\mathsf{Hom}}^{\mathrm{eff}}(\Lambda(m),\Lambda(n)))\\&\simeq\underline{\mathsf{Hom}}^{\mathrm{eff}}(M,\Lambda(n-m)).\end{aligned}$$

Therefore, we are left to check that

$$\underline{\mathsf{Hom}}^{\mathrm{eff}}(M,\Lambda(0))\in\mathbf{DM}^{\mathrm{eff}}_{\leqslant 0}(k;\Lambda)\qquad\text{and}\qquad\underline{\mathsf{Hom}}^{\mathrm{eff}}(M,\Lambda(1))\in\mathbf{DM}^{\mathrm{eff}}_{\leqslant 1}(k;\Lambda).$$

We split the argument accordingly in two parts.

Part 1: The category $\mathbf{DM}^{\mathrm{eff}}_{\leqslant 0}(k;\Lambda)$ is monoidal closed; we denote by $\underline{\mathsf{Hom}}^{\mathrm{eff}}_{\leqslant 0}(-,-)$ its internal Hom. Also, the functor $\mathrm{L}\pi_0$ of Proposition 4.15 is monoidal. Given an effective motive L, we thus have the following isomorphisms

$$\begin{aligned}&\hom_{\mathbf{DM}^{\mathrm{eff}}(k;\Lambda)}(L,\underline{\mathsf{Hom}}^{\mathrm{eff}}(M,\Lambda(0)))\\\simeq\ &\hom_{\mathbf{DM}^{\mathrm{eff}}(k;\Lambda)}(L\otimes M,\Lambda(0))\\\simeq\ &\hom_{\mathbf{DM}^{\mathrm{eff}}_{\leqslant 0}(k;\Lambda)}(\mathrm{L}\pi_0(L\otimes M),\Lambda(0))\\\simeq\ &\hom_{\mathbf{DM}^{\mathrm{eff}}_{\leqslant 0}(k;\Lambda)}(\mathrm{L}\pi_0(L)\otimes\mathrm{L}\pi_0(M),\Lambda(0))\\\simeq\ &\hom_{\mathbf{DM}^{\mathrm{eff}}_{\leqslant 0}(k;\Lambda)}(\mathrm{L}\pi_0(L),\underline{\mathsf{Hom}}^{\mathrm{eff}}_{\leqslant 0}(\mathrm{L}\pi_0(M),\Lambda(0)))\\\simeq\ &\hom_{\mathbf{DM}^{\mathrm{eff}}(k;\Lambda)}(L,\underline{\mathsf{Hom}}^{\mathrm{eff}}_{\leqslant 0}(\mathrm{L}\pi_0(M),\Lambda(0))).\end{aligned}$$

This shows that $\underline{\mathsf{Hom}}^{\mathrm{eff}}(M,\Lambda(0))$ is isomorphic to $\underline{\mathsf{Hom}}^{\mathrm{eff}}_{\leqslant 0}(\mathrm{L}\pi_0(M),\Lambda(0))$, and therefore belongs to $\mathbf{DM}^{\mathrm{eff}}_{\leqslant 0}(k;\Lambda)$ as wanted.

Part 2: The subcategory $\mathbf{DM}^{\mathrm{eff}}_{\leqslant 1}(k;\Lambda)$ is not closed under the tensor product. Nonetheless, it has a natural monoidal structure; its tensor product $-\otimes_1-$ is given by

$$M\otimes_1 N := \mathrm{LAlb}(M\otimes N)$$

where LAlb is the left adjoint to the obvious inclusion (see Proposition 4.15). With this monoidal structure, the functor $\mathrm{LAlb}:\mathbf{DM}^{\mathrm{eff}}(k;\Lambda)\longrightarrow\mathbf{DM}^{\mathrm{eff}}_{\leqslant 1}(k;\Lambda)$ is monoidal (see [14, Proposition 7.1.2]).

We denote by $\underline{\mathsf{Hom}}^{\mathrm{eff}}_{\leqslant 1}(-,-)$ the internal Hom relative to the monoidal structure on $\mathbf{DM}^{\mathrm{eff}}_{\leqslant 1}(k;\Lambda)$. A similar computation as in Part 1 shows that $\underline{\mathsf{Hom}}^{\mathrm{eff}}(M,\Lambda(1))$ is isomorphic to $\underline{\mathsf{Hom}}^{\mathrm{eff}}_{\leqslant 1}(\mathrm{LAlb}(M),\Lambda(1))$, and therefore belongs to $\mathbf{DM}^{\mathrm{eff}}_{\leqslant 1}(k;\Lambda)$ as wanted. □

Proposition 4.26 — *Assume Conjecture 4.22. Let $M\in\mathbf{DM}^{\mathrm{eff}}_{\mathrm{gm}}(k;\Lambda)$ be an effective geometric motive. Then $\underline{\mathsf{Hom}}^{\mathrm{eff}}(M,-)$ preserves $\mathbf{DM}^{\mathrm{eff}}_{\leqslant n}(k;\Lambda)$, i.e., if N belongs to $\mathbf{DM}^{\mathrm{eff}}_{\leqslant n}(k;\Lambda)$, then so is $\underline{\mathsf{Hom}}^{\mathrm{eff}}(M,N)$.*

Proof. One reduces to the case where $M=\mathrm{M}(X)$ with X smooth and projective of pure dimension d. Recall that $M^\vee=\mathrm{M}(X)(-d)[-2d]$ is the strong dual of M. We thus have the following isomorphisms

$$\underline{\mathsf{Hom}}(M,N)\simeq M^\vee\otimes N\simeq\underline{\mathsf{Hom}}(\Lambda(d),M\otimes N)[-2d].$$

Applying $\nu^{\geqslant 0}$, we deduce that

$$\underline{\mathsf{Hom}}^{\mathrm{eff}}(M,N)\simeq\underline{\mathsf{Hom}}^{\mathrm{eff}}(\Lambda(d),M\otimes N)[-2d].$$

Now, as $M\in\mathbf{DM}^{\mathrm{eff}}_{\leqslant d}(k;\Lambda)$ and $N\in\mathbf{DM}^{\mathrm{eff}}_{\leqslant n}(k;\Lambda)$, we see that

$$M\otimes N\in\mathbf{DM}^{\mathrm{eff}}_{\leqslant n+d}(k;\Lambda).$$

Applying Conjecture 4.22, the result follows. □

Our second conjecture concerns the homotopy t-structure.

Conjecture 4.27 — *The homotopy t-structure on $\mathbf{DM}^{\mathrm{eff}}(k;\Lambda)$ restricts to a t-structure on $\mathbf{DM}^{\mathrm{eff}}_{\leqslant n}(k;\Lambda)$ for any $n\in\mathbb{N}$. Moreover, the heart of the homotopy t-structure on $\mathbf{DM}^{\mathrm{eff}}_{\leqslant n}(k;\Lambda)$ is the category $\mathbf{HI}_{\leqslant n}(k;\Lambda)$ of n-presented homotopy invariant sheaves with transfers.*

Remark 4.28 — As said before, an effective motive $M\in\mathbf{DM}^{\mathrm{eff}}(k;\Lambda)$ is a complex of sheaves with transfers on Sm/k. We denote by $\mathrm{h}_r(M)$ the homology sheaves of M; these are objects of $\mathbf{HI}(k;\Lambda)$. Conversely, any object of $\mathbf{HI}(k;\Lambda)$ is itself an effective motive in an obvious way. This is said, we can rephrase the first assertion in Conjecture 4.27 as the following implication:

$$\left[M\in\mathbf{DM}^{\mathrm{eff}}_{\leqslant n}(k;\Lambda)\right]\quad\Rightarrow\quad\left[\mathrm{h}_r(M)\in\mathbf{DM}^{\mathrm{eff}}_{\leqslant n}(k;\Lambda),\ \forall r\in\mathbb{Z}\right].$$

Also, the second assertion in Conjecture 4.27 gives the following equality

$$\mathbf{HI}_{\leqslant n}(k;\Lambda) = \mathbf{HI}(k;\Lambda) \cap \mathbf{DM}^{\mathrm{eff}}_{\leqslant n}(k;\Lambda).$$

Said differently, a homotopy invariant sheaf with transfers is n-presented if and only if it belongs to $\mathbf{DM}^{\mathrm{eff}}_{\leqslant n}(k;\Lambda)$.

Lemma 4.29 — *Conjecture 4.27 holds when $n \leqslant 1$.*

Proof. This is proven in [13]. □

Proposition 4.30 — *Assume Conjectures 4.22 and 4.27. Let X be a smooth k-variety. Then, for $n \in \mathbb{N}$, $\underline{\mathrm{CH}}^n(X;\mathbb{Q})$ is n-presented.*

Proof. Let $\overline{X}$ be a smooth compactification of X whose complement is a normal crossing divisor with irreducible components D_i, for $i \in I$. There exists an exact sequence of homotopy invariant sheaves with transfers

$$\bigoplus_{i\in I} \underline{\mathrm{CH}}^{n-1}(D_i;\mathbb{Q}) \longrightarrow \underline{\mathrm{CH}}^n(\overline{X};\mathbb{Q}) \longrightarrow \underline{\mathrm{CH}}^n(X;\mathbb{Q}) \longrightarrow 0.$$

Therefore, it is enough to prove the claim under the assumption that X is smooth and projective. Also, we may assume that X has pure dimension d. Of course, we only need to consider the case $n \leqslant d$.

Using duality, we have for $U \in \mathrm{Sm}/k$:

$$\begin{aligned}
\mathrm{CH}^n(X\times U;\mathbb{Q}) &\simeq \hom_{\mathbf{DM}(k;\,\mathbb{Q})}(\mathrm{M}(X)\otimes \mathrm{M}(U), \mathbb{Q}(n)[2n])\\
&\simeq \hom_{\mathbf{DM}(k;\,\mathbb{Q})}(\mathrm{M}(U), \mathrm{M}(X)\otimes \mathbb{Q}(n-d)[2n-2d])\\
&\simeq \hom_{\mathbf{DM}(k;\,\mathbb{Q})}(\mathrm{M}(U)[2d-2n], \underline{\mathsf{Hom}}(\mathbb{Q}(d-n), \mathrm{M}(X)))\\
&\simeq \hom_{\mathbf{DM}^{\mathrm{eff}}(k;\,\mathbb{Q})}(\mathrm{M}(U)[2d-2n], \underline{\mathsf{Hom}}^{\mathrm{eff}}(\mathbb{Q}(d-n), \mathrm{M}(X))).
\end{aligned}$$

This shows that $\underline{\mathrm{CH}}^n(X;\mathbb{Q})$ coincides with $\mathrm{h}_{2d-2n}\underline{\mathsf{Hom}}^{\mathrm{eff}}(\mathbb{Q}(d-n), \mathrm{M}(X))$. Using the combination of Conjecture 4.22 and Conjecture 4.27, we are done. □

Remark 4.31 — Thanks to Lemma 4.29, the conclusion of Proposition 4.30 is known for $n = 0, 1$. (It is also known for $n = \dim(X)$ for obvious reasons.) The case $n = 2$ is already extremely interesting but also, unfortunately, completely out of reach.

References

[1] John Frank Adams, *Stable homotopy and generalised homology*, University of Chicago Press, Chicago, Ill.-London, 1974, Chicago Lectures in Mathematics. MR 0402720 (53 #6534)

[2] Yves André, *Mumford-Tate groups of mixed Hodge structures and the theorem of the fixed part*, Compositio Math. **82** (1992), no. 1, 1–24. MR 1154159 (93b:14026)

[3] Yves André and Bruno Kahn, *Nilpotence, radicaux et structures monoïdales*, Rend. Sem. Mat. Univ. Padova **108** (2002), 107–291, With an appendix by Peter O'Sullivan. MR 1956434 (2003m:18009)

[4] Joseph Ayoub, *Les six opérations de Grothendieck et le formalisme des cycles évanescents dans le monde motivique. I*, Astérisque (2007), no. 314, x+466 pp. (2008). MR 2423375 (2009h:14032)

[5] ______, *Les six opérations de Grothendieck et le formalisme des cycles évanescents dans le monde motivique. II*, Astérisque (2007), no. 315, vi+364 pp. (2008). MR 2438151 (2009m:14007)

[6] ______, *Note sur les opérations de Grothendieck et la réalisation de Betti*, J. Inst. Math. Jussieu **9** (2010), no. 2, 225–263. MR 2602027 (2011c:14061)

[7] ______, *The n-motivic t-structures for $n = 0$, 1 and 2*, Adv. Math. **226** (2011), no. 1, 111–138. MR 2735752 (2012b:14039)

[8] ______, *La réalisation étale et les opérations de Grothendieck*, Ann. Sci. Éc. Norm. Supér. (4) **47** (2014), no. 1, 1–145. MR 3205601

[9] ______, *L'algèbre de Hopf et le groupe de Galois motiviques d'un corps de caractéristique nulle, I*, J. Reine Angew. Math. **693** (2014), 1–149. MR 3259031

[10] ______, *Periods and the conjectures of Grothendieck and Kontsevich-Zagier*, Eur. Math. Soc. Newsl. (2014), no. 91, 12–18. MR 3202399

[11] ______, *Motifs des variétés analytiques rigides*, Mem. Soc. Math. France **140-141** (2015), 1–386.

[12] ______, *Une version relative de la conjecture des périodes de Kontsevich-Zagier*, Ann. of Math. (2) **181** (2015), no. 3, 905–992. MR 3296818

[13] Joseph Ayoub and Luca Barbieri-Viale, *1-motivic sheaves and the Albanese functor*, J. Pure Appl. Algebra **213** (2009), no. 5, 809–839. MR 2494373 (2010a:14025)

[14] Luca Barbieri-Viale and Bruno Kahn, *On the derived category of 1-motives*, Astérisque (2016), no. 381, xi+254. MR 3545132

[15] Ingrid Bauer, *Bloch's conjecture for Inoue surfaces with $p_g = 0$, $K^2 = 7$*, Proc. Amer. Math. Soc. **142** (2014), no. 10, 3335–3342. MR 3238411

[16] Ingrid Bauer and Davide Frapporti, *Bloch's conjecture for generalized Burniat type surfaces with $p_g = 0$*, Rend. Circ. Mat. Palermo (2) **64** (2015), no. 1, 27–42. MR 3324371

[17] Spencer Bloch, *Algebraic cycles and higher K-theory*, Adv. in Math. **61** (1986), no. 3, 267–304. MR 852815 (88f:18010)

[18] Spencer Bloch, Arnold Kas, and David I. Lieberman, *Zero cycles on surfaces with $p_g = 0$*, Compositio Math. **33** (1976), no. 2, 135–145. MR 0435073 (55 #8035)

[19] Mikhail V. Bondarko, *Differential graded motives: weight complex, weight filtrations and spectral sequences for realizations; Voevodsky versus Hanamura*, J. Inst. Math. Jussieu **8** (2009), no. 1, 39–97. MR 2461902 (2010f:14017)

[20] ______, *Weight structures vs. t-structures; weight filtrations, spectral sequences, and complexes (for motives and in general)*, J. K-Theory **6** (2010), no. 3, 387–504. MR 2746283

[21] ______, *Weight structures and 'weights' on the hearts of t-structures*, Homology Homotopy Appl. **14** (2012), no. 1, 239–261. MR 2954675

[22] Aldo Brigaglia, Ciro Ciliberto, and Claudio Pedrini, *The Italian school of algebraic geometry and Abel's legacy*, The legacy of Niels Henrik Abel, Springer, Berlin, 2004, pp. 295–347. MR 2077577 (2005e:14001)

[23] Frédéric Déglise, *Modules homotopiques*, Doc. Math. **16** (2011), 411–455. MR 2823365 (2012h:14053)

[24] Pierre Deligne, *Théorie de Hodge. II*, Inst. Hautes Études Sci. Publ. Math. (1971), no. 40, 5–57. MR 0498551 (58 #16653a)

[25] ______, *La conjecture de Weil pour les surfaces $K3$*, Invent. Math. **15** (1972), 206–226. MR 0296076 (45 #5137)

[26] Hélène Esnault and Marc Levine, *Surjectivity of cycle maps*, Astérisque (1993), no. 218, 203–226, Journées de Géométrie Algébrique d'Orsay (Orsay, 1992). MR 1265315 (94m:14008)

[27] Henri Gillet and Christophe Soulé, *Descent, motives and K-theory*, J. Reine Angew. Math. **478** (1996), 127–176. MR 1409056 (98d:14012)

[28] Vladimir Guletskiĭ, *Finite-dimensional objects in distinguished triangles*, J. Number Theory **119** (2006), no. 1, 99–127. MR 2228952 (2007k:18012)

[29] Annette Huber, *Realization of Voevodsky's motives*, J. Algebraic Geom. **9** (2000), no. 4, 755–799. MR 1775312 (2002d:14029)

[30] ______, *Slice filtration on motives and the Hodge conjecture (with an appendix by J. Ayoub)*, Math. Nachr. **281** (2008), no. 12, 1764–1776. MR 2473327 (2009i:14020)

[31] Hiroshi Inose and M. Mizukami, *Rational equivalence of 0-cycles on some surfaces of general type with $p_g = 0$*, Math. Ann. **244** (1979), no. 3, 205–217. MR 553252 (82c:14009)

[32] Uwe Jannsen, *Motives, numerical equivalence, and semi-simplicity*, Invent. Math. **107** (1992), no. 3, 447–452. MR 1150598 (93g:14009)

[33] Bruno Kahn, Jacob P. Murre, and Claudio Pedrini, *On the transcendental part of the motive of a surface*, Algebraic cycles and motives. vol. 2, London Math. Soc. Lecture Note Ser., vol. 344, Cambridge Univ. Press, Cambridge, 2007, pp. 143–202. MR 2187153 (2009c:14010)

[34] Shun-Ichi Kimura, *Chow groups are finite dimensional, in some sense*, Math. Ann. **331** (2005), no. 1, 173–201. MR 2107443 (2005j:14007)

[35] Maxim Kontsevich and Don Zagier, *Periods*, Mathematics unlimited—2001 and beyond, Springer, Berlin, 2001, pp. 771–808. MR 1852188 (2002i:11002)

[36] Carlo Mazza, Vladimir Voevodsky, and Charles Weibel, *Lecture notes on motivic cohomology*, Clay Mathematics Monographs, vol. 2, American Mathematical Society, Providence, RI; Clay Mathematics Institute, Cambridge, MA, 2006. MR 2242284 (2007e:14035)

[37] John Milnor, *Introduction to algebraic K-theory*, Princeton University Press, Princeton, N.J.; University of Tokyo Press, Tokyo, 1971, Annals of Mathematics Studies, No. 72. MR 0349811 (50 #2304)

[38] David Mumford, *Rational equivalence of 0-cycles on surfaces*, J. Math. Kyoto Univ. **9** (1968), 195–204. MR 0249428 (40 #2673)

[39] ———, *An algebraic surface with K ample, $(K^2) = 9$, $p_g = q = 0$*, Amer. J. Math. **101** (1979), no. 1, 233–244. MR 527834 (80j:14032)

[40] Jacob P. Murre, Jan Nagel, and Chris A. M. Peters, *Lectures on the theory of pure motives*, University Lecture Series, vol. 61, American Mathematical Society, Providence, RI, 2013. MR 3052734

[41] Amnon Neeman, *The Grothendieck duality theorem via Bousfield's techniques and Brown representability*, J. Amer. Math. Soc. **9** (1996), no. 1, 205–236. MR 1308405 (96c:18006)

[42] Yu. P. Nesterenko and Andrei A. Suslin, *Homology of the general linear group over a local ring, and Milnor's K-theory*, Izv. Akad. Nauk SSSR Ser. Mat. **53** (1989), no. 1, 121–146. MR 992981 (90a:20092)

[43] Fabrice Orgogozo, *Isomotifs de dimension inférieure ou égale à un*, Manuscripta Math. **115** (2004), no. 3, 339–360. MR 2102056 (2005j:14022)

[44] Dorin Popescu, *General Néron desingularization*, Nagoya Math. J. **100** (1985), 97–126. MR 818160 (87f:13019)

[45] ———, *General Néron desingularization and approximation*, Nagoya Math. J. **104** (1986), 85–115. MR 868439 (88a:14007)

[46] A. A. Rojtman, *The torsion of the group of 0-cycles modulo rational equivalence*, Ann. of Math. (2) **111** (1980), no. 3, 553–569. MR 577137 (81g:14003)

[47] Markus Rost, *Chow groups with coefficients*, Doc. Math. **1** (1996), no. 16, 319–393 (electronic). MR 1418952 (98a:14006)

[48] Andrei Suslin and Seva Joukhovitski, *Norm varieties*, J. Pure Appl. Algebra **206** (2006), no. 1-2, 245–276. MR 2220090 (2008a:14015)

[49] Burt Totaro, *Milnor K-theory is the simplest part of algebraic K-theory*, K-Theory **6** (1992), no. 2, 177–189. MR 1187705 (94d:19009)

[50] Vladimir Voevodsky, *Motivic cohomology groups are isomorphic to higher Chow groups in any characteristic*, Int. Math. Res. Not. (2002), no. 7, 351–355. MR 1883180 (2003c:14021)

[51] ______, *Open problems in the motivic stable homotopy theory. I*, Motives, polylogarithms and Hodge theory, Part I (Irvine, CA, 1998), Int. Press Lect. Ser., vol. 3, Int. Press, Somerville, MA, 2002, pp. 3–34. MR 1977582 (2005e:14030)

[52] ______, *On motivic cohomology with* $\mathbf{Z}/l$*-coefficients*, Ann. of Math. (2) **174** (2011), no. 1, 401–438. MR 2811603 (2012j:14030)

[53] Vladimir Voevodsky, Andrei Suslin, and Eric M. Friedlander, *Cycles, transfers, and motivic homology theories*, Annals of Mathematics Studies, vol. 143, Princeton University Press, Princeton, NJ, 2000. MR 1764197 (2001d:14026)

[54] Claire Voisin, *Sur les zéro-cycles de certaines hypersurfaces munies d'un automorphisme*, Ann. Scuola Norm. Sup. Pisa Cl. Sci. (4) **19** (1992), no. 4, 473–492. MR 1205880 (93m:14005)

[55] ______, *Bloch's conjecture for Catanese and Barlow surfaces*, J. Differential Geom. **97** (2014), no. 1, 149–175. MR 3229054

[56] Jörg Wildeshaus, *On the interior motive of certain Shimura varieties: the case of Picard surfaces*, Manuscripta Math. **148** (2015), no. 3-4, 351–377. MR 3414481

Hodge Theory and L^2-analysis
ALM 39, pp. 127–149

Resolution of Singularities of Differential Forms and Hsiang-Pati Coordinates*

André Belotto da Silva, Edward Bierstone†

Abstract

This is an expository article on resolution of singularities of differential forms on a complex- or real-analytic variety X_0 (or an algebraic variety X_0 over a field of characteristic zero). Given X_0, we ask whether there is a global resolution of singularities $\sigma : X \to X_0$ such that the pulled-back cotangent sheaf of X_0 is generated by differential monomials in suitable coordinates at every point of X ("Hsiang-Pati coordinates"). Existence of Hsiang-Pati coordinates (or desingularization of the cotangent sheaf) is equivalent to monomialization of Fitting ideals generated by minors of a given order of the logarithmic Jacobian matrix of σ. We can prove resolution of singularities of the cotangent sheaf in dimension up to three. It was previously known for surfaces with isolated singularities (Hsiang-Pati 1985, Pardon-Stern 2001). Consequences include monomialization of the induced Fubini-Study metric on the smooth part of a complex projective variety X_0; there have been important applications of the latter to L_2-cohomology.

2000 Mathematics Subject Classification: Primary 14E05, 14E15, 32S45; Secondary 14C30, 14F43, 32C15, 32S20, 32S35

Keywords and Phrases: Resolution of singularities, monomialization, logarithmic differential forms, Fitting ideals, Hsiang-Pati coordinates.

1 Overview

This an expository article on resolution of singularities (or monomialization) of differential forms on a singular algebraic or analytic variety, and the construction

*In honour of Steve Zucker, with warmest wishes.

†Université Paul Sabatier, Institut de Mathématiques de Toulouse, 118 route de Narbonne, F-31062 Toulouse Cedex 9, France; University of Toronto, Department of Mathematics, 40 St. George Street, Toronto, ON, Canada M5S 2E4.

E-mail: A. Belotto, andre.belotto_da_silva@math.univ-toulouse.fr; E. Bierstone, bierston@math.toronto.edu

Research supported in part by NSERC grant OGP0009070.

of associated *Hsiang-Pati coordinates*. There are several directions of possible application; in particular, the study of the L_2-cohomology of the smooth part of a singular variety, and the problem of *monomialization* or *toroidalization* of morphisms in birational geometry. Let X_0 denote either an algebraic variety over a field of characteristic zero, or a complex- or real-analytic variety. We assume that X_0 is reduced; i.e., that its structure sheaf has no nilpotents. Let $\operatorname{Sing} X_0$ denote the singular subset of X_0. The problem of desingularization of differential forms is made precise in the following conjecture.

Conjecture 1.1 (Resolution of singularities of the cotangent sheaf)**.** There is a *resolution of singularities* of X_0 (i.e., a proper birational or bimeromorphic morphism $\sigma : X \to X_0$ such that X is smooth, σ is an isomorphism over $X_0 \backslash \operatorname{Sing} X_0$, and $\sigma^{-1}(\operatorname{Sing} X_0)$ is the support of a simple normal crossings divisor E on X), such that the pulled-back cotangent sheaf of X_0 is locally generated by *differential monomials*

$$d(\boldsymbol{u}^{\boldsymbol{\alpha}_i}),\ i = 1, \ldots, s, \text{ and } d(\boldsymbol{u}^{\boldsymbol{\beta}_j} v_j),\ j = 1, \ldots, n-s, \tag{1.1}$$

where $n = \dim X_0$, $(\boldsymbol{u}, \boldsymbol{v}) = (u_1, \ldots, u_s, v_1, \ldots, v_{n-s})$ are local (analytic or étale) coordinates on X, and

1. $\operatorname{supp} E = (u_1 \cdots u_s = 0)$,
2. the multiindices $\boldsymbol{\alpha}_1, \ldots, \boldsymbol{\alpha}_s \in \mathbb{N}^s$ are linearly independent over $\mathbb{Q}$,
3. $\{\boldsymbol{\alpha}_i, \boldsymbol{\beta}_j\}$ is totally ordered (with respect to the componentwise partial ordering of $\mathbb{N}^s$).

An important consequence, for example in the case that X_0 is a complex affine (respectively, projective) variety, is that the pull-back to X of the induced Hermitian (respectively, *Fubini-Study*) metric on $X_0 \backslash \operatorname{Sing} X_0$ is locally quasi-isometric to

$$\sum_{i=1}^{s} d(\boldsymbol{u}^{\boldsymbol{\alpha}_i}) \otimes \overline{d(\boldsymbol{u}^{\boldsymbol{\alpha}_i})} + \sum_{j=1}^{n-s} d(\boldsymbol{u}^{\boldsymbol{\beta}_j} v_j) \otimes \overline{d(\boldsymbol{u}^{\boldsymbol{\beta}_j} v_j)}.$$

(Two metrics on a manifold are said to be *quasi-isometric* if the identity mapping is a quasi-isometry with respect to them. We recall that the Fubini-Study metric on complex projective space $P^n(\mathbb{C})$ is quasi-isometric to the standard Hermitian metric on a neighbourhood of the origin in an affine chart.)

The problem of desingularization of the cotangent sheaf (or the "Hsiang-Pati problem") can be reformulated in terms of principalization of logarithmic Fitting ideal sheaves (an approach suggested already by [17]); see §2.2 below. Given a resolution of singularities $\sigma : X \to X_0$, the *logarithmic Fitting ideal* $\mathcal{F}_k(\sigma)$ denotes the sheaf of ideals of $\mathcal{O}_X$ generated locally by the minors of order $n-k$ of the Jacobian matrix of σ with respect to a logarithmic basis of 1-forms on X (§2.1).

Theorem 1.2. *Let $\sigma : (X, E) \to (X_0, \operatorname{Sing} X_0)$ be a resolution of singularities of X_0. Then the following conditions are equivalent.*

1. *The logarithmic Fitting ideals $\mathcal{F}_k(\sigma)$, $k = 0, \ldots, n-1$, are all principal monomial ideals (generated locally by monomials in components of the exceptional divisor E).*

2. *The morphism σ satisfies Conjecture 1.1.*

Conjecture 1.1 can be strengthened by asking that σ be a composite of blowings-up with smooth admissible centres (see Definitions 1.8). We will give the main ideas of a recent proof of the following result [1].

Theorem 1.3. *Conjecture 1.1 (in the preceding stronger form) holds for varieties of dimension ≤ 3.*

The result was previously proved (at least locally) in the case of surfaces (2-dimensional varieties) with isolated singularities by W.-C. Hsiang and V. Pati [16, 1985], and a more conceptual proof in this case was given by W. Pardon and M. Stern [17, 2001]. Our formulation of Conjecture 1.1 is due to B. Youssin [21, 1998]. A system of coordinates as in Conjecture 1.1 will be called *Hsiang-Pati coordinates*.

One of the main interests of the Hsiang-Pati problem has been for applications to the L_2-cohomology of the smooth part of a singular variety, going back to the original ideas of Cheeger [6, 7]. The L_2-cohomology groups depend on a metric, but only up to quasi-isometry. Hsiang and Pati used their result to prove that the intersection cohomology (with the middle perversity) of a complex surface X_0 equals the L_2-cohomology of $X_0 \backslash \operatorname{Sing} X_0$ (Cheeger-Goresky-MacPherson conjecture [8]). Several controversial articles on both the Hsiang-Pati problem and the L_2-cohomology of singular varieties have perhaps discouraged work on these questions (cf. Example 1.5 below); we hope that our work will lead to a renewal of interest.

Melrose has suggested that Hsiang-Pati coordinates might be useful for an extension of his *b-calculus* to singular varieties in general. The Hsiang-Pati problem is also closely related to the problem of *monomialization* or *toroidalization* of a morphism, discussed in §1.3 below.

1.1 Overview of the article

The log Fitting ideals cannot be principalized by a simple application of resolution of singularities because $\mathcal{F}_k(\sigma)$ does not, in general, commute with pull-back (even up to multiplication by a principal monomial ideal); see Example 1.4 below.

In general, we show nevertheless that standard desingularization techniques can be used to principalize the Fitting ideal $\mathcal{F}_0(\sigma)$ of highest order minors, as well as the Fitting ideals of lower order minors. More precisely, we can reduce to the case that, if the logarithmic Jacobian matrix has rank r at $a \in X$, then $\mathcal{F}_0(\sigma)$ as well as $\mathcal{F}_{n-1}(\sigma)_a, \ldots, \mathcal{F}_{n-r-1}(\sigma)_a$ are principal, and the first $r+1$ components of σ at a (with respect to suitable local coordinates) are given by Hsiang-Pati monomials

$$\sigma_1 = v_1, \quad \ldots, \quad \sigma_r = v_r, \quad \sigma_{r+1} = \boldsymbol{u}^{\boldsymbol{\alpha}_1},$$

(where $(\boldsymbol{u}, \boldsymbol{v})$ are coordinates at a in which $\operatorname{supp} E = (u_1 \cdots u_s = 0)$; see §2.4).

An immediate consequence is that Conjecture 1.1 holds in the case $\dim X_0 \leq 2$. Moreover, to prove Theorem 1.3 (when $\dim X_0 = 3$), it remains only to principalize $\mathcal{F}_1(\sigma)$ at points of log rank 0; the image of such points in X_0 forms a discrete subset.

Principalization of $\mathcal{F}_1(\sigma)$ is technically the most difficult part of the proof. We argue by induction on the maximal value of an upper-semicontinuous local invariant ρ of $\mathcal{F}_1(\sigma)$. The invariant ρ has possible values $0, 1, \ldots, \infty$, and $\rho(a) = 0$ if and only if $\mathcal{F}_1(\sigma)_a$ is a principal monomial ideal (§3.1). Our proof of principalization of $\mathcal{F}_1(\sigma)$ has three main steps (cf. §3.2):

Step 1. Reduction to the case that $\rho(a) < \infty$, for all a. In this case, at a point a of log rank 0, we can write the components of σ (with respect to suitable local coordinates as in Conjecture 1.1) as $\sigma_1 = \boldsymbol{u}^{\boldsymbol{\alpha}}$ and $\sigma_i = g_i(\boldsymbol{u}) + \boldsymbol{u}^{\boldsymbol{\delta}} T_i$, $i > 1$, where $\boldsymbol{u}^{\boldsymbol{\alpha}}$ divides each σ_i, every dg_i is in the submodule generated by $d(\boldsymbol{u}^{\boldsymbol{\alpha}})$, and (in the case that a is a 1- or 2-point) T_2 can be written essentially in Weierstrass polynomial form with respect to a distinguished variable v (Lemma 3.3; we say that a is an *s-point* when $\operatorname{supp} E = (u_1 \cdots u_s = 0)$ at a).

Step 2. Reduction to prepared normal form (Lemma 3.5). By further blowings-up, we reduce to the case that the coefficients of the T_i as expansions in v are monomials (times units) in local coordinates as above, and the zeroth coefficients (i.e., the coefficients of v^0) are essentially in Hsiang-Pati form (as components of a morphism in dimension two).

Step 3. Decrease of ρ, by further blowings-up (Lemma 3.11).

1.2 Examples

The following examples illustrate some of the challenges in the Hsiang-Pati problem.

Example 1.4. Let $\sigma = (\sigma_1, \sigma_2, \sigma_3)$ be the morphism given by

$$\sigma_1 = u, \quad \sigma_2 = u(v + w^2), \quad \sigma_3 = u^2 w,$$

where $E = (u = 0)$. Then the logarithmic Fitting ideals $\mathcal{F}_0(\sigma) = (u^4)$, $\mathcal{F}_1(\sigma) = (u^2)$, $\mathcal{F}_2(\sigma) = (u)$ are all principal monomial ideals. Consider the blowing-up $\tau : \widetilde{X} \to X$ with centre given by the smooth curve $(u = v = 0)$; clearly, τ is admissible. We claim that $\tilde{\sigma} := \sigma \circ \tau : \widetilde{X} \to X_0$ does not satisfy the Hsiang-Pati conditions over the pre-image of the origin. Indeed, consider the "u-chart" of the blowing-up τ, with coordinates (x, y, z) in which τ is defined by $(u, v, w) = (x, xy, z)$. In this chart, $\tilde{\sigma}$ is given by

$$\tilde{\sigma}_1 = x, \quad \tilde{\sigma}_2 = x(xy + z^2), \quad \tilde{\sigma}_3 = x^2 z,$$

and $\tilde{\sigma}$ does not satisfy the Hsiang-Pati properties at the origin. The Log-Fitting ideals are $\mathcal{F}_0(\tilde{\sigma}) = (x^5)$, $\mathcal{F}_1(\tilde{\sigma}) = (x^3, x^2 z)$, $\mathcal{F}_2(\tilde{\sigma}) = (x)$; in particular, $\mathcal{F}_1(\tilde{\sigma})$ is not principal.

Example 1.5. Let $\sigma = (\sigma_1, \sigma_2, \sigma_3)$ be the morphism given by

$$\sigma_1 = u^2, \quad \sigma_2 = u^3(v^2 + uw), \quad \sigma_3 = u^4 v,$$

where $E = (u = 0)$. Then the log Fitting ideals $\mathcal{F}_0(\sigma)$, $\mathcal{F}_2(\sigma)$ are principal, and $\mathcal{F}_1(\sigma) = u^5 \cdot (u, v)$ at the 1-point $a = 0$. This is essentially the simplest example of a morphism written in prepared normal form (Lemma 3.5), that does not yet satisfy the conditions of Conjecture 1.1. In this example, σ can be reduced to Hsiang-Pati form (in fact, σ can be monomialized; cf. Section 1.3 below) by two blowings-up.

Let τ denote the blowing-up with centre $C = (u = v = 0)$ (C is the locus of points where the invariant $\rho = 1$). Then τ can be covered by two coordinate charts. Let $b \in \tau^{-1}(a)$. There are two possibilities:

1. b belongs to the u-chart, with coordinates $(x, \tilde{v}, \tilde{w})$ in which τ is given by $(u, v, w) = (x, x\tilde{v}, \tilde{w})$. Then

$$\sigma_1 \circ \tau = x^2, \quad \sigma_2 \circ \tau = x^4(x\tilde{v}^2 + \tilde{w}), \quad \sigma_3 \circ \tau = x^5 \tilde{v},$$

   and $\sigma \circ \tau$ is in Hsiang-Pati form in this chart; in fact, $\sigma_2 \circ \tau = x^4 w'$ after a coordinate change $w' = \tilde{w} + x\tilde{v}^2$, at any $b \in \tau^{-1}(a)$ in the chart.

2. b is the origin of the v-chart, with coordinates $(x, y, \tilde{w})$ in which τ is given by $(u, v, w) = (xy, y, \tilde{w})$. In this chart, $\sigma_2 \circ \tau = x^3 y^4 (y + x\tilde{w})$; b is a 2-point (with $\rho(b) = \infty$) and one more blowing-up, with centre the 2-curve $(x = y = 0)$, is needed to reduce to Hsiang-Pati form. Note that this centre is globally defined in the source of τ.

We remark that Pati [18] and Taalman [19] claim that, in dimension three, $\mathcal{F}_1(\sigma)$ is necessarily principal at a 1-point, because "since u does not divide R [where $R = T_2 = v^2 + uw$ in the example above], the 2-form $du \wedge dR$ is nowhere-vanishing [on $(u = 0)$], and thus R is a coordinate independent of u" [19, p. 258] (cf. [18, p. 443]). Example 1.5 shows this is not true.

Example 1.6. Let $\sigma = (\sigma_1, \sigma_2, \sigma_3, \sigma_4, \sigma_5)$ be the morphism given by

$$\sigma_1 = u^2, \quad \sigma_2 = u^3(v^3 + (y^2 + ux^2)uv + u^3 w), \quad \sigma_3 = u^4 v, \quad \sigma_4 = u^4 y, \quad \sigma_5 = u^4 x,$$

where $E = (u = 0)$. Then $\mathcal{F}_0(\sigma)$ and $\mathcal{F}_4(\sigma)$ are principal at 0. It might appear reasonable to blow up $(x = y = 0)$ to principalize $\mathcal{F}_3(\sigma)$, but this centre is not in $\operatorname{supp} E$.

1.3 Resolution of singularities, monomialization of differential forms and toroidalization of a morphism

Given X_0, we can ask whether there exists a resolution of singularities as in Conjecture 1.1 such that the components of σ (with respect to suitable local coordinates of a smooth local embedding variety of X_0) are themselves monomials (rather than only their differentials). This is not true, in general, because it would imply that X_0 is given locally by binomial equations; i.e, that X_0 locally has a toric structure.

It is therefore reasonable to ask whether a morphism $\sigma : X \to Y$ can be monomialized by blowings-up in both the source *and the target* — this is the problem of monomialization, or of *toroidalization* if we require, in addition, that the transformed morphism locally has components that are monomials with respect to the exceptional divisors on the source and target (see Definition 1.9).

Example 1.7. Let $X_0 \subset \mathbb{C}^3$ denote the surface

$$x_3^3 - x_1^2 x_2 x_3 - x_1^4 = 0$$

(cf. [3, Rmk. 7.3]). Then $\operatorname{Sing} X_0 = (x_3 = x_1 = 0)$. Let σ_1 denote the blowing-up of $\mathbb{C}^3$ with centre $\operatorname{Sing} X_0$. Then σ_1 is given by $x_1 = y_1$, $x_2 = y_2$, $x_3 = y_1 y_3$, in the x_1-coordinate chart, in which $\sigma_1^{-1}(X_0)$ is defined by $y_1^3(y_3^3 - y_2 y_3 - y_1) = 0$. The variety X_0 is the image of the smooth surface $X_1 : y_1 = y_3^3 - y_2 y_3$ (the *strict transform* of X_0), and the *exceptional divisor* is $(y_1 = 0)$. The latter has simple normal crossings with respect to X_1 except at the origin. The smooth surface X_1 has a system of coordinates given by (y_2, y_3).

Two more blowings-up σ_2, σ_3 are needed (provided that we blow up no simple normal crossings points) to make the transform of X_1 normal crossings with the exceptional divisor. The composite of the three blowings-up provides a resolution of singularities $\tau : X_3 \to X_0$ given by $x_1 = z_1 z_2^2$, $x_2 = z_2(z_2 - z_1)$, $x_3 = z_1 z_2^3$ in a coordinate chart, where the exceptional divisor E is $(z_1 z_2 = 0)$. The morphism τ is not in the Hsiang-Pati form of Conjecture 1.1.

After one more blowing-up σ_4 of the origin, we have $x_1 = w_1 w_2^3$, $x_2 = w_2^2(1 - w_1)$, $x_3 = w_1 w_2^4$ in the coordinate chart where σ_4 is given by $(z_1, z_2) = (w_1 w_2, w_2)$. Now, after a coordinate change $u_1 = w_1(1 - w_1)^{-3/2}$, $u_2 = w_2(1 - w_1)^{1/2}$, we have a morphism $\sigma : X \to \mathbb{C}^3$ given by $x_1 = u_1 u_2^3$, $x_2 = u_2^2$, $x_3 = u_1 u_2^4 \times \text{unit}$. Then dx_3 is in the submodule of differential forms on X generated by dx_1, dx_2, so that σ is in Hsiang-Pati form (after permuting x_1, x_2).

Of course, σ factors through the morphism $\tau : X_3 \to X_0 \subset \mathbb{C}^3$,

$$\begin{array}{ccc} X & = & X \\ {\scriptstyle\sigma_4}\downarrow & & \downarrow{\scriptstyle\sigma} \\ X_3 & \xrightarrow{\tau} & X_0 \end{array}$$

and $\sigma_4 : X \to X_3$ is trivially toroidal.

Toroidalization is a natural version of resolution of singularities for a morphism; before a precise formulation of the problem, we give several notions concerning resolution of singularities that will be used throughout the article.

Definitions 1.8 (*Admissible blowings-up* and *resolution of singularities*). A *resolution of singularities* of X_0 is a proper birational or bimeromorphic morphism $\sigma : X \to X_0$ such that X is smooth, σ is an isomorphism over $X_0 \backslash \operatorname{Sing} X_0$, and $\sigma^{-1}(\operatorname{Sing} X_0)$ is the support of a simple normal crossings divisor E on X (the *exceptional divisor*). See [4, 5]. We write $\sigma : (X, E) \to (X_0, \operatorname{Sing} X_0)$.

Given a smooth variety X_0 with a simple normal crossings divisor E_0, a blowing-up $\sigma : X \to X_0$ is called *admissible* (for E_0) if the centre of σ is smooth and has only normal crossings with E_0. We write $\sigma : (X, E) \to (X_0, E_0)$, where E denotes the birational transform of E_0 (given by the strict transform of each component of E_0) plus the exceptional divisor of σ (the inverse image of the centre of blowing up). Likewise for the composite σ of a sequence of recursively defined admissible blowings-up. (We regard E_0 as an initial "exceptional divisor" and therefore E as the *exceptional divisor* of σ.) An admissible blowing-up is called *combinatorial* if its centre is an intersection of components of E.

A singular variety X_0 locally admits an embedding $X_0|_U \hookrightarrow M_0$ in a smooth variety M_0 (U denotes an open subset of X_0). A divisor E_0 on X_0 has only *normal crossings* if E_0 is the restriction of ambient normal crossings divisors, for a suitable covering of X_0 by embeddings in smooth varieties. The notions of simple normal crossings divisor, admissible and combinatorial blowing-up all make sense in the same way.

If X_0 is an algebraic or compact analytic variety, then a resolution of singularities $(X, E) \to (X_0, \operatorname{Sing} X_0)$ can be obtained as a composite of a finite sequence of smooth admissible blowings-up (admissible, as above, relative to a local embedding in a smooth variety). In the case of a general analytic variety X_0, resolution of singularities can be obtained by a morphism which can be realized as a finite sequence of smooth admissible blowings-up over any relatively compact open subset of X_0. (We will still say, somewhat loosely, that σ is a "composite of blowings-up".)

Definition 1.9. Let $\varphi_0 : X_0 \to Y_0$ denote a proper birational or bimeromorphic morphism. A *toroidalization* of φ_0 is a commutative diagram

$$\begin{array}{ccc} X & \xrightarrow{\sigma} & X_0 \\ {\scriptstyle\varphi}\downarrow & & \downarrow{\scriptstyle\varphi_0} \\ Y & \xrightarrow{\tau} & Y_0 \end{array}$$

where σ, τ are sequences of global admissible blowings-up with the properties that, if E, D denote the exceptional divisors of σ, τ (respectively), then $\varphi^{-1}(D) = E$ (i.e., $\varphi^{-1}(\operatorname{supp} D) = \operatorname{supp} E$), φ is an isomorphism over $Y \setminus \operatorname{supp} D$, and φ can be expressed locally as

$$x_i = \boldsymbol{u}^{\boldsymbol{\alpha}_i}, \quad y_j = \boldsymbol{u}^{\boldsymbol{\beta}_j}(c_j + v_j), \quad z_k = w_k, \tag{1.2}$$

with respect to coordinates $(\boldsymbol{u}, \boldsymbol{v}, \boldsymbol{w}) = (u_1, \dots, u_p, v_1, \dots, v_q, w_1, \dots, w_r)$ and $(\boldsymbol{x}, \boldsymbol{y}, \boldsymbol{z}) = (x_1, \dots, x_p, y_1, \dots, y_q, z_1, \dots, z_r)$ in X and Y (respectively), where $\boldsymbol{u}$ and $(\boldsymbol{x}, \boldsymbol{y})$ represent the exceptional divisors, the exponents $\boldsymbol{\alpha}_i$ are $\mathbb{Q}$-linearly independent, and the c_j are nonzero constants.

The formulas (1.2) are simply what we would obtain from a monomial mapping $x_i = \boldsymbol{u}^{\boldsymbol{\alpha}_i}$ (with the $\boldsymbol{\alpha}_i$ linearly independent) with respect to coordinates $\boldsymbol{u} = (u_1, \dots, u_n)$ and $\boldsymbol{x} = (x_1, \dots, x_n)$ where the u_j and x_i all represent exceptional divisors, by translation of coordinates to a general nearby point.

In Definition 1.9, we can also assume that X_0, Y_0 are equipped with respective simple normal crossings divisors E_0, D_0, and that $E_0 = \varphi_0^{-1}(D_0)$. Then E and D mean $\sigma^{-1}(E_0)$ and $\tau^{-1}(D_0)$, respectively. The problem of toroidalization in the sense of Definition 1.9 reduces to the latter formulation with X_0, Y_0 both smooth, by resolution of singularities.

The problem of monomialization of differential forms is a natural intermediate step towards toroidalization (although Example 1.7 is too simple to illustrate this assertion) — this is indicated by our approach to Conjecture 1.1 or Theorem 1.3. By blowings-up in the source, we aim to express the successive components of σ as the Hsiang-Pati monomials $(\boldsymbol{u}^{\boldsymbol{\alpha}_i}, \boldsymbol{u}^{\boldsymbol{\beta}_j} v_j)$, in the ordering of Conjecture 1.1(3), plus additional terms whose differentials are in the submodule generated by (1.1); see §2.2 and Lemma 3.3. The passage from such a statement to monomialization of σ by blowings-up in the source and target is an interesting problem that we plan to treat in a future article.

The problem of monomialization or toroidalization of morphisms has an extensive literature (see, for example, [2, 10] and references therein), though the only general results either are of a local nature or involve generically finite rather than birational or bimeromorphic modifications. Cutkosky has proved global monomialization for algebraic morphisms in dimension up to three [10, 11]. Our proof of Theorem 1.3 is strongly influenced by [11]; in particular, the invariant ρ of Section 3.1 below is introduced in the latter. The normal forms of [11, Section 3] cannot be obtained, however, as claimed in the proof of [11, Lemma 3.6] (see [12]); we use different normal forms in our Lemma 3.5. The proofs of our results do not depend on [10, 11, 12].

Cutkosky's arguments do not extend to analytic morphisms in an evident way because they involve local blowings-up that are globalized by algebraic techniques (e.g., Bertini's theorem) that are not available in an analytic category. One of the main differences of our approach from that of [10, 11] is that we use as choices of centre of blowing up only subspaces that *a priori* have global meaning.

2 Logarithmic Fitting ideals

2.1 Logarithmic differential forms

Let X denote a smooth variety and let Ω^1_X denote the $\mathcal{O}_X$-module (i.e., the sheaf of $\mathcal{O}_X$-modules) of differential 1-forms on X; Ω^1_X is locally free of rank $n = \dim X$.

Suppose that E is a simple normal crossings divisor on X. Then we let $\Omega^1_X(\log E)$ denote the $\mathcal{O}_X$-module of *logarithmic 1-forms* on X. This can be defined as follows. If U is an étale or analytic local coordinate chart of X at a point a, with coordinates $(\boldsymbol{u}, \boldsymbol{v}) = (u_1, \dots, u_s, v_1, \dots, v_{n-s})$ such that $a = 0$ and $(u_i = 0)$, $i = 1, \dots, s$, are the components of E in U (we say the coordinates $(\boldsymbol{u}, \boldsymbol{v})$ are *adapted* to E), then $\Omega^1_X(\log E)$ is locally free at a with basis given by

$$\frac{du_i}{u_i}, \quad i = 1, \dots, s, \quad \text{and} \quad dv_j, \quad j = 1, \dots, n-s. \tag{2.1}$$

There is a natural inclusion $\Omega^1_X \hookrightarrow \Omega^1_X(\log E)$ (given by writing any 1-form in terms of a "logarithmic basis" (2.1)).

Given a singular variety X_0, we also write $\Omega^1_{X_0}$ for the *cotangent sheaf* of X_0. ($\Omega^1_{X_0}$ has stalk $\underline{m}_{X_0,a}/\underline{m}^2_{X_0,a}$ at a, where $\underline{m}_{X_0,a}$ denotes the maximal ideal of $\mathcal{O}_{X_0,a}$.) Suppose that $\sigma : (X, E) \to (X_0, \operatorname{Sing} X_0)$ is a resolution of singularities (in particular, $\operatorname{supp} E = \sigma^{-1}(\operatorname{Sing} X_0)$). Let $\sigma^*\Omega^1_{X_0}$ denote the submodule of Ω^1_X generated by the pull-back of $\Omega^1_{X_0}$.

Consider a local embedding $X_0|_V \hookrightarrow M_0$ over a neighbourhood V of $b = \sigma(a)$, where M_0 is smooth. Then $\Omega^1_{X_0}|_V$ is induced by the restriction to V of $\Omega^1_{M_0}$, and $\sigma^*(\Omega^1_{X_0}|_V) = \sigma^*\Omega^1_{M_0}$. If $\dim M_0 = N$ and σ is expressed in components $\sigma = (\sigma_1, \ldots, \sigma_N)$ with respect to local coordinates of M_0 at b, then we define the *logarithmic Jacobian matrix* of σ (with respect to the given coordinates at a and b) by

$$\log \operatorname{Jac} \sigma := \begin{pmatrix} u_1 \dfrac{\partial \sigma_1}{\partial u_1} & \cdots & u_s \dfrac{\partial \sigma_1}{\partial u_s} & \dfrac{\partial \sigma_1}{\partial v_1} & \cdots & \dfrac{\partial \sigma_1}{\partial v_{n-s}} \\ \vdots & \ddots & \vdots & \vdots & \ddots & \vdots \\ u_1 \dfrac{\partial \sigma_N}{\partial u_1} & \cdots & u_s \dfrac{\partial \sigma_N}{\partial u_s} & \dfrac{\partial \sigma_N}{\partial v_1} & \cdots & \dfrac{\partial \sigma_N}{\partial v_{n-s}} \end{pmatrix}. \tag{2.2}$$

Note that every minor of order n of $\log \operatorname{Jac} \sigma$ equals $u_1 \cdots u_s$ times the corresponding minor of order n of the standard Jacobian matrix $\operatorname{Jac} \sigma$.

The rank at a of $\log \operatorname{Jac} \sigma$ will be called the *logarithmic rank* $\log \operatorname{rk}_a \sigma$ of σ at a. It is clear from (2.2) that, if $a \in \operatorname{supp} E$, then $\log \operatorname{rk}_a \sigma = \operatorname{rk}_a(\sigma|_{E(a)})$, where $E(a)$ denotes the intersection of the components of E containing a. (If $a \notin \operatorname{supp} E$, then $\log \operatorname{rk}_a \sigma := \operatorname{rk}_a \sigma$.)

2.2 Logarithmic Fitting ideals and Theorem 1.2

For each $k = 0, \ldots, n-1$, the *logarithmic Fitting ideal* $\mathcal{F}_k = \mathcal{F}_k(\sigma)$ denotes the sheaf of ideals of $\mathcal{O}_X$ generated by the minors of order $n-k$ of $\log \operatorname{Jac} \sigma$. The Fitting ideals $\mathcal{F}_k$ are independent of the choices of adapted local coordinates of (X, E), the local embedding $X_0|_V \hookrightarrow M_0$ and the local coordinates of M_0; in fact (the transpose of) $\log \operatorname{Jac} \sigma$ is a presentation of the quotient $\mathcal{O}_X$-module $\Omega^1_X(\log E)/\sigma^*\Omega^1_{X_0}$ at a, and the $\mathcal{F}_k$ depend only on this module (cf. [13, §20.2]).

An ideal sheaf $\mathcal{F}$ will be called a *principal E-monomial ideal* (or a *principal monomial ideal* if E is clear from the context) if $\mathcal{F}$ is a principal ideal sheaf generated locally by a monomial in generators of the components of E.

The following lemma is a more precise local statement of Theorem 1.2 that is useful in our proof of Theorem 1.3; the lemma immediately implies Theorem 1.2.

Lemma 2.1. *Let $\sigma : X \to M_0$ be a morphism between smooth varieties. Say $n = \dim X$, $N = \dim M_0$. Suppose that the* critical locus *of σ (i.e., $\{a \in X : \operatorname{rk}_a \sigma < n\}$) is the support of a simple normal crossings divisor E on X. Let $a \in X$. Then, for each $k = 1, \ldots, n$, the following are equivalent:*

1. *The logarithmic Fitting ideals $\mathcal{F}_{n-1}(\sigma), \ldots, \mathcal{F}_{n-k}(\sigma)$ are all principal E-monomial ideals at a.*

2. *There are (analytic or étale) coordinates $(\boldsymbol{u},\boldsymbol{v})$ of X at a adapted to E, and coordinates $\boldsymbol{z} = (z_1,\dots,z_N)$ of M_0 at $\sigma(a)$, such that, writing $\sigma = (\sigma_1,\dots,\sigma_N)$ with respect to the coordinates $\boldsymbol{z}$,*

   (a) *the submodule $\mathcal{M}_k$ of $\Omega^1_{X,a}$ generated by the pull-backs $\sigma^* dz_m = d\sigma_m$, $m = 1,\dots,k$, is also generated by differential monomials $d(\boldsymbol{u}^{\boldsymbol{\alpha}_i})$, $i = 1,\dots,l_k$, and $d(\boldsymbol{u}^{\boldsymbol{\beta}_j} v_j)$, $j = 1,\dots,k-l_k$, for some $l_k \le k$, satisfying conditions as in Conjecture 1.1;*

   (b) *for each $m > k$, $\sigma_m = g_m + S_m$, where $dg_m \in \mathcal{M}_k$ and S_m is divisible by $\boldsymbol{u}^{\max\{\boldsymbol{\alpha}_{l_k},\boldsymbol{\beta}_{k-l_k}\}}$.*

Given (2) in Lemma 2.1, it is easy to see that, for each $m = 1,\dots,k$, the log Fitting ideal $\mathcal{F}_{n-m}$ is generated by $\boldsymbol{u}^{\boldsymbol{\gamma}_m}$, where $\boldsymbol{\gamma}_m$ is the sum of the first m elements of $\{\boldsymbol{\alpha}_i,\boldsymbol{\beta}_j\}$ (as an ordered set). The converse implication (1) $\Longrightarrow$ (2) can be proved by induction on k, by an explicit computation; details are given in [1, Section 3].

2.3 Transformation of logarithmic Fitting ideals by blowing up, and monomialization of the Fitting ideal of maximal minors

Let X be a smooth variety with simple normal crossings divisor E, and let $\beta : (X',E') \to (X,E)$ denote an admissible blowing-up with centre $C \subset \operatorname{supp} E$. With the respect to adapted local coordinates $(\boldsymbol{u},\boldsymbol{v})$ at a point $a \in C$ (see §2.1), we can assume that C is given by

$$u_1 = \cdots = u_k = 0, \text{ for some } k \ge 1, \text{ and } v_1 = \cdots = v_l = 0, \text{ for some } l \ge 0.$$

The blowing-up β is combinatorial (Definitions 1.8) precisely when $l = 0$.

Over the $(\boldsymbol{u},\boldsymbol{v})$ coordinate chart, X' can be covered by $k+l$ coordinate charts — a "u_i-chart", for each $i = 1,\dots,k$, and a "v_j-chart", for each $j = 1,\dots,l$. For example, the u_1-chart has coordinates $(\boldsymbol{u}',\boldsymbol{v}') = (u'_1,\dots,u'_s,v'_1,\dots,v'_{n-s})$ given by

$$u'_1 = u_1, \qquad u'_i = \begin{cases} u_i/u_1, & 2 \le i \le k, \\ u_i, & i > k, \end{cases} \qquad v'_j = \begin{cases} v_j/u_1, & j \le l, \\ v_j, & j > l, \end{cases}$$

and the v_1-chart has coordinates $(\boldsymbol{u}',\boldsymbol{v}') = (u'_1,\dots,u'_{s+1},v'_2,\dots,v'_{n-s})$ given by

$$u'_i = \begin{cases} u_i/v_1, & 1 \le i \le k, \\ u_i, & k < i \le s, \end{cases} \qquad u'_{s+1} = v_1, \qquad v'_j = \begin{cases} v_j/v_1, & 2 \le j \le l, \\ v_j, & j > l. \end{cases}$$

We compute

$$\log \operatorname{Jac}(\sigma \circ \beta) = ((\log \operatorname{Jac} \sigma) \circ \beta) \cdot B,$$

where B denotes the $n \times n$ matrix

$$\begin{pmatrix} A & 0 \\ 0 & I \end{pmatrix} \cdot \operatorname{Jac} \beta \cdot \begin{pmatrix} C & 0 \\ 0 & D \end{pmatrix},$$

with A (respectively, C) the $s \times s$ diagonal matrix with diagonal entries $1/u_i$ (respectively, u_i'), I the identity matrix of order $n-s$, and D a diagonal matrix of order $n-s$ with first entry 1 in the u_1-chart or u'_{s+1} in the v_1-chart, and remaining diagonal entries 1. We make the following simple but important observations.

Remarks 2.2. (1) $\det B = \text{exc}^l$, where exc denotes the exceptional divisor of σ; i.e., $\text{exc} = u_1'$ in the u_1-chart or u'_{s+1} in the v_1-chart. Therefore, $\mathcal{F}_0(\sigma \circ \beta) = \text{exc}^l \cdot \beta^* \mathcal{F}_0(\sigma)$.

(2) If β is a combinatorial blowing-up ($l = 0$), then B is invertible, so that every minor of $\log \text{Jac}\,(\sigma \circ \beta)$ is a linear combination of minors of the same order of $(\log \text{Jac}\,\sigma) \circ \beta$, and vice-versa. Therefore, $\mathcal{F}_k(\sigma \circ \beta) = \beta^* \mathcal{F}_k(\sigma)$, $k = 0, \ldots$.

Theorem 2.3. *Given a reduced variety X_0, there is a resolution of singularities $\sigma : (X, E) \to (X_0, \text{Sing}\, X_0)$ such that σ is a composite of admissible blowings-up and $\mathcal{F}_0(\sigma)$ is a principal monomial ideal.*

Moreover, if σ is a resolution of singularities such that $\mathcal{F}_0(\sigma)$ is a principal monomial ideal, and $\beta : (X', E') \to (X, E)$ is an admissible blowing-up, then $\mathcal{F}_0(\sigma \circ \beta)$ is a principal monomial ideal.

Proof. The second assertion is immediate from Remark 2.2(1). Suppose that σ is a resolution of singularities of the variety X_0 by admissible blowings-up. We can then apply resolution of singularities of an ideal to $\mathcal{F}_0(\sigma)$, and the first assertion again follows from Remark 2.2(1). $\square$

2.4 Logarithmic rank and monomialization of Fitting ideals of low-order minors

In this section, we assume that X_0 is a complex-analytic variety or an algebraic variety over an algebraically closed field of characteristic zero. The results also hold, however, for a real-analytic variety or an algebraic variety over an arbitrary field of characteristic zero (see Remark 2.6). Let $\sigma : (X, E) \to (X_0, Y_0 := \text{Sing}\, X_0)$ be a resolution of singularities of X_0. In particular, $\sigma^* \mathcal{I}_{Y_0}$ is a principal E-monomial ideal. Set

$$p := \max_E \log \text{rk}\, \sigma = \dim Y_0,$$

and, for each $k = 0, \ldots, p$, set

$$\Sigma_k := \{a \in E : \log \text{rk}_a \sigma \leq p - k\},$$
$$Y_k := \sigma(\Sigma_k).$$

Since σ is proper, each Y_k is a closed subvariety of X_0, and

$$\text{Sing}\, X_0 = Y_0 \supset Y_1 \supset \cdots \supset Y_p.$$

Define

$$\mathcal{I}_{Y_k} := \text{ideal of } Y_k \text{ in } \mathcal{O}_{X_0}, \quad k = 0, \ldots, p.$$

Lemma 2.4. *After further admissible blowings-up over Y_0 if necessary, we can assume that, for each k, $\Sigma_k = \sigma^{-1}(Y_k)$, $Y_k \backslash Y_{k+1}$ is smooth, and $\sigma^* \mathcal{I}_{Y_k}$ is a principal monomial ideal.*

Remarks 2.5. (1) It follows from the lemma that, for each k, $\log\operatorname{rk}\sigma = p-k$ on $\Sigma_k\backslash\Sigma_{k+1}$.

(2) The condition that $\sigma^*\mathcal{I}_{Y_k}$ be a principal monomial ideal is not stable under admissible blowing-up.

Proof of Lemma 2.4. $Y_0 = \operatorname{Sing} X_0$, so that $\Sigma_0 = \operatorname{supp} E = \sigma^{-1}(Y_0)$ and $\sigma^*\mathcal{I}_{Y_0}$ is a principal monomial ideal with support Σ_0.

Of course, $\Sigma_1 \subset \Sigma_0$. Set

$$Y_1' := \sigma(\Sigma_1) \cup \operatorname{Sing} Y_0.$$

Then $\dim Y_1' \leq p-1$. By resolution of singularities, after further admissible blowings-up with centres over Y_1' (i.e., in the inverse image of Y_1'), we can assume that $\sigma^*\mathcal{I}_{Y_1'}$ is a principal monomial ideal. Then $\Sigma_1 = \operatorname{supp}\sigma^*\mathcal{I}_{Y_1'} = \sigma^{-1}(Y_1')$, $Y_1 = Y_1'$ and $Y_0\backslash Y_1$ is smooth.

Again, $\Sigma_2 \subset \Sigma_1$. Let

$$Y_2' := \sigma(\Sigma_2) \cup \operatorname{Sing} Y_1.$$

Then $\dim Y_2' \leq p-2$. After further admissible blowings-up with centres over Y_2', we can assume that $\sigma^*\mathcal{I}_{Y_2'}$ is a principal monomial ideal. Then $\Sigma_2 = \operatorname{supp}\sigma^*\mathcal{I}_{Y_2'} = \sigma^{-1}(Y_2')$, $Y_2 = Y_2'$ and $Y_1\backslash Y_2$ is smooth. Moreover, it is still true that $\sigma^*\mathcal{I}_{Y_1}$ is a principal monomial ideal, $\Sigma_1 = \operatorname{supp}\sigma^*\mathcal{I}_{Y_1} = \sigma^{-1}(Y_1)$, and $Y_0\backslash Y_1$ is smooth.

We can continue in the same way to prove the lemma. □

Remark 2.6. If X_0 is a real-analytic variety, or an algebraic variety over a field that is not algebraically closed, then the $\sigma(\Sigma_k)$ need not be closed varieties. The proof of Lemma 2.4 goes through, however, provided that each $\sigma(\Sigma_k)$ lies in a closed variety of dimension $= d_k := \max\{\log\operatorname{rk}_a\sigma : a \in \Sigma_k\}$; in this case, we can simply replace each Y_k' in the proof by the smallest closed subvariety of Y_{k-1} containing $\sigma(\Sigma_k) \cup \operatorname{Sing} Y_{k-1}$. The preceding condition holds in the algebraic case, in general, by Chevalley's theorem [14, IV, 1.8.4]. It holds in the real-analytic case because X_0 has a complexification $X_0^{\mathbb{C}}$ [20], and σ is induced by a resolution of singularities $\sigma^{\mathbb{C}} : (X^{\mathbb{C}}, E^{\mathbb{C}}) \to (X_0^{\mathbb{C}}, \operatorname{Sing} X_0^{\mathbb{C}})$ of $X_0^{\mathbb{C}}$ (in particular, $\sigma^{\mathbb{C}}$ is proper). Lemma 2.4 applies to $\sigma^{\mathbb{C}}$. In the proof of Lemma 2.4 in the real case, we can take Y_k' to be the real part (i.e., the invariance space with respect to the canonical autoconjugation) of a union of components of $(Y_k^{\mathbb{C}})'$; cf. [15, Sect. 2].

Theorem 2.7. *Let $\sigma : (X,E) \to (X_0, Y_0 := \operatorname{Sing} X_0)$ denote a resolution of singularities of X_0 satisfying the conclusion of Lemma 2.4. (We use the notation at the beginning of the subsection.) Let $a \in E$, and let $r := \log\operatorname{rk}_a\sigma$. Then, for any local embedding $X_0 \hookrightarrow M_0$ (at a) in a smooth variety M_0, we can choose coordinates $(\boldsymbol{u},\boldsymbol{v}) = (u_1,\ldots,u_s,v_1,\ldots,v_{n-s})$ adapted to E for X at $a=0$, and coordinates $z = (z_1,\ldots,z_N)$ for M_0 at $\sigma(a)$, with respect to which, if $\sigma = (\sigma_1,\ldots,\sigma_N)$, then*

$$\sigma_1 = v_1,\ \ldots,\ \sigma_r = v_r,\ \sigma_{r+1} = \boldsymbol{u}^{\boldsymbol{\alpha}_1},$$

where $\boldsymbol{\alpha}_1 \in \mathbb{N}^s$, $\mathcal{I}_{Y_{p-r}}$ is generated by $z_{r+1},\ldots,z_N$ at $\sigma(a)$, and $\sigma^*\mathcal{I}_{Y_{p-r}}$ is generated by $\sigma_{r+1} = \sigma^*(z_{r+1})$ at a.

Proof. Let E_i, $i = 1, \ldots, s$, denote the components of E at a. We call $\cap E_i$ the *stratum* $E(a)$ of a (cf. §2.1). We can assume that

$$\operatorname{supp} \sigma^* \mathcal{I}_{Y_{p-r}} = \bigcup_{i=1}^{t} E_i$$

at a, where $t \leq s$. Then $\log \operatorname{rk}_b \sigma = r$, for all $b \in \cup_{i=1}^t E_i$ near a.

Let $z = (z_1, \ldots, z_N)$ denote coordinates for M_0 at $\sigma(a)$. It follows from the inverse mapping theorem that, after permuting the z_j if necessary, we can choose coordinates $(\boldsymbol{u}, \boldsymbol{v}) = (u_1, \ldots, u_s, v_1, \ldots, v_{n-s})$ for X adapted to E at $a = 0$, such that $(v_1, \ldots, v_r)$ forms part of a system of coordinates for $E(a)$ at a, and

1. $\sigma_1 = v_1, \ \ldots, \ \sigma_r = v_r$ at a;

2. for each $j > r$, $\sigma_j = \sigma_j(v_1, \ldots, v_r)$ on E_i at a, $i = 1, \ldots, t$.

Since Y_{p-r} is smooth at $\sigma(a)$, then $z_j - \sigma_j(z_1, \ldots, z_r)$, $j > r$, generate the ideal of Y_{p-r} at a. After a coordinate change

$$\bar{z}_j := z_j - \sigma_j(z_1, \ldots, z_r), \quad j > r,$$

we can therefore assume that $z_{r+1}, \ldots, z_N$ generate $\mathcal{I}_{Y_{p-r}}$ at $\sigma(a)$ (so that $\sigma_{r+1}, \ldots,$ σ_N generate $\sigma^* \mathcal{I}_{Y_{p-r}}$ at a). Since the latter is a principal monomial ideal, we can also assume that $\sigma_{r+1} = \boldsymbol{u}^{\boldsymbol{\alpha}_1}$, as required. □

The following is an immediate consequence of Theorems 2.3, 1.2 and 2.7.

Corollary 2.8. *Conjecture 1.1 holds in the case that* $\dim X_0 \leq 2$.

3 Three-dimensional case: sketch of the proof

In this section, we sketch the proof of Theorem 1.3. (Full details can be found in [1].) Assume that $\dim X_0 = 3$. By Theorems 2.3, 1.2 and 2.7, there is a resolution of singularities $\sigma : (X, E) \to (X_0, \operatorname{Sing} X_0)$ such that $\mathcal{F}_0(\sigma)$, $\mathcal{F}_2(\sigma)$ are principal monomial ideal sheaves and, moreover, if $a \in \operatorname{supp} E$ and $\log \operatorname{rk}_a \sigma > 0$, then $\mathcal{F}_1(\sigma)_a$ is also a principal monomial ideal. We will make $\mathcal{F}_1(\sigma)$ a principal monomial ideal by admissible blowings-up that preserve the preceding conditions on σ.

We argue by induction on the maximal value of an invariant ρ of $\mathcal{F}_1(\sigma)$ (§3.1 below). This invariant has the property that $\mathcal{F}_1(\sigma)_a$ is a principal monomial ideal if and only if $\rho(a) = 0$ (Lemma 3.2); in this case, we will say that a is *resolved.*

The blowings-up that we use to principalize $\mathcal{F}_1(\sigma)$ will have the additional property that $\log \operatorname{rk} \sigma$ is identically zero on the image in X of every centre (or, equivalently, that every centre of blowing up lies over $\operatorname{supp} \mathcal{F}_2(\sigma)$). Thus we will blow up only over a discrete subset of X_0. This is particularly important in the case of (not necessarily compact) analytic varieties because it means that only compact centres will be blown up and, as a consequence, that our algorithm is global.

A crucial difficulty in the problem of desingularization of the cotangent sheaf (in comparison with resolution of singularities of a variety in characteristic zero) is the lack of a hypersurface of "maximal contact" associated to the invariant ρ. An algorithm for Theorem 1.3 is described in §3.2. We first blow up to put σ in Weierstrass form (a version of Weierstrass preparation, Lemma 3.3). A second step deals with problems arising from a lack of maximal contact; Weierstrass form is used to obtain more precise local normal forms for σ at points of the maximum locus of ρ. The latter are used in a final step to decrease the maximal value of ρ.

3.1 Invariant of a logarithmic Fitting ideal

This section applies to arbitrary dimension, although we will use it in §3.2 following only in the case of dimension 3. We use the notation of §2. Let $\sigma : (X, E) \to (X_0, \operatorname{Sing} X_0)$ denote a resolution of singularities of X_0, $n = \dim X_0$.

Definition 3.1. Given $k = 0, \dots, n-1$ and $a \in X$, let $\mathcal{R}_{k,a}$ denote the *residual ideal* of $\mathcal{F}_k(\sigma)_a$ in $\mathcal{O}_{X,a}$; i.e., $\mathcal{F}_k(\sigma)_a = \prod_q \mathcal{I}_{E_q,a}^{\mu_q} \cdot \mathcal{R}_{k,a}$, with the $\mu_q \in \mathbb{N}$ as large as possible, where $\{E_q\}$ denotes the set of components of E and $\mathcal{I}_{E_q}$ is the ideal sheaf of E_q. Let $\rho_k(a)$ denote the order of $\mathcal{R}_{k,a}$ in the local ring $\mathcal{O}_{X,a}/\sum_{a\in E_q} \mathcal{I}_{E_q,a}$ (cf. [5, Section 5], [11, Section 2]; $\rho_k(a) := \infty$ if and only if $\mathcal{R}_{k,a} = 0$ in the preceding local ring).

The following lemma lists several properties of the basic invariant $\rho_k(a)$ that are all either clear or easy to prove.

Lemma 3.2. *1.* $0 \le \rho_k(a) \le \infty$.

2. $\rho_k(a) = 0$ *if and only if* $\mathcal{F}_k(\sigma)_a$ *is a principal monomial ideal.*

3. ρ_k *is upper-semicontinuous in the Zariski topology of* X.

4. *If* a *is an* n*-point, then* $\rho_k(a) = 0$ *or* ∞.

5. *If* $\beta : (X', E') \to (X, E)$ *is a combinatorial blowing-up and* $a' \in \sigma^{-1}(a)$, *then* $\rho_k(a') \le \rho_k(a)$.

We will only need ρ_k in the case that $k = n-2$ in this article; i.e., for the log Fitting ideal of 2×2 minors. Write $\rho := \rho_{n-2}$. Lemma 3.3 below extends in a straightforward way to $\rho_k(a)$, for any k, with the assumption that the Fitting ideals $\mathcal{F}_{n-1}(\sigma)_a, \mathcal{F}_{n-2}(\sigma)_a, \dots, \mathcal{F}_{k+1}(\sigma)_a$ are all principal monomial ideals. We present the lemma only in the case needed for the remainder of the paper, in part so that we can fix notation that will be used in the following sections.

Lemma 3.3 (Weierstrass form). *Let* $a \in \operatorname{supp} E$ *be an* s*-point* $(1 \le s \le n)$. *Suppose that* $\mathcal{F}_{n-1}(\sigma)_a$ *is a principal monomial ideal and that* $\log \operatorname{rk}_a \sigma = 0$. *Then:*

1. *Let* M_0 *denote a local embedding variety for* X_0 *at* $\sigma(a)$. *Then there are adapted local coordinates* $(\boldsymbol{u}, \boldsymbol{v}) = (u_1, \dots, u_s, v_1, \dots v_{n-s})$ *for* X *at* a *(where the* $(u_k = 0)$ *are the components of* E *at* a*), and local coordinates* $\boldsymbol{z} =$

$(z_1, \dots, z_N)$ for M_0 at $\sigma(a)$ with respect to which the components σ_i of σ can be written

$$\begin{aligned} \sigma_1 &= \boldsymbol{u}^{\boldsymbol{\alpha}}, \quad \boldsymbol{\alpha} \in \mathbb{N}^s, \\ \sigma_i &= g_i(\boldsymbol{u}) + \boldsymbol{u}^{\boldsymbol{\delta}} T_i, \quad i = 2, \dots, N, \end{aligned} \tag{3.1}$$

where $\boldsymbol{u}^{\boldsymbol{\alpha}}$ divides all σ_i, each g_i and T_i is analytic (or regular), each dg_i is in the submodule generated by $d\boldsymbol{u}^{\boldsymbol{\alpha}}$, the T_i are not simultaneously divisible by any u_k, and $\boldsymbol{\delta}$ is linearly independent of $\boldsymbol{\alpha}$ if some T_i is a unit.

2. *Given coordinates as above in which (3.1) satisfied, let*

$$d = d(a) := \min\{|\boldsymbol{\gamma}| : \boldsymbol{\gamma} \in \mathbb{N}^{n-s} \text{ and } \partial_{\boldsymbol{v}}^{\boldsymbol{\gamma}} T_i \text{ is a unit, for some } i\}.$$

Then

(a) *$\rho(a) < \infty$ if and only if $d(a) < \infty$;*

(b) *$\rho(a) = 0$ if and only if $d(a) = 0$ or 1;*

(c) *if $0 < \rho(a) < \infty$, then $d(a) = \rho(a) + 1$.*

3. *Suppose that $0 < \rho(a) < \infty$. Then there are adapted coordinates $(\boldsymbol{u}, v, \boldsymbol{w}) = (u_1, \dots, u_s, v, w_1, \dots w_{n-s-1})$ for X at a and coordinates $\boldsymbol{z} = (z_1, \dots, z_N)$ for M_0 at $\sigma(a)$ with respect to which the components σ_i of σ can be written as in (3.1) with*

$$T_i(\boldsymbol{u}, v, \boldsymbol{w}) = \widetilde{T}_i(\boldsymbol{u}, v, \boldsymbol{w}) v^d + \sum_{j=0}^{d-1} a_{ij}(\boldsymbol{u}, \boldsymbol{w}) v^j, \quad i = 2, \dots, N, \tag{3.2}$$

where $\widetilde{T}_2$ is a unit that we will also denote U, $a_{2,d-1} = 0$, and all monomials of order $\leq d + |\boldsymbol{\delta}|$ in the formal expansions of the $\boldsymbol{u}^{\boldsymbol{\delta}} T_i$ at a, $i = 2, \dots, N$, are "linearly independent" of $\boldsymbol{u}^{\boldsymbol{\alpha}}$ (i.e., have exponents with respect to $(\boldsymbol{u}, v, \boldsymbol{w})$ that are linearly independent of $(\boldsymbol{\alpha}, 0, \boldsymbol{0})$).

Proof. It is easy to obtain (1), where the formal expansion of each g_i at a is a sum of monomials $\boldsymbol{u}^{\boldsymbol{\beta}}$ where each $\boldsymbol{\beta}$ is a rational multiple $q\boldsymbol{\alpha}$, $q \geq 1$ (see [1, Rmks. 2.1, 3.2(1)]).

In (2), if $d = 0$, then $\boldsymbol{\delta}$ is linearly independent of $\boldsymbol{\alpha}$ and T_i is a unit, for some i; say $i = 2$. Then, after a coordinate change, we can assume $T_2 = 1$, so that $\mathcal{F}_{n-2}(\sigma)_a$ is generated by $\boldsymbol{u}^{\boldsymbol{\alpha}+\boldsymbol{\delta}}$. If $d = 1$, then, after a coordinate change, we can assume $T_2 = v_1$, and again $\mathcal{F}_{n-2}(\sigma)_a$ is generated by $\boldsymbol{u}^{\boldsymbol{\alpha}+\boldsymbol{\delta}}$. On the other hand, (a) is clear and it is easy to see directly from the log Jacobian matrix of σ that the residual ideal $\mathcal{R}_a$ of $\mathcal{F}_{n-2}(\sigma)_a$ is generated modulo $\sum_{a \in E_k} \mathcal{I}_{E_k,a}$ by the partial derivatives $\partial_{v_l} T_i$, $i = 2, \dots, N$, $l = 1, \dots, n-s$, together with $(\alpha_p \delta_q - \alpha_q \delta_p) T_i$, $i = 2, \dots, N$, $p, q = 1, \dots, s$. It follows that, if $d \geq 1$, then $\rho(a) = d - 1$.

Given (3.1), after a permutation of the coordinates $(z_1, \dots, z_N)$ and a generic linear coordinate change in $\boldsymbol{v} = (v_1, \dots, v_{n-s})$, we can write the T_i in the form (3.2), where $\widetilde{T}_2 = U$ is a unit, if we allow the sum in T_2 also to go from $j = 0$ to

$d-1$. Then we can eliminate $a_{2,d-1}$ by completing the dth power with respect to v. The final condition in (3) can be obtained by "moving" the lower order monomials of (the formal expansions of) the $\boldsymbol{u}^{\boldsymbol{\delta}}T_i$ that are rational powers of $\boldsymbol{u}^{\boldsymbol{\alpha}}$ to the g_i. □

3.2 Sketch of the proof of Theorem 1.3

Step 1. Reduction to the case that $\rho(a) < \infty$, for all a. If $0 < \rho(a) < \infty$ and $\log \operatorname{rk}_a \sigma = 0$, then the conclusions of Lemma 3.3(3) hold; in this case, we say that σ is in *Weierstrass form* at a.

Note that the set of 2-points forms a collection of curves ("2-curves") each given by (a connected component of) the intersection of precisely two components of E. A 2-curve either is closed or has limiting 3-points. By Lemma 2.4, we can assume that, throughout each 2-curve, either $\log \operatorname{rk} \sigma = 0$ or $\log \operatorname{rk} \sigma = 1$. A 2-curve on which $\log \operatorname{rk} \sigma = 0$ is relatively compact. By Theorem 2.7, $\rho = 0$ on every 2-curve where $\log \operatorname{rk} \sigma = 1$.

Lemma 3.4. *By combinatorial blowings-up (more precisely, by composing σ with a morphism $\tau : (\widetilde{X}, \widetilde{E}) \to (X, E)$ that restricts to a finite sequence of combinatorial blowings-up over any relatively compact open subset of X_0), we can reduce to the case that*

1. *$\rho(a) < \infty$ at every point a (and, therefore, by Lemma 3.3, we can choose coordinates at every nonresolved point a and its image $\sigma(a)$ in which σ has prepared Weierstrass form);*

2. *ρ is generically zero on every component of the set of 1-points, and on every 2-curve.*

In particular, in this case, the set of nonresolved points comprises isolated 2-points, isolated 1-points, and closed curves that are generically 1-points. Moreover, τ can be chosen so that no centre of blowing up includes points over a 2-curve in X where $\rho = 0$ (in particular, every centre lies over the locus $(\log \operatorname{rk} \sigma = 0)$*).*

Proof of Lemma 3.4. It is clear that $\rho(a) < \infty$ at every 1-point a. It follows from Lemma 3.3 that ρ is generically zero on every component of the set of 1-points.

Recall we can assume that $\log \operatorname{rk} \sigma$ is constant (either 0 or 1) on every 2-curve in X, and $\rho = 0$ on every 2-curve in X where $\log \operatorname{rk} \sigma = 1$.

Three-points are isolated. We can reduce to the case that $\rho(a) = 0$ at every 3-point a, by combinatorial blowings-up (principalizing a globally defined ideal sheaf generated locally by monomials in the components of the exceptional divisor). The blowings-up involved have centres that are closures of 2-curves, but it is unnecessary to blow up 2-curves that are already resolved; i.e., on which $\rho = 0$. It follows that ρ is generically zero on every 2-curve with a 3-point a as a limiting point.

Suppose that a is a 2-point. If $\rho(a) < \infty$ at a 2-point a, then ρ is generically zero on the 2-curve containing a, again by Lemma 3.3. A blowing-up with centre given by the intersection of two components of E is combinatorial. If $\rho(a) = \infty$, then we can reduce to the case that $\rho < \infty$ over a by finitely many such blowings-up (since the effect of such blowings-up is to principalize the ideal generated by the coefficients $a_{ij}(u_1, u_2)$ of formal expansions $T_i = \sum_{j=0}^{\infty} a_{ij}(u_1, u_2)v^j$ at a; cf. Lemma 3.3). It therefore follows that we can reduce to $\rho < \infty$ on every 2-curve, by a locally-finite sequence of combinatorial blowings-up. □

Step 2. Reduction to prepared normal form.

Lemma 3.5 (Prepared normal form). *Suppose that $\rho(a) < \infty$, for all $a \in X$ (and that $\rho = 0$ on every 2-curve with non-compact closure; cf. Step 1 above). Assume that ρ takes a maximum value $\rho_{\max} > 0$, and let $\Sigma \subset X$ denote the closed subset on which ρ takes the value $\rho_{\max}$ (so that $\Sigma \subset \operatorname{supp} E$). Then, by a (locally) finite sequence of admissible blowings-up over* $(\log \operatorname{rk} \sigma = 0)$, *we can reduce to the case that, for every point $a \in \Sigma$, σ has the Weierstrass form of Lemma 3.3, where the coefficients a_{ij} satisfy the following additional conditions.*

1. *At a 2-point a, with adapted coordinates* $(\boldsymbol{u}, v) = (u_1, u_2, v)$, *where* $\operatorname{supp} E = (u_1 u_2 = 0)$,

$$\begin{aligned} a_{ij} &= \boldsymbol{u}^{\boldsymbol{r}_{ij}} \tilde{a}_{ij}(\boldsymbol{u}), \quad i = 2, \ldots, N, \quad j = 1, \ldots, d-1, \\ a_{i_0,0} &= \boldsymbol{u}^{\boldsymbol{\beta}}, \qquad\qquad \textit{for some } i_0, \end{aligned} \tag{3.3}$$

   where each $\tilde{a}_{ij}$ is either zero or a unit, $\boldsymbol{\delta} + \boldsymbol{\beta}$ is linearly independent of $\boldsymbol{\alpha}$, and $\boldsymbol{u}^{\boldsymbol{\beta}}$ divides a_{i0}, for all i.

2. *At a 1-point a, with adapted coordinates* (u, v, w), *where* $\operatorname{supp} E = (u = 0)$,

$$\begin{aligned} a_{ij} &= u^{r_{ij}} w^{s_{ij}} \tilde{a}_{ij}(u, w) \quad i = 2, \ldots, N, \quad j = 1, \ldots, d-1, \\ a_{i_0,0} &= u^{\beta} w, \qquad\qquad \textit{for some } i_0, \end{aligned} \tag{3.4}$$

   where each $\tilde{a}_{ij}$ is either zero or a unit, and u^{β} divides a_{i0}, for all i.

The blowings-up involved do not increase the value of ρ over any point.

We will say that a is a *prepared* 2-point (resp., a *prepared* 1-point) if σ has Weierstrass form (3.2) at a, where the coefficients are given by (3.3) (resp., (3.4)), with respect to suitable adapted coordinates at a. In either case, we will also say that σ has *prepared normal form* at a. At a *generic* prepared 1-point, all $s_{ij} = 0$ in (3.4). Points where not all $s_{ij} = 0$ will be called *non-generic*.

The proof of Lemma 3.5 is by induction on pairs $(\rho(a), \iota(a))$ (ordered lexicographically), where $\iota(a)$ is a secondary invariant with values in $\mathbb{N}$, introduced below. If $\iota(a) = 0$, then σ has prepared normal form at a (Lemma 3.9). The proof of Lemma 3.5 is given by an algorithm organized in three distinct cases, beginning with σ in Weierstrass form:

- a is a 2-point;

- a is a 1-point with $\mathrm{ord}_a(T_i) = d(a)$ (where (T_i) denotes the ideal generated by $T_2, \ldots, T_N$ and ord_a means the order at a);
- a is a 1-point where $\mathrm{ord}_a(T_i) < d(a)$.

In each case, we find a sequence of blowings-up (over a) such that, for every point b in the preimage of a, $(\rho(b), \iota(b)) < (\rho(a), \iota(a))$. The third case is the most delicate, since the invariant ρ increases during the process. See [1, Section 6] for full details of the proof. Here we will only define the invariant ι and give its important properties.

Suppose that σ has Weierstrass form (3.2) in adapted coordinates $(\boldsymbol{u}, v)$ at a 2-point a; respectively, in adapted coordinates (u, v, w) at a 1-point a. For each $i = 2, \ldots, N$, write

$$\begin{aligned} d\sigma_1|_{(v=0)} \wedge d\sigma_i|_{(v=0)} &= d(\boldsymbol{u}^{\boldsymbol{\alpha}}) \wedge d(\boldsymbol{u}^{\boldsymbol{\delta}} a_{i0}(\boldsymbol{u})) \\ &= H_i(\boldsymbol{u}) \frac{du_1}{u_1} \wedge \frac{du_2}{u_2}; \end{aligned}$$

respectively,

$$\begin{aligned} d\sigma_1|_{(v=0)} \wedge d\sigma_i|_{(v=0)} &= d(u^{\alpha}) \wedge d(u^{\delta} a_{i0}(u, w)) \\ &= H_i(u, w) \frac{du}{u} \wedge dw. \end{aligned}$$

In either case, let $\mathcal{H}_a$ denote the ideal generated by H_i, $i = 2, \ldots, N$, in the local ring of $(v = 0)$ at a.

Remarks 3.6. $\mathcal{H}_a$ is the log Fitting ideal of 2×2 minors of the morphism $\sigma|_{(v=0)}$ at a. Blowing up of the point a in $(v = 0)$ is admissible for $E|_{(v=0)}$. If $\mathcal{H}_a$ is a principal monomial ideal (i.e., generated by a monomial in components of $E|_{(v=0)}$), then, by Theorem 2.3 and Lemma 2.1, $\sigma|_{(v=0)}$ can be written in Hsiang-Pati form $d\sigma_1|_{(v=0)} = d(\boldsymbol{u}^{\boldsymbol{\alpha}})$ (resp., $d(u^{\alpha})$), and $a_{i_0,0} = \boldsymbol{u}^{\boldsymbol{\beta}}$ (resp., $u^{\beta} w$), for some i_0, where $\boldsymbol{u}^{\boldsymbol{\beta}}$ (reap. u^{β}) satisfies the additional conditions given in Lemma 3.5.

Let $\mathcal{G}_a$ denote the ideal

$$\mathcal{G}_a := \left(\prod_{(i,j) \in J} a_{ij} \right) \cdot \mathcal{H}_a,$$

where $J := \{(i, j) : a_{ij} \neq 0,\ i = 2, \ldots, N,\ j = 1, \ldots, d - 1\}$.

Remarks 3.7. (1) Since $\mathcal{G}_a$ is an ideal of functions in two variables, it follows from resolution of singularities that, after a finite number $\iota(a; \boldsymbol{u})$ (respectively, $\iota(a; (u, w))$) of blowings-up of discrete sets (beginning with a and) lying over a, the pull-back of $\mathcal{G}_a$ is a principal ideal generated by a monomial $\tilde{\boldsymbol{u}}^{\boldsymbol{\gamma}}$ (resp., $\tilde{u}^p \tilde{w}^q$) with respect to adapted coordinates $\tilde{\boldsymbol{u}}$ (resp. $(\tilde{u}, \tilde{w})$) at any point over a. (At each step, the centre of blowing-up is the finite set of points over a at which the pull-back of $\mathcal{G}_a$ is not already a principal ideal generated by such a monomial.) Note that the blowing-up of $(v = 0)$ with centre a corresponds to the blowing-up of X with centre $(\boldsymbol{u} = \boldsymbol{0})$ (resp., $(u = w = 0)$).

(2) In particular, multiplication of $\mathcal{G}_a$ by a monomial in components of the exceptional divisor does not change the value of $\iota(a; \boldsymbol{u})$ (resp., $\iota(a; (u, w))$).

Definition 3.8. Let $\iota(a)$ denote the minimum of $\iota(a; \boldsymbol{u})$ (respectively, $\iota(a; (u, w))$) over all adapted coordinate systems $(\boldsymbol{u}, v)$ (resp. (u, v, w)) at a in which σ has Weierstrass form.

Of course, $\mathcal{H}_a$ and $\mathcal{G}_a$ themselves depend on the coordinates in Weierstrass form.

Lemma 3.9. *If $\iota(a) = 0$, then σ has prepared normal form at a.*

Proof. Assume that $\iota(a) = 0$. Then we can choose adapted local coordinates in which each coefficient a_{ij}, $j > 0$, in (3.2) is a monomial times a unit as in (3.3) (resp., (3.4)), and $\mathcal{H}_a$ is a principal ideal generated by a monomial $\boldsymbol{u}^\gamma$ (resp., $u^p w^q$) as in Remarks 3.7. In the 1-point case, necessarily $q = 0$ since $\sigma|(v = 0)$ has rank 2 outside $\operatorname{supp} E$ (i.e., since the Fitting ideal of 3×3 minors of $\log \operatorname{Jac} \sigma$ is supported in E). By Lemma 2.1, we can choose coordinates also in which the coefficients a_{i0} satisfy the conditions of Lemma 3.5. □

Lemma 3.10. *If $\iota(a) \neq 0$, then there is a neighbourhood of a in which $(\rho(b), \iota(b)) < (\rho(a), \iota(a))$, $b \neq a$.*

Proof. This is clear. □

Step 3. Further admissible blowings-up to decrease the maximal value of the invariant ρ. We use the following lemma to complete the proof of Theorem 1.3.

Lemma 3.11. *Assume that $\rho(a) < \infty$, for all $a \in X$, and that ρ takes a maximum value $\rho_{\max} > 0$ on X. Let $\Sigma \subset X$ denote the maximum locus of ρ. Suppose that σ has prepared normal form at every point $a \in \Sigma$ (see Lemma 3.5). Let A denote the discrete set of all non-generic points of Σ (i.e., all 2-points and non-generic 1-points of Σ). Then there is a morphism $\tau : (\widetilde{X}, \widetilde{E}) \to (X, E)$ given by a locally finite sequence of admissible blowings-up over Σ such that $\rho < \rho_{\max}$ throughout $\widetilde{X}$. Moreover, τ can be realized as a composite $\tau = \tau_1 \circ \tau_2 \circ \tau_3$, where*

1. *$\tau_1 : (X_1, E_1) \to (X, E)$ is a single blowing-up with centre A;*
2. *$\tau_2 : (X', E') \to (X_1, E_1)$ is the composite of a locally finite sequence of admissible blowings-up $(X_{i+1}, E_{i+1}) \to (X_i, E_i)$, $i \geq 1$, with centres $\overline{\Sigma_i \setminus A_i}$, where Σ_i is the maximum locus of ρ and A_i the preimage of A in X_i;*
3. *$\tau_3 : (\widetilde{X}, \widetilde{E}) \to (X', E')$ is the composite of a locally finite sequence of blowings-up with centres over A.*

Proof of Theorem 1.3. The theorem follows from Lemmas 3.4, 3.5 and 3.11 by induction on the maximal value of ρ, at least for X on which ρ assumes a maximum value (e.g., algebraic varieties or the restrictions of analytic varieties to relatively compact open sets). Theorem 1.3 follows for analytic varieties, in general, because the preceding lemmas show that, if we start with a resolution of singularities σ as at the beginning of this section, then ρ can be everywhere decreased to zero by finitely many blowings-up over each point of the discrete subset Γ of X_0 given by the image of $(\log \mathrm{rk}\, \sigma = 0)$. □

A sequence of blowings-up as in the conclusion of Lemma 3.11 will be called *permissible.*

To prove Lemma 3.11, we show that, in general, every prepared point $a \in \operatorname{supp} E$ admits a neighbourhood U over which there is a morphism $\tau : (\widetilde{U}, \widetilde{E}) \to (U, E|_U)$ given by a permissible finite sequence of blowings-up over the locus $(\rho = \rho(a))$, such that $\rho(b) < \rho(a)$, for all $b \in \widetilde{U}$. The blowing-up sequence depends on the nature of a (2-point, generic 1-point or non-generic 1-point). Lemma 3.11 clearly follows from this local statement, because the sequence of blowings-up in Lemma 3.11(2) is uniquely determined by the maximum value of ρ. We will present the main ideas of the proof of Lemma 3.11 here; full details can again be found in [1, Section 7].

Note that, if a is a generic prepared 1-point, then there is a neighbourhood U of a over which $A = \emptyset$, so that a permissible blowing-up sequence means a finite sequence as in Lemma 3.11(2).

Suppose that σ has prepared normal form (3.3) in adapted coordinates $(\boldsymbol{u}, v)$ at a 2-point a (respectively, prepared normal form (3.4) in adapted coordinates (u, v, w) at a 1-point a). Although v (and w) are not globally defined, there is a neighbourhood U of a in which $(v = 0)$ (or $(v = 0)$ and $(w = 0)$) are smooth hypersurfaces that we can add to E to obtain a divisor D on U. We consider D a "declared exceptional divisor". The divisor D and a corresponding monomial ideal $\mathcal{I} = \mathcal{I}(\sigma, a)$ are defined according to the nature of a, as follows.

Definitions 3.12 (*Declared exceptional divisor* and associated monomial ideal)**.** We use the notation of Lemma 3.5.

- If a is a prepared 2 point (3.3),

$$D := E|_U + (v = 0) = (u_1 u_2 v = 0),$$
$$\mathcal{I} := (v^d,\, \boldsymbol{u}^{\boldsymbol{r}_{ij}} v^j, \boldsymbol{u}^{\boldsymbol{\beta}}),$$

  where $\boldsymbol{\beta}$ will also be denoted $\boldsymbol{r}_{i_0,0}$ (for reasons evident from (3.3)).

- If a is a generic prepared 1-point ((3.4) with all $s_{ij} = 0$),

$$D := E|_U + (v = 0) = (uv = 0),$$
$$\mathcal{I} := (v^d,\, u^{r_{ij}} v^j,\, u^{\beta}),$$

  where β will also be denoted $r_{i_0,0}$.

- If a is a non-generic prepared 1-point (3.4),

$$D := E|_U + (v = 0) + (w = 0) = (uvw = 0),$$
$$\mathcal{I} := (v^d,\, u^{r_{ij}} w^{s_{ij}} v^j,\, u^{\beta} w),$$

  where again β will also be denoted $r_{i_0,0}$.

The notation for $\mathcal{I}$ in each case above is understood to mean that (i,j) runs over the index set $J := \{(i,j) : a_{ij} \neq 0,\ i = 2,\ldots,N,\ j = 1,\ldots,d-1\}$.

Remarks 3.13. (1) D and $\mathcal{I}(\sigma, a)$ depend on the adapted coordinates at a. Nevertheless, if U is a small open neighbourhood of a and $\rho(b) = \rho(a)$, where $b \in U$, then D induces a declared exceptional divisor at b, and $\mathcal{I}(\sigma, a)$ induces $\mathcal{I}(\sigma, b)$.

(2) A blowing-up that is admissible for D is also admissible for $E|_U$. Suppose that $\tau : \widetilde{U} \to U$ is given by a finite sequence of blowings-up that are admissible for D. If $\widetilde{D}$, $\widetilde{E}$ denote the transforms of the divisors D, $E|_U$, respectively, then $\widetilde{D} = \widetilde{E}$ + strict transform of $(v = 0)$ (or $\widetilde{D} = \widetilde{E}$ + strict transforms of $(v = 0)$ and $(w = 0)$ in the non-generic prepared 1-point case).

Proposition 3.14. *Let $a \in X$ and suppose that σ has prepared normal form at a. Take U, D and $\mathcal{I}$ as in Definitions 3.12. Let $\tau : (\widetilde{U}, \widetilde{E}) \to (U, E|_U)$ be a sequence of blowings-up with centres over* $\operatorname{supp} E|_U$ *that are combinatorial with respect to D, and let $b \in \tau^{-1}(a)$. If $\tau^*(\mathcal{I})$ is a principal $\widetilde{D}$-monomial ideal, then $\tau^*(\mathcal{I})$ is also $\widetilde{E}$-monomial, and $\rho(b) < \rho(a)$.*

The proof of Proposition 3.14 is an elementary but intricate case-by-case analysis (see [1, Prop. 7.3]).

Note that the hypotheses of Proposition 3.14 does not exclude $(v = w = 0)$ as centre of blowing up (in the non-generic 1-point case). Proposition 3.14 is a purely local assertion; we do not claim to principalize $\mathcal{I}(\sigma, a)$ by blowings-up that are global over X. The proposition plays an important part in the proof of Lemma 3.11 but, in the latter, we do not necessarily principalize $\mathcal{I}(\sigma, a)$ at every point b over a given $a \in \Sigma$ because ρ may decrease before $\mathcal{I}$ becomes principal (see [1, Lemma 5.4]).

References

[1] A. Belotto da Silva, E. Bierstone, V. Grandjean and P.D. Milman, *Resolution of singularities of the cotangent sheaf of a singular variety*, Adv. in Math. **307** (2017), 780–832.

[2] D. Abramovich, K. Karu, K. Matsuki and J. Włodarczyk, *Torification and factorization of birational maps*, J. Amer. Math. Soc. **15** (2002), 531–572.

[3] E. Bierstone and P.D. Milman, *Semianalytic and subanalytic sets*, Inst. Hautes Etudes Sci. Publ. Math. **67** (1988), 5–42.

[4] E. Bierstone and P.D. Milman, *Canonical desingularization in characteristic zero by blowing up the maximum strata of a local invariant*, Invent. Math. **128** (1997), 207–302.

[5] E. Bierstone and P.D. Milman, *Functoriality in resolution of singularities*, Publ. R.I.M.S. Kyoto Univ. **44** (2008), 609–639.

[6] J. Cheeger, *On the spectral geometry of spaces with cone-like singularities*, Proc. Nat. Acad. Sci. U.S.A. **76** (1979), 2103–2106.

[7] J. Cheeger, *On the Hodge theory of Riemannian pseudomanifolds*, Geometry of the Laplace operator (Proc. Sympos. Pure Math., Univ. Hawaii, Honolulu, Hawaii, 1979), pp. 91–146, Proc. Sympos. Pure Math. **XXXVI**, Amer. Math. Soc., Providence, R.I., 1980.

[8] J. Cheeger, M. Goresky and R. MacPherson, *L^2-cohomology and intersection homology of singular algebraic varieties*, Seminar on Differential Geometry, pp. 303–340, Ann. of Math. Stud. **102**, Princeton Univ. Press, Princeton, N.J., 1982.

[9] S.D. Cutkosky, *Monomialization of morphisms from 3-folds to surfaces*, Lecture Notes in Math. **1786**, Springer-Verlag, Berlin, 2002.

[10] S.D. Cutkosky, *Toroidalization of dominant morphisms of 3-folds*, Mem. Amer. Math. Soc. **190** (2007).

[11] S.D. Cutkosky, *A simpler proof of toroidalization of morphisms from 3-folds to surfaces*, Ann. Inst. Fourier (Grenoble) **63** (2013), 865–922.

[12] S.D. Cutkosky, *Errata to A simpler proof of toroidalization of morphisms from 3-folds to surfaces*, http://faculty.missouri.edu/ cutkoskys/Torerrata.pdf, University of Missouri, 2015. Web September 8, 2015.

[13] D. Eisenbud, *Commutative algebra with a view toward algebraic geometry*, Graduate Texts in Mathematics **150**, Springer, New York, 1995.

[14] A. Grothendieck, *Eléments de géométrie algébrique IV. Etude locale des schémas et des morphismes de schémas, I*, Inst. Hautes Etudes Sci. Publ. Math. **20** (1964), 5–259.

[15] H. Hironaka, *Stratification and flatness*, Real and complex singularities (Proc. Ninth Nordic Summer School/NAVF Sympos. Math., Oslo, 1976), pp. 199–265, Sijthoff and Noordhoff, Alphen aan den Rijn, 1977.

[16] W.C. Hsiang and V. Pati, *L^2-cohomology of normal algebraic surfaces. I*, Invent. Math. **81** (1985), 395–412.

[17] W. Pardon and M. Stern, *Pure Hodge structure on the L_2-cohomology of varieties with isolated singularities*, J. Reine Angew. Math. **533** (2001), 55–80.

[18] V. Pati, *The Laplacian on algebraic three folds with isolated singularities*, Proc. Indian Acad. Sci. Math. Sci. **104** (1994), 435–481.

[19] L. Taalman, *The Nash sheaf of a complete resolution*, Manuscripta Math. **106** (2001), 249–270.

[20] A. Tognoli, *Proprietà globali degli spazi analitici reali*, Ann. Mat. Pura Appl. (4) **75** (1967), 143–218.

[21] B. Youssin, *Monomial resolutions of singularities*, Abstract, Conference on Geometric Analysis and Singular Spaces, Oberwolfach, June 21-27, 1998.

Hodge Theory and L^2-analysis
ALM 39, pp. 151–205

Nilpotent Cones and Their Representation Theory*

P. Brosnan, G. Pearlstein, C. Robles†

Abstract

We describe two approaches to classifying the possible monodromy cones C arising from nilpotent orbits in Hodge theory. The first is based upon the observation that C is contained in the open orbit of any interior point $N \in C$ under an associated Levi subgroup determined by the limit mixed Hodge structure. The possible relations between the interior of C and its faces are described in terms of signed Young diagrams.

The second approach is to understand the Tannakian category of nilpotent orbits via a category D introduced by Deligne in a letter to Cattani and Kaplan. In analogy with Hodge theory, there is a functor from D to a subcategory $\hat{D} \subset D$ of SL_2-orbits. We prove that these fibers are, roughly speaking, algebraic. We also give a correction to a result [16] of K. Kato.

2000 Mathematics Subject Classification: 14D07, 32G20, 17B08, 32S35.

1 Introduction

The object of Steven Zucker's first published paper [28] was the study of normal functions arising from algebraic cycles and the Hodge conjecture. More precisely, let $X \subset \mathbb{P}^m$ be a smooth projective variety of dimension $2d$ and ζ be a primitive

---

*Dedicated to Steven Zucker in recognition of his contributions to Hodge theory, on the occasion of his 65th birthday.

†Department of Mathematics, Mathematics Building, University of Maryland, College Park, MD 20742-4015. E-mail: pbrosnan@umd.edu; Mathematics Department, Mail-stop 3368, Texas A&M University, College Station, TX 77843-3368. E-mail: gpearl@math.tamu.edu; Mathematics Department, Duke University, Box 90320, Durham, NC 27708-0320. E-mail: robles@math.duke.edu

Patrick Brosnan is partially supported by NSF grant DMS 1361159. Gregory Pearlstein is partially supported by NSF grant DMS 1361120. Colleen Robles is partially supported by NSF grants DMS 02468621 and 1361120.

Hodge class of type (d,d) on X. Let Y be a smooth hyperplane section of X. Then, the long exact sequence for relative cohomology of the pair (X,Y) gives

$$\cdots \to H^{2d-1}(X) \overset{i^*}{\hookrightarrow} H^{2d-1}(Y) \to H^{2d}(X,Y) \to H^{2d}(X) \to \cdots$$

where $i : Y \hookrightarrow X$ is inclusion. Letting $H^{2d-1}_{\mathrm{van}}(Y)$ denote the cokernel of i^* and pulling back the above sequence along the morphism of Hodge structure $\mathbb{Z}(-d) \to H^{2d}(X)$ defined by ζ determines an extension

$$0 \to H^{2d-1}_{\mathrm{van}}(Y) \to E \to \mathbb{Z}(-d) \to 0$$

in the category of mixed Hodge structures. The set of all such extensions is the intermediate Jacobian $J(H^{2d-1}_{\mathrm{van}}(Y))$.

Applying the construction of the previous paragraph to the smooth fibers of a Lefschetz pencil of hyperplane sections of X yields the prototypical example of a normal function ν_ζ. Moreover, in this context, there is a 1-1 correspondence $\zeta \leftrightarrow \nu_\zeta$ between normal functions and primitive, integral Hodge classes on X.

More precisely, Zucker proved [29] using L^2 methods that if S is a curve with smooth completion $j : S \hookrightarrow \bar{S}$ and $\mathcal{V} \to S$ is a variation of Hodge structure of weight k then $H^i(\bar{S}, j_*\mathcal{V})$ carries a functorial Hodge structure of weight $i+k$ (Theorem 7.12). Furthermore (Theorem 9.2, op. cit.), if $\mathcal{V}$ is pure of weight $2p-1$ then the cohomology classes of normal functions on S surject onto the integral (p,p) classes in $H^1(\bar{S}, j_*\mathcal{V})$.

One of the key tools in Zucker's proofs are W. Schmid's orbit theorems [24]. Roughly speaking, the nilpotent orbit theorem asserts that a variation of Hodge structure $\mathcal{V} \to S$ admits a local approximation near a point in the boundary of any normal crossing compactification $S \hookrightarrow \bar{S}$ by a nilpotent orbit

$$\theta(z_1, \ldots, z_r) = \exp(\sum_j z_j N_j) F_\infty \tag{1.1}$$

determined by the local monodromy logarithms $N_1, \ldots, N_r$ and limit Hodge filtration F_∞ of $\mathcal{V}$. The SL_2-orbit theorem [24, 4] further asserts that θ can be approximated by an auxiliary nilpotent orbit which is governed by a representation ρ of $SL_2(\mathbb{R})^r$. In this way, the norm of a flat multivalued section σ of $\mathcal{V}$ is determined by its weights with respect to ρ. The SL_2-orbit theorem also implies that the limit Hodge filtration F_∞ is part of a limit mixed Hodge structure (F_∞, W).

Moving beyond families of smooth projective varieties, Deligne conjectured that given a surjective, quasiprojective morphism $\bar{f} : \bar{X} \to \bar{S}$ there should exist a Zariski open subset $S \subset \bar{S}$ over which $\bar{f}$ restricts to give a variation of mixed Hodge structure [7] on the cohomology of the fibers. Furthermore, there should be a category of good variations of mixed Hodge structure which contains every variation of mixed Hodge structure of geometric origin, and has all of the salient features of the pure case.

In [27], Steenbrink and Zucker defined a category of admissible variations of mixed Hodge structure over a curve S, and proved that in this category one had limit mixed Hodge structures. Moreover, if $\mathcal{V} \to S$ is an admissible variation of graded-polarized mixed Hodge structure over S and $j : S \hookrightarrow \bar{S}$ is a smooth completion of S then the cohomology groups $H^i(\bar{S}, R^k j_* \mathcal{V})$ carry functorial mixed Hodge structures (cf. Theorem 4.1 [27]). In [15], Kashiwara defined a category of admissible variations of graded-polarized mixed Hodge structure in several variables using a curve test, and christened the associated nilpotent orbits *infinitesimal mixed Hodge modules.*

In particular, the category of infinitesimal mixed Hodge modules (IMHM) in a fixed number of variables is an abelian tensor category (4.3.3 and 5.2.6 [15]) which becomes a neutral Tannakian category when equipped with the functor ω which takes an IMHM to the underlying $\mathbb{R}$-vector space. The category of IMHM has a natural subcategory corresponding to nilpotent orbits with limit mixed Hodge structure which is split over $\mathbb{R}$. The Tannakian Galois group of the category of split orbits in one variable is described by the first two authors in [1].

In the case of nilpotent orbits of pure Hodge structures in one variable, a split orbit is the same thing as SL_2-orbit: If D is a period domain upon which the Lie group $G_{\mathbb{R}}$ acts transitively by automorphisms then a nilpotent orbit $\theta(z)$ with values in D is an SL_2-orbit if there exists a representation $\rho : \mathrm{SL}_2(\mathbb{R}) \to G_{\mathbb{R}}$ such that

$$\theta(g.\sqrt{-1}) = \rho(g).\theta(\sqrt{-1}) \tag{1.2}$$

for all $g \in \mathrm{SL}_2(\mathbb{R})$. A classification of such orbits may be deduced from (i) Lemma 6.24 of [24], and (ii) the classification of nilpotent $N \in \mathfrak{g}_{\mathbb{R}} = \mathrm{Aut}(V_{\mathbb{R}}, Q)$ (which is reviewed in §2.3). A full classification in the case of orbits into a Mumford–Tate domain is given in [22].

One of the intricacies of the SL_2-orbit theorem [4] in several variables is the construction of the system

$$(\hat{N}_1, H_1, \hat{N}_1^+), \ldots, (\hat{N}_r, H_r, \hat{N}_r^+) \tag{1.3}$$

of commuting SL_2-triples attached to the nilpotent orbit (1.1). In [8], Deligne gave a purely linear algebraic construction of (1.3) via an iterative construction which ensures that each N_j is a sum of $\hat{N}_j$ and a collection of highest weight vectors for $(\hat{N}_j, H_j, \hat{N}_j^+)$. Published accounts of the resulting Deligne systems appear in [25, 14, 1].

The data of an IMHM consists of a set of commuting nilpotent endomorphisms $N_1, \ldots, N_r$ together with Hodge and weight filtrations which satisfy a number of compatibility conditions. In [16], Kato observed that for each non-negative real number a, the substitution

$$N_j \mapsto \phi^a(N_j) = \sum_{k=0}^{j-1} \frac{a^k}{k!} N_{j-k}$$

defines a functor ϕ^a : IMHM $\to$ IMHM such that $\phi^a \circ \phi^b = \phi^{a+b}$ and ϕ^0 is the identity. Moreover, the functor ϕ extends to a category DH of Deligne–Hodge systems which contains IMHM as a subcategory. Kato further claimed that for any object θ of DH there exists an $a \geq 0$ such that $\phi^b \theta$ belongs to IMHM for all $b > a$.

As Kato explains in the introduction to [16], one of his motivations to study Deligne–Hodge systems was to have a framework to study degenerations of Hodge structure which are not polarizable. Such a framework could potentially be very useful in the study of degenerations of motives over non-archimedean local fields. However, in §6.2 we construct an explicit example of a two variable Deligne–Hodge system θ which violates Kato's assertion (i.e. $\phi^b \theta$ is never an IMHM). In §6.4 we show that Kato's claim is true for Deligne–Hodge systems which satisfy a suitable graded-polarization condition.

Accordingly, one can study IMHM in several variables by using the results of [22] to classify the possible several variable SL_2-orbits with data (1.3), and then impose the representation theoretic conditions required to extend $\hat{N}_j$ to a candidate N_j. This second step can be done in the category DH. Application of ϕ^b for b sufficiently large then produces the required IMHM. In §6.5 we show that the set of all Deligne systems with fixed data (1.3) forms an algebraic variety.

The approach outlined in the previous paragraph attempts to construct the monodromy cone

$$\mathcal{C} = \left\{ \sum_{j=1}^{r} a_j N_j \mid a_1, \dots, a_r > 0 \right\} \tag{1.4}$$

of an IMHM starting from the edges of the closure of $\mathcal{C}$. Alternatively, one can try to construct orbits starting from an element of the interior of $\mathcal{C}$. This second approach, *for nilpotent orbits of pure Hodge structure*, is the subject of §3–5.

A rough outline is as follows: Without loss of generality we can pass to the case where (1.1) is a nilpotent orbit with limit mixed Hodge structure split over $\mathbb{R}$. For any $N \in \mathcal{C}$, it then follows that $e^{zN} F_\infty$ is an SL_2-orbit by the results of Cattani, Kaplan and Schmid [2, 4]. The results of [22] classify the possible pairs (N, F_∞).

To reverse this process, select an SL_2-orbit $e^{zN} F_\infty$ with associated representation $\rho : \mathrm{SL}_2 \to G_\mathbb{R}$. Let $G_\mathbb{R}^{0,0}$ be the connected subgroup of $G_\mathbb{R}$ consisting of elements which preserve the limit mixed Hodge structure of $e^{zN} F_\infty$. Let $\mathcal{N}$ denote the orbit of N under the adjoint action of $G_\mathbb{R}^{0,0}$. Then, by Lemma 3.5 and Corollary 3.6 it follows that if $\mathcal{C}$ is the cone (1.4) of a several variable nilpotent orbit with limit Hodge filtration F_∞ and $N \in \mathcal{C}$ then $\mathcal{C} \subset \mathcal{N}$. Sections 4 and 5 implement this process for a number of examples related to period domains of weight 2.

Section 2 is of a different nature: The set $\mathrm{Nilp}(\mathfrak{g}_\mathbb{R})$ of nilpotent elements in a real semisimple (or reductive) Lie algebra is a classical, and very well understood, object of study in representation theory; for an excellent introduction see [6] and the references therein. In particular, there is a great deal of Hodge theoretic information that one can glean from the representation theorists' understanding of $\mathrm{Nilp}(\mathfrak{g}_\mathbb{R})$. In §2 we review the classification of nilpotent $N \in \mathfrak{g}_\mathbb{R}$ by signed Young diagrams. Particularly noteworthy here are (i) Đoković's Theorem 2.21 characterizing a partial order on the $\mathrm{Ad}(G_\mathbb{R})$ conjugacy classes of nilpotent elements $N \in \mathfrak{g}_\mathbb{R}$, and (ii) the description in §2.5 of the signed Young diagram associated to a polarized mixed Hodge structure (F, W): as illustrated in §5.2 together these provide representation theoretic constraints on the degenerations associated with the faces of a nilpotent cone underlying a nilpotent orbit on a period domain D.

Acknowledgments

In 1995, Steven Zucker was one of the instructors at a workshop for graduate students and recent PhD's held in connection with the AMS Summer Institute in Algebraic Geometry held at UC Santa Cruz. This is where the first two authors became acquainted, and we thank Steven for all of the things we have learned from him. The third author would like to thank William McGovern, Mark Green and Phillip Griffiths for helpful discussions and correspondence in connection with this work.

2 Classification of nilpotent endomorphisms

The nilpotent elements N in a classical Lie algebra $\mathfrak{g}_\mathbb{R}$ are classified by partially signed Young diagrams. This classification is up to the action of $\mathrm{Ad}(G_\mathbb{R})$; so what is really being classified are the $\mathrm{Ad}(G_\mathbb{R})$–orbits in $\mathrm{Nilp}(\mathfrak{g}_\mathbb{R})$, of which there are only finitely many.[1] Collingwood and McGovern's [6] is an excellent reference for the material in this section.

It is convenient to begin with the classification of nilpotent endomorphisms over $\mathbb{C}$ by (unsigned) Young diagrams; the basic idea is that (i) the Young diagram encodes the Jordan normal form of $N \in \mathfrak{g}_\mathbb{C}$, and (ii) the Jordan normal form determines N up to the action of $\mathrm{Ad}(G_\mathbb{C})$. Before reviewing the classifications over $\mathbb{R}/\mathbb{C}$ we need to recall the notion of a standard triple.

2.1 Standard triples

Let $\mathfrak{g}$ be a Lie algebra defined over $\mathbf{k} = \mathbb{R}$ or $\mathbb{C}$. A *standard triple* in $\mathfrak{g}$ is a set of three elements $\{N^+, Y, N\} \subset \mathfrak{g}$ such that

$$[Y, N^+] \;=\; 2N^+\,, \quad [N^+, N] \;=\; Y \quad \text{and} \quad [Y, N] \;=\; -2N\,.$$

Note that $\{N^+, Y, N\}$ span a *3–dimensional semisimple subalgebra* (TDS) of $\mathfrak{g}$ isomorphic to $\mathfrak{sl}(2, \mathbf{k})$. We call Y the *neutral element*, N the *nilnegative element* and N^+ the *nilpositive element*, respectively, of the standard triple.

Theorem 2.1 (Jacobson–Morosov). *Every nilpotent $N \in \mathfrak{g}$ can be realized as the nilnegative of a standard triple.*

Example 2.2. The matrices

$$\mathbf{n}^+ = \begin{pmatrix} 0 & 1 \\ 0 & 0 \end{pmatrix}, \quad \mathbf{y} = \begin{pmatrix} 1 & 0 \\ 0 & -1 \end{pmatrix} \quad \text{and} \quad \mathbf{n} = \begin{pmatrix} 0 & 0 \\ 1 & 0 \end{pmatrix} \tag{2.3}$$

[1] In contrast, there are infinitely many $\mathrm{Ad}(G_\mathbb{R})$ semisimple orbits in $\mathfrak{g}_\mathbb{R}$.

form a standard triple in $\mathfrak{sl}(2,\mathbb{R})$; while the matrices

$$\overline{\mathbf{e}} = \frac{1}{2}\begin{pmatrix} \mathrm{i} & 1 \\ 1 & -\mathrm{i} \end{pmatrix}, \quad \mathbf{z} = \begin{pmatrix} 0 & \mathrm{i} \\ -\mathrm{i} & 0 \end{pmatrix} \quad \text{and} \quad \mathbf{e} = \frac{1}{2}\begin{pmatrix} -\mathrm{i} & 1 \\ 1 & \mathrm{i} \end{pmatrix} \tag{2.4}$$

form a standard triple in $\mathfrak{su}(1,1)$.

2.2 Nilpotents over $\mathbb{C}$

The classification of the nilpotent elements in classical, complex, semisimple Lie algebras (which is due to Gerstenhaber [11], and Springer and Steinberg [26]) is given by partitions or, equivalently, Young diagrams.

Let $V_\mathbb{C}$ be a $\mathbb{C}$–vector space of dimension n and fix a nilpotent element $N \in \mathrm{End}(V_\mathbb{C})$. Let $\mathfrak{sl}_2\mathbb{C} \subset \mathrm{End}(V_\mathbb{C})$ be the TDS spanned by a standard triple containing N as the nilnegative element (§2.1). Let

$$V_\mathbb{C} \;=\; \bigoplus_{\ell \geq 0} V(\ell) \tag{2.5}$$

be the $\mathfrak{sl}_2\mathbb{C}$–decomposition of $V_\mathbb{C}$; here each $V(\ell) \simeq (\mathrm{Sym}^\ell \mathbb{C}^2)^{m_\ell}$ is the direct sum of m_ℓ irreducible $\mathfrak{sl}_2\mathbb{C}$–modules of dimension $\ell + 1$. In particular, $V(\ell)$ admits a basis of the form

$$\{N^a v_i \mid 1 \leq i \leq m_\ell\,,\ 0 \leq a \leq \ell\}\,.$$

Here $N^\ell v_i \neq 0$ and $N^{\ell+1} v_i = 0$. Each

$$\{N^a v_i \mid 0 \leq a \leq \ell\}$$

is an N*–string* of length $\ell + 1$ and we think of $V(\ell)$ as spanned by m_ℓ of these N–strings. Let

$$P(\ell) \;:=\; \mathrm{span}_\mathbb{C}\{v_i \mid 1 \leq i \leq m_\ell\} \;\subset\; V(\ell)$$

be the subspace of highest weight vectors. (In Hodge–theoretic language, this is the vector space of N*–primitive vectors* in $V(\ell)$.)

Note that $\sum m_\ell(\ell+1) = n$. So we may associate to the nilpotent N a partition

$$\mathbf{d} \;=\; [d_i] \;=\; [(\ell+1)^{m_\ell}]_{0 \leq \ell \in \mathbb{Z}}$$

of n; here $(\ell+1)^{m_\ell}$ indicates that the part $d_i = \ell + 1$ occurs m_ℓ times. The partition $\mathbf{d}$ is identified with the Young diagram whose i–th row contains d_i boxes. For example, the partition $\mathbf{d} = (4,2,2,1)$ of $n = 9$ is identified with the Young diagram

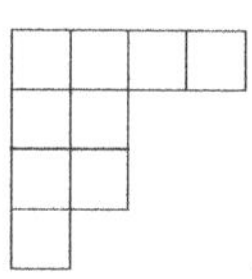
.

We think of each row of the Young diagram as representing an N–string. In this example, we have

u	Nu	N^2u	N^3u
v	Nv		
w	Nw		
x			

and $V_{\mathbb{C}} = V(3) \oplus V(1) \oplus V(0)$ with $m_3 = 1 = m_0$ and $m_2 = 2$.

2.2.1 $G_{\mathbb{C}} = \mathrm{Aut}(V_{\mathbb{C}})$

The Jordan normal form for elements of $\mathfrak{g}_{\mathbb{C}} = \mathrm{End}(V_{\mathbb{C}})$ implies that two nilpotents $N_1, N_2 \in \mathfrak{g}_{\mathbb{C}}$ lie in the same $\mathrm{Ad}(G_{\mathbb{C}})$–orbit if and only if the corresponding partitions $\mathbf{d}_1$ and $\mathbf{d}_2$ are equal. That is, the $\mathrm{Aut}(V_{\mathbb{C}})$–orbits $\mathcal{N}$ in $\mathrm{Nilp}(\mathrm{End}(V_{\mathbb{C}}))$ are indexed by partitions of $n = \dim V$; equivalently, they are indexed by Young diagrams of size n.

Example 2.6. The nilpotent $\mathrm{Aut}(V_{\mathbb{C}})$–conjugacy classes $\mathcal{N} \subset \mathrm{End}(V_{\mathbb{C}})$ for $n = 5$ are indexed by

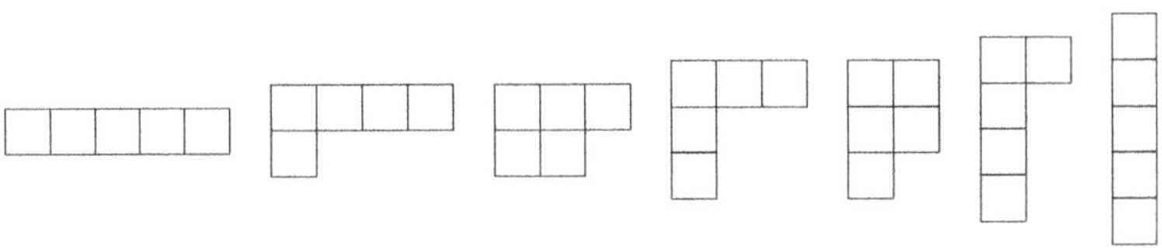

- Given a nilpotent $N \in \mathcal{N}$ in the conjugacy class indexed by the Young diagram 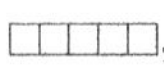, V admits a basis of the form $\{v, Nv, \ldots, N^4v\}$ with $N^5v = 0$.
- Given a nilpotent $N \in \mathcal{N}$ in the conjugacy class indexed by the Young diagram 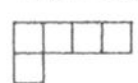, V admits a basis of the form $\{u\,;\, v, Nv, \ldots, N^3v\}$ with $Nu = 0$ and $N^4 = 0$.
- And so on.

Remark 2.7. Note that the trivial nilpotent conjugacy class $\mathcal{N} = \{0\}$ is indexed by the vertical partition $[1^n]$.

2.2.2 $G_{\mathbb{C}} = \mathrm{Aut}(V_{\mathbb{C}}, Q)$

Fix $w \in \mathbb{Z}$ and let Q be a nondegenerate bilinear form on $V_{\mathbb{C}}$ satisfying

$$Q(u,v) \;=\; (-1)^w Q(v,u)\,.$$

Set $G_{\mathbb{C}} = \mathrm{Aut}(V_{\mathbb{C}}, Q)$ and $\mathfrak{g}_{\mathbb{C}} = \mathrm{End}(V_{\mathbb{C}}, Q)$. Given a nonzero $N \in \mathrm{Nilp}(\mathfrak{g}_{\mathbb{C}})$, we may assume that the TDS is contained in $\mathfrak{g}_{\mathbb{C}}$ (cf. Jacobson and Morosov's Theorem 2.1). Then

$$Q_\ell(u,v) \;:=\; Q(u, N^\ell v)$$

defines a non–degenerate bilinear form on $P(\ell)$.

- If $w+\ell$ is even, then Q_ℓ is symmetric.
- If $w+\ell$ is odd, then Q_ℓ is skew–symmetric. This implies that m_ℓ is even. So, if w is even/odd, then the even/odd parts of $\mathbf{d}$ must occur with even multiplicity.

Theorem 2.8 (Symplectic algebras (w odd)). *Let Q be a skew-symmetric bilinear form on a complex vector space $V_{\mathbb{C}}$, and set $G_{\mathbb{C}} = \mathrm{Aut}(V_{\mathbb{C}}, Q)$ with Lie algebra $\mathfrak{g}_{\mathbb{C}} = \mathrm{End}(V_{\mathbb{C}}, Q)$. Then the $\mathrm{Ad}(G_{\mathbb{C}})$–orbits in $\mathrm{Nilp}(\mathfrak{g}_{\mathbb{C}})$ are indexed by the partitions of $2m = \dim V_{\mathbb{C}}$ in which the odd parts occur with even multiplicity.*

Example 2.9. Suppose that $G_{\mathbb{C}} = \mathrm{Sp}(6,\mathbb{C})$. The nilpotent conjugacy classes in $\mathfrak{g}_{\mathbb{C}}$ are enumerated by the partitions

$$[6]\,,\ [4,2]\,,\ [4,1^2]\,,\ [3^2]\,,\ [2^3]\,,\ [2^2,1^2]\,,\ [2,1^4]\,,\ [1^6]\,.$$

The corresponding Young diagrams are

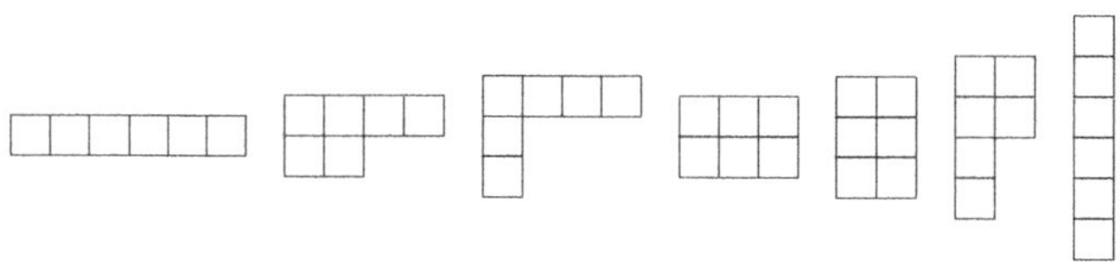

We say that a partition is *very even* if all parts d_i are even and occur with even multiplicity.

Theorem 2.10 (Orthogonal algebras (w even)). *Let Q be a symmetric bilinear form on a complex vector space $V_{\mathbb{C}}$, and set $\mathfrak{g}_{\mathbb{C}} = \mathrm{End}(V_{\mathbb{C}}, Q)$.*

(a) *Let $G_{\mathbb{C}} = \mathrm{Aut}(V_{\mathbb{C}}, Q)$. The $\mathrm{Ad}(G_{\mathbb{C}})$–orbits in $\mathrm{Nilp}(\mathfrak{g}_{\mathbb{C}})$ are indexed by the partitions of $n = \dim V_{\mathbb{C}}$ in which the even parts occur with even multiplicity.*

(b) *Let $G^\circ_{\mathbb{C}} = \mathrm{SO}(n,\mathbb{C}) \subset G_{\mathbb{C}}$. The $\mathrm{Ad}(G^\circ_{\mathbb{C}})$–orbits in $\mathrm{Nilp}(\mathfrak{g}_{\mathbb{C}})$ are indexed by partitions $\mathbf{d} = [d_i]$ of n in which the even parts occur with even multiplicity, and with the caveat that a very even partition is associated with two distinct orbits.*

Example 2.11. Suppose that $G_{\mathbb{C}} = \mathrm{O}(7,\mathbb{C})$. The nilpotent conjugacy classes in $\mathfrak{g}_{\mathbb{C}} = \mathfrak{so}(7,\mathbb{C})$ are enumerated by the partitions

$$[7]\,,\ [5,1^2]\,,\ [3^2,1]\,,\ [3,2^2]\,,\ [3,1^4]\,,\ [2^2,1^3]\,,\ [1^7]\,.$$

The corresponding Young diagrams are

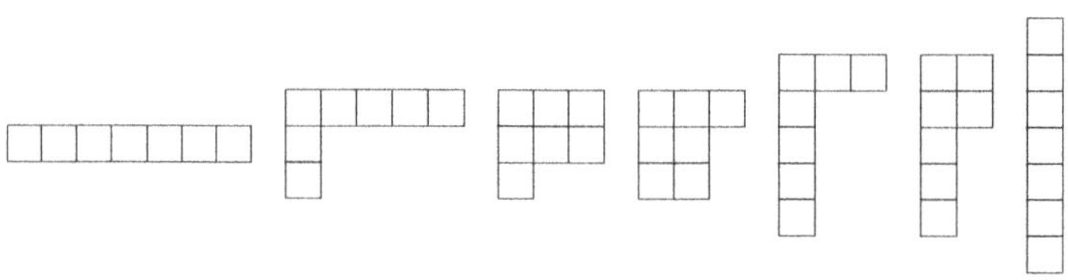

2.3 Nilpotents over $\mathbb{R}$ (signed Young diagrams)

Let $V_{\mathbb{R}}$ be a real vector space of dimension n. Given $w \in \mathbb{Z}$, fix a nondegenerate bilinear form $Q : V_{\mathbb{R}} \times V_{\mathbb{R}} \to \mathbb{R}$ satisfying $Q(u,v) = (-1)^w Q(v,u)$. Set

$$G_{\mathbb{R}} \ = \ \mathrm{Aut}(V_{\mathbb{R}}, Q)\,.$$

The classification of $\mathrm{Ad}(G_\mathbb{R})$–conjugacy classes of nilpotent $N \in \mathfrak{g}_\mathbb{R}$ is due to Springer and Steinberg [26], and is given by (partially) signed Young diagrams. A *signed Young diagram* is a Young diagram in which the boxes of a fixed row are either labeled with alternating $\pm$ signs, or are left blank. For the real forms $G_\mathbb{R}$ under consideration, the blank rows occur with even multiplicity.

Theorem 2.12 (Symplectic algebras (w odd)). *Let Q be a skew-symmetric bilinear form on a real vector space $V_\mathbb{R}$ of dimension $2m$, and set $G_\mathbb{R} = \mathrm{Aut}(V_\mathbb{R}, Q) \simeq \mathrm{Sp}(2m, \mathbb{R})$ with Lie algebra $\mathfrak{g}_\mathbb{R} = \mathrm{End}(V_\mathbb{R}, Q) \simeq \mathfrak{sp}(2m, \mathbb{R})$. Then the $\mathrm{Ad}(G_\mathbb{R})$-orbits in $\mathrm{Nilp}(\mathfrak{g}_\mathbb{R})$ are indexed by the signed Young diagrams of size $2m$ in which (i) the rows of even length are signed, and (ii) the rows of odd length are unsigned and occur with even multiplicity.*

Example 2.13. Suppose that $G_\mathbb{R} = \mathrm{Sp}(4, \mathbb{R})$. The nilpotent conjugacy classes in $\mathfrak{g}_\mathbb{R} = \mathfrak{sp}(4, \mathbb{R})$ are enumerated by the signed Young diagrams

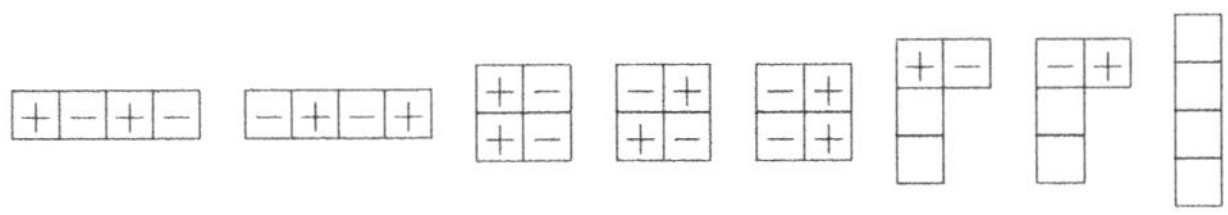

Given a signed Young diagram, let $b^\pm$ be the number of boxes labeled with a $\pm$, and let $2b^0$ be the number of unlabeled boxes. The *signature* of a signed Young diagram is $\mathrm{sig}\, Y = (s^+, s^-)$ where $s^+ = b^+ + b^0$ and $s^- = b^- + b^0$.

Remark 2.14. The signed Young diagrams of Theorem 2.12 are all of signature (m, m).

Theorem 2.15 (Orthogonal algebras (w even)). *Let Q be a symmetric bilinear form on a real vector space $V_\mathbb{R}$, and set $\mathfrak{g}_\mathbb{R} = \mathrm{End}(V_\mathbb{R}, Q)$ and $G_\mathbb{R} = \mathrm{Aut}(V_\mathbb{R}, Q) \simeq \mathrm{O}(a, b)$. The $\mathrm{Ad}(G_\mathbb{R})$–orbits in $\mathrm{Nilp}(\mathfrak{g}_\mathbb{R})$ are indexed by signed Young diagrams of size $n = \dim V_\mathbb{R}$ and signature (a, b) in which (i) the rows of odd length are signed, and (ii) the rows of even length are unsigned and occur with even multiplicity.*

Remark 2.16. For the analog of Theorem 2.10(b) with $G^\circ_\mathbb{R}$ the connected identity component of $\mathrm{SO}(a, b)$, see [6, Theorem 9.3.4]: In the case that the unsigned Young diagram characterizing the $\mathrm{Ad}(G_\mathbb{R})$–orbit $\mathcal{N}$ of N is very even, the orbit $\mathcal{N}$ decomposes into two $\mathrm{Ad}(G^\circ_\mathbb{R})$–orbits.

Example 2.17. Suppose that $G_\mathbb{R} = \mathrm{O}(3, 3)$. The nilpotent conjugacy classes in $\mathfrak{g}_\mathbb{R} = \mathfrak{so}(3, 3)$ are enumerated by the signed Young diagrams

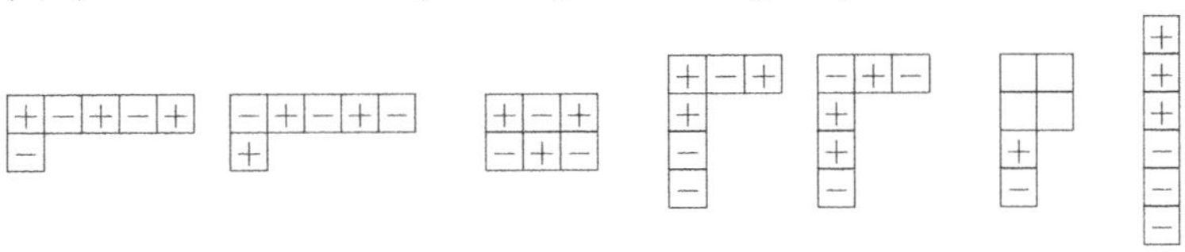

Example 2.18. Suppose that $G_\mathbb{R} = \mathrm{O}(4, 2)$. The nilpotent conjugacy classes in $\mathfrak{g}_\mathbb{R} = \mathfrak{so}(4, 2)$ are enumerated by the signed Young diagrams

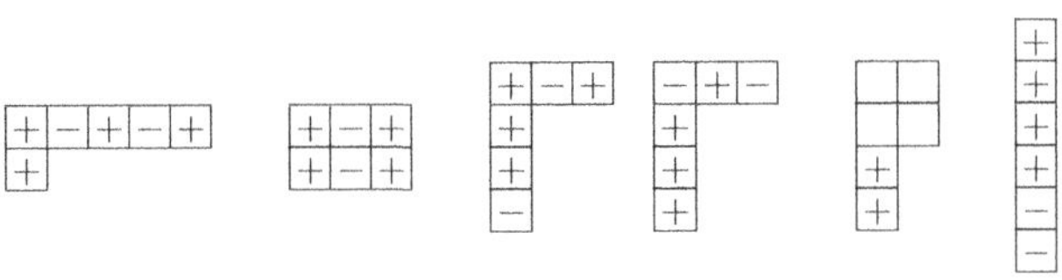

Example 2.19. Suppose that $G_{\mathbb{R}} = \mathrm{O}(5,1)$. The nilpotent conjugacy classes in $\mathfrak{g}_{\mathbb{R}} = \mathfrak{so}(5,1)$ are enumerated by the signed Young diagrams

$$\begin{array}{|c|c|c|}\hline + & - & + \\ \hline + \\ \cline{1-1} + \\ \cline{1-1} + \\ \cline{1-1} \end{array} \qquad \begin{array}{|c|}\hline + \\ \hline + \\ \hline + \\ \hline + \\ \hline + \\ \hline - \\ \hline \end{array}$$

Remark 2.20. The Lie algebras $\mathfrak{g}_{\mathbb{R}}$ of *compact* Lie groups $G_{\mathbb{R}}$ contain no nilpotent elements other than the trivial $N = 0$.

2.4 Partial order on conjugacy classes

Given two nilpotent elements $N_1, N_2 \in \mathfrak{g}_{\mathbb{R}}$, let $\mathcal{N}_i = \mathrm{Ad}(G_{\mathbb{R}})N_i$ denote the associated conjugacy classes. We define a partial order on the set of conjugacy classes by

$$\mathcal{N}_1 \leq \mathcal{N}_2 \quad \text{if} \quad \mathcal{N}_1 \subset \overline{\mathcal{N}}_2\,.$$

Đoković's Theorem 2.21 characterizes the partial order in terms of the signed Young diagram classifying the conjugacy classes.

Given a signed Young diagram Y, let Y' be the signed Young diagram obtained by removing the last (right-most) box from each row of Y. Inductively define $Y^{(k)}$ by $Y^{(0)} = Y$ and $Y^{(k+1)} = (Y^{(k)})'$. Given two signed Young diagrams Y_1 and Y_2 of signatures $s_1 = (s_1^+, s_1^-)$ and $s_2 = (s_2^+, s_2^-)$, respectively, we write $s_1 \leq s_2$ if $s_1^+ \leq s_2^+$ and $s_1^- \leq s_2^-$. Then we put a partial order the signed Young diagrams by $Y_1 \leq Y_2$ if $\mathrm{sig}\, Y_1^{(k)} \leq \mathrm{sig}\, Y_2^{(k)}$ for all k. We write $Y_1 < Y_2$ when $Y_1 \leq Y_2$ but $Y_1 \neq Y_2$. The following is [10, Theorem 5].

Theorem 2.21 (Đoković). *Let $\mathcal{N}_1, \mathcal{N}_2 \subset \mathfrak{g}_{\mathbb{R}}$ be two nilpotent $\mathrm{Ad}(G_{\mathbb{R}})$–conjugacy classes, and let Y_1 and Y_2 be the associated signed Young diagrams (§2.3). Then $\mathcal{N}_1 \subset \overline{\mathcal{N}}_2$ if and only if $Y_1 \leq Y_2$.*

Remark 2.22. The analogous partial order on $\mathrm{Ad}(G_{\mathbb{C}})$–conjugacy classes $\mathcal{N}_{\mathbb{C}} \subset \mathrm{Nilp}(\mathfrak{g}_{\mathbb{C}})$ was characterized by Gerstenhaber [11, 12].

Example 2.23. If we write $\mathcal{N}' \to \mathcal{N}$ to indicate $\mathcal{N}' < \mathcal{N}$, then the partial order on the nilpotent conjugacy classes of Example 2.13 is given by

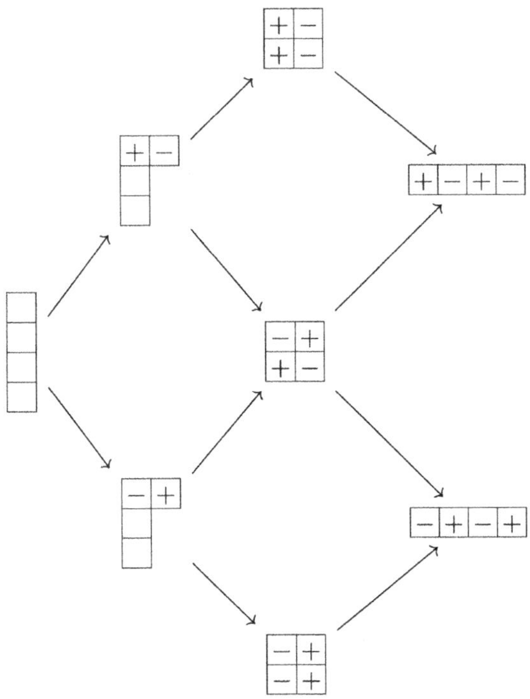

with the remaining relations given by transitivity.

Example 2.24. Likewise the partial order on the nilpotent conjugacy classes of Example 2.17 is given by

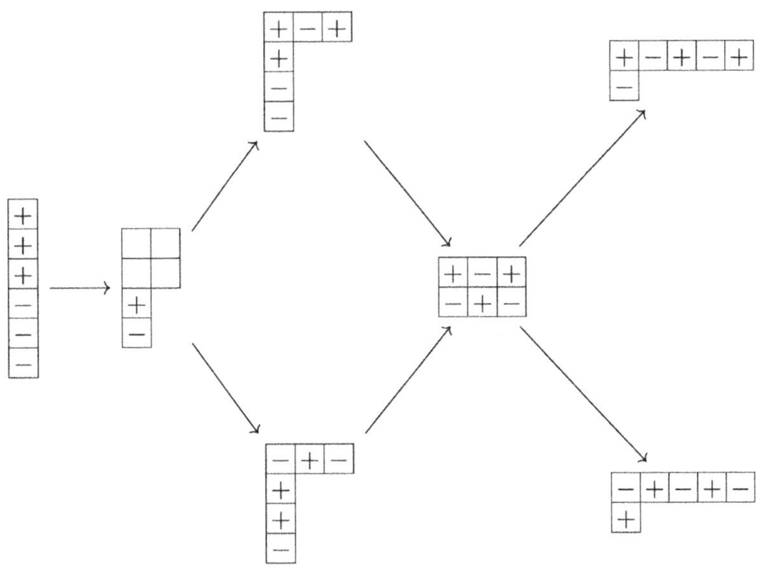

with the remaining relations given by transitivity.

Remark 2.25. In the event that $\mathcal{N}$ decomposes into two $\mathrm{Ad}(G_\mathbb{R}^\circ)$–orbits $\mathcal{N}_1$ and $\mathcal{N}_2$ (cf. Remark 2.16), the orbits are not comparable; that is, $\mathcal{N}_i \not\subset \overline{\mathcal{N}}_j$, for $i \neq j$.

2.5 Polarized mixed Hodge structures and signed Young diagrams

Let $D = G_\mathbb{R}/G_\mathbb{R}^0$ be a *period domain* parameterizing weight w, Q–polarized Hodge structures, so that

$$G_\mathbb{R} \;=\; \mathrm{Aut}(V_\mathbb{R}, Q)\,.$$

Let (F, N) be an $\mathbb{R}$–split nilpotent orbit on D, and let

$$\mathcal{N} \;:=\; \mathrm{Ad}(G_\mathbb{R})\cdot N$$

be the corresponding conjugacy class.

Let $W_\bullet(N)$ be the monodromy filtration of N; then $(F, W_\bullet(N))$ is a polarized mixed Hodge structure. Let

$$V_\mathbb{C} \;=\; \bigoplus V^{p,q}$$

be the associated Deligne bigrading. Without loss of generality we assume that $(F, W_\bullet(N))$ is $\mathbb{R}$–split. The associated *Hodge diamond* is the configuration of points in the pq–plane for which $V^{p,q} \neq 0$. In this section we explain how to construct the signed Young diagram indexing $\mathcal{N}$ from the Hodge diamond. (This, along with Đoković's Theorem 2.21, will give constraints on the degenerations associated with the faces of a nilpotent cone $\sigma \ni N$ underlying a nilpotent orbit on D. See §5 for an illustration.)

Fix p, q and define ℓ by $w + \ell = p + q$. Suppose $\ell \geq 0$ and let

$$P^{p,q} \;:=\; \ker\left\{N^{\ell+1} : V^{p,q} \to V^{p-\ell-1,q-\ell-1}\right\}$$

be the *N–primitive subspace*. (Recall that $N^\ell : V^{p,q} \to V^{p-\ell,q-\ell}$ is an isomorphism.) By our hypothesis that the polarized mixed Hodge structure is $\mathbb{R}$–split, we have $\overline{P^{p,q}} = P^{q,p}$. Moreover,

$$Q_\ell(\cdot,\cdot) \;:=\; Q(\cdot, N^\ell\cdot)$$

is a nondegenerate bilinear form on $(P^{p,q} + P^{q,p}) \cap V_\mathbb{R}$ satisfying the symmetry

$$Q_\ell(u,v) \;=\; (-1)^{w+\ell} Q_\ell(v,u)\,.$$

First suppose that $p = q$. Then $P^{p,q}$ is real and admits a basis of Q_ℓ–orthogonal real vectors. Given one such basis vector $v \in V_\mathbb{R}$,

$$v\,,\ Nv\,,\ \ldots\,,\ N^\ell v$$

is an N–string, and the polarization conditions assert

$$0 \;<\; Q(v, N^\ell v) \;=\; Q_\ell(v,v)\,.$$

In the lexicon of Đoković's [10], the "isomorphism class" of this N–string is the *rank* $\ell+1$ *gene*

$$g^+(\ell+1)\,, \quad \begin{cases} \text{if } w \text{ is even and } \ell \equiv 0 \bmod 4, \text{ or} \\ \text{if } w \text{ is odd and } \ell \equiv 1 \bmod 4; \end{cases}$$

$$g^-(\ell+1)\,, \quad \begin{cases} \text{if } w \text{ is even and } \ell \equiv 2 \bmod 4, \text{ or} \\ \text{if } w \text{ is odd and } \ell \equiv 3 \bmod 4. \end{cases}$$

Pictorially the *positive* gene $g^+(\ell+1)$ is identified with a row of $\ell+1$ boxes, labeled with alternating signs and beginning with $+$; that is, $g^+(\ell+1)$ is visualized as $\boxed{+}\boxed{-}\boxed{+}\boxed{-}\cdots$. The *negative* gene $g^-(\ell+1)$ is depicted as $\boxed{-}\boxed{+}\boxed{-}\boxed{+}\cdots$.

Next suppose that $p \neq q$. Fix $\xi = u + \mathbf{i}v \in P^{p,q}$, with $u, v \in V_{\mathbb{R}}$. The polarization conditions assert that $Q_\ell(\xi,\xi) = 0$; equivalently,

$$Q_\ell(u,u) \;=\; Q_\ell(v,v)\,,$$

and

$$Q_\ell(u,v) \;=\; 0 \quad \text{if } w+\ell \text{ is even.}$$

The polarization conditions also impose $\mathbf{i}^{p-q}Q_\ell(\xi,\bar{\xi}) > 0$ for all $0 \neq \xi$. Equivalently we have the following:

$$\begin{array}{llll} \text{If } p-q \equiv 0 \bmod 4, & \text{then } \; 0 < Q_\ell(u,u)\,; & & \\ \text{if } p-q \equiv 2 \bmod 4, & \text{then } \; 0 > Q_\ell(u,u)\,; & & \\ \text{if } p-q \equiv 1 \bmod 4, & \text{then } \; 0 = Q_\ell(u,u) & \text{and} & 0 < Q_\ell(u,v)\,; \\ \text{if } p-q \equiv 3 \bmod 4, & \text{then } \; 0 = Q_\ell(u,u) & \text{and} & 0 > Q_\ell(u,v)\,. \end{array}$$

(Note that $p-q$ is even if and only if $p+q = w+\ell$ is even.) Again, in the language of [10], a *chromosome* is a formal linear combination of genes with non-negative integral coefficients. (A chromosome is just a signed Young diagram.) The two N–strings $\{u,\ldots,N^\ell u\}$ and $\{v,\ldots,N^\ell v\}$ in $V_{\mathbb{R}}$ correspond to the chromosome

$$\begin{array}{ll} 2g(\ell+1)\,, & \text{if } w+\ell \text{ is odd;} \\ 2g^+(\ell+1)\,, & \begin{cases} \text{if } w \text{ is even and } \ell, p-q \equiv 0 \bmod 4, \text{ or} \\ \text{if } w \text{ is even and } \ell, p-q \not\equiv 0 \bmod 4, \text{ or} \\ \text{if } w \text{ is odd, } \ell \equiv 1 \bmod 4 \text{ and } p-q \equiv 0 \bmod 4, \text{ or} \\ \text{if } w \text{ is odd, } \ell \equiv 3 \bmod 4 \text{ and } p-q \not\equiv 0 \bmod 4; \end{cases} \\ 2g^-(\ell+1)\,, & \text{otherwise.} \end{array}$$

The *unpolarized* gene $g(\ell+1)$ is indicated by a row of $\ell+1$ boxes, without labels.

Definition 2.26. The *partially signed Young diagram* $Y(F,N)$ (or chromosome) *associated with the* $\mathbb{R}$*–split polarized mixed Hodge structure* (F,N) is the union of genes obtained from the N–string decomposition of the standard representation $V_{\mathbb{R}}$.

Note that $G_{\mathbb{R}}$ acts on $\mathbb{R}$–split polarized mixed Hodge structures by $g\cdot(F,N) = (gF, \mathrm{Ad}_g N)$, for $g \in G_{\mathbb{R}}$. It follows from the classification results of §2.3 that $Y(F,N)$ depends only on the $G_{\mathbb{R}}$–conjugacy class of (F,N).

Example 2.27 (Period domain for $\mathbf{h} = (3,3,3)$). This example was studied by Cattani and Kaplan in [5, §4]. We have $G_{\mathbb{R}} = \mathrm{O}(3,6)$, and there are five conjugacy classes of $\mathbb{R}$–split PMHS.

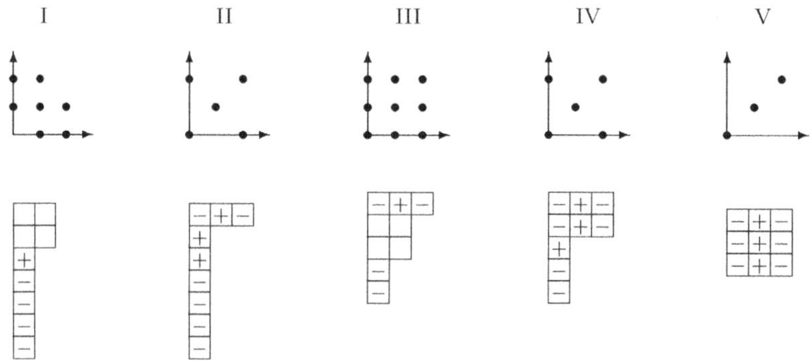

Đoković's Theorem 2.21 yields

$$\mathcal{N}_{\mathrm{I}} < \mathcal{N}_{\mathrm{II}} < \mathcal{N}_{\mathrm{III}} < \mathcal{N}_{\mathrm{IV}} < \mathcal{N}_{\mathrm{IV}} .$$

Example 2.28 (Period domain for $\mathbf{h} = (1,1,1,1,1,1)$). We have $G_{\mathbb{R}} = \mathrm{Sp}(3,\mathbb{R})$, and there are seven (conjugacy classes of) $\mathbb{R}$–split PMHS.

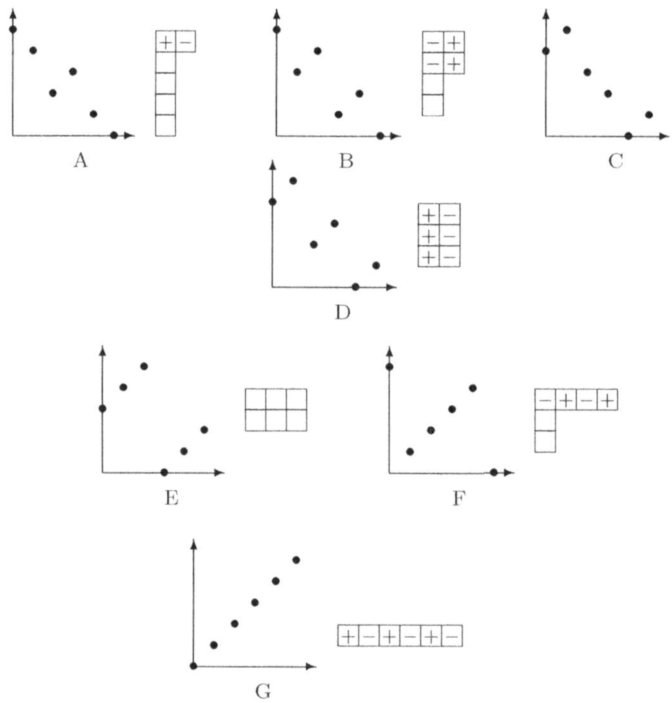

From Theorem 2.21 we find

$$\mathcal{N}_{\mathrm{A}} < \mathcal{N}_{\mathrm{C}} ,\ \mathcal{N}_{\mathrm{F}} ; \quad \mathcal{N}_{\mathrm{B}} < \mathcal{N}_{\mathrm{E}} ,\ \mathcal{N}_{\mathrm{F}} ; \quad \mathcal{N}_{\mathrm{C}} < \mathcal{N}_{\mathrm{D}} ,\ \mathcal{N}_{\mathrm{E}} ; \quad \mathcal{N}_{\mathrm{D}} ,\ \mathcal{N}_{\mathrm{E}} ,\ \mathcal{N}_{\mathrm{F}} < \mathcal{N}_{\mathrm{G}} ,$$

where the relations obtained by the transitivity of the partial order are omitted.

3 Nilpotent cones

The goal of this section is to describe one approach to identifying the nilpotent cones that underlie nilpotent orbits on a period domain (or, more generally, a Mumford–Tate domain [13]). We begin in §3.1 by reviewing the definition of nilpotent orbits; in §3.1 we outline the strategy. The approach will be worked out for weight two period domains in §4, and illustrated in the case of Hodge numbers $\mathbf{h} = (2, m, 2)$ in §5.

3.1 Nilpotent orbits

Let $V_\mathbb{R}$ be a real vector space of dimension n with a $\mathbb{Q}$–structure defined by a lattice $V_\mathbb{Z} \subset V_\mathbb{R}$. Fix $w \in \mathbb{Z}$ and let Q be a nondegenerate $(-1)^w$–symmetric bilinear form on $V_\mathbb{R}$ defined over $\mathbb{Q}$. Fix Hodge numbers $\mathbf{h} = \{h^{p,q} \mid p+q = w\}$. Let D denote the period domain parameterizing Q–polarized Hodge structures on $V_\mathbb{Q}$ with Hodge numbers $\mathbf{h}$. Let $\check{D}$ denote the compact dual of D. A (m–variable) *nilpotent orbit* on D consists of a pair $(F; N_1, \ldots, N_m)$ such that $F \in \check{D}$, the $N_i \in \mathfrak{g}_\mathbb{R}$ are commuting nilpotents and $N_i F^p \subset F^{p-1}$, and the holomorphic map $\psi : \mathbb{C}^m \to \check{D}$ defined by

$$\psi(z^1, \ldots, z^m) \ = \ \sum_i \exp(z^i N_i) F \tag{3.1}$$

has the property that $\psi(z) \in D$ for $\mathrm{Im}(z^i) \gg 0$. The associated (open) *nilpotent cone* is

$$\sigma \ = \ \{t^i N_i \mid t^i > 0\} . \tag{3.2}$$

Recall that the monodromy filtration $W_\bullet(N)$ is independent of our choice of $N \in \sigma$; so $W_\bullet(\sigma)$ is well-defined. Given a nilpotent orbit we will assume, without loss of generality, that the polarized mixed Hodge structure $(F, W_\bullet(\sigma))$ is $\mathbb{R}$–split. Let

$$V_\mathbb{C} \ = \ \oplus V^{p,q} \quad \text{and} \quad \mathfrak{g}_\mathbb{C} \ = \ \oplus \mathfrak{g}^{p,q}$$

denote the Deligne bigradings [4, (2.12)]. Recall that

$$\sigma \ \subset \ \mathfrak{g}_\mathbb{R}^{-1,-1} .$$

3.2 Identification of cones underlying nilpotent orbits

Observe that

$$\mathfrak{m}_\mathbb{C} \ := \ \oplus_p \mathfrak{g}^{p,p} \ \subset \ \mathfrak{g}_\mathbb{C}$$

is the subalgebra of $\mathfrak{g}_\mathbb{C}$ preserving the subspaces

$$V_m \ := \ \bigoplus_{q-p=m} V^{p,q} \ \subset \ V_\mathbb{C} \, , \quad m \in \mathbb{Z}. \tag{3.3}$$

Visually, $\mathfrak{m}$ corresponds to the dots on the diagonal $p = q$ in the Hodge diamond. Because the PMHS is $\mathbb{R}$–split, the subalgebra $\mathfrak{m}$ is defined over $\mathbb{R}$. Let $M^0_{\mathbb{R}} \subset G_{\mathbb{R}}$ be the *connected* Lie subgroup with (Levi) Lie algebra

$$\mathfrak{m}^0_{\mathbb{R}} := \mathfrak{g}^{0,0}_{\mathbb{R}} .$$

It will be convenient to note that

$$\text{the real form } \mathfrak{m}^0_{\mathbb{R}} = \mathfrak{g}^{0,0}_{\mathbb{R}} \text{ is the subalgebra of } \mathfrak{g}_{\mathbb{R}} \text{ preserving the } V^{p,q}. \tag{3.4}$$

Lemma 3.5. *Let (F, N) be an $\mathbb{R}$–split nilpotent orbit on $D = G_{\mathbb{R}}/G^0_{\mathbb{R}}$ with Deligne bigrading $\mathfrak{g}_{\mathbb{C}} = \oplus \mathfrak{g}^{p,q}$. Let $M^0_{\mathbb{R}} \subset M_{\mathbb{R}}$ be as defined above.*

(a) *The orbit*

$$\mathcal{N}^0 := \mathrm{Ad}(M^0_{\mathbb{R}}) \cdot N$$

is open in $\mathfrak{g}^{-1,-1}_{\mathbb{R}}$.[2]

(b) *Suppose that $\mathcal{W}^\circ_N$ is the connected component of*

$$\mathcal{W}_N := \{N' \in \mathfrak{g}^{-1,-1}_{\mathbb{R}} \mid W(N) = W(N')\}$$

containing N. Then $\mathcal{W}^\circ_N = \mathcal{N}^0$.

The lemma is well-known; the proof is included in the appendix for completeness. The key points are that (i) $\mathcal{W}_N$ is preserved under the adjoint action of $M^0_{\mathbb{R}}$, and (ii) the orbit $\mathrm{Ad}(M^0_{\mathbb{R}}) \cdot N'$ is open in $\mathfrak{g}^{-1,-1}_{\mathbb{R}}$ for every $N' \in \mathcal{W}_N$. Thus $\mathcal{W}_N$ is a disjoint union of open $\mathrm{Ad}(M^0_{\mathbb{R}})$–orbits.

From Lemma 3.5(b) we obtain:

Corollary 3.6. *Assume the hypotheses of Lemma 3.5. If a nilpotent cone σ containing N underlies a nilpotent orbit, then*

$$\sigma \subset \mathcal{N}^0 = \mathrm{Ad}(M^0_{\mathbb{R}}) \cdot N \subset \mathfrak{g}^{-1,-1} \quad \text{and} \quad N_i \in \overline{\mathcal{N}}^0 .$$

Remark 3.7. Together Theorem 2.21 and Corollary 3.6 give us representation theoretic constraints on the degenerations associated with the faces of a nilpotent cone underlying a nilpotent orbit. See §5.2 for an illustration.

For the converse to Corollary 3.6, recall Cattani and Kaplan's [3, Theorem 2.3]

Theorem 3.8 (Cattani–Kaplan). *Fix $F \in \check{D}$ and a nilpotent cone* (3.2) *with the properties that:*

(i) *$N_i F^p \subset F^{p-1}$ for every i;*

(ii) *$N_i^{k+1} = 0$, where k is the level of the Hodge structures on D; and*

(iii) *the filtration $W_\bullet(N)$ does not depend on the choice of $N \in \sigma$.*

[2] The superscript of 0 in $\mathcal{N}^0$ is meant to distinguish these nilpotent $\mathrm{Ad}(M^0_{\mathbb{R}})$–conjugacy classes from the $\mathrm{Ad}(G_{\mathbb{R}})$–conjugacy classes $\mathcal{N}$ of §2.3.

Then $(F;N)$ is a limiting mixed Hodge structure for some $N \in \sigma$, *if and only if $(F;N_1,\dots,N_m)$ is an m–variable nilpotent orbit.*

This, along with Lemma 3.5(b), yields the converse to Corollary 3.6:

Proposition 3.9. *Given an $\mathbb{R}$–split PMHS (F,N) on D, let $M^0_{\mathbb{R}}$ and $\mathcal{N}^0$ be as defined above. If $\sigma \subset \mathcal{N}^0$ is a nilpotent cone, then σ underlies a nilpotent orbit at F.*

The upshot of this discussion is

Remark 3.10. The cones σ underlying a multivariable nilpotent orbit on a domain D may be identified as follows. Begin with an $\mathbb{R}$–split PMHS $(F,W_\bullet(N))$ on D. The Deligne bigrading determines the diagonal Levi subgroup M. Any nilpotent cone $\sigma \subset \mathcal{N}^0$ will underlie a nilpotent orbit, and all such cones arise in this fashion. So to identify the nilpotent cones underlying a nilpotent orbit we must have a good enough/explicit enough geometric description of $\mathcal{N}^0$ to understand how the nilpotent cones can "fit" inside. So the strategy proceeds in three steps:

Step 1: Enumerate the $\mathbb{R}$–split PMHS (F,N) on D. For an arbitrary Mumford–Tate domain $D = G_{\mathbb{R}}/G^0_{\mathbb{R}}$, with $G_{\mathbb{R}}$ connected, these are given (up to the action of $G_{\mathbb{R}}$) by [22]. As Cattani has pointed out, if D is a period domain and $G_{\mathbb{R}}$ is the full automorphism group $\mathrm{Aut}(V_{\mathbb{R}},Q)$, then the $\mathbb{R}$–split PMHS are enumerated (again, up to the action of $G_{\mathbb{R}}$) by the Hodge diamonds.

Step 2: Determine the $\mathrm{Ad}(M^0_{\mathbb{R}})$–orbit $\mathcal{N}^0$ of $N \in \mathfrak{g}^{-1,-1}_{\mathbb{R}}$. This will require a good description of $\mathfrak{g}^{-1,-1}_{\mathbb{R}}$ as an $M^0_{\mathbb{R}}$–module. In the case of weight two period domains this description is given in §4.

Step 3: Understand the nilpotent cones $\sigma \subset \mathcal{N}^0$, by which we mean that we should have a good enough/explicit enough geometric description of $\mathcal{N}^0$ (from Step 2) to understand how the nilpotent cones can "fit" inside. In practice this involves identifying the abelian subalgebras of $\mathfrak{g}^{-1,-1}_{\mathbb{R}}$ and their intersections with $\mathcal{N}^0$.

Also, in the context of "understanding the cones", it should be noted that a theorem of Đoković's can help us understand, given a nilpotent on a period domain, constraints on the degenerations coming from the faces of the cone; this is discussed in §5.2.

3.3 The CKS commuting SL(2)'s

To a nilpotent cone σ underlying a nilpotent orbit on a Hodge domain D, Cattani, Kaplan and Schmid [4] associate a set of commuting $\mathfrak{sl}(2)$'s. When the nilpotent orbit is $\mathbb{R}$–split the $\mathfrak{sl}(2)$'s are contained in $\mathfrak{m}_{\mathbb{R}}$ by construction, and therefore in $\mathfrak{m}^{ss}_{\mathbb{R}} = [\mathfrak{m}_{\mathbb{R}},\mathfrak{m}_{\mathbb{R}}]$.

Given a Cartan decomposition $\mathfrak{m}^{ss}_{\mathbb{R}} = \mathfrak{k}\oplus\mathfrak{k}^{\perp}$, the *real rank* of $\mathfrak{m}^{ss}_{\mathbb{R}}$ is the dimension of a maximal subspace of $\mathfrak{k}^{\perp}$ consisting of commuting semisimple elements.

Lemma 3.11 (Nilpotent cones versus commuting $\mathfrak{sl}(2)$'s). *The number of (non-trivial) commuting $\mathfrak{sl}(2)$'s is bounded by the real rank of the semisimple factor $\mathfrak{m}_{\mathbb{R}}^{ss} = [\mathfrak{m}_{\mathbb{R}}, \mathfrak{m}_{\mathbb{R}}]$.*

Proof. To see this, let $Y_1, \ldots, Y_\ell$ denote the neutral elements of the commuting $\mathfrak{sl}(2)$'s. They are linearly independent and so span an ℓ–dimensional abelian subspace of $\mathfrak{m}_{\mathbb{R}}$ consisting of semisimple elements. It remains to show that we may choose a Cartan decomposition $\mathfrak{m}_{\mathbb{R}}^{ss} = \mathfrak{k} \oplus \mathfrak{k}^{\perp}$ so that $Y_i \in \mathfrak{k}^{\perp}$; for then $s \leq \mathrm{rank}_{\mathbb{R}}\, \mathfrak{m}_{\mathbb{R}}^{ss}$. To see this, let $\mathfrak{s}_i = \mathrm{span}\{N_i^+, Y_i, N_i\} \subset \mathfrak{m}_{\mathbb{R}}^{ss}$ denote the i–th $\mathfrak{sl}(2)$. Then $\mathfrak{k}_i = \mathrm{span}_{\mathbb{R}}\{N_i^+ - N_i\}$ and $\mathfrak{k}_i^{\perp} = \mathrm{span}_{\mathbb{R}}\{Y_i, N_i^+ + N_i\}$ defines a Cartan decomposition of $\mathfrak{s}_i$. Taking the sum yields a Cartan decomposition of $\mathfrak{s}_1 \oplus \cdots \oplus \mathfrak{s}_\ell \subset \mathfrak{m}_{\mathbb{R}}^{ss}$. This Cartan decomposition can be extended to one of $\mathfrak{m}_{\mathbb{R}}^{ss}$ [20, Theorem 6]. □

3.4 Two special cases

We finish §3 with discussions of two special cases that we will encounter in these notes.

3.4.1 The Hermitian case

We say that a simple factor $\mathfrak{m}' \subset \mathfrak{m}$ is *Hermitian* if any of the following equivalent conditions hold:[3]

(a) $\mathfrak{m}' = (\mathfrak{m}' \cap \mathfrak{g}^{1,1}) \oplus (\mathfrak{m}' \cap \mathfrak{g}^{0,0}) \oplus (\mathfrak{m}' \cap \mathfrak{g}^{-1,-1})$;

(b) $\mathfrak{m}' \cap \mathfrak{g}^{p,p} = 0$ when $|p| \geq 2$;

(c) $\mathfrak{m}' \cap \mathfrak{g}^{-1,-1}$ is abelian.

When $\mathfrak{m}$ is Hermitian there will exist nilpotent cones $N \in \sigma \subset \mathcal{N}^0$ underlying nilpotent orbits that are open in $\mathfrak{g}_{\mathbb{R}}^{-1,-1}$.

3.4.2 The contact case

We say that a simple factor $\mathfrak{m}' \subset \mathfrak{m}$ is *contact* if $\dim \mathfrak{m}' \cap \mathfrak{g}^{2,2} = 1$ and $\mathfrak{m}' \cap \mathfrak{g}^{p,p} = 0$ for all $|p| \geq 3$. In this case the maximal abelian subspaces of $\mathfrak{g}^{-1,-1}$ are the Lagrangian subspaces Λ of a symplectic form ν on $\mathfrak{g}^{-1,-1}$ that is invariant under the reductive (Levi) subalgebra $\mathfrak{g}^{0,0}$. The symplectic form is defined (up to scale) by choosing a nonzero z in the one–dimensional $\mathfrak{g}_{\mathbb{R}}^{-2,-2}$ and setting $[x, y] =: \nu(x, y) z$ for any $x, y \in \mathfrak{g}^{-1,-1}$.

4 Weight two period domains

Suppose that D is a period domain parameterizing effective weight two Hodge structures (Hodge numbers $\mathbf{h} = (h^{2,0}, h^{1,1}, h^{0,2})$). Our goal in this section is to address Steps 2 and 3 in the strategy (outlined in Remark 3.10) to identify the

[3] Equivalence requires that we assume the horizontal distribution is bracket–generating.

nilpotent cones underlying nilpotent orbits on D. Section 4.1 provides the necessary descriptions of both $M^0_{\mathbb{R}}$ and $\mathfrak{g}_{\mathbb{R}}^{-1,-1}$ as an $M^0_{\mathbb{R}}$–representation to explicitly describe the orbits $\mathcal{N}^0$. Section 4.2 describes a decomposition of the orbits $\mathcal{N}^0$ into simpler objects, culminating in an explicit description of $\mathcal{N}^0$ by Proposition 4.18.

4.1 The representation theory

We assume given an $\mathbb{R}$–split nilpotent orbit (F, N) on D. The Hodge diamond of the Deligne bigrading $V_{\mathbb{C}} = \oplus V^{p,q}$ is contained in

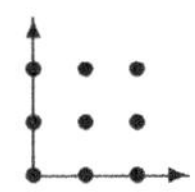

In particular,

$$V_0 = V^{0,0} \oplus V^{1,1} \oplus V^{2,2}\,, \quad V_1 = V^{1,0} \oplus V^{2,1} \quad \text{and} \quad V_2 = V^{2,0}\,.$$

Our first goal in this section is to describe the subgroup

$$M_{\mathbb{R}} = \{g \in G_{\mathbb{R}} \mid g(V_m) \subset V_m\,,\ \forall m\} \tag{4.1}$$

of $G_{\mathbb{R}}$ preserving the subspaces (3.3). Note that $\mathfrak{m}_{\mathbb{R}}$ is the Lie algebra of $M_{\mathbb{R}}$. In an abuse of notation, we let $G_{\mathbb{R}} \cap \mathrm{Aut}(V_m)$ denote the subgroup of elements $g \in G_{\mathbb{R}}$ that preserve V_m (and therefore also preserve V_{-m}) and act trivially on V_ℓ for all $\ell \neq \pm m$. Let

$$Q_0 := Q|_{V_0}$$

denote the restriction of Q to V_0. Define a nondegenerate skew-Hermitian form Q_1^* on V_1 by

$$Q_1^*(u,v) := \mathbf{i}Q(u,\bar{v})\,,$$

and a nondegenerate Hermitian form Q_2^* on V_2 by

$$Q_2^*(u,v) := -Q(u,\bar{v})\,.$$

Let

$$a := \dim V^{2,2}\,, \quad a+b := \dim V^{1,1}\,, \quad c := \dim V^{1,2} \quad \text{and} \quad d := \dim V^{0,2}\,.$$

Then Q_0 has signature $(a+b, 2a)$, so that

$$\mathrm{Aut}(V_{0,\mathbb{R}}, Q_0) \simeq \mathrm{O}(a+b, 2a)\,.$$

Next note that $g \in \mathrm{Aut}(V_1)$ stabilizes the skew-Hermitian Q_1^* if and only if it stabilizes the Hermitian $\mathbf{i}Q_1^*$; as will be noted in Remark 4.5, the latter has signature (c, c) so that

$$\mathrm{Aut}(V_1, Q_1^*) = \mathrm{Aut}(V_1, \mathbf{i}Q_1^*) \simeq \mathrm{U}(c,c)\,.$$

Finally, we note that the Hermitian form Q_2^* is positive definite, so that

$$\mathrm{Aut}(V_2, Q_2^*) \simeq \mathrm{U}(d)\,.$$

Proposition 4.2. *Let (F, N) be an $\mathbb{R}$–split nilpotent orbit on a period domain parameterizing effective weight two Hodge structures. The diagonal subgroup is*

$$\begin{aligned} M_{\mathbb{R}} &= (G_{\mathbb{R}} \cap \mathrm{Aut}(V_0)) \times (G_{\mathbb{R}} \cap \mathrm{Aut}(V_1)) \times (G_{\mathbb{R}} \cap \mathrm{Aut}(V_2)) \\ &= \mathrm{Aut}(V_{0,\mathbb{R}}, Q_0) \times \mathrm{Aut}(V_1, Q_1^*) \times \mathrm{Aut}(V_2, Q_2^*) \\ &\simeq \mathrm{O}(a+b, 2a) \times \mathrm{U}(c, c) \times \mathrm{U}(d)\,. \end{aligned} \tag{4.3}$$

Remark 4.4. For the groups $M_{\mathbb{R}}$ of Proposition 4.2, the subalgebra $\mathfrak{m}_{\mathbb{R}}^{\mathrm{ss}}$ has real rank $c + \min\{2a, a+b\}$, cf. [19].

Proof. From (4.1) and $\overline{V_m} = V_{-m}$ we see that

$$M_{\mathbb{R}} = G_{\mathbb{R}} \cap \{\mathrm{Aut}(V_0) \times \mathrm{Aut}(V_1) \times \mathrm{Aut}(V_2)\}\,.$$

From the facts that (i) $G_{\mathbb{R}} = \mathrm{Aut}(V_{\mathbb{R}}, Q)$ and (ii) the subspaces $V_m + V_{-m}$, $m = 0, 1, 2$, are all pairwise orthogonal we see that the first equation of (4.3) holds. It remains to identify the factors $G_{\mathbb{R}} \cap \mathrm{Aut}(V_m)$, $m = 0, 1, 2$. As V_0 is defined over $\mathbb{R}$, and $Q|_{V_{0,\mathbb{R}}}$ is nondegenerate of signature $(a+b, 2a)$, it follows directly that

$$G_{\mathbb{R}} \cap \mathrm{Aut}(V_0) = \mathrm{Aut}(V_{0,\mathbb{R}}, Q) \simeq \mathrm{O}(a+b, 2a)\,.$$

Next consider $G_{\mathbb{R}} \cap \mathrm{Aut}(V_2)$. From the polarization condition $0 < -Q(u, \overline{u})$ for all $0 \neq u \in V^{2,0} = V_2$, we see that we may pick a basis $\{z_s = x_s + \mathrm{i} y_s \mid s = 1, \ldots, d\}$ so that $Q(z_s, \bar{z}_t) = -\delta_{st}$ and $x_s, y_s \in V_{\mathbb{R}}$ for all s, t. Next observe that:

(i) From $-\delta_{st} = Q(z_s, \bar{z}_t)$ and $0 = Q(z_s, z_t)$ we may deduce that $-\frac{1}{2}\delta_{st} = Q(x_s, x_t) = Q(y_s, y_t)$ and $0 = Q(x_s, y_t)$ for all s, t.

(ii) Given an element $g \in G_{\mathbb{R}}$ preserving V_2, we may define a real $2d \times 2d$ matrix $A = A(g) = (A^i_j)$ by $g(x_s) =: A^r_s x_r + A^{d+r}_s y_r$ and $g(y_s) =: A^r_{d+s} x_r + A^{d+r}_{d+s} y_r$. (Here we employ the Einstein summation convention: any index appearing as a subscript and superscript is summed over. In particular, in the previous, we sum over r.) From the observations of (i) we see that g preserves Q if and only if A is orthogonal, $A^t = A^{-1}$. Moreover, g preserves V_2 if and only if $A^r_s = A^{d+r}_{d+s}$ and $A^r_{d+s} = -A^{d+r}_s$. Consequently, $g(z_s) = B^r_s z_r$, where $B^r_s := A^r_s + \mathrm{i} A^r_{d+s}$, defines a complex $d \times d$ matrix $B = (B^r_s)$ satisfying $\overline{B}^t = B^{-1}$. That is, $g \in \mathrm{Aut}(V_2, Q_2^*) \simeq \mathrm{U}(d)$.

(iii) Conversely, reversing the argument of (ii), we see that any $g \in \mathrm{Aut}(V_2, Q_2^*)$ defines an element of $G_{\mathbb{R}} \cap \mathrm{Aut}(V_2)$.

From (ii) and (iii) we deduce that

$$G_{\mathbb{R}} \cap \mathrm{Aut}(V_2) = \mathrm{Aut}(V_2, Q_2^*)\,.$$

It remains to show that $G_{\mathbb{R}} \cap \mathrm{Aut}(V_1) = \mathrm{Aut}(V_1, Q_1^*)$.

(iv) If $\{w_s\}_{s=1}^c$ is a basis of $V^{2,1}$, then $\{\overline{w}_s\}_{s=1}^c$, $\{N w_s\}_{s=1}^c$ and $\{N \overline{w}_s\}_{s=1}^c$ are bases of $V^{1,2}$, $V^{1,0}$ and $V^{0,1}$, respectively. The polarization condition $\mathrm{i}Q(u, N\bar{u}) > 0$, and the fact that $Q(V^{p,q}, V^{r,s}) = 0$ when either $p + s \neq 2$

or $q+r\neq 2$, imply that we may choose the basis $\{w_s\}_{s=1}^c$ of $V^{2,1}$ so that $Q_1^*(w_s, Nw_t) = \delta_{st}$ and $Q_1^*(w_s, w_t) = 0 = Q_1^*(Nw_s, Nw_t)$ for all s,t. That is, Q_1^* is represented by the matrix

$$\mathcal{J} := \begin{pmatrix} 0 & I_c \\ -I_c & 0 \end{pmatrix},$$

where is I_c the $c\times c$ identity matrix, with respect to the basis $\{w_s, Nw_s\}_{s=1}^c$ of V^1.

Remark 4.5. The Hermitian form $\mathbf{i}Q_1^*(\cdot,\cdot) = -Q(\cdot,\bar{\cdot})$ is represented by the matrix $2\begin{pmatrix} I_c & 0 \\ 0 & -I_c \end{pmatrix}$ with respect to the basis $\{w_s+\mathbf{i}Nw_s\,,\, w_s-\mathbf{i}Nw_s\}_{s=1}^c$ of V_1.

(v) Define $u_s, v_s \in V_{\mathbb{R}}$ by $w_s = u_s + \mathbf{i}v_s$. Then

$$\begin{aligned} 0 = Q(u_s,u_t) \;&=\; Q(u_s,v_t) \;=\; Q(v_s,v_t)\,,\\ 0 = Q(Nu_s,Nu_t) \;&=\; Q(Nu_s,Nv_t) \;=\; Q(Nv_s,Nv_t)\,,\\ 0 = Q(u_s,Nu_t) \;&=\; Q(v_s,Nv_t)\,,\\ \frac{1}{2}\delta_{st} = Q(u_s,Nv_t) \;&=\; -Q(v_s,Nu_t)\,. \end{aligned}$$

(vi) Given an element $g\in G_{\mathbb{R}}$ preserving V_1, we may define a real $4c\times 4c$ matrix $A = A(g) = (A_j^i)$ by

$$\begin{aligned} g(u_s) &=: A_s^r u_r + A_s^{c+r} v_r + A_s^{2c+r} Nu_r + A_s^{3c+r} Nv_r\,,\\ g(v_s) &=: A_{c+s}^r u_r + A_{c+s}^{c+r} v_r + A_{c+s}^{2c+r} Nu_r + A_{c+s}^{3c+r} Nv_r\,,\\ g(Nu_s) &=: A_{2c+s}^r u_r + A_{2c+s}^{c+r} v_r + A_{2c+s}^{2c+r} Nu_r + A_{2c+s}^{3c+r} Nv_r\,,\\ g(Nv_s) &=: A_{3c+s}^r u_r + A_{3c+s}^{c+r} v_r + A_{3c+s}^{2c+r} Nu_r + A_{3c+s}^{3c+r} Nv_r\,. \end{aligned}$$

(Again, the Einstein summation convention is in effect and we sum over r.) From the observations of (v) we see that g preserves Q if and only if $A^tJA = J$, where

$$J := \begin{pmatrix} 0 & 0 & 0 & I_c \\ 0 & 0 & -I_c & 0 \\ 0 & -I_c & 0 & 0 \\ I_c & 0 & 0 & 0 \end{pmatrix}$$

is the $4c\times 4c$ matrix representing $Q|_{V_1\oplus\overline{V_1}}$ with respect to the basis $\{u_s, v_s, Nu_s, Nv_s\}_{s=1}^c$; here I_c is the $c\times c$ identity matrix. As in (ii) we see that g preserves V_1 if and only if

$$\begin{aligned} A_s^r &= A_{c+s}^{c+r}, & A_{c+s}^r &= -A_s^{c+r},\\ A_s^{2c+r} &= A_{c+s}^{3c+r}, & A_{c+s}^{2c+r} &= -A_s^{3c+r},\\ A_{2c+s}^r &= A_{3c+s}^{c+r}, & A_{3c+s}^r &= -A_{2c+s}^{c+r},\\ A_{2c+s}^{2c+r} &= A_{3c+s}^{3c+r}, & A_{3c+s}^{2c+r} &= -A_{2c+s}^{3c+r}\,. \end{aligned}$$

Consequently, $g(w_s) = B^r_s w_r + B^{c+r}_s N w_r$ and $g(N w_s) = B^r_{c+s} w_r + B^{c+r}_{c+s} N w_r$, where $B^r_s = A^r_s + \mathbf{i} A^r_{c+s}$, $B^{c+r}_s = A^{2c+r}_s + \mathbf{i} A^{2c+r}_{c+s}$, $B^r_{c+s} = A^r_{2c+s} + \mathbf{i} A^r_{3c+s}$ and $B^{c+r}_{c+s} = A^{2c+r}_{2c+s} + \mathbf{i} A^{2c+r}_{3c+s}$ defines a $2c \times 2c$ complex matrix $B = (B^i_j)$ satisfying $\overline{B}^t \mathcal{J} B = \mathcal{J}$. That is, $g \in \mathrm{Aut}(V_1, Q^*_1) \simeq \mathrm{U}(c,c)$.

(vii) Conversely, reversing the computation of (vi), we see that any $g \in \mathrm{Aut}(V_1, Q^*_1)$ defines an element of $G_\mathbb{R} \cap \mathrm{Aut}(V_1)$.

From (vi) and (vii) we see that $G_\mathbb{R} \cap \mathrm{Aut}(V_1) = \mathrm{Aut}(V_1, Q^*_1)$. □

Remark 4.6. It will be helpful to select a basis for $V_{0,\mathbb{R}}$. The polarization condition $0 < Q(u, N^2 u)$ for all $0 \neq u \in V^{2,2}_\mathbb{R}$ implies that we may pick a basis $\{e_s \mid 1 \leq s \leq a\}$ of $V^{2,2}_\mathbb{R}$ so that $Q(e_s, N^2 e_t) = \delta_{st}$. Note that $\{N^2 e_s\}_{s=1}^a$ is a basis of $V^{0,0}_\mathbb{R}$, and that the $\{N e_s\}_{s=1}^a \subset V^{1,1}_\mathbb{R}$ are linearly independent. Moreover, $Q(N e_s, N e_t) = -\delta_{st}$, and we may complete the $\{N e_s\}$ to a basis $\{N e_s, f_j \mid s = 1, \ldots, a,\ j = 1, \ldots, b\}$ of $V^{1,1}_\mathbb{R}$ so that $Q(N e_s, f_j) = 0$ and $Q(f_j, f_k) = \delta_{jk}$ for all s, j, k. It then follows that, relative to the basis $\{e_s\,,\, N e_s, f_j\,,\, N^2 e_s\}$ of $V_{0,\mathbb{R}}$, the polarization Q is given by

$$Q|_{V_{0,\mathbb{R}}} = \begin{pmatrix} 0 & 0 & \mathbf{1}_a \\ 0 & \mathbf{1}_{a,b} & 0 \\ \mathbf{1}_a & 0 & 0 \end{pmatrix}, \quad \text{where} \quad \mathbf{1}_{a,b} = \begin{pmatrix} -\mathbf{1}_a & 0 \\ 0 & \mathbf{1}_b \end{pmatrix}.$$

Proposition 4.7. *Let (F, N) be an $\mathbb{R}$-split nilpotent orbit on a period domain parameterizing effective weight two Hodge structures. The subgroup $M^0_\mathbb{R} \subset M_\mathbb{R}$ is the connected identity component of*

$$\begin{aligned} G^{0,0}_\mathbb{R} &:= \{g \in G_\mathbb{R} \mid g(V^{p,q}) \subset V^{p,q}\ \forall\ p,q\} \\ &= \left(G^{0,0}_\mathbb{R} \cap \mathrm{Aut}(V_0)\right) \times \left(G^{0,0}_\mathbb{R} \cap \mathrm{Aut}(V_1)\right) \times \left(G^{0,0}_\mathbb{R} \cap \mathrm{Aut}(V_2)\right) \\ &\simeq \left(\mathrm{GL}(a,\mathbb{R}) \times \mathrm{O}(b,a)\right) \times \mathrm{GL}(c,\mathbb{C}) \times \mathrm{U}(d)\,. \end{aligned} \tag{4.8}$$

Proof. The proof is very like that of Proposition 4.2, so we will merely sketch the argument. To see that $M^0_\mathbb{R}$ is the connected identity component of $G^{0,0}_\mathbb{R}$, it suffices to observe that $G^{0,0}_\mathbb{R}$ is the maximal subgroup of $G_\mathbb{R}$ with Lie algebra $\mathfrak{m}^0_\mathbb{R} = \mathfrak{g}^{0,0}_\mathbb{R}$, and to recall that $M^0_\mathbb{R}$ is connected.

The subgroup $G^{0,0}_\mathbb{R} \subset M_\mathbb{R}$ is determined as follows: First, note that the argument establishing the first equality of (4.3) also yields the equality of (4.8). The factor $G_\mathbb{R} \cap \mathrm{Aut}(V_2) = \mathrm{Aut}(V_2, Q^*_2) \simeq \mathrm{U}(d)$ of $M_\mathbb{R}$ preserving $V_2 = V^{2,0}$ necessarily lies in $G^{0,0}_\mathbb{R}$. Next, from observations (v) and (vi) in the proof of Proposition 4.2, we see that

$$G^{0,0}_\mathbb{R} \cap \mathrm{Aut}(V_1) \simeq \left\{ \begin{pmatrix} (\overline{D}^t)^{-1} & 0 \\ 0 & D \end{pmatrix} \,\middle|\, D \in \mathrm{GL}(c,\mathbb{C}) \right\}. \tag{4.9}$$

Likewise, working with the basis of Remark 4.6 we see that

$$G^{0,0}_\mathbb{R} \cap \mathrm{Aut}(V_0) \simeq \left\{ \begin{pmatrix} E_1^{-1} & 0 & 0 \\ 0 & E_2 & 0 \\ 0 & 0 & E_1^t \end{pmatrix} \,\middle|\, \begin{array}{c} E_1 \in \mathrm{GL}(a,\mathbb{R}) \\ E_2 \in \mathrm{O}(b,a) \end{array} \right\}. \tag{4.10}$$

□

Next we describe $\mathfrak{g}_{\mathbb{R}}^{-1,-1}$ as an $M_{\mathbb{R}}^{0}$–representation.

Proposition 4.11. *Let (F, N) be an $\mathbb{R}$–split nilpotent orbit on a period domain parameterizing effective weight two Hodge structures. As a $G_{\mathbb{R}}^{0,0}$–representation*

$$\mathfrak{g}_{\mathbb{R}}^{-1,-1} \simeq \mathcal{H}_c \oplus \mathrm{Hom}_{\mathbb{R}}(\mathbb{R}^a, \mathbb{R}^{a+b}), \tag{4.12}$$

where $\mathcal{H}_c$ is the set of $c \times c$ Hermitian matrices. More precisely:

(i) *The factor $\mathrm{U}(d) \simeq G_{\mathbb{R}}^{0,0} \cap \mathrm{Aut}(V_2)$ acts trivially on $\mathfrak{g}_{\mathbb{R}}^{-1,-1}$.*

(ii) *If $g \in G_{\mathbb{R}}^{0,0} \cap (\mathrm{Aut}(V_1) \times \mathrm{Aut}(V_0)) \simeq \mathrm{GL}(c, \mathbb{C}) \times (\mathrm{GL}(a, \mathbb{R}) \times \mathrm{O}(b, a))$ is represented by $(D; E_1, E_2)$ as in (4.9) and (4.10), then the action of g on $(X, Y) \in \mathcal{H}_c \oplus \mathrm{Hom}_{\mathbb{R}}(\mathbb{R}^a, \mathbb{R}^{a+b})$ is $(X, Y) \mapsto (DX\overline{D}^t, E_2 Y E_1)$.*

Proof. The key observation is that $\mathfrak{g}_{\mathbb{R}}^{-1,-1}$ decomposes as

$$\mathfrak{g}_{\mathbb{R}}^{-1,-1} = \left(\mathfrak{g}_{\mathbb{R}}^{-1,-1} \cap \mathrm{End}(V_1)\right) \oplus \left(\mathfrak{g}_{\mathbb{R}}^{-1,-1} \cap \mathrm{End}(V_0)\right).$$

From the characterization of $G_{\mathbb{R}} \cap \mathrm{Aut}(V_1)$ in the proof of Proposition 4.2 (items (v) and (vi)) we see that the the first summand is

$$\mathfrak{g}_{\mathbb{R}}^{-1,-1} \cap \mathrm{End}(V_1) \simeq \left\{ \begin{pmatrix} 0 & 0 \\ X & 0 \end{pmatrix} \;\middle|\; X = \overline{X}^t \text{ an Hermitian } c \times c \text{ matrix} \right\}. \tag{4.13}$$

Likewise, relative to the basis of Remark 4.6, the second summand is

$$\mathfrak{g}_{\mathbb{R}}^{-1,-1} \cap \mathrm{End}(V_0) \simeq \left\{ \begin{pmatrix} 0 & 0 & 0 \\ Y & 0 & 0 \\ 0 & -Y^t \mathbf{1}_{a,b} & 0 \end{pmatrix} \;\middle|\; Y \text{ an } (a+b) \times a \text{ matrix} \right\}. \tag{4.14}$$

This establishes (4.12).

To complete the proof it remains to check that the adjoint action of $g \in G_{\mathbb{R}}^{0,0}$ on $\mathfrak{g}_{\mathbb{R}}^{-1,-1}$ is as described. Making use of the identifications (4.9) and (4.10), this is an exercise in matrix multiplication that we leave to the reader. □

4.2 Orbit decomposition

Assume the hypotheses of Proposition 4.11 and use the decomposition (4.12) to write $N = N_1 + N_0$ with

$$N_1 \in \mathrm{Sym}^2 \mathbb{R}^c \quad \text{and} \quad N_0 \in \mathrm{Hom}_{\mathbb{R}}(\mathbb{R}^a, \mathbb{R}^{a+b}).$$

Note that

- $N_0 \neq 0$ if and only if $V^{2,2} \neq 0$ (equivalently, $a \neq 0$), and
- $N_1 \neq 0$ if and only if $V^{2,1} \neq 0$ (equivalently, $c \neq 0$).

Propositions 4.7 and 4.11 imply

$$\mathcal{N}^0 \;=\; \mathcal{N}_0^0 \times \mathcal{N}_1^0 \tag{4.15}$$

where

- $\mathcal{N}_0^0$ is the orbit of N_0 under $\mathrm{GL}(a,\mathbb{R}) \times \mathrm{O}(b,a)$, and
- $\mathcal{N}_1^0$ is the orbit of N_1 under $\mathrm{GL}(c,\mathbb{C})$.

Note that the two summands in (4.12) commute. In particular, $[\mathcal{N}_0^0, \mathcal{N}_1^0] = 0$. Consequently, any maximal σ will be of the form

$$\sigma \;=\; \sigma_0 \times \sigma_1\,, \tag{4.16}$$

with $\sigma_i = \sigma \cap \mathcal{N}_i^0$. From Proposition 4.2 we see that the second and third factors, $\mathrm{U}(c,c)$ and $\mathrm{U}(d)$, of $M_\mathbb{R}$, are always Hermitian (§3.4.1). The third factor we disregard as it acts trivially on $\mathfrak{g}_\mathbb{R}^{-1,-1}$, and we have

$$\max \dim_\mathbb{R} \sigma_1 \;=\; \frac{1}{2}c(c+1)\,. \tag{4.17}$$

The first factor $\mathrm{O}(a+b, 2a)$ is Hermitian if and only if $a = 0, 1$. In this case $\max\dim_\mathbb{R} \sigma_0 = a(a+b)$. The first factor is contact (§3.4.2) if and only if $a = 2$, and in this case $\max\dim_\mathbb{R}\sigma_0 = (2+b)$. To summarize, with $\max\dim_\mathbb{R}\sigma$ denoting the maximal possible dimension of a nilpotent cone $\sigma \subset \mathcal{N}^0$ underlying a nilpotent orbit:

- If $a = 0, 1$, then $\max\dim_\mathbb{R}\sigma = a(a+b) + \frac{1}{2}c(c+1)$.
- If $a = 2$, then $\max\dim_\mathbb{R}\sigma = (2+b) + \frac{1}{2}c(c+1)$.

For the cases $a = 0, 1, 2$, the maximal nilpotent cones σ underlying a nilpotent orbit all have the same dimension. This will not be the case when $a > 2$. However, one may use [23] to identify the dimensions of the maximal cones.

It remains to describe the orbits $\mathcal{N}_0^0$ and $\mathcal{N}_1^0$.

Proposition 4.18. *Let (F, N) be an $\mathbb{R}$–split nilpotent orbit on a period domain parameterizing effective weight two Hodge structures. Let $N = N_0 + N_1$ be the decomposition given by* (4.12).

(a) *The orbit $\mathcal{N}_1^0 = \mathrm{GL}(c,\mathbb{R}) \cdot N_1$ is the set*

$$\mathcal{X} \;:=\; \left\{ X \in \mathfrak{g}_\mathbb{R}^{-1,-1} \cap \mathrm{End}(V_1) \;\middle|\; 0 < \mathbf{i}Q(u, X\bar{u}) \ \forall\ 0 \neq u \in V^{2,1} \right\}.$$

Under the identification of (4.13), *this orbit is parameterized by the positive, Hermitian $c \times c$ matrices $X = D\overline{D}^t$ with $D \in \mathrm{GL}(c,\mathbb{C})$.*

(b) *The orbit* $\mathcal{N}_0^0 = (\mathrm{GL}(a,\mathbb{R}) \times \mathrm{O}(a,b))^\circ \cdot N_0$ *is the connected component* $\mathcal{Y}^\circ$ *of*

$$\mathcal{Y} := \left\{ Y \in \mathfrak{g}_{\mathbb{R}}^{-1,-1} \cap \mathrm{End}(V_0) \;\Big|\; Q(Yu, Yu) < 0\,,\ \forall\, 0 \neq u \in V_{\mathbb{R}}^{2,2} \right\}$$

containing N_0. *Under the identification of* (4.14), $\mathcal{Y}$ *is parameterized by the* $(a+b) \times a$ *matrices* $Y = \begin{pmatrix} \alpha E_1 \\ \beta E_1 \end{pmatrix}$ *with* α *an* $a \times a$ *matrix and* β *a* $b \times a$ *matrix such that* $\alpha^t\alpha - \beta^t\beta = \mathbf{I}_a$, *and* $E_1 \in \mathrm{GL}(a,\mathbb{R})$.

The proofs of $\mathcal{N}_1^0 = \mathcal{X}$ and $\mathcal{N}_0^0 = \mathcal{Y}^\circ$ are variations on the argument establishing Lemma 3.5, and will extend to more general situations in a fairly straightforward manner.

In the proof below, it will be helpful to keep in mind that

$$\begin{aligned} \mathfrak{g}_{\mathbb{R}}^{-1,-1} \cap \mathrm{End}(V_1) &= \mathrm{End}(V_{\mathbb{R}}, Q) \cap \mathrm{Hom}(V^{2,1}, V^{1,0})\,, \\ \mathfrak{g}_{\mathbb{R}}^{-1,-1} \cap \mathrm{End}(V_0) &= \mathrm{End}(V_{\mathbb{R}}, Q) \cap \left(\mathrm{Hom}(V^{2,2}, V^{1,1}) \oplus \mathrm{Hom}(V^{1,1}, V^{0,0})\right)\,. \end{aligned}$$

Proof. By Lemma 3.5, we know that $\mathcal{N}^0$ is open in $\mathfrak{g}_{\mathbb{R}}^{-1,-1}$. From Proposition 4.11 and (4.15), we see that this is equivalent to the two conditions that $\mathcal{N}_i^0$ is open in $\mathfrak{g}_{\mathbb{R}}^{-1,-1} \cap \mathrm{Hom}(V_i)$, $i = 0, 1$.

To establish $\mathcal{N}_1^0 = \mathcal{X}$, observe that $\mathcal{X}$ is open, convex and preserved under the action of $M_{\mathbb{R}}^0$. By definition $N_1 \in \mathcal{X}$. Since $\mathcal{X}$ is preserved under the action of $M_{\mathbb{R}}^0$, it is immediate that $\mathcal{N}_1^0 \subset \mathcal{X}$. It remains to show that equality holds. An argument analogous to that establishing Lemma 3.5 shows that $\mathcal{X}$ is a union of open $M_{\mathbb{R}}^0$-orbits. The equality $\mathcal{X} = \mathcal{N}_1^0$ then follows from the convexity of $\mathcal{X}$.

Under the identification (4.13), N_1 is represented by $X = \mathbf{1}_c$. (View this as indicating that N_1 gives us a specific isomorphism $V^{2,1} \simeq V^{1,0}$.) So the action of $g = (D; E_1, E_2) \in M_{\mathbb{R}}^0$ on N_1 is $\mathbf{1}_c \mapsto D\overline{D}^t$, by Proposition 4.11.[4]

We briefly sketch the argument establishing $\mathcal{N}_0^0 = \mathcal{Y}^\circ$ which is very like that above for $\mathcal{N}_1^0 = \mathcal{X}$. Again we observe that $\mathcal{Y}$ is open and preserved under the action of $M_{\mathbb{R}}^0$ (but *not* convex[5]). As above $\mathcal{N}_0^0 \subset \mathcal{Y}$, and one may show that $\mathcal{Y}$ is a union of open $M_{\mathbb{R}}^0$-orbits.

Under the identification (4.14), N_0 is represented by $Y = (\mathbf{1}_a\ \mathbf{0}_{b,a})^t$, where $\mathbf{0}_{b,a}$ is the $b \times a$ zero matrix. Decompose $E_2 \in \mathrm{O}(b,a)$ as

$$E_2 = \begin{pmatrix} \alpha & * \\ \beta & * \end{pmatrix},$$

with α an $a \times a$ matrix and β a $b \times a$ matrix. Then Proposition 4.11 asserts that the action of $g = (D; E_1, E_2) \in M_{\mathbb{R}}^0$ on N_0 is

$$\begin{pmatrix} \mathbf{1}_a \\ \mathbf{0}_{b,a} \end{pmatrix} \mapsto \begin{pmatrix} \alpha E_1 \\ \beta E_1 \end{pmatrix}.$$

Since $E_2 \in \mathrm{O}(b,a)$, we have $\alpha^t\alpha - \beta^t\beta = \mathbf{I}_a$.

[4] The Cholesky decomposition yields a factorization of every element $X' \in \mathcal{X}$ of the form $X' = DD^t$.

[5] This is a feature of the non-classical case that D is not Hermitian symmetric.

This completes the proof of the proposition. As a final remark, and keeping the identification (4.14) in mind, we note that $(g \cdot N_0)^2 : V_{\mathbb{R}}^{2,2} \to V_{\mathbb{R}}^{0,0}$ is represented by

$$\begin{pmatrix} 0 & 0 & 0 \\ 0 & 0 & 0 \\ E_1^t(\alpha^t\alpha - \beta^t\beta)E_1 & 0 & 0 \end{pmatrix} = \begin{pmatrix} 0 & 0 & 0 \\ 0 & 0 & 0 \\ E_1^t E_1 & 0 & 0 \end{pmatrix}. \qquad \square$$

5 Example: period domain for $\mathbf{h} = (2, *, 2)$

The goal of this section is to illustrate how the material of §§2–4 can be applied to study nilpotent orbits on the period domain D for Hodge numbers

$$\mathbf{h} \;=\; (2, h^{1,1}, 2)\,.$$

In §5.1 we identify the $\mathbb{R}$–split polarized mixed Hodge structures on D ("Step 1" of Remark 3.10). Section 5.2 describes representation theoretic constraints on the degenerations coming from the faces of a cone underlying a nilpotent orbit. In §5.3 we see that these are the only constraints: one may construct nilpotent cones, that underlie nilpotent orbits, from commuting $\mathfrak{sl}(2)$s exhibiting all remaining degenerations.

5.1 The PMHS

Set

$$m + 8 \;=\; 4 \,+\, h^{1,1} \quad \text{so that} \quad V_{\mathbb{R}} \;=\; \mathbb{R}^{m+8}\,.$$

We have

$$G_{\mathbb{R}} \;=\; \mathrm{Aut}(V_{\mathbb{R}}, Q) \;\simeq\; \mathrm{O}(4, m+4) \quad \text{and} \quad \mathfrak{g}_{\mathbb{R}} \;=\; \mathrm{End}(V_{\mathbb{R}}, Q) \;\simeq\; \mathfrak{so}(4, m+4)\,.$$

Modulo the action of $G_{\mathbb{R}}$ there are at most five polarized $\mathbb{R}$–split PMHS on D. The Hodge diamonds (HD) for these PMHS are depicted in Table 5.1, where they are denoted I, …, V. As one can see from the table, we need $m \geq 0$ to get all five degenerations; we assume this to be the case.

Table 5.1: Hodge diamonds and signed Young diagrams

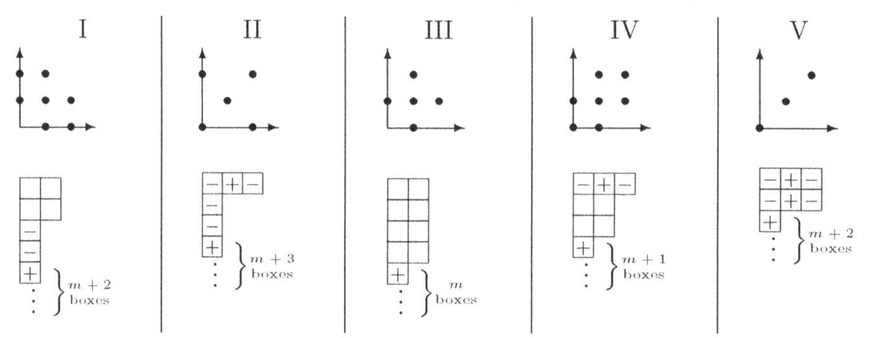

By Lemma 3.11, the number of commuting $\mathfrak{sl}(2)$'s obtained from the Cattani, Kaplan and Schmid construction [4] is bounded by the real rank of $M_{\mathbb{R}}^{ss}$. The subgroups, which are determined by Proposition 4.2, are listed in Table 5.2. In the table the maximal dimensions of the nilpotent cones $\sigma \subset \mathcal{N}^0$ underlying a nilpotent orbit are taken from §4.2.

5.2 Degenerations coming from the faces of the cone

Together Theorem 2.21 and Corollary 3.6 provide representation theoretic constraints on the degenerations associated with the faces of a nilpotent cone underlying a nilpotent orbit on a period domain D. While illustrated with examples in weight two period domains, the discussion of this section is general and applies to arbitrary period domains.

Given a polarizing nilpotent $N \in \mathfrak{g}_{\mathbb{R}}$, let

$$\mathcal{N} \;=\; \mathrm{Ad}(G_{\mathbb{R}}) \cdot N$$

denote the conjugacy class. As discussed in §2.5, these orbits are enumerated by (partially) signed Young diagrams, a.k.a. Đoković's chromosomes, and the diagrams are determined by the Hodge diamond of $V_{\mathbb{C}}$. Recall that if $\sigma = \{\lambda^i N_i \mid \lambda^i > 0\}$ is a nilpotent cone underlying a nilpotent orbit, then Corollary 3.6 yields

$$N \in \sigma \implies \sigma \subset \mathcal{N} \quad \text{and} \quad N_i \subset \overline{\mathcal{N}}. \tag{5.1}$$

In particular, every nilpotent in σ is of the same type as N. (E.g. if N is of type II, then every element of the cone is of type II.) So we can speak of the "type of σ", and there are five types of cones on the period domains for $\mathbf{h} = (2, h^{1,1}, 2)$.

Given two nilpotents N_1 and N_2 we write $\mathcal{N}_1 \leq \mathcal{N}_2$ if $\mathcal{N}_1 \subset \overline{\mathcal{N}_2}$, and $\mathcal{N}_1 < \mathcal{N}_2$ when $\mathcal{N}_1 \leq \mathcal{N}_2$ but $\mathcal{N}_1 \neq \mathcal{N}_2$. Đoković's Theorem 2.21 characterizes this partial ordering; for the conjugacy classes $\mathcal{N}$ of the polarizing N in Table 5.1 we have

$$\mathcal{N}_{\mathrm{I}} < \left\{ \begin{array}{c} \mathcal{N}_{\mathrm{II}} \\ \mathcal{N}_{\mathrm{III}} \end{array} \right\} < \mathcal{N}_{\mathrm{IV}} < \mathcal{N}_{\mathrm{V}}\,. \tag{5.2}$$

Table 5.2: The diagonal group $M_{\mathbb{R}}$

	$M_{\mathbb{R}}^{ss}$	$\mathrm{rank}_{\mathbb{R}}$	max $\dim_{\mathbb{R}}\sigma$
I	$\mathrm{O}(m+2) \times \mathrm{U}(1,1)$	1	1
II	$\mathrm{O}(m+4,2)$	2	$m+4$
III	$\mathrm{O}(m) \times \mathrm{U}(2,2)$	2	3
IV	$\mathrm{O}(m+2,2) \times \mathrm{U}(1,1)$	3	$m+3$
V	$\mathrm{O}(m+4,4)$	4	$m+4$

Given (5.1), and the dimension constraints listed in Table 5.2, this tells us something about the degenerations corresponding to the faces of the cone. For example,

(a) Any face of type I is necessarily one–dimensional.

(b) If σ is of type II, then the faces of σ are either of type I or type II — because only $\mathcal{N}_{\mathrm{I}}, \mathcal{N}_{\mathrm{II}} \le \mathcal{N}_{\mathrm{II}}$.

(c) Likewise, if σ is of type III, the faces of σ are either of type I or type III.

(d) Any face of type III is of dimension at most three.

(e) If σ is of type IV, then no face is of type V.

5.3 Commuting horizontal SL(2)s

The representation theoretic constraints (b), (c) and (e) above are the only restrictions on the degenerations coming from faces of the cone.[6] In fact, we may construct cones from commuting $\mathfrak{sl}(2)$s whose faces realize all *a priori* possible degenerations.[7] The details are as follows.

Fix a basis $\{e_1, \ldots, e_4\,,\, f_1, \ldots, f_m\,,\, e_5, \ldots, e_8\}$ of $V_{\mathbb{R}}$, so that

$$Q(e_a, e_b) = \delta^9_{a+b}\,, \quad Q(f_j, f_k) = \delta_{jk}$$

and all other pairings are zero. Then

$$H^{2,0}_\varphi \;:=\; \mathrm{span}_{\mathbb{C}}\{(e_1 - e_8) - \mathbf{i}(e_3 - e_6)\,,\, (e_2 - e_7) - \mathbf{i}(e_4 - e_5)\}$$

defines a point $\varphi \in D$. Let $\{e^1, \ldots, e^4\,,\, f^1, \ldots, f^m\,,\, e^5, \ldots, e^8\}$ denote the dual basis of $V^*_{\mathbb{R}}$, and set $e^k_\ell := e_\ell \otimes e^k \in \mathrm{End}(V_{\mathbb{R}})$. Define subspaces $\mathfrak{s}_i := \mathrm{span}_{\mathbb{R}}\{N^+_i, Y_i, N_i\} \subset \mathfrak{g}_{\mathbb{R}}$, for $i = 1, \ldots, 4$, by

$$\begin{array}{lll}
N_1 = e^6_8 - e^1_3\,, & Y_1 = (e^1_1 - e^8_8) - (e^3_3 - e^6_6)\,, & N^+_1 = e^8_6 - e^3_1, \\
N_2 = e^5_7 - e^2_4\,, & Y_2 = (e^2_2 - e^7_7) - (e^4_4 - e^5_5)\,, & N^+_2 = e^7_5 - e^4_2, \\
N_3 = e^1_6 - e^3_8\,, & Y_3 = (e^1_1 - e^8_8) + (e^3_3 - e^6_6)\,, & N^+_3 = e^6_1 - e^8_3, \\
N_4 = e^2_5 - e^4_7\,, & Y_4 = (e^2_2 - e^7_7) + (e^4_4 - e^5_5)\,, & N^+_4 = e^5_2 - e^7_4\,.
\end{array}$$

It is straightforward to confirm that the $\mathfrak{s}_i \simeq \mathfrak{sl}(2,\mathbb{R})$ are commuting $\mathfrak{sl}(2)$'s in $\mathfrak{g}_{\mathbb{R}}$, and are horizontal at the point φ. Given $\mathcal{I} = \{i_1, \ldots, i_\ell\} \subset \{1, \ldots, 4\}$, let $\mathfrak{s}_{\mathcal{I}} = \mathfrak{s}_{i_1} \oplus \cdots \oplus \mathfrak{s}_{i_s}$ denote the corresponding sum of $\mathfrak{sl}(2)$'s. Set $|\mathcal{I}| = \ell$. Each horizontal $\mathfrak{s}_{\mathcal{I}}$ determines an $\mathbb{R}$–split PMHS, and the corresponding Hodge diamonds are

$\mathcal{I}$	Hodge diamond
$\lvert\mathcal{I}\rvert = 1$	I
$\mathcal{I} = \{1,3\}, \{2,4\}$	II
$\mathcal{I} \neq \{1,3\}, \{2,4\}$	III
$\lvert\mathcal{I}\rvert = 3$	IV
$\lvert\mathcal{I}\rvert = 4$	V

In particular, $\sigma = \mathrm{span}_{\mathbb{R}_{>0}}\{N_1, \ldots, N_4\}$ is a 4–dimensional nilpotent cone whose faces realize every combination of degeneration not ruled out by §5.2.

[6]This fails for more general period domains: there are additional constraints, essentially imposed by horizontality [18].

[7]The resulting degenerations correspond to those of [17, §6] obtained from Cayley transforms $\mathbf{c}_\alpha$ in non-compact imaginary roots.

6 Deligne systems

In this section, we give a counterexample to the following assertion of Kato [16, Theorem 1.4], and provide a corrected statement:

Theorem 6.1. *Let $(V, W, N_1, \ldots, N_n, F)$ be a Deligne–Hodge system of n variables. Then for $N'_j = \sum_{k=1}^{j} a_{j,k} N_k$ $(1 \le j \le n)$ with $a_{j,k} > 0$ $(1 \le k \le j \le n)$ such that $a_{j,k}/a_{j,k+1} \gg 0$ $(1 \le k < j \le n)$, $(V, W, N'_1, \ldots, N'_n, F)$ is an* IMHM *of n variables.*

In a nutshell, the problem is that one needs a polarizability condition on the original Deligne-Hodge system to guarantee that some modification is an IMHM. More precisely, the flaw in the proof, which starts on page 857 of [16], appears to be the following: The second sentence of the third paragraph of the proof says, "on gr^W_w, put *the* bilinear form in Proposition 3.2.7." We've added the emphasis here because the key problem with the proof appears to be the word "the". To every object $\mathbf{V} = (V, W, N_1, \ldots, N_r, F)$ in DH_r, one can associate an SL_2-orbit or, equivalently, an object $\hat{\mathbf{V}}$ in $\widehat{\mathrm{DH}}_r$. Any $\hat{\mathbf{V}}$ can be polarized by a bilinear form Q. But Q is not unique. If $\hat{\mathbf{V}}$ is irreducible, then Q is unique up to non-zero scalar multiple. But, in general, there is no unique Q even up to scalar multiple. This becomes a problem because the condition that the N_i be infinitesimal isometries of Q is non-trivial, and they impose (possibly contradictory) conditions on what Q can be.

Before proceeding to give the counterexample in §6.2, we first provide a short account of Deligne systems in §6.1. In §6.3 we revisit the counterexample from a categorical point of view. In §6.4 we show that Kato's theorem holds in the presence of a suitable graded-polarization condition which can be stated in terms of the associated SL_2-orbit. In §6.5, we discuss the geometry of Deligne systems with a given underlying SL_2-orbit.

The definition of infinitesimal mixed Hodge module appears in §4 of [15], and the notion of Deligne system will be defined in the next section. For completeness, we record the definition of an IMHM here:

Definition 6.2. An infinitesimal mixed Hodge module consists of

(1) A finite dimensional real vector space $V_{\mathbb{R}}$ equipped with an increasing filtration W and a collection of non-degenerate bilinear forms $Q_k : Gr^W_k \otimes Gr^W_k \to \mathbb{R}$ of parity $(-1)^k$;

(2) A decreasing filtration F of $V_{\mathbb{C}} = V_{\mathbb{R}} \otimes \mathbb{C}$;

(3) Nilpotent endomorphisms $N_1, \ldots, N_r$ of $V_{\mathbb{R}}$ which preserve W and act by infinitesimal isometries on Gr^W,

such that

(a) $N_j(F^p) \subset F^{p-1}$ for all j and p;

(b) $e^{\sum_j z_j N_j} F$ induces a nilpotent orbit of pure Hodge structure of weight k on Gr^W_k which is polarized by Q_k;

(c) For any subset J of $\{1, \ldots, r\}$ there exists a relative weight filtration $M(J)$ such that (i) $N_j M_k(J) \subset M_{k-2}(J)$ for all $j \in J$ and (ii) $M(J)$ is the weight filtration of $\sum_{j \in J} N_j$ relative to W.

In particular, if W is pure of weight k (i.e. $Gr^W_\ell = 0$ unless $\ell = k$) then an IMHM is the same thing as a nilpotent orbit of pure Hodge structure of weight k. [Condition (c) follows from the results of Cattani and Kaplan.]

6.1 Preliminary remarks

Fix a field K of characteristic zero, and let W be an increasing filtration of a finite dimensional K-vector space V. Then, a grading of W is a semisimple endomorphism Y of V with integral eigenvalues such that

$$W_k = \bigoplus_{j \le k} E_j(Y) \tag{6.3}$$

where $E_j(Y)$ is the j-eigenspace of Y.

Let N be a nilpotent endomorphism of V which preserves W, i.e. $N(W_k) \subseteq W_k$. There exists at most one relative weight filtration $M = M(N, W)$ (cf. [27]) such that

(a) $N(M_k) \subseteq M_{k-2}$ for all k;

(b) If Gr^W_k is non-zero and $\ell \ge 0$ then the induced map

$$N^\ell : Gr^M_{k+\ell} Gr^W_k \to Gr^M_{k-\ell} Gr^W_k$$

is an isomorphism for each non-negative integer ℓ.

In the case of interest, W is the weight filtration of an admissible variation of mixed Hodge structure over the punctured disk and N is the local monodromy logarithm. In this setting N is a $(-1,-1)$-morphism of the limit mixed Hodge structure (F, M) where $M = M(N, W)$. Let

$$V = \bigoplus_{p,q} I^{p,q} \tag{6.4}$$

be the associated Deligne bigrading of (F, M) ([4, (2.12)]) and $Y = Y_{(F,M)}$ be the grading of M which acts as multiplication by $p + q$ on $I^{p,q}$. Then,

$$[Y, N] = -2N \tag{6.5}$$

since N is a $(-1,-1)$-morphism. Moreover, Y preserves W since (F, M) induces a mixed Hodge structure on each W_k by (3.13) of [27].

Definition 6.6. A 1-variable Deligne system over K consists of the following data:

- An increasing filtration W of a finite dimensional K-vector space V;

- A nilpotent endomorphism N of V which preserves W such that $M = M(N,W)$ exists;

- A grading Y of M which preserves W and satisfies $[Y,N] = -2N$.

A morphism of Deligne systems $(W,N,Y) \to (\tilde{W},\tilde{N},\tilde{Y})$ is an endomorphism T of the underlying K-vector spaces such that $T(W_i) \subset \tilde{W}_i$ and

$$\tilde{Y} \circ T - T \circ Y = 0, \qquad \tilde{N} \circ T - T \circ N = 0 .$$

Example 6.7. By the remarks of the previous paragraphs, if $(e^{zN}F,W)$ is an admissible nilpotent orbit then (W,N,Y) is a Deligne system where $Y = Y_{(F,M)}$ and $M = M(N,W)$.

To continue, we recall the following: An sl_2-pair consists of a nilpotent endomorphism N of a finite dimensional K-vector space V and grading H of the monodromy weight filtration $W(N)$ such that $[H,N] = -2N$. Moreover, there is a 1-1 correspondence between sl_2-pairs and representations ρ of $\mathrm{sl}_2(K)$ on V such that

$$N = \rho\begin{pmatrix} 0 & 0 \\ 1 & 0 \end{pmatrix}, \qquad H = \rho\begin{pmatrix} 1 & 0 \\ 0 & -1 \end{pmatrix}.$$

That such a representation determines an sl_2-pair follows from the structure of the irreducible representations of $\mathrm{sl}_2(K)$. Conversely, given an sl_2-pair, the elements of the kernel of $N : E_{-k}(H) \to E_{-k-2}(H)$ are lowest weight vectors for ρ.

In particular, given a 1-variable Deligne system (W,N,Y) let Y' be a grading of W which commutes with Y and

$$N = N_0 + N_{-1} + N_{-2} + \cdots \tag{6.8}$$

be the decomposition of N into eigencomponents relative to $\mathrm{ad}(Y')$. Then, N_0 and $H = Y - Y'$ induce the action of an sl_2-pair on each Gr^W_k. Let ρ_k be the corresponding representation of $\mathrm{sl}_2(K)$ on Gr^W_k and ρ be the representation of $\mathrm{sl}_2(K)$ which acts as ρ_k on $E_k(Y')$ via isomorphism $E_k(Y') \cong Gr^W_k$. Let

$$N_0^+ = \rho\begin{pmatrix} 0 & 1 \\ 0 & 0 \end{pmatrix}. \tag{6.9}$$

Accordingly, given a Deligne system (W,N,Y), each choice of grading Y' of W which commutes with Y determines a corresponding sl_2-triple. Moreover, a short calculation shows that since both N_0 and $H = Y - Y'$ commute with Y' so does N_0^+. Likewise, it is easy to see that each component N_{-k} appearing in (6.8) is weight -2 for $\mathrm{ad}\,Y$ since $[Y,Y'] = 0$.

Theorem 6.10 ([8]). *Let (W,N,Y) be a Deligne system. Then, there exists an unique, functorial grading $Y' = Y'(N,Y)$ of W which commutes with Y such that*

$$[N - N_0, N_0^+] = 0 \tag{6.11}$$

where N_0 is determined by $\mathrm{ad}(Y')$ *via* (6.8).

Sketch. The outline of Deligne's proof is as follows (cf. Theorem (4.4) in [21]): Let

$$\mathrm{gl}_{-r}(W) = \{ \alpha \in \mathrm{gl}(V) \mid \alpha(W_k) \subseteq W_{k-r} \}$$

and $\mathrm{gl}^Y_{-r}(W)$ be the subalgebra of elements of $\mathrm{gl}_{-r}(W)$ which commute with Y. Then, the set of all gradings of W which commute with Y is an affine space upon which the group $\exp(\mathrm{gl}^Y_{-1}(W))$ acts simply transitively via the adjoint action.

Deligne now claims by induction that it is possible construct a sequence of gradings $Y'_0, Y'_1, \ldots$ of W which commute with Y such that if $(N_0, Y - Y'_r, N_0^+)$ is sl_2-triple associated to Y'_r then

$$[N - N_0, N_0^+] \in \mathrm{gl}^Y_{-r-1}(W), \tag{6.12}$$

i.e. if $N = N_0 + N_{-1} + \cdots$ relative to $\mathrm{ad}(Y'_r)$ then $[N_0^+, N_j] = 0$ for $j < 0$.

Given the finite length of W, this process terminates in the desired grading Y'. The induction base $r = 0$ is trivial since any grading Y'_0 of W which commutes with Y will suffice. Suppose therefore that the required gradings $Y'_1, \ldots, Y'_{k-1}$ have been constructed.

Let $N_0, N_{-1}, \ldots$ be the components of N relative to $\mathrm{ad}\, Y'_{k-1}$. Then, since Y'_{k-1} commutes with $(N_0^+, Y - Y'_{k-1}, N_0)$ it follows that

$$N_{-k} = [N_0, \gamma_{-k}] + N'_{-k} \tag{6.13}$$

where N'_{-k} is highest weight $k-2$ for $(N_0^+, Y - Y'_{k-1}, N_0)$ and

$$\gamma_{-k} \in E_k(\mathrm{ad}(Y - Y'_{k-1})) \cap E_{-k}(\mathrm{ad}\, Y'_{k-1}). \tag{6.14}$$

In particular, equation (6.14) implies that $\gamma_{-k} \in E_0(\mathrm{ad}\, Y)$. A short calculation shows that

$$Y'_k = \mathrm{Ad}(1 + \gamma_{-k}) Y'_{k-1}$$

is a grading of W which commutes with Y and satisfies (6.12) for $r = k$. □

Regarding the functoriality of this construction, we recall the following from Deligne's appendix to [27]: Let $\mathfrak{n}$ be the one dimensional Lie algebra over K with generator N, and let $\mathcal{N}$ denote the category of nilpotent $\mathfrak{n}$-modules. If V_1 and $V_2 \in \mathcal{N}$ then $V_1 \otimes V_2 \in \mathcal{N}$ via the action

$$N(v_1 \otimes v_2) = N(v_1) \otimes v_2 + v_1 \otimes N(v_2). \tag{6.15}$$

Likewise, if V_1 and $V_2 \in \mathcal{N}$ then $\mathrm{Hom}_K(V_1, V_2) \in \mathcal{N}$ via

$$N(A) = N \circ A - A \circ N. \tag{6.16}$$

Theorem 6.17 (A.4 [27]). *Suppose that V_1, $V_2 \in \mathcal{N}$ carry finite increasing filtrations W such that N acts as an admissible endomorphism of (V_i, W) for $i = 1, 2$. Define*

$$W_k(V_1 \otimes V_2) = \sum_{i+j=k} (W_i V_1 \otimes W_j V_2).$$

Then, $N\otimes 1+1\otimes N$ is an admissible endomorphism of $V_1\otimes V_2$ with relative weight filtration

$$M_k(N, W(V_1\otimes V_2)) = \sum_{i+j=k} (M_iV_1\otimes M_jV_2).$$

Lemma 6.18. *A choice of Deligne system data on finite dimensional K-vector spaces V_1 and V_2 determines a Deligne system on $V_1\otimes V_2$ with weight filtration $W(V_1\otimes V_2)$, nilpotent endomorphism*

$$N(v_1\otimes v_2) = N(v_1)\otimes v_2 + v_1\otimes N(v_2)$$

and grading

$$Y(v_1\otimes v_2) = Y(v_1)\otimes v_2 + v_1\otimes Y(v_2)$$

of $M(N, V_1\otimes V_2)$. Moreover, the corresponding grading $Y'(N,Y)$ of $W(V_1\otimes V_2)$ is given by

$$Y'(v_1\otimes v_2) = Y'(v_1)\otimes v_2 + v_1\otimes Y'(v_2). \tag{6.19}$$

Proof. Deligne's Theorem A.4 asserts the existence of the relative weight filtration of N and W. A short calculation shows that $[Y,N]=-2N$. Likewise, the definition of $W(V_1\otimes V_2)$ coupled with the fact that action of Y on V_j preserves $W(V_j)$ implies that the action of Y on $V_1\otimes V_2$ preserves $W(V_1\otimes V_2)$. Therefore, the induced actions of Y and N determine a Deligne system on $V_1\otimes V_2$ with weight filtration $W(V_1\otimes V_2)$.

To complete the proof, we need to show that (6.19) is a grading of $W(V_1\otimes V_2)$ which satisfies the conditions of Theorem 6.10. To this end, we note that

$$N = \left(\sum_{a\leq 0} N_a\right)\otimes 1 + 1\otimes\left(\sum_{b\leq 0} N_b\right)$$

relative to the action of adY' on gl(V_1) and gl(V_2). Direct calculation shows that

$$[Y', N_a\otimes 1 + 1\otimes N_b] = (a+b)(N_a\otimes 1 + 1\otimes N_b)$$

and hence $N_0 = N_0\otimes 1 + 1\otimes N_0$ is the zero eigencomponent of the action of N on $V_1\otimes V_2$ relative to adY'.

Setting $N_0^+ = N_0^+\otimes 1 + 1\otimes N_0^+$ and $H = (Y-Y')\otimes 1 + 1\otimes(Y-Y')$ one easily checks that (N_0, H, N_0^+) is an sl$_2$-triple for $V_1\otimes V_2$. Direct computation shows that $[Y,Y']=0$ on $V_1\otimes V_2$ and that

$$[N-N_0, N_0^+] = 0$$

on $V_1\otimes V_2$. □

An analogous argument proves that a Deligne system on V induces a Deligne system on $V^* = \mathrm{Hom}(V,K)$ via (6.16) where K is equipped with the trivial Deligne system $W_0 = K$, $N=0$ and $Y=0$. Accordingly, a choice of Deligne systems on V_1 and V_2 determines a Deligne system on $\mathrm{Hom}(V_1,V_2)$.

Lemma 6.20 ([8]). *If $T : V_1 \to V_2$ is a morphism of Deligne systems then T commutes with the associated grading Y' of the Deligne system* $\mathrm{Hom}(V_1, V_2)$. *Equivalently, if T is a morphism of Deligne systems $(W, N, Y) \to (\tilde{W}, \tilde{N}, \tilde{Y})$ then*

$$T \circ Y' = \tilde{Y}' \circ T \tag{6.21}$$

where $Y' = Y'(N, Y)$ and $\tilde{Y}' = Y'(\tilde{N}, \tilde{Y})$.

Proof. Since T preserves $W(V_1 \otimes V_2)$ it follows that

$$T = \sum_{j \leq 0} T_j, \qquad [Y', T_j] = jT_j. \tag{6.22}$$

Moreover, since $[Y, Y'] = 0$ and $[Y, T] = 0$ it follows from the Jacobi identity that $[Y, T_j] = 0$. In particular, $[H, T_0] = [Y - Y', T_0] = 0$. Likewise, $[N, T] = 0$ and the fact that N and T preserve $W\mathrm{Hom}(V_1, V_2)$ implies $[N_0, T_0] = 0$. Therefore, $[N_0^+, T_0] = 0$.

By equation (6.22), $[H, T_j] = [Y - Y', T_j] = -jT_j$. Moreover, $[N, T] = 0 \implies [N_0, T_{-1}] = 0$ since $N_{-1} = 0$. But, T_{-1} has weight 1 with respect to $\mathrm{ad}(H)$, and hence $T_{-1} = 0$.

Assume by induction that $T_j = 0$ for $j = -2, \ldots, 1-k$. Then,

$$[N, T] = 0 \implies [N_0, T_{-k}] + [N_{-k}, T_0] = 0.$$

Accordingly, by the Jacobi identity,

$$[N_0^+, [N_0, T_{-k}]] = -[N_0^+, [N_{-k}, T_0]] = -[[N_0^+, N_{-k}], T_0] - [N_{-k}, [N_0^+, T_0]] = 0.$$

Once again, this forces $T_{-k} = 0$ since T_{-k} is weight $k > 0$ for $\mathrm{ad}(H)$. Consequently, $T = T_0$. □

Remark 6.23. The definition of Deligne system given above is due to Kato. The original (and equivalent) formulation in Christine Schwarz's paper [25] is that a Deligne system is given by the data (W, N, Y', Y). Deligne's theorem (6.10) is then stated as the assertion that given (W, N, Y) as above there is a unique choice of grading Y' of W which completes (W, N, Y) to a Deligne system. As outlined in (6.26) below, Deligne also considered the several variable case. Axioms for the several variable Deligne systems are given below in Definition 6.25 (see also [25]).

The origin of Deligne's letter to Cattani and Kaplan is a question related to a mysterious splitting operation which arises in Schmid's SL_2-orbit theorem. Namely (Prop. 2.20, [4]), given any mixed Hodge structure (F, W) on V there exists a unique, real element

$$\delta \in \bigoplus_{p,q<0} \mathrm{gl}(V)^{p,q}$$

such that $(e^{-i\delta}F, W)$ is a mixed Hodge structure which is split over $\mathbb{R}$. On the other hand (Lemma 6.60, [4]), by the SL_2-orbit theorem, if $e^{zN}F$ is a nilpotent orbit of pure Hodge structure then

$$e^{iyN}F = e^{\zeta}\left(1 + \sum_{k>0} g_k y^{-k}\right) e^{iyN} e^{-i\delta} F$$

where ζ is given by universal Lie polynomials in the Hodge components of δ. In particular, since ζ is real and depends only on the Hodge components of δ, it follows that

$$\hat{F} = e^{\zeta}e^{-i\delta}F$$

is another split mixed Hodge structure attached to an arbitrary mixed Hodge structure (F, W). In [1] this operation $(F, W) \mapsto (\hat{F}, W)$ is called the sl_2-splitting of (F, W). In the work of Kato and Usui, this operation is called the canonical splitting. For future reference we let $\hat{Y}_{(F,W)} = Y_{(\hat{F},W)}$.

In [8], Deligne asserts that if $(e^{zN}F, W)$ is an admissible nilpotent orbit with limit mixed Hodge structure (F, M) which is split over $\mathbb{R}$ then

$$\hat{Y}_{(e^{iN}F,W)} = Y'(N, Y_{(F,M)}). \tag{6.24}$$

A published proof was given by the first two authors in [1].

In the second part of his letter, Deligne focuses on applying his construction to several variable systems.

Definition 6.25. An r-variable Deligne system consists of data

$$\begin{pmatrix} W^0 & W^1 & \cdots & W^{r-1} & Y^r \\ & N_1 & \cdots & N_{r-1} & N_r \end{pmatrix} \tag{6.26}$$

where

- $W^0, \ldots, W^r$ are increasing filtrations of a finite dimensional K-vector space V;
- Y^r is a grading of W^r;

such that

(a) $N_1, \ldots, N_r$ are commuting nilpotent endomorphisms of V which preserve W^0;

(b) $M(N_j, W^{j-1})$ exists and equals W^j for $j = 1, \ldots, r$;

(c) Let $1 \leq j \leq r$, $0 \leq k < j-1$, $\ell \in \mathbb{Z}$, and let $U = W^k_\ell$, then the restriction $W^j|_U$ of W^j to U is the relative monodromy filtration of $N_j|_U$ with respect to $W^{j-1}|_U$;

(d) $N_j(W^k_\ell) \subseteq W^k_\ell$ for any j, k, ℓ, and $N_j(W^k_\ell) \subseteq W^k_{\ell-2}$ if $k \geq j$;

(e) Y^r preserves each W^j and $[Y^r, N_j] = -2N_j$ for all j.

A morphism of Deligne systems

$$T : (W, N_1, \ldots, N_r; Y) \to (\tilde{W}, \tilde{N}_1, \ldots, \tilde{N}_r, \tilde{Y}) \tag{6.27}$$

is a homomorphism of the underlying vector spaces such that

$$T(W^j_i) \subset \tilde{W}^j_i, \qquad T \circ N_j = \tilde{N}_j \circ T$$

for all j (and i), and $T \circ Y^r = \tilde{Y}^r \circ T$.

Theorem 6.28 (Deligne [8]). *Starting from Y^r, the iterative application of the construction $Y^{j-1} = Y'(N_j, Y^j)$ to a system* (6.26) *satisfying the conditions* (a)–(e) *yields a system of commuting gradings such that if $\hat{N}_j$ is the degree zero part of N_j with respect to* $\operatorname{ad} Y^{j-1}$ *and $H_j = Y^j - Y^{j-1}$ then*

$$(\hat{N}_1, H_1), \ldots, (\hat{N}_r, H_r) \tag{6.29}$$

are commuting sl_2*-pairs.*

By condition (b), the weight filtrations $W^1, \ldots, W^r$ are determined by W^0 and $N_1, \ldots, N_r$. Therefore, following [25], we abuse notation and refer to any tuple of data $(W, N_1, \ldots, N_r; Y^r)$ which generates a system (6.26) satisfying conditions (a)–(e) with $W^0 = W$ as a *Deligne system.* For future use, we define a *pre-Deligne system* to consist of data $(W, N_1, \ldots, N_r)$ as above which satisfy conditions (a)–(d).

Lemma 6.30. *If $(W^0, N_1, \ldots, N_r; Y^r)$ is a Deligne system with associated gradings Y^i then $(W^0, N_1, \ldots, N_j; Y^j)$ is also a Deligne system.*

Corollary 6.31. *Let T be a morphism of Deligne systems* (6.27) *with associated grading Y^i and $\tilde{Y}^i$. Then,*

$$\tilde{Y}^i \circ T = T \circ Y^i. \tag{6.32}$$

Proof. Equation (6.32) is true for $i = r$ by hypothesis. Likewise, we know that T is a morphism of the Deligne systems

$$(W^{r-1}, N_r, Y^r) \to (\tilde{W}^{r-1}, \tilde{N}_r, \tilde{Y}^r).$$

By Lemma (6.20), this implies (6.32) for $i = r - 1$. Accordingly, we have a morphism of Deligne systems

$$(W^0, N_1, \ldots, N_{r-1}; Y^{r-1}) \to (\tilde{W}^0, \tilde{N}_1, \ldots, \tilde{N}_{r-1}; \tilde{Y}^{r-1})$$

and hence (6.32) holds by downward induction. □

Lemma 6.33 (Deligne [8]). *The set of systems* (6.26) *satisfying conditions* (a)–(e) *forms an abelian category.*

Sketch. Let T be a morphism of Deligne systems (6.27). Then, equation (6.32) holds for all i by the previous Corollary. This forces T to be compatible with all of the associated filtrations and representations of sl_2. We leave the details to the reader. □

Returning to the splitting operation (6.24), suppose that $\theta(z) = \exp(\sum_j z_j N_j)F$ is a polarizable nilpotent orbit of pure Hodge structure of weight k on V with limit mixed Hodge structure (F, W^r). Then, $(W, N_1, \ldots, N_r; Y^r)$ is a Deligne system where $Y^r = Y_{(F,W^r)}$ and W pure of weight k on V. As shown in [1], in the case where (F, W^r) is split over $\mathbb{R}$, the resulting sl_2-pairs (6.29) generate the representation of $\mathrm{SL}_2^r(\mathbb{R})$ occurring in the SL_2-orbit theorem of Cattani, Kaplan and Schmid [4].

The construction of the previous paragraph motivates the following:

Definition 6.34 (cf. [16]). A Deligne–Hodge system $(W, N_1, \dots, N_r; F)$ consists of a pre-Deligne system $(W, N_1, \dots, N_r)$ over $K = \mathbb{C}$ equipped with:

(i) A real structure $V = V_{\mathbb{R}} \otimes \mathbb{C}$ to which W and $N_1, \dots, N_r$ descend;

(ii) A decreasing filtration F of V;

such that

(f1) $N_j F^p \subset F^{p-1}$ for any $1 \leq j \leq n$ and $p \in \mathbb{Z}$;

(f2) (F, W^r) is a mixed Hodge structure. Furthermore, for $1 \leq k < n$, $w \in \mathbb{Z}$ and for $U = W^k_w$, $(F|_U, W^r|_U)$ is a mixed Hodge structure.

Then, by the results of Kashiwara we have a forgetful functor

$$\mathbf{V} \to (W, N_1, \dots, N_r; F)$$

from IMHM to the category of Deligne–Hodge systems.

Let D_r denote the category of Deligne systems in r-variables over $K = \mathbb{C}$ and DH_r denote the category of Deligne–Hodge systems in r-variables. Then, by setting $Y^r := Y_{(F, W^r)}$ (the Deligne splitting), we have a forgetful functor

$$(V_{\mathbb{R}}, W, N_1, \dots, N_r; F) \mapsto (W, N_1, \dots, N_r; Y^r) \tag{6.35}$$

from DH_r to D_r.

In Proposition 5.6.2 of [15], Kashiwara shows that the category of IMHM is abelian, and W, $M(N_1, W)$, Gr^W, $Gr^{M(N_1, W)}$, etc. are exact functors. In Proposition 1.8 of [16], Kato asserts that the category of Deligne–Hodge systems is abelian via the embedding of Theorem 6.1 of Deligne–Hodge systems into the category of IMHM. Since Theorem 6.1 is false, this invalidates Kato's proof.

In §3.2.1 of [16], Kato defines a category of SL_2-orbits. An alternative description of this category is as follows: Let $DH_r \to DH_r$ be the functor defined by the rule

$$(V_{\mathbb{R}}, W, N_1, \dots, N_r; F) \mapsto (V_{\mathbb{R}}, W, \hat{N}_1, \dots, \hat{N}_r; \hat{F}) \tag{6.36}$$

where the $\hat{N}_j$ are the degree zero part of N_j with respect to $\operatorname{ad} Y^{j-1}$ and $(\hat{F}, W^r)$ is the sl_2-splitting of (F, W^r). An object of DH_r is an SL_2-orbit if it is fixed by this functor. The set $\widehat{DH}_r$ of all SL_2-orbits in DH_r is a full subcategory.

Example 6.37. $\widehat{DH}_0$ is the category of mixed Hodge structures which are split over $\mathbb{R}$. In the case where W is pure, $\widehat{DH}_1$ is the set of nilpotent orbits with limit mixed Hodge structure split over $\mathbb{R}$. In the case where W is mixed, $\widehat{DH}_1$ consists of admissible nilpotent orbits with limit mixed Hodge structures which are split over $\mathbb{R}$ and $N = N_0$.

Lemma 6.38. *The composite functor* **IMHM** $\to \widehat{DH}$ *is essentially surjective, i.e. for any object* $\hat{\mathbf{V}} \in \widehat{DH}_r$ *there exists a choice of graded-polarizations relative to which* $\hat{\mathbf{V}}$ *is an admissible nilpotent orbit with limit mixed Hodge structure split over* $\mathbb{R}$ *and* $N_j = \hat{N}_j$ *for each* j.

Sketch. This is Proposition 3.2.7 in Kato. For $\widehat{DH}_0$, this is just the statement that any $\mathbb{R}$-split mixed Hodge structure admits a graded-polarization. For $\widehat{DH}_1$ this statement follows from Lemma 6.24 of [24] which asserts that a horizontal $\mathrm{sl}_2(\mathbb{R})$ representation determines a polarizable nilpotent orbit of Hodge structure (the choice of polarizing form is defined up to scale on each irreducible Hodge subrepresentation of $\mathrm{sl}_2(\mathbb{R})$). See Kato paper for the several variable case. □

Definition 6.39. Let $\mathbf{V} = (N_1, \dots, N_r; W, F)$ be an object of DH_r with underlying vector space V. If $\mathbf{V}$ is pure of weight w then $Q : V \otimes V \to \mathbb{R}(w)$ polarizes $\mathbf{V}$ if

(a) Each N_j is an infinitesimal isometry of Q;

(b) Q polarizes the associated SL_2-orbit $\hat{\mathbf{V}}$ obtained by the application of (6.36) to $\mathbf{V}$, i.e. $\hat{\mathbf{V}}$ satisfies the axioms of an IMHM with Q as the polarizing form.

If $\mathbf{V}$ is not pure then $\mathbf{V}$ is said to be graded-polarizable if there exists a polarization for each of the induced Deligne–Hodge systems on Gr^W.

6.2 A counterexample

In this section we construct an explicit counterexample to Kato's theorem 6.1 in the case where $\mathbf{V}$ is not graded-polarizable.

Define

$$(V, W, N_1, N_2; F) \tag{6.40}$$

as follows:

— V is the four dimensional real vector space with ordered basis (e_1, f_1, e_2, f_2);

— N_1 and N_2 are the nilpotent endomorphisms

$$\begin{array}{llll} N_1(e_1) = f_1, & N_1(f_1) = 0, & N_1(e_2) = f_2, & N_1(f_2) = 0, \\ N_2(e_1) = f_2, & N_2(f_1) = 0, & N_2(e_2) = 0, & N_2(f_2) = 0, \end{array}$$

— W is the increasing filtration on V, with $\mathrm{gr}^W_k V = 0$ for $k \neq 1$;

— F is the decreasing filtration on $V_{\mathbb{C}}$ with $\mathrm{gr}^k_F V = 0$ for $k \notin [0,1]$ and with $F^1 = \langle e_1, e_2 \rangle$.

Proposition 6.41. *The data* (6.40) *defines an object* $\mathbf{V}$ *of* DH_2.

Proof. We need to check conditions (a)–(d) and (f).

(a) Clearly $N_1 N_2 = N_2 N_1 = N_1^2 = N_2^2 = 0$. So the the N_i commute and are nilpotent. They also obviously respect W because W is trivial.

(b) Set $W^0 = W$. Then $W^1 := M(N_1, W^0)$ exists and is split by the endomorphism

$$Y^1(e_1) = 2e_1, \qquad Y^1(f_1) = 0, \qquad Y^1(e_2) = 2e_2, \qquad Y^1(f_2) = 0. \tag{6.42}$$

Since $N_2(W_k^1) \subset W_{k-2}^1$, the filtration $W^2 := M(N_2, W^1)$ exists and is equal to W^1. In fact, the mixed Hodge structure, (V, W^2, F) is split over $\mathbb{R}$ with $Y^1 = Y^2 = Y_{(W^2,F)}$ the Deligne splitting. ($Y^2 v = (p+q)v$ for $v \in I_{(W^2,F)}^{(p,q)}$.) For this Hodge structure, we have $I^{(1,1)} = \langle e_1, e_2 \rangle$ and $I^{(0,0)} = \langle f_1, f_2 \rangle$.

(c) This is the condition that, if $0 \leq k < j - 1 \leq n - 1$ and $U = W_w^{(k)}$ for some w, then $W_U^{(j)}$ is the relative monodromy filtration of $N_j|U$ with respect to $W^{(j-1)}$. That is trivial in this case, because we have to have $k = 0$ and $j = 2$, and $W^{(0)}$ is trivial.

(d) Also trivial in this case.

(f.1) This is the horizontality condition which is trivial for the given F.

(f.2) The requirement here is that, for $0 \leq k < n, w \in \mathbb{Z}$ and $U := W_w^{(k)}$, $(W^{(n)}|_U, F|_U)$ is a mixed Hodge structure. When $k = 0$ we just have to check that $(V, W^{(n)}, F)$ is a mixed Hodge structure. That is fairly obvious. It is also clear when $k = 1$, because $W^1 = W^2$. □

Suppose now that Theorem 6.1 holds for (6.40). Then there exists an $a \in \mathbb{R}_+$ such that upon setting $N_1' = N_1$ and $N_2' = aN_1 + N_2$ the data

$$(V, N_1', N_2', W, F) \tag{6.43}$$

underlies an IMHM.

For $z = (z_1, z_2) \in \mathbb{C}^2$, set $N'(z) = \sum_{i=1}^2 z_i N_i'$ and $F'(z) = e^{N'(z)}F$. For $i = 1, 2$, set $e_i(z) = \exp(N'(z))e_i$. Then $F'^1(z) = \langle e_1(z), e_2(z) \rangle$.

Since (V, W, N_1', N_2', F) is an IMHM, there exists a skew-symmetric form

$$Q : V \otimes V \to \mathbb{R}(-1)$$

respecting the N_i' and polarizing $(V, F'(y))$ for $y = (y_1, y_2) \in \mathbb{R}^2$ with $y_1, y_2 \gg 0$. Fix this Q. (Here, when we say that Q respects the N_i', we mean that the N_i' are infinitesimal isometries of Q.)

Lemma 6.44. *Both N_1 and N_2 respect Q.*

Proof. Since $N_1 = N_1'$, and N_1' respects Q, N_1 respects Q. So $N_2 = N_2' - aN_1'$ also respects Q. □

Lemma 6.45. *We have*

(a) $Q(f_1, f_2) = 0$.

(b) $Q(e_1, e_2) = 0$.

(c) $Q(e_2, f_2) = 0$.

Proof. (a) Since N_1 respects Q,

$$0 = Q(N_1e_1, f_2) + Q(e_1, N_1f_2) = Q(f_1, f_2) + 0 = Q(f_1, f_2).$$

(b) Set $q(z) := Q(e_1(z), e_2(z))$ and note that $q(z)$ is a polynomial in the variables z_1, z_2. Since Q polarizes $F'(y)$ for $y = (y_1, y_2)$ with $y_1, y_2 \gg 0$, $q(y) = 0$ for $y_1, y_2 \gg 0$. But, since q is a polynomial, this implies that $q = 0$ identically. So $q(0) = Q(e_1, e_2) = 0$.

(c) Since N_2 respects Q, we have $0 = Q(N_2e_1, e_2) + Q(e_1, N_2e_2) = Q(f_2, e_2)$. So $Q(e_2, f_2) = 0$ as well. □

Proposition 6.46. *Contrary to Lemma* (6.45) (c), $Q(f_2, e_2) > 0$. *Consequently, the Deligne system* (V, N_1', N_2', W, F) *from* (6.43) *does not admit a form* Q *making it into a* IMHM.

Proof. Suppose (V, N_1', N_2', W, F) is an IMHM polarized by Q. Assume $y_1, y_2 \gg 0$ so that $(V, F(y))$ is a pure Hodge structure of weight 1 polarized by Q. Moreover, assume y_1 and y_2 are positive.

Let $C(y)$ denote the Weil operator on the pure Hodge structure $(V, F(y))$. It is given on V^{pq} by multiplication by i^{p-q}. (We are using the sign conventions from §1.2 of Kashiwara [15].)

Since Q is a polarization on $(V, F(y))$, the form $Q(Cu, \bar{v})$ is positive definite. It follows that $Q(Ce_2(y), \bar{e}_2(y)) > 0$. So we compute

$$\begin{aligned}
0 < Q(Ce_2(y), \bar{e}_2(y)) &= Q(C(e_2 + i(y_1 + ay_2)f_2), e_2 - i(y_1 + ay_2)f_2) \\
&= iQ(e_2 + i(y_1 + ay_2)f_2, e_2 - i(y_1 + ay_2)f_2) \\
&= 2(y_1 + ay_2)Q(e_2, f_2).
\end{aligned}$$

Since y_1, y_2 and a are positive, it follows that $Q(e_2, f_2) > 0$. □

6.3 Categorical comments

An IMHM $\mathbf{V}$ is split if $\mathbf{V} \cong \oplus Gr_k^W \mathbf{V}$. The category of split infinitesimal Hodge modules is, more or less by definition, polarizable Tannakian in the sense of Saavedra-Rivano (see, for example, page 169 of [9]). It follows that the category IMHM^s of split infinitesimal mixed Hodge modules is semi-simple (by Proposition 4.11 on page 169 of [9]).

However, we have the following.

Proposition 6.47. *The example in* (6.40) *is not a semi-simple object.*

Proof. Let $H = \langle e_2, f_2 \rangle$. By restriction of (W, N_1, N_2, F) to H we obtain a sub-object $\mathbf{H}$ of $\mathbf{V}$. (The restriction of N_2 to H is zero.) But this sub-object is clearly not a direct summand. □

The proposition gives another way to see that Theorem 6.1 fails for $\mathbf{V}$: Since split objects in IMHM are semi-simple, the theorem would imply that split objects in DH_r are also semi-simple.

6.4 Graded polarizability

On the other hand, the result of Kato's paper [16] holds once the polarization is added in. In other words, every graded-polarizable Deligne–Hodge system gives rise to an IMHM:

Theorem 6.48. *Let* $(V, W, N_1, \ldots, N_n, F)$ *be a graded polarizable* DH *system of* n *variables. Then for* $N'_j = \sum_{k=1}^{j} a_{j,k} N_k$ $(1 \le j \le n)$ *with* $a_{j,k} > 0$ $(1 \le k \le j \le n)$ *such that* $a_{j,k}/a_{j,k+1} \gg 0$ $(1 \le k < j \le n)$, $(V, W, N'_1, \ldots, N'_n, F)$ *is an IMHM of* n *variables.*

Essentially, Kato's proof goes though word for word upon addition of this polarizability hypothesis: Suppose (Q_w) is a fixed polarization of the graded of $\mathbf{V}$. Write D for the mixed period domain corresponding to Q_w. It sits in the so-called compact dual $\check{D}$. Write G for the group of isometries of Q preserving W. Then $G(\mathbb{C})$ acts on the algebraic variety $\check{D}$, while $G(\mathbb{R})$ acts on the open complex submanifold D. The N_i and H_i are all in the Lie algebra $\mathfrak{g}$ of G and, most importantly, the function β constructed by Kato lies in $G(\mathbb{R})$. On the other hand, note that, without Q, there really is no period domain D (or $\check{D}$).

Now, Kato's proof shows that $\beta(y)F(y)$ converges to

$$I = \exp(\sum_j i\hat{N}_j)\hat{F}$$

as long as $y_i/y_{i+1} \gg 0$ for all i. In particular, $\beta(y)F(y)$ lies in D for such y. Since $\beta(y) \in G(\mathbb{R})$, it follows that $F(y)$ lies in D as well.

6.5 Geometric structure

By Theorem 6.48, the question of when a given Deligne–Hodge system gives rise to an IMHM for an appropriate substitution $N_j \to N'_j$ reduces to a question about the polarizability of the underlying SL_2-orbit. Accordingly, we make the following definition:

Definition 6.49. The sl_2-type of a Deligne system $(W, N_1, \ldots, N_r, Y^r)$ consists of the weight filtration W and the associated sl_2-pairs $(\hat{N}_1, H_1), \ldots, (\hat{N}_r, H_r)$ of (6.29).

Remark 6.50. In the case where W is pure of weight k the sum

$$Y^r = k\,\mathrm{Id} + H_1 + \cdots + H_r \tag{6.51}$$

for any possible associated Deligne system. In particular, we can then recover the intermediate gradings Y^j and hence the elements $H_j = Y^j - Y^{j-1}$ via the iterated application of Deligne construction $Y^j = Y'(\hat{N}_j, Y^{j+1})$. Thus, in the pure case, an sl_2-type is equivalent to $(W, \hat{N}_1, \ldots, \hat{N}_r, Y^r)$. In [16] (cf. Prop. 3.3.2), Kato calls such Deligne systems an associated SL(2)-orbit.

The remainder of this section is devoted to proving that the set of all Deligne systems with a given sl_2-type forms an algebraic variety. We start with a series of lemmata:

Lemma 6.52. *Let $(W, \hat{N}, H)$ be an sl_2-type. Let $\mathcal{Y}(W, \hat{N}, H)$ be the set of all gradings of W which commute with $\hat{N}$ and H. Define $\mathrm{gl}_{-1}(W, \hat{N}, H)$ to be the subalgebra of $\mathrm{gl}_{-1}(W)$ consisting of elements which commute with $\hat{N}$ and H. Then, $\mathcal{Y}(W, \hat{N}, H)$ is an affine space upon which the subgroup $\exp(\mathrm{gl}_{-1}(W, \hat{N}, H))$ acts simply transitively.*

Proof. The set $\mathcal{Y}(W)$ of all gradings of W is an affine space upon which the subalgebra $W_{-1}\mathrm{gl}(V)$ acts simply transitively by $Y \mapsto Y + \beta$. Accordingly, let $Y \in \mathcal{Y}(W, \hat{N}, H)$ and $\beta \in W_{-1}\mathrm{gl}(V)$. Then, clearly $Y + \beta \in \mathcal{Y}(W, \hat{N}, H)$ if and only if $\beta \in \mathrm{gl}_{-1}(W, \hat{N}, H)$. Thus, $\mathcal{Y}(W, \hat{N}, H)$ is an affine space upon which $\mathrm{gl}_{-1}(W, \hat{N}, H)$ acts simply transitively by $Y \mapsto Y + \beta$.

On the other hand, as discussed in Proposition 2.2 of [4] the group $\exp(W_{-1}\mathrm{gl}(V))$ also acts simply transitively on $\mathcal{Y}(W)$ via the adjoint action. Moreover, the relation between α, $\beta \in W_{-1}\mathrm{gl}(V)$ defined by the equation

$$e^{\mathrm{ad}\,\alpha}Y = Y + \beta$$

is given by universal Lie polynomials in the eigencomponents of α and β with respect to $\mathrm{ad}\,Y$. In particular, since Y commutes with $\hat{N}$ and H, if β also commutes with $\hat{N}$ and H then so do all of its eigencomponents, and hence α also has this property. □

Let $(W, \hat{N}, H)$ be an sl_2-type. Define $\mathcal{U}(W, \hat{N}, H)$ to be the set of pairs (Y^0, N_-) where $Y^0 \in \mathcal{Y}(W, \hat{N}, H)$ and

$$N_- = \sum_{k \geq 2} N_{-k}$$

where N_{-k} is either 0 or an element of $E_{-k}(\mathrm{ad}\,Y^0)$ which is of highest weight $k-2$ for the representation of sl_2 generated by $(\hat{N}, H)$. Let

$$\pi : \mathcal{U}(W, \hat{N}, H) \to \mathcal{Y}(W, \hat{N}, H)$$

denote projection $(Y^0, N_-) \to Y^0$.

Lemma 6.53. *$\pi : \mathcal{U}(W, \hat{N}, H) \to \mathcal{Y}(W, \hat{N}, H)$ is an equivariant vector bundle in the sense that for any $\alpha \in \mathrm{gl}_{-1}(W, \hat{N}, H)$:*

$$e^{\alpha} : \pi^{-1}(Y^0) \to \pi^{-1}(e^{\alpha}.Y^0)$$

is a linear isomorphism on the fibers, and the group $\exp(\mathrm{gl}_{-1}(W, \hat{N}, H))$ acts transitively on the base.

For future use, we record the following:

Lemma 6.54. *Let W and W' be increasing gradings of a finite dimensional vector space V over a field of characteristic zero, with respective gradings Y and Y'. If $[Y, Y'] = 0$ then Y preserves W'.*

Proof. Mutually commuting semisimple endomorphisms can be simultaneously diagonalized over any field which contains all the eigenvalues of both endomorphisms. Since Y and Y' have integral eigenvalues

$$V = \bigoplus_{p,q} V(p,q), \qquad V(p,q) = E_p(Y) \cap E_q(Y').$$

Therefore, Y preserves $W'_k = \oplus_{q \leq k} V(p,q)$. □

Lemma 6.55. *Let $\mathcal{S}$ denote the set of all Deligne systems with* sl_2*-type* $(W, \hat{N}, H)$. *Then, the map $\Psi : \mathcal{S} \to \mathcal{U}(W, \hat{N}, H)$ which sends the Deligne system (W, N, Y^1) to (Y^0, N_-) where $Y^0 = Y'(N, Y^1)$ and $N_- = \sum_{k \geq 2} N_{-k}$ is the sum of components of N of negative weight with respect to* $\mathrm{ad}\, Y^0$ *is a bijection.*

Proof. A Deligne system with sl_2-type (W, N, H) produces an element of $\mathcal{U}(W, \hat{N}, H)$ via Ψ. Moreover, Ψ is injective since a Deligne system with sl_2-type $(W, \hat{N}, H)$ is recovered from its image under Ψ by the rule:

$$N = \hat{N} + N_-, \qquad Y^1 = Y^0 + H. \tag{6.56}$$

It remains to prove that Ψ is surjective. Let (Y^0, N_-) be an element of $\mathcal{U}(W, \hat{N}, H)$, and define N and Y^1 by (6.56). It is sufficient to show that (W, N, Y^1) is a Deligne system, since by construction $\Psi(W, N, Y^1) = (Y^0, N_-)$.

Accordingly, let W^1 be the weight filtration determined by Y^1. We need to check that (W, N, Y^1) satisfies Deligne system axioms (a)–(e).

(e) By construction, $[Y^1, Y^0] = 0$ and hence Y^1 preserves W^1 and W^0 by the previous Lemma. Likewise,

$$[Y^1, N] = [Y^1, \hat{N}] + \sum_{k \geq 2} [Y^0 + H, N_{-k}] = -2N. \tag{6.57}$$

(a) By (e), N is nilpotent. Similarly, since N has weights less than or equal to zero with respect to $\mathrm{ad}\, Y^0$, it preserves W^0.

(b) To verify that $W^1 = M(N, W^0)$, we begin with the assertion that $W^1 = M(\hat{N}, W^0)$: By construction,

$$[Y^1, \hat{N}] = [Y^0 + H, \hat{N}] = -2\hat{N} \tag{6.58}$$

so $\hat{N}$ lowers W^1 by 2. It remains to show that

$$\hat{N}^\ell : Gr^{W^1}_{k+\ell} Gr^{W^0}_k \to Gr^{W^1}_{k-\ell} Gr^{W^0}_k \tag{6.59}$$

is an isomorphism. However, since $H = Y^1 - Y^0$ and $[Y^1, Y^0] = 0$ it follows that

$$Gr^{W^1}_{k+\ell} Gr^{W^0}_k \cong E_{k+\ell}(Y^1) \cap E_k(Y^0) = E_\ell(H) \cap E_k(Y^0).$$

Accordingly, (6.59) is an isomorphism. Consequently, $M(N, W) = M(\hat{N}, W)$ since $[Y^1, N] = -2N$ and and changing $\hat{N} \to N$ does not change the induced action of N on Gr^W.

(c) Since $[Y^1, Y^0] = 0$, it follows that the argument of (b) above holds for the restriction of N to W^0_ℓ.

(d) By (a)+(e) above, N preserves W^0 and lowers the weights of W^1 by 2. □

Theorem 6.60. *The set $\mathcal{S}$ of all r-variable Deligne systems $(W, N_1, \dots, N_r, Y^r)$ of* sl_2*-type*

$$(W, \hat{N}_1, H_1, \dots, \hat{N}_r, H_r) \tag{6.61}$$

is an algebraic variety.

Proof. We induct on the number of variables. The case $r = 1$ is covered by the previous Lemma.

To continue, we observe that if $S = (W, N_1, \dots, N_r; Y^r)$ is a point of $\mathcal{S}$ then

$$W^j = M(\hat{N}_j, W^{j-1}), \qquad j = 1, \dots, r$$

and hence the weight filtrations W^j attached to S are determined by the sl_2-type (6.61). Let $\mathcal{Y}$ denote the affine variety consisting of gradings of W^{r-1}.

Let $\mathcal{S}'$ denote the set of Deligne systems with sl_2-type$(W, \hat{N}_1, H_1, \dots, \hat{N}_{r-1}, H_{r-1})$. Let $\mathcal{S}''$ denote the set of Deligne systems with sl_2-type $(W^{r-1}, \hat{N}_r, H_r)$. By the induction hypothesis, both $\mathcal{S}'$ and $\mathcal{S}''$ are algebraic varieties, and hence so is the product $\mathcal{S}' \times \mathcal{S}''$. Let $\mathcal{P} \subset \mathcal{S}' \times \mathcal{S}''$ be the fiber product over $\mathcal{Y}$ defined by the maps

$$S' = (W, N_1, \dots, N_{r-1}, Y^{r-1}) \mapsto Y^{r-1},$$
$$S'' = (W^{r-1}, N_r, Y^r) \mapsto Y'(N_r, Y^r).$$

A point $S \in \mathcal{S}$ determines a point $(S', S'') \in \mathcal{P}$ by the rule

$$S' = (W, N_1, \dots, N_{r-1}, Y^{r-1}), \qquad S'' = (W^{r-1}, N_r, Y^r) \tag{6.62}$$

such that:

(I) Y^r commutes with the gradings $Y^{r-1}, \dots, Y^0$ attached to S';

(II) N_r has only non-positive eigenvalues relative to $\mathrm{ad}\, Y^j$ for $0 \le j \le r-1$;

(III) $[N_r, N_j] = 0$ for $1 \le j \le r-1$;

(IV) $[Y^r, N_j] = -2N_j$ for $1 \le j \le r$.

Conversely, let $(S', S'') \in \mathcal{P}$ be a point of the form (6.62) which satisfies conditions (I)–(IV). Then,

$$S = (W, N_1, \dots, N_r; Y^r) \tag{6.63}$$

is a Deligne system:

(a) Since each N_j is part of a Deligne system, it is nilpotent. Similarly, since $(N_1, \dots, N_{r-1})$ are part of a Deligne system, they mutually commute. Condition (III) implies the remaining commutativity conditions. Likewise, by hypothesis, $N_1, \dots, N_{r-1}$ preserve W^0. Condition (II) implies that N_r preserves W^0.

(b) The fact that S' and S'' are Deligne systems implies the existence of all required relative weight filtrations.

(c) Since S' and S'' are Deligne systems, this condition is automatic except for the extremal case: W^r restricted to $U = W^k_\ell$ for $k < r-1$ is the relative weight filtration of $W^{r-1}|_U$ and $N_r|_U$. This follows from properties (I) and (II).

(d) Since S' and S'' are Deligne systems, the only unresolved cases are $N_r(W^k_\ell) \subset W^k_\ell$ for $k < r-1$ (which follows from (II)) and $N_j(W^r_\ell) \subset W^r_{\ell-2}$ which follows from (IV).

(e) $[Y^r, N_j] = -2N_j$ is property (IV). Property (I) implies that Y^r preserves each W^j.

Let $\mathcal{A}$ be the algebraic subvariety of $\mathcal{P}$ defined by properties (I)–(IV). Given $(S', S'') \in \mathcal{A}$, the corresponding Deligne system S has the same sl_2-type as S' and S'', i.e. $S \in \mathcal{S}$. A simple check shows that the maps

$$\mathcal{S} \to \mathcal{A} \to \mathcal{S}, \qquad \mathcal{A} \to \mathcal{S} \to \mathcal{A}$$

are the identity, and hence $\mathcal{S}$ is isomorphic to the algebraic variety $\mathcal{A}$. □

In the case where $(\hat{N}_1, H_1), \ldots, (\hat{N}_r, H_r)$ are all infinitesimal isometries of bilinear forms on Gr^W, we can ask that all N_j's appearing above are also infinitesimal isometries. This is again an algebraic condition. More generally, if the initial sl_2-type belongs a Mumford–Tate Lie algebra $\mathfrak{m}$ then it is an algebraic condition for all the N_j's to belong to $\mathfrak{m}$. Likewise, the condition that each N_j be horizontal with respect to a given Hodge filtration F is also algebraic.

As independent check of the compatibility of the next few examples with the results of the earlier sections of this paper, we develop our examples starting from Lemma (6.24) in Schmid [24]. Namely, if $H_{\mathbb{C}} = H_{\mathbb{R}} \otimes \mathbb{C}$ is Hodge structure of weight k equipped with a horizontal action of $\mathrm{sl}_2(\mathbb{C})$ then $H_{\mathbb{C}}$ is a direct sum of irreducible representations of the form $S(n) \otimes \mathbb{R}(m)$ and $S(n) \otimes E(p,q)$ where

— $S(n) = \mathrm{Sym}^n(\mathbb{C}^2)$ where $\mathbb{C}^2$ is equipped with the standard matrix action of sl_2 with respect to the basis $e = (1,0)$ and $f = (0,1)$ of $\mathbb{C}^2$. Relative to the limit mixed Hodge structure e is type $(1,1)$ and f is type $(0,0)$. The polarization is defined by

$$Q(e^j f^{n-j}, e^{n-j} f^j) = (-1)^n (-1)^j j!(n-j)!;$$

— $E(p,q) = \mathbb{C}^2$ equipped with the trivial action of sl_2 and $e + if$ of type (p,q);

— $\mathbb{R}(m)$ is $\mathbb{C}$ equipped with a pure Hodge structure of type $(-m,-m)$.

Before proceeding with the examples, we note that in the case where W is pure of weight k, equation (6.51) applies and we shall omit W from the data of the Deligne system. If W is pure of weight k, we also always have $N_1 = \hat{N}_1$.

Example 6.64. Let $(N_1, N_2; F)$ generate a pure nilpotent orbit of odd weight $2m+1$ of "vanishing cycle" type, i.e. there exist linearly independent elements α_1 and α_2 of the underlying real vector space $V_{\mathbb{R}}$ such that $Q(\alpha_1, \alpha_2) = 0$ and

$$N_j(\gamma) = Q(\gamma, \alpha_j)\alpha_j.$$

Let $W^j = W(\sum_{i\leq j} N_i)[-(2m+1)]$ and assume without loss of generality that (F, W) is split over $\mathbb{R}$. In Schmid's terminology, the corresponding horizontal sl_2 action generated by $N = N_1 + N_2$ and the limit mixed Hodge structure (F, W^2) decomposes the underlying vector space as

$$[S(1) \otimes \mathbb{R}(-m)] \oplus [S(1) \otimes \mathbb{R}(-m)] \oplus K$$

where K is a pure Hodge structure of weight $2m + 1$ with trivial sl_2-action. The two $S(1)$ factors correspond to the isotypical components of highest weight 1 for (N_1, H_1) and (N_2, H_2), i.e. $\hat{N}_j = N_j$.

More concretely, there exist dual elements α_1' and α_2' of type $(m+1, m+1)$ with respect to (F, W^2) such that $Q(\alpha_j', \alpha_k) = \delta_{jk}$, and (F, W^2) has Deligne bigrading

$$I^{m+1,m+1} = \mathrm{span}(\alpha_1', \alpha_2'), \qquad \bigoplus_p I^{p,2m+1-p} = K, \qquad I^{m,m} = \mathrm{span}(\alpha_1, \alpha_2).$$

The associated sl_2-type is given by (N_1, H_1) and (N_2, H_2) where

$$\begin{array}{lll} E_1(H_1) = \mathrm{span}(\alpha_1'), & E_0(H_1) = K \oplus \mathrm{span}(\alpha_2, \alpha_2'), & E_{-1}(H_1) = \mathrm{span}(\alpha_1), \\ E_1(H_2) = \mathrm{span}(\alpha_2'), & E_0(H_2) = K \oplus \mathrm{span}(\alpha_1, \alpha_1'), & E_{-1}(H_2) = \mathrm{span}(\alpha_2). \end{array}$$

To analyze the corresponding variety of Deligne systems via the process described above, we start with the set $\mathcal{S}$ attached to W^0 and (N_1, H_1). Since W^0 is pure of weight k, $\mathcal{S}$ is a point. This becomes the set $\mathcal{S}'$ at the next step, and we have to consider the set $\mathcal{S}''$ attached to W^1 and (N_2, H_2). By (6.51), $Y^1 = Y^0 + H_1$ which eliminates the freedom to pick a point in $\mathcal{Y}(W^1, N_2, H_2)$. The remaining freedom in $\mathcal{U}(W^1, N_2, H_2)$ is to select an element which is weight -2 for $\mathrm{ad}\, Y^1$ and highest weight 0 for (N_2, H_2). This is exactly the space spanned by N_1, and so the possible set of Deligne systems consists of the triples

$$(N_1, N_2 + aN_1, Y^2) \tag{6.65}$$

where a is an arbitrary scalar.

Example 6.66. More generally, let $(N_1, N_2; F)$ generate a two variable sl_2-orbit of weight k polarized by Q, i.e. the limit mixed Hodge structure (F, W^2) split over $\mathbb{R}$ and $\hat{N}_j = N_j$. Let $Y^1 = k\,\mathrm{Id} + H_1$ and $Y^2 = Y^1 + H_2$. Then, the set of Deligne systems with the same sl_2-type consists of all triples

$$(N_1, N_2 + \eta, Y^2)$$

where $\eta = \sum_{\ell\geq 2} \eta_{-\ell}$ and $\eta_{-\ell}$ is weight $-\ell$ with respect to $\mathrm{ad}(Y^1)$ and highest weight $\ell - 2$ for (N_2, H_2).

In particular, let Ω denote the set of all $(-1,-1)$-morphisms of (F, W^2) which are lowest weight -2 for (N_1, H_1) and highest weight zero for (N_2, H_2). Note that $\Omega \neq 0$ since $N_1 \in \Omega$. Fix an inner product on Ω. Recall that $F_o = e^{iN_1+iN_2}F$ belongs to the classifying space D. Moreover, since D is an open subset of the compact dual $\check{D}$ in the analytic topology, there exists an real number $t > 0$ such that if $\omega \in \Omega$ has norm 1 and $|\tau| < t$ then $e^{i\tau\omega}F_o$ also belongs to D. The following argument shows that $(N_1, N_1 + t\omega, N_2; F)$ generates a nilpotent orbit: By assumption, the map

$$\theta(z_1, z_2, z_3) = \exp(z_1 N_1 + z_2(N_1 + t\omega) + z_3 N_2)F \tag{6.67}$$

is horizontal. Moreover ω commutes with N_1 and N_2 since it is lowest weight -2 for (N_1, H_1) and highest weight zero for (N_2, H_2).

To see that θ takes values in D when the imaginary parts of $z_j = x_j + \sqrt{-1}y_j$ are sufficiently large, observe that

$$y_1 N_1 + y_2(N_1 + t\omega) + y_3 N_2 = (y_1 + y_2)\,(N_1 + \tau\omega) + y_3 N_2$$

where $\tau = (y_2 t)/(y_1 + y_2)$ is positive and less than t. Therefore, since ω is lowest weight -2 for (N_1, H_1) and highest weight zero for (N_2, H_2) it follows that

$$\begin{aligned}\exp(iy_1 N_1 + iy_2(N_1 + t\omega) + iy_3 N_2)F &= \exp(i(y_1 + y_2)(N_1 + \tau\omega) + iy_3 N_2)F \\ &= (y_1 + y_2)^{-H_1/2} y_3^{-H_2/2} e^{i\tau\omega} F_o.\end{aligned}$$

Accordingly, since $(y_1 + y_2)^{-H_1/2} y_3^{-H_2/2}$ is a real automorphism of Q and $e^{i\tau\omega}F_o \in D$ it follows that (6.67) takes values in D provided the imaginary parts of z_1, z_2 and z_3 are positive. Likewise, if $\mathfrak{a}$ is an abelian subalgebra of Ω and $\omega_1, \dots, \omega_\ell$ are an orthonormal basis of $\mathfrak{a}$ then a similar type of argument shows that

$$(N_1, N_1 + t_1\omega_1, \dots, N_1 + t_\ell\omega_\ell, N_2; F)$$

generates a nilpotent orbit, provided $|t_j| < t$ for all j.

Example 6.68. Example (6.64) can be generalized to the weight $2m + 1$ case where there exist pairwise orthogonal, linearly independent sets of vanishing cycles $\{\alpha_{11}, \dots, \alpha_{1p}\}$ and $\{\alpha_{21}, \dots, \alpha_{2q}\}$ such that $N_i = \frac{1}{2}\sum_\ell \omega_i^{\ell\ell}$ where

$$\omega_i^{jk}(u) = Q(u, \alpha_{ij})\alpha_{ik} + Q(u, \alpha_{ik})\alpha_{ij}$$

for $j \leq k$. Then, the associated set of Deligne systems is

$$\mathcal{S}(\mathbb{F}) = \{(N_1, N_2 + \sum_{j \leq k} c_{jk}\omega_1^{jk}, Y^2) \mid c_{jk} \in \mathbb{F}\}$$

where $\mathbb{F} \subseteq \mathbb{C}$ is the field of interest.

In Schmid's terminology, the associated representation of sl_2 for $N = N_1 + N_2$ is a sum

$$[S(1) \otimes \mathbb{R}(-m)]^p \oplus [S(1) \otimes \mathbb{R}(-m)]^q \oplus K$$

where two groupings of $S(1)$ factors corresponding to the isotypical components of highest weight 1 for (N_1, H_1) and (N_2, H_2), and K is a pure Hodge structure of weight $2m+1$ with trivial sl_2-action. As in Example (6.64), there are dual elements

$$\alpha'_{11}, \dots, \alpha'_{1p}, \alpha'_{21}, \dots, \alpha'_{2q}$$

of type $(m+1, m+1)$ such that $Q(\alpha'_{ij}, \alpha_{k\ell}) = \delta_{ik}\delta_{j\ell}$. Let $A_i = \mathrm{span}(\alpha_{ij})$ and $A'_i = \mathrm{span}(\alpha'_{ij})$. Then, Deligne bigrading of (F, W^2) is given by

$$I^{m+1,m+1} = A'_1 \oplus A'_2, \qquad \bigoplus_p I^{p,2m+1-p} = K, \qquad I^{m,m} = A_1 \oplus A_2.$$

The corresponding sl_2-type is (N_1, H_1) and (N_2, H_2) where

$$\begin{aligned} E_1(H_1) &= A'_1, & E_0(H_1) &= K \oplus A_2 \oplus A'_2, & E_{-1}(H_1) &= A_1, \\ E_1(H_2) &= A'_2, & E_0(H_2) &= K \oplus A_1 \oplus A'_1, & E_{-1}(H_2) &= A_2. \end{aligned}$$

Due to the short length of the monodromy weight filtrations, it follows by Example 6.66 we are looking for infinitesimal isometries which are lowest weight -2 for (N_1, H_1) and highest weight zero for (N_2, H_2). This subspace is spanned by the elements ω_1^{ij} for $i \le j$.[8]

The short length of the monodromy weight filtrations also forces commutativity of all ω_1^{ij}, so we can form a new monodromy cone using the techniques described at the end of Example 6.66 – just pick an inner product which makes all the ω_1^{ij} orthonormal and find the appropriate value of t.

Example 6.69. Turning now to §5, we consider a two variable example of the form

$$S(2) \oplus S(2) \oplus \mathbb{R}(-1)$$

where the two $S(2)$ factors are the isotypical components of highest weight 2 for (N_1, H_1) and (N_2, H_2), and $\mathbb{R}(-1)$ is a factor of Hodge type $(1,1)$ on which both copies of sl_2 act trivially. The corresponding period domain has Hodge numbers $h^{2,0} = 2$ and $h^{1,1} = 3$.

More concretely, we start with a real vector space $V_{\mathbb{R}}$ with basis

$$\{\alpha_2, \alpha_1, \alpha_0, \beta_2, \beta_1, \beta_0, \gamma\}$$

in which we think of $\alpha_j = e^j f^{2-j}$ and $\beta_j = e^j f^{2-j}$ under the identification with $S(2) = \mathrm{Sym}^2(\mathbb{C}^2)$, and γ is the generator of $\mathbb{R}(-1)$. Accordingly, N_1 annihilates $\{\beta_2, \beta_1, \beta_0, \gamma\}$ and acts on $\alpha_j = e^j f^{2-j}$ by the rule $N_1(\alpha_j) = j\alpha_{j-1}$. Likewise, N_2 annihilates $\{\alpha_2, \alpha_1, \alpha_0, \gamma\}$ and acts on $\beta_j = e^j f^{2-j}$ by the rule $N_2(\beta_j) = j\beta_{j-1}$. The polarizing form is given by

$$Q(\alpha_j, \alpha_{2-j}) = Q(\beta_j, \beta_{2-j}) = (-1)^j j!(2-j)!, \qquad Q(\gamma, \gamma) = 1$$

[8]Sketch: One first checks that the stated conditions imply that such an element γ vanishes on $A'_2 \oplus K \oplus A_1 \oplus A_2$ and $\gamma(A'_1) \subset A_1$. Let $\gamma(\alpha'_{1\ell}) = \sum_k \gamma_\ell^k \alpha_{1k}$. Then, the infinitesimal isometry condition forces $\gamma_j^i = \gamma_i^j$.

and all other pairings zero.

The limit Hodge filtration of (F, W^2) is

$$I^{2,2} = \mathrm{span}(\alpha_2, \beta_2), \qquad I^{1,1} = \mathrm{span}(\alpha_1, \beta_1, \gamma), \qquad I^{0,0} = \mathrm{span}(\alpha_0, \beta_0)$$

which is type (V) in the setting of §5. The nilpotent orbit $\theta(z_1) = e^{z_1 N_1} e^{iN_2} F$ is of type (II) with limit Hodge numbers $h^{2,2} = h^{2,0} = h^{0,2} = 1$, $h^{1,1} = 3$ and $h^{0,0} = 1$.

The corresponding elements H_1 and H_2 are given by

$$E_2(H_2) = \mathrm{span}(\beta_2), \quad E_0(H_2) = \mathrm{span}(\alpha_2, \alpha_1, \alpha_0, \beta_1, \gamma), \quad E_{-2}(H_2) = \mathrm{span}(\beta_0),$$
$$E_2(H_1) = \mathrm{span}(\alpha_2), \quad E_0(H_1) = \mathrm{span}(\beta_2, \beta_1, \beta_0, \alpha_1, \gamma), \quad E_{-2}(H_1) = \mathrm{span}(\alpha_0).$$

To find the possible candidate deformations $N_2 \mapsto N_2 + \eta$ which preserve the underlying sl_2-orbit structure, we can start by first identifying all morphisms of type $(-1,-1)$ of (F, W^2) which are infinitesimal isometries. This space has basis $\{\eta_1, \eta_2, \eta_3, \eta_4, N_1, N_2\}$ where

$$\eta_1(\alpha_2) = \beta_1, \qquad \eta_1(\beta_1) = \frac{1}{2}\alpha_0,$$
$$\eta_2(\alpha_2) = \gamma, \qquad \eta_2(\gamma) = -\frac{1}{2}\alpha_0,$$
$$\eta_3(\beta_2) = \alpha_1, \qquad \eta_3(\alpha_1) = \frac{1}{2}\beta_0,$$
$$\eta_4(\beta_2) = \gamma, \qquad \eta_4(\gamma) = -\frac{1}{2}\beta_0$$

and annihilate all other basis elements. A short calculation shows that only η_2 and η_4 commute with N_1 and N_2 which is required for $(N_1, N_2 + \eta, Y^2)$ to be a Deligne system. However, η_4 turns out to be a lowest weight vector of weight -2 for (N_2, H_2) whereas for a deformation of Deligne systems $(N_1, N_2 + \eta, Y^2)$, we would need η to be a sum of highest weight vectors for (N_2, H_2). On the other hand, η_2 is lowest weight -2 for (N_1, H_1) and highest weight zero for (N_2, H_2). Thus, the set of all Deligne systems with given sl_2-type in this case is

$$(N_1, N_2 + a\eta_2 + bN_1, Y^2), \qquad Y^2 = 2\mathrm{Id} + H_1 + H_2$$

where a and b are arbitrary scalars.

As in Example 6.66, we obtain an associated 3 variable nilpotent orbit with data

$$(N_1, N_1 + t\eta_2, N_2; F).$$

Looking back to §5, it was predicted that for a degeneration of type (V) with Hodge numbers $h^{2,0} = 2$ and $h^{1,1} = 3$ the maximum possible dimension of a nilpotent cone is 3, which is realized by this example.

Example 6.70. As a "degenerate" case of the above, observe that we can always augment a pure, 1-variable Deligne system (N, Y^1) of weight k to a 2-variable Deligne system (N_1, N_2, Y^2) by setting $N_1 = N$, $N_2 = 0$ and $Y^2 = Y^1$. In this case, one finds that any Deligne system deformation

$$(N_1, N_2 + \eta, Y^2)$$

of (N_1, N_2, Y^2) must have η of lowest weight -2 for the representation (N_1, H_1).

In the case where (N, Y^1) arises from a pure nilpotent orbit $\theta(z) = e^{zN}F$ with limit mixed Hodge structure (F, W^1) split over $\mathbb{R}$, the method of Example 6.66 shows that if η is also $(-1,-1)$-morphism of (F, W^1) then there exists a positive number t such that $(N, N + t\eta, F)$ generates a pure nilpotent orbit.

A simple example of this type occurs when we start with a several variable pure nilpotent orbit generated by $(N_1, \dots, N_r; F)$ with limit mixed Hodge structure split over $\mathbb{R}$, and consider the 1-variable orbit θ generated by $N = \sum_j N_j$ and F. Then, all of the N_j's are lowest weight -2 for the sl_2-representation attached to θ.

In particular, let $\theta(z) = e^{zN}F$ be a 1-variable pure nilpotent orbit with limit mixed Hodge structure (F, W^1) split over $\mathbb{R}$. Let $G_{\mathbb{R}} = \mathrm{Aut}_{\mathbb{R}}(Q)$ and $G_{\mathbb{R}}^{0,0}$ be the subset of elements which preserve (F, W^1). Let $\mathcal{N}^0$ be the orbit of N under the adjoint action of $G_{\mathbb{R}}^{0,0}$. Then, any element of $\mathcal{N}$ which commutes with N will be of lowest weight -2 for the representation attached to θ. More generally, we can construct several variable nilpotent orbits in this way.

Example 6.71. Let D be a period domain with Hodge numbers $h^{2,0} = 1$ and $h^{1,1} = m + 1$. Then, one possible type of 1-variable sl_2-orbit $\theta(z) = e^{zN}F$ corresponds to the Schmid form

$$S(2) \oplus [\mathbb{R}(-1)]^m$$

with $\alpha_j = e^j f^{2-j}$ for $j = 0, 1, 2$ spanning $S(2)$ and $[\mathbb{R}(-1)]^m$ generated by elements $\gamma_1, \dots, \gamma_m$. The polarization is

$$Q(\alpha_j, \alpha_{2-j}) = (-1)^j j!(2-j)!, \qquad Q(\gamma_i, \gamma_j) = \delta_{ij}$$

and all other pairings zero. The subspace of infinitesimal isometries which are $(-1,-1)$-morphisms and lowest weight -2 for (N, H) is spanned by $\eta_1, \dots, \eta_m$ where

$$\eta_i(\alpha_2) = \gamma_i, \qquad \eta_i(\gamma_i) = -\frac{1}{2}\alpha_0$$

and η_i annihilates all other basis elements. It is easy to see that $[\eta_i, \eta_j] = 0$ for all i and j. Accordingly, there is a positive real number t such that

$$(N, N + t_1\eta_1, \dots, N + t_m\eta_m; F)$$

generates a pure nilpotent orbit provided $|t_j| < t$ for all j.

Example 6.72. To obtain an example where the deformation of a pure sl_2-orbit $(N_1, N_2; F)$ with a Deligne system deformation $(N_1, N_2 + \eta, Y^2)$ with η of weight -3 with respect to Y^1, we consider an sl_2-orbit of Schmid type

$$S(3) \oplus [S(1) \otimes \mathbb{R}(-1)]$$

for $N = N_1 + N_2$ where $S(3)$ is the isotypical component of highest weight 3 for N_1 and $S(1) \otimes \mathbb{R}(-1)$ is the isotypical component of highest weight 1 for N_2.

More concretely, we have a 6 dimensional real vector space $V_{\mathbb{R}}$ with basis $\{\alpha_3, \alpha_2, \alpha_1, \alpha_0, \beta_1, \beta_0\}$ and sl_2 action in which N_1 annihilates $\{\beta_1, \beta_0\}$ and acts on

$\alpha_j = e^j f^{3-j}$ by the rule $N_1(\alpha_j) = j\alpha_{j-1}$. Similarly, N_2 annihilates $\{\alpha_3, \alpha_2, \alpha_1, \alpha_0\}$ and acts on $\beta_j = e^j f^{1-j}$ by the rule $N_2(\beta_j) = j\beta_{j-1}$. The polarization is given by

$$Q(\alpha_j, \alpha_{3-j}) = -(-1)^j j!(3-j)!, \qquad Q(\beta_1, \beta_0) = 1$$

and all other pairings equal to zero.

In this setting, the limit mixed Hodge structure of (F, W^2) is of Hodge–Tate type with α_j of type (j, j) and β_j of type $(j+1, j+1)$. For the mixed Hodge structure $(e^{iN_2}F, W^1)$ the Hodge type of α_j remains unchanged but $\beta_1 + i\beta_0$ is now of Hodge type $(2, 1)$. The linear map η which annihilates $\{\alpha_2, \alpha_1, \alpha_0, \beta_1\}$ and acts by

$$\eta(\alpha_3) = \beta_1, \qquad \eta(\beta_0) = -\frac{1}{6}\alpha_0$$

is an infinitesimal isometry which is a morphism of type $(-1, -1)$ for (F, W^2). It is weight -3 with respect to $\operatorname{ad} Y^1$ and commutes with N_1. A bit more calculation shows that it is lowest weight -3 for (N_1, H_1) and highest weight 1 for (N_2, H_2). So $(N_1, N_2 + \eta, Y^2)$ is a deformation of the Deligne system (N_1, N_2, Y^2) with the same sl_2-type. Since it is also polarizable, it follows that there is an $a > 0$ such that

$$(N_1, N_2 + \eta + aN_1; F)$$

generates a polarizable nilpotent orbit. We also note that unlike the original orbit (N_1, N_2, F), the new orbit has maximal unipotent monodromy since now

$$Gr_4^{W^2} = N_1(Gr_6^{W^2}) + (N_2 + \eta + aN_1)(Gr_6^{W^2}).$$

Let $(N_1, \ldots, N_r; F)$ generate a pure nilpotent of weight k which is polarized by Q. Let $W^j = W(\sum_{i \leq j} N_i)[-k]$ and assume that (F, W^r) is split over $\mathbb{R}$. Let $(\hat{N}_j, H_j)$ be the associated sl_2-type, and $G_{\mathbb{R}} = \mathrm{Aut}_{\mathbb{R}}(Q)$. Then, for any $g \in G_{\mathbb{R}}$ the data

$$(\mathrm{Ad}(g)N_1, \ldots, \mathrm{Ad}(g)N_r; g(F)) \tag{6.73}$$

generates a nilpotent orbit of weight k which is polarized by Q.

Theorem 6.74. *The nilpotent orbit generated by* (6.73) *has the same limit mixed Hodge structure and* sl_2*-type as the orbit generated by* $(N_1, \ldots, N_r; F)$ *if and only if* g *preserves* F *and each* sl_2*-pair* $(\hat{N}_j, H_j)$.

Proof. Suppose that $g \in G_{\mathbb{R}}$ preserves F and each sl_2-pair. Then, g preserves

$$Y^j = k\,\mathrm{Id} + H_1 + \cdots + H_j \tag{6.75}$$

and hence g preserves the weight filtrations W^j. In particular, g preserves the mixed Hodge structure (F, W^r) and hence g preserves $Y^r = Y_{(F,W^r)}$. By the functoriality of Deligne's construction, it then follows that

$$Y(\mathrm{Ad}(g)N_j, \mathrm{Ad}(g)Y^j) = \mathrm{Ad}(g)Y(N_j, Y_j) = \mathrm{Ad}(g)Y^{j-1} = Y^{j-1}$$

and so the corresponding chain of gradings remains the same. Likewise, the component of $\mathrm{Ad}(g)N_j$ of degree 0 with respect to $Y^{j-1} = \mathrm{Ad}(g)Y^{j-1}$ is just $\mathrm{Ad}(g)\hat{N}_j$.

By assumption $\mathrm{Ad}(g)\hat{N}_j = \hat{N}_j$. Therefore (6.73) is a nilpotent orbit with the same sl_2-type as $(N_1, \ldots, N_r; F)$.

Conversely, suppose that (6.73) defines a pure nilpotent orbit with the same sl_2-type as the orbit generated by $(N_1, \ldots, N_r; F)$. Then, we must have the same associated weight filtrations

$$\begin{aligned} W(\mathrm{Ad}(g)(N_1 + \cdots + N_j))[-k] &= g(W(N_1 + \cdots + N_j)[-k]) \\ &= W(N_1 + \cdots + N_j)[-k] \end{aligned}$$

since we have the same associated set of gradings (6.75). As in the previous paragraph, g must preserve the mixed Hodge structure (F, W^r) and hence $\mathrm{Ad}(g)Y^r = Y^r$. By the functoriality of Deligne systems,

$$Y(\mathrm{Ad}(g)N_r, Y^r) = Y(\mathrm{Ad}(g)N_r, \mathrm{Ad}(g)(Y^r)) = \mathrm{Ad}(g)Y^{r-1}.$$

Therefore

$$\mathrm{Ad}(g)Y^{r-1} - Y^r = H_r = Y^r - Y^{r-1}$$

which implies that $\mathrm{Ad}(g)Y^{r-1} = Y^{r-1}$. Repeating this argument down the chain of gradings shows that g preserves each H_j. As in the first paragraph of the proof, the component of $\mathrm{Ad}(g)N_j$ of degree 0 with respect to $Y^{j-1} = \mathrm{Ad}(g)Y^{j-1}$ is just $\mathrm{Ad}(g)\hat{N}_j$. To obtain the same sl_2-type, we must therefore have $\mathrm{Ad}(g)\hat{N}_j = \hat{N}_j$ □

In summary, the set of Deligne systems with given sl_2-type forms an algebraic variety, and this remains so once we layer on the existence of a polarization and/or a filtration F with respect to which all of the N_j's are horizontal. Starting from a pure nilpotent orbit with limit mixed Hodge structure which is split over $\mathbb{R}$, the real points of the algebraic group $\mathcal{G}$ are consisting of isometries which preserve the limit Hodge filtration and associated sl_2-pairs acts upon the set of nilpotent orbits with these properties.

A Proof of Lemma 3.5

First note that $\mathcal{N}^0 \subset \mathcal{W}_N^\circ$.

Note that $Y \in \mathfrak{m}_{\mathbb{R}}^{0,\mathrm{ss}}$ acts on $\mathfrak{g}_{\mathbb{R}}^{\ell,\ell} \subset \mathfrak{m}_{\mathbb{R}}$ by the scalar 2ℓ.

Claim. Each $N' \in \mathcal{W}_N$ may be realized as the nilnegative element of a standard triple in $\mathfrak{m}_{\mathbb{R}}$ containing Y as the neutral element.

Remark. Malcev's Theorem implies that N and N' are conjugate under the action of $\mathrm{Ad}(M_{\mathbb{C}}^0)$. Unfortunately, Malcev's Theorem does not hold over $\mathbb{R}$.[9]

To prove the claim it suffices to construct $N'_+ \in \mathfrak{g}_{\mathbb{R}}^{1,1}$ with the property that $[N'_+, N'] = Y$. Given $\ell \geq 0$, it follows from the definition of $\mathcal{W}(N)$ and properties of the filtration $W(N)$ that

[9] As an example notice that $N = \begin{pmatrix} 0 & 1 \\ 0 & 0 \end{pmatrix}$ is not conjugate to $-N$ in $\mathrm{SL}_2\mathbb{R}$ although they can both be completed to a standard triple containing $Y = \begin{pmatrix} 1 & 0 \\ 0 & -1 \end{pmatrix}$ as the neutral element.

$$(N')^{2\ell} : \mathfrak{g}^{\ell,\ell} \to \mathfrak{g}^{-\ell,-\ell} \text{ is an isomorphism.}$$

Let

$$P_{2\ell}(N') \;:=\; \ker\{(N')^{2\ell+1} : \mathfrak{g}_{\mathbb{R}}^{\ell,\ell} \to \mathfrak{g}_{\mathbb{R}}^{-\ell-1,-\ell-1}\}.$$

Fix a basis $\{v_\ell^1, \ldots, v_\ell^{d_\ell}\}$ of $P_{2\ell}(N')$. Then

$$\bigcup_{\ell \geq 0} \{(N')^k v_\ell^i \mid 1 \leq i \leq d_\ell\,,\; 0 \leq k \leq 2\ell\}$$

is a basis of $\mathfrak{m}_{\mathbb{R}}$, and we may define $N'_+ \in \mathrm{End}(\mathfrak{m}_{\mathbb{R}})$ by

$$(N')^k v_\ell^i \;\mapsto\; k(2\ell - k + 1)(N')^{k-1} v_\ell^i\,.$$

Then $\{N'_+, Y, N'\}$ is a standard triple in $\mathrm{End}(\mathfrak{m}_{\mathbb{R}})$. Since both $Y, N' \in \mathfrak{m}_{\mathbb{R}} \subset \mathrm{End}(\mathfrak{m}_{\mathbb{R}})$, we necessarily have $N'_+ \in \mathfrak{m}_{\mathbb{R}}$. This proves the claim. ◇

From the proof of the claim we see that

$$\mathfrak{m}_{\mathbb{R}}^0 \;=\; \bigoplus_{\ell \geq 0} (N')^\ell P_{2\ell}(N')\,,$$

and we deduce that the orbit

$$\mathcal{N}' \;:=\; \mathrm{Ad}(M_{\mathbb{R}}^0) \cdot N' \;\subset\; \mathfrak{g}_{\mathbb{R}}^{-1,-1}$$

has

$$\dim \mathcal{N}' \;=\; \dim_{\mathbb{R}} \mathfrak{m}_{\mathbb{R}}^0 \;-\; \dim_{\mathbb{R}} P_0 \;=\; \dim_{\mathbb{R}} \mathfrak{g}_{\mathbb{R}}^{-1,-1}\,.$$

In particular, $\mathcal{N}'$ is open in $\mathfrak{g}_{\mathbb{R}}^{-1,-1}$, and this statement is independent of our choice of $N' \in \mathcal{W}_N$. The lemma now follows from the connectedness of $\mathcal{W}_N^\circ$.

References

[1] Patrick Brosnan and Gregory Pearlstein. On the algebraicity of the zero locus of an admissible normal function. *Compos. Math.*, 149(11):1913–1962, 2013.

[2] Eduardo Cattani and Aroldo Kaplan. Polarized mixed Hodge structures and the local monodromy of a variation of Hodge structure. *Invent. Math.*, 67(1):101–115, 1982.

[3] Eduardo Cattani and Aroldo Kaplan. Degenerating variations of Hodge structure. *Astérisque*, (179-180):9, 67–96, 1989. Actes du Colloque de Théorie de Hodge (Luminy, 1987).

[4] Eduardo Cattani, Aroldo Kaplan and Wilfried Schmid. Degeneration of Hodge structures. *Ann. of Math. (2)*, 123(3):457–535, 1986.

[5] Eduardo H. Cattani and Aroldo G. Kaplan. Horizontal SL_2-orbits in flag domains. *Math. Ann.*, 235(1):17–35, 1978.

[6] David H. Collingwood and William M. McGovern. *Nilpotent orbits in semisimple Lie algebras*. Van Nostrand Reinhold Mathematics Series. Van Nostrand Reinhold Co., New York, 1993.

[7] Pierre Deligne. La conjecture de Weil. I. *Inst. Hautes Études Sci. Publ. Math.*, (43):273–307, 1974.

[8] Pierre Deligne. Personal letter to E. Cattani and A. Kaplan. 1993.

[9] Pierre Deligne, James S. Milne, Arthur Ogus and Shih Kuang-Yen. *Hodge cycles, motives, and Shimura varieties*. Lecture Notes in Mathematics, 900, Springer, 1982.

[10] Dragomir Ž. Djoković. Closures of conjugacy classes in classical real linear Lie groups. In *Algebra, Carbondale 1980 (Proc. Conf., Southern Illinois Univ., Carbondale, Ill., 1980)*, volume 848 of *Lecture Notes in Mathematics*, pages 63–83. Springer, Berlin, 1981.

[11] Murray Gerstenhaber. Dominance over the classical groups. *Ann. of Math. (2)*, 74:532–569, 1961.

[12] Murray Gerstenhaber. On dominance and varieties of commuting matrices. *Ann. of Math. (2)*, 73:324–348, 1961.

[13] Mark Green, Phillip Griffiths and Matt Kerr. *Mumford-Tate groups and domains: their geometry and arithmetic*, volume 183 of *Annals of Mathematics Studies*. Princeton University Press, Princeton, NJ, 2012.

[14] Aroldo Kaplan and Gregory Pearlstein. Singularities of variations of mixed Hodge structure. *Asian J. Math.*, 7(3):307–336, 2003.

[15] Masaki Kashiwara. A study of variation of mixed Hodge structure. *Publ. Res. Inst. Math. Sci.*, 22(5):991–1024, 1986.

[16] Kazuya Kato. On SL(2)-orbit theorems. *Kyoto J. Math.*, 54(4):841–861, 2014.

[17] Matt Kerr and Colleen Robles. Hodge theory and real orbits in flag varieties. arXiv:1407.4507, 2014.

[18] Matt Kerr, Gregory Pearlstein and Colleen Robles. Polarized relations on horizontal SL(2)'s. arXiv = 1705.03117.

[19] Anthony W. Knapp. *Lie groups beyond an introduction*, volume 140 of *Progress in Mathematics*. Birkhäuser Boston Inc., Boston, MA, second edition, 2002.

[20] G. D. Mostow. Some new decomposition theorems for semi-simple groups. *Mem. Amer. Math. Soc.*, 14: 31–44, 1955.

[21] Gregory J. Pearlstein. Degenerations of mixed Hodge structure. *Duke Math. J.*, 110(2):217–251, 2001.

[22] Colleen Robles. Classification of horizontal SL(2)s. *Compos. Math.*, 152(5):918–954, 2016.

[23] Colleen Robles. Schubert varieties as variations of Hodge structure. *Selecta Math. (N.S.)*, 20(3):719–768, 2014.

[24] Wilfried Schmid. Variation of Hodge structure: the singularities of the period mapping. *Invent. Math.*, 22:211–319, 1973.

[25] Christine Schwarz. Relative monodromy weight filtrations. *Math. Z.*, 236(1):11–21, 2001.

[26] T. A. Springer and R. Steinberg. Conjugacy classes. In *Seminar on Algebraic Groups and Related Finite Groups (The Institute for Advanced Study, Princeton, N.J., 1968/69)*, volume 131 of *Lecture Notes in Mathematics*, pages 167–266. Springer, Berlin, 1970.

[27] Joseph Steenbrink and Steven Zucker. Variation of mixed Hodge structure. I. *Invent. Math.*, 80(3):489–542, 1985.

[28] Steven Zucker. Generalized intermediate Jacobians and the theorem on normal functions. *Invent. Math.*, 33(3):185–222, 1976.

[29] Steven Zucker. Hodge theory with degenerating coefficients. L_2 cohomology in the Poincaré metric. *Ann. of Math. (2)*, 109(3):415–476, 1979.

Hodge Theory and L^2-analysis
ALM 39, pp. 207–243

Recent Results on Cohomology Jump Loci*

Nero Budur, Botong Wang†

Abstract

This is a survey of recent results related to cohomology jump loci. It emphasizes connections with deformations with cohomology constraints, global structural results for rank one local systems and line bundles, some connections with restrictions on fundamental groups and homotopy types, and connections with classical singularity theory and Bernstein-Sato ideals.

1 Introduction

Let $\mathcal{M}$ be a moduli space parametrizing objects P that come with $\mathbb{C}$ vector spaces

$$H^i(P) \quad (i \in \mathbb{Z}).$$

For example, some cohomology theory. The **cohomology jump loci** of $\mathcal{M}$ are defined set-theoretically as

$$\mathcal{V}^i_k = \{P \in \mathcal{M} \mid \dim H^i(P) \geq k\}.$$

For every i, there is a chain of inclusions

$$\mathcal{M} = \mathcal{V}^i_0 \supset \mathcal{V}^i_1 \supset \mathcal{V}^i_2 \supset \cdots .$$

In many situations, the cohomology jump loci form closed subsets of the moduli space. When the moduli is an algebraic variety or a scheme, the cohomology jump loci are usually closed subschemes. Cohomology jump loci are at least as complex

*Dedicated to Steve Zucker's 65th birthday.

†KU Leuven, Department of Mathematics, Celestijnenlaan 200B, B-3001 Leuven, Belgium. E-mail: Nero.Budur@wis.kuleuven.be; KU Leuven, Department of Mathematics, Celestijnenlaan 200B, B-3001 Leuven, Belgium. E-mail: wang_botong@hotmail.com

as the moduli, and in general everything can be a moduli space. Hence the study of the cohomology jump loci can be described as finding a balance between general setups one can focus on and interesting properties one can draw.

In this article one finds a survey of the recent results related to cohomology jump loci. It emphasizes connections with deformations with cohomology constraints, global structural results for rank one local systems and line bundles, some connections with restrictions on fundamental groups and homotopy types, and connections with classical singularity theory and Bernstein-Sato ideals.

There are many results left out. For example, we do not survey all the implications on quasi-projective groups one can draw from the global structure of cohomology jump loci of rank one local systems. We also do not survey the large body of work dedicated to complements of hyperplane arrangements. One can find these two topics in the surveys by Suciu [Su-2011, Su-2015].

Acknowledgements. The first author thanks Steve Zucker and the Department of Mathematics of the Johns Hopkins University for the hospitality during the writing of this survey. The first author was partly sponsored by NSA, FWO, a KU Leuven OT grant, and a Flemish Methusalem grant.

2 Cohomology jump ideals

2.1 Old algebraic example: Fitting ideals

Let R be a commutative ring and M a finitely generated R-module. Consider a free finite rank presentation of M

$$R^n \xrightarrow{d} R^m \longrightarrow M \longrightarrow 0.$$

Define the **Fitting ideals** of M

$$J_k(M) \subset R$$

as the ideals

$$J_k(M) = I_{m-k+1}(d)$$

generated by the minors of size $m-k+1$ of the matrix d. We set $J_k(M)=0$ for $k \leq 0$, and $J_k(M) = R$ for $k > m$. One has the classical result of Alexander and Fitting, see [Ei-1995, §20.2]:

Theorem 2.1. *Let R be a commutative ring and M a finitely generated R-module. The Fitting ideal $J_k(M)$ depends only on M and k, and not on the presentation of M.*

Let $\mathcal{M} = Spec(R)$. The ideals $J_k(M)$ define closed subschemes

$$\mathcal{V}_k(M) \subset \mathcal{M}.$$

There is a chain of closed embeddings

$$\mathcal{M} = \mathcal{V}_0(M) \supset \mathcal{V}_1(M) \supset \cdots \supset \mathcal{V}_{m+1}(M).$$

Thinking of $\mathcal{M}$ as the moduli space of its closed points, each closed point comes with an assigned vector space

$$(m = \text{maximal ideal of } R) \mapsto M \otimes_R R/m.$$

Moreover, $\mathcal{V}_k(M)$ is set-theoretically

$$\mathcal{V}_k(M) = \{m \subset R \text{ maximal ideal } \mid \dim_{R/m}(M \otimes_R R/m) \geq k\},$$

hence it is a cohomology jump locus.

2.2 Cohomology jump ideals are generalized Fitting ideals

In [BW-2013], the construction and the key properties of the Fitting ideals were generalized simultaneously in two directions: to define invariants of a module from a whole free resolution, not just from a free presentation, and to define these invariants for complexes of modules.

Let R be a noetherian commutative ring, and let $M^\bullet$ be a complex of R-modules, bounded above. Suppose $H^i(M^\bullet)$ is a finitely generated R-module. There always exists a bounded above complex $F^\bullet$ of finitely generated free R-modules and a morphism of complexes

$$g : F^\bullet \xrightarrow{\sim} M^\bullet$$

which is a quasi-isomorphism. We define the **cohomology jump ideals** of $M^\bullet$ to be

$$J^i_k(M^\bullet) = I_{rank(F^i)-k+1}(d^{i-1} \oplus d^i)$$

where I denotes the determinantal ideal as above, and

$$d^{i-1} : F^{i-1} \to F^i \quad \text{and} \quad d^i : F^i \to F^{i+1}$$

are the differentials of the complex $F^\bullet$. The same conventions for ideals generated by minors as in the previous subsection apply when $rank(F^i) - k + 1$ falls outside the range of the size of minors.

The **cohomology jump loci** of a bounded above complex $M^\bullet$ of R-modules with finitely generated cohomology are the closed subschemes

$$\mathcal{V}^i_k(M^\bullet) \subset Spec(R)$$

defined by the cohomology jump ideals $J^i_k(M^\bullet)$. In some sense, this is the mother of all examples of cohomology jump loci.

Remark 2.2. If M is a finitely generated R-module, then the Fitting ideals of M coincide with the 0-cohomology jump ideals of M seen as a complex

$$M[0] = [\cdots \to 0 \to M \to 0 \to \cdots]$$

with trivial terms outside degree 0:

$$J_k(M) = J^0_k(M[0]) \quad \text{and} \quad \mathcal{V}_k(M) = \mathcal{V}^0_k(M[0]).$$

Theorem 2.3. ([BW-2013]) *Let R be a noetherian commutative ring. Let $M^\bullet$ be a complex of R-modules, bounded above, with finitely generated cohomology. Then $J_k^i(M^\bullet)$ depends only on M, i, and k, and not on the choice of resolution $F^\bullet$ of $M^\bullet$.*

An important first property is:

Corollary 2.4. *If $M^\bullet$ is quasi-isomorphic to $N^\bullet$, then $J_k^i(M^\bullet) = J_k^i(N^\bullet)$.*

Since taking determinantal ideals commutes with base change, one has the following base change property of the cohomology jump ideals:

Corollary 2.5. *Let R and $M^\bullet$ be defined as above, and let S be a noetherian R-algebra. Moreover, suppose $M^\bullet$ is a complex of flat R-modules, then*

$$J_k^i(M^\bullet) \cdot S = J_k^i(M^\bullet \otimes_R S),$$

where we regard $M^\bullet \otimes_R S$ as a complex of S modules.

When R is a field, by definition $J_k^i(M^\bullet) = 0$ if $\dim H^i(M^\bullet) \geq k$ and $J_k^i(M^\bullet) = R$ if $\dim H^i(M^\bullet) < k$. Thus, we have the following:

Corollary 2.6. *Suppose $M^\bullet$ is a complex of flat R-modules. Then for any maximal ideal m of R, $J_k^i(M^\bullet) \subset m$ if and only if*

$$\dim_{R/m} H^i(M^\bullet \otimes_R R/m) \geq k.$$

This says that the cohomology jump ideals deserve their name.

Remark 2.7. (Tor jumping loci) If M is a finitely generated R-module, the ideals $J_k^i(M)$ put a closed subscheme structure on the Tor jumping loci:

$$\mathcal{V}_k^i(M)_{red} = \{P \in Spec(R) \text{ closed point } \mid \dim_{R/m_P} \mathrm{Tor}_{-i}^R(M, R/m_P) \geq k\}.$$

Remark 2.8. (Syzygies) If M is a finitely generated R-module and $F^\bullet$ is a free resolution, the syzygies of M are defined as the kernels of the differentials of $F^\bullet$. Hence the cohomology jump loci of a syzygy of M coincide with the far-enough-to-the-left cohomology jump loci of M.

2.3 Dimension bounds

Recall the following classical homological algebra statement from [Ei-1995, Theorem 20.9]:

Theorem 2.9. *Let R be a ring. Let*

$$0 \longrightarrow F^0 \xrightarrow{d^0} F^1 \longrightarrow \cdots \longrightarrow F^{n-1} \xrightarrow{d^{n-1}} F^n$$

be a complex of free finitely generated R-modules. Then this complex is exact if and only if

$$rank(F^i) = rank(d^{i-1} \oplus d^i)$$

and

$$depth\ I_{rank(d^i)}(d^i) \geq n - i.$$

By applying this statement to the exact complex $(F^\bullet[-1] \oplus F^\bullet, d[-1] \oplus d)$, one has the following universal bound for cohomology support loci:

Theorem 2.10. *With the same assumptions as in the previous theorem, if $F^\bullet$ is exact, then*

$$\mathit{depth}\ J_1^i(F^\bullet) \geq n - i.$$

3 Deformation theory

3.1 Local study

By a **deformation problem** we mean describing the formal scheme or the analytic germ $\mathcal{M}_{(P)}$ at some object P of some moduli space $\mathcal{M}$. A deformation problem is equivalent to describing the corresponding functor, which we denote also by $\mathcal{M}_{(P)}$:

$$\mathcal{M}_{(P)} : \mathsf{ART} \longrightarrow \mathsf{SET},$$

where ART is the category of Artinian local algebras of finite type over k, and SET is the category of sets. Assuming $\mathcal{M}$ is a k-scheme, on objects the functor is defined as

$$A \mapsto \mathrm{Hom}_{k-\mathrm{sch}}(\mathit{Spec}(A), \mathcal{M}_{(P)}) = \mathrm{Hom}_{k-\mathrm{alg}}(\mathcal{O}_{\mathcal{M}_{(P)}}, A).$$

This functor is usually well-defined even if the moduli space is not. That is, by passing to the categorical language, one can still ask about the deformations of some object P inside a category. For example, one can ask about the deformations of a coherent sheaf on an algebraic variety, regardless of any stability conditions necessary for the construction of moduli spaces.

Suppose now the objects parametrized by $\mathcal{M}$ come with a cohomology theory, so that one has the stratification of $\mathcal{M}$ by the cohomology jump loci $\mathcal{V}_k^i$. The study of the formal or analytic germs of the cohomology jump loci at an object P amounts to studying the associated functor on Artinian local algebras as a subfunctor

$$\mathcal{V}^i_{k,(P)} \subset \mathcal{M}_{(P)}.$$

Even if the moduli space does not exist, the corresponding deformation problem with cohomology constraints is usually well-defined.

3.2 Differential graded Lie algebras

In a letter written by P. Deligne to J. Millson in 1986 in connection with the paper [GM-1988] by Goldman-Millson, a principle of sweeping generality was stated. The letter is available on J. Millson's webpage [De-1986].

Deligne's Principle. *Every deformation problem over a field of characteristic zero is controlled by a differential graded Lie algebra, with two quasi-isomorphic differential graded Lie algebras giving the same deformation theory.*

We explain next the meaning of Deligne's Principle.

From now on, $k = \mathbb{C}$. We use the abbreviation DGLA for a **differential graded Lie algebra**.

Definition 3.1. A **DGLA** consists of the following set of data:

1. a graded vector space $C = \bigoplus_{i\in\mathbb{N}} C^i$ over $\mathbb{C}$,
2. a Lie bracket which is bilinear, graded skew-commutative, and satisfies the graded Jacobi identity, i.e., for any $\alpha \in C^i, \beta \in C^j$ and $\gamma \in C^k$,
$$[\alpha, \beta] + (-1)^{ij}[\beta, \alpha] = 0$$
and
$$(-1)^{ki}[\alpha, [\beta, \gamma]] + (-1)^{ij}[\beta, [\gamma, \alpha]] + (-1)^{jk}[\gamma, [\alpha, \beta]] = 0,$$
3. a family of linear maps, called the differential maps, $d^i : C^i \to C^{i+1}$, satisfying $d^{i+1}d^i = 0$ and the Leibniz rule, i.e., for $\alpha \in C^i$ and $\beta \in C$
$$d[\alpha, \beta] = [d\alpha, \beta] + (-1)^i[\alpha, d\beta]$$
where $d = \sum d^i : C \to C$.

We denote this DGLA by (C, d), or C when there is no risk of confusion. A homomorphism of DGLAs is a linear map which preserves the grading, Lie bracket, and the differential maps.

Definition 3.2. A homomorphism of DGLAs $g : C \to D$ is **1-equivalent** if it induces an isomorphism on cohomology up to degree 1 and a monomorphism at degree 2. Two DGLAs are of the same 1-**homotopy type**, if they can be connected by a zig-zag of 1-equivalences.

The nicest DGLAs form a particular class:

Definition 3.3. For a DGLA (C, d), the cohomology forms a natural DGLA $(H^\bullet(C), 0)$ with zero differential. We say the DGLA C is 1-**formal** if C is of the same 1-homotopy type as $(H^\bullet(C), 0)$.

Given a DGLA (C, d), one can attach naturally a functor, called the **deformation functor**,
$$\mathrm{Def}(C, d) : \mathsf{ART} \longrightarrow \mathsf{SET}$$
defined on Artinian local rings by
$$A \mapsto \frac{\text{Maurer-Cartan elements of } C \otimes_{\mathbb{C}} m_A}{\text{gauge}},$$
that is,
$$A \mapsto \frac{\{\omega \in C^1 \otimes_{\mathbb{C}} m_A \mid d\omega + \frac{1}{2}[\omega, \omega] = 0\}}{C^0 \otimes_{\mathbb{C}} m_A}.$$
Here m_A is the maximal ideal of A, and $C \otimes_{\mathbb{C}} m_A$ is naturally a DGLA by letting
$$[\alpha \otimes a, \beta \otimes b] = [\alpha, \beta] \otimes ab,$$
$$d(\alpha \otimes a) = d\alpha \otimes a.$$

Since $(C \otimes_{\mathbb{C}} m)^0 = C^0 \otimes_{\mathbb{C}} m$ is a nilpotent Lie algebra, the Campbell-Hausdorff multiplication defines a nilpotent Lie group structure on the space $C^0 \otimes m$. The gauge action of an element $\lambda \in C^0 \otimes m$ on $C^1 \otimes m$ is given by

$$\alpha \mapsto \exp(\operatorname{ad}\lambda)\alpha + \frac{1 - \exp(\operatorname{ad}\lambda)}{\operatorname{ad}\lambda}(d\lambda)$$

in terms of power series.

Suppose now one has a deformation problem $\mathcal{M}_{(P)}$ over $\mathbb{C}$ as in §3.1. To say that a DGLA C controls this deformation problem means that the two functors

$$\mathcal{M}_{(P)}) : \mathsf{ART} \to \mathsf{SET} \qquad \text{and} \qquad \mathrm{Def}(C) : \mathsf{ART} \to \mathsf{SET}$$

are isomorphic:

$$\mathcal{M}_{(P)} = \mathrm{Def}(C).$$

Constructing a DGLA responsible for a deformation problem is an ad-hoc process with no all-encompassing recipe. Typically the DGLAs one first constructs in deformation theory are infinite-dimensional, and there might be multiple choices of such constructions. The second statement of Deligne's Principle is then crucial:

Theorem 3.4 ([GM-1988]). *The deformation functor* $\mathrm{Def}(C)$ *only depends on the 1-homotopy type of C. More precisely, if a morphism of DGLA $f : C \to D$ is 1-equivalent, then the induced transformation on functors $f_* : \mathrm{Def}(C) \to \mathrm{Def}(D)$ is an isomorphism.*

This finishes the explanation of Deligne's Principle.

Example 3.5. ([FIM-2012]) For a complex manifold or complex algebraic variety X, the infinitesimal deformations of an $\mathcal{O}_X$-coherent sheaf E are controlled by the DGLA of global sections

$$\Gamma(X, \mathcal{A}^{\bullet}(\mathcal{E}nd^{\bullet}(\tilde{E}^{\bullet})))$$

of any acyclic resolution $\mathcal{A}^{\bullet}$ of the sheaf of DGLAs $\mathcal{E}nd^{\bullet}(\tilde{E}^{\bullet})$ of a locally free resolution $\tilde{E}^{\bullet}$ of E. If X is a complex manifold, then $\mathcal{A}^{\bullet}$ can be chosen to be the Dolbeault resolution.

Remark 3.6. A theorem due to Lurie [Lu-2011] and Pridham [Pr-2011] in the framework of derived algebraic geometry states that, with the appropriate axiomatization, Deligne's Principle is a theorem. However, in this axiomatization the notion of a deformation problem might be different than the classical notion we have defined, and a classical deformation problem might not come from any new type of deformation problem.

3.3 Deformations with constraints

By a **deformation problem with cohomology constraints** we mean the study of formal germs of cohomology jump loci

$$\mathcal{V}^i_{k,(P)} \subset \mathcal{M}_{(P)}$$

for all i and k at once. This is equivalent to describing the corresponding functors on local Artin rings which we denote also by $\mathcal{V}^i_{k,(P)}$, and which in fact are usually well-defined even if the formal germs are not.

In [BW-2013], a generalization of Deligne's Principle has been proposed and illustrated in several examples:

New Principle. *Every deformation problem with cohomology constraints over a field of characteristic zero is controlled by a pair consisting of a DGLA together with a DGL module, with two quasi-isomorphic pairs giving the same deformation theory with cohomology constraints.*

Let us explain what this means. The deformation problem without constraints $\mathcal{M}_{(P)}$ comes, as claimed by Deligne's Principle, with a DGLA C. Typically this is some kind of endomorphism object. That is, C does not come alone, but it comes with a natural module M on which it acts. Given a pair (C, M) consisting of a DGLA and DGL module, [BW-2013] defines canonically **cohomology jump deformation functors**

$$\mathrm{Def}^i_k(C, M) : \mathsf{ART} \to \mathsf{SET}.$$

Then

$$\mathcal{V}^i_{k,(P)} = \mathrm{Def}^i_k(C, M) \quad \text{for all } i, k$$

as subfunctors of $\mathcal{M}_{(P)} = \mathrm{Def}(C)$. We recall now the basic details.

3.4 Deformation theory via DGLA pairs

Definition 3.7. Given a DGLA (C, d_C), we define a **module** over (C, d_C) to be the following set of data:

1. a graded vector space $M = \bigoplus_{i \in \mathbb{N}} M^i$ together with a bilinear multiplication map $C \times M \to M$, $(\alpha, \xi) \mapsto \alpha\xi$, such that for any $\alpha \in C^i$ and $\xi \in M^j$, $\alpha\xi \in M^{i+j}$. And furthermore, for any $\alpha \in C^i, \beta \in C^j$ and $\zeta \in M$, we require
$$[\alpha, \beta]\zeta = \alpha(\beta\zeta) - (-1)^{ij}\beta(\alpha\zeta),$$

2. a family of linear maps $d^i_M : M^i \to M^{i+1}$ (write $d_M = \sum_{i \in \mathbb{Z}} d^i_M : M \to M$), satisfying $d^{i+1}_M d^i_M = 0$. And we require it to be compatible with the differential on C, i.e., for any $\alpha \in C^i$,
$$d_M(\alpha\xi) = (d_C\alpha)\xi + (-1)^i\alpha(d_M\xi).$$

Definition 3.8. A **homomorphism** of (C, d_C)-modules $f : (M, d_M) \to (N, d_N)$ is a linear map $f : M \to N$ such that

1. f preserves the grading, i.e., $f(M^i) \subset N^i$,

2. f is compatible with multiplication by elements in C, i.e., $f(\alpha\xi) = \alpha f(\xi)$, for any $\alpha \in C$ and $\xi \in M$,

3. f is compatible with the differentials, i.e., $f(d_M\alpha) = d_N f(\alpha)$.

Fixing a DGLA (C, d_C), the category of C-modules is an abelian category.

Definition 3.9. A **DGLA pair** is a DGLA (C, d_C) together with a (C, d_C)-module (M, d_M). Usually, we denote such a pair simply by (C, M). A homomorphism of DGLA pairs $g : (C, M) \to (D, N)$ consists of a map $g_1 : C \to D$ of DGLA and a C-module homomorphism $g_2 : M \to N$, considering N as a C-module induced by g_1. For $q \in \mathbb{N} \cup \{\infty\}$, we call g a q-**equivalence** if g_1 is 1-equivalent and g_2 is q-equivalent. Here a morphism of complexes is q-equivalent if it induces an isomorphism on cohomology up to degree q and a monomorphism at degree $q + 1$ (for example, ∞-equivalent means quasi-isomorphic, which also means homotopy equivalence as complexes since these are complexes of vector spaces). Moreover, we define two DGLA pairs to be of the same q-**homotopy type**, if they can be connected by a zig-zag of q-equivalences. Two DGLA pairs have the same **homotopy type** if they have the same ∞-homotopy type.

By keeping the 1-equivalence for the DGLAs in the definition of the q-equivalence for DGLA pairs, we are using in an economical manner the fact that the Maurer-Cartan set (to be introduced below) of a DGLA is independent under 1-equivalences.

Definition 3.10. Let (C, M) be a DGLA pair. Then $(H^\bullet(C), d = 0)$, the cohomology of C with zero differentials, is a DGLA, and $(H^\bullet(M), d = 0)$, the cohomology of M with zero differentials, is an $H^\bullet(C)$-module. We call the DGLA pair

$$(H^\bullet(C), H^\bullet(M))$$

the **cohomology DGLA pair** of (C, M).

Definition 3.11. We say the DGLA pair (C, M) is q-**formal** if (C, M) is of the same q-homotopy type as $(H^\bullet(C), H^\bullet(M))$. A pair is **formal** if it is ∞-formal.

From now, for a DGLA pair (C, M) we always assume that M is bounded above as a complex and $H^j(M)$ is a finite dimensional $\mathbb{C}$-vector space for every $j \in \mathbb{Z}$. This is needed in order to apply the results about cohomology jump ideals from §2.2.

Lemma 3.12. *Let (C, M) be a DGLA pair. Let $A \in \mathsf{ART}$. Given any Maurer-Cartan element ω of $C \otimes_{\mathbb{C}} m_A$, that is*

$$\omega \in C^1 \otimes_{\mathbb{C}} m_A \text{ with } d\omega + \frac{1}{2}[\omega, \omega] = 0,$$

there is a complex of A-modules

$$(M \otimes_{\mathbb{C}} A, d_\omega) \tag{1}$$

with

$$d_\omega := d \otimes \mathrm{id}_A + \omega.$$

Moreover, $(M \otimes_{\mathbb{C}} A, d_\omega)$ has finitely generated cohomology over A.

In particular, one can apply the construction of cohomology jump ideals from §2.2 to obtain ideals

$$J^i_k(M\otimes_{\mathbb{C}} A, d_\omega)\subset A.$$

The **cohomology jump deformation functors** of the DGLA pair (C, M)

$$\mathrm{Def}^i_k(C,M):\mathsf{ART}\to\mathsf{SET}$$

are defined by

$$A\mapsto\frac{\{\omega\in C^1\otimes_{\mathbb{C}} m_A\mid d\omega+\frac{1}{2}[\omega,\omega]=0\text{ and }J^i_k(M\otimes_{\mathbb{C}} A,d_\omega)=0\}}{C^0\otimes_{\mathbb{C}} m_A}.$$

This gives a well-defined subfunctor of Def(C), see [BW-2013]. Theorem 3.4 admits the following refinement:

Theorem 3.13. *The cohomology jump functor* $\mathrm{Def}^i_k(C,M)$ *only depends on the number k and on the i-homotopy type of (C, M).*

3.5 Finite-dimensional DGLA pairs

In the presence of finite dimensional DGLA pairs, the cohomology jump deformation functors are pro-represented by schemes with explicit equations, up to the gauge action. We explain what this means.

Let (C, M) be a finite dimensional DGLA pair. Fixing a base $e_1,\ldots,e_b$ of C^1, denote by $x_1,\ldots,x_b$ the dual base. We identify the affine coordinate ring $\mathcal{O}_{C^1}$ with the polynomial ring $\mathbb{C}[x_1,\ldots,x_b]$. The universal element of C^1 is defined as the element

$$w_{univ}=\sum_{i=1}^{b}e_i\otimes x_i\quad\text{in}\quad C^1\otimes_{\mathbb{C}}\mathcal{O}_{C_1}.$$

The universal Maurer-Cartan equations are obtained by setting in $C^2\otimes_{\mathbb{C}}\mathcal{O}_{C^1}$,

$$(d_C\otimes id_{\mathcal{O}_{C^1}})w_{univ}+\frac{1}{2}[w_{univ},w_{univ}]=0,$$

where the Lie bracket is extended by multiplication to $C\otimes_{\mathbb{C}}\mathcal{O}_{C^1}$. The universal Maurer-Cartan equations define a closed affine subscheme

$$\mathcal{F}(C)\subset C^1\cong\mathbb{A}^b_{\mathbb{C}}$$

containing the origin. This is also called the *versal deformation, or Kuranishi space* of C.

Def(C) is pro-represented by $\mathcal{F}(C)_{(0)}$ modulo the gauge action when this quotient exists in the category of schemes. This issue does not arise when, for example, the gauge action of C^0 on C^1 is trivial. In that case,

$$\mathrm{Def}(C)\cong\mathcal{F}(C)_{(0)}.$$

If C is 1-formal, then the universal Maurer-Cartan equations become the quadratic (that is, degree at most two) equations

$$[w_{univ}, w_{univ}] = 0.$$

In this case, $\mathcal{F}(C)$ is a quadratic cone and we denote it by $\mathcal{Q}(C)$.

On $\mathcal{F}(C)$, one can define the $\mathcal{O}_{\mathcal{F}(C)}$-linear *universal Aomoto complex*

$$(M^{\bullet} \otimes_{\mathbb{C}} \mathcal{O}_{\mathcal{F}(C)}, d_M \otimes id + w_{univ}).$$

Taking the cohomology jump ideals

$$J^i_k(C, M) \subset \mathcal{O}_{\mathcal{F}(C)}$$

of this complex, one obtains conic affine closed subschemes

$$\mathcal{F}^i_k(C, M) \subset \mathcal{F}(C).$$

Now, the cohomology jump deformation functors $\mathrm{Def}^i_k(C, M)$ are pro-represented by $\mathcal{F}^i_k(C, M)$ modulo the gauge action, when this quotient exists as a scheme. When the action of C^0 on C^1 is trivial,

$$\mathrm{Def}^i_k(C, M) \cong \mathcal{F}^i_k(C, M)_{(0)}.$$

When the DGLA pair is formal, the maps in the universal Aomoto complex are given just by multiplication with the image of w_{univ} in $\mathcal{O}_{\mathcal{Q}(C)}$. Hence the entries of the matrices representing these maps are images of linear forms from $\mathcal{O}_{C^1} = \mathbb{C}[x_1, \dots, x_b]$. The $J^i_k(C, M)$ are generated by minors of these matrices, as explained in 2. We denote the loci $\mathcal{F}^i_k(C, M)$ in this case by

$$\mathcal{R}^i_k(C, M),$$

and they are classically called *resonance varieties.*

Remark 3.14. One can define the quadratic cones and the resonance varieties from cohomology in the non-formal cases as well, but the web of connections with original problem is in general largely lost and their geometric meaning is harder to grasp.

3.6 Dimension bounds

We address the question of providing bounds for the dimensions of cohomology jump loci

$$\mathcal{V}^i_k \subset \mathcal{M}$$

locally around the point P. Assume for simplicity that the controlling DGLA pair (C, M) for these loci is a formal DGLA pair with trivial action of H^0C on H^1C. Recall from the last subsection that it is enough to bound the dimension the resonance varieties of the cohomology pair,

$$\mathcal{R}^i_k = \mathcal{R}^i_k(HC, HM) \subset \mathcal{Q}(HC) = \mathcal{Q}.$$

Define the **exactness degree**

$$\mathrm{exdeg}(C, M)$$

to be the smallest degree at which the universal Aomoto complex for the cohomology pair

$$(H^{\bullet}M \otimes_{\mathbb{C}} \mathcal{O}_{\mathcal{Q}}, w_{univ}\cdot)$$

is not exact. By Theorem 2.10, the depth of the ideal defining $\mathcal{R}^i_1$ in $\mathcal{Q}$ is $\geq$ $\mathrm{exdeg}(C, M) - i$. Suppose in addition that $\mathcal{Q}$ is nice enough so that the depth of ideals in $\mathcal{O}_{\mathcal{Q}}$ is the same as codimension, for example, that $\mathcal{O}_{\mathcal{Q}}$ is Cohen-Macaulay. Then locally at P one has

$$\operatorname{codim} \mathcal{V}^i_1 \geq \mathrm{exdeg}(C, M) - i. \tag{2}$$

4 Deformations of holomorphic vector bundles

Let X be compact complex manifold. A holomorphic vector bundle on X has non-interesting cohomology if X is not compact. The cohomology jump loci of line bundles on a smooth projective curve are classically called Brill-Nother loci, see [A+1985]. Cohomology jump loci of vector bundles can be thus considered as generalized Brill-Nother loci.

Let E be a locally free $\mathcal{O}_X$-coherent sheaf on X. By Example 3.5, the deformations of E are governed by the **Dolbeault complex** of E,

$$(A^{0,\bullet}_{\mathrm{Dol}}(\mathcal{E}nd(E))) = (\Gamma(X, \mathcal{E}nd(E) \otimes_{\mathcal{O}_X} \Omega^{0,\bullet}_X), \bar{\partial}),$$

with the natural DGLA structure. The deformations of E with cohomology constraints are governed by the DGLA pair

$$(A^{0,\bullet}_{\mathrm{Dol}}(\mathcal{E}nd(E)), A^{0,\bullet}_{\mathrm{Dol}}(E)).$$

The main issue arising when applying the deformation techniques of §§3.1—3.4 is that the DGLA pairs involved are usually infinite-dimensional and not formal, even in the case of smooth projective curves. We will focus here on cases when formality holds.

Consider the moduli space $\mathcal{M}$ of stable rank n holomorphic vector bundles E with vanishing Chern classes on a compact Kähler manifold X. We consider the Hodge-theoretic flavored cohomology jump loci

$$\mathcal{V}^{pq}_k(F) \subset \mathcal{M}$$

given by

$$\mathcal{V}^{pq}_k(F) = \{E \in \mathcal{M} \mid \dim H^q(X, E \otimes_{\mathcal{O}_X} F \otimes_{\mathcal{O}_X} \Omega^p_X) \geq k\}$$

with the natural scheme structure, for fixed p and fixed poly-stable bundle F with vanishing Chern classes. In this case, the deformation problem with cohomology constraints is governed by the DGLA pair

$$(A^{0,\bullet}_{\mathrm{Dol}}(\mathcal{E}nd(E)), A^{p,\bullet}_{\mathrm{Dol}}(E \otimes F)).$$

Theorem 4.1 ([D+1975], [Si-1992]). *Let X be a compact Kähler manifold. For any $E \in \mathcal{M}$, the DGLA pair*

$$(A^{0,\bullet}_{\mathrm{Dol}}(\mathcal{E}nd(E)), A^{p,\bullet}_{\mathrm{Dol}}(E \otimes F))$$

is formal.

Let

$$\mathcal{Q}(E) = \{\eta \in H^1(X, \mathcal{E}nd(E)) \mid \eta \wedge \eta = 0 \in H^2(X, \mathcal{E}nd(E))\},$$
$$\mathcal{R}^{pq}_k(E; F) = \{\eta \in \mathcal{Q}(E) \mid \dim H^q(H^\bullet(X, E \otimes F \otimes \Omega^p_X), \eta \wedge \cdot) \geq k\}.$$

The set $\mathcal{Q}(E)$ is endowed with a natural closed scheme structure defined using the corresponding universal quadratic equations on the affine space $H^1(X, \mathcal{E}nd(E))$. In particular, $\mathcal{Q}(E)$ has quadratic singularities. The set $\mathcal{R}^{pq}_k(E; F)$ has a natural closed subscheme structure of $\mathcal{Q}(E)$ defined using the cohomology jump ideals as in §3.5 of the universal complex over $\mathcal{Q}(E)$ with $\mathcal{O}_{\mathcal{Q}(E)}$-linear maps given by the cup-product:

$$(H^\bullet(X, E \otimes F \otimes \Omega^p_X) \otimes_{\mathbb{C}} \mathcal{O}_{\mathcal{Q}(E)}, \eta_{univ} \wedge \cdot).$$

Note that this is a complex of free finitely generated $\mathcal{O}_{\mathcal{Q}(E)}$-modules such that the entries of the matrices representing the maps in the complex come from linear polynomials in $\mathbb{C}[H^1(X, \mathcal{E}nd(E))]$. More precisely, $\mathcal{Q}(E)$ and $\mathcal{R}^{pq}_k(E; F)$ are the quadratic cone and, respectively, the resonance varieties as defined in §3.5 of the cohomology DGLA pair with trivial differentials

$$(H^\bullet(X, \mathcal{E}nd(E)), H^\bullet(X, E \otimes F \otimes \Omega^p_X)). \tag{3}$$

Formality of the DGLA pair implies:

Theorem 4.2. ([BW-2013]) *Let X be a compact Kähler manifold. Let E and F be a stable and, respectively, a poly-stable holomorphic vector bundle with vanishing Chern classes on X. Then there is an isomorphism of formal schemes*

$$\mathcal{V}^{pq}_k(F)_{(E)} \cong \mathcal{R}^{pq}_k(E; F)_{(0)}.$$

This isomorphim means that one has explicit simple equations describing locally at E the cohomology jump loci $\mathcal{V}^{pq}_k(F)$, via the universal equations mentioned above for $\mathcal{R}^{pq}_k(E; F)$. Note that although the equations are simple, the outcome is not necessarily so.

The isomorphism

$$(\mathcal{M})_{(E)} \cong \mathcal{Q}(E)_{(0)},$$

which is a particular case of Theorem 4.2, was shown by Nadel [Na-1988] and Goldman-Millson [GM-1988]. For rank $n = 1$ bundles, the result is due to Green-Lazarsfeld [GL-1987, GL-1991].

By fixing the attention on a generic vector bundle E laying in $\mathcal{V}^{pq}_k(F)$, that is, if

$$\dim H^q(X, E \otimes F \otimes \Omega^p_X) = k,$$

then the generators of the cohomology jump ideals defining $\mathcal{R}^{pq}_k(E; F)$ are precisely all the entries of the matrices defining the cohomology jump ideals. Since these entries are linear forms, one obtains:

Corollary 4.3. *With the same assumptions as in Theorem 4.2, if*

$$\dim H^q(X, E \otimes F \otimes \Omega_X^p) = k,$$

then $\mathcal{V}_k^{pq}(F)$ has quadratic algebraic singularities at E.

This was also showed for $F \otimes \Omega_X^p = \mathcal{O}_X$ by Martinengo [Mar-2009] and B. Wang [Wa-2012].

Remark 4.4. The assumptions in the theorem are made such that the controlling DGLA pair is formal and the gauge action of H^0 is trivial on H^1. As explained in §3.5, whenever this is the case, the quadratic cone and the resonance varieties of the cohomology DGLA pair describe locally the moduli space and the cohomology jump loci. We will present repetitions of this type of thinking to obtain similar results in completely different contexts. This is how one puts the New Principle of §3.3 to work.

5 Deformations of representations and local systems

To a group G, one can associate the set of rank n complex linear representations,

$$\mathrm{Hom}(G, GL(n, \mathbb{C})).$$

When G is finitely presented, this is an affine scheme. Any finite presentation provides equations for this scheme. These are in general hard to compute and to extract information from. We will consider groups G in a more geometric context, namely as fundamental groups of topological spaces.

Let X be a smooth manifold which is of the homotopy type of a finite type CW-complex, and let $x \in X$ be a base point. Then the fundamental group $\pi_1(X, x)$ is finitely presented. We will use the notation

$$\mathbf{R}(X, n) = \mathrm{Hom}(\pi_1(X, x), GL(n, \mathbb{C}))$$

for the moduli space of representations.

Recall that a (complex) local system of rank n on a topological space is a locally constant sheaf of $\mathbb{C}$-vector spaces of rank n. Every closed point $\rho \in \mathbf{R}(X, n)$ corresponds to a rank n local system L_ρ on X. Conversely, a local system L_ρ on X defines a representation ρ uniquely up to conjugation. This corresponds to the fact that local systems do not come with a choice of base point x.

Let W be a local system of any rank on X. One can define the cohomology jump loci

$$\mathcal{V}_k^i(W) \subset \mathbf{R}(X, n)$$

given by

$$\mathcal{V}_k^i(W) = \{\rho \in \mathbf{R}(X, n) \mid \dim H^i(X, L_\rho \otimes_{\mathbb{C}} W) \geq k\}$$

with a natural closed scheme structure. For this structure, one uses the cohomology jump ideals of appropriately chosen universal complexes: either formed via a universal twisted cochain complex on the universal cover of X, or by using the universal rank n local system on $\mathbf{R}(X,n)$. Both universal complexes lead to isomorphic cohomology jump loci due to invariance of the latter under quasi-isomorphisms, e.g. [DP-2014, BW-2013].

Remark 5.1. (1) For a group G, one can define the cohomology jump loci of G by setting $X = K(G,1)$.

(2) For a smooth manifold X, it is known that the cohomology jump loci $\mathcal{V}^1_k(\mathbb{C}_X)$ of rank one representations, which we shall denote by Σ^1_k later on, are the same as the those of its fundamental group. In other words, Σ^1_k are invariants of $\pi_1(X)$, and can be computed from a finite presentation of $\pi_1(X)$.

We address the local structure of the cohomology jump loci $\mathcal{V}^i_k(W)$ at a representation $\rho \in \mathbf{R}(X,n)$. Equivalently, we ask how one can describe all the infinitesimal deformations of ρ constrained by the condition that the degree i cohomology of the corresponding local system L_ρ has dimension $\geq k$.

It is almost true that this infinitesimal deformation problem with cohomology constraints is governed by the DGLA pair given by de Rham complexes

$$(A^\bullet_{\mathrm{DR}}(\mathcal{E}nd(L_\rho)), A^\bullet_{\mathrm{DR}}(L_\rho \otimes_{\mathbb{C}} W)).$$

Here, for a local system L on X, the de Rham complex $A^\bullet_{\mathrm{DR}}(L)$ is the complex of global L-valued C^∞ forms with the usual differential. Also, $\mathcal{E}nd(L_\rho)$ is the local system of endomorphisms of L_ρ.

To make this completely true, one needs to reintroduce in the picture the choice of base point x. Let

$$\varepsilon : A^\bullet_{\mathrm{DR}}(\mathcal{E}nd(L_\rho)) \to \mathfrak{g} = \mathcal{E}nd(L_\rho)|_x$$

be the restriction map. This is a DGLA map, hence we can also define

$$\mathfrak{h} = \varepsilon(H^0(X, \mathcal{E}nd(L_\rho))).$$

Goldman-Millson [GM-1988] attach to an augmented DGLA $(A^\bullet_{\mathrm{DR}}(\mathcal{E}nd(L_\rho)); \varepsilon)$ a deformation functor

$$\mathrm{Def}(A^\bullet_{\mathrm{DR}}(\mathcal{E}nd(L_\rho)); \varepsilon),$$

defined as the deformation functor of the kernel of ε. They show that the formal scheme of $\mathbf{R}(X,n)$ at ρ pro-represents this functor. Similarly, one can define cohomology jump loci sub-deformation functors attached to an augmented DGLA pair

$$(A^\bullet_{\mathrm{DR}}(\mathcal{E}nd(L_\rho)), A^\bullet_{\mathrm{DR}}(L_\rho \otimes_{\mathbb{C}} W); \varepsilon)$$

as in §3.4, such that they describe the local structure of $\mathcal{V}^i_k(W)$ at ρ, [BW-2013].

Remark 5.2. In the eye-opening paper [DP-2014], Dimca and Papadima have addressed the reduced local structure at the trivial representation of cohomology jump loci of representations of fundamental groups. In this case, the de Rham complex $A^\bullet_{\mathrm{DR}}(X)$ of X plays the controlling role, as it is a module over itself. The results in [BW-2013] were the outcome of the desire of the authors to extend the results of [DP-2014] beyond the reduced structure and beyond the trivial representation. The solution in terms of DGLA pairs to infinitesimal deformation problems with cohomology constraints seemes natural to us only after reading [DP-2014].

In cases where formality of the (augmented) DGLA pair is present, one can write explicit local equations for $\mathcal{V}^i_k(W)$ as explained in §3.5. Let

$$\mathcal{Q}(\rho) = \{\eta \in Z^1(\pi_1(X), \mathfrak{gl}(n,\mathbb{C})_{\mathrm{ad}\,\rho}) \mid \bar\eta \wedge \bar\eta = 0 \in H^2(X, \mathcal{E}nd(L_\rho))\},$$

$$\mathcal{R}^i_k(\rho, W) = \{\eta \in \mathcal{Q}(\rho) \mid \dim H^i(H^\bullet(X, L_\rho \otimes_{\mathbb{C}} W), \bar\eta \wedge \cdot) \geq k\},$$

with the natural scheme structures, where Z^1 stands for the vector space of 1-cocycles, and ρ is the adjoint representation, and $\bar\eta$ is the image of η in cohomology.

Similarly, for a local system L on X, define

$$\mathcal{Q}(L) = \{\eta \in H^1(X, \mathcal{E}nd(L)) \mid \eta \wedge \eta = 0 \in H^2(X, \mathcal{E}nd(L))\},$$

$$\mathcal{R}^i_k(L, W) = \{\eta \in \mathcal{Q}(L) \mid \dim H^i(H^\bullet(X, L \otimes W), \eta \wedge \cdot) \geq k\},$$

with the natural scheme structures.

Theorem 5.3. ([BW-2013]) *Let X be a compact Kähler manifold, $\rho \in \mathbf{R}(X,n)$ be a semi-simple representation, and W be a semi-simple local system on X. Then there is an isomorphism of formal schemes*

$$\mathcal{V}^i_k(W)_{(\rho)} \cong \mathcal{R}^i_k(\rho, W)_{(0)} \cong (\mathcal{R}^i_k(L_\rho, W) \times \mathfrak{g}/\mathfrak{h})_{(0)}.$$

Remark 5.4. In particular, one has that

$$\mathbf{R}(X,n)_{(\rho)} \cong \mathcal{Q}(\rho)_{(0)} \cong (\mathcal{Q}(L) \times \mathfrak{g}/\mathfrak{h})_{(0)},$$

a result due to Simpson [Si-1992].

Remark 5.5. With the same assumptions one obtains as before that, if in addition $k = \dim H^i(X, L_\rho)$, then $\mathcal{V}^i_k(W)$ has quadratic singularities at ρ.

Remark 5.6. If ρ is a rank one representation, then the affine space $\mathfrak{g}/\mathfrak{h} = 0$. In fact, a more general statement holds as we now describe.

Let $\mathcal{M}_{\mathrm{B}}(X,n)$ be the moduli space of irreducible local systems on X of rank n, see [Si-1994], and let W another local system. Let

$$\Sigma^i_k(W) \subset \mathcal{M}_{\mathrm{B}}(X,n)$$

be the cohomology jump loci of irreducible local systems. The controlling DGLA pair in this case is the non-augmented pair

$$(A^\bullet_{\mathrm{DR}}(\mathcal{E}nd(L)), A^\bullet_{\mathrm{DR}}(L \otimes W)).$$

From the formality of this pair one obtains:

Theorem 5.7. ([BW-2013]) *Let X be a compact Kähler manifold. Let L be an irreducible local system on X, and let W be a semi-simple local system. There is an isomorphism of formal schemes*

$$\mathcal{M}_B(X,n)_{(L)} \cong \mathcal{Q}(L)_{(0)}$$

inducing an isomorphism

$$\Sigma^i_k(W)_{(L)} \cong \mathcal{R}^i_k(L,W)_{(0)}.$$

This has also been proved by a different method by Popa-Schnell [PSc-2013] for rank one local systems L, $W = \mathbb{C}_X$, and X is a smooth projective complex variety.

Remark 5.8. Due to Simpson's non-abelian Hodge theory [Si-1992], a similar statement holds for Higgs bundles on smooth projective varieties, via a DGLA pair arising from the Higgs complex, see [BW-2013].

6 Line bundles and rank one local systems

In this section we specialize to line bundles and rank one local systems. For the space of rank one local systems on a topological space X we use the notation

$$\mathcal{M}_{\mathrm{B}}(X) = \mathrm{Hom}(\pi_1(X), \mathbb{C}^*) = \mathrm{Hom}(H_1(X,\mathbb{Z}), \mathbb{C}^*) = H^1(X, \mathbb{C}^*).$$

We will use the following notation for the cohomology jump loci of rank one local systems:

$$\Sigma^i_k = \{L \in \mathcal{M}_{\mathrm{B}}(X) \mid \dim H^i(X,L) \geq k\} \subset \mathcal{M}_{\mathrm{B}}(X),$$

and identify a representation $\rho \in \mathrm{Hom}(\pi_1(X), \mathbb{C}^*)$ with the associated local system L_ρ. We first recall how the local systems and line bundles are related in the compact case.

6.1 Local systems vs. Higgs line bundles

We recall the following from [Si-1992]. Let X be a compact Kähler manifold. Let

$$\mathcal{M}_{\mathrm{Dol}}(X) = \{(E,\theta) \mid \text{Higgs line bundle}\} = \mathrm{Pic}^\tau(X) \times H^0(X, \Omega^1_X),$$

where

$$\mathrm{Pic}^\tau(X) = \ker\{c_1 : \mathrm{Pic}(X) \to H^2(X,\mathbb{R})\}.$$

The space $\mathcal{M}_{\mathrm{Dol}}(X)$ is endowed with a natural complex structure. There is an isomorphism of real analytic groups

$$\mathcal{M}_{\mathrm{B}}(X) \xrightarrow{\sim} \mathcal{M}_{\mathrm{Dol}}(X),$$
$$L_\rho \mapsto (E_\rho, \theta_\rho),$$

which can be thought of as the analytification-tropicalization of $\mathcal{M}_{\mathrm{B}}(X)$. More precisely, E_ρ is the analytification

$$E_\rho = L_\rho \otimes_{\mathbb{C}} \mathcal{O}_X,$$

and θ_ρ is the tropicalization

$$\theta_\rho = (1,0)\text{-part of } \log ||\rho|| \in H^1(X,\mathbb{R}) = \mathrm{Hom}(\pi_1(X),\mathbb{R}).$$

Let

$$\mathcal{U}_B(X) \subset \mathcal{M}_{\mathrm{B}}(X)$$

be the unitary local systems. We have

$$\mathcal{U}_B(X) = \mathrm{Hom}(H_1(X,\mathbb{Z}), S^1),$$

where $S^1 \subset \mathbb{C}^*$ is the unit circle, and

$$\mathcal{U}_{Dol}(X) = \mathrm{Pic}^\tau(X).$$

$U_B(X)$ is a totally real subgroup of $\mathcal{M}_{\mathrm{B}}(X)$, whereas $\mathcal{U}_{Dol}(X)$ is a finite disjoint union of copies of the Picard variety $\mathrm{Pic}^0(X)$. For example, for a smooth projective curve of genus g, we have $\mathcal{U}_B(X) = (S^1)^{2g}$ and $\mathcal{U}_{Dol}(X) = \mathrm{Jac}(X)$, the Jacobian of X.

For a Higgs line bundle (E,θ), let

$$H^{pq}(E,\theta) = H^p(H^q(X, E\otimes\Omega_X^\bullet),\theta).$$

The cohomologies of rank one local systems and of Higgs line bundles are related via:

Theorem 6.1. ([GL-1987, Be-1988, GL-1991, Si-1992]) *Let X be a compact Kähler manifold. Let $L \in \mathcal{M}_B(X)$ be a rank one local system with associated Higgs line bundle $(E,\theta) \in \mathcal{M}_{Dol}(X)$. Then*

$$H^i(X,L) \cong \bigoplus_{p+q=i} H^{pq}(E,\theta).$$

This has been extended by Simpson beyond the rank one case in [Si-1992, Si-1994].

We fix in this section the following notation about cohomology jump loci:

$$\begin{aligned}
\Sigma_k^i &= \{L \in \mathcal{M}_{\mathrm{B}}(X) \mid \dim H^i(X,L) \geq k\} && \subset \mathcal{M}_{\mathrm{B}}(X),\\
\mathcal{V}_k^{pq} &= \{E \in \mathrm{Pic}^\tau(X) \mid \dim H^q(X, E\otimes\Omega_X^p) \geq k\} && \subset \mathrm{Pic}^\tau(X),\\
\mathcal{V}_k^{i} &= \{E \in \mathrm{Pic}^\tau(X) \mid \dim H^i(X, E) \geq k\} && \subset \mathrm{Pic}^\tau(X),\\
\mathcal{H}_k^{pq} &= \{(E,\theta) \in \mathcal{M}_{\mathrm{Dol}}(X) \mid \dim H^{p,q}(E,\theta) \geq k\} && \subset \mathcal{M}_{\mathrm{Dol}}(X).
\end{aligned}$$

Remark 6.2. Since $\mathrm{Pic}^\tau(X) = \mathcal{U}_{Dol}(X) \subset \mathcal{M}_{\mathrm{Dol}}(X)$, one has that

$$\mathcal{V}_k^{pq} = \mathcal{H}_k^{pq} \cap \mathrm{Pic}^\tau(X).$$

Arapura's trick [Ar-1992] shows, via the theorem above, that every irreducible component of an $\mathcal{H}_k^{pq}$ comes from an irreducible component of Σ_k^i.

Therefore we can take the point of view that studying line bundles with torsion first Chern class is a particular case of studying rank one local systems.

6.2 Local study

Let X be a complex manifold. Consider the local study of the loci $\Sigma_k^i \subset \mathcal{M}_{\mathrm{B}}(X)$ at a rank one representation ρ. The DGLA pair describing the formal schemes

$$\Sigma^i_{k,(\rho)}$$

is

$$(A^{\bullet}_{\mathrm{DR}}(X), A^{\bullet}_{\mathrm{DR}}(L_\rho)),$$

where we denote by $A^{\bullet}_{\mathrm{DR}}(X)$ the de Rham complex of X.

Although $A^{\bullet}_{\mathrm{DR}}(X)$ might not be formal, from the definition of the deformation functor $\mathrm{Def}(A^{\bullet}_{\mathrm{DR}}(X))$ one easily computes that

$$\mathcal{M}_{\mathrm{B}}(X)_{(\rho)} \cong H^1(X,\mathbb{C})_{(0)}.$$

This is of course not necessary, since we have the description

$$\mathcal{M}_{\mathrm{B}}(X) \cong H^1(X,\mathbb{C}^*).$$

However, the de Rham complex plays the crucial role in understanding further the properties of the cohomology jump loci.

In particular, $\Sigma^i_{k,(\mathbf{1})}$, the formal schemes at the trivial representation, are described by the DGLA pair consisting of $A^{\bullet}_{\mathrm{DR}}(X)$ acting on itself. The Lie structure is trivial here, $A^{\bullet}_{\mathrm{DR}}(X)$ is actually a commutative differential graded algebra (CDGA). Using the deformation theory of DGLA pairs as in §3.5, one has the following description, due for the reduced structure of Σ^1_k to Dimca, Papadima, and Suciu:

Theorem 6.3. ([DPS-2009]) *Let X be a 1-formal space (e.g. a compact Kähler manifold, or the complement of a hyperplane arrangement), that is, the de Rham complex is 1-formal. Then*

$$\Sigma^1_{k,(\mathbf{1})} \cong \mathcal{R}^1_{k,(0)},$$

and $\mathcal{R}^1_k$ is the tangent cone at $\mathbf{1} \in \mathcal{M}_B(X)$ of Σ^1_k. In particular, the irreducible components of Σ^1_k passing though $\mathbf{1}$ are subtori.

Here, $\mathcal{R}^i_k$ are the cohomology jump loci of the universal Aomoto complex

$$(H^{\bullet}(X,\mathbb{C}) \otimes_{\mathbb{C}} \mathbb{C}[H^1(X,\mathbb{C})], w_{univ} \cup .).$$

Set-theoretically,

$$\mathcal{R}^i_k = \{w \in H^1(X,\mathbb{C}) \mid \dim H^i(H^{\bullet}(X,\mathbb{C}), w \cup .) \geq k\}.$$

This follows from making explicit for this case the Section 3.5.

Remark 6.4. (Hyperplane arrangements I.) This result, for the reduced germs of Σ^i_k at the trivial representation, for the case when X is the complement of a hyperplane arrangement, had been obtained earlier by Esnault, Schechtman, and Viehweg [ESV-1992]. By Orlik-Solomon [OS-1980], the cohomology ring of X is a combinatorial invariant of the hyperplane arrangement, that is, it only depends on the lattice of intersection of the hyperplanes and not on their positions. Hence, the components through $\mathbf{1}$ of Σ_k^i can be detected combinatorially in this case from the resonance varieties $\mathcal{R}_k^i$. In the context of hyperplane arrangements, the resonance varieties have been introduced by Falk, see the survey articles by Suciu referred to in the Introduction.

Remark 6.5. (Hyperplane arrangements II.) A very difficult folklore conjecture states that all the cohomology jump loci Σ_k^i of the complement of a hyperplane arrangement are combinatorial invariants. A particular case of this conjecture states that the eigenvalues of the monodromy on the Milnor fiber of a hyperplane arrangement are determined combinatorially. The best result so far in this direction is the recent paper of Papadima and Suciu [PaS-2014]. Surprisingly, they use the cohomology jump loci of rank two local systems and resonance varieties defined over finite fields to address the rank one case.

Remark 6.6. The de Rham complex of X encodes the homotopy type of the topological space X. A consequence is that the formal scheme

$$\Sigma_k^i(X)_{(\mathbf{1})}$$

at the trivial representation only depends on the i-homotopy type of X. For the underlying reduced germs, this is due to Dimca-Papadima [DP-2014]. The cohomology jump deformation functors from §3.4 give the statement for the non-reduced structure as well.

Remark 6.7. (Gysin model) Typically $A^{\bullet}_{\mathrm{DR}}(X)$ is not formal. Hence the explicit description of $\Sigma_k^i(X)_{(\mathbf{1})}$ depends on finding at least a finite model for $A^{\bullet}_{\mathrm{DR}}(X)$, that is, a finite dimensional CDGA quasi-isomorphic with $A^{\bullet}_{\mathrm{DR}}(X)$. The crucial observation of Dimca-Papadima [DP-2014] is that, if X is a smooth quasi-projective variety, there is always such a model. This is the *Gysin model* of Morgan [Mo-1978], defined with respect to any smooth projective compactification

$$j : X \to Y$$

with a simple normal crossings divisor $Y \setminus X = \bigcup_{i \in S} D_i$ as boundary. Here D_i are the irreducible components. For $I \subset S$, let $D_I = \cap_{i \in I} D_i$. The Gysin CDGA $\mathcal{A}^{\bullet}$ is defined by setting

$$\mathcal{A}^i = \bigoplus_{p+q=i} \mathcal{A}^{p,q}, \qquad \mathcal{A}^{p,q} = \bigoplus_{|I|=q} H^p(D_I, \mathbb{C}).$$

The multiplication is the cup-product

$$\mathcal{A}^{p,q} \otimes \mathcal{A}^{p',q'} \xrightarrow{\cup} \mathcal{A}^{p+p',q+q'}.$$

The differential on $\mathcal{A}^{\bullet}$ is defined using the Gysin maps from intersections of the boundary divisors and it satisfies

$$\mathcal{A}^{p,q} \longrightarrow \mathcal{A}^{p+2,q-1}.$$

The Gysin model $\mathcal{A}^\bullet$ is in fact the (total complex of the) E_2 page of the Leray spectral sequence for the open embedding j:

$$\mathcal{A}^{p,q} = E_2^{p,q} = H^p(Y, R^q j_* \mathbb{C}_X) \Rightarrow \mathbb{H}^i(Y, Rj_*\mathbb{C}_X) = H^i(X, \mathbb{C}).$$

It is well-known that this spectral sequence degenerates at the E_3 page, [PS-2008]. In other words, the cohomology CDGA of $\mathcal{A}^\bullet$ is the cohomology of X. The more general statement, showing by §3.4 that the infinitesimal deformations of the trivial rank one local system on X with cohomology constraints are governed by the Gysin model, is the following:

Theorem 6.8. ([Mo-1978]) *Let X be a smooth complex quasi-projective variety. Then there is a quasi-isomorphism of CDGAs between the de Rham complex and the Gysin model,*

$$A^\bullet_{\mathrm{DR}}(X) \cong \mathcal{A}^\bullet.$$

Hence, as explained in §3.5, the resonance varieties of the Gysin model describe the local structure of Σ^i_k at the trivial representation.

Using the positive grading on the Gysin model, Dimca and Papadima draw global consequences from local information, generalizing Theorem 6.3:

Theorem 6.9. ([DP-2014]) *Let X be a smooth quasi-projective complex variety. Then the components of Σ^i_k containing the trivial representation are subtori of $\mathcal{M}_B(X)$.*

Further improvements of this result are described next.

6.3 Structure of the cohomology jump loci

We describe now in more detail the global structural results on cohomology jump loci of rank one local systems and line bundles. The proofs of the structure theorems for cohomology jump loci are all similar in plan: a combination of local study, information from mixed Hodge structure theory, and arithmetic.

Theorem 6.10. ([Ar-1997]) *Let X be the complement in a compact Kähler manifold $\bar{X}$ of a divisor with simple normal crossings singularities. Assume that $H^1(\bar{X}, \mathbb{C}) = 0$. Then Σ^i_k are finite union of unitary translates of subtori of $\mathcal{M}_B(X)$. The same holds for Σ^1_1 with no assumption on $\bar{X}$, with the extra feature that the positive dimensional components are torsion translates.*

Theorem 6.11. ([BW-2015]) *Let X be a smooth complex quasi-projective variety. Then the cohomology jump loci of rank one local systems Σ^i_k are a finite unions of torsion translates of subtori of $\mathcal{M}_B(X)$.*

Theorem 6.12. ([Wa-2013]) *Let X be a compact Kähler manifold. Then the cohomology jump loci of line bundles Σ^i_k and $\mathcal{V}^{pq}_k$ are finite unions of torsion translates of subtori of $\mathcal{M}_B(X)$ and, respectively,* $\mathrm{Pic}^\tau(X)$.

Here, by subtori of $\mathcal{M}_{\mathrm{B}}(X)$ we mean *affine tori* $(\mathbb{C}^*)^s$. By subtori of $\mathrm{Pic}^\tau(X)$ we mean *compact complex subtori*, that is, connected compact complex Lie subgroups; their underlying topological spaces are real tori.

By Remark 6.2, the statement about $\mathcal{V}_k^{pq}$ in the third theorem follows from the statement about Σ_k^i.

Remark 6.13. The theorems build upon a long list of partial results. The chronology is the following:

- Beauville [Be-1988] showed the linearity property of $\mathcal{V}_1^1$ for a compact Kähler manifold and conjectured the third theorem for $\mathcal{V}_k^{pq}$.

- Green-Lazarsfeld [GL-1991] showed the linearity of $\mathcal{V}_k^{pq}$ and Σ_k^1 for a compact Kähler manifold.

- Arapura [Ar-1992] showed that Σ_k^i are finite unions of unitary translated subtori for a compact Kähler manifold. He also showed that $\mathcal{H}_k^{pq}$ are finite unions of unitary translated subtori of $\mathrm{Pic}^\tau(X)$ times vector subspaces of $H^0(X, \Omega_X^1)$, recovering thus the result of [GL-1991].

- Simpson [Si-1993] proved the theorems for X smooth projective, namely by showing that the components of Σ_k^i, and hence of $\mathcal{V}_k^{pq}$, are torsion translated subtori.

- Arapura [Ar-1997] showed the first theorem. A small gap in the proof was filled later in [AB+2013].

- Campana [Cam-2001] showed the third theorem for Σ_k^1.

- Pink and Roessler [PR-2004] reproved the statement about $\mathcal{V}_k^{pq}$ for a smooth projective variety via reduction modulo p.

- In [Bu-2009] we showed that $\Sigma_k^i \cap \mathcal{U}_B(X)$, the cohomology jump loci of unitary rank one local systems on a smooth quasi-projective variety X, and their Hodge theoretic refinements, are finite unions of torsion translated subtori of $\mathrm{Pic}^\tau(X)$ times rational convex polytopes. Combined with Theorem 6.10, this gives Theorem 6.11 under the assumptions on the compactification as in Theorem 6.10, as remarked by Dimca and Libgober.

- Dimca-Papadima [DP-2014] showed Theorem 6.9.

- The final statements of the last two theorems were after this just highly nontrivial statements about existence of torsion points in each irreducible component of Σ_k^i, this being shown recently in [BW-2015] and, respectively, [Wa-2013].

The three theorems suggest that the following question might have a positive answer:

Question 6.14. *Let X be the complement of a simple normal crossings divisor in a compact Kähler manifold. Are the cohomology jump loci Σ_k^i always finite unions of torsion translated subtori?*

6.4 Fibrations

The structure theorem is more refined in some cases. For a compact Kähler manifold X, let

$$a_X : X \to Alb(X)$$

be the Albanese map. For a compact analytic variety Y, we say that Y has *maximal Albanese dimension* if $\dim Y$ equals the dimension of the Albanese of any smooth model of Y.

Theorem 6.15. ([GL-1991]) *Let X be a compact Kähler manifold. Let Z be a positive dimensional irreducible component of $\mathcal{V}^i_k \subset \mathrm{Pic}^\tau(X)$. Then there exists an analytic dominant map*

$$f : X \to Y$$

with connected fibers to an analytic variety Y of maximal Albanese dimension and

$$\dim Y \leq i,$$

such that Z is a translate of $f^*\mathrm{Pic}^0(Y)$.

As an application, a more careful analysis of the reasons behind Theorem 6.15 allowed Pareschi and Popa to generalize the classical Castelnuovo - de Franchis theorem from surfaces to other dimensions:

Theorem 6.16. ([PP-2009]) *Let X be a compact Kähler manifold of maximal Albanese dimension and irregularity $q(X)$. If*

$$\chi(\omega_X) < q(X) - \dim X$$

then X admits a surjective morphism with connected fibers onto a normal compact analytic variety Y with

$$0 < \dim Y < \dim X$$

and with any smooth model $\tilde{Y}$ of Albanese general type and with $\chi(\omega_{\tilde{Y}}) > 0$.

In the non-compact case, one has the following:

Theorem 6.17. ([Ar-1997]) *Let X be the complement in a compact Kähler manifold $\bar{X}$ of a divisor with simple normal crossings singularities. Assume that $H^1(\bar{X}, \mathbb{C}) = 0$. Let W be a positive dimension irreducible component of Σ^i_k. Then there exists an analytic map*

$$f : X \to T$$

to an extension T of a compact complex torus by an affine torus, such that W is a translate of $f^*\mathcal{M}_B(T)$.

In Theorems 6.15 and 6.17, the translates are by torsion elements whenever the assumptions of Theorems 6.10—6.12 hold.

The components of the support loci Σ^1_1 of the first cohomology are completely understood in terms of *fibrations*, i.e. surjective analytic maps with connected general fiber, onto curves. Generalizing previous results of [GL-1987, Be-1988, Cat-1991, GL-1991, Ar-1992] from the compact case to the non-compact case, Arapura proved:

Theorem 6.18. ([Ar-1997]) *Let X be the complement in a compact Kähler manifold of a divisor with simple normal crossings singularities. Then there is a bijection between:*

- *the set equivalence classes of fibrations*
$$f : X \to C,$$
where C is a smooth Riemann surfaces minus finitely many points with $\chi(C) < 0$, under the equivalence relation $f \sim f'$ if there exists an isomorphism $g : C \to C'$ with $g \circ f = f'$, and

- *the irreducible positive-dimensional components of Σ_1^1 containing the trivial local system.*

To a fibration f corresponds the component $f^*\mathcal{M}_B(C) = f^*\Sigma_1^1(C)$.

Remark 6.19. This result has recently been generalized to normal complex varieties by Arapura, Dimca, and Hain [ADH-2015]. In this paper, they show that Σ_k^1 for a normal variety behaves as for a smooth variety. This opens up the question of how much of what is known about cohomology jump loci of smooth varieties goes through to varieties with singularities.

Remark 6.20. ([Ar-1997]) An interesting application is that the cardinality $N_b(X)$ of the set of equivalence classes of fibrations onto curves as above with fixed Betti number $b_1(C) = b$ depends only on the fundamental group $\pi_1(X)$. Indeed, it is known that the support of Σ_k^1 can be computed from $\pi_1(X)$. By the theorem, $N_b(X)$ is the number of irreducible components of Σ_1^1 of dimension b.

By the torsion translate property, the components of Σ_1^1 not passing through the trivial local system, must contain a torsion point. Hence, one can account for these components by applying the theorem to a finite cover of X. Or, equivalently, one can use the orbifold language, to which we refer to [AB+2013]. Moreover, in [Di-2007a, Di-2007b, AB+2013], all the positive dimensional components of Σ_k^1 for $k > 1$ are also accounted for. The structure of Σ_k^1 forces constraints on fundamental groups $\pi_1(X)$, in light of Remark 5.1. Here are some of the additional properties:

Theorem 6.21. *Let X be a smooth quasi-projective complex variety. Then:*

- ([Di-2007b]) *Any irreducible component of Σ_k^1 is also one of Σ_l^1 for any $1 \leq l \leq k$.*

- ([DPS-2008])*Any two distinct irreducible components of Σ_k^1 intersect in at most finitely many points, and the intersection points are torsion.*

- ([AB+2013]) *If Z and Z' are distinct irreducible components of Σ_k^1 and, respectively, $\Sigma_{k'}^1$, and if $\rho \in Z \cap Z'$, then $\rho \in \Sigma_{k+k'}^1$.*

6.5 Dimension bounds

Theorem 6.15 implies a dimension bound:

Theorem 6.22. ([GL-1987, GL-1991]) *Let X be a compact Kähler manifold. The cohomology support loci $\mathcal{V}_1^i \subset \mathrm{Pic}^\tau(X)$ satisfy*
$$\operatorname{codim} \mathcal{V}_1^i \geq \dim a_X(X) - i.$$

Remark 6.23. This codimension bound can also be thought of as a special case of the codimension bound (2) for DGLA pairs. Indeed, consider $\mathcal{V}_1^i$ locally at trivial line bundle $\mathcal{O}_X$. This is described by the DGLA pair (3) which simplifies in this case to

$$(H^\bullet(X, \mathcal{O}_X), H^\bullet(X, \mathcal{O}_X)).$$

Let $\mathrm{exdeg}(X)$ be the exactness degree of this pair as defined in §3.6. Here the quadratic cone of the pair is the affine space

$$\mathcal{Q} = H^1(X, \mathcal{O}_X).$$

The universal Aomoto complex in this case is

$$(H^\bullet(X, \mathcal{O}_X) \otimes_{\mathbb{C}} \mathcal{O}_{\mathcal{Q}}, w_{univ}).$$

Applying (2), one has that

$$\mathrm{codim}\, \mathcal{V}_1^i \geq \mathrm{exdeg}(X) - i.$$

Then the theorem follows after identifying

$$\mathrm{exdeg}(X) = \dim a_X(X),$$

this equality being explicitly stated in [LP-2010, Theorem A].

The bound on the codimension of $\mathcal{V}_1^i$ implies the generic vanishing:

Theorem 6.24. ([GL-1987]) *Let X be a compact Kähler manifold. Then for a general line bundle $L \in \mathrm{Pic}^\tau(X)$,*

$$H^i(X, L) = 0 \qquad \textit{for} \qquad i < \dim a_X(X).$$

Theorem 6.22 has been generalized to Hodge pieces and to local systems by Popa and Schnell:

Theorem 6.25. ([PSc-2013]) *Let X be a smooth complex projective variety of dimension n.*

- *The cohomology support loci $\mathcal{V}_1^{pq} \subset \mathrm{Pic}^\tau(X)$ satisfy*

$$\mathrm{codim}\, \mathcal{V}_1^{pq} \geq |p + q - n| - \delta(a_X).$$

- *The cohomology support loci $\Sigma_1^i \subset \mathcal{M}_B(X)$ satisfy*

$$\mathrm{codim}\, \Sigma_1^i \geq 2(|i - n| - \delta(a_X)).$$

The first part in the theorem implies the second part by the correspondence between local systems and Higgs line bundles. Here $\delta(a_X)$ is the defect of semi-smallness of the Albanese map,

$$\delta(a_X) = \max_{l \in \mathbb{N}}(2l - \dim X + \dim A_l),$$

where

$$A_l = \{y \in Alb(X) \mid \dim f^{-1}(y) \geq l\}.$$

A weaker statement for abelian varieties was also proved by Krämer-Weissauer [KW-2015].

Remark 6.26. We like to think of Theorem 6.25 as, again, a special case of the codimension bound (2). In this case, the controlling DGLA pair for $\mathcal{V}^{pq}$ locally at the trivial line bundle is

$$(H^{\bullet}(X,\mathcal{O}_X), H^{\bullet}(X,\Omega_X^p)),$$

according to (3). The quadratic cone of this pair is $\mathcal{Q} = H^1(X,\mathcal{O}_X)$ again. The universal Aomoto complex of this pair is

$$(H^{\bullet}(X,\Omega_X^p)\otimes\mathcal{O}_{\mathcal{Q}}, w_{univ}).$$

Let $\operatorname{exdeg}(X,p)$ be the exactness degree of this complex. By (2),

$$\operatorname{codim}\mathcal{V}_1^{pq} \geq \operatorname{exdeg}(X,p) - q.$$

By Hodge symmetry, it is enough to consider the case $p+q\leq n$. Then the result follows from the bound

$$\operatorname{exdeg}(X,p)\geq n-p-\delta(a_X),$$

shown implicitly in [PSc-2013].

The proof in [PSc-2013] of Theorem 6.25 goes in two steps. The first step is to identify a rather natural class of coherent sheaves $\mathcal{F}$ on an abelian variety A of dimension n such that the relative cohomology jump loci

$$\mathcal{V}_k^i(\mathcal{F}) = \{L\in\operatorname{Pic}^0(A)\mid \dim H^i(A,\mathcal{F}\otimes L)\geq k\}\subset\operatorname{Pic}^0(A)$$

satisfy automatically the bound

$$\operatorname{codim}\mathcal{V}_1^i(\mathcal{F})\geq i.$$

These are called *GV sheaves.* This bound is by duality related to the bound (2) obtained by classical homological algebra. The second step is to note that the derived direct image of the trivial line bundle under the Albanese map of X decomposes into special sheaves as above by the Decomposition Theorem for proper maps. The special sheaves are:

Theorem 6.27. ([PSc-2013]) *Let A be a complex abelian variety. Let M be a mixed Hodge module on A with underlying filtered $\mathcal{D}$-module $(\mathcal{M},F)$. Then*

$$\operatorname{codim}\mathcal{V}_1^i(\operatorname{gr}_k^F\mathcal{M})\geq i$$

for all k and i.

6.6 Application to abelian varieties and compact complex tori

([PPS-2015]) For a stronger statement and a generalization of the previous theorem to compact complex tori, see the recent article of Pareschi, Popa, and Schnell

[PPS-2015]. The GV sheaves, and their related cousins the M-regular sheaves, that is, coherent sheaves $\mathcal{F}$ satisfying the stronger inequality

$$\operatorname{codim} \mathcal{V}_1^i(\mathcal{F}) > i,$$

have been used to prove the following numerical characterizations in terms of plurigenera:

Theorem 6.28.

- ([CH-1999]) *A smooth complex projective variety X is birational to an abelian variety if and only if $P_1(X) = P_2(X) = 1$ and $q(X) = \dim X$.*
- ([PPS-2015]) *A compact Kähler manifold X is bimeromorphic to a compact complex torus if and only if $\dim H^1(X, \mathbb{C}) = 2\dim X$ and $P_1(X) = P_2(X) = 1$.*

6.7 Constructible complexes on tori

Let $\mathcal{K}$ be a $\mathbb{C}$-constructible complex on a complex manifold. Consider the relative hypercohomology jump loci of rank one local systems

$$\Sigma_k^i(\mathcal{K}) \subset \mathcal{M}_{\mathrm{B}}(X)$$

given by

$$\Sigma_k^i(\mathcal{K}) = \{L \in \mathcal{M}_{\mathrm{B}}(X) \mid \dim \mathbb{H}^i(X, \mathcal{K}^\bullet \otimes_{\mathbb{C}} L) \geq k\}.$$

Theorem 6.29. ([GaL-1996]) *Let $\mathcal{K}$ be a perverse sheaf on the affine torus $X = (\mathbb{C}^*)^n$. Then $\Sigma_k^i(\mathcal{K})$ is a finite unions of translates of subtori of $\mathcal{M}_B(X)$ and*

$$\operatorname{codim} \Sigma_1^i(\mathcal{K}) \geq i.$$

We expect the translates to be by torsion elements.

As a corollary, one has the following generic vanishing, the perverse sheaf counterpart of Theorem 6.24.

Theorem 6.30. *Let $X \subset (\mathbb{C}^*)^n$ be a closed subvariety of dimension d such that the shifted constant sheaf $\mathbb{C}_X[d]$ is a perverse sheaf on $(\mathbb{C}^*)^n$ (e.g. a locally complete intersection). Then*

$$(-1)^d \chi(X) \geq 0.$$

Remark 6.31. This result had appeared already in [LS-1991] and was attributed to Laumon in [GaL-1996]. The positivity of the signed Euler characteristic has been later conjectured for any closed subvariety of $(\mathbb{C}^*)^n$ in [HS-2014]. Counterexamples to this conjecture were recently constructed in [BW-2014].

In the compact complex torus case, one has:

Theorem 6.32. ([Wa-2013, PPS-2015]) *Let X be a compact complex torus. Let $\mathcal{K}$ be a perverse sheaf on X coming from a polarizable real Hodge module. Then $\Sigma_k^i(\mathcal{K})$ is a finite unions of torsion translates of subtori of $\mathcal{M}_B(X)$.*

Here a polarizable real Hodge module is a generalization of unitary local systems, as opposed to the usual polarizable Hodge modules being generalizations of quasi-unipotent local systems of geometric origin. The result of [Wa-2013] deals with polarizable Hodge modules, the extension to real polarizable Hodge modules is due to [PPS-2015].

7 Fundamental groups and homotopy type

7.1 Fundamental groups

Let X be a connected smooth manifold. As pointed out in Remark 5.1, the moduli space of rank one local systems $\mathcal{M}_{\mathrm{B}}(X)$ and the cohomology jump loci Σ^1_k are determined by the fundamental group $\pi_1(X)$. The statement that each irreducible component of Σ^1_k is a torsion translated subtorus puts nontrivial conditions on the fundamental group of X.

First, let us recall some notations in group theory. Given a group G, we denote its derived subgroup $[G, G]$ by DG or D^1G. Let $D^nG = D(D^{n-1}G)$. $\mathrm{Rank}(G/D^nG)$ is defined to be

$$\sum_{1\leq j\leq n} \dim_{\mathbb{Q}}(D^{j-1}G/D^jG)\otimes_{\mathbb{Z}}\mathbb{Q}.$$

$\mathrm{Rank}(G/D^nG)$ can be a natural number or $+\infty$.

A group G is called *virtually nilpotent* if it has a finite index subgroup G' which is nilpotent. A group G is called *polycyclic* if there is a sequence of subgroups $G = G_0 \triangleright G_1 \triangleright \cdots \triangleright G_n = \{0\}$ such that each G_j is a normal subgroup of G_{j-1} and each G_{j-1}/G_j is cyclic.

The following theorem is proved using Simpson's result about Σ^i_k of smooth projective varieties.

Theorem 7.1. ([AN-1999]) *Let X be a smooth projective variety, and let $G = \pi_1(X)$. Suppose that DG is finitely generated and G/D^nG is solvable of finite rank for some n. Then there are normal subgroups $P \supseteq Q \supseteq D^nG$ so that*

1. *G/P is finite,*
2. *P/Q is nilpotent,*
3. *Q/D^nG is a torsion group.*

When G is polycyclic, $D^nG = 0$ for sufficiently large n. Moreover the assumption of the theorem is automatically satisfied. Thus we have the following consequence.

Corollary 7.2. ([AN-1999]) *Let X be a smooth projective variety, and let $G = \pi_1(X)$. Suppose G is polycyclic. Then G is virtually nilpotent.*

By extending Simpson's result to Σ_k^1 of compact Kähler manifolds, in [Cam-2001] Campana generalized the above results to the case when X is a compact Kähler manifold.

Let now $G_n = [G_{n-1}, G]$ denote the *lower central series* of a group G. The *Chen groups* of a group G are the successive quotients in the lower central series of $[G, D^2G]$. A remarkable result of Cohen-Schenck [CS-2013], conjectured by Suciu, is that the resonance varieties $\mathcal{R}_1^1$, as in Remark 6.4, determine combinatorially the ranks of the Chen groups of the fundamental group of a complement of hyperplane arrangements:

Theorem 7.3. ([CS-2013]) *Let $\theta_k(G)$ be the rank of the k-th Chen group of the fundamental group G of a complement U of a hyperplane arrangement. Let h_k be the number of k-dimensional irreducible components of the resonance variety*

$$\mathcal{R}_1^1(U) = \{w \in H^1(U,\mathbb{C}) \mid H^1(H^\bullet(U,\mathbb{C}), w \cup .) \neq 0\}.$$

Then

$$\theta_k(G) = (k-1) \sum_{m \geq 2} h_m \binom{m+k-2}{k}.$$

This is a consequence of 1-formality, reducedness of the resonance varieties, and some other special properties that fundamental groups of complements of hyperplane arrangements enjoy. Beyond such groups, it remains to be determined what is the precise relation between the possibly-nonreduced cohomology jump loci and Chen groups.

7.2 Homotopy type of smooth projective varieties and compact Kähler manifolds

As we have discussed in Theorem 6.10, Theorem 6.11 and Theorem 6.12 and thereafter, the special geometry of smooth quasi-projective varieties and compact Kähler manifolds has strong implications on the cohomology jump loci of rank one local systems Σ_k^i. Next, we will show some examples of how to use cohomology jump loci to detect the non-algebraicity of compact Kähler manifolds.

Every one dimensional compact complex manifold is projective hence Kähler. In dimension two, there are compact complex manifolds that are not Kähler, for example the Hopf surface or more generally any compact complex surface X with odd $b_1(X)$. There are also compact Kähler surfaces that are not projective, for example a general 2-dimensional complex torus. However, using the classification of complex surfaces, Kodaira proved that every compact Kähler surface can be deformed to a smooth projective surface. In higher dimensions, Kodaira problem asks whether a higher dimensional compact Kähler manifold can always be deformed to a smooth projective variety.

In [Vo-2004], Voisin gave a negative answer to Kodaira problem by a topological method. More precisely, Voisin constructed compact Kähler manifolds that are not of the same homotopy type as any smooth projective variety. She also showed how to modify the example to obtain compact Kähler manifolds that are not of the same rational (or real) homotopy type as any smooth projective variety. Let us review Voisin's construction first.

Given a compact complex torus T of dimension at least two, suppose there is a torus endomorphism ϕ of T whose induced endomorphism $H^1(\phi)$ on $H^1(T,\mathbb{Z})$ satisfies some special properties. Voisin proved that the special endomorphism forces that the rational Neron-Severi group $NS_{\mathbb{Q}}(T) = H^{1,1}(T,\mathbb{C}) \cap H^2(T,\mathbb{Q})$ is trivial. Hence T is not an abelian variety.

Fix a compact complex torus T with the special endomorphism ϕ as above. Voisin constructed a 4-dimensional compact Kähler manifold X by blowing up a sequence of points and surfaces on $T \times T$. Voisin managed to encode the information of the endomorphism ϕ into the cohomology ring $H^\bullet(X,\mathbb{Z})$. She showed that if another compact Kähler manifold Y has the same cohomology ring, i.e., $H^\bullet(Y,\mathbb{Z}) \cong H^\bullet(X,\mathbb{Z})$ as graded commutative rings, then $Alb(Y)$ is not an abelian variety, and hence Y is not projective.

Voisin also gave other constructions so she can encode the information of ϕ in the $\mathbb{R}$-coefficient cohomology ring $H^\bullet(X,\mathbb{R})$. Thus the optimal result of [Vo-2004] is the following.

Theorem 7.4. [Vo-2004] *There exists a compact Kähler manifold X which is not of the real homotopy type of any smooth projective variety.*

In [Wa-2013], Voisin's theorem is reproved using cohomology jump loci. A compact Kähler manifold X is constructed which is very similar to Voisin's example. Now, the information of the endomorphism ϕ is now encoded in the cohomology jump locus $\Sigma^2_1(X)$ of rank one local systems. Thus, by Voisin's argument using rational Neron-Severi group, we obtain the following theorem.

Theorem 7.5. [Wa-2013] *There exists a compact Kähler manifold X with the following property. Suppose Y is another compact Kähler manifold with $H^1(Y,\mathbb{C}) \cong H^1(X,\mathbb{C})$, or equivalently there is an isomorphism of germs $\mathcal{M}_B(Y)_{(0)} \cong \mathcal{M}_B(X)_{(0)}$. If the isomorphism $\mathcal{M}_B(Y)_{(\mathbf{1})} \cong \mathcal{M}_B(X)_{(\mathbf{1})}$ induces an isomorphism $\Sigma^2_1(Y)_{(\mathbf{1})} \cong \Sigma^2_1(X)_{(\mathbf{1})}$, then Y is not projective.*

As mentioned in Remark 6.6, the pairs of germs $\Sigma^i_k(X)_{(\mathbf{1})} \subset \mathcal{M}_{\mathrm{B}}(X)_{(\mathbf{1})}$ are determined by the real i-homotopy type of X for all i and k. Thus we can slightly improve Voisin's result to the following.

Corollary 7.6. *There exists a compact Kähler manifold X that is not of real 2-homotopy type of any smooth projective variety.*

8 Small ball complements and Bernstein-Sato ideals

We now let $X = \mathbb{C}^n$ and let

$$f : X \to \mathbb{C}$$

be a holomorphic function. For a point $x \in f^{-1}(0)$, let

$$U_{f,x} = B_x \cap (X \setminus f^{-1}(0)),$$

where B_x is a small ball centered at x in X.

Conjecture 8.1. *Let $U_{f,x}$ be the complement of a hypersurface singularity germ as above. Then the cohomology jump loci of rank one local systems $\Sigma^i_k \subset \mathcal{M}_B(U_{f,x})$ are finite unions of torsion translates of subtori.*

Remark 8.2. It is known that the small ball complements $U_{f,x}$ are 1-formal spaces, see [DH-1988]. Therefore Theorem 6.3 applies, and hence the conjecture is true for those irreducible components of $\Sigma^1_k(U_{f,x})$ containing the trivial rank one local system, that is, those components are subtori.

Remark 8.3. This conjecture is stated erroneously as a theorem in [Li-2009]. We have also believed it to be proved in *loc. cit.* until the writing of [BW-2013] when we realized that the rather difficult Theorem 6.11 should in principle be easier to prove. A red flag about [Li-2009] was raised then. Scouting for the details, we remarked the following. Let L be a rank one local system on a smooth quasi-projective variety X. It is known that if L is unitary, then the cohomology $H^\bullet(X, L)$ admits a mixed Hodge structure [Ti-1987]. The attached theory of weights leads, via a well-known argument of Deligne, to a technique for degeneration of associated spectral sequences, cf. Remark 6.7. In a more generalized context, this is also the reason why Theorem 6.32 works for polarizable real Hodge modules. However, if L is not unitary, it is not clear how to construct mixed Hodge structures nor how to attach weights. This leaves unproved beyond the unitary case Lemma 3.5 of [Li-2009] on the degeneration of a certain spectral sequence. At least for this reason, the main result of [Li-2009] is unfortunately still open, and we stated this as the above conjecture.

The above conjecture generalizes a classical result in singularity theory. Recall that, by Milnor, the small ball complement $U_{f,x}$ is a smooth locally trivial fibration over a small punctured disc Δ^* in $\mathbb{C}$ via the map f. The homotopy class of the fiber is called *the Minor fiber of f at x* and it is denoted by $F_{f,x}$. The conjecture implies the well-known:

Theorem 8.4. (Monodromy Theorem) *Let $f : (\mathbb{C}^n, x) \to (\mathbb{C}, 0)$ be the germ of a holomorphic function. Then the eigenvalues of the monodromy on the cohomology $H^i(F_{f,x}, \mathbb{C})$ of the Milnor fiber are roots of unity for every i.*

The precise relation between the conjecture and the Monodromy Theorem is the following. Assume, for simplicity, that there is locally at x a splitting

$$f = \prod_{1 \leq i \leq r} f_i$$

such that

$$\text{the } f_i \text{ define mutually distinct, reduced, irreducible analytic branches of } f. \quad (4)$$

In this case,

$$\mathcal{M}_B(U_{f,x}) = (\mathbb{C}^*)^r,$$

and a rank one local system L on $U_{f,x}$ is identified with the r-tuple $\lambda \in (\mathbb{C}^*)^r$ given by the monodromies around each of the branches.

Lemma 8.5. *Let f and x be as above, satisfying (4). Let*

$$diag : \mathbb{C}^* \longrightarrow \mathcal{M}_B(U_{f,x}) = (\mathbb{C}^*)^r$$

be the diagonal embedding $\gamma \mapsto (\gamma, \dots, \gamma)$. Then intersection of the cohomology support loci of rank one local systems

$$\bigcup_i \Sigma_1^i(U_{f,x}) \subset \mathcal{M}_B(U_{f,x})$$

with the image of the diagonal embedding is the set of eigenvalues of the Milnor monodromy on the cohomology of $F_{f,x}$.

Remark 8.6. When (4) does not hold, one has a similar statement after adjusting the definition of the diagonal embedding. The lemma follows immediately from [Bu-2015, Prop. 3.31, Thm. 4]. Note that, as pointed out by Liu-Maxim [LM-2014], in all the statements in [Bu-2015] where the uniform support $Supp^{unif}\,\psi_F\mathbb{C}_X$ of the Sabbah specialization complex appears, the *unif* should be dropped to conform to what is proven in [Bu-2015]. There are two other different proofs of this lemma known to us: one using the Leray spectral sequence for the Milnor fibration, the other one using perverse sheaves related to the open embedding of $U_{f,x}$ in B_x.

A different approach to the Monodromy Theorem was given by Malgrange and Kashiwara. For a holomorphic function $f : X \to \mathbb{C}$, one defines the *Bernstein-Sato polynomial* $b_f(s)$ to be the monic generator of the ideal of polynomials $b(s) \in \mathbb{C}[s]$ satisfying

$$b(s)f^s = Pf^{s+1},$$

for some operator

$$P \in \mathcal{O}_X[\partial_1, \dots, \partial_n, s],$$

where $\mathcal{O}_X$ are the holomorphic functions on X, and $\partial_i = \partial/\partial x_i$ are the usual derivations with respect to a choice of coordinates on $X = \mathbb{C}^n$. A non-trivial result of Bernstein and Sato guarantees the existence of a non-zero $b_f(s)$ for any f. If f is a polynomial, then $\mathcal{O}_X$ can be replaced with the polynomial ring in n variables.

One has the classical main result about Bernstein-Sato polynomials, letting $\mathrm{Exp}(\alpha) = \exp(2\pi i\alpha)$:

Theorem 8.7. *Let $f : \mathbb{C}^n \to \mathbb{C}$ be a holomorphic function. Then:*

- ([Ka-1976]) *The roots of $b_f(s)$ are negative rational numbers.*
- ([Mal-1983, Ka-1982]) *The set*

  $$\{\mathrm{Exp}(\alpha) \mid \alpha \text{ is a root of } b_f(s)\} \subset \mathbb{C}^*$$

  is the set of all eigenvalues of the Milnor monodromy on the cohomologies of Milnor fibers $F_{f,x}$ of f at points $x \in f^{-1}(0)$.

We have taken in [Bu-2015] the point of view that this result of Malgrange and Kashiwara hides a more general statement about rank one local systems and we tried to uncover what this statement is. Our answer is in terms of the *Bernstein-Sato ideals*

$$B_{F,x} \subset \mathbb{C}[s_1, \ldots, s_r]$$

attached to the collection of functions

$$F = (f_1, \ldots, f_r),$$

with, for simplicity, $f = \prod_{i=1}^{r} f_i$ as in (4). The ideal $B_{F,x}$ is defined as the the ideal generated by the polynomials $b \in \mathbb{C}[s_1, \ldots, s_r]$ such that

$$b(s_1, \ldots, s_r) \prod_{i=1}^{r} f_i^{s_i} = P \prod_{i=1}^{r} f_i^{s_i+1}$$

for some operator

$$P \in \mathcal{O}_{X,x}[\partial_1, \ldots, \partial_n, s_1, \ldots, s_r],$$

where x_i are local coordinates at x.

For $y \in f^{-1}(0)$ close to x, one has the cohomology support loci $\Sigma_1^i(U_{f,y})$ in $\mathcal{M}_{\mathrm{B}}(U_{f,y})$. Since now there might be branches f_i with $f_i(y) \neq 0$, it is possible that $\mathcal{M}_{\mathrm{B}}(U_{f,y})$ is a smaller affine torus than $(\mathbb{C}^*)^r$. Fixing the coordinates $t_1, \ldots, t_r$ on $(\mathbb{C}^*)^r$, the smaller affine torus is cut out by the equations $t_i = 1$ for those i with $f_i(y) \neq 0$. By eliminating these equations, we define the *uniform cohomology support loci*

$$\Sigma_1^i(U_{f,y})^{unif} \subset (\mathbb{C}^*)^r,$$

all sitting inside the same affine torus for all y close to x. Consider the zero locus

$$Z(B_{F,x}) \subset \mathbb{C}^r$$

of the Bernstein-Sato ideal.

Theorem 8.8. ([Bu-2015]) *With the notation as above,*

$$\mathrm{Exp}(Z(B_{F,x})) \supset \bigcup_{\substack{i \in \mathbb{Z} \\ y \in f^{-1}(0) \text{ close to } x}} \Sigma_1^i(U_{f,y})^{unif},$$

and we conjectured that equality holds.

References

[Ar-1992] D. Arapura, *Higgs line bundles, Green-Lazarsfeld sets, and maps of Kähler manifolds to curves.* Bull. Amer. Math. Soc. 26 (1992), no. 2, 310–314.

[Ar-1997] D. Arapura, *Geometry of cohomology support loci for local systems. I.* J. Algebraic Geom. 6 (1997), no. 3, 563–597.

[AN-1999] D. Arapura, M. Nori, *Solvable fundamental groups of algebraic varieties and Kähler manifolds.* Compositio Math. 116 (1999), no. 2, 173–188.

[A+1985] E. Arbarello, M. Cornalba, P.A. Griffiths, J. Harris, *Geometry of algebraic curves. Vol. I.* Grundlehren der Mathematischen Wissenschaften 267. Springer-Verlag, New York, 1985.

[AB+2013] E. Artal Bartolo, J.I. Cogolludo-Agustín, D. Matei, *Characteristic varieties of quasi-projective manifolds and orbifolds.* Geom. Topol. 17 (2013), no. 1, 273–309.

[ADH-2015] D. Arapura, A. Dimca, R. Hain, On the fundamental groups of normal varieties. arXiv:1412.1483.

[Be-1988] A. Beauville, *Annulation du H^1 et systémes paracanoniques sur les surfaces.* J. Reine Angew. Math. 388 (1988), 149–157.

[Bu-2009] N. Budur, *Unitary local systems, multiplier ideals, and polynomial periodicity of Hodge numbers.* Adv. Math. 221 (2009), 217–250.

[Bu-2015] N. Budur, *Bernstein-Sato ideals and local systems.* Ann. Inst. Fourier 64 (2015).

[BW-2013] N. Budur, B. Wang, *Cohomology jump loci of differential graded Lie algebras.* arXiv:1309.2264. To appear in Compositio Math.

[BW-2014] N. Budur, B. Wang, *The signed Euler characteristic of very affine varieties.* arXiv:1403.4371. To appear in Int. Math. Res. Notices.

[BW-2015] N. Budur, B. Wang, *Cohomology jump loci of quasi-projective varieties.* Ann. Sci. École Norm. Sup. 48 (2015), no. 1.

[Cam-2001] F. Campana, *Ensembles de Green-Lazarsfeld et quotients résolubles des groupes de Kähler.* J. Algebraic Geom. 10 (2001), no. 4, 599–622.

[Cat-1991] F. Catanese, *Moduli and classification of irregular Kähler manifolds (and algebraic varieties) with Albanese general type fibrations.* Inv. Math. 104 (1991), 263–289.

[CH-1999] J. Chen, C. Hacon, *Characterization of abelian varieties.* Invent. Math. 143 (2001), no. 2, 435–447.

[CS-2013] D.Cohen, H. Schenck, *Chen ranks and resonance.* arXiv:1312.3652.

[D+1975] P. Deligne, P. Griffiths, J. Morgan, D. Sullivan, *Real homotopy theory of Kähler manifolds.* Invent. Math. 29 (1975), no. 3, 245–274.

[De-1986] P. Deligne, *Letter to J. Millson*, April 24, 1986. `http://www.math.umd.edu/~millson/papers/deligne.pdf`

[Di-2007a] A. Dimca, *Characteristic varieties and constructible sheaves.* Rend. Lincei Mat. Appl. 18 (2007), 365–389.

[Di-2007b] A. Dimca, *On the irreducible components of characteristic varieties,* An. Ştiinţ. Univ. "Ovidius" Constanţa Ser. Mat. 15 (2007), 67–73.

[DP-2014] A. Dimca, S. Papadima, *Nonabelian cohomology jump loci from an analytic viewpoint.* Commun. Contemp. Math. 16 (2014), no. 4, 1350025, 47 pp.

[DPS-2008] A. Dimca, S. Papadima, A.I. Suciu, *Alexander polynomials: essential variables and multiplicities.* Int. Math. Res. Not. 2008.

[DPS-2009] A. Dimca, S. Papadima, A.I. Suciu, *Topology and geometry of cohomology jump loci.* Duke Math. J. 148 (2009), no. 3, 405–457.

[DH-1988] A. Durfee, R. Hain, *Mixed Hodge structures on the homotopy of links.* Math. Ann. 280 (1988), no. 1, 69–83.

[Ei-1995] D. Eisenbud, *Commutative algebra. With a view toward algebraic geometry.* Graduate Texts in Mathematics, 150. Springer-Verlag, New York, 1995.

[ESV-1992] H. Esnault, V. Schechtman, E. Viehweg, *Cohomology of local systems of the complements of hyperplanes.* Invent. Math. 109 (1992), no. 3, 557–561; *Erratum,* ibid. 112 (1993), no. 2, 447.

[FIM-2012] D. Fiorenza, D. Iacono, E. Martinengo, *Differential graded Lie algebras controlling infinitesimal deformations of coherent sheaves.* J. Eur. Math. Soc. (JEMS) 14 (2012), no. 2, 521–540.

[GaL-1996] O. Gabber, F. Loeser, *Faisceaux pervers l-adiques sur un tore.* Duke Math. J. 83 (1996), no. 3, 501–606.

[GL-1987] M. Green, R. Lazarsfeld, *Deformation theory, generic vanishing theorems, and some conjectures of Enriques, Catanese and Beauville.* Invent. Math. 90 (1987), no. 2, 389–407.

[GL-1991] M. Green, R. Lazarsfeld, *Higher obstructions to deforming cohomology groups of line bundles.* J. Amer. Math. Soc. 4 (1991), no. 1, 87–103.

[GM-1988] W. Goldman, J. Millson, *Deformations of flat bundles over Kähler manifolds.* Inst. Hautes Études Sci. Publ. Math. No. 67 (1988), 43–96.

[HS-2014] J. Huh, B. Sturmfels, *Likelihood Geometry.* Combinatorial Algebraic Geometry, Lecture Notes in Mathematics 2108, Springer, 2014.

[Ka-1976] M. Kashiwara, *B-functions and holonomic systems. Rationality of roots of B-functions.* Invent. Math. 38 (1976/77), no. 1, 33–53.

[Ka-1982] M. Kashiwara, *Vanishing cycle sheaves and holonomic systems of differential equations.* Algebraic geometry (Tokyo/Kyoto, 1982), 134–142, Lecture Notes in Math., 1016, Springer, Berlin, 1983.

[KW-2015] T. Krämer, R. Weissauer, *Vanishing theorems for constructible sheaves on abelian varieties.* J. Alg. Geom. 24 (2015), 531–568.

[Li-2009] A. Libgober, *Non-vanishing loci of Hodge numbers of local systems.* Manuscripta Math. 128 (2009), no. 1, 1–31.

[Lu-2011] J. Lurie, *Formal Moduli Problems.* 2011. `http://www.math.harvard.edu/~lurie/papers/DAG-X.pdf`

[LM-2014] Y. Liu, L. Maxim, *Characteristic varieties of hypersurface complements.* arXiv:1411.7360.

[LP-2010] R. Lazarsfeld, M. Popa, *Derivative complex, BGG correspondence, and numerical inequalities for compact Káhler manifolds.* Invent. Math. 182 (2010), no. 3, 605–633.

[LS-1991] F. Loeser, C. Sabbah, *Caractérisation des D-modules hypergéométriques irréductibles sur le tore.* C. R. Acad. Sci. Paris Sér. I Math. 312 (1991), no. 10, 735–738. *Caractérisation des D-modules hypergéométriques irréductibles sur le tore. II.* C. R. Acad. Sci. Paris Sér. I Math. 315 (1992), no. 12, 1263–1264.

[Mal-1983] B. Malgrange, *Polynômes de Bernstein-Sato et cohomologie évanescente.* Analysis and topology on singular spaces, II, III (Luminy, 1981), 243–267, Astérisque, 101–102, Soc. Math. France, Paris, 1983.

[Mar-2009] E. Martinengo, *Local structure of Brill-Noether strata in the moduli space of flat stable bundles.* Rend. Semin. Mat. Univ. Padova 121 (2009), 259–280.

[Mar-2012] E. Martinengo, *Infinitesimal deformations of Hitchin pairs and Hitchin map.* Internat. J. Math. 23 (2012), no. 7, 125–153.

[Mo-1978] J.W. Morgan, *The algebraic topology of smooth algebraic varieties.* Inst. Hautes Études Sci. Publ. Math. 48 (1978), 137–204. *Correction to: "The algebraic topology of smooth algebraic varieties"*, Inst. Hautes Études Sci. Publ. Math. 64 (1986), 185.

[Na-1988] A. Nadel, *Singularities and Kodaira dimension of the moduli space of flat Hermitian-Yang-Mills connections.* Compositio Math. 67 (1988), no. 2, 121-128.

[OS-1980] P. Orlik, L. Solomon, *Combinatorics and topology of complements of hyperplanes.* Invent. Math. 56 (1980), no. 2, 167–189.

[PaS-2014] S. Papadima, A.I. Suciu, *The Milnor fibration of a hyperplane arrangement: from modular resonance to algebraic monodromy.* arXiv:1401.0868.

[Pr-2011] J.P. Pridham, *Unifying derived deformation theories.* Adv. Math. 224 (2010), no. 3, 772–826. *Corrigendum.* Adv. Math. 228 (2011), no. 4, 2554–2556.

[PP-2009] G. Pareschi, M. Popa, *Strong generic vanishing and a higher-dimensional Castelnuovo-de Franchis inequality.* Duke Math. J. 150 (2009), no. 2, 269–285.

[PPS-2015] G. Pareschi, M. Popa, C. Schnell, *Hodge modules on complex tori and generic vanishing for compact Kähler manifolds.* arXiv:1505.00635.

[PR-2004] R. Pink, D. Roessler, *A conjecture of Beauville and revisited.* Math. Ann. 330 (2004), no. 2, 293–308.

[PS-2008] C. Peters, J. Steenbrink, *Mixed Hodge structures.* Ergebnisse der Mathematik und ihrer Grenzgebiete. 3. Folge. A Series of Modern Surveys in Mathematics, 52. Springer-Verlag, Berlin, 2008.

[PSc-2013] M. Popa, C. Schnell, *Generic vanishing theory via mixed Hodge modules.* Forum Math. Sigma 1 (2013), e1, 60 pp.

[Si-1992] C. Simpson, *Higgs bundles and local systems.* Inst. Hautes Études Sci. Publ. Math. 75 (1992), 5–95.

[Si-1993] C. Simpson, *Subspaces of moduli spaces of rank one local systems.* Ann. Sci. École Norm. Sup. (4) 26 (1993), no. 3, 361–401.

[Si-1994] C. Simpson, *Moduli of representations of the fundamental group of a smooth variety. II.* Inst. Hautes Études Sci. Publ. Math. 80 (1994), 5–79.

[Su-2011] A.I. Suciu, *Fundamental groups, Alexander invariants, and cohomology jumping loci.* Topology of algebraic varieties and singularities, 179–223, Contemp. Math., 538, Amer. Math. Soc., Providence, RI, 2011.

[Su-2015] A.I. Suciu, *Around the tangent cone theorem.* arXiv:1502.02279.

[Ti-1987] K. Timmerschdidt, *Mixed Hodge theory for unitary local systems.* J. reine angew. Math. 379 (1987), 152–171.

[Vo-2004] C. Voisin, *On the homotopy types of compact Khler and complex projective manifolds.* Invent. Math. 157 (2004), no. 2, 329–343.

[Wa-2012] B. Wang, *Cohomology jump loci in the moduli spaces of vector bundles.* arXiv:1210.1487.

[Wa-2013] B. Wang, *Torsion points on the cohomology jump loci of compact Kähler manifolds.* arXiv:1312.6619.

Hodge Theory and L^2-analysis
ALM 39, pp. 245–282

On Semipositivity, Injectivity and Vanishing Theorems*

Osamu Fujino†

Abstract

This is a survey article on the recent developments of semipositivity, injectivity, and vanishing theorems for higher-dimensional complex projective varieties.

2010 Mathematics Subject Classification: Primary 14F17; Secondary 14E30, 14D07.

Keywords and Phrases: Semipositivity, nefness, vanishing theorem, injectivity theorem, canonical ring, pluricanonical divisor.

1 Introduction

This paper is a survey article on the recent developments of semipositivity, injectivity, and vanishing theorems for higher-dimensional complex projective varieties (see, for example, [Fn7], [Fn9], [Fn11], [FF], and [FFS]).

We know that many important generalizations of the Kodaira vanishing theorem, for example, the Kawamata–Viehweg vanishing theorem, Kollár's injectivity, torsion-free, and vanishing theorems, the Nadel vanishing theorem, and so on, were obtained in 1980s. They have already played crucial roles in the study of higher-dimensional complex projective varieties. We note that the Fujita–Zucker–Kawamata semipositivity theorem for direct images of relative canonical bundles has also played important roles. One of my main motivations was to establish a more general cohomological package based on the theory of mixed Hodge structures on cohomology with compact support. Now I think that our new results are almost satisfactory (see Theorems 2.12, 2.13, and 3.6). They are waiting for

*Dedicated to Professor Steven Zucker on the occasion of his 65th birthday.

†Department of Mathematics, Graduate School of Science, Osaka University, Toyonaka, Osaka 560-0043, Japan. E-mail: fujino@math.sci.osaka-u.ac.jp.

applications. I hope that the reader would find various applications of our semipositivity, injectivity, and vanishing theorems.

Let us see the contents of this paper. In Section 2, we first discuss the Hodge theoretic aspect of Kodaira-type vanishing theorems (see, for example, [EV], [Ko4, Part III], [Fn7], [Fn9], and [Fn11]). I emphasize the importance of Kollár's injectivity theorem and its generalizations. I think that one of the most important recent developments is the introduction of mixed Hodge structures on cohomology with compact support in order to generalize Kollár's injectivity theorem (see, for example, [Fn3], [Fn7], [Fn9], and [Fn11]). Next we discuss Enoki's injectivity theorem, which is an analytic counterpart of Kollár's injectivity theorem. I like Enoki's idea since it is very simple and powerful. Enoki's proof only uses the standard results of the theory of harmonic forms on compact Kähler manifolds. Although I obtained some generalizations of Enoki's injectivity theorem and their applications (see [Fn4] and [Fn5]), I think that they are not satisfactory for most geometric applications.

In Section 3, we treat several semipositivity theorems for direct images of relative (log) canonical bundles and relative pluricanonical bundles. The (numerical) semipositivity of direct images of relative (log) canonical bundles discussed in this paper is more or less Hodge theoretic. Note that mixed Hodge structures on cohomology with compact support are also very useful for semipositivity theorems. By considering their variations, we can prove a powerful semipositivity theorem by the theory of gradedly polarizable admissible variation of mixed Hodge structure (see [FF] and [FFS]). Unfortunately, since I am not familiar with the recent developments of semipositivity theorems by L^2 methods, I do not discuss the analytic aspect of semipositivity theorems in this paper. In Subsection 3.1, we explain new semipositivity theorems for direct images of relative pluricanonical bundles with the aid of the minimal model program (see [Fn12]). I think that it is highly desirable to recover them without using the minimal model program.

In Section 4, we will see that pluricanonical divisors sometimes behave much better than canonical divisors. We discuss two different topics. In Subsection 4.1, we explain Kollár's famous result on plurigenera in étale covers of smooth projective varieties of general type. We give Lazarsfeld's proof using the theory of asymptotic multiplier ideal sheaves for the reader's convenience and also a proof based on the minimal model program. The proof based on the minimal model program is harder than Lazarsfeld's proof but is interesting and natural from the minimal model theoretic viewpoint. In Subsection 4.2, we explain Viehweg's ampleness theorem for direct images of relative pluricanonical bundles, which is buried in Viehweg's papers. I think that these results may help the reader to understand the reason why we should consider pluricanonical divisors for the study of higher-dimensional algebraic varieties.

In Section 5, we quickly review the finite generation of (log) canonical rings due to Birkar–Cascini–Hacon–McKernan. I want to emphasize that we need the semipositivity theorem discussed in Section 3 when we treat (log) canonical rings for varieties which are not of (log) general type (see [FMo] and [Fn10]). We also explain the nonvanishing conjecture, which is one of the most important conjectures for higher-dimensional complex projective varieties.

Section 6 is an appendix, where we collect some definitions with the intention of helping the reader to understand this paper. The reader can read each section separately.

Acknowledgments. The author was partially supported by Grant-in-Aid for Young Scientists (A) 24684002 and Grant-in-Aid for Scientific Research (S) 24224001 from JSPS. He thanks Professor Steven Zucker for useful comments and advice. He also thanks Professor Fabrizio Catanese for answering his questions and Yoshinori Gongyo for various discussions. Finally, he thanks Shin-ichi Matsumura for sending his preprints.

We will work over $\mathbb{C}$, the complex number field, throughout this paper. In this paper, a scheme means a separated scheme of finite type over $\mathbb{C}$.

2 On injectivity theorems and vanishing theorems

I think that one of the most fundamental results for complex projective varieties is Kollár's injectivity theorem (see Theorem 2.1). The importance of the Kawamata–Viehweg (or Nadel) vanishing theorem for the study of higher-dimensional complex algebraic varieties is repeatedly emphasized in many papers and textbooks (see, for example, [KoM] and [Laz]). On the other hand, I think that the importance of Kollár's injectivity theorem has not been emphasized so far in the standard literature.

Let us recall Kollár's injectivity theorem.

Theorem 2.1 ([Ko1, Theorem 2.2]). *Let X be a smooth projective variety and let L be a semiample Cartier divisor on X, that is, the complete linear system $|mL|$ has no base points for some positive integer m. Let D be a member of $|kL|$ for some positive integer k. Then*

$$H^i(X, \mathcal{O}_X(K_X + lL)) \to H^i(X, \mathcal{O}_X(K_X + (l+k)L)),$$

induced by the natural inclusion $\mathcal{O}_X \hookrightarrow \mathcal{O}_X(D) \simeq \mathcal{O}_X(kL)$, is injective for every i and every positive integer l.

Remark 2.2. If we assume that L is ample, $l = 1$, and k is sufficiently large in Theorem 2.1, then we obtain that

$$H^i(X, \mathcal{O}_X(K_X + L)) \hookrightarrow H^i(X, \mathcal{O}_X(K_X + (1+k)L)) = 0$$

for every $i > 0$ by Serre's vanishing theorem. Therefore, Theorem 2.1 quickly recovers the Kodaira vanishing theorem for projective varieties (see Theorem 2.3 below).

For the reader's convenience, we recall:

Theorem 2.3 (Kodaira vanishing theorem for projective varieties). *Let X be a smooth projective variety and let L be an ample Cartier divisor on X. Then we have*

$$H^i(X, \mathcal{O}_X(K_X + L)) = 0$$

for every $i > 0$.

We will give a proof of Theorem 2.3 after we discuss E_1-degenerations of Hodge to de Rham type spectral sequences.

Note that Theorem 2.1 is obviously a generalization of Tankeev's pioneering result.

Theorem 2.4 ([Tan, Proposition 1]). *Let X be a smooth projective variety with $\dim X \geq 2$. Assume that the complete linear system $|L|$ has no base points and determines a morphism $\Phi_{|L|} : X \to Y$ onto a variety Y with $\dim Y \geq 2$. Then*

$$H^0(X, \mathcal{O}_X(K_X + 2D)) \to H^0(D, \mathcal{O}_D((K_X + 2D)|_D))$$

is surjective for almost all divisors $D \in |L|$. Equivalently,

$$H^1(X, \mathcal{O}_X(K_X + D)) \to H^1(X, \mathcal{O}_X(K_X + 2D))$$

is injective for almost all divisors $D \in |L|$.

By Theorem 2.1, we can prove:

Theorem 2.5 ([Ko1, Theorem 2.1]). *Let X be a smooth projective variety, let Y be an arbitrary projective variety, and let $f : X \to Y$ be a surjective morphism. Then we have the following properties.*

(i) $R^i f_* \mathcal{O}_X(K_X)$ is torsion-free for every i.

(ii) Let H be an ample Cartier divisor on Y, then

$$H^j(Y, \mathcal{O}_Y(H) \otimes R^i f_* \mathcal{O}_X(K_X)) = 0$$

for every $j > 0$ and every i.

Theorem 2.5 (i) and (ii) are called Kollár's torsion-freeness and the Kollár vanishing theorem respectively. We give a small remark on Theorem 2.5.

Remark 2.6. If $f = id_X : X \to X$ in Theorem 2.5 (ii), then we have $H^i(X, \mathcal{O}_X(K_X + H)) = 0$ for every $i > 0$ and every ample Cartier divisor H on X. This is nothing but the Kodaira vanishing theorem for projective varieties (see Theorem 2.3). If f is birational in Theorem 2.5 (i), then $R^i f_* \mathcal{O}_X(K_X) = 0$ for every $i > 0$ since $R^i f_* \mathcal{O}_X(K_X)$ is a torsion sheaf for every $i > 0$. This is the Grauert–Riemenschneider vanishing theorem for birational morphisms between projective varieties.

In [Ko1], Kollár proved Theorem 2.1 and Theorem 2.5 simultaneously. Therefore, the relationship between Theorem 2.1 and Theorem 2.5 is not clear in [Ko1]. Now it is well-known that Theorem 2.1 and Theorem 2.5 are equivalent by the works of Kollár himself and Esnault–Viehweg (see, for example, [Ko4, Chapter 9] and [EV]). We note that Theorem 2.1 follows from the E_1-degeneration of Hodge to de Rham spectral sequence.

2.7 (E_1-degeneration of Hodge to de Rham spectral sequence). Let V be a smooth projective variety. Then the spectral sequence

$$E_1^{p,q} = H^q(V, \Omega_V^p) \Rightarrow H^{p+q}(V, \mathbb{C})$$

degenerates at E_1. This is a direct consequence of the Hodge decomposition for compact Kähler manifolds.

Therefore, we can see that Theorem 2.1 is a result of the theory of *pure* Hodge structures. Thus, it is natural to consider *mixed* generalizations of Theorem 2.1.

We do not repeat the proof of Theorem 2.1 depending on the E_1-degeneration of Hodge to de Rham spectral sequence in 2.7 here. For the details, see, for example, [Ko4, Chapter 9] and [EV].

2.8. We note Deligne's famous generalization of the E_1-degeneration in 2.7. Let V be a smooth projective variety and let Δ be a simple normal crossing divisor on V. Then the spectral sequence

$$E_1^{p,q} = H^q(V, \Omega_V^p(\log \Delta)) \Rightarrow H^{p+q}(V \setminus \Delta, \mathbb{C})$$

degenerates at E_1 by Deligne's theory of mixed Hodge structures for smooth noncompact algebraic varieties (see [Del]).

Unfortunately, the E_1-degeneration in 2.8 seems to produce no useful generalizations of Theorem 2.1. We think that the following E_1-degeneration is a correct ingredient for mixed generalizations of Theorem 2.1.

2.9. Let V and Δ be as in 2.8. Then the spectral sequence

$$E_1^{p,q} = H^q(V, \Omega_V^p(\log \Delta) \otimes \mathcal{O}_V(-\Delta)) \Rightarrow H_c^{p+q}(V \setminus \Delta, \mathbb{C})$$

degenerates at E_1. This is a consequence of mixed Hodge structures on cohomology with compact support $H_c^\bullet(V \setminus \Delta, \mathbb{C})$.

Remark 2.10. In 2.9, we see that $H^q(V, \Omega_V^p(\log \Delta) \otimes \mathcal{O}_V(-\Delta))$ is dual to $H^{n-q}(V, \Omega_V^{n-p}(\log \Delta))$ by Serre duality, where $n = \dim V$. Moreover, $H_c^{p+q}(V \setminus \Delta, \mathbb{C})$ is dual to $H^{2n-(p+q)}(V \setminus \Delta, \mathbb{C})$ by Poincaré duality. Therefore, we can check the E_1-degeneration in 2.9 by the E_1-degeneration in 2.8. However, it is better to discuss mixed Hodge structures on cohomology with compact support in order to treat more general situations below (see Theorem 2.12, Theorem 2.13, Theorem 3.6, and so on).

We give a proof of the Kodaira vanishing theorem for projective varieties by using the E_1-degeneration in 2.9 in order to get the reader to grow more comfortable with the E_1-degeneration in 2.9.

Proof of Theorem 2.3. By the standard covering trick (see, for example, Step 1 in the proof of [Ko1, Theorem 2.2]), we can reduce Theorem 2.3 to the case when the complete linear system $|L|$ has no base points. So, we assume that $|L|$ has no base points for simplicity. We take a smooth member D of $|L|$ by Bertini's theorem. We put $\iota : X \setminus D \hookrightarrow X$. By the E_1-degeneration of

$$E_1^{p,q} = H^q(X, \Omega_X^p(\log D) \otimes \mathcal{O}_X(-D)) \Rightarrow H_c^{p+q}(X \setminus D, \mathbb{C}),$$

we obtain that the natural map

$$\pi : H^j(X, \iota_! \mathbb{C}_{X \setminus D}) \to H^j(X, \mathcal{O}_X(-D))$$

induced by $\iota_! \mathbb{C}_{X \setminus D} \subset \mathcal{O}_X(-D)$ is surjective for every j. Since

$$\iota_! \mathbb{C}_{X \setminus D} \subset \mathcal{O}_X(-mD) \subset \mathcal{O}_X(-D)$$

for every $m \geq 1$, we obtain that

$$\pi : H^j(X, \iota_! \mathbb{C}_{X \setminus D}) \to H^j(X, \mathcal{O}_X(-mD)) \xrightarrow{p} H^j(X, \mathcal{O}_X(-D))$$

and that p is surjective for every j. Note that $H^j(X, \mathcal{O}_X(-mD)) = 0$ for $j < \dim X$ and for $m \gg 0$ by the Serre vanishing theorem. Thus we obtain that $H^j(X, \mathcal{O}_X(-D)) = 0$ for $j < \dim X$. By Serre duality, we have $H^i(X, \mathcal{O}_X(K_X + D)) = 0$ for every $i > 0$. □

We next give a remark on [EV].

Remark 2.11. Let V be a smooth projective variety and let $A + B$ be a simple normal crossing divisor on V such that A and B have no common irreducible components. In [EV], Esnault–Viehweg discussed the E_1-degeneration of

$$\begin{aligned} E_1^{p,q} &= H^q(V, \Omega_V^p(\log(A+B)) \otimes \mathcal{O}_V(-B)) \\ &\Rightarrow \mathbb{H}^{p+q}(V, \Omega_V^\bullet(\log(A+B)) \otimes \mathcal{O}_V(-B)) \end{aligned}$$

(see also [DI]). This E_1-degeneration contains the E_1-degenerations in 2.8 and in 2.9 as special cases. However, they did not pursue geometric applications of the E_1-degeneration in 2.9, that is, in the case when $A = 0$.

By using the E_1-degeneration in 2.9 and some more general E_1-degenerations arising from mixed Hodge structures on cohomology with compact support, we can obtain various generalizations of Theorem 2.1 and Theorem 2.5. We write the following useful generalizations without explaining the precise definitions and the notation here (see 6.6, 6.7, 6.8, 6.9 in Section 6).

Theorem 2.12 (Injectivity theorem for simple normal crossing pairs)**.** *Let (X, Δ) be a simple normal crossing pair such that Δ is an $\mathbb{R}$-divisor on X whose coefficients are in $[0, 1]$, and let $\pi : X \to V$ be a proper morphism between schemes. Let L be a Cartier divisor on X and let D be an effective Cartier divisor that is permissible with respect to (X, Δ). Assume the following conditions:*

(i) $L \sim_{\mathbb{R}, \pi} K_X + \Delta + H$,

(ii) H is a π-semiample $\mathbb{R}$-divisor, and

(iii) $tH \sim_{\mathbb{R},\pi} D + D'$ *for some positive real number* t*, where* D' *is an effective* $\mathbb{R}$*-Cartier* $\mathbb{R}$*-divisor that is permissible with respect to* (X, Δ).

Then the homomorphisms

$$R^q\pi_*\mathcal{O}_X(L) \to R^q\pi_*\mathcal{O}_X(L + D),$$

which are induced by the natural inclusion $\mathcal{O}_X \hookrightarrow \mathcal{O}_X(D)$*, are injective for all* q.

Theorem 2.12 is a generalization of Theorem 2.1.

Theorem 2.13. *Let* $f : (Y, \Delta) \to X$ *be a proper morphism from an embedded simple normal crossing pair* (Y, Δ) *to a scheme* X *such that* Δ *is an* $\mathbb{R}$*-divisor whose coefficients are in* $[0, 1]$*. Let* L *be a Cartier divisor on* Y *and let* q *be an arbitrary nonnegative integer. Then we have the following properties.*

(i) Assume that $L - (K_Y + \Delta)$ *is* f*-semi-ample. Then every associated prime of* $R^qf_*\mathcal{O}_Y(L)$ *is the generic point of the* f*-image of some stratum of* (Y, Δ).

(ii) Let $\pi : X \to V$ *be a proper morphism between schemes. Assume that*

$$f^*H \sim_{\mathbb{R}} L - (K_Y + \Delta),$$

where H *is nef and log big over* V *with respect to* $f : (Y, \Delta) \to X$*. Then we have*

$$R^p\pi_*R^qf_*\mathcal{O}_Y(L) = 0$$

for every $p > 0$.

Theorem 2.13 (i) and (ii) are generalizations of Theorem 2.5 (i) and (ii) respectively. For the details, see, for example, [Fn3, Sections 5 and 6], [Fn7, Theorem 1.1], [Fn9, Theorem 1.1], and [Fn11, Chapter 5]. Note that Theorem 2.12 and Theorem 2.13 have already played crucial roles in the proof of the fundamental theorems for log canonical pairs and semi-log canonical pairs (see, for example, [Fn3], [Fn6], and [Fn11]).

Anyway, the formulation of Theorem 2.12 and Theorem 2.13 is natural and useful from the minimal model theoretic viewpoint, although it may look unduly technical and artificial.

Remark 2.14. Let V and Δ be as in 2.8. In the traditional framework, $\mathcal{O}_V(K_V + \Delta)$ was recognized to be $\det \Omega^1_V(\log \Delta)$. On the other hand, in our new framework for vanishing theorems, we see $\mathcal{O}_V(K_V + \Delta)$ as

$$\mathcal{H}om_{\mathcal{O}_V}(\mathcal{O}_V(-\Delta), \mathcal{O}_V(K_V))$$

and $\mathcal{O}_V(-\Delta)$ as the 0th term of $\Omega^\bullet_V(\log \Delta) \otimes \mathcal{O}_V(-\Delta)$.

I think that it is not so easy to understand the statements of Theorem 2.12 and Theorem 2.13. So we give a very special case of Theorem 2.12 to clarify the main difference between Theorem 2.1 and Theorem 2.12.

Theorem 2.15. *Let X be a smooth projective variety and let Δ be a simple normal crossing divisor on X. Let L be a semiample Cartier divisor on X and let D be a member of $|kL|$ for some positive integer k such that D contains no strata of Δ. Then the homomorphism*

$$H^i(X, \mathcal{O}_X(K_X + \Delta + lL)) \to H^i(X, \mathcal{O}_X(K_X + \Delta + (l+k)L))$$

induced by the natural inclusion $\mathcal{O}_X \hookrightarrow \mathcal{O}_X(D)$ is injective for every positive integer l and every i.

If $\Delta = 0$ in the above, then Theorem 2.15 is nothing but Kollár's original injectivity theorem (Theorem 2.1).

Remark 2.16. Let Δ be a simple normal crossing divisor on a smooth variety X. Let $\Delta = \sum_{i \in I} \Delta_i$ be the irreducible decomposition of Δ. Then a closed subset W of X is called a stratum of Δ if W is an irreducible component of $\Delta_{i_1} \cap \cdots \cap \Delta_{i_k}$ for some $\{i_1, \ldots, i_k\} \subset I$.

Remark 2.17. Let Δ be a simple normal crossing divisor on a smooth variety V. Then W is a stratum of Δ if and only if W is a log canonical center of (V, Δ) (see 6.5 in Section 6).

We have discussed the Hodge theoretic aspect of Kodaira-type vanishing theorems. For the details and various related topics, see [EV], [Ko4, Part III], [Fn11], and references therein.

2.1 Complex analytic setting

After Kollár obtained Theorem 2.1, Enoki (see [Eno, Theorem 0.2]) proved:

Theorem 2.18 (Enoki's injectivity theorem)**.** *Let X be a compact Kähler manifold and let $\mathcal{L}$ be a semipositive line bundle on X. Then, for any nonzero holomorphic section s of $\mathcal{L}^{\otimes k}$ with some positive integer k, the multiplication homomorphism*

$$\times s : H^i(X, \omega_X \otimes \mathcal{L}^{\otimes l}) \longrightarrow H^i(X, \omega_X \otimes \mathcal{L}^{\otimes (l+k)}),$$

induced by $\otimes s$, is injective for every i and every positive integer l.

Remark 2.19. Let $\mathcal{L}$ be a holomorphic line bundle on a compact Kähler manifold X. We say that $\mathcal{L}$ is semipositive if there exists a smooth hermitian metric h on $\mathcal{L}$ such that $\sqrt{-1}\Theta_h(\mathcal{L})$ is a semipositive $(1,1)$-form on X, where $\Theta_h(\mathcal{L}) = D^2_{(\mathcal{L},h)}$ is the curvature form and $D_{(\mathcal{L},h)}$ is the Chern connection of $(\mathcal{L}, h)$.

Remark 2.20. Let X be a smooth projective variety and let $\mathcal{L}$ be a line bundle on X. If $\mathcal{L}$ is semiample, that is, $|\mathcal{L}^{\otimes k}|$ has no base points for some positive integer k, then $\mathcal{L}$ is semipositive in the sense of Remark 2.19.

Enoki's proof in [Eno] is arguably simpler than the proof of Theorem 2.1 based on Hodge theory. It only uses the standard results in the theory of harmonic forms on compact Kähler manifolds. Let us see the ideas of Enoki's proof of Theorem 2.18.

Idea of Proof of Theorem 2.18. We put $n = \dim X$. Let $\mathcal{H}^{n,i}(X, \mathcal{L}^{\otimes l})$ (resp. $\mathcal{H}^{n,i}(X, \mathcal{L}^{\otimes(l+k)})$ be the space of $\mathcal{L}^{\otimes l}$-valued (resp. $\mathcal{L}^{\otimes(l+k)}$-valued) harmonic (n,i)-forms on X. By using the Nakano identity and the semipositivity of $\mathcal{L}$, we can easily check that $s \otimes \varphi$ is harmonic for every $\varphi \in \mathcal{H}^{n,i}(X, \mathcal{L}^{\otimes l})$. Therefore,

$$\times s : H^i(X, \omega_X \otimes \mathcal{L}^{\otimes l}) \longrightarrow H^i(X, \omega_X \otimes \mathcal{L}^{\otimes(l+k)})$$

is nothing but $\otimes s : \mathcal{H}^{n,i}(X, \mathcal{L}^{\otimes l}) \to \mathcal{H}^{n,i}(X, \mathcal{L}^{\otimes(l+k)}) : \varphi \mapsto s \otimes \varphi$, which is obviously injective. □

We note that Theorem 2.18 is better than Theorem 2.1 by Remark 2.20. Unfortunately, I do not know how to generalize Enoki's theorem appropriately for various geometric applications. Although I obtained some generalizations of Theorem 2.18 and their applications in [Fn4] and [Fn5], they are not so useful in the minimal model program compared with Theorem 2.12 and Theorem 2.13. Related to Theorem 2.15, we have:

Conjecture 2.21. *Let X be a compact Kähler manifold and let Δ be a simple normal crossing divisor on X. Let $\mathcal{L}$ be a semipositive line bundle on X and let s be a nonzero holomorphic section of $\mathcal{L}^{\otimes k}$ on X for some positive integer k. Assume that $(s = 0)$ contains no strata of Δ. Then the multiplication homomorphism*

$$\times s : H^i(X, \omega_X \otimes \mathcal{O}_X(\Delta) \otimes \mathcal{L}^{\otimes l}) \to H^i(X, \omega_X \otimes \mathcal{O}_X(\Delta) \otimes \mathcal{L}^{\otimes(l+k)}),$$

induced by $\otimes s$, is injective for every positive integer l and every i.

I do not know the precise relationship between the injectivity theorems of Kollár and Enoki. Thus I pose:

Problem 2.22. Clarify the relationship between Kollár's injectivity theorem (Theorem 2.1) and Enoki's injectivity theorem (Theorem 2.18).

For almost all geometric applications, we use Theorem 2.5 (ii) for $i = 0$. Theorem 2.5 (ii) for $i = 0$ is sufficient for Viehweg's theory of weak positivity (see [Vie1], [Vie2], and [Fn14]). See also Subsection 4.2 below. Note that Theorem 2.5 (ii) for $i = 0$ is a special case of Ohsawa's vanishing theorem.

Theorem 2.23 ([Oh1, Theorem 3.1]). *Let X be a compact Kähler manifold, let $f : X \to Y$ be a holomorphic map to an analytic space Y with a Kähler form σ, and let (E, h) be a holomorphic vector bundle on X with a smooth hermitian metric h. Assume that $\sqrt{-1}\Theta_h(E) \geq_{\mathrm{Nak}} \mathrm{Id}_E \otimes f^*\sigma$, that is, $\sqrt{-1}\Theta_h(E) - \mathrm{Id}_E \otimes f^*\sigma$ is semipositive in the sense of Nakano, where $\Theta_h(E)$ is the curvature form of (E, h). Then*

$$H^j(Y, f_*(\omega_X \otimes E)) = 0$$

for every $j > 0$.

For the proof of Theorem 2.23, see [Oh1]. I am not so familiar with Theorem 2.23 and do not know if the formulation of Theorem 2.23 is natural or not.

Remark 2.24. For the details of σ and $f^*\sigma$ in Theorem 2.23, see [Oh1, §3]. Note that Y is permitted to have singularities.

By comparing Theorem 2.23 with Theorem 2.5 (ii), it is natural to consider:

Conjecture 2.25. *With the same assumptions as in Theorem 2.23, we have*

$$H^j(Y, R^i f_*(\omega_X \otimes E)) = 0$$

for every i and every positive integer j.

We close this section with:

Problem 2.26. Clarify the relationship between Kollár's vanishing theorem (Theorem 2.5 (ii)) and Ohsawa's vanishing theorem (Theorem 2.23).

For Enoki-type injectivity theorems, see, [Eno], [Take], [Oh2], [Fn4], [Fn5], [Ma1], [Ma2], [Ma3], [Ma4], [Ma5], [GM], etc.

Remark 2.27 (Added in September 2016). After I wrote this paper, there are some developments. In [Ma8], Shin-ichi Matsumura proved Conjecture 2.21 under the extra assumption that Δ is smooth (see [Ma8, Theorem 1.3 and Corollary 1.4]). He also proved Conjecture 2.25 completely in [Ma6]. For the precise statement, see [Ma6, Theorem 1.3]. In [FMa], he and I obtained a generalization of Enoki's injectivity theorem, which can be seen as a generalization of Nedel's vanishing theorem (see [FMa, Theorem A]). For some further developments, see [Fn16] and [Ma7].

3 On local freeness and semipositivity theorems

Let us start with Fujita's semipositivity theorem in [Fujita].

Theorem 3.1 ([Fujita, (0.6) Main Theorem]). *Let $f : M \to C$ be a surjective morphism from a compact Kähler manifold onto a smooth projective curve C with connected fibers. Then $f_*\omega_{M/C}$ is nef.*

Before we go further, let us recall the definition of nef locally free sheaves.

Definition 3.2 (Nef locally free sheaves). Let $\mathcal{E}$ be a locally free sheaf of finite rank on a complete algebraic variety V. Then $\mathcal{E}$ is called nef if $\mathcal{E} = 0$ or $\mathcal{O}_{\mathbb{P}_V(\mathcal{E})}(1)$ is nef on $\mathbb{P}_V(\mathcal{E})$, that is, $\mathcal{O}_{\mathbb{P}_V(\mathcal{E})}(1) \cdot C \geq 0$ for every curve C on $\mathbb{P}_V(\mathcal{E})$. A nef locally free sheaf $\mathcal{E}$ was originally called a (numerically) semipositive locally free sheaf in the literature.

Remark 3.3. Assume that X is a smooth projective variety for simplicity. Let $\mathcal{L}$ be a line bundle on X. Then $\mathcal{L}$ is nef in the sense of Definition 3.2 if and only if $\mathcal{L}$ is nef in the usual sense. If $\mathcal{L}$ is semipositive in the sense of Remark 2.19, then $\mathcal{L}$ is nef. However, a nef line bundle $\mathcal{L}$ is not necessarily semipositive in the sense of Remark 2.19.

Remark 3.4. Note that f is not necessarily smooth in Theorem 3.1. If f is smooth in Theorem 3.1, then the nefness of $f_*\omega_{M/C}$ follows directly from Griffiths's calculations of connections and curvatures in [Gri].

Although Fujita's theorem was inspired by Griffiths's paper [Gri] (see Remark 3.4), Fujita's original proof of Theorem 3.1 in [Fujita] is not so Hodge theoretic. In the introduction of [Fujita], Fujita wrote:

> The method looks rather elementary and purely computational, but it depends deeply (often implicitly) on the theory of variation of Hodge structures.

Professor Steven Zucker informed me that he read Fujita's article [Fujita] when it appeared in 1978 and reproved Fujita's theorem from rather basic Hodge theory that appeals to Steenbrink's work [St]. It is clear that he had already been very familiar with Schmid's result (see [Sc]) on asymptotic behavior of Hodge metrics (see [Zuc1] and [Zuc2]). I think that he could write [Zuc3] without any difficulties. It seems that he is the first one who directly applies Hodge theory to obtain semipositivity results like Theorem 3.1, that is, semipositivity results for nonsmooth morphisms.

Independently, Kawamata obtained the following semipositivity theorem in [Kaw1] by using Schmid's paper [Sc]. His result is:

Theorem 3.5 ([Kaw1, Theorem 5]). *Let $f : X \to Y$ be a surjective morphism between smooth projective varieties with connected fibers which satisfies the following conditions:*

(i) There is a Zariski open dense subset Y_0 of Y such that $D = Y \setminus Y_0$ is a simple normal crossing divisor on Y.

(ii) Put $X_0 = f^{-1}(Y_0)$ and $f_0 = f|_{X_0}$. Then f_0 is smooth.

(iii) The local monodromies of $R^n f_{0*}\mathbb{C}_{X_0}$ around D are unipotent, where $n = \dim X - \dim Y$.

Then $f_*\omega_{X/Y}$ is a locally free sheaf and nef.

However, the proof of Theorem 3.5 in [Kaw1] seems to be insufficient when $\dim Y \geq 2$. We do not repeat the comments on the troubles in [Kaw1] here. For the details, see the comments in [FFS, 4.6. Remarks]. Fortunately, we have some generalizations of Theorem 3.5 in [Fn2], [FF], and [FFS] (see, for example, Theorem 3.6 below). The proofs in [FF] and [FFS] are independent of Kawamata's arguments in [Kaw1]. Note that our arguments in [FF] and [FFS] need some results on Hodge theory obtained after the publication of Kawamata's paper [Kaw1] (see, for example, [CK], and [CKS]). Kawamata could and did use only [Del], [Gri], and [Sc] on Hodge theory when he wrote [Kaw1]. Although I sometimes called Theorem 3.5 the Fujita–Kawamata semipositivity theorem (see, for example, [FF]), it is probably not accurate. It is more appropriate to call it the Fujita–Zucker–Kawamata semipositivity theorem. I apologize for suppressing Zucker's contribution [Zuc3].

Theorem 3.5 follows from the theory of polarizable variation of *pure* Hodge structure. It is natural to consider *mixed* generalizations of Theorem 3.5. We have already seen that mixed Hodge structures on cohomology with compact support are very useful (see Section 2). So, we consider their variations and prove some powerful generalizations of Theorem 3.5, which depend on the theory of gradedly polarizable admissible variation of *mixed* Hodge structure (see, for example, [SZ], and [Kas]). We have:

Theorem 3.6 (Semipositivity theorem). *Let (X, D) be a simple normal crossing pair such that D is reduced and let $f : X \to Y$ be a projective surjective morphism onto a smooth complete algebraic variety Y. Assume that every stratum of (X, D) is dominant onto Y. Let Σ be a simple normal crossing divisor on Y such that every stratum of (X, D) is smooth over $Y^* = Y \setminus \Sigma$. Then $R^p f_*\omega_{X/Y}(D)$ is locally free for every p. We put $X^* = f^{-1}(Y^*)$, $D^* = D|_{X^*}$, and $d = \dim X - \dim Y$. We further assume that all the local monodromies on $R^{d-i}(f|_{X^*\setminus D^*})_!\mathbb{Q}_{X^*\setminus D^*}$ around Σ are unipotent. Then we obtain that $R^i f_*\omega_{X/Y}(D)$ is a nef locally free sheaf on Y.*

For the definitions and the notation used in Theorem 3.6, see 6.6 and 6.7 in Section 6. Theorem 3.6 was first obtained in [FF]. Then, we gave an alternative proof of Theorem 3.6 based on Saito's theory of mixed Hodge modules (see [Sa1], [Sa2], and [Sa3]) in [FFS]. As an application of Theorem 3.6, we establish the projectivity of various moduli spaces (for the details, see [Fn6], [Fn8], [KvP], etc.). For a new approach, see [Fn15]. In the introduction of [Fujita], Fujita wrote:

> Perhaps our result is closely related with the problem about the (quasi-) projectivity of moduli spaces. Of course, however, the relation will not be simple.

Now we know that generalizations of Fujita's semipositivity theorem (see Theorem 3.1 and Theorem 3.6) with Viehweg's covering arguments (see [Vie1] and [Vie2]) are useful for the projectivity of coarse moduli spaces of stable (log-)varieties (see, for example, [Ko2], [Fn8], and [KvP]).

Anyway, by Theorem 3.5, we have:

Theorem 3.7 (Fujita, Zucker, Kawamata, $\cdots$). *Let $f : X \to Y$ be a surjective morphism between smooth projective varieties with connected fibers. Then there exists a generically finite morphism $\tau : Y' \to Y$ from a smooth projective variety Y' with the following property. Let X' be any resolution of the main component of $X \times_Y Y'$. Then $f'_*\omega_{X'/Y'}$ is a nef locally free sheaf, where f' is the composite $X' \to X \times_Y Y' \to Y'$.*

Theorem 3.7 has already played crucial roles in the study of higher-dimensional algebraic varieties. For some geometric applications, we have to treat $f_*\omega_{X/Y}^{\otimes m}$ or $f'_*\omega_{X'/Y'}^{\otimes m}$ with $m \geq 2$ (see Section 4). Thus we have:

Conjecture 3.8 (Semipositivity of direct images of relative pluricanonical bundles). *Let $f : X \to Y$ be a surjective morphism between smooth projective varieties* *with connected fibers. Then there exists a generically finite morphism $\tau : Y' \to Y$ from a smooth projective variety Y' with the following property. Let X' be any resolution of the main component of $X \times_Y Y'$ sitting in the following commutative diagram:*

$$\begin{array}{ccc} X' & \longrightarrow & X \\ {\scriptstyle f'}\downarrow & & \downarrow{\scriptstyle f} \\ Y' & \xrightarrow[\tau]{} & Y. \end{array}$$

Then $f'_*\omega_{X'/Y'}^{\otimes m}$ is a nef locally free sheaf for every positive integer m.

Note that the local freeness of $f'_*\omega_{X'/Y'}^{\otimes m}$ for $m \geq 2$ in Conjecture 3.8 is highly nontrivial even when f' is a smooth projective morphism. The following theorem by Siu (see [Siu2]) is nontrivial for $m \geq 2$ and can be proved only by using L^2 methods. For a simpler proof, see [Pă1].

Theorem 3.9 (Siu). *Let $f : X \to Y$ be a smooth projective morphism between smooth quasiprojective varieties with connected fibers. Then $f_*\omega_{X/Y}^{\otimes m}$ is locally free for every nonnegative integer m.*

Theorem 3.9 is a clever application of the Ohsawa–Takegoshi L^2 extension theorem. We have no Hodge theoretic proofs of Theorem 3.9. Therefore, we ask:

Problem 3.10. Find a Hodge theoretic proof or an algebraic proof of Theorem 3.9.

We note:

Remark 3.11. If Y is projective in Theorem 3.9, then $f_*\omega_{X/Y}^{\otimes m}$ is nef for every positive integer m by Theorem 3.17 below. Therefore, Conjecture 3.8 holds true when $f : X \to Y$ is smooth.

We recommend the reader to see [FF] and [FFS] for the Hodge theoretic aspect of semipositivity theorems discussed in this section. Note that the style of [FF] is the same as my other papers, whereas, [FFS] is written in the language of Saito's theory of mixed Hodge modules.

3.1 New semipositivity theorems using MMP

In this subsection, we discuss new semipositivity theorems obtained through the use of the minimal model program, following [Fn12] and [Fn13].

Let us start with the definition of (good) minimal models. We recommend the reader to see 6.2 and 6.5 in Section 6 if he is not familiar with the minimal model program.

Definition 3.12 (Good minimal models). Let $f : X \to Y$ be a projective morphism between normal quasiprojective varieties. Let Δ be an effective $\mathbb{Q}$-divisor on X such that (X, Δ) is kawamata log terminal. A pair (X', Δ') sitting in a diagram

$$\begin{array}{ccc} X & \overset{\phi}{\dashrightarrow} & X' \\ & {\scriptstyle f}\searrow \quad \swarrow{\scriptstyle f'} & \\ & Y & \end{array}$$

is called a minimal model of (X, Δ) over Y if

(i) X' is $\mathbb{Q}$-factorial,

(ii) f' is projective,

(iii) ϕ is birational and ϕ^{-1} has no exceptional divisors,

(iv) $\phi_* \Delta = \Delta'$,

(v) $K_{X'} + \Delta'$ is f'-nef, and

(vi) $a(E, X, \Delta) < a(E, X', \Delta')$ for every ϕ-exceptional divisor E contained in X.

Furthermore, if $K_{X'} + \Delta'$ is f'-semiample, then (X', Δ') is called a *good* minimal model of (X, Δ) over Y. When Y is a point, we usually omit "over Y" in the above definitions; we sometimes simply say that (X', Δ') is a relative (good) minimal model of (X, Δ).

We also need the notion of weakly semistable morphisms introduced by Abramovich–Karu (see [AK]).

Definition 3.13 (Weakly semistable morphisms). Let $f : X \to Y$ be a projective surjective morphism between quasiprojective varieties. Then $f : X \to Y$ is called weakly semistable if

(i) the varieties X and Y admit toroidal structures $(U_X \subset X)$ and $(U_Y \subset Y)$ with $U_X = f^{-1}(U_Y)$,

(ii) with this structure, the morphism f is toroidal,

(iii) the morphism f is equidimensional,

(iv) all the fibers of the morphism f are reduced, and

(v) Y is smooth.

It then follows that X has only rational Gorenstein singularities (see [AK, Lemma 6.1]). Both $(U_X \subset X)$ and $(U_Y \subset Y)$ are toroidal embeddings without self-intersection in the sense of [KKMS, Chapter II, §1]. For the details, see [AK].

We propose the following conjecture.

Conjecture 3.14. *Let $f : X \to Y$ be a weakly semistable morphism with connected fibers. Then $f_* \omega_{X/Y}^{\otimes m}$ is locally free for every $m \geq 1$.*

By the argument in [Fn12, Section 4] (see also [Fn13]), we have:

Theorem 3.15 (Local freeness). *Let $f : X \to Y$ be a weakly semistable morphism with connected fibers. Assume that the geometric generic fiber $X_{\overline{\eta}}$ of $f : X \to Y$ has a good minimal model. Then $f_*\omega_{X/Y}^{\otimes m}$ is locally free for every $m \geq 1$.*

Sketch of Proof of Theorem 3.15. Let us consider the following commutative diagram:

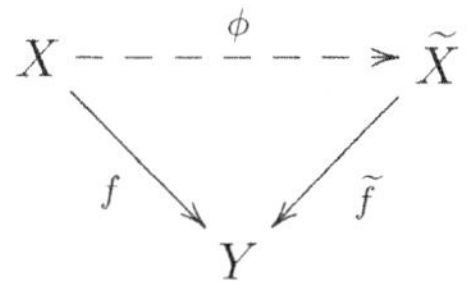

where $\widetilde{f} : \widetilde{X} \to Y$ is a relative good minimal model of $f : X \to Y$. We can always construct a relative good minimal model by the assumption that the geometric generic fiber of f has a good minimal model. Then we have

$$f_*\omega_{X/Y}^{\otimes m} \simeq \widetilde{f}_*\mathcal{O}_{\widetilde{X}}(mK_{\widetilde{X}/Y}) \qquad (\heartsuit)$$

for every positive integer m. We note that X has only rational Gorenstein singularities.

The following lemma due to Nakayama is a variant of Kollár's torsion-freeness: Theorem 2.5 (i). It is a key ingredient of the proof of Theorem 3.15.

Lemma 3.16 ([Nak, Corollary 3]). *Let $g : V \to C$ be a projective surjective morphism from a normal quasiprojective variety V to a smooth quasiprojective curve C. Assume that V has only canonical singularities and that K_V is g-semiample. Then $R^ig_*\mathcal{O}_V(mK_V)$ is locally free for every i and every positive integer m.*

By the above isomorphism ($\heartsuit$), it is sufficient to prove the local freeness of $\widetilde{f}_*\mathcal{O}_{\widetilde{X}}(mK_{\widetilde{X}/Y})$. Let P be an arbitrary closed point of Y. Since $f : X \to Y$ is weakly semistable, we can prove that the diagram

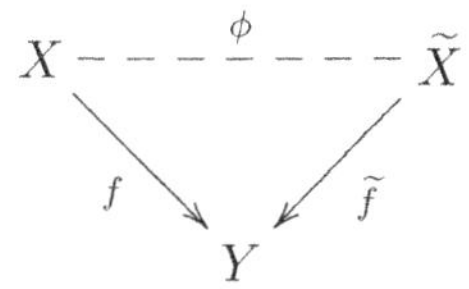

behaves well under the base change by $C \hookrightarrow Y$, where C is a general smooth curve on Y passing through P. Roughly speaking, by this observation, we can reduce the problem to the case when Y is a smooth projective curve. Note that f and $\widetilde{f}$ are both flat. By Lemma 3.16, we see that $\dim H^0(\widetilde{X}_y, \mathcal{O}_{\widetilde{X}}(mK_{\widetilde{X}/Y})|_{\widetilde{X}_y})$ is independent of $y \in Y$. Therefore, $\widetilde{f}_*\mathcal{O}_{\widetilde{X}}(mK_{\widetilde{X}/Y})$ is locally free by the flat base change theorem. Thus, we obtain that $f_*\omega_{X/Y}^{\otimes m}$ is locally free. For the details, see [Fn12, Section 4] and [Fn13]. □

By the argument in [Fn12, Section 5], we can prove:

Theorem 3.17 (Semipositivity). *Let $f : X \to Y$ be a weakly semistable morphism between projective varieties with connected fibers. Let $m \geq 1$ be fixed. Assume that $f_*\omega_{X/Y}^{\otimes m}$ is locally free. Then $f_*\omega_{X/Y}^{\otimes m}$ is nef.*

Idea of Proof of Theorem 3.17. The following theorem by Popa–Schnell is a clever and interesting application of the Kollár vanishing theorem: Theorem 2.5 (ii).

Theorem 3.18 ([PoSc, Theorem 1.4]). *Let $f : V \to W$ be a surjective morphism from a smooth projective variety V onto a projective variety W with $\dim W = n$. Let $\mathcal{L}$ be an ample line bundle on W such that $|\mathcal{L}|$ has no base points. Let k be a positive integer. Then*

$$f_*\omega_V^{\otimes k} \otimes \mathcal{L}^{\otimes l}$$

is generated by global sections for every $l \geq k(n+1)$.

By Viehweg's fiber product trick and the local freeness of $f_*\omega_{X/Y}^{\otimes m}$, we can prove that there exists an ample line bundle $\mathcal{A}$ on Y such that

$$\left(\bigotimes^{s} f_*\omega_{X/Y}^{\otimes m}\right) \otimes \mathcal{A}$$

is generated by global sections for every positive integer s by Theorem 3.18. Here, we used the fact that weakly semistable morphisms behave well by taking fiber products. This implies that $f_*\omega_{X/Y}^{\otimes m}$ is nef. For the details, see [Fn12, Section 5]. □

As we saw above, a key ingredient of Theorem 3.15 (resp. Theorem 3.17) is Kollár's torsion-freeness (resp. Kollár's vanishing theorem). Of course, the existence of relative good minimal models plays a crucial role in the proof of Theorem 3.15.

Remark 3.19. In the proof of Theorem 3.15, we need the finite generation of relative canonical ring

$$R(X/Y) = \bigoplus_{m}^{\infty} f_*\mathcal{O}_X(mK_X)$$

from [BCHM] to construct a relative good minimal model of $f : X \to Y$. Note that the finite generation of $R(X/Y)$ is more or less Hodge theoretic when $X_{\overline{\eta}}$ is not of general type. This is because the reduction argument due to Fujino–Mori (see Theorem 5.4 below and [FMo]) uses Theorem 3.5.

Remark 3.20. Let V be a smooth projective variety. It is well-known that V has a good minimal model when $\dim V - \kappa(V) \leq 3$.

By combining Theorem 3.15 with Theorem 3.17, we obtain:

Theorem 3.21. *Let $f : X \to Y$ be a surjective morphism between smooth projective varieties with connected fibers. Assume that*

$$f : X \xrightarrow{\delta} X^{\dagger} \xrightarrow{f^{\dagger}} Y$$

such that $f^{\dagger} : X^{\dagger} \to Y$ is weakly semistable and that δ is a resolution of singularities. We further assume that the geometric generic fiber $X_{\overline{\eta}}$ of f has a good minimal model. Then $f_*\omega_{X/Y}^{\otimes m}$ is a nef locally free sheaf for every positive integer m.

By the weak semistable reduction theorem due to Abramovich–Karu (see [AK]) and Theorem 3.21, we have:

Theorem 3.22. *Let $f : X \to Y$ be a surjective morphism between smooth projective varieties with connected fibers. Assume that the geometric generic fiber $X_{\overline{\eta}}$ of $f : X \to Y$ has a good minimal model. Then there exists a generically finite morphism $\tau : Y' \to Y$ from a smooth projective variety Y' with the following property. Let X' be any resolution of the main component of $X \times_Y Y'$ sitting in the following commutative diagram:*

$$\begin{array}{ccc} X' & \longrightarrow & X \\ {\scriptstyle f'}\downarrow & & \downarrow{\scriptstyle f} \\ Y' & \xrightarrow[\tau]{} & Y. \end{array}$$

Then $f'_*\omega_{X'/Y'}^{\otimes m}$ is a nef locally free sheaf for every positive integer m.

This means that Conjecture 3.8 holds true under the assumption that the geometric generic fiber of f has a good minimal model. More precisely, Conjecture 3.8 follows from Conjecture 3.14 by the weak semistable reduction theorem due to Abramovich–Karu (see [AK]) and Theorem 3.17. Moreover, Conjecture 3.14 holds under the assumption that the geometric generic fiber has a good minimal model (see Theorem 3.15).

We close this section with Takayama's result. Using complex analytic methods, Takayama in [Taka] strengthened Theorem 3.21 as follows.

Theorem 3.23 (Takayama). *In Theorem 3.21, for every positive integer m, the m-th Narasimhan–Simha Hermitian metric g_m on the locally free sheaf $\mathcal{E}_m = f_*\omega_{X/Y}^{\otimes m}$ has Griffiths semipositive curvature, the induced singular Hermitian metric $h = e^{-\varphi}$ on $\mathcal{O}_{\mathbb{P}_X(\mathcal{E}_m)}(1)$ of $\mathbb{P}_X(\mathcal{E}_m)$ has semipositive curvature, and the Lelong number of the local weight φ is zero everywhere on $\mathbb{P}_X(\mathcal{E}_m)$. In particular, $\mathcal{O}_{\mathbb{P}_X(\mathcal{E}_m)}(1)$ is nef.*

For the definition of the Narasimhan–Simha Hermitian metric and the details of Theorem 3.23, see the original paper [Taka]. We point out that Theorem 3.23 is based on the arguments in [Fn12] and [Fn13].

I am not so familiar with the analytic aspect of semipositivity theorems. For the details, see [Ber], [BP1], [BP2], [Mou], [MT1], [MT2], [MT3], [PăT], [Taka], etc.

4 Canonical divisors versus pluricanonical divisors

In this section, let us see that mK_X with $m \geq 2$ sometimes behaves much better than K_X. We will discuss two different topics: Kollár's result on plurigenera in étale covers of smooth projective varieties of general type and Viehweg's ampleness theorem on direct images of relative pluricanonical bundles of semistable families of projective varieties. I was very much impressed by these results.

4.1 Plurigenera in étale covers

Let us recall Kollár's famous result on plurigenera in étale covers of smooth projective varieties of general type (see [Ko3]). For the details and some related topics, see also [Ko5, 2. Vanishing Theorems] and [Ko4, Chapter 15].

Theorem 4.1 (Kollár). *Let X be a smooth projective variety of general type. Let $f : Y \to X$ be an étale morphism from a smooth projective variety Y. Then we have*

$$h^0(Y, \mathcal{O}_Y(mK_Y)) = \deg f \cdot h^0(X, \mathcal{O}_X(mK_X))$$

for every positive integer $m \geq 2$.

Here, we will present Lazarsfeld's proof of Theorem 4.1 following [Laz, Theorem 11.2.23]. It is actually an easy application of the theory of asymptotic multiplier ideal sheaves. We will give an alternative proof of Theorem 4.1 after we discuss canonical models of smooth projective varieties of general type in Theorem 4.5.

Proof. Let D be a big Cartier divisor on X. Then $\mathcal{J}(X, \|D\|)$ denotes the asymptotic multiplier ideal sheaf associated to the complete linear systems $|mD|$ for all $m \gg 0$. For the details of $\mathcal{J}(X, \|D\|)$, see [Laz, Chapter 11]. By the Nadel vanishing theorem (see [Laz, Theorem 11.2.12 (ii)]),

$$H^i(X, \mathcal{O}_X(mK_X) \otimes \mathcal{J}(X, \|(m-1)K_X\|)) = 0$$

for every $i > 0$ and every $m \geq 2$. Therefore, we have

$$\begin{aligned}&h^0(X, \mathcal{O}_X(mK_X) \otimes \mathcal{J}(X, \|(m-1)K_X\|)) \\ &= \chi(X, \mathcal{O}_X(mK_X) \otimes \mathcal{J}(X, \|(m-1)K_X\|))\end{aligned}$$

for every $m \geq 2$. Since $\mathcal{J}(X, \|mK_X\|) \subset \mathcal{J}(X, \|(m-1)K_X\|)$ (see [Laz, Theorem 11.1.8 (ii)]), we have

$$\begin{aligned}&H^0(X, \mathcal{O}_X(mK_X) \otimes \mathcal{J}(X, \|mK_X\|)) \\ &= H^0(X, \mathcal{O}_X(mK_X) \otimes \mathcal{J}(X, \|(m-1)K_X\|)) \\ &= H^0(X, \mathcal{O}_X(mK_X))\end{aligned}$$

for every $m \geq 1$ by [Laz, Proposition 11.2.10]. Thus, we obtain

$$h^0(X, \mathcal{O}_X(mK_X)) = \chi(X, \mathcal{O}_X(mK_X) \otimes \mathcal{J}(X, \|(m-1)K_X\|))$$

for every $m \geq 2$. Similarly, we have

$$h^0(Y, \mathcal{O}_Y(mK_Y)) = \chi(Y, \mathcal{O}_Y(mK_Y) \otimes \mathcal{J}(Y, \|(m-1)K_Y\|))$$

for $m \geq 2$. Since f is étale, $K_Y = f^*K_X$ and

$$\mathcal{J}(Y, \|(m-1)K_Y\|) = f^*\mathcal{J}(X, \|(m-1)K_X\|)$$

by [Laz, Theorem 11.2.16]. Thus we have

$$\begin{aligned}&\chi(Y, \mathcal{O}_Y(mK_Y) \otimes \mathcal{J}(Y, \|(m-1)K_Y\|))\\ &= \chi(Y, f^*(\mathcal{O}_X(mK_X) \otimes \mathcal{J}(X, \|(m-1)K_X\|)))\\ &= \deg f \cdot \chi(X, \mathcal{O}_X(mK_X) \otimes \mathcal{J}(X, \|(m-1)K_X\|))\end{aligned}$$

for $m \geq 2$. Therefore, we obtain the desired equality $h^0(Y, \mathcal{O}_Y(mK_Y)) = \deg f \cdot h^0(X, \mathcal{O}_X(mK_X))$ for every $m \geq 2$. □

The proof of Theorem 4.1 says that mK_X with $m \geq 2$ should be seen as $K_X + (m-1)K_X$. Since $m \geq 2$, $(m-1)K_X$ is big. Therefore, we can apply the Nadel vanishing theorem to

$$\begin{aligned}&\mathcal{O}_X(mK_X) \otimes \mathcal{J}(X, \|(m-1)K_X\|)\\ &= \mathcal{O}_X(K_X + (m-1)K_X) \otimes \mathcal{J}(X, \|(m-1)K_X\|).\end{aligned}$$

Obviously, the equality in Theorem 4.1 does not hold for $m = 1$.

Example 4.2. Let C be a smooth projective curve with the genus $g(C) \geq 2$. Let $f : \widetilde{C} \to C$ be an étale cover with $\deg f = n \geq 2$. Then we have

$$2g(\widetilde{C}) - 2 = n(2g(C) - 2)$$

by Hurwitz's formula. This implies that $g(\widetilde{C}) = n(g(C) - 1) + 1$. Thus we have

$$h^0(\widetilde{C}, \mathcal{O}_{\widetilde{C}}(K_{\widetilde{C}})) \neq n \cdot h^0(C, \mathcal{O}_C(K_C)).$$

The following example also shows that mK_X with $m \geq 2$ sometimes contains much more information than K_X.

Example 4.3 (Godeaux surface)**.** We put

$$Y = (Z_0^5 + Z_1^5 + Z_2^5 + Z_3^5 = 0) \subset \mathbb{P}^3.$$

Then Y is a smooth projective surface such that

$$\mathcal{O}_Y(K_Y) = \mathcal{O}_{\mathbb{P}^3}(-4+5)|_Y = \mathcal{O}_Y(1)$$

is very ample. Therefore, Y is of general type,

$$h^0(Y, \mathcal{O}_Y(K_Y)) = h^0(\mathbb{P}^3, \mathcal{O}_{\mathbb{P}^3}(1)) = 4,$$

and $q(Y) = h^1(Y, \mathcal{O}_Y) = 0$. We put $G = \mathbb{Z}/5\mathbb{Z}$. Then G acts freely on Y by

$$[Z_0 : Z_1 : Z_2 : Z_3] \mapsto [Z_0 : \zeta Z_1 : \zeta^2 Z_2 : \zeta^3 Z_3]$$

where $\zeta = \exp\left(\frac{2\pi\sqrt{-1}}{5}\right)$. We put $X = Y/G$. Then X is a smooth projective surface with ample canonical divisor. Let $f : Y \to X$ be the natural map. Then f is a finite étale morphism. We can directly check that $h^0(X, \mathcal{O}_X(K_X)) = 0$ by $H^0(X, \mathcal{O}_X(K_X)) = H^0(Y, \mathcal{O}_Y(K_Y))^G$. Note that $q(X) = h^1(X, \mathcal{O}_X) = 0$, $K_Y^2 = 5$, and $K_X^2 = 1$. We also note that X is known as a Godeaux surface. It is well-known that the linear system $|mK_X|$ gives an embedding into a projective space for every $m \geq 5$. Note that

$$4 = h^0(Y, \mathcal{O}_Y(K_Y)) \neq \deg f \cdot h^0(X, \mathcal{O}_X(K_X)) = 0.$$

4.4 (Canonical models)**.** Let us discuss canonical models of finite étale covers of smooth projective varieties of general type. The existence of canonical models was unknown when [Ko3] was written. Let $\pi : V \to W$ be a projective surjective morphism from a smooth quasiprojective variety V onto a quasiprojective variety W. Assume that K_V is π-big. Then, by [BCHM], the (relative) canonical ring

$$R(V/W) = \bigoplus_{m=0}^{\infty} \pi_* \mathcal{O}_V(mK_V)$$

is a finitely generated $\mathcal{O}_W$-algebra. We put

$$V_c = \mathrm{Proj}_W R(V/W)$$

and call it the canonical model of V over W or the relative canonical model of $\pi : V \to W$. It is well-known that V_c is birationally equivalent to V over W, V_c has only canonical singularities, and K_{V_c} is ample over W.

A finite étale morphism between smooth projective varieties of general type induces a natural finite étale morphism between their canonical models:

Theorem 4.5. *Let $f : Y \to X$ be a finite étale morphism between smooth projective varieties of general type. Let X_c be the canonical model of X and let Y_c be the canonical model of Y. Then there exists a finite étale morphism $f_c : Y_c \to X_c$ such that*

$$\begin{array}{ccc} Y & \dashrightarrow & Y_c \\ {\scriptstyle f}\downarrow & & \downarrow{\scriptstyle f_c} \\ X & \dashrightarrow & X_c \end{array}$$

is commutative.

The following proof was suggested by Yoshinori Gongyo.

Proof. By taking an elimination of indeterminacy of $X \dashrightarrow X_c$ and the base change of Y, we may assume that $g : X \to X_c$ is a morphism. Let $Y' \to X_c$ be the relative canonical model of $g \circ f : Y \to X_c$ (see [BCHM] and 4.4). Then, by using the negativity lemma (see, for example, [KoM, Lemma 3.39]), we see that $K_{Y'} = f'^* K_{X_c}$ and that $f' : Y' \to X_c$ is finite since $K_{Y'}$ is f'-ample. Therefore, $K_{Y'}$ is ample since K_{X_c} is ample. This implies that $Y' = Y_c$, that is, Y' is the canonical model of Y. Note that Y is smooth and is finite over X. Therefore, Y is the normalization of the main component of $X \times_{X_c} Y_c$. Thus we obtain the following commutative diagram:

$$\begin{array}{ccc} Y & \longrightarrow & Y_c \\ {\scriptstyle f}\downarrow & & \downarrow{\scriptstyle f_c} \\ X & \underset{g}{\longrightarrow} & X_c. \end{array}$$

By Lemma 4.8 below, we obtain that f_c is a finite étale morphism. □

By the proof of Theorem 4.5, we have:

Theorem 4.6. *Let $f : Y \to X$ be a finite étale morphism between smooth projective varieties. Let X_m be a minimal model of X. Then we can construct a commutative diagram:*

$$\begin{array}{ccc} Y & \overset{\varphi}{\dashrightarrow} & \widetilde{Y} \\ {\scriptstyle f}\downarrow & & \downarrow{\scriptstyle \widetilde{f}} \\ X & \dashrightarrow & X_m \end{array}$$

such that $\widetilde{f} : \widetilde{Y} \to X_m$ is a finite étale morphism and that φ is birational.

For the definition of minimal models, recall Definition 3.12.

Proof. As in the proof of Theorem 4.5, we may assume that $g : X \to X_m$ is a morphism. Let $\widetilde{f} : \widetilde{Y} \to X_m$ be the relative canonical model of $g \circ f : Y \to X_m$. Then, by the proof of Theorem 4.5, $\varphi : Y \dashrightarrow \widetilde{Y}$ and $\widetilde{f} : \widetilde{Y} \to X_m$ satisfy the desired properties. □

Related to Theorem 4.5 and Theorem 4.6, we have:

Problem 4.7. Find a projective variety X such that X has only $\mathbb{Q}$-factorial terminal (or canonical) singularities with nef (or ample) canonical divisor and a finite étale morphism $f : Y \to X$ such that Y is not $\mathbb{Q}$-factorial.

The following lemma is a special case of [NZ, Lemma 3.9].

Lemma 4.8. *We consider the following commutative diagram of quasiprojective varieties:*

$$\begin{array}{ccc} Y & \overset{q}{\longrightarrow} & W \\ {\scriptstyle f}\downarrow & & \downarrow{\scriptstyle h} \\ X & \underset{p}{\longrightarrow} & V \end{array}$$

such that

(i) X and Y are smooth,

(ii) p and q are projective birational morphisms,

(iii) V and W are normal and have only rational singularities,

(iv) f is a finite étale morphism, and

(v) h is finite.

Then h is an étale morphism.

We give a proof for the reader's convenience. It is interesting for me that the proof of Lemma 4.8 below uses the E_1-degeneration of Hodge to de Rham type spectral sequences for projective simple normal crossing varieties.

Proof. We take an arbitrary point $P \in V$. By taking a birational modification of X and the base change of Y, we may assume that $E = \mathrm{Supp}(p^{-1}(P))$ is a simple normal crossing divisor on X. Since f is étale, $f^{-1}(E)$ is a simple normal crossing divisor on Y. Note that $f^{-1}(E)$ is the disjoint union of $E_Q = \mathrm{Supp}(q^{-1}(Q))$ for points $Q \in h^{-1}(P)$. Since V has only rational singularities, $R^i p_* \mathcal{O}_X = 0$ for every $i > 0$.

Claim. *The natural map*

$$\pi : H^i(E, \mathbb{C}) \to H^i(E, \mathcal{O}_E)$$

induced by the inclusion $\mathbb{C}_E \hookrightarrow \mathcal{O}_E$ is surjective for every i.

Proof of Claim. By using the Mayer–Vietoris simplicial resolution of a projective simple normal crossing variety E, we can construct a cohomological mixed Hodge complex $(K_{\mathbb{Z}}, (K_{\mathbb{Q}}, W_{\mathbb{Q}}), (K_{\mathbb{C}}, W, F))$ which induces a natural mixed Hodge structure on $H^\bullet(E, \mathbb{Z})$. By the theory of mixed Hodge structures, we see that

$$E_1^{p,q} = \mathbb{H}^{p+q}(E, \mathrm{Gr}_F^p K_{\mathbb{C}}) \Rightarrow H^{p+q}(E, \mathbb{C})$$

degenerates at E_1. We can directly check that $\mathrm{Gr}_F^0 K_{\mathbb{C}}$ is quasi-isomorphic to $\mathcal{O}_E$ by using the Mayer–Vietoris simplicial resolution of E. Thus, we obtain that

$$\pi : H^i(E, \mathbb{C}) \to H^i(E, \mathcal{O}_E)$$

is surjective for every i. □

By the following commutative diagram:

$$\begin{array}{ccc} (R^i p_* \mathbb{C}_X)_P & \xrightarrow{\simeq} & H^i(E, \mathbb{C}) \\ \downarrow & & \downarrow \pi \\ (R^i p_* \mathcal{O}_X)_P & \longrightarrow & H^i(E, \mathcal{O}_E), \end{array}$$

we obtain that $H^i(E, \mathcal{O}_E) = 0$ for every $i > 0$, as $R^i p_* \mathcal{O}_X = 0$ for every $i > 0$. Thus, we obtain $\chi(E, \mathcal{O}_E) = 1$. By the same argument, we have $\chi(E_Q, \mathcal{O}_{E_Q}) = 1$. On the other hand,

$$\chi(f^{-1}(E), \mathcal{O}_{f^{-1}(E)}) = \deg f \cdot \chi(E, \mathcal{O}_E) = \deg f$$

since f is étale. Therefore, we have

$$\sharp h^{-1}(P) = \sum_{Q \in h^{-1}(P)} \chi(E_Q, \mathcal{O}_{E_Q}) = \chi(f^{-1}(E), \mathcal{O}_{f^{-1}(E)}) = \deg f.$$

This implies that $f : E_Q \to E$ is an isomorphism for every $Q \in h^{-1}(P)$. Thus, by the theorem on formal functions, we have

$$\widehat{\mathcal{O}}_{V,P} = (p_* \mathcal{O}_X)^{\wedge}_P \simeq (q_* \mathcal{O}_Y)^{\wedge}_Q = \widehat{\mathcal{O}}_{W,Q}$$

for every $Q \in h^{-1}(P)$. Thus, h is an étale morphism. □

We give an alternative proof of Theorem 4.1, one that is natural from the minimal model theoretic viewpoint. Unfortunately, it may be more complicated than Lazarsfeld's proof given before.

Proof of Theorem 4.1. By Theorem 4.5, we may replace $f : Y \to X$ with $f_c : Y_c \to X_c$. Note that

$$h^0(X_c, \mathcal{O}_{X_c}(mK_{X_c})) = h^0(X, \mathcal{O}_X(mK_X))$$

and

$$h^0(Y_c, \mathcal{O}_{Y_c}(mK_{Y_c})) = h^0(Y, \mathcal{O}_Y(mK_Y))$$

for every $m \geq 1$. By the Kawamata–Viehweg vanishing theorem for singular varieties (see, for example, [Fn11, Corollary 5.7.7]), we have

$$H^i(X_c, \mathcal{O}_{X_c}(mK_{X_c})) = H^i(Y_c, \mathcal{O}_{Y_c}(mK_{Y_c})) = 0$$

for every $i > 0$ and every $m \geq 2$. We also note that $f_c^* \mathcal{O}_{X_c}(mK_{X_c}) = \mathcal{O}_{Y_c}(mK_{Y_c})$ for every m since f_c is étale. Therefore, we obtain

$$\begin{aligned} h^0(Y_c, \mathcal{O}_{Y_c}(mK_{Y_c})) &= \chi(Y_c, \mathcal{O}_{Y_c}(mK_{Y_c})) \\ &= \deg f_c \cdot \chi(X_c, \mathcal{O}_{X_c}(mK_{X_c})) \\ &= \deg f_c \cdot h^0(X_c, \mathcal{O}_{X_c}(mK_{X_c})) \end{aligned}$$

for every $m \geq 2$. This implies the desired equality. □

Remark 4.9. In Lemma 4.8, the assumption that V and W have only rational singularities (see (iii) in Lemma 4.8) is indispensable.

Consider the following example. Let $C \subset \mathbb{P}^2$ be an elliptic curve and let $V \subset \mathbb{P}^3$ be a cone over $C \subset \mathbb{P}^2$. Let $p : X \to V$ be the blow-up at the vertex P of V and let E be the p-exceptional divisor on X. Note that there is a natural $\mathbb{P}^1$-bundle structure $\pi : X \to C$ and E is a section of π. We take a nontrivial finite étale cover $D \to C$. We put $Y = X \times_C D$ and $F = E \times_C D$. Let H be an ample Cartier divisor on V. We consider $q = \Phi_{|mf^*p^*H|} : Y \to W$ for a sufficiently large positive integer m. Note that q contracts F to an isolated normal singular point Q of W. Then we have the following commutative diagram

$$\begin{array}{ccc} Y & \xrightarrow{q} & W \\ {\scriptstyle f}\downarrow & & \downarrow{\scriptstyle h} \\ X & \xrightarrow[p]{} & V \end{array}$$

such that f is étale, h is finite, but h is not étale. Note that $h^{-1}(P) = Q$ since $f^{-1}(E) = F$. It is also false that the singularities of V and W are rational.

4.2 Viehweg's ampleness theorem

We treat direct images of relative pluricanonical bundles. The following theorem is buried in Viehweg's papers (see [Vie1] and [Vie2]). The statement seems to be magical.

Theorem 4.10 (Viehweg). *Let $f : X \to Y$ be a surjective morphism from a smooth projective variety X onto a smooth projective curve Y with connected fibers. Then $f_*\omega_{X/Y}^{\otimes m}$ is nef for every positive integer m. In particular, we have $\deg\det f_*\omega_{X/Y}^{\otimes m} \geq 0$ for every positive integer m. Assume that f is semistable. If $\deg\det f_*\omega_{X/Y}^{\otimes k} > 0$, that is, $\det f_*\omega_{X/Y}^{\otimes k}$ is ample, for some positive integer k, then $f_*\omega_{X/Y}^{\otimes k'}$ is ample, where k' is any multiple of k with $k' \geq 2$.*

Proof. From Kawamata (see [Kaw2, Theorem 1]), we have that $f_*\omega_{X/Y}^{\otimes m}$ is nef for every $m \geq 1$. Or, by Viehweg's weak positivity: [Vie1, Theorem III] (see also [Fn14, Theorem 4.3 and Theorem 5.5] and Remark 4.14 below), $f_*\omega_{X/Y}^{\otimes m}$ is weakly positive for every $m \geq 1$. Since Y is a smooth projective curve, the weak positivity implies that $f_*\omega_{X/Y}^{\otimes m}$ is nef for every $m \geq 1$. By [Vie2, Theorem 3.5] (see also [Fn14, Theorem 5.11]), $\deg\det f_*\omega_{X/Y}^{\otimes k} > 0$ implies that $f_*\omega_{X/Y}^{\otimes k'}$ is big in the sense of Viehweg (see [Fn14, Definition 3.1]) when f is semistable. Then, by [Vie2, Lemma 3.6] (see also [Fn14, Lemma 3.7]), there is a generically isomorphic injection

$$\bigoplus_r \mathcal{A} \to S^\nu(f_*\omega_{X/Y}^{\otimes k'})$$

for some ample invertible sheaf $\mathcal{A}$ on Y and some positive integer ν, where $r = \operatorname{rank} S^\nu(f_*\omega_{X/Y}^{\otimes k'})$. This implies that $S^\nu(f_*\omega_{X/Y}^{\otimes k'})$ is ample by [Laz, Theorem 6.4.15]. Therefore, $f_*\omega_{X/Y}^{\otimes k'}$ is ample by [Har, Proposition 2.4]. □

Roughly speaking, Theorem 4.10 says if $f_*\omega_{X/Y}$ is a little bit positive then $f_*\omega_{X/Y}^{\otimes m}$ is very positive for $m \geq 2$. Viehweg's arguments in [Vie1] and [Vie2] (see also [Fn14]) use his clever covering trick and fiber product trick. They are geometric. It seems to be very important to find a more direct approach to Theorem 4.10. Thus, we have:

Problem 4.11. Find an analytic (and more direct) proof of Theorem 4.10.

The example by Catanese–Dettweiler (see [CD1], [CD2], and [CD3]) below says that the condition $k' \geq 2$ in Theorem 4.10 is indispensable.

Example 4.12 (Catanese–Dettweiler). There exist a smooth projective surface X of general type and a smooth projective curve Y such that $f : X \to Y$ is semistable and that $f_*\omega_{X/Y} = A \oplus Q$, where A is an ample vector bundle of rank 2 and Q is a unitary flat vector bundle of rank 4. Moreover, Q is not semiample. In this case, $\deg\det f_*\omega_{X/Y} > 0$. By Theorem 4.10, $f_*\omega_{X/Y}^{\otimes m}$ is ample for every $m \geq 2$. However, $f_*\omega_{X/Y}$ is not ample.

Note that the construction of Example 4.12 in [CD2] depends on the theory of variation of Hodge structure. For the details, see [CD1], [CD2], and [CD3].

Problem 4.13. Find similar examples to Example 4.12 that do not use the theory of variation of Hodge structure.

Although we do not discuss Viehweg's weak positivity in this paper, we give a remark on the weak positivity of $f_*\omega_{X/Y}^{\otimes m}$ for the interested reader:

Remark 4.14. Let $f : X \to Y$ be a surjective morphism between smooth projective varieties with connected fibers. Then $f_*\omega_{X/Y}^{\otimes m}$ is weakly positive for every positive integer m by Viehweg (see [Vie1]). Viehweg's original proof of his weak positivity uses Theorem 3.5. Given what we know now, we can prove the weak positivity of $f_*\omega_{X/Y}^{\otimes m}$ by Kollár's vanishing theorem: Theorem 2.5 (ii). Moreover, Theorem 3.18 drastically simplifies the proof of Viehweg's weak positivity of $f_*\omega_{X/Y}^{\otimes m}$. For the details, see [Fn14].

Anyway, we should consider not only K_X but also mK_X with $m \geq 2$ in order to understand complex projective varieties much better.

5 On finite generation of (log) canonical rings

In this section, we quickly discuss the finite generation of (log) canonical rings due to Birkar–Cascini–Hacon–McKernan (see [BCHM]) and some related topics (see [FMo] and [Fn10]). For simplicity, we only treat the absolute setting in this section. However, the relative setting is very important and is indispensable for some applications (cf. Remark 3.19).

Theorem 5.1. *Let X be a smooth projective variety and let Δ be an effective $\mathbb{Q}$-divisor on X such that* SuppΔ *is a simple normal crossing divisor and that the coefficients of Δ are less than one. Assume that $K_X + \Delta$ is big. Then the log canonical ring*

$$R(X, K_X + \Delta) = \bigoplus_{m=0}^{\infty} H^0(X, \mathcal{O}_X(\lfloor m(K_X + \Delta)\rfloor))$$

is a finitely generated $\mathbb{C}$-algebra.

Theorem 5.1 was first obtained in [BCHM]. For the proof of Theorem 5.1, see also [CL] and [Pă2]. We know that the proof of Theorem 5.1 was greatly influenced by Siu's extension argument (see [Siu1]) based on the Ohsawa–Takegoshi L^2 extension theorem (see [HM], [CL], and [Pă2]) although [HM] and [CL] only use the Kawamata–Viehweg vanishing theorem and do not use L^2 methods. I feel that Theorem 5.1 is not Hodge theoretic. It is natural to see that Theorem 5.1 is more closely related to L^2 methods than to Hodge theory.

By combining Theorem 5.1 with the result in [FMo], we have:

Theorem 5.2. *Let X be a smooth projective variety and let Δ be an effective $\mathbb{Q}$-divisor on X such that* $\mathrm{Supp}\Delta$ *is a simple normal crossing divisor and that the coefficients of Δ are less than one. Then the log canonical ring*

$$R(X, K_X + \Delta) = \bigoplus_{m=0}^{\infty} H^0(X, \mathcal{O}_X(\lfloor m(K_X + \Delta)\rfloor))$$

is a finitely generated $\mathbb{C}$-algebra.

Note that we do not assume that $K_X + \Delta$ is big in Theorem 5.2. As a corollary of Theorem 5.2, we get when $\Delta = 0$:

Corollary 5.3. *Let X be a smooth projective variety. Then the canonical ring*

$$R(X) = \bigoplus_{m=0}^{\infty} H^0(X, \mathcal{O}_X(mK_X))$$

is a finitely generated $\mathbb{C}$-algebra.

We note that the formulation of Theorem 5.1, which may look artificial, is indispensable for the proof of Corollary 5.3. We will see how to reduce Theorem 5.2 to Theorem 5.1. I think that this reduction step is more or less Hodge theoretic. I do not know how to prove Corollary 5.3 without using this reduction argument based on the Fujino–Mori canonical bundle formula (see [FMo]). We can prove the following theorem from [FMo].

Theorem 5.4 (Fujino–Mori). *Let X be a smooth projective variety, let Δ be an effective $\mathbb{Q}$-divisor on X such that* $\mathrm{Supp}\Delta$ *is a simple normal crossing divisor with $\lfloor\Delta\rfloor = 0$. Let $f : X \to Y$ be a surjective morphism between smooth projective varieties with connected fibers. Assume that $\kappa(X_{\overline{\eta}}, K_{X_{\overline{\eta}}} + \Delta|_{X_{\overline{\eta}}}) = 0$ where $X_{\overline{\eta}}$ is the geometric generic fiber of $f : X \to Y$. Then we can construct a commutative diagram:*

$$\begin{array}{ccc} X & \xleftarrow{\ p\ } & X' \\ {\scriptstyle f}\downarrow & & \downarrow{\scriptstyle f'} \\ Y & \xleftarrow[\ q\ ]{} & Y' \end{array}$$

with the following properties.

(i) p and q are projective birational morphisms.

(ii) *X' and Y' are smooth projective varieties.*

(iii) *There exists an effective $\mathbb{Q}$-divisor Δ' on X' such that* SuppΔ' *is a simple normal crossing divisor on X' with $\lfloor \Delta' \rfloor = 0$ and that*

$$H^0(X, \mathcal{O}_X(\lfloor m(K_X + \Delta)\rfloor)) \simeq H^0(X', \mathcal{O}_{X'}(\lfloor m(K_{X'} + \Delta')\rfloor))$$

for every positive integer m.

(iv) *There exist a positive integer k, a nef $\mathbb{Q}$-divisor M on Y', and an effective $\mathbb{Q}$-divisor D on Y' such that* Supp$D$ *is a simple normal crossing divisor with $\lfloor D \rfloor = 0$ and that*

$$\begin{aligned} &H^0(X', \mathcal{O}_{X'}(\lfloor mk(K_{X'} + \Delta')\rfloor)) \\ &\simeq H^0(Y', \mathcal{O}_{Y'}(\lfloor mk(K_{Y'} + M + D)\rfloor)) \end{aligned}$$

for every positive integer m.

We do not give a proof of Theorem 5.4 here. For the details, see [FMo, Theorem 4.5].

Remark 5.5. Theorem 5.4 is an application of the Fujino–Mori canonical bundle formula discussed in [FMo]. The $\mathbb{Q}$-divisor M is called the semistable part in [FMo] and now is usually called the moduli part in the literature. Note that the nefness of M comes from Theorem 3.5. Therefore, Theorem 5.4 is more or less Hodge theoretic.

Using Kodaira's lemma and Hironaka's resolution theorem, we can prove:

Proposition 5.6. *If $K_{Y'} + M + D$ is big in Theorem 5.4, then there exist a birational morphism $r : Z \to Y'$ from a smooth projective variety Z, an effective $\mathbb{Q}$-divisor Δ_Z on Z, and positive integers a and b such that* SuppΔ_Z *is a simple normal crossing divisor with $\lfloor \Delta_Z \rfloor = 0$ and that*

$$\begin{aligned} &H^0(Y', \mathcal{O}_{Y'}(\lfloor ma(K_{Y'} + M + D)\rfloor)) \\ &\simeq H^0(Z, \mathcal{O}_Z(\lfloor mb(K_Z + \Delta_Z)\rfloor)) \end{aligned}$$

for every positive integer m.

Proof. This is an easy consequence of Kodaira's lemma on big $\mathbb{Q}$-divisors and Hironaka's resolution of singularities. We note that we can choose Δ_Z with $\lfloor \Delta_Z \rfloor = 0$ since M is nef and $\lfloor D \rfloor = 0$. More precisely, by Kodaira's lemma, we have $K_{Y'} + M + D \sim_{\mathbb{Q}} A + E$, where A is an ample $\mathbb{Q}$-divisor and E is an effective $\mathbb{Q}$-divisor. Thus we have $(1+\varepsilon)(K_{Y'} + M + D) \sim_{\mathbb{Q}} K_{Y'} + (M + \varepsilon A) + D + \varepsilon E$ for every positive rational number ε. If ε is sufficiently small, then $(Y', D + \varepsilon E)$ is kawamata log terminal. Since $M + \varepsilon A$ is ample, we can take an effective $\mathbb{Q}$-divisor B such that $B \sim_{\mathbb{Q}} M + \varepsilon A$ and that $(Y', \Delta_{Y'})$ is still kawamata log terminal, where $\Delta_{Y'} = B + D + \varepsilon E$. Therefore, we can find positive integers a and b such that $a(K_{Y'} + M + D) \sim b(K_{Y'} + \Delta_{Y'})$. By Hironaka's resolution theorem, we can take $r : Z \to Y'$ and Δ_Z such that SuppΔ_Z is a simple normal crossing divisor with $\lfloor \Delta_Z \rfloor = 0$ and that $H^0(Z, \mathcal{O}_Z(\lfloor m(K_Z + \Delta_Z)\rfloor)) \simeq H^0(Y', \mathcal{O}_{Y'}(\lfloor m(K_{Y'} + \Delta_{Y'})\rfloor))$ for every positive integer m. Thus we have the desired properties. □

Let us see how to prove Theorem 5.2 by using Theorem 5.1, Theorem 5.4, and Proposition 5.6.

Proof of Theorem 5.2. Let X and Δ be as in Theorem 5.2. Assume that $\kappa(X, K_X+\Delta) \geq 0$ and that $K_X + \Delta$ is not big. We consider the Iitaka fibration $\Phi_{|m(K_X+\Delta)|} : X \dashrightarrow Y$ for some large and divisible positive integer m. By taking a suitable birational modification, we may assume that $f : X \to Y$ is a surjective morphism between smooth projective varieties with connected fibers. Then we apply Theorem 5.4 and Proposition 5.6 to $f : X \to Y$. Thus, we see that $R(X, K_X + \Delta)$ is a finitely generated $\mathbb{C}$-algebra if and only if $R(Z, K_Z + \Delta_Z)$ is a finitely generated $\mathbb{C}$-algebra. By Theorem 5.1, we know that $R(Z, K_Z + \Delta_Z)$ is a finitely generated $\mathbb{C}$-algebra. Therefore, we obtain Theorem 5.2. □

As I explained in Remark 5.5, Theorem 5.4 is more or less Hodge theoretic. So, we pose:

Problem 5.7. Prove Corollary 5.3 without using Hodge theory.

Remark 5.8. Theorem 3.5 holds true under the assumption that X is only a compact Kähler manifold. This is because we can use the theory of variation of Hodge structure even for compact Kähler manifolds. Therefore, Theorem 5.4 holds true under the assumption that X is a compact complex manifold in Fujiki's class $\mathcal{C}$. As a consequence, we see that Theorem 5.2 holds for compact complex manifolds in Fujiki's class $\mathcal{C}$. For the details, see [Fn10, Section 5]. As a special case, we have the corollary below.

Corollary 5.9. *Let X be a compact Kähler manifold. Then the canonical ring*

$$R(X) = \bigoplus_{m=0}^{\infty} H^0(X, \omega_X^{\otimes m})$$

is a finitely generated $\mathbb{C}$-algebra.

Remark 5.10. There exists a compact complex non-Kähler manifold whose canonical ring is not a finitely generated $\mathbb{C}$-algebra (see [Fn10, Example 6.4], which is essentially due to Wilson [Wil]). This means that Corollary 5.9 does not hold for compact complex non-Kähler manifolds. The reader can find a smooth morphism $f : X \to Y$ from a compact complex non-Kähler manifold X onto $Y = \mathbb{P}^1$ with connected fibers such that $f_*\omega_{X/Y} \simeq \mathcal{O}_{\mathbb{P}^1}(-2)$ (see [Fn10, Example 6.1]). This example is due to Atiyah. Therefore, Theorem 3.5 does not hold for compact non-Kähler manifolds. Of course, by this example, we see that Theorem 5.4 does not hold true without assuming that X is a compact complex manifold in Fujiki's class $\mathcal{C}$. For the details, see [Fn10].

As a generalization of Theorem 5.2, we have Conjecture 5.11, where we are allowing the coefficients of Δ to equal one.

Conjecture 5.11. *Let X be a smooth projective variety and let Δ be an effective $\mathbb{Q}$-divisor on X such that* $\mathrm{Supp}\Delta$ *is a simple normal crossing divisor and that the coefficients of Δ are less than or equal to one. Then the log canonical ring*

$$R(X, K_X + \Delta) = \bigoplus_{m=0}^{\infty} H^0(X, \mathcal{O}_X(\lfloor m(K_X + \Delta)\rfloor))$$

is a finitely generated $\mathbb{C}$-algebra.

Remark 5.12. Of course, we think that Conjecture 5.11 also holds for compact complex manifolds in Fujiki's class $\mathcal{C}$. However, I could not reduce Conjecture 5.11 for compact complex manifolds in Fujiki's class $\mathcal{C}$ to Conjecture 5.11 for projective varieties. Therefore, Conjecture 5.11 for compact complex manifolds in Fujiki's class $\mathcal{C}$ may be much harder than for projective varieties. For the details, see [Fn10].

I have already discussed Conjecture 5.11 in detail in a joint paper with Yoshinori Gongyo (see [FG]). In [FG], we clarified the relationship among various conjectures in the minimal model program related to Conjecture 5.11. So we do not repeat the discussions about Conjecture 5.11 here. We strongly recommend the interested reader to see [FG].

In order to prove Conjecture 5.11, the following famous conjecture seems to be unavoidable. I think that it is a very difficult open problem for higher-dimensional algebraic varieties.

Conjecture 5.13 (Nonvanishing conjecture). *Let X be a smooth projective variety such that the canonical divisor K_X is pseudoeffective. Then we have $H^0(X, \mathcal{O}_X(mK_X)) \neq 0$ for some positive integer m, equivalently, $\kappa(X) \geq 0$.*

For the reader's convenience, let us recall the definition of pseudoeffective divisors.

Definition 5.14. Let X be a smooth projective variety and let D be a Cartier divisor on X. Then D is pseudoeffective if $D + A$ is big for every ample $\mathbb{Q}$-divisor A on X.

The characterization of pseudoeffective divisors via singular hermitian metrics may be helpful.

Remark 5.15. Let X be a smooth projective variety and let D be a Cartier divisor on X. Then D is pseudoeffective if and only if $\mathcal{O}_X(D)$ has a singular hermitian metric h with $\sqrt{-1}\Theta_h \geq 0$ in the sense of currents.

The characterization of uniruled varieties due to Boucksom–Demailly–Păun–Peternell (see [BDPP]) is important and helps us understand Conjecture 5.13.

Theorem 5.16. *Let X be a smooth projective variety. Then X is uniruled if and only if K_X is not pseudoeffective.*

For the reader's convenience, we recall the definition of uniruled varieties.

Definition 5.17. Let X be a smooth projective variety with $\dim X = n$. Then X is uniruled if there exist a smooth projective variety Y with $\dim Y = n-1$ and a dominant rational map $Y \times \mathbb{P}^1 \dashrightarrow X$.

Therefore, Conjecture 5.13 says that the Kodaira dimension of X is nonnegative if X is not covered by rational curves. Thus, Conjecture 5.13 looks very reasonable. However, I do not know how to attack it.

Remark 5.18 (Added in September 2016). By Kenta Hashizume's recent result, Conjecture 5.13 implies that any projective log canonical pair (X, Δ) such that $K_X + \Delta$ is pseudoeffective has a minimal model. For the details, see [Has].

6 Appendix

In this appendix, we collect some definitions, which may help us understand this paper, for the reader's convenience. For the details, see [Fn3] and [Fn11]. Recall that a scheme means a separated scheme of finite type over $\mathbb{C}$ in this paper.

6.1 (Iitaka dimension and Kodaira dimension). Let X be a normal projective variety and let D be a $\mathbb{Q}$-Cartier divisor on X. Then $\kappa(X, D)$ denotes the Iitaka dimension of D. More precisely,

$$\kappa(X, D) = \limsup_{m \to \infty} \frac{\log \dim H^0(X, \mathcal{O}_X(\lfloor mD \rfloor))}{\log m}.$$

For the definition of $\lfloor mD \rfloor$, see 6.4 below. Let X be a smooth projective variety. Then we put $\kappa(X) = \kappa(X, K_X)$ and call it the Kodaira dimension of X.

6.2 ($\mathbb{Q}$-factorial). Let X be a normal variety. Then X is called $\mathbb{Q}$-factorial if every prime divisor D on X is $\mathbb{Q}$-Cartier.

Note that $\mathbb{Q}$-factoriality sometimes plays crucial roles in the minimal model program. The notion of terminal singularities and canonical singularities is indispensable in the minimal model program.

6.3 (Terminal singularities and canonical singularities). Let X be a normal variety such that K_X is $\mathbb{Q}$-Cartier. If there exists a projective birational morphism $f : Y \to X$ from a smooth variety Y such that the exceptional locus $\mathrm{Exc}(f) = \sum_{i \in I} E_i$ is a simple normal crossing divisor on Y. In this situation, we can write

$$K_Y = f^* K_X + \sum_{i \in I} a_i E_i.$$

If $a_i > 0$ for every $i \in I$, then we say that X has only terminal singularities. If $a_i \geq 0$ for every $i \in I$, then we say that X has only canonical singularities. It is well-known that X has only rational singularities when X has only canonical singularities.

6.4 (Round-down of $\mathbb{Q}$-divisors). Let $D = \sum a_i D_i$ be a $\mathbb{Q}$-divisor on a normal variety X. Note that D_i is a prime divisor for every i and that $D_i \neq D_j$ for $i \neq j$. Of course, $a_i \in \mathbb{Q}$ for every i. We put $\lfloor D \rfloor = \sum \lfloor a_i \rfloor D_i$ and call it the round-down of D. For every rational number x, $\lfloor x \rfloor$ is the integer defined by $x - 1 < \lfloor x \rfloor \leq x$.

In the minimal model program, we usually use the notion of pairs.

6.5 (Singularities of pairs). A pair (X, Δ) consists of a normal variety X and an effective $\mathbb{R}$-divisor Δ on X such that $K_X + \Delta$ is $\mathbb{R}$-Cartier. A pair (X, Δ) is called kawamata log terminal (resp. log canonical) if for any projective birational morphism $f : Y \to X$ from a normal variety Y, $a(E, X, \Delta) > -1$ (resp. ≥ -1) for every E, where

$$K_Y = f^*(K_X + \Delta) + \sum_E a(E, X, \Delta)E$$

defines $a(E, X, \Delta)$. Let (X, Δ) be a log canonical pair and let W be a closed subset of X. Then W is called a log canonical center of (X, Δ) if there are a projective birational morphism $f : Y \to X$ from a normal variety Y and a prime divisor E on Y such that $a(E, X, \Delta) = -1$ and that $f(E) = W$.

To help understand Theorem 2.12, Theorem 2.13, and Theorem 3.6, we quickly explain the notion of simple normal crossing pairs and some related topics.

6.6 (Simple normal crossing pairs). Let Z be a simple normal crossing divisor on a smooth variety M and let B be an $\mathbb{R}$-divisor on M such that $\mathrm{Supp}(B + Z)$ is a simple normal crossing divisor and that B and Z have no common irreducible components. We put $\Delta_Z = B|_Z$ and consider the pair (Z, Δ_Z). We call (Z, Δ_Z) a *globally embedded* simple normal crossing pair. A pair (X, Δ) is called a simple normal crossing pair if it is Zariski locally isomorphic to a globally embedded simple normal crossing pair. If (X, Δ) is a simple normal crossing pair and X is a divisor on a smooth variety M, then (X, Δ) is called an embedded simple normal crossing pair. Of course, a globally embedded simple normal crossing pair is automatically an embedded simple normal crossing pair.

6.7 (Strata and permissibility). Let (X, Δ) be a simple normal crossing pair. Assume that the coefficients of Δ are in $[0, 1]$. Let $\nu : X^\nu \to X$ be the normalization. We put

$$K_{X^\nu} + \Theta = \nu^*(K_X + \Delta),$$

that is, Θ is the sum of the inverse images of Δ and the singular locus of X. Then we see that (X^ν, Θ) is log canonical. Let W be a closed subset of X. Then W is called a stratum of (X, Δ) if W is an irreducible component of X or the ν-image of some log canonical center of (X^ν, Θ). A Cartier divisor D on X is permissible with respect to (X, Δ) if D contains no strata of (X, Δ) in its support. A finite $\mathbb{R}$-linear combination of permissible Cartier divisors with respect to (X, Δ) is called a permissible $\mathbb{R}$-divisor with respect to (X, Δ).

We need the notion of nef and log big divisors for Theorem 2.13.

6.8 (Nef and log big divisors). Let $f : (Y, \Delta) \to X$ be a proper morphism from a simple normal crossing pair (Y, Δ) to a scheme X. Assume that the coefficients of Δ are in $[0, 1]$. Let $\pi : X \to V$ be a proper morphism between schemes. Let H be an $\mathbb{R}$-Cartier divisor on X. Then H is nef and log big over V with respect to $f : (Y, \Delta) \to X$ if H is nef over V and $H|_{f(W)}$ is big over $\pi \circ f(W)$ for every stratum W of (Y, Δ). Note that if H is ample over V then H is nef and log big over V with respect to $f : (Y, \Delta) \to X$.

For Theorem 2.12 and Theorem 2.13, let us recall the definition of $\mathbb{R}$-linear equivalence.

6.9 ($\mathbb{R}$-divisors). Let B_1 and B_2 be two $\mathbb{R}$-Cartier divisors on a scheme X. Then B_1 is linearly (resp. $\mathbb{Q}$-linearly, or $\mathbb{R}$-linearly) equivalent to B_2, denoted by $B_1 \sim B_2$ (resp. $B_1 \sim_{\mathbb{Q}} B_2$, or $B_1 \sim_{\mathbb{R}} B_2$) if

$$B_1 = B_2 + \sum_{i=1}^{k} r_i(f_i)$$

such that $f_i \in \Gamma(X, \mathcal{K}_X^*)$ and $r_i \in \mathbb{Z}$ (resp. $r_i \in \mathbb{Q}$, or $r_i \in \mathbb{R}$) for every i. Here, $\mathcal{K}_X$ is the sheaf of total quotient rings of $\mathcal{O}_X$ and $\mathcal{K}_X^*$ is the sheaf of invertible elements in the sheaf of rings $\mathcal{K}_X$. We note that (f_i) is a principal Cartier divisor associated to f_i, that is, the image of f_i by $\Gamma(X, \mathcal{K}_X^*) \to \Gamma(X, \mathcal{K}_X^*/\mathcal{O}_X^*)$, where $\mathcal{O}_X^*$ is the sheaf of invertible elements in $\mathcal{O}_X$.

Let $f : X \to Y$ be a morphism between schemes. If there is an $\mathbb{R}$-Cartier divisor B on Y such that

$$B_1 \sim_{\mathbb{R}} B_2 + f^*B,$$

then B_1 is said to be relatively $\mathbb{R}$-linearly equivalent to B_2. It is denoted by $B_1 \sim_{\mathbb{R}, f} B_2$ or $B_1 \sim_{\mathbb{R}, Y} B_2$.

References

[AK] D. Abramovich, K. Karu, Weak semistable reduction in characteristic 0, Invent. Math. **139** (2000), no. 2, 241–273.

[Ber] B. Berndtsson, Curvature of vector bundles associated to holomorphic fibrations, Ann. of Math. (2) **169** (2009), no. 2, 531–560.

[BP1] B. Berndtsson, M. Păun, Bergman kernels and the pseudoeffectivity of relative canonical bundles, Duke Math. J. **145** (2008), no. 2, 341–378.

[BP2] B. Berndtsson, M. Păun, A Bergman kernel proof of the Kawamata subadjunction theorem, preprint (2008). arXiv:0804.3884 [math.AG]

[BCHM] C. Birkar, P. Cascini, C. D. Hacon, J. McKernan, Existence of minimal models for varieties of log general type, J. Amer. Math. Soc. **23** (2010), no. 2, 405–468.

[BDPP] S. Boucksom, J.-P. Demailly, M. Păun, T. Peternell, The pseudo-effective cone of a compact Kähler manifold and varieties of negative Kodaira dimension, J. Algebraic Geom. **22** (2013), no. 2, 201–248.

[CL] P. Cascini, V. Lazić, New outlook on the minimal model program, I, Duke Math. J. **161** (2012), no. 12, 2415–2467.

[CD1] F. Catanese, M. Dettweiler, The direct image of the relative dualizing sheaf needs not be semiample, C. R. Math. Acad. Sci. Paris **352** (2014), no. 3, 241–244.

[CD2] F. Catanese, M. Dettweiler, Answer to a question by Fujita on variation of Hodge structures, to appear in Adv. Stud. Pure Math. arXiv:1311.3232 [math.AG]

[CD3] F. Catanese, M. Dettweiler, Vector bundles on curves coming from variation of Hodge structures, Internat. J. Math. **27** (2016), no. 7, 1640001, 25 pp.

[CK] E. Cattani, A. Kaplan, Polarized mixed Hodge structures and the local monodromy of a variation of Hodge structure, Invent. Math. **67** (1982), no. 1, 101–115.

[CKS] E. Cattani, A. Kaplan, W. Schmid, Degeneration of Hodge structures, Ann. of Math. (2) **123** (1986), no. 3, 457–535.

[Del] P. Deligne, Théorie de Hodge. II, Inst. Hautes Études Sci. Publ. Math. No. **40** (1971), 5–57.

[DI] P. Deligne, L. Illusie, Relèvements modulo p^2 et décomposition du complexe de de Rham, Invent. Math. **89** (1987), no. 2, 247–270.

[Eno] I. Enoki, Kawamata–Viehweg vanishing theorem for compact Kähler manifolds, *Einstein metrics and Yang–Mills connections (Sanda, 1990)*, 59–68, Lecture Notes in Pure and Appl. Math., **145**, Dekker, New York, 1993.

[EV] H. Esnault, E. Viehweg, *Lectures on vanishing theorems*, DMV Seminar, **20**, Birkhäuser Verlag, Basel, 1992.

[Fn1] O. Fujino, Remarks on algebraic fiber spaces, J. Math. Kyoto Univ. **45** (2005), no. 4, 683–699.

[Fn2] O. Fujino, Higher direct images of log canonical divisors, J. Differential Geom. **66** (2004), no. 3, 453–479.

[Fn3] O. Fujino, Fundamental theorems for the log minimal model program, Publ. Res. Inst. Math. Sci. **47** (2011), no. 3, 727–789.

[Fn4] O. Fujino, A transcendental approach to Kollár's injectivity theorem, Osaka J. Math. **49** (2012), no. 3, 833–852.

[Fn5] O. Fujino, A transcendental approach to Kollár's injectivity theorem II, J. Reine Angew. Math. **681** (2013), 149–174.

[Fn6] O. Fujino, Fundamental theorems for semi log canonical pairs, Algebr. Geom. **1** (2014), no. 2, 194–228.

[Fn7] O. Fujino, Vanishing theorems, *Minimal models and Extremal Rays (Kyoto, 2011)*, 299–321, Adv. Stud. Pure Math., **70**, Math. Soc. Japan, Tokyo, 2016.

[Fn8] O. Fujino, Semipositivity theorems for moduli problems, preprint (2012). arXiv:1210.5784 [math.AG]

[Fn9] O. Fujino, Injectivity theorems, to appear in Adv. Stud. Pure Math. arXiv:1303.2404 [math.AG]

[Fn10] O. Fujino, Some remarks on the minimal model program for log canonical pairs, J. Math. Sci, Univ. Tokyo **22** (2015), no. 1, 149–192.

[Fn11] O. Fujino, Foundations of the minimal model program, MSJ Memoirs, **35**, Math. Soc. Japan, Tokyo, 2017.

[Fn12] O. Fujino, Direct images of relative pluricanonical bundles, Algebr. Geom. **3** (2016), no. 1, 50–62.

[Fn13] O. Fujino, Corrigendum: Direct images of relative pluricanonical bundles (Algebraic Geometry **3**, no. 1, (2016), 50–62), Algebr. Geom. **3** (2016), no. 2, 261–263.

[Fn14] O. Fujino, Subadditivity of the logarithmic Kodaira dimension for morphisms of relative dimension one revisited, preprint (2015). Kyoto-Math 2015-02

[Fn15] O. Fujino, Vanishing and semipositivity theorems for semi-log canonical pairs, preprint (2015). arXiv:1509.00531 [math.AG]

[Fn16] O. Fujino, Kollár–Nadel type vanishing theorem, preprint (2016). arXiv:1607.04684 [math.AG]

[FF] O. Fujino, T. Fujisawa, Variations of mixed Hodge structure and semipositivity theorems, Publ. Res. Inst. Math. Sci. **50** (2014), no. 4, 589–661.

[FFS] O. Fujino, T. Fujisawa, M. Saito, Some remarks on the semipositivity theorems, Publ. Res. Inst. Math. Sci. **50** (2014), no. 1, 85–112.

[FG] O. Fujino, Y. Gongyo, On log canonical rings, to appear in Adv. Stud. Pure Math. arXiv:1302.5194 [math.AG]

[FMa] O. Fujino, S. Matsumura, Injectivity theorem for pseudo-effective line bundles and its applications, preprint (2016). arXiv:1605.02284 [math.CV]

[FMo] O. Fujino, S. Mori, A canonical bundle formula, J. Differential Geom. **56** (2000), no. 1, 167–188.

[Fujita] T. Fujita, On Kähler fiber spaces over curves, J. Math. Soc. Japan **30** (1978), no. 4, 779–794.

[GM] Y. Gongyo, S. Matsumura, Versions of injectivity and extension theorems, Ann. Sci. Éc. Norm. Supér. (4) **50** (2017), no. 2, 479–502.

[Gri] P. Griffiths, Periods of integrals on algebraic manifolds. III. Some global differential-geometric properties of the period mapping, Inst. Hautes Études Sci. Publ. Math. No. **38** (1970), 125–180.

[HM] C. D. Hacon, J. McKernan, Extension theorems and the existence of flips, *Flips for 3-folds and 4-folds*, 76–110, Oxford Lecture Ser. Math. Appl., **35**, Oxford Univ. Press, Oxford, 2007.

[Har] R. Hartshorne, Ample vector bundles, Inst. Hautes Études Sci. Publ. Math. No. **29** (1966), 63–94.

[Has] K. Hashizume, On the non-vanishing conjecture and existence of log minimal models, preprint (2016). arXiv:1609.00121 [math.AG]

[Kas] M. Kashiwara, A study of variation of mixed Hodge structure, Publ. Res. Inst. Math. Sci. **22** (1986), no. 5, 991–1024.

[Kaw1] Y. Kawamata, Characterization of abelian varieties, Compositio Math. **43** (1981), no. 2, 253–276.

[Kaw2] Y. Kawamata, Kodaira dimension of algebraic fiber spaces over curves, Invent. Math. **66** (1982), no. 1, 57–71.

[Kaw3] Y. Kawamata, Kodaira dimension of certain algebraic fiber spaces, J. Fac. Sci. Univ. Tokyo Sect. IA Math. **30** (1983), no. 1, 1–24.

[KKMS] G. Kempf, F. F. Knudsen, D. Mumford, B. Saint-Donat, *Toroidal embeddings. I*, Lecture Notes in Mathematics, Vol. **339**, Springer-Verlag, Berlin-New York, 1973.

[Ko1] J. Kollár, Higher direct images of dualizing sheaves. I, Ann. of Math. (2) **123** (1986), no. 1, 11–42.

[Ko2] J. Kollár, Projectivity of complete moduli, J. Differential Geom. **32** (1990), no. 1, 235–268.

[Ko3] J. Kollár, Shafarevich maps and plurigenera of algebraic varieties, Invent. Math. **113** (1993), no. 1, 177–215.

[Ko4] J. Kollár, *Shafarevich maps and automorphic forms*, M. B. Porter Lectures, Princeton University Press, Princeton, NJ, 1995.

[Ko5] J. Kollár, Singularities of pairs, *Algebraic geometry—Santa Cruz 1995*, 221–287, Proc. Sympos. Pure Math., **62**, Part 1, Amer. Math. Soc., Providence, RI, 1997.

[KoM] J. Kollár, S. Mori, *Birational geometry of algebraic varieties*, Cambridge Tracts in Mathematics, **134**, Cambridge University Press, Cambridge, 1998.

[KvP] S. Kovács, Z. Patakfalvi, Projectivity of the moduli space of stable log-varieties and subadditivity of log-Kodaira dimension, to appear in J. Amer. Math. Soc. arXiv:1503.02952 [math.AG]

[Laz] R. Lazarsfeld, *Positivity in algebraic geometry. II. Positivity for vector bundles, and multiplier ideals*, Results in Mathematics and Related Areas. 3rd Series. A Series of Modern Surveys in Mathematics, **49**, Springer-Verlag, Berlin, 2004.

[Ma1] S. Matsumura, A Nadel vanishing theorem via injectivity theorems, Math. Ann. **359** (2014), no. 3-4, 785–802.

[Ma2] S. Matsumura, A Nadel vanishing theorem for metrics with minimal singularities on big line bundles, Adv. Math. **280** (2015), 188–207.

[Ma3] S. Matsumura, An injectivity theorem with multiplier ideal sheaves of singular metrics with transcendental singularities, to appear in J. Algebraic Geom. arXiv:1308.2033 [math.CV]

[Ma4] S. Matsumura, Some applications of the theory of harmonic integrals, Complex Manifolds **2** (2015), 16–25.

[Ma5] S. Matsumura, Injectivity theorems with multiplier ideal sheaves and their applications, *Complex analysis and geometry*, 241–255, Springer Proc. Math. Stat., **144**, Springer, Tokyo, 2015.

[Ma6] S. Matsumura, A vanishing theorem of Kollár–Ohsawa type, Math. Ann. **366** (2016), no. 3-4, 1451–1465.

[Ma7] S. Matsumura, Injectivity theorems with multiplier ideal sheaves for higher direct images under Kähler morphisms, preprint (2016). arXiv:1607.05554 [math.CV]

[Ma8] S. Matsumura, A transcendental approach to injectivity theorem for log canonical pairs, preprint (2016). arXiv:1607.07213 [math.CV]

[Mou] C. Mourougane, Images directes de fibrés en droites adjoints, Publ. Res. Inst. Math. Sci. **33** (1997), no. 6, 893–916.

[MT1] C. Mourougane, S. Takayama, Hodge metrics and positivity of direct images, J. Reine Angew. Math. **606** (2007), 167–178.

[MT2] C. Mourougane, S. Takayama, Hodge metrics and the curvature of higher direct images, Ann. Sci. Éc. Norm. Supér. (4) **41** (2008), no. 6, 905–924.

[MT3] C. Mourougane, S. Takayama, Extension of twisted Hodge metrics for Kähler morphisms, J. Differential Geom. **83** (2009), no. 1, 131–161.

[Nak] N. Nakayama, Invariance of the plurigenera of algebraic varieties under minimal model conjectures, Topology **25** (1986), no. 2, 237–251.

[NZ] N. Nakayama, D.-Q. Zhang, Building blocks of étale endomorphisms of complex projective manifolds, Proc. Lond. Math. Soc. (3) **99** (2009), no. 3, 725–756.

[Oh1] T. Ohsawa, Vanishing theorems on complete Kähler manifolds, Publ. Res. Inst. Math. Sci. **20** (1984), no. 1, 21–38.

[Oh2] T. Ohsawa, On a curvature condition that implies a cohomology injectivity theorem of Kollár–Skoda type, Publ. Res. Inst. Math. Sci. **41** (2005), no. 3, 565–577.

[Pă1] M. Păun, Siu's invariance of plurigenera: a one-tower proof, J. Differential Geom. **76** (2007), no. 3, 485–493.

[Pă2] M. Păun, Lecture notes on rational polytopes and finite generation, *Analytic and algebraic geometry*, 371–403, IAS/Park City Math. Ser., **17**, Amer. Math. Soc., Providence, RI, 2010.

[PăT] M. Păun, S. Takayama, Positivity of twisted relative pluricanonical bundles and their direct images, preprint (2014). arXiv:1409.5504 [math.AG]

[PoSc] M. Popa, C. Schnell, On direct images of pluricanonical bundles, Algebra & Number Theory **8** (2014), no. 9, 2273–2295.

[Sa1] M. Saito, Modules de Hodge polarisables, Publ. Res. Inst. Math. Sci. **24** (1988), no. 6, 849–995 (1989).

[Sa2] M. Saito, Mixed Hodge modules, Publ. Res. Inst. Math. Sci. **26** (1990), no. 2, 221–333.

[Sa3] M. Saito, On Kollár's conjecture, *Several complex variables and complex geometry, Part 2 (Santa Cruz, CA, 1989)*, 509–517, Proc. Sympos. Pure Math., **52**, Part 2, Amer. Math. Soc., Providence, RI, 1991.

[Sc] W. Schmid, Variation of Hodge structure: the singularities of the period mapping, Invent. Math. **22** (1973), 211–319.

[Siu1] Y.-T. Siu, Invariance of plurigenera, Invent. Math. **134** (1998), no. 3, 661–673.

[Siu2] Y.-T, Siu, Extension of twisted pluricanonical sections with plurisubharmonic weight and invariance of semipositively twisted plurigenera for manifolds not necessarily of general type, *Complex geometry (Göttingen, 2000)*, 223–277, Springer, Berlin, 2002.

[St] J. Steenbrink, Limits of Hodge structures, Invent. Math. **31** (1975/76), no. 3, 229–257.

[SZ] J. Steenbrink, S. Zucker, Variation of mixed Hodge structure. I, Invent. Math. **80** (1985), no. 3, 489–542.

[Taka] S. Takayama, Singularities of Narasimhan–Simha type metrics on direct images of relative pluricanonical bundles, Ann. Inst. Fourier (Grenoble) **66** (2016), no. 2, 753–783.

[Take] K. Takegoshi, Higher direct images of canonical sheaves tensorized with semi-positive vector bundles by proper Kähler morphisms, Math. Ann. **303** (1995), no. 3, 389–416.

[Tan] S. G. Tankeev, n-dimensional canonically polarized varieties, and varieties of basic type, Izv. Akad. Nauk SSSR Ser. Mat. **35** (1971), 31–44.

[Vie1] E. Viehweg, Weak positivity and the additivity of the Kodaira dimension for certain fibre spaces, *Algebraic varieties and analytic varieties (Tokyo, 1981)*, 329–353, Adv. Stud. Pure Math., **1**, North-Holland, Amsterdam, 1983.

[Vie2] E. Viehweg, Weak positivity and the additivity of the Kodaira dimension. II. The local Torelli map, *Classification of algebraic and analytic manifolds (Katata, 1982)*, 567–589, Progr. Math., **39**, Birkhäuser Boston, Boston, MA, 1983.

[Wil] P. M. H. Wilson, On the canonical ring of algebraic varieties, Compositio Math. **43** (1981), no. 3, 365–385.

[Zuc1] S. Zucker, Generalized intermediate Jacobians and the theorem on normal functions, Invent. Math. **33** (1976), no. 3, 185–222.

[Zuc2] S. Zucker, Hodge theory with degenerating coefficients. L^2 cohomology in the Poincaré metric, Ann. of Math. (2) **109** (1979), no. 3, 415–476.

[Zuc3] S. Zucker, Remarks on a theorem of Fujita, J. Math. Soc. Japan **34** (1982), no. 1, 47–54.

Hodge Theory and L^2-analysis
ALM 39, pp. 283–319

The Business of Height Pairings*

Souvik Goswami, James D. Lewis†

Abstract

In algebraic geometry there is the notion of a height pairing of algebraic cycles, which lies at the confluence of arithmetic, Hodge theory and topology. After explaining a motivating example situation, we introduce new directions in this subject.

2000 Mathematics Subject Classification: 14C25, 14C30, 14C35.

Keywords and Phrases: Height pairing, Abel-Jacobi map, regulator, Deligne cohomology, Chow group.

1 Introduction

From topology one has the notion of the local linking number (or index) of two curves in 3-space. Basically this determines locally how many times a given curve winds around another (with orientation). If one of curves bounds a membrane (we think of that membrane as a precycle, in the sense that it's boundary is not zero), then the sum of these local links can be interpreted as an intersection pairing. Paragraph 2.1 in [Be3] comes to mind about this. The height pairing of two algebraic cycles is an algebraic generalization of this. Here is an example (see [C-L]) of how we view a classical algebraic cycle as bounding a precycle. Let X be a projective algebraic manifold of dimension d and $\{Z_\alpha\}$ a finite collection of irreducible subvarieties of codimension $r-1$ in X. Let $f_\alpha \in \mathbb{C}(Z_\alpha)^\times$, and consider the precycle

$$\xi_1' := \sum_\alpha (f_\alpha, Z_\alpha).$$

*In celebration of Steven M. Zucker's 65th birthday. A true pioneer in Hodge theory!

†632 Central Academic Building, University of Alberta, Edmonton, Alberta T6G 2G1, CANADA. E-mail: souvik@ualberta.ca; 632 Central Academic Building, University of Alberta, Edmonton, Alberta T6G 2G1, CANADA. E-mail: lewisjd@ualberta.ca; Second author partially supported by a grant from the Natural Sciences and Engineering Research Council of Canada.

Put

$$\xi_1 := \sum_\alpha \mathrm{div}_{Z_\alpha}(f_\alpha) \in z^r(X),$$

where $z^r(X)$ are the cycles of codimension r in X. Note that by definition $\xi_1 \in z^r_{\mathrm{rat}}(X)$, the subgroup of cycles in $z^r(X)$ rationally equivalent to zero. Alternate take: Let $\square := \mathbb{P}^1\backslash\{1\}$. Then one can interpret

$$\xi_1' = \sum_\alpha \mathrm{graph}_{Z_\alpha\times\square}(f_\alpha) \in z^r(X\times\square),$$

with

$$\partial(\xi_1') = \partial_0(\xi_1') - \partial_\infty(\xi_1') = \xi_1.$$

If $\xi_2 \in z^{d-r+1}(X)$ is in general position with respect to ξ_1 (and ξ_1'), then $|\xi_1|\cap|\xi_2| = \emptyset$; moreover

$$\sum_\alpha \int_{Z_\alpha\cap\xi_2} \log|f_\alpha| \in \mathbb{R}$$

becomes the analog of the total linking index of ξ_1 and ξ_2. Now suppose that $\xi_1 = 0$, i.e. $\partial\xi_1' = 0$. Then the real regulator of the "$K_1^{(r)}(X)$" cycle ξ_1', given by the formula,

$$R_{r,1}(\xi_1') \in H^{r-1,r-1}(X,\mathbb{R}) \simeq H^{d-r+1,d-r+1}(X,\mathbb{R})^\vee, \tag{1}$$

$$\omega \in H^{d-r+1,d-r+1}(X,\mathbb{R}) \mapsto \sum_\alpha \int_{Z_\alpha} \log|f_\alpha|\omega \in \mathbb{R},$$

is well-defined (see [Ja1], or [KLM] and the references cited there). If $\omega = [\xi_2]$ is algebraic, then

$$R_{r,1}(\xi_1')(\omega) = \sum_\alpha \int_{Z_\alpha\cap\xi_2} \log|f_\alpha|.$$

Finally, if $\omega = 0$, e.g. $\omega = [\xi_2]$ where $\xi_2 \in z^{d-r+1}_{\mathrm{rat}}(X)$, then $R_{r,1}(\xi_1')(\omega) = 0$. We deduce:

Proposition 1.1. *We have a pairing*

$$\langle\ ,\ \rangle : z^r_{\mathrm{rat}}(X) \times z^{d+1-r}_{\mathrm{rat}}(X) \to \mathbb{R}$$

given by

$$\langle\xi_1,\xi_2\rangle = R_{r,1}(\xi_1')(\xi_2) = \sum_\alpha \int_{Z_\alpha\cap\xi_2} \log|f_\alpha| \in \mathbb{R},$$

where $\xi_1 \in z^r_{\mathrm{rat}}(X)$, $\xi_2 \in z^{d+1-r}_{\mathrm{rat}}(X)$ and $\xi_1' = \sum(f_\alpha, Z_\alpha)$ is a higher Chow precycle whose divisor (boundary) is ξ_1. It is easy to see that the pairing is well-defined, i.e., it is independent of the exact choice of ξ_1', since if $\mathrm{div}(\xi_1' - \xi_1'') = 0$, *then*

$$R_{r,1}(\xi_1' - \xi_1'')(\xi_2) = 0$$

as $\xi_2 \sim_{\mathrm{rat}} 0$.

The projection formula holds trivially from the definition. That is, we have

Proposition 1.2. *Let $\pi : X \to Y$ be a flat surjective morphism between two smooth projective varieties X and Y. Then $\langle \xi_1, \pi^*\xi_2 \rangle = \langle \pi_*\xi_1, \xi_2 \rangle$ for all $\xi_1 \in z^r_{rat}(X)$ and $\xi_2 \in z^{d-r+1}_{rat}(Y)$ with $|\pi_*\xi_1| \cap |\xi_2| = \emptyset$.*

A little less obvious fact is that this pairing is symmetric. That is, it has the following property which we will call the *reciprocity property* of the pairing.

Proposition 1.3. *For all $\xi_1 \in z^r_{\mathrm{rat}}(X)$, $\xi_2 \in z^{d-r+1}_{\mathrm{rat}}(X)$ with $|\xi_1| \cap |\xi_2| = \emptyset$, $\langle \xi_1, \xi_2 \rangle = \langle \xi_2, \xi_1 \rangle$.*

Proof. Let (f, D) and (g, E) be the higher Chow precycles such that $\xi_1 = \mathrm{div}(f)$ and $\xi_2 = \mathrm{div}(g)$. We can assume, using some additional machinery [Blo1] (Lemma 4.2), that with regard to the pairs (f, D), (g, E), everything is in "general" position. For notational simplicity, let us assume that D and E are irreducible and meet properly along an irreducible curve C. Let

$$f_c := f\big|_C \in \mathbb{C}(C)^\times, \quad g_c := g\big|_C \in \mathbb{C}(C)^\times.$$

For every point $p \in C$, put

$$T_p\{f_c, g_c\} = (-1)^{\nu_p(f_c)\nu_p(g_c)} \left(\frac{f_c^{\nu_p(g_c)}}{g_c^{\nu_p(f_c)}} \right)_p, \tag{2}$$

where $\nu_p(h)$ is the vanishing order of a function h at p. Since $|\xi_1| \cap |\xi_2| = \emptyset$, it follows that

$$T_p\{f_c, g_c\} = \begin{cases} f_c^{\nu_p(g_c)}(p), & \text{if } \nu_p(g_c) \neq 0, \\ g_c^{-\nu_p(f_c)}(p), & \text{if } \nu_p(f_c) \neq 0, \\ 1, & \text{otherwise.} \end{cases}$$

Then it is a consequence of Weil reciprocity:

$$\prod_{p \in C} T_p\{f_c, g_c\} = 1,$$

that

$$\int_{D \cap \mathrm{div}(g)} \log|f| = \int_{E \cap \mathrm{div}(f)} \log|g|.$$

Obviously, this is equivalent to $\langle \xi_1, \xi_2 \rangle = \langle \xi_2, \xi_1 \rangle$. The reader can consult Chapters 2.2.2 and 2.2.3 of [GS] for a generalization of this. □

In addition, this pairing is also non-degenerate in the sense of detecting rational equivalence (see [C-L] for details). This pairing is a special case of the complex Archimedean height pairing, well presented in [MS1], and plays a role at "infinity" in §4. We will return to a generalization of this Archimedean height pairing in §9.

2 Notation and a breezy review of background material

• Unless otherwise specified, X is a smooth projective variety of dimension d defined over a subfield $k \subseteq \mathbb{C}$, and $H^\bullet(X(\mathbb{C}))$ is singular cohomology, treating X as a complex analytic space.
• For a quasi-projective variety W over a field (or more generally a noetherian and separated scheme W), $z^r(W)$ is the free abelian group generated by subvarieties of codimension r in W. The Chow group of W is defined as $\mathrm{CH}^r(W) = z^r(W)/z^r_{rat}(W)$, where $z^r_{rat}(W)$ is the subgroup of cycles rationally equivalent to zero. The rational Chow groups will be denoted by $\mathrm{CH}^r(W;\mathbb{Q}) := \mathrm{CH}^r(W)\otimes_{\mathbb{Z}}\mathbb{Q}$.
• Let $\mathbb{A}\subseteq\mathbb{R}$ be a subring. The reader is assumed to have some familiarity with the abelian category of $\mathbb{A}$-MHS (mixed Hodge structures). Two excellent reference sources are [B-Z] and [Ja1]. If $r\in\mathbb{Z}$, then the Tate twist $\mathbb{A}(r) := (2\pi \mathrm{i})^r\mathbb{A}$ is the (pure) Hodge structure with weight $-2r$ and Hodge type $(-r,-r)$. It is customary to make the further assumption that $\mathbb{A}\otimes_{\mathbb{Z}}\mathbb{Q}$ is a field, and we will assume this. The reasons have to do with Deligne's observation (his $I^{p,q}$ decomposition theorem — a user-friendly explanation provided in [St]) that the weight functor $W_\bullet$ is exact (same for the Hodge filtration functor $F^\bullet$)[1]. Let V_1, V_2 be $\mathbb{A}$-MHS. Carlson [Ca] was the first to give an explicit description of $\mathrm{Ext}^1_{\mathbb{A}-\mathrm{MHS}}(V_1,V_2)$ in terms of a "torus", with the consequence that $\mathrm{Ext}^1_{\mathbb{A}-\mathrm{MHS}}(V_1,-)$ is a right exact functor. If we assume for the moment that $\mathbb{A}$-MHS has enough injectives, then it is clear from formal homological algebra arguments that $\mathrm{Ext}^n_{\mathbb{A}-\mathrm{MHS}}(V_1,V_2)=0$ form $n\geq 2$. In general, one uses an Yoneda-Ext argument. The vanishing of the higher Ext's was first proven by Beilinson [Be1].
• Let's fix $\mathbb{A}$ as per the above paragraph, and put, for V an $\mathbb{A}$-MHS, $\Gamma(V) = \hom_{\mathbb{A}-\mathrm{MHS}}(\mathbb{A}(0),V)$, $J(V) = \mathrm{Ext}^1_{\mathbb{A}-\mathrm{MHS}}(\mathbb{A}(0),V)$. For instance, if $\mathbb{A}=\mathbb{Q}$, then the classical Hodge conjecture asserts that $\Gamma\big(H^{2r}(X(\mathbb{C}),\mathbb{Q}(r))\big)$ is generated by the fundamental classes of cycles $z^r(X;\mathbb{Q}) := z^r(X)\otimes\mathbb{Q}$. The space $\Gamma\big(H^{2r}(X(\mathbb{C}), \mathbb{Q}(r))\big)$, of dimension M say, in untwisted form is precisely $F^rH^{2r}(X(\mathbb{C}),\mathbb{C})\cap H^{2r}(X(\mathbb{C}),\mathbb{Q})\simeq\bigoplus_1^M\mathbb{Q}(-r)$. In general, $F^rH^i(X(\mathbb{C}),\mathbb{C})\cap H^i(X(\mathbb{C}),\mathbb{Q})$ need not be a Hodge structure, as first observed by Grothendieck (see [Lew1], §7)). The (unique) largest Hodge structure in $F^rH^i(X(\mathbb{C}),\mathbb{C})\cap H^i(X(\mathbb{C}),\mathbb{Q})$ is denoted by $N^r_HH^i(X(\mathbb{C}),\mathbb{Q})$. There is also a filtration by coniveau, denoted by

$$N^rH^i(X,\mathbb{Q})\subseteq N^r_HH^i(X(\mathbb{C}),\mathbb{Q}).$$

The (Grothendieck amended) general Hodge conjecture (GHC) asserts that the aforementioned inclusion is an equality (the reader can again consult [Lew1](§7) for details).

[1] Exactness is implied by strict compatibility which means that $h(F^rV_{1,\mathbb{C}}) = h(V_{1,\mathbb{C}})\cap F^rV_{2,\mathbb{C}}$ and $h(W_\ell V_{1,\mathbb{A}\otimes\mathbb{Q}}) = h(V_{1,\mathbb{A}\otimes\mathbb{Q}})\cap W_\ell V_{2,\mathbb{A}\otimes\mathbb{Q}}$ for all r and ℓ. The idea is this. For any $\mathbb{A}$-MHS V, $V_{\mathbb{C}}$ has a $\mathbb{C}$-splitting into a bigraded direct sum of complex vector spaces $I^{p,q} := F^p\cap W_{p+q}\cap\big[\overline{F^q}\cap W_{p+q}+\sum_{i\geq 2}\overline{F^{q-i+1}}\cap W_{p+q-i}\big]$, where one shows that $F^rV_{\mathbb{C}} = \oplus_{p\geq r}\oplus_q I^{p,q}$ and $W_\ell V_{\mathbb{C}} = \oplus_{p+q\leq\ell}I^{p,q}$. Then by construction of $I^{p,q}$, one has $h(I^{p,q}(V_{1,\mathbb{C}})\subseteq I^{p,q}(V_{2,\mathbb{C}})$. Hence h preserves both the Hodge and complexified weight filtrations. Now use the fact that $\mathbb{A}\otimes\mathbb{Q}$ is a field to deduce that h preserves the weight filtration over $\mathbb{A}\otimes\mathbb{Q}$.

• If V is an $\mathbb{A}$-MHS, then by [Ca], [Ja2],

$$J(V) \simeq \frac{W_0 V_{\mathbb{C}}}{F^0 W_0 V_{\mathbb{C}} + W_0 V}.$$

As an example for $X/\mathbb{C}$, $J\big(H^{2r-1}(X(\mathbb{C}), \mathbb{Z}(r))\big)$ denotes the r-th Griffiths jacobian, and $J\big(H^{2r-2}(X(\mathbb{C}), \mathbb{R}(r))\big) \simeq H^{r-1,r-1}(X(\mathbb{C}), \mathbb{R}(r-1))$, the target space (after incorporating twists, viz., after multiplication by $(2\pi \mathrm{i})^{r-1}$) of $R_{r,1}$ in (1). Indeed, $J\big(H^{2r-2}(X(\mathbb{C}), \mathbb{R}(r-1))\big)$ is a version of real Deligne cohomology $H^{2r-1}_{\mathcal{D}}(X(\mathbb{C}), \mathbb{R}(r))$, where we consider $X/\mathbb{C}$ as a real variety via $X \to \mathrm{Spec}(\mathbb{C}) \to \mathrm{Spec}(\mathbb{R})$ (see [Ja1]).

3 Intermezzo I

Steven Zucker's seminal work [Z], the L_2-cohomology in the Poincaré metric associated to a polarizable variation of Hodge structure over a base curve, turned out to provide one instance of an L_2-cohomology interpretation of a corresponding intersection cohomology, the coincidence in the general situation over an arbitrary base manifold S with $\overline{S}$ Kähler, conjectured by Deligne, and settled by the works of W. Schmid, A. Kaplan, and E. Cattani, following the development of Schmid's sl_2-orbit theorem to several variables.[2] In this part, we are interested in a lesser known result of Zucker's work, as it relates to a global function field height pairing due to Beilinson [Be3], albeit in characteristic zero. We wish to make it clear that the construction here is simply an interpretation of Section 1 in [Be3], from the point of view of the L_2-cohomology in [Z]. Start off with a diagram

$$\begin{array}{ccc} \mathscr{X} & \hookrightarrow & \overline{\mathscr{X}} \\ \rho\downarrow & & \downarrow\overline{\rho} \\ C & \stackrel{j}{\hookrightarrow} & \overline{C} \end{array}$$

where $\overline{C}$ is a smooth projective curve, C affine, $\overline{\rho}$ is proper, ρ is smooth and proper, and all varieties are smooth, defined over a field $k \subseteq \mathbb{C}$. Let $K = k(C) = k(\eta)$, $\eta \in C/k$ the generic point, and set $X_K = \mathscr{X}_\eta$, the generic fiber. Note that

$$\mathrm{CH}^r(\mathscr{X}_\eta) = \varinjlim_{U \subset C/k} \mathrm{CH}^r(\rho^{-1}(U/k)),$$

and that the cycle class map

$$\mathrm{CH}^r(\mathscr{X}_\eta; \mathbb{Q}) \to H^{2r}(\mathscr{X}_\eta(\mathbb{C}), \mathbb{Q}(r)) := \varinjlim_{U \subset C/k} H^{2r}(\rho^{-1}(U(\mathbb{C}))), \mathbb{Q}(r))$$

[2] The reader is encouraged to consult [C-K-S] for more precise details concerning this discussion.

is induced by

$$\lim_{\substack{\longrightarrow \\ U\subset C/k}} \left(\mathrm{CH}^r(\rho^{-1}(U/k);\mathbb{Q}) \to H^{2r}(\rho^{-1}(U(\mathbb{C}))),\mathbb{Q}(r))\right).$$

Warning. The definition of $H^{2r}(X_K,\mathbb{Q}(r))$, which is commonly interpreted as $H^{2r}(X_K(\mathbb{C}),\mathbb{Q}(r))$, should *not* be misconstrued as the same object as $H^{2r}(\mathscr{X}_\eta(\mathbb{C}), \mathbb{Q}(r))$, the latter defined by a limit process.

The affine Lefschetz theorem, the fact that C is a curve, together with the (known degeneration of the) Leray spectral sequence (Deligne, but see [Z](§15) and the references cited there), tells us that the Leray filtration

$$H^{2r}(\mathscr{X}(\mathbb{C}),\mathbb{Q}(r)) = L_0 \supset L_1 \supset L_2 \supset \{0\},$$

satisfies

$$L_0/L_1 = H^0(C, R^{2r}\rho_*\mathbb{Q}(r)),$$
$$L_1/L_2 = H^1(C, R^{2r-1}\rho_*\mathbb{Q}(r)),$$
$$L_2H^2(C, R^{2r-2}\rho_*\mathbb{Q}(r)) = 0,$$

with same story for $H^{2r}(\mathscr{X}_\eta(\mathbb{C}),\mathbb{Q}(r))$, where we replace C by η. It is clear then that $\xi \in \mathrm{CH}^r_{\mathrm{hom}}(X_K;\mathbb{Q}) = \mathrm{CH}^r_{\mathrm{hom}}(\mathscr{X}_\eta;\mathbb{Q})$ maps to zero in $H^0(\eta, R^{2r}\rho_*\mathbb{Q}(r))$ by the Leray spectral sequence associated to ρ. Indeed, ξ will have a spread cycle $\tilde{\xi} \in \mathrm{CH}^r(\overline{\mathscr{X}};\mathbb{Q})$, with $\tilde{\xi}\big|_{\mathscr{X}} \mapsto 0 \in H^0(C, R^{2r}\rho_*\mathbb{Q}(r))$. Thus $\tilde{\xi}\big|_{\mathscr{X}} \in H^1(C, R^{2r-1}\rho_*\mathbb{Q}(r))$. Let $d = \dim X_K$, which is the relative dimension of the flat morphism $\overline{\rho}$. Observe that the product $H^1(C, R^{2r-1}\rho_*\mathbb{Q}(r)) \otimes H^1(C, R^{2d-2r+1}\rho_*\mathbb{Q}(d-r+1)) \xrightarrow{\cup} H^2(C, R^{2d}\rho_*\mathbb{Q}(d)) = 0$, is zero. Indeed, to re-iterate, this is due to the affine Lefschetz theorem applied to a smooth affine C with cohomological degree $2 > 1 = \dim C$. Notice however that $\overline{\mathscr{X}}$ is complete (and smooth), and hence $H^{2r}(\overline{\mathscr{X}},\mathbb{Q}(r))$ is a pure Hodge structure of weight zero, viz., $H^{2r}(\overline{\mathscr{X}},\mathbb{Q}(r)) = W_0H^{2r}(\overline{\mathscr{X}},\mathbb{Q}(r))$. Furthermore $[\xi]$ is the restriction of $[\tilde{\xi}] \in H^{2r}(\overline{\mathscr{X}},\mathbb{Q}(r))$. Thus it is well known (rather implicit after reading [Z](§14) and more to the point in [PM]), that for $\xi \in \mathrm{CH}^r_{\mathrm{hom}}(X_K;\mathbb{Q})$,

$$[\xi] \in W_0H^1(\eta, R^{2r-1}\rho_*\mathbb{Q}(r)) = W_0H^1(C, R^{2r-1}\rho_*\mathbb{Q}(r))$$
$$= H^1(\overline{C}, j_*R^{2r-1}\rho_*\mathbb{Q}(r)),$$

the latter being the object of interest in [Z]. Note that the pairing (also being a pairing of intersection cohomologies)

$$H^1(\overline{C}, j_*R^{2r-1}\rho_*\mathbb{Q}(r)) \otimes H^1(\overline{C}, j_*R^{2d-2r+1}\rho_*\mathbb{Q}(d-r+1))$$
$$\xrightarrow{\cup} H^2(\overline{C}, R^{2d}\rho_*\mathbb{Q}(d)) \simeq \mathbb{Q},$$

is non-degenerate [Z]. Now for $\xi_1 \in \mathrm{CH}^r_{\mathrm{hom}}(X_K;\mathbb{Q})$, $\xi_2 \in \mathrm{CH}^{d-r+1}_{\mathrm{hom}}(X_K;\mathbb{Q})$, we arrive at Beilinson's (global case) height pairing over the function field of a curve [Be3], viz., $\langle\ ,\ \rangle : \mathrm{CH}^r_{\mathrm{hom}}(X_K;\mathbb{Q}) \otimes \mathrm{CH}^{d-r+1}_{\mathrm{hom}}(X_K;\mathbb{Q}) \to$

$$H^1(\overline{C}, j_*R^{2r-1}\rho_*\mathbb{Q}(r)) \otimes H^1(\overline{C}, j_*R^{2d-2r+1}\rho_*\mathbb{Q}(d-r+1)) \simeq \mathbb{Q}.$$

Remark 3.1. *For each closed point $v \in C$, Beilinson [Be3] attached a local linking number $\langle\ ,\ \rangle_v$, and showed that the global height pairing is the sum of local ones.*

4 The arithmetic scenario

Now let X be a smooth projective variety of dimension d, defined over a number field k (i.e., $[k:\mathbb{Q}]<\infty$). Denote by $z^\bullet_{\text{hom}}(X)$ the nullhomologous cycles. Under some assumptions, Beilinson in [Be3] defined a height pairing

$$z^r_{\text{hom}}(X;\mathbb{Q})\times z^{d-r+1}_{\text{hom}}(X;\mathbb{Q})\to\mathbb{R},$$

which factors through $z^\bullet_{\text{rat}}(X;\mathbb{Q})$, viz., an induced height pairing

$$\langle\ ,\ \rangle_{\text{HT}}:\text{CH}^r_{\text{hom}}(X;\mathbb{Q})\times\text{CH}^{d-r+1}_{\text{hom}}(X;\mathbb{Q})\to\mathbb{R}. \tag{3}$$

This pairing should have a number of conjectural properties, for example

Conjecture 4.1 (Conjecture 5.4 (a) of [Be3]). *The height pairing is non-degenerate.*

Conjecture 4.2 (Hodge-index Conjecture 5.5 of [Be3]). *Assume that a hard Lefschetz conjecture holds on null-homologous cycles (Conjecture 5.3 of [Be3]) and consider the primitive cycle decomposition. Let $L_X\in CH^1(X)$ be the class of a hyperplane section. Then the form $\langle\cdot,L^{d-2r+1}\cdot\rangle_{\text{HT}}$ is definite of sign $(-1)^r$ on the primitive r-cycles for $r\le\frac{d+1}{2}$.*

These conjectures seem to mimick the nondegeneracy and the polarizing properties of the cohomology of X. For example, Conjecture 4.1 is an analog of the non-degeneracy of

$$H^{2r-1}(X(\mathbb{C}),\mathbb{Q})\times H^{2d-2r+1}(X(\mathbb{C}),\mathbb{Q})\to\mathbb{Q}.$$

It is instructive to explain the idea behind the pairing: The ingredient comes from

4.3 Arithmetic Chow groups. References for this section are [GS], [BGS], and [Ku]. We will only provide a brief glimpse into this fascinating theory. Interested readers may consult the references cited above for details (especially [BGS]). We begin with a motivating example (Chapter III of [Neu]):

Example 4.4. *Consider a number field k with the number ring $\mathcal{O}_k$. A prime $\wp$ of k is a class of equivalent valuations of k. The non-Archimedean equivalence classes are called* ***finite*** *primes and accordingly the Archimedean ones are* ***infinite*** *primes. The infinite primes $\wp$ are obtained from the embeddings $\tau:k\hookrightarrow\mathbb{C}$. There are two sorts of these: the real primes, corresponding to the real embeddings, and the complex primes corresponding to the pairs of complex conjugate non-real embeddings. The finite primes will be denoted formally by $\wp$ and the infinite primes by $\wp_\infty$.*

To each prime $\wp\in k$ (finite or infinite), we associate a canonical homomorphism

$$v_\wp:k^*\to\mathbb{R}$$

from the multiplicative group k^* of k. If $\wp$ is a finite prime, then $v_\wp$ is a $\wp$-adic exponential valuation which is normalized by the condition $v_\wp(k^*) = \mathbb{Z}$. If $\wp$ is infinite, then $v_\wp$ is given by $v_\wp(a) = -\ln|\tau a|$, where τ is an embedding which defines $\wp$.

Arakelov class group of $\mathcal{O}_k$: *The group $\widehat{\mathrm{Div}}(\mathcal{O}_k)$- of (Arakelov) divisors is defined by elements of the form*

$$D := \sum_{\wp} m_\wp \wp + \sum_{\wp_\infty} \lambda_\infty \wp_\infty,$$

where $m_\wp \in \mathbb{Z}$ and $\lambda_\infty \in \mathbb{R}$, respectively. The principal divisors $\widehat{\mathcal{P}}(\mathcal{O}_k)$ are of the form

$$\sum_{\wp} v_\wp(\alpha)\wp + \sum_{\wp_\infty}(-\log|\alpha|_{\wp_\infty}),$$

where $|\alpha|_{\wp_\infty} = |\tau\alpha|$ if $\wp_\infty$ is real, and $|\alpha|_{\wp_\infty} = |\tau\alpha|^2$ if $\wp_\infty$ is complex. We define the Arakelov class group of $\mathcal{O}_k$ as the quotient

$$\widehat{\mathrm{Cl}}(\mathcal{O}_k) := \widehat{\mathrm{Div}}(\mathcal{O}_k)/\widehat{\mathcal{P}}(\mathcal{O}_k).$$

One can define a real number, called the degree of a divisor D as

$$\widehat{deg}(D) := \sum_{\wp} m_\wp \log N_\wp + \sum_{\wp_\infty} \lambda_\infty,$$

with $N_\wp = |\mathcal{O}_k/\wp|$. The degree of a principal divisor is zero by the product formula. Hence we get a well-defined (continuous) homomorphism

$$\widehat{deg} : \widehat{\mathrm{Cl}}(\mathcal{O}_k) \to \mathbb{R}.$$

More generally, consider a regular, projective and flat scheme $\widetilde{X} \to S = \mathrm{Spec}(\mathcal{O}_k)$ of absolute dimension $d+1$. Such a scheme will be referred to as a regular arithmetic variety. Note that, from the definition $\widetilde{X}$ can be seen as a projective and flat scheme over $\mathrm{Spec}(\mathbb{Z})$.

For a regular arithmetic variety $\widetilde{X}$, and any integer $r \geq 0$ we let $Z^r(\widetilde{X})$ be the free abelian group of cycles of codimension p over $\widetilde{X}$. The set of complex points $\widetilde{X}(\mathbb{C})$ of $\widetilde{X}$ can be identified with the disjoint union $\coprod_{\sigma:k\hookrightarrow\mathbb{C}} \widetilde{X}_\sigma(\mathbb{C})$. Let $F_\infty : \widetilde{X}(\mathbb{C}) \to \widetilde{X}(\mathbb{C})$ be the antiholomorphic involution coming from complex conjugation. We denote by $D^{r,r}(\widetilde{X}_\mathbb{R})$ the set of real currents in $D^{r,r}(\widetilde{X}(\mathbb{C}))$ (with respect to a suitable action F_∞^* of F_∞ on $D^{r,r}(\widetilde{X}(\mathbb{C})))$[3].

Now, any cycle $Z \in Z^r(\widetilde{X})$ defines a current $\delta_Z \in D^{r,r}(\widetilde{X}_\mathbb{R})$ by integration on its set of complex points. A *Green current* for Z is any current $g \in D^{r-1,r-1}(\widetilde{X}_\mathbb{R})$ such that $dd^c g + \delta_Z$ is smooth (see §1 of [BGS] for details and notations). Denote by

[3] A comment is in order here: In Arakelov setting, a current $\alpha \in D^{r,r}(\widetilde{X}(\mathbb{C}))$ is called real if $F_\infty^*(\alpha) = (-1)^r\alpha$ (see either Section 3.2 of [GS] or 2.1 of [BGS]).

$\widehat{Z}^r(\widetilde{X})$- the group of pairs (Z, g_Z) where $Z \in Z^r(\widetilde{X})$ and g_Z is a Green current for Z, with addition defined component-wise. Let $\widehat{R}^r(\widetilde{X}) \subset \widehat{Z}^r(\widetilde{X})$ be the subgroup generated by pairs of the form $(0, \partial u + \overline{\partial} v)$, where u and v are currents of type $(r-2, r-1)$ and $(r-1, r-2)$ respectively (∂ and $\overline{\partial}$ being suitable operations on the space of currents), and $(\operatorname{div}(f), -\log|f|^2)$, where f is a rational function on an integral subscheme $\widetilde{Y} \subset \widetilde{X}$ of codimension $r-1$, and $-\log|f|^2$ is the current on $\widetilde{X}(\mathbb{C})$ obtained by restricting forms to the smooth part of $\widetilde{Y}(\mathbb{C})$ and integrating against the L^1 function $-\log|f|^2$. Now define the *arithmetic Chow group* of codimension r as

$$\widehat{\mathrm{CH}}^r(\widetilde{X}) = \widehat{Z}^r(\widetilde{X})/\widehat{R}^r(\widetilde{X}).$$

Remark 4.5. *Arithmetic Chow groups can be defined for more general types of arithmetic varieties assuming only that the generic fibre is smooth (refer to §3.2 of [GS] for details). In case our arithmetic variety is $S = \operatorname{Spec}(\mathcal{O}_k)$ for the number ring $\mathcal{O}_k$ of k, the arithmetic Chow group $\widehat{CH}^1(S) \cong \widehat{\mathrm{Cl}}(\mathcal{O}_k)$.*[4]

We note down some crucial properties of arithmetic Chow groups:

- (Theorem 4.2.3 of [GS]) There is a cup product of arithmetic Chow groups

$$\widehat{\mathrm{CH}}^r(\widetilde{X}) \otimes \widehat{\mathrm{CH}}^s(\widetilde{X}) \to \widehat{\mathrm{CH}}^{r+s}(\widetilde{X}; \mathbb{Q}),$$

  formally defined by the formula $[(Z_1, g_{Z_1})] \cdot [(Z_2, g_{Z_2})] = [(Z_1 \cdot Z_2, g_{Z_1} \star g_{Z_2})]$, where $\star$ denotes the star product of Green currents (§1 of [GS]).

- (Theorems 3.6.1 and 4.2.3 of [GS]) Let $f : \widetilde{X} \to \widetilde{Y}$ be a morphism of regular arithmetic varieties. Then there is a pull-back homomorphism $f^* : \widehat{\mathrm{CH}}^r(\widetilde{Y}) \to \widehat{\mathrm{CH}}^r(\widetilde{X}; \mathbb{Q})$. It is multiplicative, i.e., given $\alpha \in \widehat{\mathrm{CH}}^r(\widetilde{Y})$ and $\beta \in \widehat{\mathrm{CH}}^s(\widetilde{Y})$, we have

$$f^*(\alpha \cdot \beta) = f^*(\alpha) \cdot f^*(\beta).$$

  Further if f is proper, $f_k : \widetilde{X}_k \to \widetilde{Y}_k$ is smooth and $\widetilde{X}, \widetilde{Y}$ are equidimensional, then there is a push-forward homomorphism

$$f_* : \widehat{\mathrm{CH}}^r(X) \to \widehat{\mathrm{CH}}^{r-\delta}(Y) \ (\delta := \dim(X) - \dim(Y))$$

  satisfying the projection formula

$$f_*(f^*(\alpha) \cdot \beta) = \alpha \cdot f_*(\beta).$$

[4] One uses a special case of Theorem 3.3.5, exact sequence (i) of [GS], setting $\widetilde{X} = S = \operatorname{Spec}(\mathcal{O}_k)$ (see Section 3.4 of [GS] for details).

4.6 Arithmetic height pairing. Let $\widehat{\mathrm{CH}}^*(\widetilde{X})$ be the arithmetic Chow theory as defined above. One can define an *arithmetic degree* map as a push-forward

$$\widehat{deg}_{\widetilde{X}} : \widehat{\mathrm{CH}}^{d+1}(\widetilde{X}) \to \widehat{\mathrm{CH}}^1(S) \cong \widehat{\mathrm{Cl}}(\mathcal{O}_k) \to \mathbb{R},$$

where $\widehat{\mathrm{Cl}}(\mathcal{O}_k) \to \mathbb{R}$ is the $\widehat{deg}$ map of Example 4.4.[5] Together with the arithmetic intersection, it defines a pairing

$$\widehat{\mathrm{CH}}^r(\widetilde{X};\mathbb{Q}) \otimes \widehat{\mathrm{CH}}^{d-r+1}(\widetilde{X};\mathbb{Q}) \to \widehat{\mathrm{CH}}^{d+1}(\widetilde{X};\mathbb{Q}) \to \mathbb{R}.$$

For a smooth projective variety X/k, assume that it has a *regular model* $\widetilde{X}$, i.e., a regular arithmetic variety which is projective and flat over $S := \mathrm{Spec}(\mathcal{O}_k)$, together with an isomorphism $\widetilde{X}_k \cong X/k$. Let's explore this a little bit further. We are considering a family $\widetilde{X} \to S$, where the generic fibre is isomorphic to X/k. This resembles the situation $\rho : \mathscr{X} \to C$ of §3. Now for each finite prime $\wp \in S$, we get a finite fibre $\widetilde{X}_\wp := \widetilde{X} \times_S \mathrm{Spec}(\mathcal{O}_k/\wp\mathcal{O}_k)$, and for each embedding $\sigma : k \hookrightarrow \mathbb{C}$, we get a "fibre at infinity" $\widetilde{X}_\sigma := \widetilde{X} \times_\sigma \mathbb{C}$. We think of the whole family (fibres over finite and infinite primes/embeddings) as a completion of $\widetilde{X}$ over $\overline{S} := S \cup \{\sigma : k \hookrightarrow \mathbb{C}\}_\sigma$, resembling the situation $\overline{\rho} : \overline{\mathscr{X}} \to \overline{C}$ in §3.

Remark 4.7. *The existence of a regular model for X is a highly non-trivial problem. As a basic example, smooth projective curves have regular models (after possibly extending the base field). Apart from that, triple product of curves (Gross-Schoen) and abelian varieties (Künnemann) provides us with a large class of examples.*

With this set up and under a further assumption ((17) of [Ku]), Beilinson's height pairing can be interpreted in light of arithmetic intersection

$$\widehat{\mathrm{CH}}^r(\widetilde{X};\mathbb{Q}) \otimes \widehat{\mathrm{CH}}^{d-r+1}(\widetilde{X};\mathbb{Q}) \to \widehat{\mathrm{CH}}^{d+1}(\widetilde{X};\mathbb{Q}) \to \mathbb{R}.$$

This pairing may a priori depend on the choice of $\widetilde{X}$. Since our primary aim is to detect non-trivial cycles, this choice is not a hinderance, once we have one. To get a more earthly description, we stretch the analogy with Example 4.4 even further. Think of $\widetilde{X}$ as a projective scheme $\mathrm{Proj}\,(\mathcal{O}_k[X_0, \ldots, X_N]/I) \to S$, for some homogeneous ideal I. The height pairing then is a sum of non-Archimedean and Archimedean parts

$$\langle \xi_1, \xi_2 \rangle_{\mathrm{HT}} := \sum_{\wp} \langle \xi_1, \xi_2 \rangle_\wp + \sum_{\wp_\infty} \langle \xi_1, \xi_2 \rangle_{\wp_\infty},$$

where (at least for finite primes $\wp$ of good reduction) $\langle \xi_1, \xi_2 \rangle_\wp$ is given by

$$\log N_\wp \times \left(\#|(\xi_1 \cap \xi_2)_\wp|, \text{counting multiplicities}\right), \text{ albeit heuristically!}$$

[5] We remark here that this $\widehat{deg}$ is really the composition of the push-forward morphism $\widehat{\mathrm{CH}}^1(S) \to \widehat{\mathrm{CH}}^1(\mathrm{Spec}(\mathbb{Z}))$ attached to the unique morphism $S \to \mathrm{Spec}(\mathbb{Z})$, and the isomorphism $\widehat{\mathrm{CH}}^1(\mathrm{Spec}(\mathbb{Z})) \cong \mathbb{R}$ (see 2.1.3 of [BGS] for a detailed discussion of arithmetic degree maps).

On the subgroup of cycles algebraically equivalent to zero, the height pairing is given by the Néron-Tate pairing

$$J^r_{\text{alg}}(X)(\overline{k})_{\mathbb{Q}} \times J^{d-r+1}_{\text{alg}}(X)(\overline{k})_{\mathbb{Q}} \to \mathbb{R},$$

where

$$J^r_{\text{alg}}(X)(\overline{k})_{\mathbb{Q}} := \Phi_r(\text{CH}^r_{\text{alg}}(X/\overline{k};\mathbb{Q})),$$

and

$$\Phi_r : \text{CH}^r_{\text{hom}}(X/k) \to J\big(H^{2r-1}(X(\mathbb{C}), \mathbb{Z}(r))\big)$$

is the Griffiths Abel-Jacobi map.

Remark 4.8. *Much like the notion of a height function, one can extend the height pairing for a smooth and projective X defined over $\overline{\mathbb{Q}}$ (see 4.0.6 of [Be3]).*

Remark 4.9. *The Archimedean part of the height pairing $\sum_{\wp_\infty}\langle \xi_1, \xi_2\rangle_{\wp_\infty}$ is given by the star product of green currents g_{ξ_1} and g_{ξ_2} (see §1 of [MS1] for details). Restricting further to the subgroups $z^r_{rat}(X)$ and $z^{d-r+1}_{rat}(X)$, this Archimedean part of the height pairing is the one defined in §1 (up to factors).*

5 Bloch-Beilinson filtration

It was first indicated in [Blo2], and later fortified by Beilinson, that for X smooth and projective over a field k, there should be a descending filtration

$$F^0 := CH^r(X;\mathbb{Q}) \supset F^1 = CH^r_{\text{hom}}(X;\mathbb{Q}) \supset \cdots \supset F^r \supset \{0\},$$

satisfying

$$Gr^\nu_F \text{CH}^r(X;\mathbb{Q}) := \frac{F^\nu \text{CH}^r(X;\mathbb{Q})}{F^{\nu+1}\text{CH}^r(X;\mathbb{Q})} \simeq \text{Ext}^\nu_{\mathcal{MM}}\big(\text{Spec}(k), h^{2r-\nu}(X)(r)\big),$$

where $\mathcal{MM}$ is the conjectural category of mixed motives over k. A number of candidate filtrations have been proposed (names will suffice) by Jannsen [Ja3], S. Saito [SSa], M. Saito/M. Asakura [A], Murre [Mu], Griffiths-Green [G-G], Lewis [Lew2], Lewis/Kerr [K-L], Raskind [Ra], and so forth.... In the case $k = \mathbb{C}$ we have seen that in the category of MHS, $\text{Ext}^{\bullet\geq 2}_{\text{MHS}} = 0$, and yet $F^2\text{CH}^r(X;\mathbb{Q})$ need not be zero (Mumford [M], Bloch [Blo2], Lewis (op. cit. and [Lew4]), Schoen (see [Ja2]), Roitman [Ro], Griffiths-Green [G-G-P],...). Even in the case where $\text{trdeg}_{\mathbb{Q}} k = 1$, there are examples from some of the references (op. cit.) that $F^2\text{CH}^r(X_k;\mathbb{Q}) \neq 0$. Indeed Beilinson and Bloch have independently conjectured the following:

Conjecture 5.1. *Let $X/\overline{\mathbb{Q}}$ be smooth and projective. Then the Griffiths Abel-Jacobi map*

$$\Phi_r : CH^r_{\text{hom}}(X/\overline{\mathbb{Q}};\mathbb{Q}) \to J(H^{2r-1}(X(\mathbb{C}), \mathbb{Q}(r))),$$

is injective.

Remark 5.2. *Assuming the classical Hodge conjecture, one can argue that X in the conjecture can be replaced by a smooth quasi-projective variety over $\overline{\mathbb{Q}}$. This follows from a weight filtered spectral sequence argument [K-L](p. 371).*

For $k = \mathbb{C}$, the following theorem best summarizes one's expectations: First consider fields $\overline{\mathbb{Q}} \subset K \subset \mathbb{C}$, where $K/\overline{\mathbb{Q}}$ is finitely generated. One first constructs a filtration on $\mathrm{CH}^r(X_K;\mathbb{Q})$. The "lift" from K to $\mathbb{C}$ follows from:

$$F^\nu \mathrm{CH}^r(X_{\mathbb{C}};\mathbb{Q}) = \varinjlim_{K\subset\mathbb{C}} F^\nu \mathrm{CH}^r(X_K;\mathbb{Q})$$

Theorem 5.3 ([Lew2]). *Let X/K be smooth projective of dimension d. Then for all r, there is a filtration,*

$$CH^r(X_K;\mathbb{Q}) = F^0 \supset F^1 \supset \cdots \supset F^\nu \supset F^{\nu+1} \supset \\ \cdots \supset F^r \supset F^{r+1} = F^{r+2} = \cdots,$$

which satisfies the following

(i) $F^1 = CH^r_{\mathrm{hom}}(X_K;\mathbb{Q})$.

(ii) $F^2 \subseteq \big(\ker \Phi_r : CH^r_{\mathrm{hom}}(X_K;\mathbb{Q}) \to J\big(H^{2r-1}(X_K(\mathbb{C}),\mathbb{Q}(r))\big)\big)$.

(iii) $F^{\nu_1} CH^{r_1}(X_K;\mathbb{Q}) \bullet F^{\nu_2} CH^{r_2}(X_K;\mathbb{Q}) \subset F^{\nu_1+\nu_2} CH^{r_1+r_2}(X_K;\mathbb{Q})$, *where $\bullet$ is the intersection product.*

(iv) *F^ν is preserved under the action of correspondences between smooth projective varieties over K.*

(v) *Assume that the Künneth components of the diagonal class*

$$[\Delta_X] = \oplus_{p+q=2d}[\Delta_X(p,q)] \in H^{2d}(X\times X,\mathbb{Q}(d)))$$

are algebraic and defined over K. Then

$$\Delta_X(2d-2r+\ell+m, 2r-\ell-m)_*\big|_{Gr_F^\nu CH^r(X;\mathbb{Q})} = \delta_{\ell,\nu}\cdot \mathrm{Identity}.$$

[If we assume the conjecture that homological and numerical equivalence coincide, then (v) says that Gr_F^ν factors through the Grothendieck motive.]

(vi) *Let $D^r(X_K) := \bigcap_\nu F^\nu$. If Conjecture 5.1 holds for smooth quasi-projective varieties defined over $\overline{\mathbb{Q}}$ (vis-à-vis Remark 5.2), then $D^r(X_K) = 0$ (hence $D^r(X_{\mathbb{C}}) = 0$).*

It is instructive to briefly explain how this filtration comes about. For X/K smooth projective, one can find a smooth quasi-projective $\mathcal{S}/\overline{\mathbb{Q}}$ such that $\overline{\mathbb{Q}}(\mathcal{S})$ is identified with K. One can then spread out X/K to a family $\rho : \mathscr{X} \to \mathcal{S}$, where ρ is a smooth and proper morphism of smooth quasi-projective varieties over $\overline{\mathbb{Q}}$, and X/K is the generic fiber. As a momentary digression, we offer the reader an illuminating illustration of the notion of spreads:

Example 5.4. *Let*

$$Y/\mathbb{C} = \mathrm{Spec}\left\{\frac{\mathbb{C}[x,y]}{(\pi y^2 + (\sqrt{\pi}+4)x^3 + \mathrm{e}x)}\right\},$$

$$\mathcal{S}/\mathbb{Q} = \mathrm{Spec}\left\{\frac{\mathbb{Q}[u,v,w]}{(u-v^2)}\right\}.$$

Set

$$\mathcal{Y}_{\mathcal{S}} = \mathrm{Spec}\left\{\frac{\mathbb{Q}[x,y,u,v,w]}{(uy^2 + (v+4)x^3 + wx, u-v^2)}\right\}.$$

The inclusion

$$\frac{\mathbb{Q}[u,v,w]}{(u-v^2)} \subset \frac{\mathbb{Q}[x,y,u,v,w]}{(uy^2 + (v+4)x^3 + wx, u-v^2)}$$

defines a morphism $\mathcal{Y}_{\mathcal{S}} \to \mathcal{S}$, as varieties over $\mathbb{Q}$. Let $\eta \in \mathcal{S}$, be the generic point. Then

$$\mathbb{Q}(\eta) = \mathrm{Quot}\left(\frac{\mathbb{Q}[u,v,w]}{(u-v^2)}\right).$$

Note that the embedding

$$\mathbb{Q}(\eta) \hookrightarrow \mathbb{C}, \quad (u,v,w) \mapsto (\pi, \sqrt{\pi}, \mathrm{e}), \ \Rightarrow \mathcal{Y}_{\mathcal{S},\eta} \times \mathbb{C} = Y/\mathbb{C}.$$

We will have more to say about this in the next section.

Now here is the key point. Beilinson's absolute Hodge cohomology $H_{\mathcal{H}}$ [Be1], is a highly sophisticated cohomology theory with a number of similar properties to Deligne-Beilinson cohomology, with the advantage of incorporating weights. For our purposes here, we need the short exact sequence (p. 2 of [Be1]):

$$J\big(H^{2r-1}(\mathscr{X}(\mathbb{C});\mathbb{Q}(r))\big) \hookrightarrow H^{2r}_{\mathcal{H}}(\mathscr{X}(\mathbb{C}),\mathbb{Q}(r)) \twoheadrightarrow \Gamma\big(H^{2r}(\mathscr{X}(\mathbb{C}),\mathbb{Q}(r))\big). \tag{4}$$

There is a cycle class map $\mathrm{cl}_r : \mathrm{CH}^r(\mathscr{X}/\overline{\mathbb{Q}};\mathbb{Q}) \to H^{2r}_{\mathcal{H}}(\mathscr{X}(\mathbb{C}),\mathbb{Q}(r))$, and according to Conjecture 5.1 and Remark 5.2, one anticipates that cl_r is injective. The lowest weight part, $\underline{H}^{2r}_{\mathcal{H}}(\mathscr{X}(\mathbb{C}),\mathbb{Q}(r)) \subset H^{2r}_{\mathcal{H}}(\mathscr{X}(\mathbb{C}),\mathbb{Q}(r))$ is given by the image $H^{2r}_{\mathcal{H}}(\overline{\mathscr{X}}(\mathbb{C}),\mathbb{Q}(r)) \to H^{2r}_{\mathcal{H}}(\mathscr{X}(\mathbb{C}),\mathbb{Q}(r))$, where $\overline{\mathscr{X}}/\overline{\mathbb{Q}}$ is a smooth compactification of $\mathscr{X}/\overline{\mathbb{Q}}$. Note that $\mathrm{CH}^r(\overline{\mathscr{X}}/\overline{\mathbb{Q}}) \to \mathrm{CH}^r(\mathscr{X}/\overline{\mathbb{Q}})$ is surjective; likewise there is a cycle class map $\mathrm{CH}^r(\overline{\mathscr{X}};\mathbb{Q}) \to H^{2r}_{\mathcal{H}}(\overline{\mathscr{X}},\mathbb{Q}(r))$. Thus we conjecturally have an injection

$$\mathrm{cl}_r : \mathrm{CH}^r(\mathscr{X}/\overline{\mathbb{Q}};\mathbb{Q}) \to \underline{H}^{2r}_{\mathcal{H}}(\mathscr{X}(\mathbb{C}),\mathbb{Q}(r)).$$

The filtration $\mathcal{F}^{\nu}\mathrm{CH}^r(\mathscr{X}/\overline{\mathbb{Q}};\mathbb{Q})$ is given by the pullback of the ν-th Leray filtration of ρ on $\underline{H}^{2r}_{\mathcal{H}}(\mathscr{X}(\mathbb{C}),\mathbb{Q}(r))$, to $\mathrm{CH}^r(\mathscr{X}/\overline{\mathbb{Q}};\mathbb{Q})$. (For an excellent motivic description of the Leray filtration, the reader should consult [Ar].) Let $\eta_{\mathcal{S}}$ be the generic point of $\mathcal{S}$, and put $K := \overline{\mathbb{Q}}(\eta_{\mathcal{S}})$, and note that the sequence in (4) remains exact at the generic point, by properties of direct limits. Write $X_K := \mathscr{X}_{\eta_{\mathcal{S}}}$. The injectivity of cl_r passes to the generic point of $\mathcal{S}$, viz., $\mathrm{cl}_r : \mathrm{CH}^r(\mathscr{X}_{\eta_{\mathcal{S}}};\mathbb{Q}) \hookrightarrow \underline{H}^{2r}_{\mathcal{H}}(\mathscr{X}_{\eta_{\mathcal{S}}};(\mathbb{C}),\mathbb{Q}(r))$, leading to a filtration $\{F^{\nu}\mathrm{CH}^r(X_K;\mathbb{Q})\}_{\nu\geq 0}$. Thus following [Lew2] we introduced a decreasing filtration $F^{\nu}\mathrm{CH}^r(X_K;\mathbb{Q})$, with the property that $Gr_F^{\nu}\mathrm{CH}^r(X_K;\mathbb{Q}) \hookrightarrow E_\infty^{\nu,2r-\nu}(\eta_S)$, where $E_\infty^{\nu,2r-\nu}(\eta_S)$ is the ν-th graded piece of the Leray filtration associated to ρ on $\underline{H}_{\mathcal{H}}^{2r}(\mathscr{X}_{\eta_S},\mathbb{Q}(r))$ It is proven in [Lew2] that the term $E_\infty^{\nu,2r-\nu}(\eta_S)$ fits in a short exact sequence:

$$0 \to \underline{E}_\infty^{\nu,2r-\nu}(\eta_S) \to E_\infty^{\nu,2r-\nu}(\eta_S) \to \underline{\underline{E}}_\infty^{\nu,2r-\nu}(\eta_S) \to 0,$$

where

$$\underline{\underline{E}}_\infty^{\nu,2r-\nu}(\eta_S) = \Gamma(H^\nu(\eta_S, R^{2r-\nu}\rho_*\mathbb{Q}(r))),$$

$$\underline{E}_\infty^{\nu,2r-\nu}(\eta_S) = \frac{J(W_{-1}H^{\nu-1}(\eta_S, R^{2r-\nu}\rho_*\mathbb{Q}(r)))}{\Gamma(Gr_W^0 H^{\nu-1}(\eta_S, R^{2r-\nu}\rho_*\mathbb{Q}(r)))} \subset J(H^{\nu-1}(\eta_S, R^{2r-\nu}\rho_*\mathbb{Q}(r))).$$

Here the latter inclusion is a result of the short exact sequence:

$$0 \to W_{-1}H^{\nu-1}(\eta_S, R^{2r-\nu}\rho_*\mathbb{Q}(r)) \to W_0H^{\nu-1}(\eta_S, R^{2r-\nu}\rho_*\mathbb{Q}(r))$$

$$\to Gr_W^0 H^{\nu-1}(\eta_S, R^{2r-\nu}\rho_*\mathbb{Q}(r)) \to 0.$$

We attend to (vi). The idea comes from [Ra]; however, as noted in [Ra], it goes back to a hard Lefschetz argument due to Beauville (see [Lew2]). This will imply that $D^r(X) := D^r(X_K) = 0$ under Conjecture 5.1 for smooth quasi-projective varieties. It is instructive to the reader to explain this argument. It suffices to show that

$$\lim_{\substack{\longrightarrow \\ U\subset \mathcal{S}/\overline{\mathbb{Q}}}} F^{r+1}\mathrm{CH}^r(\rho^{-1}(U);\mathbb{Q})) =: F^{r+1}\mathrm{CH}^r(X_K;\mathbb{Q})) \tag{5}$$

$$= \lim_{\substack{\longrightarrow \\ U\subset \mathcal{S}/\overline{\mathbb{Q}}}} F^{r+j}\mathrm{CH}^r(\rho^{-1}(U);\mathbb{Q})) =: F^{r+j}\mathrm{CH}^r(X_K;\mathbb{Q}), \text{ for all } j \geq 1.$$

Let L_X be the operation of cupping with the hyperplane class of the fibers of ρ, and

$$\psi_{r+j} : F^{r+j}\mathrm{CH}^r(\rho^{-1}(U);\mathbb{Q})) \to E_\infty^{r+j,r-j},$$

the natural map. There is a commutative diagram

$$\begin{array}{ccc} F^{r+j}\mathrm{CH}^r(\rho^{-1}(U);\mathbb{Q})) & \xrightarrow{\psi_{r+j}} & E_\infty^{r+j,r-j} \\ \Big\downarrow L_X^{d-r+j} & & \wr\Big\downarrow L_X^{d-r+j} \\ F^{r+j}\mathrm{CH}^{d+j}(\rho^{-1}(U;\mathbb{Q})) & \to & E_\infty^{r+j,2d-r+j} \end{array}$$

Since $F^{r+j+1}\mathrm{CH}^{r+j}(\rho^{-1}(U);\mathbb{Q})) = \ker\psi_{r+j}$, and that

$$\lim_{\substack{\longrightarrow \\ U\subset \mathcal{S}/\overline{\mathbb{Q}}}} CH^{d+j}(\rho^{-1}(U);\mathbb{Q})) = \mathrm{CH}^{d+j}(X_K;\mathbb{Q}) = 0,$$

for $j \geq 1$, as $\dim X_K = d$, it follows that (5) holds, and we're done.

6 The business of spreads

Consider the smooth elliptic curve

$$E_{\mathbb{C}} := \mathrm{Proj}\left(\frac{\mathbb{C}[z_0, z_1, z_2]}{(2\mathrm{e}z_0z_2^2 - \pi z_1^3 + \sqrt{\pi}z_1z_0^2 + \sqrt{3}\mathrm{i}z_0^3)}\right) \subset \mathbb{P}^2(\mathbb{C}).$$

An analytic geometer may view this as a compact Riemann surface endowed with the analytic topology. If for the moment we view E as a prototypical projective algebraic manifold, then one key distinguishing feature that E (or for that matter any complex algebraic variety) has over general complex manifolds, is that it can be arrived at via base extension from a smaller subfield $K \subset \mathbb{C}$. There are lots of choices for K, but it is customary to think of it as finitely generated over $\mathbb{Q}$. To muddy the water a bit, let's consider $\xi = \{\sqrt{5}z_0^4z_1 + \mathrm{i}z_2^5 = 0\} \cap E_{\mathbb{C}} \in \mathrm{CH}^1(E_{\mathbb{C}})$. In this case, let's choose $K = \mathbb{Q}(\sqrt{\pi}, \mathrm{e}, \mathrm{i}, \sqrt{3}, \sqrt{5})$, and define

$$E_K := \mathrm{Proj}\left(\frac{K[z_0, z_1, z_2]}{(2^{-1}\mathrm{e}z_0z_2^2 - \pi z_1^3 + \sqrt{\pi}z_1z_0^2 + \sqrt{3}\mathrm{i}z_0^3)}\right) \subset \mathbb{P}_K^2.$$

By base change, we have $E_{\mathbb{C}} = E_K \times_K \mathbb{C} := E_K \times_{\mathrm{Spec}(K)} \mathrm{Spec}(\mathbb{C})$. Now K itself can be representative of the process of evaluation of a general point over $\mathbb{Q}$. Let

$$\mathcal{S} = \mathrm{Spec}\left(\frac{\mathbb{Q}[u, v, w, t, s]}{(w^2 + 1, t^2 - 3, s^2 - 5)}\right).$$

Let $\eta_{\mathcal{S}} \in \mathcal{S}/\mathbb{Q}$ be the generic point. Note that by definition $\mathbb{Q}(\eta_{\mathcal{S}}) = \mathbb{Q}(\mathcal{S})$ and that the evaluation map

$$\mathbb{Q}(\mathcal{S}) \hookrightarrow \mathbb{C},\ (u, v, w, t, s) \mapsto p := (\mathrm{e}, \sqrt{\pi}, \mathrm{i}, \sqrt{3}, \sqrt{5}) \in \mathcal{S}(\mathbb{C}), \tag{6}$$

identifies $\mathbb{Q}(\mathcal{S})$ with K. Any other point $q \in \mathcal{S}(\mathbb{C})$ for which evaluation defines an embedding $K \hookrightarrow \mathbb{C}$ is called a general point of $\mathcal{S}$. Now consider the quasi-projective variety $\mathcal{E}_{\mathbb{Q}}$ defined by

$$\left\{\begin{matrix} 2^{-1}uz_0z_2^2 - v^2z_1^3 + vz_1z_0^2 + wtz_0^3 = 0 \\ w^2 + 1 = t^2 - 3 = s^2 - 5 = 0 \end{matrix}\right\} \subset \mathbb{P}_{\mathbb{Q}}^2 \times_{\mathbb{Q}} \mathrm{Spec}(\mathbb{Q}[u, v, w, t, s]).$$

Likewise, ξ has an obvious spread $\tilde{\xi}$ given by

$$\left\{\begin{matrix} 2^{-1}uz_0z_2^2 - v^2z_1^3 + vz_1z_0^2 + wtz_0^3 = 0 \\ sz_0^4z_1 + wz_2^5 = 0 \\ w^2 + 1 = t^2 - 3 = s^2 - 5 = 0 \end{matrix}\right\}.$$

As in Example 5.4, there is a morphism $\rho : \mathcal{E}_{\mathbb{Q}} \to \mathcal{S}_{\mathbb{Q}}$. Then $\tilde{\xi} \in \mathrm{CH}^1(\mathcal{E}_{\mathbb{Q}})$. Indeed $\tilde{\xi}_{\eta_{\mathcal{S}}} = \xi \in \mathrm{CH}^1(E_K; \mathbb{Q})$, where $\mathcal{E}_{\eta_{\mathcal{S}}}$ is identified with E_K, under the embedding given in (6). Let us also view $\rho : \mathcal{E}(\mathbb{C}) \to \mathcal{S}(\mathbb{C})$ as the induced map of complex spaces. The datum associated to the Leray sheaf $R^{\bullet}\rho_*\mathbb{Q}$ amounts to an arithmetic variation of Hodge structure, and these ideas have played a big role in constructing algebraic invariants associated to Chow groups of algebraic cycles, as for example seen in the previous section. The reader should also consult [A], [G-G] and [Lew4] as further exploitation of these ideas. A different line of enquiry involving spreads can be found in [V]. Finally one can also spread $\mathcal{E}$ over $\mathbb{Z}$, by including the equation $2x-1=0$. This leads to an arithmetic scheme over $\mathbb{Z}$ where the business of height pairings can be addressed.

7 Intermezzo II

At this point, it should be reasonably clear to the reader that the notion of a height pairing of the form

$$F^{\nu}\mathrm{CH}^{r}(X_K;\mathbb{Q})\times F^{\nu}\mathrm{CH}^{d-r+\nu}(X_K;\mathbb{Q})\to\mathbb{R}, \tag{7}$$

generalizing (3), and providing a "polarization" on "primitive" pieces of $Gr_F^{\nu}\mathrm{CH}^r(X;\mathbb{Q})$, much the same way as with the Hodge-Riemann bilinear relations on the primitive cohomology of a projective algebraic manifold, should exist. Here K is finitely generated over $\overline{\mathbb{Q}}$. As we will see below, there is the technical requirement that K has transcendence degree $\nu-1$ over $\mathbb{Q}$, $\nu\geq 1$. Unfortunately, a proof of such a pairing seems elusive at this given time, and so we were forced to make further concessions (§8).

The relevance of these ideas should be clear. The idea of attaching a conjecturally non-degenerate pairing on graded pieces of the Bloch-Beilinson filtration is a unique new idea that is at the cross roads of arithmetic, Arakelov geometry and Hodge theory. At the heart of the notion of a height pairing of two cycles, is the idea of "spreading" a cycle out so as to form an intersection pairing, very similar to the aforementioned idea of defining a linking number of two disjoint curves in 3-space, where one curve bounds a membrane, thus creating an intersection number with the other curve. In (co-)homology theory, it is often the case that to determine whether a (co-)cycle is non-zero, is via an intersection/cup product with a complementary dimensional cycle. The "definite" properties of the Néron-Tate pairing (and conjecturally that of the Beilinson pairing) should convince one that this technology may lead to similar role in detecting the non-triviality of a specific "interesting" algebraic cycle.

8 A new pairing

Throughout this section, we will assume Conjecture 5.1 and the GHC.

If $\overline{\mathbb{Q}}$ of the previous section is replaced by K, a field of finite transcendence degree over $\overline{\mathbb{Q}}$, then as alluded to earlier, Conjecture 5.1 is false. However as indicated in §5, the notion of a conjectural Bloch-Beilinson filtration involves spreads, which is key to a generalized pairing. We will continue with the notation of §5, with $K=\overline{\mathbb{Q}}(\mathcal{S})$ finitely generated over $\overline{\mathbb{Q}}$. Let $Gr_F^{\nu}CH^r(X_K;\mathbb{Q})$ denote the graded pieces of the filtration, we have a non-canonical motivic decomposition (albeit

$Gr_F^\nu \mathrm{CH}^r(X_K;\mathbb{Q})$ is unique)

$$\begin{aligned}\mathrm{CH}^r(X_K;\mathbb{Q}) &= \bigoplus_{\nu\geq 0} \Delta_{X_K}(2d-2r+\nu, 2r-\nu)_*\mathrm{CH}^r(X_K;\mathbb{Q}) \\ &\simeq \bigoplus_{\nu\geq 0} Gr_F^\nu \mathrm{CH}^r(X_K;\mathbb{Q}),\end{aligned}$$

much like the Hodge decomposition of the de Rham cohomology of X.

Now if $X/\overline{\mathbb{Q}}$ is smooth projective, in light of Conjecture 5.1, Beilinson's height pairing could be interpreted as a pairing on Gr_F^1. In [S-G], we obtained the following extension of Beilinson's pairing for higher graded pieces:

Theorem 8.1. *Let $X/\overline{\mathbb{Q}}$ be a smooth projective variety of dimension d and let $K/\overline{\mathbb{Q}}$ be a finitely generated overfield of transcendence degree $\nu - 1$, where $\nu \geq 1$ is an integer. Then there exists a pairing*

$$\langle\ ,\ \rangle_{\mathrm{HT}}^\nu : Gr_F^\nu CH^r(X_K;\mathbb{Q}) \times Gr_F^\nu CH^{d-r+\nu}(X_K;\mathbb{Q}) \to \mathbb{R},$$

extending Beilinson's height pairing.

Proof. (Sketch only.) First note that $K \cong \overline{\mathbb{Q}}(S)$ where $S/\overline{\mathbb{Q}}$ is a smooth projective variety of dimension $\nu - 1$ and let η_S be the generic point of S. In this case $\rho : \mathscr{X} \to \mathcal{S}$ is given by $Pr_S : S \times X \to S$. We have the short exact sequence at the generic point

$$0 \to \underline{E}_\infty^{\nu,2r-\nu}(\eta_S) \to E_\infty^{\nu,2r-\nu}(\eta_S) \to \underline{\underline{E}}_\infty^{\nu,2r-\nu}(\eta_S) \to 0,$$

where

$$\underline{\underline{E}}_\infty^{\nu,2r-\nu}(\eta_S) = \Gamma\left(H^\nu(\eta_S, R^{2r-\nu}\rho_*\mathbb{Q}(r))\right) = 0$$

by the affine Lefschetz theorem and

$$\underline{E}_\infty^{\nu,2r-\nu}(\eta_S) = \frac{J\left(W_{-1}\left(H^{\nu-1}(\eta_S,\mathbb{Q}) \otimes H^{2r-\nu}(X,\mathbb{Q})\right)(r)\right)}{\Gamma\left(Gr_W^0\left(H^{\nu-1}(\eta_S,\mathbb{Q}) \otimes H^{2r-\nu}(X,\mathbb{Q})\right)(r)\right)}.$$

We also have $Gr_F^\nu CH^r(X_K;\mathbb{Q}) \hookrightarrow \underline{E}_\infty^{\nu,2r-\nu}(\eta_S)$.

The following two propositions are key to the proof.

Proposition 8.2 ([Lew3]). *There is an injective map*

$$Gr_F^\nu CH^r(X_K;\mathbb{Q}) \hookrightarrow J(H_0).$$

Here $J(H_0)$ denotes the jacobian of the pure Hodge structure H_0 defined by

$$H_0 := \left(\frac{H^{\nu-1}(S,\mathbb{Q})}{N_{\overline{\mathbb{Q}}}^1 H^{\nu-1}(S,\mathbb{Q})} \otimes \frac{H^{2r-\nu}(X,\mathbb{Q})}{N_H^{r-\nu+1}H^{2r-\nu}(X,\mathbb{Q})}\right)(r).$$

Rather than explain the details of the proof of Proposition 8.2, the main philosophical point is the expectation[6] that

$$\mathrm{Ext}^{\nu}_{\mathcal{MM}}\big(\mathrm{Spec}(K), h^{2r-\nu}(X_K)(r)\big) \simeq \mathrm{Ext}^{\nu}_{\mathcal{MM}}\left(\mathrm{Spec}(K), \frac{h^{2r-\nu}(X_K)(r)}{N_K^{r-\nu+1}}\right).$$

Unfortunately, any attempt to extend Proposition 8.2 beyond $X = X_{\overline{\mathbb{Q}}}$, viz., to X_K, involving a twisted spread $\mathscr{X} \to \mathcal{S}$, $\overline{\mathbb{Q}}(\mathcal{S}) = K$, seems highly non-trivial. Next,

Proposition 8.3 (Lewis). *There is a surjective map*

$$CH^r_{\mathrm{hom}}((S \times X)_{\overline{\mathbb{Q}}}; \mathbb{Q}) \twoheadrightarrow Gr^{\nu}_F CH^r(X_K; \mathbb{Q})$$

given by the projector $\Delta_S \otimes \Delta_X(2d - 2r + \nu, 2r - \nu)$.

Proof. First of all we observe that $\mathrm{CH}^r(S \times_{\overline{\mathbb{Q}}} X) \twoheadrightarrow \mathrm{CH}^r(X_K)$ is surjective. Therefore by Theorem 5.3(v), the composite involving the full Chow group:

$$\mathrm{CH}^r(S \times_{\overline{\mathbb{Q}}} X; \mathbb{Q}) \xrightarrow{(\mathrm{Id}\otimes\Delta_{X/\overline{\mathbb{Q}}}(2d-2r+\nu,2r-\nu))_*}$$

$$\Delta_{X_K}(2d - 2r + \nu, 2r - \nu)_*\mathrm{CH}^r(X_K; \mathbb{Q}) \simeq Gr^{\nu}_F\mathrm{CH}^r(X_K; \mathbb{Q}),$$

is surjective. Now for any smooth affine subvariety $U \subset S/\overline{\mathbb{Q}}$ of dimension $< \nu$, the affine Lefschetz theorem implies that $H^{\nu}(U, \mathbb{Q}) = 0$. Applying the Künneth formula to $H^{2r}(U \times X, \mathbb{Q}(r))$, it follows that

$$(\mathrm{Id} \otimes \Delta_{X/\overline{\mathbb{Q}}}(2d - 2r + \nu, 2r - \nu))_*\mathrm{CH}^r(U \times_{\overline{\mathbb{Q}}} X; \mathbb{Q}) \mapsto 0 \in H^{2r}(U \times X, \mathbb{Q}(r)),$$

and hence accordingly,

$$\mathrm{CH}^r_{\mathrm{hom}}(U \times_{\overline{\mathbb{Q}}} X; \mathbb{Q}) \xrightarrow{(\mathrm{Id}\otimes\Delta_{X/\overline{\mathbb{Q}}}(2d-2r+\nu,2r-\nu))_*}$$

$$\Delta_{X_K}(2d - 2r + \nu, 2r - \nu)_*\mathrm{CH}^r(X_K; \mathbb{Q}) \simeq Gr^{\nu}_F\mathrm{CH}^r(X_K; \mathbb{Q})$$

is surjective. Finally

$$\mathrm{CH}^r_{\mathrm{hom}}(S \times_{\overline{\mathbb{Q}}} X; \mathbb{Q}) \twoheadrightarrow \mathrm{CH}^r_{\mathrm{hom}}(U \times_{\overline{\mathbb{Q}}} X; \mathbb{Q})$$

is surjective by the Hodge conjecture, and the proposition follows. □

By our assumptions,

$$\Phi_r : \mathrm{CH}^r_{\mathrm{hom}}((S \times X)_{\overline{\mathbb{Q}}}; \mathbb{Q}) \hookrightarrow J\big(H^{2r-1}(S \times X, \mathbb{Q}(r))\big).$$

We have the following decomposition at the level of jacobians:

$$J(H^{2r-1}(S \times X, \mathbb{Q}(r))) \cong J(H_0) \oplus J(H_0^{\perp}),$$

[6]This is also apparent in the work of Shuji Saito [SSa].

where $H_0^{\perp}$ arises due to polarization. Let P_1 be the projector

$$H^{2r-1}(S \times X, \mathbb{Q}(r)) \twoheadrightarrow H_0,$$

and w_1 be an algebraic cycle lying in the Künneth component

$$H^{2(d+\nu-r)-1}(S \times X, \mathbb{Q}(d+\nu-r)) \otimes H^{2r-1}(S \times X, \mathbb{Q}(r))$$

corresponding to it. Let

$$\Xi_1 := w_{1,*}(\mathrm{CH}^r_{\mathrm{hom}}((S \times X)_{\overline{\mathbb{Q}}}; \mathbb{Q})).$$

Since we are assuming Conjecture 5.1, $F^2CH^r((S\times X)_{\overline{\mathbb{Q}}}; \mathbb{Q}) = 0$ and Ξ_1 is independent of the choice of algebraic cycle representative corresponding to P_1. Viewing everything inside the jacobian, we get

$$P_{1,*} : \mathrm{CH}^r_{\mathrm{hom}}((S \times X)_{\overline{\mathbb{Q}}}; \mathbb{Q}) \twoheadrightarrow Gr^{\nu}_F\mathrm{CH}^r(X_K; \mathbb{Q})$$

and

$$\Phi_r|_{\Xi_1} : \Xi_1 \cong Gr^{\nu}_F\mathrm{CH}^r(X_K; \mathbb{Q}).$$

By a similar procedure, we get $\Xi_2 := w_{2,*}(\mathrm{CH}^{d-r+\nu}_{\mathrm{hom}}((S\times X)_{\overline{\mathbb{Q}}}; \mathbb{Q}))$, for an algebraic cycle w_2 (similar to w_1), and an isomorphism

$$\Phi_{d-r+\nu}|_{\Xi_2} : \Xi_2 \cong Gr^{\nu}_F\mathrm{CH}^{d-r+\nu}(X_K; \mathbb{Q}).$$

Note that $d - r + \nu = (d + \nu - 1) - r + 1$ and we have Beilinson's height pairing

$$\mathrm{CH}^r_{\mathrm{hom}}((S \times X)_{\overline{\mathbb{Q}}}; \mathbb{Q}) \times \mathrm{CH}^{d-r+\nu}_{\mathrm{hom}}((S \times X)_{\overline{\mathbb{Q}}}; \mathbb{Q}) \to \mathbb{R},$$

and hence between Ξ_1 and Ξ_2. The desired pairing $\langle\ ,\ \rangle^{\nu}_{\mathrm{HT}}$ between the spaces $Gr^{\nu}_F\mathrm{CH}^r(X_K; \mathbb{Q})$ and $Gr^{\nu}_F\mathrm{CH}^{d-r+\nu}(X_K; \mathbb{Q})$ is now obtained through the isomorphisms above. □

Remark 8.4. *One can show that the height pairing above is independent of the choice of smooth projective variety $S/\overline{\mathbb{Q}}$ with $\overline{\mathbb{Q}}(S) \cong K$.*

Since our height pairing $\langle\ ,\ \rangle^{\nu}_{\mathrm{HT}}$ is given by the one developed by Beilinson, it is only natural that the conjectures in [Be3] have a natural extension for graded pieces. As an example:

Proposition 8.5. *Assume Conjecture 4.2 (Hodge-index conjecture) and let L_{X_K} denote the operation of intersecting with a hyperplane section. Then for $x \neq 0 \in Gr^{\nu}_F CH^r(X_K; \mathbb{Q})$ such that $L^{d-2r+\nu+1}_{X_K}(x) = 0$, the height pairing*

$$(-1)^r\langle x\ ,\ L^{d-2r+\nu}_{X_K}(x)\ \rangle^{\nu}_{\mathrm{HT}} > 0,$$

when $r \leq (d+\nu)/2$.

Proof. First note that the filtration developed in [Lew2] already has the property that $L_{X_K}^{d-2r+\nu}$ defines an isomorphism between

$$Gr_F^{\nu}\mathrm{CH}^r(X_K;\mathbb{Q}) \text{ and } Gr_F^{\nu}\mathrm{CH}^{d-r+\nu}(X_K;\mathbb{Q}),$$

so Proposition 8.5 makes sense. Now for any $x \in \Xi_1$

$$\begin{aligned}&\Phi_{d-r+\nu}\left(L_{S\times X}^{d-2r+\nu}(x) - w_{2,*}\circ L_{S\times X}^{d-2r+\nu}(x)\right)\\ &= [L_{S\times X}]^{d-2r+\nu}(\Phi_r(x)) - [w_2]_*\circ[L_{S\times X}]^{d-2r+\nu}(\Phi_r(x))\\ &= [L_{S\times X}]^{d-2r+\nu}(\Phi_r(x)) - [L_{S\times X}]^{d-2r+\nu}(\Phi_r(x)) = 0.\end{aligned}$$

Since we are assuming Conjecture 5.1, we get

$$L_{S\times X}^{d-2r+\nu}(x) = w_{2,*}\circ L_{S\times X}^{d-2r+\nu}(x),$$

which shows that $L_{S\times X}^{d-2r+\nu}$ maps Ξ_1 to Ξ_2, isomorphically. Further, let $\Xi_2' \subset \mathrm{CH}_{\mathrm{hom}}^{d-r+\nu+1}(S\times X;\mathbb{Q})$ be such that $\Xi_2' \cong Gr_F^{\nu}\mathrm{CH}^{d-r+\nu+1}(X_K;\mathbb{Q})$.

For $x' \in \Xi_1$,
$\Phi_r(x') = x \in Gr_F^{\nu}\mathrm{CH}^r(X_K;\mathbb{Q}) \implies \Phi_{d-r+\nu+1}(L_{S\times X}^{d-2r+\nu+1}(x')) = L_{X_K}^{d-2r+\nu+1}(x)$.
So, $L_{X_K}^{d-2r+\nu+1}(x) = 0 \implies L_{S\times X}^{d-2r+\nu+1}(x') = 0$. We also have

$$(-1)^r\langle x\, ,\, L_{X_K}^{d-2r+\nu}(x)\rangle_{\mathrm{HT}}^{\nu} = (-1)^r\langle x'\, ,\, L_{S\times X}^{d-2r+\nu}(x')\rangle_{\mathrm{HT}}\, .$$

Note that $x' \in \Xi_1 \subset \mathrm{CH}_{\mathrm{hom}}^r((S\times X)_{\overline{\mathbb{Q}}};\mathbb{Q})$ and $L_{S\times X}^{d-2r+\nu+1}(x') = 0$. Now assuming Conjecture 4.2, we conclude

$$(-1)^r\langle x'\, ,\, L_{S\times X}^{d-2r+\nu}(x')\rangle_{\mathrm{HT}} > 0\, ,$$

and Proposition 8.5 follows immediately. □

We study the following subspace of $Gr_F^{\nu}\mathrm{CH}^r(X_K;\mathbb{Q})$:

Definition 8.6. *Let* $F^{\nu}\underline{CH}_{\mathrm{alg}}^r(X_K;\mathbb{Q}) :=$

$$F^{\nu}CH^r(X_K;\mathbb{Q})\bigcap\left[\mathrm{Im}(CH_{\mathrm{alg}}^r((S\times X)_{\overline{\mathbb{Q}}};\mathbb{Q}) \to CH^r(X_K;\mathbb{Q}))\right].$$

Then we define

$$Gr_F^{\nu}\underline{CH}_{\mathrm{alg}}^r(X_K;\mathbb{Q}) := \mathrm{Im}\left(F^{\nu}\underline{CH}_{alg}^r(X_K;\mathbb{Q}) \to Gr_F^{\nu}CH^r(X_K;\mathbb{Q})\right).$$

There is one remark in order: If S' is another such variety, then we can dominate both S and S' by a desingularization of a third $S'' \hookrightarrow S\times S'$. From this, and the fact that the rational Chow group of cycles algebraically equivalent to zero being a $\mathbb{Q}$ vector space, one can show

$$\mathrm{Im}\left(\mathrm{CH}_{\mathrm{alg}}^r((S\times X)_{\overline{\mathbb{Q}}};\mathbb{Q}) \to \mathrm{CH}^r(X_K;\mathbb{Q})\right)$$

and

$$\mathrm{Im}\left(\mathrm{CH}_{\mathrm{alg}}^r((S'\times X)_{\overline{\mathbb{Q}}};\mathbb{Q}) \to CH^r(X_K;\mathbb{Q})\right)$$

are the same. Thus the definition of $Gr_F^{\nu}\underline{\mathrm{CH}}_{\mathrm{alg}}^r(X_K;\mathbb{Q})$ is independent of the choice of S. Now we have the following

Theorem 8.7. *Under the same set up as in Theorem 8.1 , we have the height pairing*

$$\langle\ ,\ \rangle^{\nu}_{\mathrm{HT,alg}} : Gr^{\nu}_{F}\underline{CH}^{r}_{\mathrm{alg}}(X_K;\mathbb{Q}) \times Gr^{\nu}_{F}\underline{CH}^{d-r+\nu}_{\mathrm{alg}}(X_K;\mathbb{Q}) \to \mathbb{R},$$

extending the Néron-Tate pairing.

Proof. Assuming Conjecture 5.1, we get that

$$\mathrm{CH}^{r}_{\mathrm{alg}}((S\times X)_{\overline{\mathbb{Q}}};\mathbb{Q}) \hookrightarrow J^{r}_{\mathrm{alg}}(S\times X)_{\mathbb{Q}}$$

and

$$\mathrm{CH}^{d-r+\nu}_{\mathrm{alg}}((S\times X)_{\overline{\mathbb{Q}}};\mathbb{Q}) \hookrightarrow J^{d-r+\nu}_{\mathrm{alg}}(S\times X)_{\mathbb{Q}}.$$

The proof now goes exactly in the same way as Theorem 8.1, if we replace $\mathrm{CH}^{r}_{\mathrm{hom}}((S\times X)_{\overline{\mathbb{Q}}};\mathbb{Q})$ (resp. $\mathrm{CH}^{d-r+\nu}_{\mathrm{hom}}((S\times X)_{\overline{\mathbb{Q}}};\mathbb{Q})$) with $\mathrm{CH}^{r}_{\mathrm{alg}}((S\times X)_{\overline{\mathbb{Q}}};\mathbb{Q})$ (resp. $\mathrm{CH}^{d-r+\nu}_{\mathrm{alg}}((S\times X)_{\overline{\mathbb{Q}}};\mathbb{Q})$). We obtain $\Xi_{1,\mathrm{alg}} \subset \mathrm{CH}^{r}_{\mathrm{alg}}((S\times X)_{\overline{\mathbb{Q}}};\mathbb{Q})$ (respectively $\Xi_{2,\mathrm{alg}} \subset \mathrm{CH}^{d-r+\nu}_{\mathrm{alg}}((S\times X)_{\overline{\mathbb{Q}}};\mathbb{Q})$), such that

$$\Xi_{1,\mathrm{alg}} \cong Gr^{\nu}_{F}\underline{\mathrm{CH}}^{r}_{\mathrm{alg}}(X_K;\mathbb{Q}), \quad \Xi_{2,\mathrm{alg}} \cong Gr^{\nu}_{F}\underline{\mathrm{CH}}^{d-r+\nu}_{\mathrm{alg}}(X_K;\mathbb{Q}).$$

The height pairing $\langle\ ,\ \rangle^{\nu}_{\mathrm{HT,alg}}$ is now given as the pairing between $\Xi_{1,\mathrm{alg}}$ and $\Xi_{2,\mathrm{alg}}$. □

Remark 8.8. *The reasons for restricting to this particular subspace are the following:*

1. *Since the height pairing for cycles algebraically equivalent to zero is given by the Néron-Tate pairing ([Be3], Remark 4.0.8), one can work without the assumption of Conjecture 5.1.*
2. *Further for cycles algebraically equivalent to zero, assumption (17) of [Ku] is no longer necessary ([Ku], §8).*

So in effect, one can freely use the machineries available from arithmetic intersection theory to compute the height pairing, albeit (GHC). We will illustrate this with an example.

8.9 An example computation. Using the formalism of arithmetic intersection theory discussed in §4, we present here a computation related to the theory developed so far.

Example 8.10. *Let* $X = C_1 \times C_2$ *be the product of smooth projective curves* C_1 *and* C_2*, defined over* $\overline{\mathbb{Q}}$*. For* $\nu = 2$*, we fix an embedding* $K = \overline{\mathbb{Q}}(C_2) \hookrightarrow \mathbb{C}$ *(so naturally* $S = C_2$ *following the set up of Theorem 8.1), and let* $p = \eta_2 \in C_2(\mathbb{C})$ *be a very general point corresponding to this embedding. To be more precise,* η_2 *is regarded as the generic point of* $\overline{\mathbb{Q}}(C_2)$ *(so* $\overline{\mathbb{Q}}(C_2) = \overline{\mathbb{Q}}(\eta_2)$*), and recall that any point* $p \in S(\mathbb{C})$ *for which evaluation at* p *defines an embedding* $\overline{\mathbb{Q}}(\eta_2) \hookrightarrow \mathbb{C}$ *is defined to be a very general point. Although this notation is a bit slang, we write* $p = \eta_2$*. We fix* $e_2 \in C_2(\overline{\mathbb{Q}})$*. For distinct points* $p_1, q_1, p_2, q_2 \in C_1(\overline{\mathbb{Q}})$*, let*

$$\xi_1 := (p_1 - q_1) \times (\eta_2 - e_2) \in Gr_F^2 \underline{CH}^2_{\text{alg}}(X_K; \mathbb{Q}),$$

$$\xi_2 := (p_2 - q_2) \times (\eta_2 - e_2) \in Gr_F^2 \underline{CH}^2_{\text{alg}}(X_K; \mathbb{Q}).$$

Assume also

$$N_H^1 \left(H^1(C_1, \mathbb{Q}) \otimes H^1(C_2, \mathbb{Q}) \right) = 0. \tag{8}$$

Then

$$\langle \xi_1, \xi_2 \rangle^2_{\text{HT,alg}} = deg \left(\Delta^2_{C_2}(1,1) \right) \langle p_1 - q_1, p_2 - q_2 \rangle_{\text{NT}},$$

where on the RHS we have Néron-Tate pairing.

We add a remark before we begin the proof:

Remark 8.11. *The assumption in (8) holds for example if we take $X = E_1 \times E_2$, a product of two non-isogenous elliptic curves; for here $N_H^1(H^1(E_1, \mathbb{Q}) \otimes H^1(E_2, \mathbb{Q})) = H^2_{alg}(E_1 \times E_2, \mathbb{Q}) \cap (H^1(E_1, \mathbb{Q}) \otimes H^1(E_2, \mathbb{Q})) = 0$ follows from the fact that any non-zero element*

$$[\xi] \in H^2_{\text{alg}}(E_1 \times E_2, \mathbb{Q}) \cap (H^1(E_1, \mathbb{Q}) \otimes H^1(E_2, \mathbb{Q}))$$

will in turn define an isogeny between E_1 and E_2.

Since Example 8.10 illustrates the potential of our theory so effectively, we will provide a more detailed proof.

Proof. From the assumptions of Example 8.10 we get that $S \times X = C_2 \times (C_1 \times C_2) \cong C_1 \times (C_2 \times C_2)$. We have the Chow-Künneth decomposition for smooth curves

$$\Delta_{C_2}(1,1) = \Delta_{C_2} - e_2 \times C_2 - C_2 \times e_2.$$

Now put

$$\tilde{\xi}_1 := (p_1 - q_1) \times \Delta_{C_2}(1,1) \in \text{CH}^2_{\text{alg}}(C_1 \times (C_2 \times C_2); \mathbb{Q}),$$

$$\tilde{\xi}_2 := (p_2 - q_2) \times \Delta_{C_2}(1,1) \in \text{CH}^2_{\text{alg}}(C_1 \times (C_2 \times C_2); \mathbb{Q}).$$

Using the assumption in (8) and basic intersection theory, one can show $\tilde{\xi}_i \in \Xi_{alg} \mapsto \xi_i, i = 1,2$ under the isomorphism $\Xi_{alg} \cong Gr_F^2 \underline{\text{CH}^2}_{alg}(X_K; \mathbb{Q})$. Here $\Xi_{alg} \subset \text{CH}^2_{\text{alg}}(C_1 \times (C_2 \times C_2); \mathbb{Q})$ is the suitable subspace (see Theorem 8.7 for details). Thus, the height pairing $\langle \xi_1, \xi_2 \rangle^2_{\text{HT,alg}}$ is given by $\langle \tilde{\xi}_1, \tilde{\xi}_2 \rangle_{\text{HT}}$. We provide a general computation in the formalism of arithmetic intersection theory, for which $\langle \tilde{\xi}_1, \tilde{\xi}_2 \rangle_{\text{HT}}$ is a particular case.

Lemma 8.12. *Let C be smooth projective curve and X be a smooth projective variety of dimension $d-1$, both defined over a number field k. Let $\alpha_1, \alpha_2 \in CH^1_{\text{alg}}(C; \mathbb{Q})$ and $\pi_1 : C \times X \to C$ and $\pi_2 : C \times X \to X$ are the projections. Given $w_1 \in CH^{r-1}(X; \mathbb{Q})$ and $w_2 \in CH^{(d-1)-(r-1)}(X; \mathbb{Q}) = CH^{d-r}(X; \mathbb{Q})$ and the cycles*

$$a_1 := \pi_1^*(\alpha_1) \cdot \pi_2^*(w_1) \in CH^r_{\text{alg}}(C \times X; \mathbb{Q}),$$

$$a_2 := \pi_1^*(\alpha_2) \cdot \pi_2^*(w_2) \in CH_{\mathrm{alg}}^{d-r+1}(C \times X; \mathbb{Q}).$$

We get the following height pairing relation :

$$\langle a_1, a_2 \rangle_{\mathrm{HT}} = (\deg(w_1 \cdot w_2)_X) \langle \alpha_1, \alpha_2 \rangle_{\mathrm{NT}},$$

where $(w_1 \cdot w_2)_X$ *is the usual intersection pairing on* X.

Proof. Let $\widetilde{C}$ be a regular semi-stable model for C over $\mathrm{Spec}(\mathcal{O}_{k'})$ (after a finite extension k' of the ground field k). Choose Z_i, $i = 1, 2$ cycles on $\widetilde{C}$ of codimension 1 such that

1. $Z_i|_C = \alpha_i$.
2. $Z_i \cdot V = 0$ for any vertical cycle V.

One can arrange the above situation by Th. 1.3 of [Hr]. Choose g_i, $i = 1, 2$, Green's functions for Z_i such that $dd^c g_i + \delta_{Z_i} = 0$ (since α_i is null-homologous, the cohomology class $[\omega_{Z_i}] = 0$). We have

$$[(Z_i, g_i)] \in \widehat{\mathrm{CH}}^1(\widetilde{C}), \ i = 1, 2.$$

Then,

$$\langle \alpha_1, \alpha_2 \rangle_{\mathrm{NT}} = \widehat{deg}_{\widetilde{C}}([(Z_1, g_1)] \cdot [(Z_2, g_2)]) \in \widehat{\mathrm{CH}}^1(\mathrm{Spec}(\mathbb{Z})) \otimes \mathbb{Q} \cong \mathbb{R}$$

is independent of the choices of Z_i, g_i.

Now, for any projective and flat model $\widetilde{X}'$ over $\mathrm{Spec}(\mathcal{O}_{k'})$ of X, by de Jong's alteration ([dJo], Theorem 8.2) we get a projective, flat and regular scheme $\widetilde{X}$ over a finite extension of k' (in turn a finite extension of k), with a finite and surjective morphism to $\widetilde{X}'$. In particular $\dim(\widetilde{X}') = \dim(\widetilde{X})$. Let W_i, $i = 1, 2$ be cycles on $\widetilde{X}$ of codimensions $r - 1$ and $d - r$ respectively such that

$$W_i|_X = w_i, \ \ i = 1, 2.$$

Let g_{W_1} (resp. g_{W_2}) be a Green current for W_1 (resp. W_2). Then

$$[(W_1, g_{W_1})] \in \widehat{\mathrm{CH}}^{r-1}(\widetilde{X}),$$

$$[(W_2, g_{W_2})] \in \widehat{\mathrm{CH}}^{d-r}(\widetilde{X}).$$

For the scheme $\widetilde{C} \times_{\mathrm{Spec}(\mathcal{O}_{k'})} \widetilde{X}$, we can use the alteration trick once more to obtain a regular flat and projective scheme Z over $\mathrm{Spec}(\mathcal{O}_{k''})$, where k'' is a finite extension of k and a dominant and finite morphism $f : Z \to \widetilde{C} \times_{\mathrm{Spec}(\mathcal{O}_{k'})} \widetilde{X}$. In particular $\dim(Z) = \dim(\widetilde{C} \times_{\mathrm{Spec}(\mathcal{O}_{k'})} \widetilde{X}) = d + 1$. For the projections

$$\pi_{\widetilde{C}} : \widetilde{C} \times_{\mathrm{Spec}(\mathcal{O}_{k'})} \widetilde{X} \to \widetilde{C},$$

$$\pi_{\widetilde{X}} : \widetilde{C} \times_{\mathrm{Spec}(\mathcal{O}_{k'})} \widetilde{X} \to \widetilde{X},$$

consider

$$f_{\widetilde{C}} := \pi_{\widetilde{C}} \circ f,$$
$$f_{\widetilde{X}} := \pi_{\widetilde{X}} \circ f,$$

and the cycles

$$\tilde{a}_1 := f^*_{\widetilde{C}}([(Z_1, g_{Z_1})]) \cdot f^*_{\widetilde{X}}([(W_1, g_{W_1})]),$$
$$\tilde{a}_2 := f^*_{\widetilde{C}}([(Z_2, g_{Z_2})]) \cdot f^*_{\widetilde{X}}([(W_2, g_{W_2})]).$$

Then (up to rational multiples, which will arise since we are using alterations and extensions of the base field k)

$$\langle a_1, a_2 \rangle_{\mathrm{HT}} = \widehat{deg}_Z\,(\tilde{a}_1 \cdot \tilde{a}_2) \in \widehat{\mathrm{CH}}^1(\mathrm{Spec}(\mathbb{Z})) \otimes \mathbb{Q} \cong \mathbb{R}.$$

Since $f^*_{\widetilde{C}}$ and $f^*_{\widetilde{X}}$ are morphisms of rings ([GS], 4.4.3 (5)),

$$\tilde{a}_1 \cdot \tilde{a}_2 = f^*_{\widetilde{C}}\left([(Z_1, g_1)] \cdot [(Z_2, g_2)]\right) \cdot f^*_{\widetilde{X}}\left([(W_1, g_{W_1})] \cdot [(W_2, g_{W_2})]\right).$$

By the projection formula for arithmetic intersection pairing ([GS], 4.4.3 (7))

$$f_{\widetilde{C},*}\,(\tilde{a}_1 \cdot \tilde{a}_2) = \underbrace{[(Z_1, g_1)] \cdot [(Z_2, g_2)]}_{\in \widehat{\mathrm{CH}}^2(\widetilde{C};\mathbb{Q})} \cdot \underbrace{f_{\widetilde{C},*}\left[f^*_{\widetilde{X}}\left([(W_1, g_{W_1})] \cdot [(W_2, g_{W_2})]\right)\right]}_{\in \widehat{\mathrm{CH}}^0(\widetilde{C};\mathbb{Q}))}.$$

Since

$$\widehat{deg}_Z\,(\tilde{a}_1 \cdot \tilde{a}_2) = \widehat{deg}_{\widetilde{C}}\left(f_{\widetilde{C},*}(\tilde{a}_1 \cdot \tilde{a}_2)\right)$$

and

$$f_{\widetilde{C},*}\left[f^*_{\widetilde{X}}\left([(W_1, g_{W_1})] \cdot [(W_2, g_{W_2})]\right)\right] = \deg(w_1 \cdot w_2)_X,$$

we obtain our desired result. We note here that since we are using $\mathbb{Q}$-valued intersection pairing, the relations among various height-pairings won't change. □

Quite generally, one can also prove the following:

Theorem 8.13 ([S-G]). *Given smooth projective curves $C_1, \ldots, C_d$ over $\overline{\mathbb{Q}}$, let $X = C_1 \times \cdots \times C_d$. For $\nu \geq 2$, we fix an embedding $K = \overline{\mathbb{Q}}(C_2 \times \cdots \times C_\nu) \hookrightarrow \mathbb{C}$, and let $p = (\eta_2, \ldots, \eta_\nu) \in C_2(\mathbb{C}) \times \cdots \times C_\nu(\mathbb{C})$ be a very general point corresponding to this embedding (see Example 8.10 for a clarification of "general"). We fix $e_j \in C_j(\overline{\mathbb{Q}})$, $j = 2, \ldots, d$. For distinct points $p_1, q_1, p_2, q_2 \in C_1(\overline{\mathbb{Q}})$ and $\nu \leq r \leq d$, let*

$$\xi_1 := Pr^*_{1,\ldots,\nu}((p_1 - q_1) \times (\eta_2 - e_2) \times \cdots \times (\eta_\nu - e_\nu)) \bigcap$$
$$Pr^*_{\nu+1,\ldots,r}(e_{\nu+1}, \ldots, e_r) \in Gr^\nu_F \underline{CH}^r_{alg}(X_K; \mathbb{Q}),$$

$$\xi_2 := Pr^*_{1,\ldots,\nu}((p_2 - q_2) \times (\eta_2 - e_2) \times \cdots \times (\eta_\nu - e_\nu)) \bigcap$$
$$Pr^*_{r+1,\ldots,d}(e_{r+1}, \ldots, e_d) \in Gr^\nu_F \underline{CH}^{d-r+\nu}_{alg}(X_K; \mathbb{Q}).$$

Assume also

$$N_H^1\left(H^1(C_1,\mathbb{Q})\otimes\cdots\otimes H^1(C_\nu,\mathbb{Q})\right)=0,$$

and

$$N_{\overline{\mathbb{Q}}}^1\left(H^1(C_2,\mathbb{Q})\otimes\cdots\otimes H^1(C_\nu,\mathbb{Q})\right)=0.$$

Then, $\langle\xi_1,\xi_2\rangle^\nu_{\mathrm{HT,alg}} = \left(\prod_{j=2}^\nu[deg(\Delta^2_{C_j}(1,1))_{C_j\times C_j}]\right)\langle p_1-q_1,p_2-q_2\rangle_{\mathrm{NT}}$, *where* $\langle\ ,\ \rangle_{\mathrm{NT}}$ *is the Néron-Tate pairing on* $(J^1(C_1)(\overline{\mathbb{Q}}))\otimes\mathbb{Q}$.

Remark 8.14. *For a self-product of a CM-elliptic curve ([S-G], §8.2), we were able to eliminate the assumption in (8) altogether.*

9 An Archimedean pairing involving the equivalence relation defining higher Chow groups

In this section, and for each $m\geq 0$, we construct a pairing on the cycle level, involving the equivalence relation in the definition of Bloch's higher Chow groups $\mathrm{CH}^r(X,m)$ defined below. The case when $m=0$ has already been defined in §1, and the nature of this pairing is more akin to the Archimedean height pairing defined in the literature. It was first discussed in [C-L]; however presentation here is intended to be more user friendly. A general construction of this pairing for all m is in order. We first recall that two subvarieties V_1, V_2 of a given variety intersect properly if $\operatorname{codim}\{V_1\cap V_2\}\geq\operatorname{codim}V_1+\operatorname{codim}V_2$. This notion naturally extends to algebraic cycles.

(i) *Higher Chow groups.* Let W/k be a quasi-projective variety over a field k. Put $z^r(W)$ = free abelian group generated by subvarieties of codimension r in W,

$$\Delta^m := \mathrm{Spec}\left(\frac{k[t_0,...,t_m]}{1-\sum_{j=0}^m t_j}\right)$$

the standard m-simplex, and $z^r(W,m)=\left\{\xi\in z^k(W\times\Delta^m)\ \middle|\ \xi \text{ meets all faces } \{j_1=\cdots=j_\ell=0\mid \ell=1,...,m\}\text{ properly}\right\}$.

Definition 9.1 ([Blo1]). *$CH^\bullet(W,\bullet)$ = homology of $\{z^\bullet(W,\bullet),\partial\}$. We put* $\mathrm{CH}^k(W)$ $:=\mathrm{CH}^k(W,0)$.

(ii) *Cubical version.* Let $\square^m:=(\mathbb{P}^1\backslash\{1\})^m$ with coordinates z_i and 2^m codimension one faces obtained by setting $z_i=0,\infty$, and boundary maps $\partial=\sum(-1)^{i-1}(\partial_i^0-\partial_i^\infty)$, where ∂_i^0, ∂_i^∞ denote the restriction maps to the faces $z_i=0$, $z_i=\infty$ respectively. The rest of the definition is completely analogous for $z^r(W,m)\subset z^r(W\times\square^m)$, except that one has to quotient out by the subgroup $z^r_{\mathrm{dgt}}(W,m)\subset z^r(W,m)$ of degenerate cycles obtained via pullbacks $\sum_{j=1}^m\mathrm{pr}_j^*: z^r(W,m-1)\to z^r(W,m)$, $\mathrm{pr}_j: W\times\square^m\to W\times\square^{m-1}$ the j-th canonical projection. It is known that both complexes are quasi-isomorphic (Bloch (unpublished)/Levine [Lv]; independently).

9.2 A quick detour via Milnor K-theory. An excellent reference for this part is [B-T]. Let $\mathbb{F}$ be a field with multiplicative group $\mathbb{F}^\times \subset \mathbb{F}$. Consider the graded tensor algebra

$$T(\mathbb{F}) := \bigoplus_{r=0}^{\infty} \{\mathbb{F}^\times\}^{\otimes_{\mathbb{Z}} m} = \mathbb{Z} \oplus \mathbb{F}^\times \oplus \cdots,$$

and let $R(\mathbb{F})$ be the graded 2-sided ideal generated by

$$\big\{\tau \otimes (1-\tau) \ \big|\ \tau \in \mathbb{F}^\times \backslash \{1\}\big\}.$$

Recall that the Milnor K-theory of $\mathbb{F}$ is given by

$$K_\bullet^M(\mathbb{F}) := T(\mathbb{F})/R(\mathbb{F}) = \bigoplus_{r=0}^{\infty} K_m^M(\mathbb{F}).$$

Further, recall that $K_r^M(\mathbb{F}) \simeq \mathrm{CH}^r(\mathrm{Spec}(\mathbb{F}), r)$ (Nesterenko/Suslin (1990), Totaro (1992)). Now let W/k be a smooth scheme over a field k. If one replaces $\mathbb{F}$ by $\mathcal{O}_W^\times$, then we arrive at the sheaf $\mathcal{K}_{r,W}^M$ of Milnor K-groups. To be more precise, let $\mathcal{O}_W$ be the sheaf of regular functions on X, with sheaf of units $\mathcal{O}_W^\times$. As in [Ka], we put

$$\mathcal{K}_{r,W}^M := \big(\underbrace{\mathcal{O}_W^\times \otimes \cdots \otimes \mathcal{O}_W^\times}_{r \text{ times}}\big)/\mathcal{J} \quad \text{(Milnor sheaf)},$$

where $\mathcal{J}$ is the subsheaf of the tensor product generated by sections of the form:

$$\big\{\tau_1 \otimes \cdots \otimes \tau_r \ \big|\ \tau_i + \tau_j = 1, \quad \text{for some } i \text{ and } j,\ i \neq j\big\}.$$

For example, $\mathcal{K}_{1,W}^M = \mathcal{O}_X^\times$. The higher Chow groups $\mathrm{CH}^r(W, m)$ come naturally equipped with a coniveau filtration involving codimension of cycles when projected to W, whose graded pieces can be computed via a local-to-global spectral sequence ([BO], [Blo1]), involving flasque resolutions of certain sheaves. Via the works of Elbaz-Vincent/Müller-Stach (1998), and Gabber (1992) (see [MS2], together with [Ke]), one of those sheaves is $\mathcal{K}_{r,W}^M$. This, together with partial degeneration of the aforementioned spectral sequence leads to:

Theorem 9.3 (See [MS2]). *For $0 \le m \le 2$, there is an isomorphism*

$$H_{\mathrm{Zar}}^{r-m}(W, \mathcal{K}_{r,W}^M) \simeq \mathit{CH}^r(W, m).$$

In the context of Milnor K-theory, the last 3 terms of the flasque resolution of $\mathcal{K}_{r,W}^M$ are

$$\bigoplus_{\mathrm{cd}_W V = r-2} K_2^M(\mathbb{C}(V)) \xrightarrow{\text{Tame}} \bigoplus_{\mathrm{cd}_W V = r-1} K_1^M(\mathbb{C}(V)) \xrightarrow{\text{div}} \bigoplus_{\mathrm{cd}_W V = r} K_0^M(\mathbb{C}(V)).$$

If we interpret this in terms of global sections, this leads to a complex whose last three terms and corresponding homologies (norm/graph maps, indicated at $\updownarrow$)

for $0 \le m \le 2$ are:

$$\begin{array}{ccccc} \bigoplus_{\mathrm{cd}_W Z=r-2} K_2^M(\mathbb{C}(Z)) & \xrightarrow{T} & \bigoplus_{\mathrm{cd}_W Z=r-1} \mathbb{C}(Z)^\times & \xrightarrow{\mathrm{div}} & \bigoplus_{\mathrm{cd}_W Z=r} \mathbb{Z} \\ \updownarrow & & \updownarrow & & \updownarrow \\ \mathrm{CH}^r(W,2) & & \mathrm{CH}^r(W,1) & & \mathrm{CH}^r(W,0) \end{array} \tag{9}$$

where as a reminder, div is the divisor map of zeros minus poles of a rational function, and T is the Tame symbol map. Again as a reminder, the Tame symbol map

$$T : \bigoplus_{\mathrm{cd}_X Z=r-2} K_2^M(\mathbb{C}(Z)) \to \bigoplus_{\mathrm{cd}_X D=r-1} K_1^M(\mathbb{C}(D))$$

is defined as follows. First $K_2^M(\mathbb{C}(Z))$ is generated by symbols $\{f, g\}$, f, $g \in \mathbb{C}(Z)^\times$, under $f \otimes g \mapsto \{f, g\}$.

For f, $g \in \mathbb{C}(Z)^\times$,

$$T(\{f,g\}) = \sum_D (-1)^{\nu_D(f)\nu_D(g)} \left(\frac{f^{\nu_D(g)}}{g^{\nu_D(f)}}\right)_D,$$

where $(\cdots)_D$ means restriction to the generic point of D, and ν_D represents order of a zero or pole along an irreducible divisor $D \subset Z$.

Example 9.4. Taking cohomologies of the complex in (9), we have:

(i) $\mathrm{CH}^r(W,0) = z^r(W)/z^r_{\mathrm{rat}}(W) =: \mathrm{CH}^r(W)$.

(ii) $\mathrm{CH}^r(W,1)$ is represented by classes of the form $\xi = \sum_j (f_j, D_j)$, where $\mathrm{codim}_X D_j = r-1$, $f_j \in \mathbb{C}(D_j)^\times$, and $\sum \mathrm{div}(f_j) = 0$; modulo the image of the Tame symbol.

(iii) $\mathrm{CH}^r(W,2)$ is represented by classes in the kernel of the Tame symbol; modulo the image of a higher Tame symbol.

In this section we will adopt the cubical version of $\mathrm{CH}^\bullet(W,\bullet)$, albeit a simplicial version can also be arranged [KLL]. The intersection product for cycles in the cubical version is easy to define. On the level of cycles, and in $W \times W \times \square^{m+n}$, one has

$$z^r(W,m) \times z^k(W,n) \to z^{r+k}(W \times W, m+n);$$

however the pullback along the diagonal

$$z^{r+k}(W \times W, m+n) \to z^{r+k}(W, m+n)$$

is not well-defined, even for smooth W. In particular, for smooth W, the issue of when an intersection product is defined, which is a general position statement involving proper intersections, has to be addressed since we will be working on the level of cycles. On the level of Chow groups, a moving lemma of Bloch (adapted to the cubical situation) guarantees a pullback for smooth W:

$$\mathrm{CH}^\bullet(W \times W, \bullet) \to \mathrm{CH}^\bullet(W, \bullet),$$

and hence an intersection product for smooth W.

Let us return to the situation where X be a projective algebraic manifold of dimension d, and let $z^r_{\mathrm{rat}}(X, m) := \partial\big(z^r(X, m+1)\big) \subset z^r(X, m)$ be the equivalence relation subgroup defining the higher Chow groups $\mathrm{CH}^r(X, m)$. As in [KLM], we will need to restrict ourselves to those precycles[7] $z^r(X, \bullet)$ that are in general position with respect to the real subsets $X \times [-\infty, 0]^\bullet$, and we will denote this by $z^\bullet_{\mathbb{R}}(X, \bullet)$. Now introduce

$$\Lambda^0(r, m, X) = \left\{ (\xi_1, \xi_2) \in z^r_{\mathbb{R},\mathrm{rat}}(X, m) \times z^{d-r+m+1}_{\mathbb{R},\mathrm{rat}}(X, m) \;\middle|\; \begin{array}{c} |\xi_1| \cap |\xi_2| = \emptyset \\ \text{in } X \times \square^m \end{array} \right\},$$

$$\Lambda^+(r, m, X) = \left\{ (\xi_1, \xi_2) \in \Lambda^0(r, m, X) \;\middle|\; \begin{array}{c} \xi_1 = \partial\xi_1', \text{ where} \\ \xi_1' \cap \xi_2 \text{ is defined} \\ \text{in } z^{d+m+1}_{\mathbb{R}}(X, 2m+1) \end{array} \right\},$$

$$\Lambda^-(r, m, X) = \left\{ (\xi_1, \xi_2) \in \Lambda^0(r, m, X) \;\middle|\; \begin{array}{c} \xi_2 = \partial\xi_2', \text{ where} \\ \xi_1 \cap \xi_2' \text{ is defined} \\ \text{in } z^{d+m+1}_{\mathbb{R}}(X, 2m+1) \end{array} \right\},$$

$$\Lambda(r, m, X) \subset \Lambda^+(r, m, X) \bigcap \Lambda^-(r, m, X)$$

is characterized by the requirement that $\xi_1' \cap \xi_2'$ is defined, viz,

$$\xi_1' \cap \xi_2' \in z^{d+m+1}_{\mathbb{R}}(X, 2m+2).$$

Theorem 9.5. *There are natural pairings*

$$\langle\ ,\ \rangle^+_m : \Lambda^+(r, m, X) \to \mathbb{C}/\mathbb{Z}(1),$$
$$\langle\ ,\ \rangle^-_m : \Lambda^-(r, m, X) \to \mathbb{C}/\mathbb{Z}(1),$$

which satisfy the following:

(i) *(Reciprocity) On* $\Lambda(r, m, X)$, $\langle\ ,\ \rangle^+_m = (-1)^m \langle\ ,\ \rangle^-_m$.

(ii) *(Bilinearity) If* $(\xi_1^{(1)}, \xi_2)$, $(\xi_1^{(2)}, \xi_2) \in \Lambda^+(r, m, X)$, *then*

$$\langle \xi_1^{(1)} + \xi_1^{(2)}, \xi_2 \rangle^+_m = \langle \xi_1^{(1)}, \xi_2 \rangle^+_m + \langle \xi_1^{(2)}, \xi_2 \rangle^+_m.$$

If $(\xi_1, \xi_2^{(1)})$, $(\xi_1, \xi_2^{(2)}) \in \Lambda^-(r, m, X)$, *then*

$$\langle \xi_1, \xi_2^{(1)} + \xi_2^{(2)} \rangle^-_m = \langle \xi_1, \xi_2^{(1)} \rangle^-_m + \langle \xi_1, \xi_2^{(2)} \rangle^-_m.$$

(iii) *(Projection formula) Let* $\pi : X \to Y$ *be a flat surjective morphism between two smooth projective varieties, with* $\dim X = d$. *Then* $\langle \xi_1, \pi^*\xi_2 \rangle^\pm_m = \langle \pi_*\xi_1, \xi_2 \rangle^\pm_m$ *for all* $\xi_1 \in z^r_{\mathrm{rat}}(X, m)$ *and* $\xi_2 \in z^{d-r+m+1}_{\mathrm{rat}}(Y, m)$ *with* $(\pi_*\xi_1, \xi_2) \in \Lambda^\pm(r + s - d, m, Y)$, *where* $s := \dim Y$.

[7] As a reminder, for $\xi \in z^r(X, \bullet)$ to be a cycle, we require $\partial\xi = 0$.

Proof. We first recall the definition of Deligne cohomology. Good sources for this are [Ki], [Ja1] and [KLM]. Let $\mathcal{D}_X^\bullet$ be the (fine) sheaf of complex-valued currents acting on C^∞ complex-valued compactly supported $(2d-\bullet)$-forms, where we recall $\dim X = d$. One has a decomposition into Hodge type:

$$\mathcal{D}_X^\bullet = \bigoplus_{p+q=\bullet} \mathcal{D}_X^{p,q},$$

where $\mathcal{D}_X^{p,q}$ acts on $(d-p, d-q)$ forms, with Hodge filtration,

$$F^r\mathcal{D}_X^\bullet = \bigoplus_{p+q=\bullet, p\geq r} \mathcal{D}_X^{p,q}.$$

Likewise, for a subring $\mathbb{A} \subseteq \mathbb{C}$, there is the (soft)[8] sheaf subcomplex $\mathcal{C}_X^\bullet(\mathbb{A}) \subset \mathcal{D}_X^\bullet$ of $\mathbb{A}$-coefficient Borel-Moore chains on X. The global sections of a given sheaf $\mathcal{S}$ over X will be denoted by $\mathcal{S}(X)$. Next, for a morphism of complexes $\lambda : A^\bullet \to C^\bullet$, we recall the cone complex:

$$\mathrm{Cone}(A^\bullet \xrightarrow{\lambda} B^\bullet) = A^\bullet[1] \oplus B^\bullet,$$

with differential

$$\delta_D : A^{q+1} \oplus B^q \to A^{q+2} \oplus B^{q+1}, \quad (a,b) \overset{\delta_D}{\mapsto} (-da, \lambda(a) + db).$$

Definition 9.6. *Fix a subring $\mathbb{A} \subseteq \mathbb{R}$. The Deligne cohomology of X is given by*

$$H^i_{\mathcal{D}}(X, \mathbb{A}(j)) := H^i\big(\mathrm{Cone}\big(\mathcal{C}_X^\bullet(X, \mathbb{A}(j)) \bigoplus F^j\mathcal{D}_X^\bullet(X) \xrightarrow{\varepsilon - l} \mathcal{D}_X^\bullet(X)\big)[-1]\big).$$

It is customary of thinking of currents as associated to homology. Note that by simply regarding $\mathcal{C}_X^\bullet(\mathbb{A}(j))$, $F^j\mathcal{D}_X^\bullet$ as acyclic resolutions of the respective sheaves $\mathbb{A}(r)$ and $\Omega^j_{X,\mathrm{d-closed}}$, with quasi-isomomorphisms, $\{\mathbb{A}(j) \to 0 \to \cdots\} \approx \mathcal{C}_X^\bullet(\mathbb{A}(j))$, $\Omega_X^{\bullet \geq j} \approx F^j\mathcal{D}_X^\bullet$, the above definition, when compared with the one in [EV], already incorporates Poincaré duality.

Remark 9.7. *Generally speaking, one thinks of currents as well behaved under proper push-forwards, albeit with no defined pull-back. However, the rules can be broken here if one replaces the sheaf complex of currents on a given manifold with another which is quasi-isomorphic and having better properties with respect to pull-backs and multiplication. The situation is well documented in [Ki](§4) and [K-L](§8). The reader should keep this in mind in the discussion below. To simplify our notation, we will use the notation "·" to refer to multiplication of currents. Also we use the principal branch of the* log *function below.*

Continuing with the proof of Theorem 9.5, we now recall the description of the regulator on the level of complexes [KLM].

$$\mathrm{cl}_{r,m,X} : \mathrm{CH}^r(X,m) \to H_{\mathcal{D}}^{2r-m}(X, \mathbb{Z}(r)), \text{ viz., } \mathbb{A} = \mathbb{Z}.$$

[8] In the end, acyclicity is all that matters here.

Consider $\square^m$ with affine coordinates $(z_1, ..., z_m)$ and introduce the currents: (δ_V means integration over V)

$$\Omega_m := \left(\bigwedge_{j=1}^{m} d\log z_j\right) \cdot \delta_{\square^m},$$

$$T_{z_1} = \delta_{[-\infty,0]\times\square^{m-1}}, ..., T_m := T_{z_1} \cap \cdots \cap T_{z_m} = \delta_{[-\infty,0]^m} := \int_{[-\infty,0]^m} (-),$$

$$\begin{aligned} R_m := {} & \log z_1 d\log z_2 \wedge \cdots \wedge d\log z_m \cdot \delta_{\square^m} \\ & - (2\pi \mathrm{i}) \log z_2 d\log z_3 \wedge \cdots \wedge d\log z_m \cdot T_{z_1} + \cdots \\ & + \ (-1)^{m-1}(2\pi\mathrm{i})^{m-1} \log z_m \cdot T_{z_1} \cap \cdots \cap T_{z_{m-1}}. \end{aligned}$$

For $\xi \in z^r_{\mathbb{R}}(X, m)$, and if we let $\mathrm{pr}_{\square} : X^m \to \square^m$, $\mathrm{pr}_X : X \times \square^m \to X$ be the obvious projections, we consider the currents on X:

$$T_m(\xi) := \int_{\xi} \mathrm{pr}_{\square}^*(T_m) \wedge \mathrm{pr}_X^*(-),$$

$$\Omega_m(\xi) := \int_{\xi} \mathrm{pr}_{\square}^*(\Omega_m) \wedge \mathrm{pr}_X^*(-),$$

$$R_m(\xi) := \int_{\xi} \mathrm{pr}_{\square}^*(R_m) \wedge \mathrm{pr}_X^*(-).$$

One has the following identities [KLM]:

$$\partial T_m(\xi) = T_{m-1}(\partial\xi), \ d[\Omega_m(\xi)] = 2\pi\mathrm{i}\Omega_{m-1}(\partial\xi), \tag{10}$$

$$d[R_m(\xi)] = \Omega_m(\xi) - (2\pi\mathrm{i})^m T_m(\xi) - 2\pi\mathrm{i} R_{m-1}(\partial\xi).$$

The map $\mathrm{cl}_{r,m,X}$ is induced (up to the normalizing twist $(2\pi\mathrm{i})^{r-m}$) by

$$\xi \in z^r(X, m) \mapsto \big((2\pi\mathrm{i})^m T_m(\xi), \Omega_m(\xi), R_m(\xi)\big), \tag{11}$$

with the following caveat. One expects a quasi-isomorphism $z^r_{\mathbb{R}}(X, \bullet) \approx z^r(X, \bullet)$, which certainly holds after tensoring with $\mathbb{Q}$ [K-L]. Having said this, by the very definition of $\Lambda^{\pm}$, we can drop the $\mathbb{Q}$-coefficients from this discussion without compromising the theorem. It is easy to check that

$$\big((2\pi\mathrm{i})^m T_m(\xi), \Omega_m(\xi), R_m(\xi)\big) = (0, 0, 0) \text{ for } \xi \in z^r_{\mathrm{dgt}}(X, m).$$

For $m = 0$, note that $(T_0(\xi), \Omega_0(\xi), (2\pi\mathrm{i})^m R_0(\xi)) = (\xi, \delta_\xi, 0)$. First an observation. For precycles $\alpha \in z^p(X, \ell)$ and $\beta \in z^q(X, n)$ (in general position), one has the relation [KLM]:

$$R_{\ell+n}(\alpha \cup \beta) = (-2\pi\mathrm{i})^{\ell} T_\ell(\alpha) \cdot R_n(\beta) + R_\ell(\alpha) \cdot \Omega_n(\beta). \tag{12}$$

9.8 The pairings. For $(\xi_1, \xi_2) \in \Lambda^+(r, m, X)$, we put

$$\langle \xi_1, \xi_2 \rangle_m^+ := (-2\pi \mathrm{i})^{m+1} T_{m+1}(\xi_1{}') \cdot R_m(\xi_2) + R_{m+1}(\xi_1{}') \cdot \Omega_m(\xi_2) \qquad (13)$$

$$\in \mathbb{C}/\mathbb{Z}(2m+1) \simeq \mathbb{C}/\mathbb{Z}(1),$$

where the latter $\simeq$ is given by multiplication by $(-4\pi^2)^{-m}$, and for $(\xi_1, \xi_2) \in \Lambda^-(r, m, X)$, we put (under $\mathbb{C}/\mathbb{Z}(2m+1) \simeq \mathbb{C}/\mathbb{Z}(1)$),

$$\langle \xi_1, \xi_2 \rangle_m^- := (-2\pi \mathrm{i})^m T_m(\xi_1) \cdot R_{m+1}(\xi_2') + R_m(\xi_1) \cdot \Omega_{m+1}(\xi_2') \in \mathbb{C}/\mathbb{Z}(1).$$

Note that for dimension and general position reasons alone,

$$T_{m+1}(\xi_1') \cdot T_m(\xi_2) = 0 = T_m(\xi_1) \cdot T_{m+1}(\xi_2') \in z_{\mathbb{R}}^{d+m+1}(X, 2m+1), \qquad (14)$$

and likewise over $|\xi_1 \cap \xi_2'|$ or $|\xi_1' \cap \xi_2|$,

$$\Omega_{m+1}(\xi_1') = 0 = \Omega_{m+1}(\xi_2'), \qquad (15)$$

using the fact that $\dim |\xi_1 \cap \xi_2'|$, $\dim |\xi_1' \cap \xi_2| \leq m$ and that $\Omega_{m+1}(\xi_1')$, $\Omega_{m+1}(\xi_2')$ are meromorphic currents involving $m+1$ holomorphic differentials. This, together with

$$R_m(\xi_1) \wedge \Omega_{m+1}(\xi_2') = \mathrm{pr}_{\square}^*(R_m \wedge \Omega_{m+1}) \cdot \delta_{\xi_1 \cap \xi_2'} \text{ (by fiberwise Fubini)},$$

implies (using (12)) the simpler expression:

$$\langle \xi_1, \xi_2 \rangle_m^- := (-2\pi \mathrm{i})^m T_m(\xi_1) \cdot R_{m+1}(\xi_2').$$

Furthermore, the vanishing relations in (14) and (15) imply that the pairings $\langle \xi_1, \xi_2 \rangle_m^\pm$ correspond (up to twist) to $(*)$ in a Deligne complex triple of the form $(0, 0, *)$ (see the RHS of (11)). Note that if either $\partial \xi_1' = 0$ or $\partial \xi_2' = 0$, then the pairings $\langle \xi_1, \xi_2 \rangle_m^\pm$ amount to a cup product in Deligne cohomology of the regulator of a higher Chow cycle, together with one which is nullhomologous (in Deligne cohomology), which is zero in:

$$H_{\mathcal{D}}^{2d+1}(X, \mathbb{Z}(d+m+1)) \simeq \mathbb{C}/\mathbb{Z}(1),$$

where firstly after incorporating the normalizing twist (just preceding (11)), and in our setting, we arrive at the isomorphisms:

$$H_{\mathcal{D}}^{2d+1}(X, \mathbb{Z}(d+m+1)) \simeq \mathbb{C}/\mathbb{Z}(d+m+1) \overset{\times (2\pi \mathrm{i})^{d+m}}{\simeq} \mathbb{C}/\mathbb{Z}(1).$$

Hence the pairings $\langle \xi_1, \xi_2 \rangle_m^\pm$ do not depend on the choices of the ξ_j''s. For simplicity, we will assume given $(\xi_1, \xi_2) \in \Lambda(r, m, X)$. By definition, this implies that

$$\xi_1' \cap \xi_2,\ \xi_1 \cap \xi_2' \in z_{\mathbb{R}}^{d+m+1}(X, 2m+1),\ \xi_1' \cap \xi_2' \in z_{\mathbb{R}}^{d+m+1}(X, 2m+2),$$

which is important in ensuring that the currents above are defined. Next, the relations

$$R_{2m+1}(\partial\{\xi_1' \cup \xi_2'\}) = R_{2m+1}(\xi_1 \cup \xi_2') + (-1)^{m+1} R_{2m+1}(\xi_1' \cup \xi_2)$$

imply that

$$\langle \xi_1, \xi_2 \rangle_m^+ = (-1)^m \langle \xi_1, \xi_2 \rangle_m^-. \tag{16}$$

We remark in passing that in the case $m = 0$, and after taking real parts, equation (16) implies the reciprocity result in Proposition 1.3. The remaining claims in Theorem 9.5 are left to the reader. □

Remark 9.9. *We can pass to a real-valued height pairing using the composite* $\mathbb{C}/\mathbb{Z}(1) \to \mathbb{C}/\mathbb{R}(1) \simeq \mathbb{R}$.

We put

$$\langle\ ,\ \rangle_m := \langle\ ,\ \rangle_m^+\big|_{\Lambda(r,m,X)} = (-1)^m \langle\ ,\ \rangle_m^-\big|_{\Lambda(r,m,X)}.$$

We will denote by $\langle\ ,\ \rangle_m^{\mathbb{R}}$ the corresponding real pairing.

In the case $m = 0$, we have $\Lambda^0(r,0,X) \subset z^r_{\mathrm{rat}}(X) \times z^{d-r+1}_{\mathrm{rat}}(X)$. Let $(\xi_1, \xi_2) \in z^r_{\mathrm{rat}}(X) \times z^{d-r+1}_{\mathrm{rat}}(X)$. By considering the cases where $y = |\xi_j|$, $j = 1, 2$, and regarding the real pairing $\langle\ ,\ \rangle_0^{\mathbb{R}}$ below, we may assume that the domain is given by

$$\Lambda := \big\{(\xi_1, \xi_2) \in z^r_{\mathrm{rat}}(X) \times z^{d-r+1}_{\mathrm{rat}}(X) \ \big|\ |\xi_1| \cap |\xi_2| = \emptyset\big\},$$

and thus we have pairings

$$\langle\ ,\ \rangle_0 : \Lambda^0(r,0,X) \to \mathbb{C}/\mathbb{Z}(1),$$

$$\langle\ ,\ \rangle^{\mathbb{R}} := \langle\ ,\ \rangle_0^{\mathbb{R}} : \Lambda \to \mathbb{R}.$$

Let $\xi_1 := \operatorname{div}(f, D) \in z^r_{\mathrm{rat}}(X, 0)$, $\xi_2 := \operatorname{div}(g, E) \in z^{d-r+1}_{\mathrm{rat}}(X, 0)$ be given. In this case D and E are irreducible subvarieties of X of $\operatorname{codim}_X D = r - 1$ and $\operatorname{codim}_X E = d - r$, and $f \in \mathbb{C}(D)^\times$, $g \in \mathbb{C}(E)^\times$. Then

$$\langle \xi_1, \xi_2 \rangle_0 = \int_{\{D \setminus f^{-1}[-\infty,0]\} \cap \xi_2} \log f \in \mathbb{C}/\mathbb{Z}(1).$$

Similarly,

$$\langle \xi_1, \xi_2 \rangle_0^{\mathbb{R}} = \int_{D \cap \xi_2} \log |f|.$$

Remark 9.10. *It is instructive to work out the case* $m = 1$. *Let*

$$\xi_1 := \sum_j (g_j, D_j) \in z^r_{\mathrm{rat}}(X, 1)$$

and

$$\xi_2 := T(\{f_1, f_2\}, E) \in z^{d-r+2}_{\mathrm{rat}}(X, 1)$$

be given, where T is the Tame symbol. (We will also be working under the assumption that $(\xi_1, \xi_2) \in \Lambda(r, 1, X)$.) In this case E and D_j are irreducible subvarieties of X of $\operatorname{codim}_X D_j = r - 1$ and $\operatorname{codim}_X E_j = d - r$, and $f_i \in \mathbb{C}(E)^\times$, $i = 1, 2$, and $g_j \in \mathbb{C}(D_j)^\times$. Set $C_j = D_j \cap E$, which is a curve in X, and put

$$g_{j,C_j} := g\big|_{C_j}, \quad f_{i,C_j} = f_i\big|_{C_j}.$$

Then using the identification $\langle\xi_1,\xi_2\rangle_1 = -\langle\xi_1,\xi_2\rangle_1^-$, *a simple computation yields:*

$$\langle\xi_1,\xi_2\rangle_1 = \sum_j \Bigg[(2\pi \mathrm{i}) \int_{g_{j,C_j}^{-1}[-\infty,0]} \log f_1 d\log f_2$$

$$- (2\pi\mathrm{i})^2 \int_{(f_{1,C_j}\times g_{j,C_j})^{-1}[-\infty,0]^2} \log f_2\Bigg] \in \mathbb{C}/\mathbb{Z}(3) \simeq \mathbb{C}/\mathbb{Z}(1).$$

(Recall in equation (13) the identification $\mathbb{C}/\mathbb{Z}(2m+1) \simeq \mathbb{C}/\mathbb{Z}(1)$, *which explains the need for the identification* $\mathbb{C}/\mathbb{Z}(3) \simeq \mathbb{C}/\mathbb{Z}(1)$ *in the case* $m = 1$.*) Let* γ *be the (closed) curve given by*

$$\gamma := \sum_j g_{j,C_j}^{-1}[-\infty,0].$$

Taking the real part of $\langle\xi_1,\xi_2\rangle_1$ *and applying a Stokes' theorem argument, one can show that:*

$$\langle\xi_1,\xi_2\rangle_1^{\mathbb{R}} = -2\pi \int_\gamma \big[\log|f_1| d\arg f_2 - \log|f_2| d\arg f_1\big]. \tag{17}$$

Equation (17) is easily seen to be non-trivial. [Take for example $X := E = \mathbb{P}^2 \ni [z_0,z_1,z_2]$, *and consider* $\mathbb{P}^1 = \ell_j := D_j := V(z_j)$, $g_0 = -z_1/z_2$, $g_1 = -z_2/z_0$, $g_2 = -z_0/z_1$. *Note that* $g_j \in \mathbb{C}(\ell_j)^\times$ *and that* $\sum_{j=0}^2 \mathrm{div}_{\ell_j}(g_j) = 0$. *Put* $L := z_0+z_1+z_2$, *and* $t_j = z_j/L$, *hence* $t_0+t_1+t_2 = 1$. *Thus in affine coordinates,* $g_0 = -t_1/t_2$, $g_1 = -t_2/t_0$, $g_2 = -t_0/t_1$, *and* $\ell_j = V(t_j)$. *Now let* γ *be the corresponding* 1*-cycle, which is the boundary of the real simplex* $\{t_0+t_1+t_2 = 1 \mid t_j \in [0,1]\}$. *Let* $\nu : \mathbb{P}^2 \to \mathbb{P}^1$ *be the projection from* $[0,0,1]$ *(explicit:* $\nu([z_0,z_1,z_2]) = [z_0,z_1]$*)*. *Then* $\nu(t_0,t_1,t_2) = (t_0,t_1)$ *and the aforementioned real simplex becomes* $\{t_0+t_1 \le 1 \mid t_j \in [0,1]\}$, *and* $\nu_*(\gamma)$ *is the obvious boundary. Consider* p *in the interior of* $\nu_*(\gamma)$, *and in terms of the coordinate* $w = t_0 + \mathrm{i}t_1$, $t_j \in \mathbb{R}$, *set* $h(w) = w - p$. *Observe that* $\int_{\nu_*(\gamma)} d\log h \neq 0$. *Now put*

$$f_2 := \nu^*(h(w)) = \frac{z_0 + \mathrm{i}z_1 + p\cdot L}{L},$$

and choose $f_1 \in \mathbb{R}^\times\backslash\{\pm 1\}$.*]*

Acknowledgements: The authors would like to thank Vincent Maillot for suggesting the idea of a height pairing on graded pieces of a Bloch-Beilinson filtration, and to José Burgos Gil for providing us with the idea of the proof of Lemma 8.12. We are also grateful from correspondence with Klaus Künnemann. A warm thanks goes to Lizhen Ji for his superb logistics in bringing this conference to fruition, and of course, to our esteemed colleague Steven Zucker, for making this wonderful event possible. Finally, the referee deserves a warm thanks for doing a splendid job!

References

[Ar] Arapura, D., *The Leray spectral sequence is motivic,* Invent. Math. **160** (2005), no. 3, 567–589.

[A] Asakura, M., *Motives and algebraic de Rham cohomology,* in: The Arithmetic and Geometry of Algebraic Cycles, Proceedings of the CRM Summer School, June 7-19, 1998, Banff, Alberta, Canada (Editors: B. Gordon, J. Lewis, S. Müller-Stach, S. Saito and N. Yui), NATO Science Series **548**, Kluwer Academic Publishers, 2000, 133–154.

[B-T] Bass, H., Tate, J., *The Milnor ring of a global field,* in Algebraic K-Theory II, Lecture Notes in Mathematics **342**, Springer-Verlag, 1972, 349–446.

[Be1] Beilinson, A., *Notes on absolute Hodge cohomology,* in Applications of Algebraic K-Theory to Algebraic Geometry and Number Theory, Contemporary Mathematics **55**, part 1, AMS, Providence 1986, 35–68.

[Be2] ——, *Higher regulators and values of L-functions,* J. Sov. Math. **30** (1985), 2036–2070.

[Be3] ——, *Height pairing between algebraic cycles,* Contemporary Mathematics. **37** (1987), 1–24.

[Blo1] Bloch, S., *Algebraic cycles and higher K-theory,* Adv. Math. **61** (1986), 267–304.

[Blo2] ——, *Lectures on algebraic cycles,* Duke University Mathematics Series, IV, Duke University, Mathematics Department, Durham, N.C., 1980. 182 pp.

[Blo3] ——, *Height pairing between algebraic cycles,* K-Theory, Arithmetic and Geometry, Lecture Notes in Mathematics **1289**, Springer-Verlag, 2006, 1–26.

[Blo4] ——, *The moving lemma for higher Chow groups,* J. Algebraic Geom. **3**(3) (1994), 493–535.

[BO] Bloch, S., Ogus, A., *Gersten's conjecture and the homology of schemes,* Annales scientifiques de l'É.N.S. 4e série, **7** (1974), no. 2, 181–201.

[BGS] Bost, J.-B., Gillet, H., Soulé, C., *Heights of projective varieties and positive Green forms,* J. Amer. Math. Soc. **7** (1994), no. 4, 903–1027.

[B-Z] Brylinski, J.-L., Zucker, S., *An overview of recent advances in Hodge theory,* Complex analysis - several variables **6**, Itogi Nauki i Tekhniki. Ser. Sovrem. Probl. Mat. Fund. Napr., **69**, VINITI, Moscow, 1991, 48–165.

[KLL] Kerr, M., Lewis, J. D., Lopatto, P., (with an appendix by Burgos, J.), *Simplicial Abel-Jacobi maps and reciprocity laws.* In progress.

[C-L] Chen, X., Lewis, J. D., *Real regulators on self-products of K3 surfaces,* J. Algebraic Geom. **20** (2011), no. 1, 101–125.

[C-K-S] Cattani, E., Kaplan, A., Schmid, W., *L_2 and intersection cohomologies for a polarizable variation of Hodge structure,* Invent. Math. **87** (1987), no. 2, 217–252.

[Ca] Carlson, J., *Extension of mixed Hodge structures*, In Journées de Géométrie Algébrique d'Angers 1979, Sijhoff and Nordhoff (1980), 107–127.

[dJo] de Jong, A. J., *Smoothness, semi-stability and alterations,* I.H.E.S. Publ. Math. **83** (1996), 51–93.

[De] Deligne, P., *Théorie de Hodge, II, III,* Inst. Hautes Études Sci. Publ. Math. No. **40** (1971), 5–57; No. **44** (1974), 5–77.

[EV] Esnault, H., Viehweg, E., *Deligne-Beilinson cohomology*, in Beilinson's Conjectures on Special Values of L-Functions (Rapoport, Schappacher, Schneider, eds.), Perspect. Math. **4**, Academic Press, 1988, 43–91.

[G-G] Green, M., Griffiths, P. A., *Hodge-theoretic invariants for algebraic cycles,* Int. Math. Res. Notices (2003), no. 9, 477–510.

[G-G-P] Green, M., Griffiths, P. A., Paranjape, K., *Cycles over fields of transcendence degree* 1, Michigan Math. J. **52** (2004), 181–187.

[GS] Gillet, H., Soulé, C., *Arithmetic intersection theory,* Publ. Math. I.H.E.S., (1990), 94–174.

[Hr] Hriljac, P., *Heights and Arakelov's intersection theory,* Amer. J. Math. **107** (1985), 23–38.

[Ja1] Jannsen, U., *Deligne cohomology, Hodge-$\mathcal{D}$-conjecture, and motives,* in Beilinson's Conjectures on Special Values of L-Functions (Rapoport, Schappacher, Schneider, eds.), Perspect. Math. **4**, Academic Press, 1988, 305–372.

[Ja2] ——, *Mixed Motives and Algebraic K-Theory,* Lecture Notes in Mathematics **1000**, Springer-Verlag, 1990.

[Ja3] ——, *Equivalence relations on algebraic cycles,* in: The Arithmetic and Geometry of Algebraic Cycles, Proceedings of the CRM Summer School, June 7-19, 1998, Banff, Alberta, Canada (Editors: B. Gordon, J. Lewis, S. Müller-Stach, S. Saito and N. Yui), NATO Science Series **548**, Kluwer Academic Publishers, 2000, 225–260.

[Ka] Kato, K., *Milnor K-theory and the Chow group of zero cycles,* Applications of algebraic K-theory to algebraic geometry and number theory, Part I, II, 1986, 241–253.

[KLM] Kerr, M., Lewis, J. D., Müller-Stach, S., *The Abel-Jacobi map for higher Chow groups,* Compos. Math. **142** (2006), no. 2, 374–396.

[K-L] Kerr, M., Lewis, J. D., *The Abel-Jacobi map for higher Chow groups, II,* Invent. Math. **170** (2007), no. 2, 355–420.

[Ke] Kerz, M., *The Gersten conjecture for Milnor K-theory,* Invent. Math. **175** (2009), no. 1, 1–33.

[Ki] King, J., *Log complexes of currents and functorial properties of the AbelJacobi map,* Duke Math. J. **50** (1983), no, 1, 1–53.

[Ku] Künnemann, K., *Higher Picard varieties and the height pairing,* Amer. J. Math. **118** (1996), 781–797.

[Lv] Levine, M., *Bloch's higher Chow groups revisited,* Asterisque **226** (1994), 235–320.

[Lew1] Lewis, J. D., *A Survey of the Hodge Conjecture. Second edition. Appendix B by B. Brent Gordon,* CRM Monograph Series **10**, American Mathematical Society, Providence, RI, 1999.

[Lew2] ——, *A filtration on the Chow groups of a complex projective variety,* Compositio Math. **128** (2001), no. 3, 299–322.

[Lew3] ——, *Arithmetic normal functions and filtrations on Chow groups,* Proc. Amer. Math. Soc. **140** (2012), no. 8, 2663–2670.

[Lew4] ——, *Cycles on varieties over subfields of* $\mathbb{C}$ *and cubic equivalence,* Motives and algebraic cycles, Fields Inst. Commun. **56**, Amer. Math. Soc., Providence, RI, 2009, 233–247.

[M] Mumford, D., *Rational equivalence of 0-cycles on surfaces,* J. Math. Kyoto Univ. **9** (1968), 195–204.

[Mu] Murre, J. P., *On a conjectural filtration on the Chow groups of an algebraic variety, Parts I and II,* Indag. Math. **4** (1993), 177–201.

[MS1] Müller-Stach, S., *A remark on height pairings,* In Algebraic cycles and Hodge theory (Torino, 1993), Lecture Notes in Mathematics **1594**, Springer, Berlin, 1994, 253–259.

[MS2] ——, *Algebraic cycle complexes,* in Proceedings of the NATO Advanced Study Institute on the Arithmetic and Geometry of Algebraic Cycles **548** (Lewis, Yui, Gordon, Müller-Stach, S. Saito, eds.), Kluwer Academic Publishers, Dordrecht, The Netherlands, 2000, 285–305.

[Neu] Neukirch, J., *Algebraic number theory,* Grundlehren der Mathematischen Wissenschaften [Fundamental Principles of Mathematical Sciences] **322**, Springer-Verlag, Berlin, 1999.

[PM] Peters, C., Saito, M., *Lowest weights in cohomology of variations of Hodge structure,* Nagoya Math. J. **206** (2012), 1–24.

[Ra] Raskind, W., *Higher l-adic Abel-Jacobi mappings and filtrations on Chow groups,* Duke Math. J. **78** (1995), no. 1, 33–57.

[Ro] Roitman, A. A., *Rational equivalence of zero-dimensional cycles,* (Russian) Mat. Zametki **28** (1980), no. 1, 85–90.

[SSa] Saito, S., *Motives and filtrations on Chow groups,* Invent. Math. **125** (1996), 149–196.

[S] Schoen, C., *On Hodge structures and nonrepresentability of Chow groups,* Compositio Math. **88** (1993), no. 3, 285–316.

[S-G] Goswami, S., *Height pairing on graded pieces of a Bloch-Beilinson Filtration,* PhD Thesis, 2015.

[St] Steenbrink, J. H. M., *A summary of mixed Hodge theory,* Motives (Seattle, WA, 1991), Proc. Sympos. Pure Math. **55**, Part 1, Amer. Math. Soc., Providence, RI, 1994, 31–41.

[V] Voisin, C., *Hodge loci and absolute Hodge classes,* Compositio Math. **113** (2007), 945–958.

[Z] Zucker, S., *Hodge theory with degenerating coefficients: L_2-cohomology in the Poincaré metric,* Annals of Math. **109** (1979), 415–476.

Hodge Theory and L^2-analysis
ALM 39, pp. 321–376

Extremal Degenerations of Polarized Hodge Structures*

Mark Green, Phillip Griffiths, Colleen Robles†

Abstract

The aim of the paper is to provide a Hodge theoretic answer to the question: what is the least/most singular variety to which a smooth projective variety can degenerate?

2010 Mathematics Subject Classification: 14D07, 32G20, 14M15, 14M17.

Keywords and Phrases: Variation of Hodge structure, reduced limit period mapping, flag variety, flag domain.

I Introduction

An interesting question in algebraic geometry is: *In what ways can a smooth projective variety X degenerate?* Here one imagines a situation

$$\mathcal{X} \xrightarrow{\pi} S \tag{I.1}$$

where $\mathcal{X}$ and S are complex manifolds with $\mathcal{X} \subset \mathbb{P}^N \times S$ and where π is a proper holomorphic mapping with X a smooth fibre. Then π is a holomorphic submersion over a Zariski open set $S^* \subset S$, and one is interested in which varieties $X_s = \pi^{-1}(s)$ can arise when $s \in S \backslash S^*$. The question of course needs refinement; e.g., by assuming some sort of semi–stable reduction for (I.1) (cf. [1]).

*Dedicated to Steven Zucker in recognition of his contributions to Hodge theory, on the occasion of his 65th birthday.

†Robles is partially supported by the National Science Foundation through grants DMS-1309238 and DMS-1361120. A portion of this work was undertaken while Robles was a member of the Institute for Advanced Study; she thanks the institute for a wonderful working environment and the Robert and Luisa Fernholz Foundation for financial support.

Hodge theory provides an invariant associated to (I.1). Namely there is a period mapping

$$\Phi : S^* \to \Gamma\backslash D \tag{I.2}$$

where $D = G_{\mathbb{R}}/R$ is a Hodge domain and for $s \in S^*$,

$$\Phi(s) \text{ is the polarized Hodge structure on } H^n(X_s, \mathbb{Q})_{\rm prim}.$$

The ambiguity in the identification of $H^n(X_s, \mathbb{Q})_{\rm prim}$ with a fixed vector space V is given by the image of the monodromy representation

$$\rho : \pi_1(S^*) \to \Gamma \subset G\,.$$

We note that the invariant (I.2) of (I.1) only depends on the family over S^*; it does not depend on the generally non-unique semi-stable reduction, although as we shall see it strongly limits what the singular fibres can be.

There are two ways of attaching Hodge–theoretic data to the limits

$$\lim_{s\to s_o} \Phi(s)\,; \tag{I.3}$$

these data will then reflect the specialization $X_s \to X_{s_o}$. The first, and traditional, way is to think of (I.3) as giving a *limiting mixed Hodge structure.* Specifically, following [8] and [16] and taking $S\backslash S^*$ to a be a local normal crossing divisor, one attaches to D a set $B(\Gamma)$ of equivalence classes of limiting mixed Hodge structures and extends (I.2) to an *extended period mapping*

$$\Phi_e : S \to \Gamma\backslash\left(D \cup B(\Gamma)\right)\,.$$

One may roughly think of $\Phi_e(s_o)$ as containing a *maximal* amount of Hodge–theoretic information in the limit.

To explain the second, more recent method we assume that $S = \Delta$ is the unit disc and $S^* = \Delta^*$ the punctured disc so that (I.1) becomes

$$\Phi : \Delta^* \to \Gamma_T\backslash D \tag{I.4}$$

where T is the monodromy transformation and $\Gamma_T = \{T^k\}_{k\in\mathbb{N}}$. It is known that T is quasi-unipotent. Since, in this paper, the essential information in the limit $\lim_{s\to s_0}\Phi(s)$ is invariant under base change, we will assume that T is unipotent with logarithm N. Then (I.4) may be lifted to a mapping of the upper–half plane

$$\tilde\Phi : \mathbb{H} \to D\,,$$

and following [15, Appendix to Lecture 10] and [18] we define[1] the *reduced limit period mapping* associated to (I.4) by

$$\lim_{\substack{\operatorname{Im} z \to \infty \\ \operatorname{Re} z \text{ bounded}}} \tilde\Phi(z) \in \partial D \tag{I.5}$$

where D is embedded in its compact dual $\check D$ and $\partial D \subset \check D$, cf. Section II.G.

[1]Precise definitions of all notions discussed in this introduction are given in later sections.

To explain this a bit more, in the situation (I.4) the boundary component $B(\Gamma)$ referred to above becomes

$$B(N) = \left\{ \begin{array}{c} \text{equivalence classes of limiting mixed Hodge structures} \\ (V, W_\bullet(N), F^\bullet) \text{ with monodromy weight filtration } W_\bullet(N) \end{array} \right\} .$$

There is then a mapping

$$\Phi_\infty : B(N) \to \partial D ,$$

and the reduced limit period mapping (I.5) is the composition of this mapping with $\Phi_e(s_o)$, where $s_o = 0 \in \Delta$. We will abbreviate it by $\Phi_\infty(s_o)$. It is well–defined since (I.5) is a fixed point of T. We may roughly think of $\Phi_\infty(s_o)$ as containing the *minimal* amount of Hodge–theoretic information in the limit (I.3). The extended period mapping Φ_e determines the limit in all torodial compactifications as in [2]. For all Hermitian symmetric domains D, the mapping Φ_∞ corresponds to the Satake–Bailey–Borel compactification [3, 25].

One advantage of the reduced limit period mapping is that it maps to a space on which the group $G_\mathbb{R}$ acts. One may then use the rich and well understood structure of the partially ordered set of $G_\mathbb{R}$–orbits in ∂D to define what is meant by extremal degenerations of a polarized Hodge structure, cf. Section III.A. Specifically, a $G_\mathbb{R}$–orbit $\mathcal{O} \subset \partial D$ is said to be *polarized relative to the Hodge domain D* in case there is a period mapping (I.4) whose reduced limit period lies in $\mathcal{O}$, cf. Definition II.31. When the infinitesimal period relation is bracket–generating[2] (cf. §II.D.2), all orbits in ∂D of real codimension–one in $\check{D}$ are polarizable relative to some Hodge domain structure on the open $G_\mathbb{R}$–orbit D, cf. [15, 18] or Section III.C. The unique closed orbit in ∂D is sometimes, but not always, polarizable. The general question of polarizability is discussed in [18, 19].

A degeneration (I.4) of a polarized Hodge structure $\Phi(s)$ is said to be *minimal* if the reduced limit period lies in a $G_\mathbb{R}$–orbit of real codimension–one in $\check{D}$; it is said to be *maximal* if its reduced limit lies in an orbit whose closure does not lie in a proper sub-orbit that is polarizable relative to D, cf. Definition III.4. One way to think of this is the following: Points of D are given by polarized Hodge structures $(V, Q, F^\bullet)$. Then arbitrary period mappings may be well–approximated by nilpotent orbits with the same limit period mapping. Thus a minimal degeneration of $F^\bullet \in D$ is given by a nilpotent orbit such that

$$\lim_{\substack{\operatorname{Im} z \to \infty \\ \operatorname{Re} z \text{ bounded}}} e^{zN} \cdot F^\bullet \quad \text{lies in a } G_\mathbb{R}\text{–orbit of real codimension–one in } \check{D}.$$

Intuitively these are the *least* degenerate limiting mixed Hodge structures that $F^\bullet \in D$ can specialize to. (For brevity "$G_\mathbb{R}$–orbit of real codimension–one in $\check{D}$" will henceforth be expressed as "codimension–one $G_\mathbb{R}$–orbit".) Similarly, maximal degenerations are the *most* degenerate that $F^\bullet$ can specialize to.

The main results of this paper will describe the *extremal* — the minimal and maximal — degenerations of polarized Hodge structures in a number of cases. These will be described in terms of the types of limiting mixed Hodge structure

[2] If D is a period domain, then the IPR is bracket–generating if and only if there are no 'gaps' in the Hodge numbers [7].

that maps to the reduced limit period point in ∂D. For example, we will prove the following theorem on minimal degenerations for a period domain.

Theorem I.6. *Given a period domain D parameterizing polarized Hodge structures of weight n, the minimal degenerations have either*

$$\begin{array}{l} N^2 = 0 \quad \textit{and} \quad \operatorname{rank} N \in \{1,2\}\,, \quad \textit{or} \\ N^2 \neq 0\ ,\ N^3 = 0 \quad \textit{and} \quad \operatorname{rank} N = 2\,. \end{array}$$

(In place of Theorem I.6 we will prove the more precise Theorem IV.1.) The Hodge diamonds for the corresponding limiting mixed Hodge structures are pictured in Figure I.1, and the algebro-geometric interpretation of the theorem is discussed in §I.B.

At the other extreme, the maximal degenerations polarizing the closed $G_\mathbb{R}$–orbit are described by the following two results.

Proposition I.7. *If the limiting mixed Hodge structure is of Hodge–Tate type, then the limit period is a point in the closed orbit.*

Theorem I.8. *Let D be a period domain parameterizing weight n Hodge structures. If there exists a limiting mixed Hodge structure $(V, W_\bullet(N), F^\bullet)$ that maps to the closed $G_\mathbb{R}$–orbit in $\check{D}$, but is* not *of Hodge–Tate type, then $n = 2m$ is even and:*

(a) *For $k \neq 0$, $\mathrm{Gr}^{W_\bullet(N)}_{n+k,\mathrm{prim}}$ is of Hodge–Tate type.* (Thus k is even.)

(b) *For $k \neq 0$, $\mathrm{Gr}^{W_\bullet(N)}_{n+k,\mathrm{prim}} \neq 0$ implies $k \equiv 2$ (mod) 4.*

(c) *$\mathrm{Gr}^{W_\bullet(N)}_{n,\mathrm{prim}} \neq 0$, and the only nonzero $I^{p,q}_{\mathrm{prim}}$, with $p+q=n$, are*

$$I^{m+1,m-1}_{\mathrm{prim}} \quad \textit{and} \quad I^{m-1,m+1}_{\mathrm{prim}}\,.$$

The proposition and theorem are proved in Sections V.B and V.D, respectively. See §I.C for discussion of these, and related, results.

Remark I.9. The (polarizable) $G_\mathbb{R}$–orbits in $\overline{D}$ form a poset (partially ordered set) whose structure [20] is expected to have algebro–geometric applications [13]. In the present paper, we consider only the extremal elements in the poset.

Throughout the paper we shall deal with two types of polarized Hodge structures.

Type I. These are effective polarized Hodge structures $(V, Q, F^\bullet)$ of weight $n \geq 0$ that we think of as $H^n(X,\mathbb{C})_{\mathrm{prim}}$ for a smooth algebraic variety X of dimension n.

Type II. These are polarized Hodge structures $(\mathfrak{g}, Q_\mathfrak{g}, F^\bullet_\mathfrak{g})$ of weight $n = 0$ and where, unless otherwise mentioned, $-Q_\mathfrak{g}$ is the Cartan–Killing form.

We think of polarized Hodge structures of Type I as directly related to algebraic geometry. Limiting mixed Hodge structures $(V, W_\bullet(N), F^\bullet)$ arising from polarized

Hodge structures of Type I may be pictured in the first quadrant of the (p,q)–plane in terms of the Deligne splitting

$$V_{\mathbb{C}} = \bigoplus_{\substack{0 \le p,q \\ 0 \le p+q \le 2n}} I^{p,q}$$

where dots indicate a possibly nonzero $I^{p,q}$. For example, a pure Hodge structure of weight $n = 5$ is depicted as

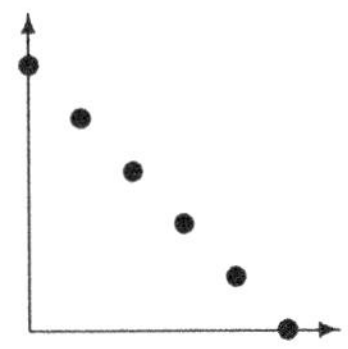

A polarized Hodge structure of Type I gives one of Type II with $\mathfrak{g} \subset \mathrm{End}(V, Q)$. In this case the corresponding Hodge domains are the same. The polarized Hodge structures of Type II are especially convenient when studying the geometry of the $G_{\mathbb{R}}$–orbits $\mathcal{O}$ in $\check{D}$. For example, suppose that the limiting mixed Hodge structure $(V, W_{\bullet}(N), F^{\bullet})$ is $\mathbb{R}$–split. The induced adjoint limiting mixed Hodge structure $(\mathfrak{g}, W_{\bullet}(N)_{\mathfrak{g}}, F^{\bullet}_{\mathfrak{g}})$ is also $\mathbb{R}$–split. Let $\mathfrak{g}_{\mathbb{C}} = \oplus I^{p,q}_{\mathfrak{g}}$ be the Deligne splitting. Let

$$F^{\bullet}_{\infty} = \lim_{z \to \infty} e^{zN} F^{\bullet} \in \mathcal{O}\,.$$

Then the tangent and normal spaces are naturally identified with

$$T_{F^{\bullet}_{\infty}} \mathcal{O} = \bigoplus_{p>0 \text{ or } q>0} \left(I^{p,q}_{\mathfrak{g}} + I^{q,p}_{\mathfrak{g}} \right)_{\mathbb{R}}\,, \tag{I.10a}$$

$$N_{F^{\bullet}_{\infty}} \mathcal{O} = \mathbf{i} \bigoplus_{p,q>0} \left(I^{p,q}_{\mathfrak{g}} + I^{q,p}_{\mathfrak{g}} \right)_{\mathbb{R}}\,, \tag{I.10b}$$

cf. Section III.B. In particular, much of the geometry (such as dimension and codimension, CR–tangent space, and intrinsic Levi form) associated with the $G_{\mathbb{R}}$–orbit $\mathcal{O} \subset \check{D}$ can be "read off" from the Deligne splitting. For example, (I.10b) implies that the (real) codimension of $\mathcal{O} \subset \check{D}$ is

$$\mathrm{codim}_{\check{D}}\, \mathcal{O} = \sum_{p,q>0} \dim_{\mathbb{C}} I^{p,q}_{\mathfrak{g}}\,. \tag{I.11}$$

Moreover, each $I^{p,q}_{\mathfrak{g}}$ may be realized as a direct sum of root spaces (and a Cartan subalgebra if $p = q = 0$), and this Lie theoretic structure plays an essential rôle in the analysis.

I.A Setting and assumptions

We will be concerned only with limiting mixed Hodge structures corresponding to unipotent local monodromy. This will often rule out "generic" specializations.

For example the "general" degeneration of a surface is nodal (and has finite order monodromy). From the Hodge–theoretic perspective these are not boundary components, but rather correspond to Hodge subdomains; such objects are of interest, but are not among those that we are concerned with here.

Throughout we take $G_\mathbb{R}$ to be the connected identity component of the real Lie group that is the set of $\mathbb{R}$–points of a $\mathbb{Q}$–algebraic group. For example, if D is a period domain, then $G_\mathbb{R}$ is the connected identity component of $\mathrm{Aut}(V, Q)$. For simplicity we will assume that the Lie algebra $\mathfrak{g}_\mathbb{R}$ is semisimple. This means that important reductive examples, such as $G_\mathbb{R} = \mathrm{U}(p,q)$, are not considered. This involves no loss of generality to the extent that the results of the paper are concerned with weight zero Hodge structures on $\mathfrak{g}_\mathbb{R}$, the $G_\mathbb{R}$–orbit structure of $\check{D} = G_\mathbb{C}/P$, and nilpotent orbits on D; however, some care must be taken when generalizing the results involving Hodge structures on V. More precisely, most of our results hold for reductive $G_\mathbb{R}$; the exceptions are Proposition I.15 (cf [24, Proposition 4.1]), and then by extension the last statement of Theorem I.16 and Corollary V.5.

We work entirely over $\mathbb{R}$ and do not consider lattices $\Gamma \subset G_\mathbb{Q}$. This means the resulting analysis will be a bit coarse as the rational structure will constrain the degenerations that may arise geometrically. This omission will be addressed in [13].

In fact, [13] will be concerned with potential algebro-geometric applications of the methods and results of this paper and the related works [17, 18, 19, 24]. In an important series of papers applying the minimal model program for threefolds, the KSBA (Kollár–Sheperd-Barron–Alexeev) moduli space $\mathcal{M}^{\mathrm{KSBA}}$ has been constructed for surfaces X of general type, cf. [21] and the references therein. This space has a canonical compactification $\overline{\mathcal{M}}^{\mathrm{KSBA}}$ which is a projective algebraic variety whose boundary components correspond to certain singular surfaces X_0. The local structure of the singularities of the X_0's is understood (loc. cit.), but we are not aware of any nonclassical[3] examples in which the global structure of the boundary components of $\mathcal{M}^{\mathrm{KSBA}}$ is understood. By constructing a suitable partial compactification $\Gamma\backslash D^*$ of $\Gamma\backslash D$ and an extension $\overline{\mathcal{M}}^{\mathrm{KSBA}} \to \Gamma\backslash D^*$ of the period mapping one may hope to use Hodge theory as reflected in the boundary structure of $\Gamma\backslash D^*$, as a guide to understand the boundary structure of $\overline{\mathcal{M}}^{\mathrm{KSBA}}$. In analogy with the Satake–Baily–Borel compactification, the boundary components attached to $\Gamma\backslash D$ are arithmetic quotients of complex submanifolds of the $G_\mathbb{R}$–orbits $\mathcal{O} \subset \overline{D}$; consequently the boundary structure of $\Gamma\backslash D^*$ will be inherited from that of D. In the work in progress [13] this program is being explored in the case when X is a minimal, regular surface of general type with $p_g(X) = 2$ and $K_X^2 = 2$.

I.B Minimal degenerations

In this section we illustrate the minimal degenerations of Theorem I.6. We shall describe the nonzero $I^{p,q}$. Figure I.1 illustrates the possibilities for weights one

[3] *Nonclassical* in the sense that the corresponding period domain is not a Hermitian symmetric domain.

through four; from these the reader will easily guess (correctly) what the general case will be.

Remark I.12. In both Figure I.1 and the discussion below the nilpotents N with $N^2 \neq 0$ polarize a codimension one orbit if and only if $n = 2m$ is even, $h^{m,m} = 1$ and $\dim V \geq 5$. (When these constraints fail the degeneration is still minimal in a Lie theoretic sense; see Remark I.14.)

The general rule is this: Let $h^{p,q}$ be the Hodge numbers of the polarized Hodge structure $(V, Q, F^\bullet)$ and

$$i^{p,q} = \dim I^{p,q}.$$

If $N \neq 0$ and $N^2 = 0$, then for one pair $p_o < q_o$ with $p_o + q_o = n$ we have:

- i^{p_o+1,q_o}, $i^{p_o,q_o-1} = 1$;
- $i^{p_o,q_o} = h^{p_o,q_o} - 1$ and $i^{p_o+1,q_o-1} = \begin{cases} h^{p_o+1,q_o-1} - 1, & p_o + 1 \neq q_o - 1, \\ h^{p_o+1,q_o-1} - 2, & p_o + 1 = q_o - 1; \end{cases}$
- for all other $p < q$, $i^{p,q} = h^{p,q}$.

Put another way, for some $p_o < q_o$ one class in V^{p_o,q_o} and one class in V^{p_o+1,q_o-1} disappear in $\mathrm{Gr}_n^{W(N)}$ and reappear as classes

$$\begin{aligned} \alpha &\in I^{p_o+1,q_o}, \\ N\alpha &\in I^{p_o,q_o-1}. \end{aligned}$$

If $N^2 \neq 0$ and $N^3 = 0$, then $n = 2m$ is even and we have:

- $i^{m-1,m-1}$, $i^{m+1,m+1} = 1$;
- $i^{m-1,m+1} = h^{m-1,m+1} - 1$ and $i^{m+1,m-1} = h^{m+1,m-1} - 1$;
- for all other $p < q$, $i^{p,q} = h^{p,q}$.

In this case, one class in $V^{m-1,m+1}$ and one class in $V^{m+1,m-1}$ disappear in $\mathrm{Gr}_n^{W(N)}$ and reappear as classes

$$\begin{aligned} \alpha &\in I^{m+1,m+1}, \\ N^2\alpha &\in I^{m-1,m-1}. \end{aligned}$$

Moreover, the degeneration is minimal if and only if $h^{m,m} = 1$ (Remark I.12).

The pictures above are particularly revealing when

$$h^{p,q} = \begin{cases} 1 & \text{for all } p \neq q, \\ 2 & \text{if } p = q. \end{cases} \tag{I.13}$$

In this case they are as pictured in Figure I.2; in these figures a uncircled node indicates $i^{p,q} = 1$, a circled node indicates $i^{p,q} = 2$.

From the algebro–geometric perspective, Theorem I.6 may at first glance seem surprising. For example, for a smooth threefold X a "generic" specialization might be thought to be $X \to X_o$ where X_o has a node. In the case that (I.13) holds, this is the right most picture of Figure I.2 for $n = 3$. The middle picture would arise

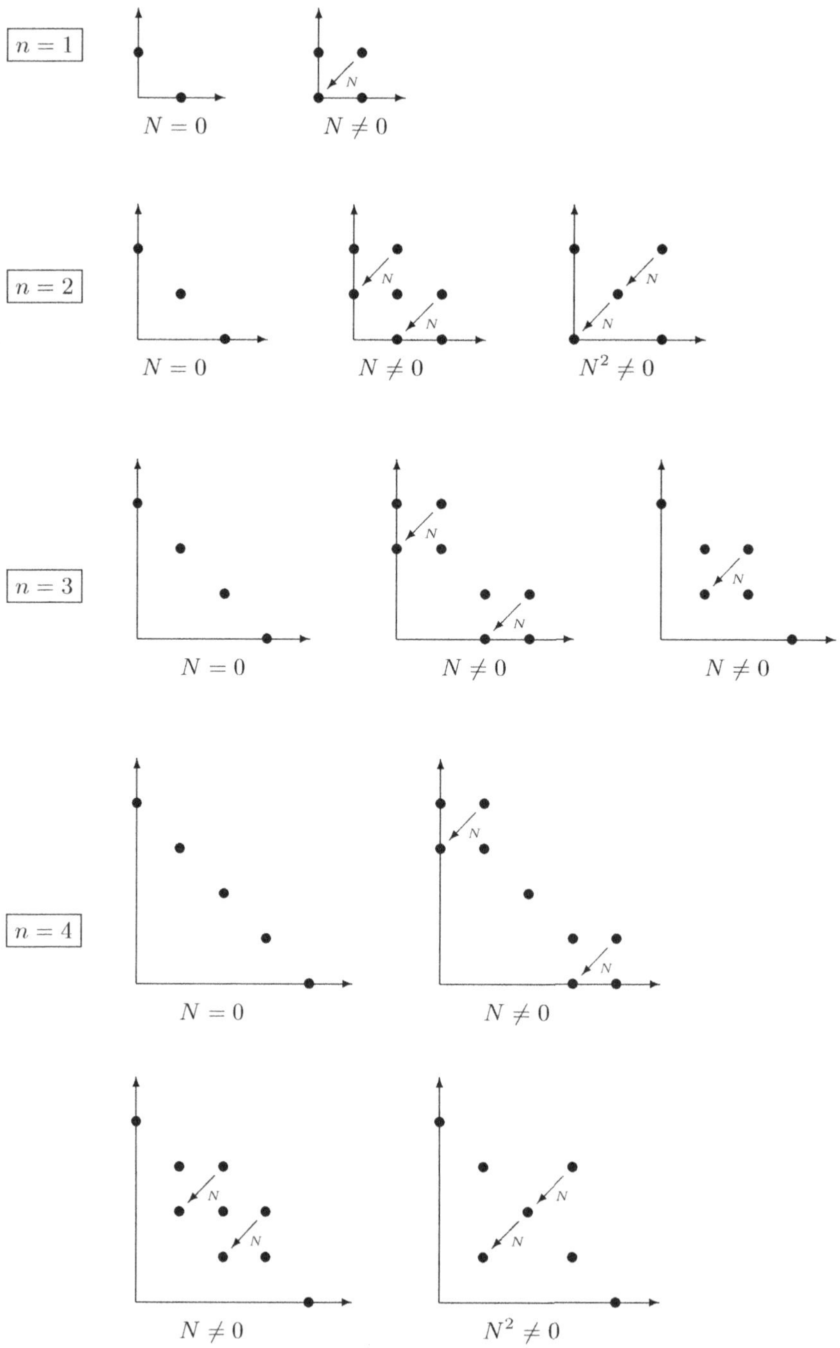

Figure I.1: Minimal degenerations for weight n period domains

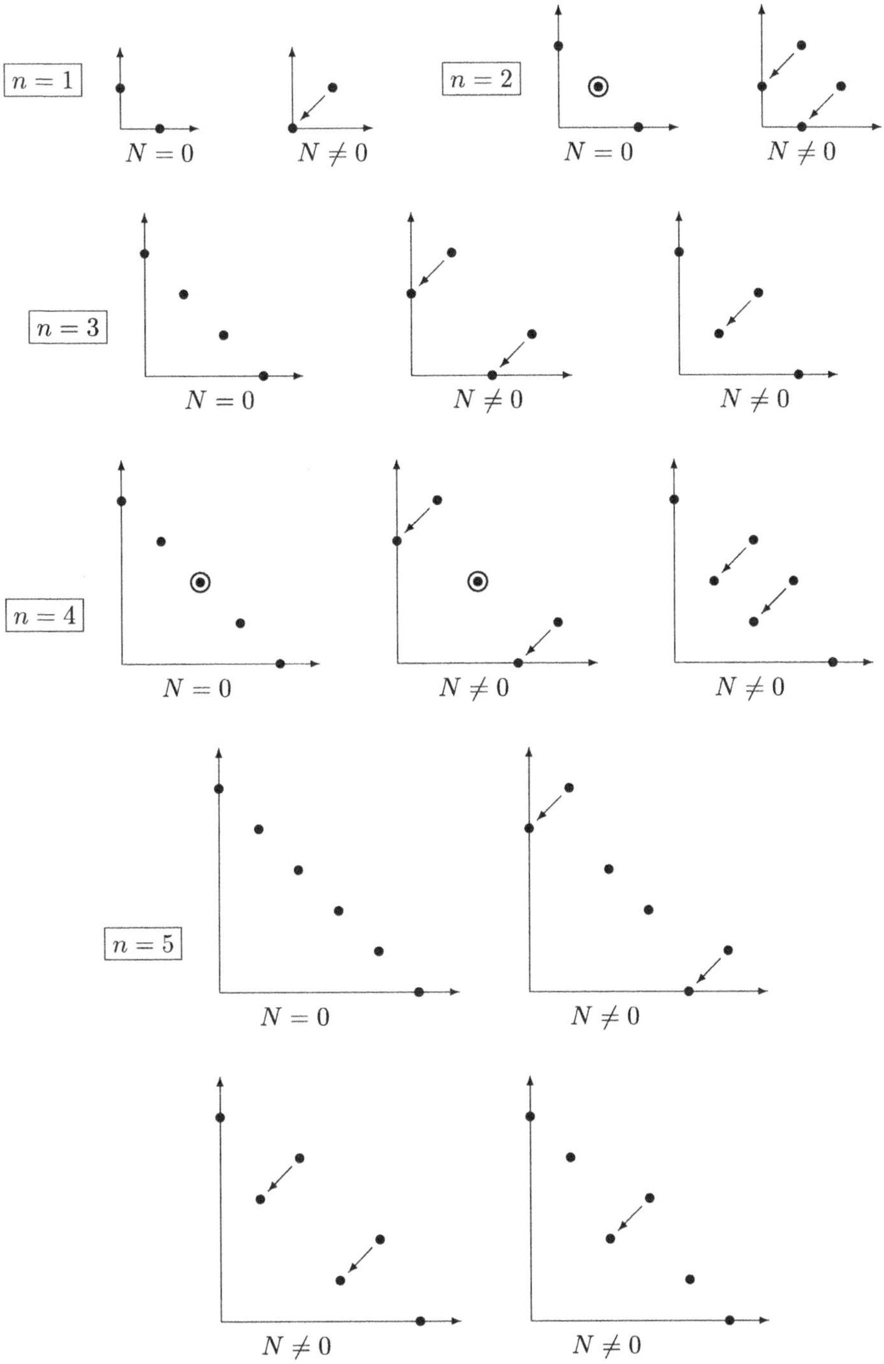

Figure I.2: Minimal degenerations for period domains with (I.13)

from X_o having a smooth double surface D, where there is an $\omega \in H^0(\Omega^3_X)$ that specializes to $\omega_o \in H^0(\Omega^3_{X_o}(\log D))$ whose residue goes to the class in $I^{2,0}$.

In the case that $n = 5$, the second and fourth picture of Figure I.2 may be interpreted as in the $n = 3$ case. For the third we may think of a five-fold X specializing to X_o with the local equation $\{x_1x_2 + x_3x_4 = 0\}$ in $\mathbb{C}^6$; that is, X_o has a double point along a surface. Then the $h^{4,1}$ drops by one. In summary, the three degenerations (in the $n = 5$ case) correspond to the local equations

$$\begin{aligned} &x_1x_2 = 0\,, && \text{(double four-fold)},\\ &x_1x_2 + x_3x_4 = 0\,, && \text{(double surface)},\\ &x_1x_2 + x_3x_4 + x_5x_6 = 0\,, && \text{(double point)}. \end{aligned}$$

In the more general setting of Hodge domains, the $I^{p,q}$ may be more complicated than those of Figure I.1: there may be more N–strings, and they may be have length greater than two. What we can say, in terms of generalizing Theorem I.6, is Proposition IV.5. Nonetheless, from the Lie theoretic perspective, the codimension–one orbits all possess a uniform structure in the following sense: for an appropriate choice of Cartan subalgebra $\mathfrak{h}_\mathbb{R}$ (essentially one may think of this as reflecting a "good choice" of basis of V), the nilpotent N will be a root vector and the normal space may be identified with a real root space [15, 18]. That is,

$$N \in \mathfrak{g}^\alpha_\mathbb{R}\,,$$

where $\alpha \in \mathfrak{h}^*_\mathbb{C}$ is a root and the root space $\mathfrak{g}^\alpha \subset \mathfrak{g}_\mathbb{C}$ is defined over $\mathbb{R}$, and (I.10b) becomes

$$N_{F^\bullet_\infty}\mathcal{O} = \mathbf{i}\, I^{1,1}_\mathfrak{g}(\mathbb{R}) = \mathbf{i}\,\mathfrak{g}^\alpha_\mathbb{R}\,.$$

Remark I.14. When the constraints of Remark I.12 fail, the nilpotents with $N^2 \neq 0$ that are described above will polarize $G_\mathbb{R}$–orbits $\mathcal{O} \subset \partial D$ of codimension greater than one. So these degenerations are not minimal in the sense of this paper. However, they have the property that N may still be realized as an element of a root space $\mathfrak{g}^\alpha_\mathbb{R}$ (with α short). So one may think of these degenerations as "close to minimal" from this Lie theoretic perspective.

I.C Maximal degenerations

We recall that mixed Hodge structure $(V, W_\bullet, F^\bullet)$ is *of Hodge–Tate type* if the $I^{p,q} = 0$ for all $p \neq q$. (The Hodge structures in row 4 of Figure II.1 are of Hodge–Tate type.) For limiting mixed Hodge structures there is the general

Proposition I.15. *The limiting mixed Hodge structure $(V, W_\bullet(N), F^\bullet)$ is of Hodge–Tate type if and only if the associated adjoint limiting mixed Hodge structure $(\mathfrak{g}, W_\bullet(N)_\mathfrak{g}, F^\bullet_\mathfrak{g})$ is of Hodge–Tate type.*

The proposition is proved in Section V.B.

For the description here of the maximal degenerations we shall make the assumption *the closed $G_{\mathbb{R}}$–orbit $\mathcal{O}_{\rm cl}$ is polarizable relative to D.*

Thus the image (I.5) of (I.4) under the reduced limit period map is a point $F^\bullet_\infty \in \mathcal{O}_{\rm cl}$.

Theorem I.16. *The following are equivalent:*

(a) *The orbit $\mathcal{O}_{\rm cl}$ is totally real; i.e., the Cauchy–Riemann tangent space $T^{\rm CR}\mathcal{O}_{\rm cl} = 0$.*

(b) *The real dimension of the closed orbit is the complex dimension of the compact dual:* $\dim_{\mathbb{R}}\mathcal{O}_{\rm cl} = \dim_{\mathbb{C}}\check{D}$.

(c) *The stabilizer* $P = \mathrm{Stab}_{G_{\mathbb{C}}}(F^\bullet_\infty)$ *is $\mathbb{R}$–split and* $\mathcal{O}_{\rm cl} = G_{\mathbb{R}}/P_{\mathbb{R}}$.

Any of these imply that the limiting mixed Hodge structures $(V, W_\bullet(N), F^\bullet)$ and $(\mathfrak{g}, W_\bullet(N)_{\mathfrak{g}}, F^\bullet_{\mathfrak{g}})$ are of Hodge–Tate type.

The theorem is proved as Theorem III.22 and Corollary V.5.

From an algebro–geometric perspective it is not surprising that the most degenerate limiting mixed Hodge structure is one of Hodge–Tate type. More interesting is that there are both Hodge theoretic (Lemma V.7) and Lie theoretic (Lemma V.10 and Remark V.12) obstructions to a given type of polarized Hodge structure being able to degenerate to one of Hodge–Tate type.

In general there is the following

Theorem I.17. *Suppose that the limiting mixed Hodge structure $(V, W_\bullet(N), F^\bullet)$ is sent to the closed orbit under the reduced limit period mapping* (I.5). *Then Deligne splitting $\mathfrak{g}_{\mathbb{C}} = \oplus I^{p,q}_{\mathfrak{g}}$ associated with the induced limiting mixed Hodge structure $(\mathfrak{g}, W_\bullet(N)_{\mathfrak{g}}, F^\bullet_{\mathfrak{g}})$ satisfies:*

$$\begin{aligned} I^{p,-q}_{\mathfrak{g}} &= 0 \quad \textit{for all} \quad p \neq q > 0\,, \\ I^{p,-p}_{\mathfrak{g}} &= 0 \quad \textit{for all} \quad \textit{odd} \ \ p \geq 3\,, \\ I^{p,q}_{\mathfrak{g}} &= 0 \quad \textit{for all} \quad p+q \neq 0 \quad \textit{with} \quad |p-q| > 2\,. \end{aligned}$$

The constraints of Theorem I.17 are illustrated in Figure III.3.b. We will prove the slightly stronger Theorem V.1.

For period domains one may reconstruct the limiting mixed Hodge structure $(V, W_\bullet(N), F^\bullet)$ from $(\mathfrak{g}, W_\bullet(N)_{\mathfrak{g}}, F^\bullet_{\mathfrak{g}})$; doing so, Theorem I.17 yields Theorem I.8 (which is is proved in Section V.D).

A convenient schematic to picture a limiting mixed Hodge structures is its decomposition into N–strings

$$\begin{array}{ccccccccc} H^0(n) & \xrightarrow{N} & H^0(n-1) & \xrightarrow{N} \cdots \xrightarrow{N} & H^0(1) & \xrightarrow{N} & H^0 \\ & & H^1(n-1) & \xrightarrow{N} \cdots \xrightarrow{N} & H^1 & & \\ & & & \vdots & & & \\ & & & H^{n-1}(1) \xrightarrow{N} H^{n-1} & & & \\ & & & H^n & & & \end{array} \tag{I.18}$$

where

$$H^k = N^{n-k}\operatorname{Gr}^{W_\bullet(N)}_{2n-k,\mathrm{prim}}$$

is a polarized Hodge structure of weight k. (It may happen that H^k is a Tate twist of a lower weight Hodge structure.) Under the schematic (I.18), the possibilities in Theorem I.8 are:

$\boxed{n=2}$ We have $H^1=0$, $H^0\neq 0$ and H^2 has type $(*,0,*)$. In terms of the $i^{p,q}_{\mathrm{prim}}$ the Hodge numbers are

$$h^{2,0} = i^{2,0}_{\mathrm{prim}} + i^{2,2}_{\mathrm{prim}}\,, \quad h^{1,1} = i^{2,2}_{\mathrm{prim}}\,.$$

$\boxed{n=4}$ We have $H^0=H^1=H^3=0$, $H^2\neq 0$ is of Hodge–Tate type and H^4 has type $(0,*,0,*,0)$.

$\boxed{n=6}$ We have $H^1=H^2=H^3=H^5=0$, at least one of H^0 and H^4 is nonzero and of Hodge–Tate type, and H^6 has type $(0,0,*,0,*,0,0)$.

Corollary I.19. *If the limiting mixed Hodge structure corresponds to a point in the closed orbit, then* $\operatorname{Gr}^{W_\bullet(N)}_*$ *is rigid in the sense that it does not admit a non-trivial variation of Hodge structure.*

The reason is that the infinitesimal period relation is trivial for the $\operatorname{Gr}^{W_\bullet(N)}_*$.

The above results were under the assumption that the limit period map sends (I.4) to a point in the closed orbit. As we have discussed, in some, but not all, cases the limiting mixed Hodge structure is of Hodge–Tate type. In the other direction we have Proposition I.7.

Given a degeneration of polarized Hodge structures (I.4) whose reduced limit mapping (I.5) goes to a point in a $G_{\mathbb{R}}$–orbit $\mathcal{O}\subset\partial D$, for the limiting mixed Hodge structure $(V,W_\bullet(N),F^\bullet)$ the above may be summarized as follows:

(a) If $\mathcal{O}$ is the closed orbit, and is totally real, then $(V,W_\bullet(N),F^\bullet)$ is Hodge–Tate.

(b) If $(V,W_\bullet(N),F^\bullet)$ is Hodge–Tate, then $\mathcal{O}$ is the closed orbit.

(c) If $\mathcal{O}$ is the closed orbit and $(V,W_\bullet(N),F^\bullet)$ is *not* Hodge–Tate, then the nonzero $\operatorname{Gr}^{W_\bullet(N)}_{n+k,\mathrm{prim}}$ have $k\equiv 2$ (mod) 4, and are Hodge–Tate, and $\operatorname{Gr}^{W_\bullet(N)}_{n,\mathrm{prim}}$ is as close to being Hodge–Tate as the Hodge numbers of $(V,Q,F^\bullet)$ will allow.

I.D Notation

- $\Delta=\{t\in\mathbb{C} : |t|<1\}$ is the unit disc, $\Delta^*=\{t\in\Delta : t\neq 0\}$ is the punctured unit disc, and $\mathbb{H}=\{z\in\mathbb{C} : \operatorname{Im}(z)>0\}$ is the upper–half plane.
- $\mathcal{X}\to\Delta$ is a semi–stable reduction.
- $\check{D}=G_{\mathbb{C}}/P$ is a generalized flag variety containing $D=G_{\mathbb{R}}/R$ as an open $G_{\mathbb{R}}$–orbit with a compact isotropy group $R=G_{\mathbb{R}}\cap P$.

- $\Phi : \Delta^* \to \Gamma_T \backslash D$ is a period mapping with lift $\tilde{\Phi} : \mathbb{H} \to D$ and $\tilde{\Phi}_* : T\mathbb{H} \to I \subset T\check{D}$, where I denotes the infinitesimal period relation.
- $\tilde{B}(N) \subset \check{D}$ is the set of N–polarized limiting mixed Hodge structures, and $B(N)$ the boundary component consisting of equivalence classes of limiting mixed Hodge structures.
- $\Phi_\infty : B(N) \to \partial D$ is the reduced limit period map.
- $F^\bullet_{\lim} = \lim_{t\to 0}\left(e^{-\ell(t)N}F^\bullet_t\right) = \lim_{\mathrm{Im}(z)\to\infty} e^{-zN}\tilde{\Phi}(z)$ where $\ell(t) = \frac{1}{2\pi\mathbf{i}}\log t$ and $t = e^{2\pi\mathbf{i}z} \in \Delta^*$ for $z \in \mathbb{H}$.
- $F^\bullet_\infty = \lim_{\mathrm{Im}(z)\to\infty}\tilde{\Phi}(z)$.
- $\mathcal{T} \subset R$ is a compact maximal torus of $G_\mathbb{R}$.
- $\mathfrak{g}_\mathbb{C}$ and $\mathfrak{g}_\mathbb{R}$ are the Lie algebras of $G_\mathbb{C}$ and $G_\mathbb{R}$.
- $(V, W_\bullet(N), F^\bullet)$ a limiting mixed Hodge structure with $F^\bullet = F^\bullet_{\lim}$, $(V, W_\bullet, \tilde{F}^\bullet)$ is the associated $\mathbb{R}$–split limiting mixed Hodge structure associated, and $(\mathfrak{g}, W_\bullet(N)_\mathfrak{g}, F^\bullet_\mathfrak{g})$ the induced adjoint limiting mixed Hodge structures on $\mathfrak{g}$.
- $V_\mathbb{C} = \oplus I^{p,q}$ and $\mathfrak{g}_\mathbb{C} = \oplus I^{p,q}_\mathfrak{g}$ are the Deligne splittings of a limiting mixed Hodge structure $(V, W_\bullet(N), F^\bullet)$ and the induced adjoint limiting mixed Hodge structure $(\mathfrak{g}, W_\bullet(N)_\mathfrak{g}, F^\bullet_\mathfrak{g})$.
- $-Q_\mathfrak{g}$ is the Killing form on $\mathfrak{g}$.
- $\mathfrak{h} \subset \mathfrak{g}_\mathbb{C}$ is a Cartan subalgebra with roots $\Delta \subset \mathfrak{h}^*$.[4]
- Given $x \in \check{D}$, $\mathcal{O}_x = G_\mathbb{R} \cdot x$ is the $G_\mathbb{R}$–orbit.

II Reduced limit period mappings

II.A Generalized flag varieties

A *generalized flag variety* is a homogeneous complex manifold

$$\check{D} = G_\mathbb{C}/P$$

where $G_\mathbb{C}$ is a complex, semi–simple Lie group and P is a parabolic subgroup.[5] When $P = B$ is a Borel subgroup we shall use the term *flag variety*. At the reference point $x_o = P\mathbb{1}$ there is a natural identification of the holomorphic tangent space

$$T_{x_o}\check{D} = \mathfrak{g}_\mathbb{C}/\mathfrak{p}\,. \tag{II.1}$$

[4]We shall use the notation Δ for both the set of roots of $(\mathfrak{g}_\mathbb{C}, \mathfrak{h})$ and for the unit disc in $\mathbb{C}$; the context should make it clear which use of Δ is being made.

[5]We use the notation $\check{D}$ because we shall mainly think of it as the compact dual of a generalized flag domain.

We shall describe $\mathfrak{p}$ in terms of the roots associated to a Cartan subalgebra $\mathfrak{h} \subset \mathfrak{p}$, which always exists. The root space decomposition of $(\mathfrak{g}_{\mathbb{C}}, \mathfrak{h})$ is

$$\mathfrak{g}_{\mathbb{C}} = \mathfrak{h} \oplus \bigoplus_{\alpha\in\Delta} \mathfrak{g}^{\alpha}$$

where $\Delta \subset \mathfrak{h}^*$ is the set of *roots* and $\mathfrak{g}^{\alpha}$ is the one–dimensional *root space.* We let $\Delta^+ \subset \Delta$ denote a choice of *positive roots.* This determines a Borel subalgebra $\mathfrak{b} \supset \mathfrak{h}$ by

$$\mathfrak{b} = \mathfrak{h} \oplus \bigoplus_{\alpha\in\Delta^+} \mathfrak{g}^{-\alpha} .$$

Conversely, a choice of Borel $\mathfrak{b} \supset \mathfrak{h}$ determines the positive roots by

$$\Delta^+ = \{\alpha \in \Delta \ : \ \mathfrak{g}^{-\alpha} \subset \mathfrak{b}\} .$$

Let $\Delta^+_s \subset \Delta^+$ denote the set of *simple roots* relative to that choice.

The first description of parabolic subalgebras $\mathfrak{p} \subset \mathfrak{g}$ is in terms of subsets $\Sigma \subset \Delta^+_s$. Denoting by $\langle\Sigma\rangle := \Delta \cap \mathrm{span}_{\mathbb{Z}} \Sigma$ the roots spanned by Σ, we set

$$\mathfrak{p}_\Sigma = \underbrace{\mathfrak{h} \oplus \Big(\bigoplus_{\alpha\in\langle\Sigma\rangle} \mathfrak{g}^{\alpha}\Big)}_{\mathfrak{p}_r} \oplus \underbrace{\Big(\bigoplus_{\beta\in\Delta^-\setminus\langle\Sigma\rangle^-} \mathfrak{g}^{\beta}\Big)}_{\mathfrak{p}_n}$$

where $\Delta^- = -\Delta^+$ and $\langle\Sigma\rangle^- = \langle\Sigma\rangle \cap \Delta^-$. Then $\mathfrak{p}_\Sigma$ is a parabolic subalgebra with reductive Levi factor $\mathfrak{p}_r$ and nilpotent radical $\mathfrak{p}_n$. Note that

$$\mathfrak{h} \subset \mathfrak{b} \subset \mathfrak{p}_\Sigma .$$

When a choice of Cartan and Borel has been made, any parabolic of the form $\mathfrak{p}_\Sigma$ is a *standard parabolic.* Every parabolic subalgebra $\mathfrak{p}$ of $\mathfrak{g}$ is $\mathrm{Ad}(G_{\mathbb{C}})$–conjugate to a standard parabolic. Using the identification (II.1) of the holomorphic tangent space we have

$$T_{x_o}\check{D} = \bigoplus_{\beta\in\Delta^+\setminus\langle\Sigma\rangle^+} \mathfrak{g}^{\beta} =: \mathfrak{p}^+_n . \tag{II.2}$$

The second description of parabolic subalgebras $\mathfrak{p} \subset \mathfrak{g}$ is in terms of the set $\mathrm{Hom}(\Lambda_{\mathrm{rt}}, \mathbb{Z})$ of *grading elements* where $\Lambda_{\mathrm{rt}} = \mathrm{span}_{\mathbb{Z}} \Delta \subset \mathfrak{h}^*$ is the root lattice. If $\Delta^+_s = \{\alpha_1, \ldots, \alpha_r\}$, then there is a dual basis $\{L_1, \ldots, L_r\}$ for the set of grading elements given by

$$\alpha_j(L_i) = \delta_{ij} . \tag{II.3}$$

Given a grading element L, under the action of $\mathfrak{h}$ on $\mathfrak{g}_{\mathbb{C}}$ we have an eigenspace decomposition

$$\begin{aligned} \mathfrak{g}_{\mathbb{C}} &= \mathfrak{g}_{-k} \oplus \cdots \oplus \mathfrak{g}_0 \oplus \cdots \oplus \mathfrak{g}_k , \\ \text{with} \quad \mathfrak{g}_\ell &= \{\xi \in \mathfrak{g} \ : \ [L, \xi] = \ell\xi\} . \end{aligned} \tag{II.4}$$

Because the roots are integral linear combination of the simple roots, we see from (II.3) that the eigenvalues of L are integers $\ell \in \mathbb{Z}$. We note that

- Each $\mathfrak{g}_\ell$ is a direct sum of root spaces (and $\mathfrak{h}$ when $\ell = 0$). Explicitly,

$$\begin{aligned} \mathfrak{g}_\ell &= \bigoplus_{\alpha(L)=\ell} \mathfrak{g}^\alpha \quad \text{if } \ell \neq 0\,, \\ \mathfrak{g}_0 &= \mathfrak{h} \oplus \bigoplus_{\alpha(L)=0} \mathfrak{g}^\alpha\,. \end{aligned} \tag{II.5}$$

- The eigenspaces $\mathfrak{g}_\ell$ and $\mathfrak{g}_{-\ell}$ pair non–degenerately under the Cartan–Killing form.
- The Jacobi identity yields

$$[\mathfrak{g}_\ell, \mathfrak{g}_m] \subset \mathfrak{g}_{\ell+m}\,. \tag{II.6}$$

- The Cartan subalgebra $\mathfrak{h} \subset \mathfrak{g}_0$ and $\mathfrak{g}_0$ is a reductive subalgebra. (Indeed, $\mathfrak{g}_0$ is a Levi factor the parabolic subalgebra (II.8).) By (II.6), each $\mathfrak{g}_\ell$ is a $\mathfrak{g}_0$–module.
- More generally, any representation U of $\mathfrak{g}_\mathbb{C}$ admits an L–eigenspace decomposition. Since the weights of $\mathfrak{g}_\mathbb{C}$ are rational linear combinations of the roots, the eigenvalues are rational $U = \oplus_{m\in\mathbb{Q}} U_m$.

Remark II.7. Grading elements may be defined without reference to a choice of positive roots (equivalently, a choice of Cartan and Borel $\mathfrak{h} \subset \mathfrak{b}$). In general, a grading element is any semisimple endomorphism of $\mathfrak{g}_\mathbb{C}$ with integer eigenvalues and the property that the eigenspace decomposition (II.4) satisfies (II.6). If $\mathfrak{g}_\mathbb{C}$ is semisimple, any such endomorphism is necessarily a grading element [6, Proposition 3.1.2(1)].[6]

Given L we may define the parabolic subalgebra

$$\mathfrak{p}_L = \mathfrak{g}_0 \oplus \mathfrak{g}_- \tag{II.8}$$

where $\mathfrak{g}_- = \mathfrak{g}_{-1} \oplus \cdots \oplus \mathfrak{g}_{-k}$. Then (II.8) is a Levi decomposition of $\mathfrak{p}_L$. This associates to every grading element a parabolic subalgebra containing $\mathfrak{h}$. The parabolic subalgebra $\mathfrak{p}_L$ is standard if and only if $\alpha_i(L) \geq 0$ for all simple roots α_i. Conversely, given a standard parabolic $\mathfrak{p}_\Sigma$, there is a canonically associated grading element

$$L_\Sigma = \sum_{\alpha_i \notin \Sigma} L_i\,. \tag{II.9}$$

In particular, the relationship between the root and grading element descriptions is: Given L define

$$\Sigma_L = \{\alpha \in \Delta_s^+ \ :\ \alpha(L) = 0\} \subset \Delta_s^+\,.$$

With the grading element notation we note that (II.2) is

$$T_{x_o}\check{D} = \mathfrak{g}_+ = \mathfrak{g}_1 \oplus \cdots \oplus \mathfrak{g}_k\,.$$

[6]The definition of grading element in [6] is more restrictive than ours: it imposes the condition that $\mathfrak{g}_1$ generate the Lie subalgebra $\mathfrak{g}_+$. Nonetheless the proof of [6, Proposition 3.1.2(1)] applies to our looser notion.

II.B Generalized flag domains

A real form $\mathfrak{g}_{\mathbb{R}}$ of $\mathfrak{g}_{\mathbb{C}}$ is the set of fixed points of a conjugation

$$\sigma : \mathfrak{g}_{\mathbb{C}} \to \mathfrak{g}_{\mathbb{C}} \quad \text{satisfying} \quad \sigma(\lambda X) = \bar{\lambda}(X) \text{ for all } \lambda \in \mathbb{C} \text{ and } X \in \mathfrak{g}_{\mathbb{C}} .$$

We denote by $G_{\mathbb{R}} \subset G_{\mathbb{C}}$ the corresponding connected real Lie group. A *generalized flag domain* is defined to be an open $G_{\mathbb{R}}$–orbit $D \subset \check{D}$ whose isotropy group is compact. A period domain is an example of a generalized flag domain. If $D = G_{\mathbb{R}} \cdot x_o \subset \check{D} = G_{\mathbb{C}}/P$, then the isotropy group is

$$R = G_{\mathbb{R}} \cap P .$$

Conversely, given a homogeneous complex manifold $D = G_{\mathbb{R}}/R$ with $R \subset G_{\mathbb{R}}$ the compact centralizer of a torus, its compact dual is a generalized flag variety $\check{D} = G_{\mathbb{C}}/P$ as above in which D is an open $G_{\mathbb{R}}$–orbit. It is known that R contains a compact maximal torus $\mathcal{T}$ whose complexified Lie algebra $\mathfrak{t}_{\mathbb{C}}$ is a Cartan subalgebra $\mathfrak{h}$ of $\mathfrak{g}$. The roots Δ of $\mathfrak{h}$ take purely imaginary values on $\mathfrak{t}$, which gives that

$$\overline{\mathfrak{g}^{\alpha}} = \mathfrak{g}^{-\alpha} .$$

We have the identification of the complexified (real) tangent space

$$T_{x_o,\mathbb{C}} D = \bigoplus_{\alpha \in \Delta \setminus \langle \Sigma \rangle} \mathfrak{g}^{\alpha} = \mathfrak{g}_{+} \oplus \mathfrak{g}_{-} ,$$

and with a suitable choice of positive roots we have

$$T^{1,0}_{x_o} D = \bigoplus_{\alpha \in \Delta^{+} \setminus \langle \Sigma \rangle^{+}} \mathfrak{g}^{\alpha} = T_{x_o} D .$$

If we specify $\check{D}$ (equivalently, $P \subset G_{\mathbb{C}}$) by a grading element L as above, then we have

$$\overline{L} = -L . \tag{II.10}$$

II.C Polarized Hodge structures

II.C.1 Definition

Let $(V, Q, F^{\bullet})$ denote a polarized Hodge structure of weight n on a real vector space V. That is, the *polarization* $Q : V \times V \to \mathbb{R}$ is a nondegenerate bilinear form such that

$$Q(u, v) = (-1)^n Q(v, u) \quad \text{for all } \ u, v \in V ,$$

and the *Hodge filtration* $F^n \subset F^{n-1} \subset \cdots \subset F^1 \subset F^0 = V_{\mathbb{C}}$ is a decreasing filtration on $V_{\mathbb{C}}$ with the properties that

$$F^p \oplus \overline{F^{n-p+1}} \ \overset{\simeq}{\to} \ V_{\mathbb{C}} ,$$

(HR1) $$Q(F^p, F^{n-p+1}) = 0 ,$$

(HR2) $$Q(Cv, \bar{v}) \ > \ 0 \quad \text{for all } 0 \neq v \in V_{\mathbb{C}}$$

where C denotes the Weil operator. The equation (HR1) is the *first Hodge–Riemann bilinear relation* (HR1); the inequality (HR2) is the *second Hodge–Riemann bilinear relation.*

A *Q–isotropic flag* is any filtration $F^\bullet$ of $V_\mathbb{C}$ satisfying (HR1). Let

$$\mathbf{f} = (f^p = \dim F^p)$$

denote the *Hodge numbers.* We may regard the Hodge structure as a point in the $\mathrm{Aut}(V_\mathbb{C}, Q)$–homogeneous generalized flag variety $\mathrm{Flag}^Q_\mathbf{f}(V_\mathbb{C})$ of Q–isotropic filtrations $F^\bullet$ of $V_\mathbb{C}$ with dimension $\dim F^p = f^p$. The $\mathrm{Aut}(V,Q)$–orbit of any one of these Hodge structures is the *period domain* of Q–polarized Hodge structures with Hodge numbers $\mathbf{f}$. It is an open subset of $\mathrm{Flag}^Q_\mathbf{f}(V_\mathbb{C})$; the latter is the *compact dual* of the period domain.

The *Hodge decomposition* is

$$V_\mathbb{C} = \bigoplus_{p+q=n} V^{p,q} \quad \text{with} \quad V^{p,q} = F^p \cap \overline{F^q}$$

and satisfies

$$\begin{aligned} Q(V^{p,q}, V^{r,s}) \ &\neq \ 0 \quad \text{only if} \quad (p,q) = (s,r)\,, \\ C|_{V^{p,q}} &= \mathbf{i}^{p-q}\mathbb{1}\,. \end{aligned}$$

We will also refer to

$$\mathbf{h} = (h^{p,q} = \dim V^{p,q})$$

as the *Hodge numbers* of the Hodge structure. Note that $f^p = \sum_{q\geq p} h^{q,n-q}$.

A polarized Hodge structure $(V, Q, F^\bullet)$ of weight n is equivalent to a homomorphism

$$\varphi : S^1 \to \mathrm{Aut}(V,Q)$$

of real algebraic groups with the properties that $\varphi(-1) = (-1)^n\mathbb{1}$ and $Q(\varphi(\mathbf{i})v, \bar{v}) > 0$ for all $0 \neq v \in V_\mathbb{C}$. The Hodge decomposition is the φ–eigenspace decomposition

$$V^{p,q} = \{v \in V_\mathbb{C} \ : \ \varphi(z)v = z^{p-q}v\,,\ \forall\, z \in S^1\}\,.$$

II.C.2 Induced PHS on End(V,Q)

There is an induced Hodge structure on the Lie algebra $\mathrm{End}(V,Q)$ of $\mathrm{Aut}(V,Q)$ defined by

$$F^p\,\mathrm{End}(V_\mathbb{C}, Q) = \{\xi \in \mathrm{End}(V_\mathbb{C}, Q) \ : \ \xi(F^q) \subset F^{p+q}\ \forall\, q\}\,.$$

Equivalently,

$$\mathrm{End}(V_\mathbb{C}, Q)^{p,q} = \{\xi \in \mathrm{End}(V_\mathbb{C}, Q) \ : \ \xi(V^{r,s}) \subset V^{p+r,q+s}\ \forall\, r,s\}\,.$$

Note that $\mathrm{End}(V_\mathbb{C}, Q)^{p,q} = 0$ if $p + q \neq 0$, so that

$$\mathrm{End}(V_\mathbb{C}, Q) = \bigoplus_{p\in\mathbb{Z}} \mathrm{End}(V_\mathbb{C}, Q)^{p,-p}$$

is a weight zero Hodge structure.

II.C.3 PHS in terms of grading elements

We may view grading elements as "infinitesimal Hodge structures" as follows.[7] Note that the induced Hodge structure on $\mathrm{End}(V, Q)$ satisfies

$$\left[\mathrm{End}(V_{\mathbb{C}}, Q)^{p,-p}, \mathrm{End}(V_{\mathbb{C}}, Q)^{q,-q}\right] \subset \mathrm{End}(V_{\mathbb{C}}, Q)^{p+q,-p-q}.$$

So, if we define L to be the semisimple endomorphism of $\mathrm{End}(V_{\mathbb{C}}, Q)$ that acts on $\mathrm{End}(V_{\mathbb{C}}, Q)^{p,-p}$ by the eigenvalue $-p$, then the discussion of Remark II.7 implies that $L \in \mathrm{End}(V_{\mathbb{C}}, Q)$ is a grading element.

The grading element L also induces the original Hodge structure on V. In particular, the standard representation $V_{\mathbb{C}}$ of $\mathrm{Aut}(V_{\mathbb{C}}, Q)$ decomposes into a direct sum $\oplus_m V_m$ of L–eigenvalues with rational eigenvalues (Remark II.7). This eigenspace decomposition is the Hodge decomposition

$$V^{p,q} = V_{(q-p)/2} = \{v \in V_{\mathbb{C}} \ : \ L(v) = \frac{1}{2}(q-p)v\}. \tag{II.11}$$

We say that *the grading element defines a polarized Hodge structure on V.*

II.D Hodge groups and domains

II.D.1 Definition

Let V be a $\mathbb{R}$-vector space, and let Q be a non-degenerate bilinear form $Q : V \times V \to \mathbb{R}$ satisfying $Q(u, v) = (-1)^n Q(v, u)$ for some $n \in \mathbb{Z}$. A reductive $\mathbb{R}$-algebraic group G is a *Hodge group* if there exists a representation $\rho : G_{\mathbb{R}} \to \mathrm{Aut}(V, Q)$ and a nonconstant homomorphism $\varphi : \mathbb{U}_{\mathbb{R}} \to G_{\mathbb{R}}$ such that $(V, Q, \rho \circ \varphi)$ is a (real) polarized Hodge structure of weight n. Let F_φ denote the corresponding Hodge flag. The triple (G, ρ, φ) is a *Hodge representation.*

Given a Hodge representation, the associated *Hodge domain* D is the set of polarized Hodge structures $\{(V, Q, \rho(g)F^\bullet_\varphi) \ : \ g \in G_{\mathbb{R}}\}$.[8] That is, D is the $G_{\mathbb{R}}$–orbit of $F^\bullet_\varphi$ in the period domain containing the polarized Hodge structure $(V, Q, F^\bullet_\varphi)$. We let $R \subset G_{\mathbb{R}}$ denote the centralizer of the circle φ, and write $D = G_{\mathbb{R}}/R$.

Example II.12. A special case of a Hodge domain is the period domain of Section II.C.1. In this case $G_{\mathbb{R}} = \mathrm{Aut}(V, Q)$ is $\mathrm{SO}(2a, b)$ for even weight, and $\mathrm{Sp}(2g, \mathbb{R})$ for odd weight.

A Hodge domain structure gives an embedding of the compact dual

$$\check{D} \hookrightarrow \mathrm{Flag}^Q_{\mathbf{f}}(V_{\mathbb{C}})$$

as the $G_{\mathbb{C}}$–orbit of any $F^\bullet \in D$. Specifically, $x \in \check{D}$ gives a flag $F^\bullet_x$ that satisfies the first Hodge–Riemann bilinear relation $Q(F^p_x, F^{n-p+1}_x) = 0$. The second Hodge–Riemann bilinear relation defines the open $G_{\mathbb{R}}$–orbit $D \subset \check{D}$.

---

[7] "Infinitesimal" because, appropriately rescaled, $\varphi'(1)$ is a grading element, and conversely every grading element may be realized as $\varphi'(1)$, cf. [23].

[8] A more precise term would be a *Hodge domain structure on the generalized flag domain D.*

In the case when V is an irreducible G–module another way of think of a Hodge domain is that it is given by a pair (φ, χ) consisting of a co-character φ and a character χ of $\mathcal{T} \subset G_\mathbb{R}$. Specifically, for $S^1 = \mathbb{R}/2\pi\mathbf{i}\mathbb{Z}$ and

$$\varphi : S^1 \to \mathcal{T}$$

the isotropy subgroup $R = Z_{G_\mathbb{R}}(\varphi(S^1))$ is the centralizer in $G_\mathbb{R}$ of the circle $\varphi(S^1)$. The $G_\mathbb{R}$–invariant complex structure on $D = G_\mathbb{R}/R$ is given by

$$\mathrm{Ad} \circ \varphi : S^1 \to \mathfrak{g}_\mathbb{R}/\mathfrak{r} \simeq T_{x_o,\mathbb{R}} G_\mathbb{R}/R\,.$$

The character χ is the highest weight of the representation $\rho : G \to \mathrm{Aut}(V, Q)$. There are conditions, not spelled out here, on the pair (φ, χ). See [14] for details.

II.D.2 Induced PHS on $\mathfrak{g}$

From the description of Section II.D.1 it is clear that a given homogeneous complex manifold D, corresponding to φ above, may be realized as a Hodge domain in multiple ways, corresponding to the χ's above (see Section II.D.4). For this work a particularly important pair of such realizations is the following: Given a generalized flag domain D realized as a Hodge domain for polarized Hodge structure $(V, Q, F^\bullet)$, another realization is as induced polarized Hodge structures on $\mathfrak{g} \subset \mathrm{End}(V, Q)$. Generalizing Section II.C.2 from $\mathrm{Aut}(V, Q)$ to the more general Hodge groups G, these are defined by

$$F^p_\mathfrak{g} = \{\xi \in \mathfrak{g}_\mathbb{C} \ : \ \xi(F^q) \subset F^{p+q} \ \forall\ q\}\,.$$

Equivalently,

$$\mathfrak{g}^{p,q} = \{\xi \in \mathfrak{g}_\mathbb{C} \ : \ \xi(V^{r,s}) \subset V^{p+r,q+s} \ \forall\ r,s\}\,.$$

Note that $\mathfrak{g}^{p,q} = 0$ if $p + q \neq 0$, so that

$$\mathfrak{g}_\mathbb{C} = \mathfrak{g}^{-k,k} \oplus \cdots \oplus \mathfrak{g}^{0,0} \oplus \cdots \oplus \mathfrak{g}^{k,-k} \tag{II.13}$$

is a weight zero Hodge structure. The decomposition (II.13) satisfies $[\mathfrak{g}^{p,-p}, \mathfrak{g}^{q,-q}] \subset \mathfrak{g}^{p+q,-p-q}$ for all p, q. Consequently $\mathfrak{g}^- := \oplus_{p>0} \mathfrak{g}^{-p,p}$ is a subalgebra. The IPR is *bracket-generating* if $\mathfrak{g}^-$ is generated by $\mathfrak{g}^{-1,1}$ as a Lie algebra. (If D is a period domain, then the IPR is bracket-generating if and only if there are no "gaps" in the Hodge numbers [7]. See [23] for further discussion of bracket-generation.)

The polarization Q on V induces a polarization $Q_\mathfrak{g}$ on $\mathfrak{g}$. The latter is invariant under G. Therefore, if $\mathfrak{g}$ is simple, then $-Q_\mathfrak{g}$ is necessarily a positive multiple of the Killing form. Unless otherwise stated,

$-Q_\mathfrak{g}$ will denote the Killing form throughout.

Notice that $F^0_\mathfrak{g}$ is the Lie algebra of the stabilizer $P \subset G_\mathbb{C}$ of *both* the Hodge structure $F^\bullet$ on V and the induced Hodge structure $F^\bullet_\mathfrak{g}$ on $\mathfrak{g}$. In particular,

the $G_\mathbb{C}$-orbits of $F^\bullet \in \mathrm{Flag}^Q_\mathbf{f}(V_\mathbb{C})$ and $F^\bullet_\mathfrak{g} \in \mathrm{Flag}^{Q_\mathfrak{g}}_{\mathbf{f}_\mathfrak{g}}(\mathfrak{g}_\mathbb{C})$ both realize the generalized flag variety $\check{D} = G_\mathbb{C}/P$ as a projective variety.

Moreover, the two infinitesimal period relations agree under this identification, cf. [14]. Likewise, if

$$R = G_{\mathbb{R}} \cap P\,,$$

then

the $G_{\mathbb{R}}$–orbits of $F^\bullet \in \mathrm{Flag}^Q_{\mathbf{f}}(V_{\mathbb{C}})$ and $F^\bullet_{\mathfrak{g}} \in \mathrm{Flag}^{Q_{\mathfrak{g}}}_{\mathbf{f}_{\mathfrak{g}}}(\mathfrak{g}_{\mathbb{C}})$ both realize the homogeneous manifold $D = G_{\mathbb{R}}/R$ as a Hodge domain.

By slight abuse of terminology we refer to the $G_{\mathbb{R}}$–orbit of $F^\bullet_{\mathfrak{g}}$ as the *adjoint Hodge domain* $D_{\mathfrak{g}}$ associated to D, where the latter is viewed as the Hodge domain for the Hodge structure on V. The reason for doing this is that coming from algebraic geometry one thinks of $(V, Q, F^\bullet)$ as arising from $H^n(X, Q)_{\mathrm{prim}}$ where X is a smooth projective variety. However, in order to study (i) the geometry of the $G_{\mathbb{R}}$–orbits in $\check{D}$ and (ii) Lie–theoretic aspects of the Hodge structure, it is necessary to work with the polarized Hodge structures $(\mathfrak{g}, Q_{\mathfrak{g}}, F^\bullet_{\mathfrak{g}})$.

II.D.3 PHS in terms of grading elements

In analogy with Section II.C.3, given a Hodge structure on V with Hodge group G, there is a canonical choice of grading element. To be precise, given the induced Hodge structure (II.13) on $\mathfrak{g}$, define L by

$$L|_{\mathfrak{g}^{-p,p}} = p\mathbb{1}\,. \tag{II.14}$$

Since L is a derivation and $\mathfrak{g}_{\mathbb{C}}$ is semisimple, L is necessarily an element of $\mathfrak{g}_{\mathbb{C}}$. Moreover, since L is semisimple, it is necessarily contained in a Cartan subalgebra. And since the eigenvalues of L on $\mathfrak{g}_{\mathbb{C}}$ are integers, L is necessarily a grading element (Remark II.7).

Conversely, given a complex semisimple Lie algebra $\mathfrak{g}_{\mathbb{C}}$ and a grading element L, there is a canonical choice of (isomorphism class of) real form $\mathfrak{g}_{\mathbb{R}}$ (and we may take the representative $\mathfrak{g}_{\mathbb{R}}$ to be defined over $\mathbb{Z}$ [23, Proposition 2.14]) with the property that the L–eigenspace decomposition (II.4) of $\mathfrak{g}_{\mathbb{C}}$ defines a weight zero Hodge structure (II.13) by the assignment

$$\mathfrak{g}^{-p,p} = \mathfrak{g}_p\,, \tag{II.15}$$

cf. [23, Proposition 2.36]. This Hodge structure on $\mathfrak{g}$ is related to the initial Hodge structure on V by the grading element: the subspaces $V^{p,q}$ are also L–eigenspaces; that is, (II.11) holds. Observe, that while the L–eigenvalues on $\mathfrak{g}$ are integers, on V they lie in $\frac{1}{2}\mathbb{Z}$.

The Hodge structure (II.15) on $\mathfrak{g}$ is polarized by $Q_{\mathfrak{g}}$; equivalently, if

$$\mathfrak{k}_{\mathbb{C}} = \bigoplus_{\ell} \mathfrak{g}_{2\ell} \quad \text{and} \quad \mathfrak{k}^\perp_{\mathbb{C}} = \bigoplus_{\ell} \mathfrak{g}_{2\ell+1}\,,$$

then $\mathfrak{k}_{\mathbb{R}} = \mathfrak{g}_{\mathbb{R}} \cap \mathfrak{k}_{\mathbb{C}}$ and $\mathfrak{k}^\perp_{\mathbb{R}} = \mathfrak{g}_{\mathbb{R}} \cap \mathfrak{k}^\perp_{\mathbb{C}}$ define a *Cartan decomposition*

$$\mathfrak{g}_{\mathbb{R}} = \mathfrak{k}_{\mathbb{R}} \oplus \mathfrak{k}^\perp_{\mathbb{R}}\,.$$

We see from (II.5) and (II.15) that each $\mathfrak{g}^{p,-p}$ is a direct sum of root spaces (and $\mathfrak{h}$ if $p = 0$). Define *compact* and *noncompact roots* by

$$\begin{aligned} \Delta_{\rm cpt} &= \{\alpha \in \Delta \ : \ \mathfrak{g}^\alpha \subset \mathfrak{k}_{\mathbb{C}}\} = \{\alpha \in \Delta \ : \ \alpha(L) \text{ is even}\}\,, \\ \Delta_{nc} &= \{\alpha \in \Delta \ : \ \mathfrak{g}^\alpha \subset \mathfrak{k}_{\mathbb{C}}^\perp\} = \{\alpha \in \Delta \ : \ \alpha(L) \text{ is odd}\}\,. \end{aligned}$$

Note that

$$\Delta = \Delta_{\rm cpt} \cup \Delta_{nc}\,.$$

II.D.4 Realizations of D

In Section II.D.3 we observed that both the Hodge structure on V and the induced Hodge structure on $\mathfrak{g}$ are given by a common grading element L. This fact may be used to deduce that the two Hodge domains D and $D_{\mathfrak{g}}$ parameterizing Hodge structures of these types are isomorphic as $G_{\mathbb{R}}$–homogeneous complex submanifolds $G_{\mathbb{R}}/R \subset G_{\mathbb{C}}/P$, cf. [14]. This is one example of a general method to realize $G_{\mathbb{R}}/R$ as a Hodge domain, which we now outline.

Given a generalized flag variety $G_{\mathbb{C}}/P$ there are canonically defined Hodge structures that realize the variety as a compact dual. Let $L = L_{\mathfrak{p}}$ be the grading element (II.9) associated with $\mathfrak{p}$. Let $\Lambda_{\rm wt} \subset \mathfrak{h}^*$ be the weight lattice of $\mathfrak{g}_{\mathbb{C}}$, and let $\{\omega_1, \ldots, \omega_r\}$ be the *fundamental weights* with respect to the simple roots $\{\alpha_1, \ldots, \alpha_r\}$. Given any dominant integral weight $\lambda \in \Lambda_{\rm wt}$, let U^λ denote the corresponding irreducible representation of $\mathfrak{g}_{\mathbb{C}}$ with highest weight λ. If λ is a weight of $G_{\mathbb{C}}$,[9] then the parabolic P is the stabilizer of the highest weight line in U^λ if and only if

$$\lambda = \sum_{\alpha_i \notin \Sigma} \lambda^i \omega_i \quad \text{with} \quad 0 < \lambda^i \in \mathbb{Z}\,. \tag{II.16}$$

In this case, the $G_{\mathbb{C}}$–orbit of the highest weight line is a homogeneous embedding $G_{\mathbb{C}}/P \hookrightarrow \mathbb{P}U^\lambda$ that realizes the homogeneous complex manifold $G_{\mathbb{C}}/P$ as a homogeneous projective variety.

Let $\mathfrak{g}_{\mathbb{R}}$ be the real form determined by L (Section II.D.3). Given an irreducible representation V of $\mathfrak{g}_{\mathbb{R}}$ there exists an irreducible representation U of $\mathfrak{g}_{\mathbb{C}}$ such that one of the following holds:

- $V_{\mathbb{C}} = U$, in which case U is *real*;
- $V_{\mathbb{C}} = U \oplus U^*$ and $U \simeq U^*$, in which case U is *quaternionic*;
- $V_{\mathbb{C}} = U \oplus U^*$ and $U \not\simeq U^*$, in which case U is *complex*.

In the case that λ is the highest weight of U, the representation V admits a polarized Hodge structure $(F^\bullet, Q)$ with Hodge group G if and only if

$$L(\lambda) \in \frac{1}{2}\mathbb{Z}\,, \tag{II.17}$$

cf. [14]. (A priori, we have only $L(\lambda) \in \mathbb{Q}$.) In this case, the Hodge structure $(V, Q, F^\bullet)$ is given by the L–eigenspace decomposition (II.11). In particular,

[9] This will always be the case if $G_{\mathbb{C}}$ is simply connected.

> *$G_{\mathbb{C}}/P$ is realized as the compact dual for any of these polarized Hodge structures.*

From this perspective, a very natural realization is given by any λ that minimizes the coefficients λ^i of (II.16) subject to the constraint (II.17). In many cases it is possible to take $\lambda^i = 1$;[10] this corresponds to the minimal homogeneous embedding $G/P \hookrightarrow \mathbb{P}U^\lambda$ of G/P as a rational homogeneous variety.

Likewise,

> *the $G_{\mathbb{R}}$–orbit of any of these Hodge filtrations $F^\bullet$ realizes $G_{\mathbb{R}}/R$ as a Hodge domain.*

Remark II.18. As the highest weight of the adjoint representation, the highest root $\tilde\alpha$ is a dominant integral weight. By definition the grading element is integer–valued on $\tilde\alpha$. So (II.17) holds with $\lambda = \tilde\alpha$. Whence, the construction above yields a polarized Hodge structure $(\mathfrak{g}, \tilde{Q}_{\mathfrak{g}}, \tilde{F}^\bullet_{\mathfrak{g}})$. This agrees with the polarized Hodge structure $(\mathfrak{g}, Q_{\mathfrak{g}}, F^\bullet_{\mathfrak{g}})$ induced (as in Section II.D.2) from any of the $(V, Q, F^\bullet)$ constructed in this section.

II.E Period mappings and nilpotent orbits

We shall only consider period mappings corresponding to a one–parameter family of degenerating polarized Hodge structures. Such is given by a Hodge domain D, a unipotent monodromy transformation $T \in G$ and a locally liftable holomorphic mapping

$$\Phi : \Delta^* \to \Gamma_T \backslash D . \tag{II.19}$$

which satisfies the infinitesimal period relation. Here $\Gamma_T = \{T^k \ : \ k \in \mathbb{Z}\}$.

Denoting by $\mathbb{H} = \{z \in \mathbb{C} \ : \ \mathrm{Im}(z) > 0\}$ the upper–half plane with covering map $\mathbb{H} \to \Delta^*$ given by

$$t = e^{2\pi \mathrm{i} z} ,$$

we may lift (II.19) to give

$$\tilde\Phi : \mathbb{H} \to D ,$$
$$\tilde\Phi(z+1) = T \cdot \tilde\Phi(z) .$$

Setting $N = \log(T) \in \mathfrak{g}^{\mathrm{nilp}}$, we may then "unwind" Φ to

$$\tilde\Psi : \mathbb{H} \to \check{D}$$

by defining

$$\tilde\Psi(z) = e^{-zN} \cdot \tilde\Phi(z) .$$

[10] The issue is the following: In the case that $\mathfrak{g}_{\mathbb{C}}$ is simple, the weights of $\mathfrak{g}_{\mathbb{C}}$ lie in the $\frac{1}{d}\mathbb{Z}$–span of the simple roots, where $1 \le d \in \mathbb{Z}$ is the determinant of the Cartan matrix. So, when $d \in \{1,2\}$ (II.17) will hold with $\lambda^i = 1$, $\alpha_i \notin \Sigma$. We have $d \in \{1,2\}$ when $\mathfrak{g}_{\mathbb{C}}$ is one of $\mathfrak{so}_{2r+1}\mathbb{C}$, $\mathfrak{sp}_{2r}\mathbb{C}$, $\mathfrak{e}_7$, $\mathfrak{e}_8$, $\mathfrak{f}_4$ or $\mathfrak{g}_2$. In the case that $\mathfrak{g}_{\mathbb{C}} = \mathfrak{sl}_n\mathbb{C}$, the determinant is $n+1$ and we will be able to satisfy (II.17) with values $\lambda^i \in \{1, \ldots, n+1\}$; in the case that $\mathfrak{g}_{\mathbb{C}} = \mathfrak{so}_{2r}\mathbb{C}$, we have $d = 4$ and will be able to satisfy (II.17) with $\lambda^i \in \{1,2\}$; in the case that $\mathfrak{g}_{\mathbb{C}} = \mathfrak{e}_6$, we have $d = 3$ and will be able to satisfy (II.17) with $\lambda^i \in \{1,2,3\}$.

Then $\tilde{\Psi}(z+1) = \tilde{\Psi}(z)$ so that there is an induced map

$$\Psi : \Delta^* \to \check{D} .$$

A basic result is that Ψ extends across the origin $t = 0$. Then setting $\ell(t) = \log(t)/2\pi\mathbf{i}$, the original period mapping is well approximated by the nilpotent orbit

$$t \mapsto e^{\ell(t)N} \cdot \Psi(0) , \tag{II.20}$$

see [26]. Implicit here is the statement that $e^{\ell(t)N} \cdot \Psi(0) \in D$ for $0 < |t| < \varepsilon$. We shall further explain this below.

We shall sometimes write $\Phi(t) = F^\bullet_t$ for the multi-valued filtration on $V_{\mathbb{C}}$. The lift $\tilde{\Phi}$ to $\mathbb{H} \to D$ will be denoted by

$$z \mapsto F^\bullet_z$$

where $F^\bullet_{z+1} = T \cdot F^\bullet_z$.

Because of the strong approximation of (II.19) by a nilpotent orbit, we shall replace Φ by the nilpotent orbit (II.20). For this we set

$$\Psi(0) = F^\bullet_{\mathrm{lim}} .$$

Note that $F^\bullet_{\mathrm{lim}}$ is defined only up to the action of Γ_T and a choice of coordinate t. Rescaling t by $t \mapsto e^{2\pi\mathbf{i}\lambda} t$ induces the change

$$F^\bullet_{\mathrm{lim}} \to e^{\lambda N} \cdot F^\bullet_{\mathrm{lim}} .$$

Thus what is well–defined is a map

$$\{\text{period mappings (II.19)}\} \times T^*_0\Delta \to \text{nilpotent orbits.}$$

The conditions that Φ define a period mapping translate into

$$\begin{aligned} e^{zN} \cdot F^\bullet_{\mathrm{lim}} &\in D \quad \text{for } \mathrm{Im}(z) \gg 0 , \\ N \cdot F^p_{\mathrm{lim}} &\subset F^{p-1}_{\mathrm{lim}} \qquad \text{(infinitesimal period relation).} \end{aligned} \tag{II.21}$$

Definition II.22. A *nilpotent orbit* is given by $(F^\bullet, N)$ where $F^\bullet \in \check{D}$, $N \in \mathfrak{g}^{\mathrm{nilp}}_{\mathbb{R}}$ and where the conditions (II.21) are satisfied with $F^\bullet$ in place of $F^\bullet_{\mathrm{lim}}$.

Two nilpotent orbits $(F^\bullet, N)$ and $('F^\bullet, N)$ are *equivalent* if

$$'F^\bullet = e^{\lambda N} F^\bullet$$

for some $\lambda \in \mathbb{C}$. We set

- o $\tilde{B}(N) = \{\text{nilpotent orbits } (F^\bullet, N)\}$,
- o and let $B(N) = e^{\mathbb{C}N} \backslash \tilde{B}(N)$ denote the set of equivalence classes of nilpotent orbits.

We next set

$$D_N = D \cup B(N)$$

and observe that the action of Γ_T extends naturally to D_N. In [16] the structure of a log–analytic variety with slits is defined on D_N, and a basic result is that the period mapping (II.19) extends to

$$\Phi_e : \Delta \to \Gamma_T \backslash D_N \tag{II.23}$$

where the origin is mapped to the equivalence class of $(F^{\bullet}_{\mathrm{lim}}, N)$. We shall refer to (II.23) as the *extended period mapping.*

Definition II.24. The mapping

$$\{\text{period mappings (II.19)}\} \;\to\; \Phi_e(0) \;\in\; \Gamma_T \backslash D_N$$

will be called the *limit period mapping.*

II.F Nilpotent orbits and limiting mixed Hodge structures

Let D be a Hodge domain parameterizing weight n, Q–polarized Hodge structures on V with Hodge group G. Associated to a nilpotent orbit $(F^{\bullet}, N)$ there is a special type of mixed Hodge structure $(V, W_{\bullet}(N), F^{\bullet})$ called a *limiting mixed Hodge structure* (limiting mixed Hodge structures). There will then be a bijection of sets

$$\{\text{nilpotent orbits}\} \overset{\text{(II.27)}}{\longleftrightarrow} \{\text{limiting mixed Hodge structures}\},$$

which will pass to the quotient by taking equivalence classes.

Since N is nilpotent, there exists $0 \le m \le n$ such that $N^m \neq 0$ and $N^{m+1} = 0$. Then the *weight filtration*

$$W_{-m}(N) \subset \cdots \subset W_0(N) \subset \cdots \subset W_m(N) = V$$

is the unique filtration that satisfies

$$\begin{aligned} N : W_k(N) &\longrightarrow W_{k-2}(N)\,, \\ N^k : \mathrm{Gr}_k^{W_\bullet(N)} &\xrightarrow{\simeq} \mathrm{Gr}_{-k}^{W_\bullet(N)}\,, \quad k \ge 0\,. \end{aligned}$$

It is always possible to complete N to a $\mathfrak{sl}_2$–*triple*

$$\{N, Y, N^+\} \subset \mathfrak{g} \tag{II.25}$$

where

$$\begin{aligned} [Y, N] &= -2N\,, \\ [Y, N^+] &= 2N^+\,, \\ [N^+, N] &= Y\,. \end{aligned}$$

The span $\mathfrak{a}$ of (II.25) is a three dimensional semisimple subalgebra (TDS) of $\mathfrak{g}$ that is isomorphic to $\mathfrak{sl}_2$. Denoting by $V_\ell = \{v \in V \ : \ Yv = \ell v\}$ the ℓ–weight space for the semisimple action of Y, we have

$$W_k(N) = \bigoplus_{\ell \leq k} V_\ell \quad \text{and} \quad \mathrm{Gr}_k^{W_\bullet(N)} \simeq V_k\,. \tag{II.26}$$

The *primitive spaces* are defined, for $k \geq 0$, by

$$\mathrm{Gr}_{k,\mathrm{prim}}^{W_\bullet(N)} = \ker\left\{N^{k+1} : \mathrm{Gr}_k^{W_\bullet(N)} \to \mathrm{Gr}_{-k-2}^{W_\bullet(N)}\right\}.$$

Decomposing V under the action of the TDS, the primitive spaces are the highest weight spaces. There are nondegenerate bilinear forms, of parity k,

$$Q_k : \mathrm{Gr}_k^{W_\bullet(N)} \times \mathrm{Gr}_k^{W_\bullet(N)} \to \mathbb{R}$$

defined by

$$Q_k(v,w) = \epsilon_k Q(v, N^k w)$$

where $\epsilon_k = \pm 1$.

The basic result [8] is

$$\begin{array}{c} (F^\bullet, N) \textit{ is a nilpotent orbit if and only if} \\ (V, W_\bullet(N), F^\bullet) \textit{ is a polarized mixed Hodge structure.} \end{array} \tag{II.27}$$

The filtration $F^\bullet$ induces a Hodge structure of weight k on $\mathrm{Gr}_k^{W_\bullet(N)}$ and N is of Hodge type $(-1,-1)$. The polarization condition means that the summand $\mathrm{Gr}_{k,\mathrm{prim}}^{W_\bullet(N)}$ of $\mathrm{Gr}_k^{W_\bullet(N)}$ is polarized by the form Q_k. We shall generally suppress mention of the polarization conditions, which we always take to be understood.

The equivalence relation on limiting mixed Hodge structures is induced by rescaling $F^\bullet \mapsto e^{\lambda N} F^\bullet$. The Hodge structures on $\mathrm{Gr}_k^{W_\bullet(N)}$ are unchanged, but some of the extension data in the mixed Hodge structure will be altered.

II.G Reduced limit period mapping

This section is summarizes material developed in [15, Appendix to Lecture 10] and [18]; the interested reader should consult those references for additional detail and proofs.

Given a period mapping (II.19) with extension (II.23) the limiting mixed Hodge structure given by the point

$$\Phi_e(0) \in \Gamma_T \backslash B(N)$$

represents, in a precise sense, the maximal amount of information in the limit of a degenerating family of polarized Hodge structures. For some purposes the other extreme of describing a minimal amount of information in the limit is useful. In that direction we will consider two notions of a *reduced limit period mapping*. For the first we lift (II.19) to

$$\tilde{\Phi} : \mathbb{H} \to D\,.$$

Definition II.28 (Version 1). The *reduced limit period mapping* is defined by

$$\Phi \ \mapsto \ \lim_{\mathrm{Im}(z)\to\infty} \tilde{\Phi}(z) \in \partial D\,.$$

The map was introduced in [18] under the term *naïve limit.* It has been further discussed in [15, Appendix to Lecture 10] and [18]. If we think of $\tilde{\Phi}(z) = F^\bullet_z$ as a filtration on $V_\mathbb{C}$, then we will set

$$\lim_{\mathrm{Im}(z)\to\infty} F^\bullet_z = F^\bullet_\infty\,.$$

The reduced limit period mapping has the properties:

(a) It is the same for Φ and for the approximating nilpotent orbit (II.20). So, without loss of generality, we shall assume that $\tilde{\Phi}(z) = e^{zN}\cdot F^\bullet$.

(b) It is independent of the lifting $\tilde{\Phi}$. In fact, $F^\bullet_\infty \in \partial D$ is a fixed point of T and the differential

$$T_* : T_{F^\bullet_\infty}\check{D} \to T_{F^\bullet_\infty}\check{D}$$

is the identity. Equivalently, the vector field on $\check{D} = G_\mathbb{C}/P$ defined by $N \in \mathfrak{g}$ vanishes to second order at $F^\bullet_\infty$.

(c) The mapping

$$\Phi_\infty : B(N) \to \partial D \tag{II.29a}$$

defined on $\tilde{B}(N)$ by

$$\Phi_\infty(F^\bullet, N) = \lim_{\mathrm{Im}(z)\to\infty} \tilde{\Phi}(z) = F^\bullet_\infty \tag{II.29b}$$

is well–defined on $B(N)$, and it is $Z(N)_\mathbb{R}$–equivariant. The image of $B(N)$ lies in a $G_\mathbb{R}$–orbit in ∂D

$$\Phi_\infty\left(B(N)\right) \subset \mathcal{O}_{F^\bullet_\infty} = G_\mathbb{R}\cdot F^\bullet_\infty\,.$$

(d) If $\mathfrak{g}_\mathbb{C} = \oplus I^{p,q}_\mathfrak{g}$ is the Deligne splitting associated with $(W(N), F^\bullet)$, then

$$F^a_\infty = \bigoplus_{q\le n-a} I^{\bullet,q}$$

and

$$\mathrm{codim}\,\mathcal{O}_{F^\bullet_\infty} = \sum_{p,q>0} \dim_\mathbb{C} I^{p,g}_\mathfrak{g}\,.$$

Definition II.30 (Version 2). Given a Hodge domain $D \subset \check{D}$ with boundary component $B(N)$, the *reduced limit period mapping* is (II.29).

Note that the first notion (Definition II.28) is the composition $\Phi_\infty \circ \Phi_e(0)$.

Definition II.31 ([17]). A $G_\mathbb{R}$–orbit $\mathcal{O} \subset \partial D$ for a generalized flag domain D is *polarizable relative to* D if there is a Hodge domain structure on D and a nilpotent orbit $(F^\bullet, N)$ such that $F^\bullet_\infty \in \mathcal{O}$. The orbit $\mathcal{O}$ is *polarizable* if there is a Hodge domain D with $\mathcal{O} \subset \partial D$ relative to which $\mathcal{O}$ is polarizable.

There are examples of generalized flag domains D and D' in a generalized flag variety $\check{D}$ and a $G_{\mathbb{R}}$–orbit $\mathcal{O} \subset \partial D \cap \partial D'$ such that $\mathcal{O}$ is polarizable relative to D, but not relative to D', cf. [15, Appendix to Lecture 10].

Let $\mathcal{N} \subset \mathfrak{g}_{\mathbb{R}}$ denote the $\mathrm{Ad}(G_{\mathbb{R}})$–orbit of N. Then

$$\mathcal{O}_{F^\bullet_\infty} = \bigcup_{N' \in \mathcal{N}} \Phi_\infty(B(N')) .$$

In this case we say that *$\mathcal{O}_{F^\bullet_\infty}$ is polarized by $\mathcal{N}$ relative to D.* In this sense, the $G_{\mathbb{R}}$–orbits in $\check{D}$ separate into those that have Hodge–theoretic significance, meaning that over $\mathbb{R}$ every point is realized as a reduced limit for some Hodge domain structure, and those that don't have Hodge–theoretic significance in this sense.

In light of the equivalence (II.27) between nilpotent orbits and limiting mixed Hodge structures, one may ask: *What point of ∂D does a limiting mixed Hodge structure map to?* To answer this we first need two general facts about mixed Hodge structures. Given a mixed Hodge structure $(V, W_\bullet, F^\bullet)$ there is the canonical *Deligne splitting*

$$V_{\mathbb{C}} = \bigoplus I^{p,q}$$

satisfying

$$F^p = \bigoplus_{q \geq p} I^{q,\bullet} \quad \text{and} \quad W_k = \bigoplus_{p+q \leq k} I^{p,q} \tag{II.32}$$

and

$$\overline{I^{p,q}} \equiv I^{q,p} \mod \bigoplus_{r<p,s<q} I^{r,s} .$$

The mixed Hodge structure is split over $\mathbb{R}$, or *$\mathbb{R}$–split*, if

$$\overline{I^{p,q}} = I^{q,p} .$$

In this case $(V, W_\bullet, F^\bullet)$ is a direct sum over $\mathbb{R}$ of pure Hodge structures $\oplus_{p+q=k} I^{p,q}$ of weight k.

For the second property, canonically associated to a mixed Hodge structure $(V, W_\bullet, F^\bullet)$ there is an $\mathbb{R}$–split mixed Hodge structure $(V, W_\bullet, \tilde{F}^\bullet)$ given by

$$\tilde{F}^\bullet = e^{-2\mathrm{i}\delta} \cdot F^\bullet$$

where $\delta \in \oplus_{p,q<0} I^{p,q}_{\mathfrak{g}}$. Here, $I^{p,q}_{\mathfrak{g}}$ is the Deligne splitting of the induced limiting mixed Hodge structure $(\mathfrak{g}, W_{\bullet,\mathfrak{g}}, F^\bullet_{\mathfrak{g}})$.

If $(V, W_\bullet(N), F^\bullet)$ is a limiting mixed Hodge structure, then so is $(V, W_\bullet(N), \tilde{F}^\bullet)$, and conversely. Moreover, we have

$$F^\bullet_\infty = \tilde{F}^\bullet_\infty ; \tag{II.33}$$

that is, denoting by $B(N)_{\mathbb{R}}$ the equivalence classes containing an $\mathbb{R}$–split limiting mixed Hodge structure, the reduced limit period mapping factors

$$\begin{array}{ccc} B(N) & \xrightarrow{\ \Phi_\infty\ } & \\ \downarrow & & \partial D \\ B(N)_{\mathbb{R}} & \longrightarrow & \end{array} \tag{II.34}$$

Next, in terms of the Deligne splitting

$$V_{\mathbb{C}} = \bigoplus \tilde{I}^{p,q}$$

associated with $(V, W_\bullet(N), \tilde{F}^\bullet)$ we have

$$F^p_\infty = \bigoplus_{q \le n-p} \tilde{I}^{\bullet,q}\,. \tag{II.35}$$

The picture is this

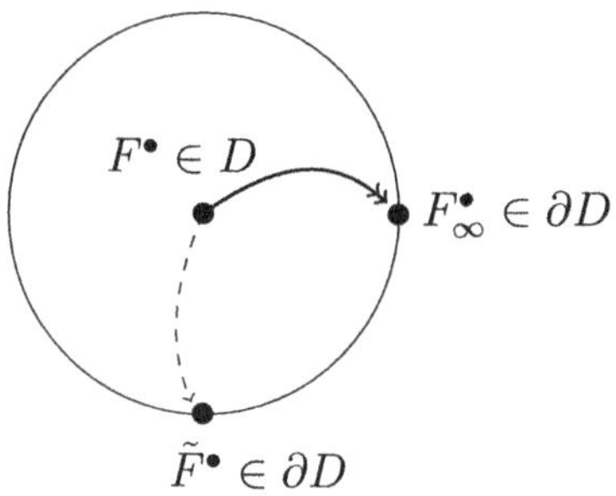

Starting with $F^\bullet \in D$ such that $(V, W_\bullet(N), F^\bullet)$ is a limiting mixed Hodge structure, we have for the reduced limit period mapping (the solid arrow emanating from $F^\bullet \in D$)

$$\lim_{y\to\infty} e^{\mathrm{i}yN} F^\bullet = F^\bullet_\infty\,.$$

We also have the map $F^\bullet \to \tilde{F}^\bullet$ (the dashed arrow), and then

$$\lim_{y\to\infty} e^{yN} \tilde{F}^\bullet = F^\bullet_\infty\,.$$

Thus, $F^\bullet_\infty$ is reached from $\tilde{F}^\bullet$ by traveling along the real one–parameter subgroup $\exp(\mathbb{R}N)$ in a $G_{\mathbb{R}}$–orbit in ∂D.

Because of (II.33) and the subsequent factorization (II.34), henceforth, unless mentioned otherwise, we shall adopt the

Convention. We shall assume that a limiting mixed Hodge structure is $\mathbb{R}$–split.

Because of this we may drop the tilde in (II.35) to have

$$F^p_\infty = \bigoplus_{q \le n-p} I^{\bullet,q}\,. \tag{II.36}$$

II.H Reduced limit period mapping for $(\mathfrak{g}, Q_{\mathfrak{g}}, F^\bullet_{\mathfrak{g}})$

The limiting mixed Hodge structure $(V, W_\bullet(N), F^\bullet)$ determines a limiting mixed Hodge structure $(\mathfrak{g}, W_\bullet(N)_{\mathfrak{g}}, F^\bullet_{\mathfrak{g}})$ by

$$\begin{aligned} F^p_{\mathfrak{g}} &= \{\xi \in \mathfrak{g}_{\mathbb{C}} \ :\ \xi(F^q) \subset F^{p+q}\ \forall\, q\}\,, \\ W_\ell(N)_{\mathfrak{g}} &= \{\xi \in \mathfrak{g}_{\mathbb{R}} \ :\ \xi(W_m(N)) \subset W_{m+\ell}(N)\ \forall\, m\}\,. \end{aligned}$$

As above $V_{\mathbb{C}} = \oplus I^{p,q}$ will denote the Deligne splitting of $V_{\mathbb{C}}$, and

$$\mathfrak{g}_{\mathbb{C}} = \bigoplus I^{p,q}_{\mathfrak{g}} \tag{II.37}$$

will denote the Deligne splitting on $\mathfrak{g}_{\mathbb{C}}$. If the initial limiting mixed Hodge structure $(V, W_\bullet(N), F^\bullet)$ is $\mathbb{R}$–split, so is the induced limiting mixed Hodge structure $(\mathfrak{g}, W_\bullet(N)_{\mathfrak{g}}, F^\bullet_{\mathfrak{g}})$ on $\mathfrak{g}$; indeed,

$$I^{p,q}_{\mathfrak{g}} = \{\xi \in \mathfrak{g}_{\mathbb{C}} \ : \ \xi(I^{r,s}) \subset I^{p+r,q+s} \ \forall \ r,s\}. \tag{II.38}$$

In particular, (II.36) implies

$$F^p_{\mathfrak{g},\infty} = \bigoplus_{q \le -p} I^{\bullet,q}_{\mathfrak{g}}.$$

Example II.39 (A G_2 Hodge domain). The exceptional simple Lie group G_2 of rank two may be realized as the Hodge group of a weight 6 Hodge structure with Hodge numbers $\mathbf{h} = (1,1,1,1,1,1,1) = (1^7)$, [18, Section 6.1.3]. The associated Hodge domain D is an open $G_2(\mathbb{R})$–orbit in the flag variety $G_2(\mathbb{C})/B$. Including D (which is polarized by $N = 0$) there are four $G_2(\mathbb{R})$–orbits that are polarized relative to D. The Deligne splittings $V_{\mathbb{C}} = \oplus I^{p,q}$ (left column) and $\mathfrak{g}_{\mathbb{C}} = \oplus I^{p,q}_{\mathfrak{g}}$ (right column) for each of these orbits are pictured in Figure II.1. A circled node indicates $i^{p,q} = 2$, and an uncircled node indicates $i^{p,q} = 1$.

The boundary ∂D contains seven $G_2(\mathbb{R})$–orbits, four of them are *not* polarized relative to D [18, Section 6.1.3].

Remark II.40 (Jacobson–Morosov parabolics). Let Y be the neutral element of an $\mathfrak{sl}_2$–triple (II.25) containing the given nilpotent N. In analogy with (II.26) we have

$$W_k(N)_{\mathfrak{g}} = \bigoplus_{\ell \le k} \mathfrak{g}_\ell, \quad \text{where} \quad \mathfrak{g}_\ell = \{\xi \in \mathfrak{g} \ : \ [Y,\xi] = \ell\xi\}$$

is the ℓ–eigenspace of Y. In particular,

$$W_0(N)_{\mathfrak{g}} = \bigoplus_{\ell \ge 0} \mathfrak{g}_{-\ell}$$

is a parabolic subalgebra of $\mathfrak{g}$. Not every parabolic $\mathfrak{p} \subset \mathfrak{g}$ may be realized as $\mathfrak{p} = W_0(N)_{\mathfrak{g}}$ for some N;[11] those that can are *Jacobson–Morosov parabolics*. When the eigenvalues of Y are even, that is $\mathfrak{g}_{2k+1} = 0$ for all $k \in \mathbb{Z}$, the we say that $W_0(N)$ is an *even Jacobson–Morosov parabolic*. A Borel subalgebra $\mathfrak{b} \subset \mathfrak{g}$ may always be realized as an even Jacobson–Morosov parabolic by taking N to be a principal nilpotent [9].

Recall that, up to the action of $\mathrm{Ad}(G_{\mathbb{C}})$, the parabolic subalgebras $\mathfrak{p} \subset \mathfrak{g}$ are indexed by (possibly empty) subsets $\Sigma \subset \Delta^+_s$ of the simple roots (Section II.A). The subset Σ indexing a Jacobson–Morosov parabolic $W_0(N)$ may be determined

[11]For example, if $n > 2$, then the parabolic $P \subset \mathrm{SL}_n\mathbb{C}$ stabilizing a line in $\mathbb{C}^n$ is not a Jacobson–Morosov parabolic, cf. [9].

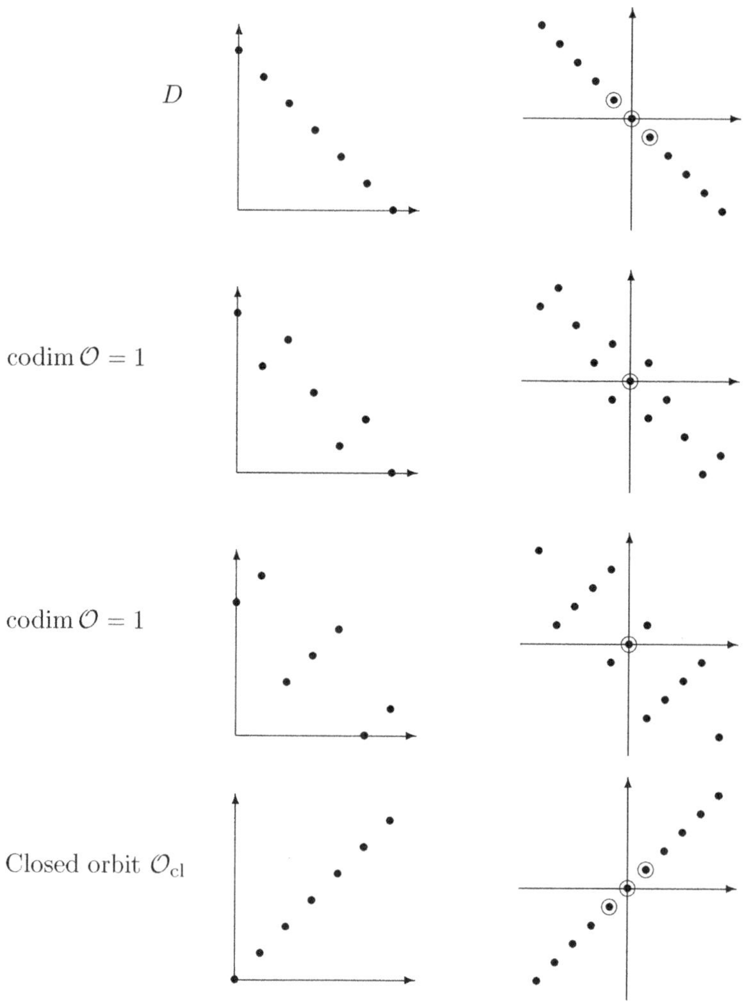

Figure II.1: Deligne splittings for polarized orbits in $G_2(\mathbb{C})/B$

as follows. The fact that the eigenvalues of Y are integers implies that Y is a grading element. (The Jacobson–Morosov parabolic $W_0(N)$ is even if and only if $\alpha_i(Y)$ is even for all simple roots.) We may choose $\mathfrak{h} \subset \mathfrak{b} \subset W_0(N)$ so that $\alpha_i(Y) \geq 0$ for all simple roots $\alpha_i \in \Delta_s^+$. (Alternatively, 'conjugate' Y so that it lies in a chosen positive Weyl chamber.) Then

$$W_0(N)_{\mathfrak{g}} = \mathfrak{p}_\Sigma\,, \quad \text{where} \quad \Sigma = \{\alpha_i \ : \ \alpha_i(Y) = 0\}\,.$$

In fact, given the normalization $\alpha_i(Y) \geq 0$, we have

$$\alpha_i(Y) \in \{0, 1, 2\}\,,$$

and the $\mathrm{Ad}(G_{\mathbb{C}})$–orbits of nilpotent $N \in \mathfrak{g}_{\mathbb{C}}$ are indexed by the *characteristic vectors* $(\alpha_1(Y), \ldots, \alpha_r(Y))$, cf. [9, 11, 22].

III Extremal degenerations of polarized Hodge structures

III.A Precise definition of extremal degenerations

Definition III.1. Let $(V, Q, F^\bullet)$ be a polarized Hodge structure. The following data constitute a *degeneration of the polarized Hodge structure* $(V, Q, F^\bullet)$:

(i) A Hodge domain D such that $F^\bullet = F^\bullet_x$ for some $x \in D$.

(ii) A period mapping

$$\Phi : \Delta^* \to \Gamma_T \backslash D \tag{III.2}$$

such that for the lift $\tilde{\Phi} : \mathbb{H} \to D$ we have

$$\tilde{\Phi}(z) = x \quad \text{for some} \quad z \in \mathbb{H}\,.$$

For $t = e^{2\pi \mathrm{i} z} \in \Delta^*$, we think of $\Phi(t) = F^\bullet_t \in \Gamma_T \backslash D$ as defining a one–parameter family of Γ_T–equivalence classes of polarized Hodge structures which tend to a "singular point" as $t \to 0$. Because of the constraints on the Hodge groups, this is a more refined notion than just a family of equivalence classes of polarized Hodge structures over the punctured disc.

In this paper two types of limits associated to (III.2) have been discussed. One is the equivalence class of limiting mixed Hodge structures $(V, W_\bullet(N), F^\bullet_{\mathrm{lim}})$ viewed as the image of the origin $\Phi_e(0)$ under the extended period map

$$\Phi_e : \Delta \to \Gamma_T \backslash \left(D \cup B(N) \right)$$

(Section II.E). The other is the image

$$F^\bullet_\infty \in \partial D$$

of the reduced limit period mapping (Section II.G)

$$\Phi_\infty : B(N) \to \partial D$$

applied to $\Phi_e(0) \in B(N)$. As noted above, Φ_∞ is invariant under T so that $F^\infty \in \partial D$ is well–defined.

The orbits $\mathcal{O}_x = G_\mathbb{R} \cdot x$ of the action of $G_\mathbb{R}$ on the generalized flag variety form a partially ordered set by the relation "contained in the closure of"

$$\mathcal{O}' \prec \mathcal{O} \quad \text{if} \quad \mathcal{O}' \subset \overline{\mathcal{O}}\,.$$

Aside from the open orbits, at one extreme are the real codimension–one orbits; at the other extreme is the unique closed orbit $\mathcal{O}_{\mathrm{cl}}$. In first approximation, an extremal degeneration of a polarized Hodge structure is a degeneration whose reduced limit period $F^\bullet_\infty$ lies in one of these two extremes. However, since there may be no degeneration with $F^\bullet_\infty$ in the closed orbit, a refinement of this notion is required.

Definition III.3. We shall say that a $G_{\mathbb{R}}$–orbit $\mathcal{O} \subset \partial D$ is *maximal relative to the Hodge domain structure on the generalized flag domain* D if it is polarizable relative to D and if there exists no other orbit $\mathcal{O}' \precneqq \mathcal{O}$ that is polarizable relative to D.

Definition III.4. A degeneration of a polarized Hodge structure is *extremal* if the reduced limit period lies in either a codimension–one orbit (in which case the degeneration is *minimal*), or in an orbit that is maximal relative to the Hodge domain structure on D (in which case the degeneration is *maximal*).

Thus extremal degenerations of a polarized Hodge structure represented by a point $x \in D$ are the least and most degenerate the reduced limit period map can realize for a period map (III.2) with $\Phi(z) = x$ for some $z \in \mathbb{H}$. There are a few subtleties in the concept.

(a) We cannot talk of the degenerations of a generic $x \in D$. As a trivial example, the infinitesimal period relation $I \subset TD$ may be zero. Even if we make the reasonable assumption that I is bracket–generating, we know of no result that ensures the existence of a period mapping (III.2).

(b) Given (III.2), we may replace it by the corresponding nilpotent orbit without changing the limit $F^{\bullet}_{\infty}$. If (III.2) arises from a family of algebraic varieties, the corresponding nilpotent orbit usually does not; however, this doesn't matter in the sense that the limiting mixed Hodge structure, constructed algebro–geometrically in [27], will be the same as that constructed analytically from the nilpotent orbit. Thus we may speak unambiguously of the degeneration of the Hodge structure on $H^n(X_t, \mathbb{Q})_{\text{prim}}$ for $X_t = \pi^{-1}(t)$, $t \neq 0$, in a family $\mathcal{X} \xrightarrow{\pi} \Delta$.

The study of extremal degenerations requires that we understand the $G_{\mathbb{R}}$–orbit structure of $\check{D} = G_{\mathbb{C}}/P$. The necessary material is presented in Sections III.B–III.E.

III.B $G_{\mathbb{R}}$–orbit structure

The representation theory of Lie groups and Lie algebras plays an essential rôle in the analysis of the geometry of the compact dual $G_{\mathbb{C}}/P$ and its $G_{\mathbb{R}}$–orbits. In this section we outline that structure.

Fix a generalized flag variety $\check{D} = G_{\mathbb{C}}/P$ and a $G_{\mathbb{R}}$–orbit $\mathcal{O}$. Given a point $x \in \mathcal{O}$ the Lie algebra $\mathfrak{p}_x$ of the stabilizer $P_x = \mathrm{Stab}_{G_{\mathbb{C}}}(x)$ contains a Cartan subalgebra $\mathfrak{h}_x$ that is stable under conjugation [12, Corollary 2.1.3]. Hence $\mathfrak{h}_x$ is defined over $\mathbb{R}$; let $\mathfrak{h}_x(\mathbb{R})$ denote the real form.

Set $\mathfrak{p} = \mathfrak{p}_x$ and $\mathfrak{h} = \mathfrak{h}_x$. Choose a Borel subalgebra $\mathfrak{h} \subset \mathfrak{b} \subset \mathfrak{p}$, and let $L \in \mathfrak{h}$ be the grading element (II.9) associated with this triple. The fact that $\mathfrak{h}$ is closed under conjugation implies that $\mathfrak{h}(\mathbb{R}) = \mathfrak{h} \cap \mathfrak{g}_{\mathbb{R}}$ is Cartan subalgebra of $\mathfrak{g}_{\mathbb{R}}$. It also implies that $\bar{\alpha}$ is a root whenever α is a root. It follows that

$\bar{L}$ *is a grading element.*[12]

[12]We may assume that $\mathfrak{h} \subset \mathfrak{k}$ if and only if x lies in an open $G_{\mathbb{R}}$–orbit. When the orbit is not open, (II.10) will fail.

Additionally, the $\mathfrak{h}$–roots of $\mathfrak{g}$ decompose into three types: we say α is *real* if $\bar\alpha = \alpha$; we say α is *imaginary* if $\bar\alpha = -\alpha$; and we say α is *complex* otherwise.

As elements of the Cartan subalgebra L and $\overline{L}$ define a bigraded eigenspace decomposition

$$\mathfrak{g} = \oplus\, \mathfrak{g}^{p,q} \tag{III.5a}$$

given by

$$\mathfrak{g}^{p,q} = \{\xi \in \mathfrak{g} \ : \ [L,\xi] = p\xi\,,\ [\overline{L},\xi] = q\xi\}\,. \tag{III.5b}$$

This is the bigrading of [18, Lemma 3.2]. If the infinitesimal period relation is bracket–generating, then the filtration corresponding to x is

$$F^p_{\mathfrak{g},x} = \bigoplus_{q\le -p} \mathfrak{g}^{q,\bullet}\,. \tag{III.6}$$

From this point on:

We assume that the infinitesimal period relation is bracket–generating.[13] (III.7)

In the event that $F^\bullet_{\mathfrak{g},x} = F^\bullet_{\mathfrak{g},\infty}$ lies in the image of a reduced limit period mapping, the bigrading (III.5) is related to the Deligne splitting (II.37) by

$$I^{p,q}_{\mathfrak{g}} = \mathfrak{g}^{q,p}\,, \tag{III.8}$$

cf. [15, 18]. If $\mathcal{O}$ is polarized by N relative to D (Definition II.31), then (III.8) implies

$$N \in I^{-1,-1}_{\mathfrak{g}} = \mathfrak{g}^{-1,-1}\,. \tag{III.9}$$

Note that $\mathfrak{g}^{p,q}$ is a direct sum of root spaces (and $\mathfrak{h}$ if $p,q=0$). In particular,

$$\begin{aligned} \mathfrak{g}^{p,q} &= \bigoplus_{\substack{\alpha(L)=p\\ \bar\alpha(L)=q}} \mathfrak{g}^\alpha\,, \quad \text{if } (p,q)\neq(0,0)\,,\\ \mathfrak{g}^{0,0} &= \mathfrak{h} \oplus \bigoplus_{\substack{\alpha(L)=0\\ \bar\alpha(L)=0}} \mathfrak{g}^\alpha\,. \end{aligned} \tag{III.10}$$

Observe that

$$\overline{\mathfrak{g}^{p,q}} = \mathfrak{g}^{q,p}\,; \tag{III.11}$$

that is,

complex conjugation corresponds to reflection about the line $p = q$.

From (III.11) we see that $\mathfrak{g}^{p,q} \oplus \mathfrak{g}^{q,p}$ is defined over $\mathbb{R}$. Let $(\mathfrak{g}^{p,q} \oplus \mathfrak{g}^{q,p})_\mathbb{R} = (\mathfrak{g}^{p,q} \oplus \mathfrak{g}^{q,p}) \cap \mathfrak{g}_\mathbb{R}$ denote the real form. Then the tangent, CR–tangent and normal

[13] In the event that the IPR is not bracket–generating one may either modify the definition of L so that (III.6) holds, as in [18], or reduce to the case that the IPR is bracket–generating as in [23, Section 3.3].

spaces are given by

$$\begin{aligned} T_x\mathcal{O} &= \bigoplus_{p>0 \text{ or } q>0} (\mathfrak{g}^{p,q} \oplus \mathfrak{g}^{q,p})_{\mathbb{R}}, \\ T_x^{\mathrm{CR}}\mathcal{O} &= \bigoplus_{\substack{p>0\\ q\geq 0}} (\mathfrak{g}^{p,-q} \oplus \mathfrak{g}^{-q,p})_{\mathbb{R}}, \\ N_x\mathcal{O} &= \mathbf{i} \bigoplus_{p,q>0} (\mathfrak{g}^{p,q} \oplus \mathfrak{g}^{q,p})_{\mathbb{R}}. \end{aligned} \tag{III.12}$$

In the case that $F^\bullet_{\mathfrak{g},x} = F^\bullet_{\mathfrak{g},\infty}$, equations (III.8) and (III.12) yield (I.10). Note also that Lie algebra of the stabilizer $R \subset G_{\mathbb{R}}$ is

$$\mathfrak{r} = (\mathfrak{p} \oplus \bar{\mathfrak{p}})_{\mathbb{R}} = \bigoplus_{p,q\geq 0} (\mathfrak{g}^{-p,-q} \oplus \mathfrak{g}^{-q,-p})_{\mathbb{R}}.$$

It is convenient to visualize this structure in the (p,q)–plane: see Figure III.1.

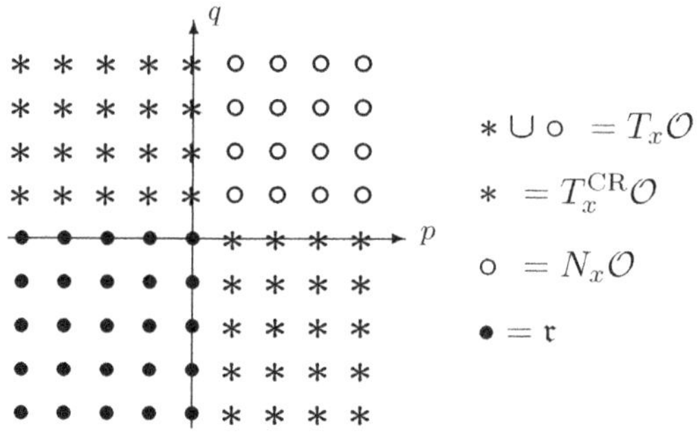

Figure III.1: Visualization of tangent and normal spaces

III.C Codimension–one orbits

Note that, if $\mathcal{O}$ is of codimension–one, then (III.10) and (III.12) imply that there exists a real root α such that

$$N_x\mathcal{O} = \mathbf{i}\, I^{p,p}_{\mathbb{R}} = \mathbf{i}\, \mathfrak{g}^{\alpha}_{\mathbb{R}}, \quad \text{with} \quad p = \alpha(L).$$

The assumption (III.7) forces $p = 1$ [18, Corollary 4.4].

Suppose that $\mathcal{O} \subset \partial D$. Then for a suitably scaled root vector $N^+ \in \mathfrak{g}^{\alpha}_{\mathbb{R}}$ (which is necessarily nilpotent) we have

$$e^{\mathbf{i}N^+} \cdot F^\bullet_x \in D.$$

We may complete N^+ to an $\mathfrak{sl}_2$–triple (II.25) with $N \in \mathfrak{g}^{-\alpha}_{\mathbb{R}}$. Note that the $\mathfrak{sl}_2 = \mathrm{span}\{N, Y, N^+\}$ gives an $\mathrm{SL}_2(\mathbb{C})$–homogeneous embedding of $\mathbb{P}^1 = \mathbb{CP}^1$ into $\check{D}$ with $\mathbb{H} = \mathbb{P}^1 \cap D$ and $\mathbb{RP}^1 = \mathbb{P}^1 \cap \mathcal{O}$. In particular,

$$\lim_{z\to\infty} e^{zN} \cdot \left(e^{\mathbf{i}N^+} \cdot F^\bullet_x\right) \in \mathcal{O}.$$

It now follows from Definition II.31 that we have recovered [18, Proposition 5.16]:

Every codimension–one orbit $\mathcal{O} \subset \partial D$ is polarizable relative to D by a root vector $N \in \mathfrak{g}_{\mathbb{R}}^{-\alpha}$.

Moreover, in this case

$$L + \overline{L} = Y ,$$

and the Deligne splitting $\mathfrak{g}_{\mathbb{C}} = \oplus I_{\mathfrak{g}}^{p,q}$ is *explicitly* given by (III.8) as an eigenspace decomposition

$$I_{\mathfrak{g}}^{p,q} = \{\xi \in \mathfrak{g}_{\mathbb{C}} \ : \ [L,\xi] = q\xi \, , \ [Y,\xi] = (p+q)\xi\} \, . \tag{III.13}$$

Moreover, in the event that the Deligne splitting on $\mathfrak{g}_{\mathbb{C}}$ is induced from one on $V_{\mathbb{C}}$ as in (II.38), we have

$$I^{p,q} = \{v \in V_{\mathbb{C}} \ : \ L(v) = qv \, , \ Y(v) = (p+q)v\} \, .$$

These decompositions are in practice straightforward to compute, cf. [19] or Example IV.6.

Remark III.14. Given a maximal compact subgroup $K_{\mathbb{R}} \subset G_{\mathbb{R}}$, each $G_{\mathbb{R}}$–orbit in $\check{D}$ contains a distinguished set of "Matsuki points" (Section III.D.1). The open orbit D is related to the codimension–one orbits $\mathcal{O} \subset \partial D$ by an application of a Cayley transform to a Matsuki point: Given a Matsuki point $x_0 \in D$ one may apply a Cayley transform $\mathbf{c}_\alpha \in G_{\mathbb{C}}$ to obtain a Matsuki point $x = \mathbf{c}_\alpha(x_0)$ in the codimension–one orbit $\mathcal{O}$, [15, 18, 19].

III.D Orbit dimensions

Fix a maximal, connected compact Lie subgroup $K_{\mathbb{R}} \subset G_{\mathbb{R}}$. Let $\mathfrak{k}_{\mathbb{R}} \subset \mathfrak{g}_{\mathbb{R}}$ denote the Lie algebra of $K_{\mathbb{R}}$, and let

$$\mathfrak{g}_{\mathbb{R}} = \mathfrak{k}_{\mathbb{R}} \oplus \mathfrak{k}_{\mathbb{R}}^{\perp} \, .$$

denote the Cartan decomposition. The complexification will be written as

$$\mathfrak{g}_{\mathbb{C}} = \mathfrak{k}_{\mathbb{C}} \oplus \mathfrak{k}_{\mathbb{C}}^{\perp} \, .$$

Let θ denote the Cartan involution

$$\theta|_{\mathfrak{k}_{\mathbb{C}}} = \mathbb{1} \quad \text{and} \quad \theta|_{\mathfrak{k}_{\mathbb{C}}^{\perp}} = -\mathbb{1} \, . \tag{III.15}$$

The dimensions of the $G_{\mathbb{R}}$ and $K_{\mathbb{R}}$–orbits through x are given by certain $\mathfrak{h}$–root "counts," cf. (III.18) and (III.19) below. A comparison of these root counts yields a characterization (Lemma III.20) of the unique closed $G_{\mathbb{R}}$–orbit in $\check{D}$.

III.D.1 Matsuki points

We say that a point $x \in \check{D} = G/P$ is a *Matsuki point* if the Lie algebra $\mathfrak{p}_x$ of the stabilizer $P_x = \mathrm{Stab}_{G_\mathbb{C}}(x)$ contains a Cartan subalgebra $\mathfrak{h}_x$ that is stable under conjugation and the Cartan involution (III.15). This implies that the $K_\mathbb{R}$–orbit is equal to the intersection of the $K_\mathbb{C}$–orbit with the $G_\mathbb{R}$–orbit: $K_\mathbb{R} \cdot x = (G_\mathbb{R} \cdot x) \cap (K_\mathbb{C} \cdot x)$, cf. [12, Chapter 8]. That is, the $G_\mathbb{R}$–orbit and the $K_\mathbb{C}$–orbit are Matsuki dual. Every $G_\mathbb{R}$–orbit (resp. $K_\mathbb{C}$–orbit) contains a Matsuki point.

The fact that $\mathfrak{h}$ is closed under the Cartan involution implies that $\theta\alpha$ is a root whenever α is a root. Moreover,

$$-\alpha = \theta\bar{\alpha} = \overline{\theta\alpha}\,, \tag{III.16}$$

and the complex roots appear in quartets

$$\{\alpha\,,\ \bar{\alpha}\,,\ \theta\alpha\,,\ \theta\bar{\alpha} = \overline{\theta\alpha}\} = \{\pm\alpha\,,\ \pm\bar{\alpha}\}\,.$$

Note that (III.16) implies that the bigrading (III.5) satisfies

$$\theta(\mathfrak{g}^{p,q}) = \mathfrak{g}^{-q,-p}\,; \tag{III.17}$$

that is,

the Cartan involution corresponds to reflection about the line $p = -q$.

III.D.2 The $G_\mathbb{R}$–orbit

Let $\mathcal{O} = G_\mathbb{R} \cdot x$ be the $G_\mathbb{R}$–orbit of x. Then

$$\dim_\mathbb{R} \mathcal{O} = \dim_\mathbb{R} \mathfrak{g}_\mathbb{R} \ -\ \dim_\mathbb{R} \mathfrak{g}_\mathbb{R} \cap \mathfrak{p}\,.$$

Note that

$$\mathfrak{g}_\mathbb{R} \cap \mathfrak{p} = (\mathfrak{p} \cap \bar{\mathfrak{p}})_\mathbb{R} = \mathfrak{h}(\mathbb{R}) \ \oplus \sum_{\alpha(L),L(\bar{\alpha})\leq 0} (\mathfrak{g}^\alpha + \mathfrak{g}^{\bar{\alpha}})_\mathbb{R}\,.$$

The sum above is over roots in (closed) lower–left quadrant

$$\Delta(\leq 0, \leq 0) = \{\alpha \in \Delta \ :\ \alpha(L), \bar{\alpha}(L) \leq 0\}\,,$$

cf. Figure III.2. Set

$$\Delta(\mathcal{O}) = \Delta \backslash \Delta(\leq 0, \leq 0)\,.$$

Then the (real) tangent space

$$T_x\mathcal{O} \ \simeq\ \mathfrak{g}_\mathbb{R}/\mathfrak{p} \ \simeq \sum_{\alpha\in\Delta(\mathcal{O})} (\mathfrak{g}^\alpha + \mathfrak{g}^{\bar{\alpha}})_\mathbb{R}$$

as a vector space, and we conclude that

$$\dim_\mathbb{R}\mathcal{O} = \frac{1}{2}\ |\{\alpha \in \Delta(\mathcal{O}) \mid \alpha \neq \bar{\alpha}\}| \ +\ |\{\alpha \in \Delta(\mathcal{O}) \mid \alpha = \bar{\alpha}\}|\,. \tag{III.18}$$

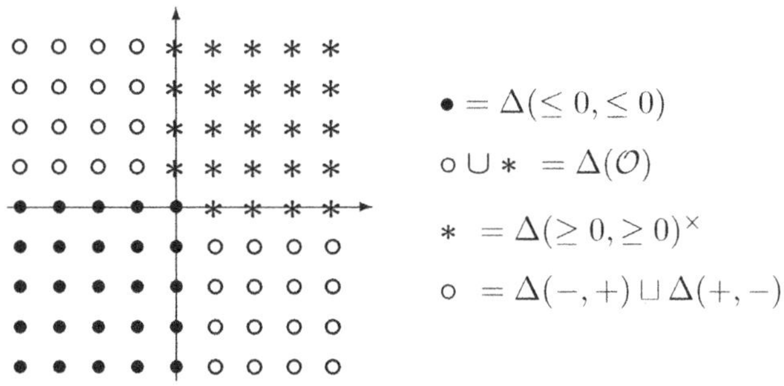

Figure III.2: Visualization of root sets

III.D.3 The $K_\mathbb{R}$–orbit

Likewise, we may compute the dimension of the $K_\mathbb{R}$–orbit

$$\mathcal{O}^{K_\mathbb{R}} = K_\mathbb{R} \cdot x$$

as follows. Decompose

$$\Delta(\mathcal{O}) = \Delta(\geq 0, \geq 0)^\times \sqcup \Delta(-,+) \sqcup \Delta(+,-)$$

into three disjoint sets by

$$\begin{aligned} \Delta(\geq 0, \geq 0)^\times &= \{\alpha \in \Delta(\mathcal{O}) \ : \ -\alpha \in \Delta(\leq 0, \leq 0)\} = -\Delta(\geq 0, \geq 0) \backslash \{(0,0)\}, \\ \Delta(+,-) &= \{\alpha \in \Delta \ : \ \alpha(L) > 0 \,,\ L(\bar\alpha) < 0\}, \\ \Delta(-,+) &= \{\alpha \in \Delta \ : \ \alpha(L) < 0 \,,\ L(\bar\alpha) > 0\}. \end{aligned}$$

We may visualize $\Delta(+,-)$ and $\Delta(-,+)$ as open quadrants in the (p,q)–plane, and $\Delta(\geq 0, \geq 0)^\times$ as the closed upper–right quadrant minus the origin, cf. Figure III.2.

We have an identification $T_o\mathcal{O}^{K_\mathbb{R}} \simeq \mathfrak{k}_\mathbb{R}/\mathfrak{p}$, as vector spaces. So the dimension of $\mathcal{O}^{K_\mathbb{R}}$ is equal to the dimension of $\mathfrak{k}_\mathbb{R}/\mathfrak{p}$. Recall that $\mathfrak{k}_\mathbb{C} \subset \mathfrak{g}_\mathbb{C}$ is the fix point locus of the Cartan involution $\theta : \mathfrak{g}_\mathbb{C} \to \mathfrak{g}_\mathbb{C}$. In particular, $\mathfrak{k}_\mathbb{C} = \{x + \theta x \ : \ x \in \mathfrak{g}_\mathbb{C}\}$. Then, making use of (III.11) and (III.17), we have

$$\begin{aligned} \dim \mathcal{O}^{K_\mathbb{R}} = &\ |\{\alpha \in \Delta(\mathcal{O}) \ : \ \alpha = \bar\alpha\}| + |\{\alpha \in \Delta(-,+) \mid \alpha = \theta\alpha\}| \\ &+ \frac{1}{2} |\{\alpha \in \Delta(-,+) \mid \alpha \neq \theta\alpha\}| \\ &+ \frac{1}{2} \left|\{\alpha \in \Delta(\leq 0, \leq 0)^\times \ \middle|\ \alpha \neq \bar\alpha\}\right| . \end{aligned} \tag{III.19}$$

III.E Characterization of closed orbits

The flag manifold $\check{D} = G/P$ contains a unique closed $G_\mathbb{R}$–orbit [28, Theorem 3.3], which is contained in the closure of every other $G_\mathbb{R}$–orbit [28, Corollary 3.4]. It follows that the closed $G_\mathbb{R}$–orbit is the unique orbit for which $\mathcal{O} = \mathcal{O}^{K_\mathbb{R}}$. Therefore

$\mathcal{O}$ is closed if and only if the two dimensions (III.18) and (III.19) are equal. Whence we obtain the following characterization of the closed $G_{\mathbb{R}}$–orbit.

Lemma III.20. *Let $x \in \check{D}$ be a Matsuki point, and let $\mathfrak{h}$ be a Cartan subalgebra of $\mathfrak{g}$ contained in the Lie algebra $\mathfrak{p}$ of the stabilizer $P = \mathrm{Stab}_G(x)$ that is stable under complex conjugation and the Cartan involution. Then the $G_{\mathbb{R}}$–orbit $\mathcal{O}$ through x is closed (equivalently, $\mathcal{O} = \mathcal{O}^{K_{\mathbb{R}}}$) if and only if the $\mathfrak{h}$–roots satisfy*

$$\theta\alpha = \alpha \quad \textit{for all} \quad \alpha \in \Delta(-,+)\,. \tag{III.21}$$

That is, by (III.16), *all the roots of $\Delta(-,+)$ are imaginary and compact.*

Keeping in mind that $\alpha \mapsto \bar{\alpha}$ is a bijection between $\Delta(+,-)$ and $\Delta(-,+)$, we see that (III.21) holds if and only if

$$\theta\alpha = \alpha \quad \text{for all} \quad \alpha \in \Delta(+,-)\,.$$

Whence the lemma may be visualized as in Figure III.3.a, where the potentially nonzero $\mathfrak{g}^{p,q}$ are indicated by a node at the point (p,q). Equivalently, we will say that the bigraded $(L,\overline{L})$–eigenspace decomposition (III.5) must be of the form depicted in Figure III.3.a.

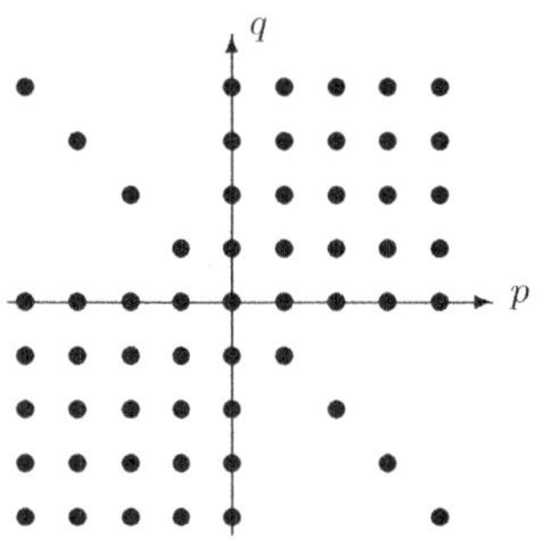

(III.3.a)

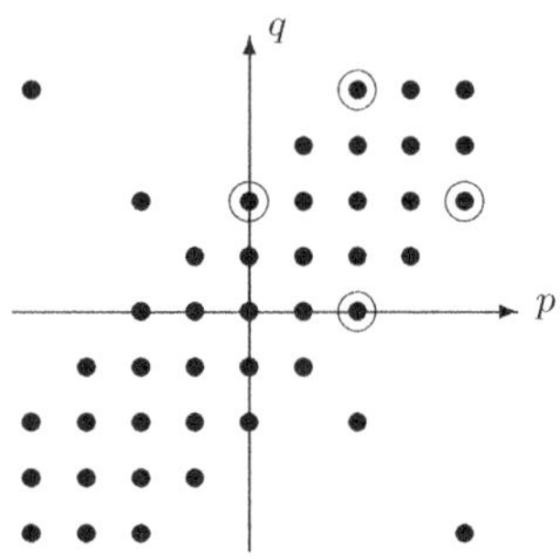

(III.3.b)

Figure III.3: Visualization of (polarized) closed $G_{\mathbb{R}}$–orbit

Remark. We remark that Figure III.3.a is necessary, but not sufficient, for an orbit to be closed. That is, there are examples in which the bigrading takes this form, but (III.21) fails.

III.F Totally real orbits

A $G_{\mathbb{R}}$–orbit $\mathcal{O} \subset \check{D}$ is *totally real* if the Cauchy–Riemannian tangent bundle $T^{\mathrm{CR}}\mathcal{O}$ is trivial. Visually, the orbit is totally real if there are no $*$'s in Figure III.1. By (III.12) and Lemma III.20 any totally real orbit is necessarily the unique closed orbit. Moreover, recalling that

$$\mathfrak{p} = \bigoplus_{p \geq 0} \mathfrak{g}^{-p,\bullet} = \mathfrak{g}^{\leq 0,\bullet}\,,$$

the first half of Theorem I.16 is now evident:

Theorem III.22. *The following are equivalent:*

(i) *The orbit $\mathcal{O}_{\rm cl}$ is totally real.*

(ii) *The real dimension of the closed orbit is the complex dimension of the compact dual:* $\dim_{\mathbb{R}}\mathcal{O}_{\rm cl} = \dim_{\mathbb{C}}\check{D}$.

(iii) *The stabilizer P is $\mathbb{R}$-split and $\mathcal{O}_{\rm cl} = G_{\mathbb{R}}/P_{\mathbb{R}}$.*

IV Minimal degenerations of polarized Hodge structures

Let D be a Hodge domain. In Section III.C we saw that every codimension-one $G_{\mathbb{R}}$-orbit $\mathcal{O} \subset \partial D$ is polarizable relative to D, and gave explicit descriptions of Deligne splittings as eigenspace decompositions. In Section IV.A we specialize to the case that D is a period domain and give a refined description. In Example IV.6 we give an example illustrating the (in general) more complicated structure in the Hodge domain case.

IV.A Minimal degenerations in period domains

Definition. A *type I basic boundary component* for weight n consists of limiting mixed Hodge structures that are of the form

$$\begin{array}{ccc} H^{n-1-2k}(k+1) & \xrightarrow{N} & H^{n-1-2k}(k) \\ & H^n & \end{array}$$

for some k with $0 \le 2k \le n-1$. A *type II basic boundary component* for weight $n = 2m$ consists of limiting mixed Hodge structures that are of the form

$$\begin{array}{ccccc} H^0(m+1) & \xrightarrow{N} & H^0(m) & \xrightarrow{N} & H^0(m-1) \\ & & H^n\,. & & \end{array}$$

Any $G_{\mathbb{R}}$-orbit $\mathcal{O} \subset \partial D$ containing the image of a basic boundary component $B(N)$ under the reduced limit period mapping is a *basic boundary orbit.*

The Deligne splittings of the limiting mixed Hodge structures in a basic boundary component are illustrated in Figure IV.1.

Theorem IV.1. *Given a period domain $D \subset \check{D}$, the codimension-one boundary orbits are basic. Additionally, the nonzero off-diagonal $I^{p,q}$ all have dimension one. Moreover, the type II basic boundary components occur only for even weight Hodge structures with $h^{m,m} = 1$.*

Proof. Any codimension-one orbit $\mathcal{O} \subset \mathrm{bd}(D)$ contains the image under the reduced limit period mapping of a boundary component $B(N)$; moreover we may

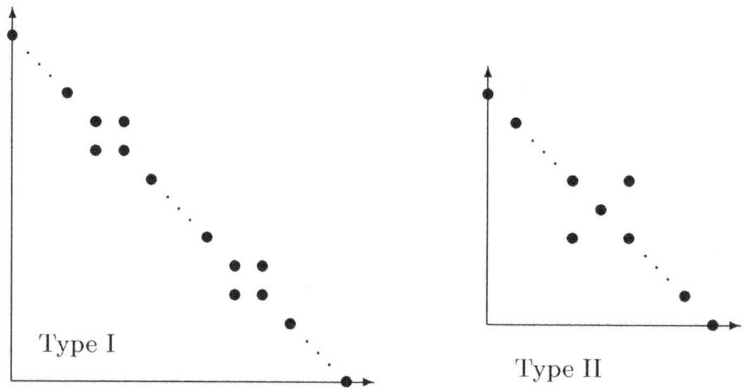

Figure IV.1: Limiting mixed Hodge structures in basic boundary component

take N to be a root vector (Section III.C). Therefore, for an appropriate choice of basis $\{e_i\}_{i=1}^d$ of V we may take N to be of one of the following normal forms. Let $\{e^i\}_{i=1}^d$ be the dual basis of V^*, and let $\{e^i_j = e_j \otimes e^i\}_{i,j=1}^d$ be the corresponding basis of $\mathrm{End}(V)$.

Suppose the weight n is odd, so that $d = 2c$. The nilpotent N is of one of the following forms

$$e^i_j - e^{c+j}_{c+i}\,,\quad e^{c+i}_j + e^{c+j}_i\,,\quad e^i_{c+j} + e^j_{c+i}\,,\quad e^i_{c+i}\,,\quad e^{c+i}_i\,,\quad \text{with}\quad i,j \le c\,. \tag{IV.2}$$

If the weight n is even, and the dimension $d = 2c$ of V is even, then the nilpotent N is of one of the following forms

$$e^i_j - e^{c+j}_{c+i}\,,\quad e^{c+i}_j - e^{c+j}_i\,,\quad e^i_{c+j} - e^j_{c+i}\,,\quad \text{with}\quad i,j \le c\,. \tag{IV.3}$$

In the event that the dimension $d = 2c+1$ is odd, then the nilpotent N is either of one of the three forms in (IV.3), or of one of the following forms

$$e^d_i - e^{c+i}_d\,,\quad e^d_{c+i} - e^i_d\,,\quad \text{with}\quad i \le c\,. \tag{IV.4}$$

It is clear from these explicit expressions that:

- If N is of one of the forms in (IV.2) or (IV.3), then $N^2 = 0$ and N has rank at most two, so that the boundary component is type I basic.
- If N if of one of the forms in (IV.4), then $N^3 = 0$ while $N^2 \neq 0$, and N has rank two, so that the boundary component is type II basic. Moreover, in this case the weight $n = 2m$ is even, and $d = 2c+1 = \sum h^{p,q}$ forces $h^{m,m}$ to be odd. Finally, from (I.11) and $\dim I^{1,1}_{\mathfrak{g}} = i^{m,m} = h^{m,m}$ we see $\mathrm{codim}_{\check{D}}\, \mathcal{O} = 1$ if and only if $h^{m,m} = 1$. □

IV.B Minimal degenerations in Hodge domains

In the case of Hodge domains the $I^{p,q}$ corresponding to a minimal degeneration may be more complicated that those arising in the case of period domains (Figure IV.1). Illustrations of this are given in Examples II.39 (see, in particular, the rows 2 and 3 of Figure II.1) and IV.6. What we can say in general is

Proposition IV.5. *Fix a flag domain $G_\mathbb{R}/R$ endowed with the structure of a Hodge domain D. Then there exists a Hodge representation (V, Q) of G that realizes $G_\mathbb{R}/R$ as Hodge domain D and with the property that each codimension–one $G_\mathbb{R}$–orbit $\mathcal{O} \subset \partial D \subset \check{D}$ is polarized by a limiting mixed Hodge structure $(V, W_\bullet(N), F^\bullet)$ such that*

$$N^3 = 0 \quad \text{as an element of} \quad \mathfrak{g}_\mathbb{R} \subset \operatorname{End}(V, Q)\,.$$

In fact, if $G_\mathbb{C} \neq G_2$ is simple, then we may take V to be the adjoint representation $\mathfrak{g}$; if $G_\mathbb{C} = G_2(\mathbb{C})$, then we may take V to be the standard representation $V = \mathbb{R}^7$.

The proof follows Example IV.6.

Example IV.6 (An F_4 Hodge domain). Consider the case that the Hodge group is the exceptional simple Lie group of rank four. The smallest representation $V_\mathbb{C}$ of $G_\mathbb{C} = F_4(\mathbb{C})$ is of dimension 26. There exist real forms V and $G_\mathbb{R} = \mathrm{F\,I}$, the latter of real rank four and with maximal compact subalgebra $\mathfrak{k}_\mathbb{R} = \mathfrak{sp}(3) \oplus \mathfrak{su}(2) \subset \mathfrak{g}_\mathbb{R}$, with the property that V admits the structure of a $G_\mathbb{R}$–Hodge representation with Hodge numbers

$$\mathbf{h} = (1^4, 2^9, 1^4)\,.$$

In this case the compact dual $G_\mathbb{C}/B$ is the flag variety and has dimension 24. The Deligne splittings associated with the four codimension–one boundary orbits are depicted in Figure IV.2; these are computed by (III.13). Each circled node has dimension two, the remainder have dimension one. Note that in the first row of Figure IV.2 we have $N^2 = 0$; in the second row $N^2 \neq 0$ and $N^3 = 0$.

Proof of Proposition IV.5. Given a polarized Hodge structure $(V, Q, F^\bullet) \in D$, let $(\mathfrak{g}, Q_\mathfrak{g}, F^\bullet_\mathfrak{g})$ denote the induced polarized Hodge structure on $\mathfrak{g}_\mathbb{R}$ (Section II.D.2). Any codimension–one boundary $G_\mathbb{R}$–orbit $\mathcal{O} \subset \partial D$ is polarized by a root vector $N \subset \mathfrak{g}_\mathbb{R}^{-\alpha}$. It is well-known that root vectors satisfy $N^4 - 0$ as elements of $\operatorname{End}(\mathfrak{g})$. (Equivalently, every α–string has length at most four.)

- If α is *not* a root of the exceptional Lie algebra $\mathfrak{g}_2$, then $N^3 = 0$ on $\mathfrak{g}_\mathbb{C}$ (every α–string has length at most three).
- If α is a root of $\mathfrak{g}_2$, then it is possible that $N^3 \neq 0$ on $\mathfrak{g}_\mathbb{C}$; this is illustrated in Figure II.1 (row 3, column 2). However, as is also illustrated in Figure II.1 (rows 2 and 3, column 1), any root vector will satisfy $N^3 = 0$ as an element of $\mathfrak{g}_\mathbb{C} \subset \operatorname{End}(\mathbb{C}^7)$ acting on the standard representation. □

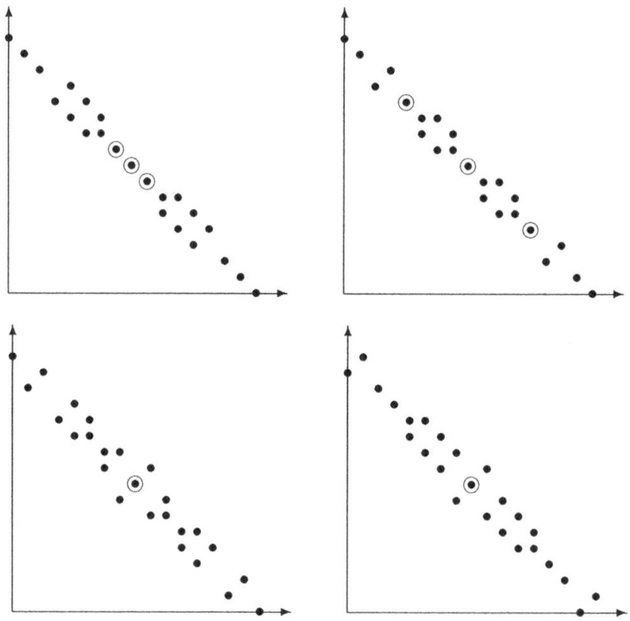

Figure IV.2: Deligne splittings associated with codimension–one boundary orbits for $D \subset F_4(\mathbb{C})/B$.

V Maximal degenerations of polarized Hodge structures

V.A Polarizable closed orbits

In the case that the closed orbit is polarizable, Lemma III.20 and Figure III.3.a may be refined. We now recall Theorem I.17 and prove the slightly stronger

Theorem V.1. *Suppose that the limiting mixed Hodge structure* $(\mathfrak{g}, W_\bullet(N), F^\bullet)$ *is sent to the closed orbit under the reduced limit period mapping. Then Deligne splitting* $\mathfrak{g}_\mathbb{C} = \oplus I^{p,q}_\mathfrak{g}$ *satisfies:*

$$I^{p,-q}_\mathfrak{g} = 0 \quad \text{for all} \quad p \neq q > 0\,, \tag{V.2a}$$

$$I^{p,-p}_\mathfrak{g} = 0 \quad \text{for all} \quad \text{odd} \;\; p \geq 3\,, \tag{V.2b}$$

$$I^{p,q}_\mathfrak{g} = 0 \quad \text{for all} \quad p+q \neq 0 \quad \text{with} \quad |p-q| > 2\,. \tag{V.2c}$$

Moreover, any N*–string in* $\oplus_{q-p=2}\,\mathfrak{g}^{p,q}$ *has length* $\equiv 3$ *mod* 4*. In particular, if* $\mathrm{Gr}^{W_\bullet(N)_\mathfrak{g}}_{k,\mathrm{prim}} \neq 0$*, then* k *is even.*

The Deligne splitting is pictured in Figure III.3.b. The circled nodes indicate those $I^{p,q}_\mathfrak{g}$ with $p-q = \pm 2$ that *may* (but need not) have a non–trivial primitive component. (There are no constraints on the primitive components with $p - q \neq \pm 2$, aside from the obvious $I^{p,-p}_\mathfrak{g} = I^{p,-p}_{\mathfrak{g},\mathrm{prim}}$ for all $|p| \geq 2$.)

Proof. Lemma III.20 implies (V.2a). Assertion (V.2c) follows from the the property $N : \mathfrak{g}^{p,q} \to \mathfrak{g}^{p-1,q-1}$ and the fact that the N–strings are uninterrupted. Given this, it follows that

$$\Gamma = \bigoplus_{p \geq 2} (\mathfrak{g}^{-p,p} \oplus \mathfrak{g}^{p,-p}) \subset \mathrm{Gr}^{W_\bullet(N)_\mathfrak{g}}_{0,\mathrm{prim}} .$$

Recall (Section II.F) that $\mathrm{Gr}^{W_\bullet(N)_\mathfrak{g}}_{0,\mathrm{prim}}$ carries a weight zero Hodge structure polarized by $Q_\mathfrak{g}$. So given a nonzero $v \in \mathfrak{g}^{-p,p} \subset \Gamma$, we have $(-1)^p Q_\mathfrak{g}(v, \bar{v}) > 0$. By (III.10), Γ is a direct sum of root spaces. Moreover, Lemma III.20 asserts that the roots are imaginary and compact. Recall that $-Q_\mathfrak{g}$ is the Killing form. Therefore, given a nonzero root vector $v \in \mathfrak{g}^\alpha \subset \mathfrak{g}^{-p,p}$, the inequality $(-1)^p Q_\mathfrak{g}(v, \bar{v}) > 0$ implies that α is compact if p is even and noncompact if p is odd. Therefore, p is necessarily even; this establishes (V.2b).

Any N–string in $\oplus_{q-p=2}\, \mathfrak{g}^{p,q}$ is necessarily of the form

$$u\,,\ Nu\,,\ N^2u\,, \ldots,\ N^{2k}u\,,$$

with $0 \neq u \in \mathfrak{g}^{k-1,k+1}$, $N^k u \in \mathfrak{g}^{-1,1}$ and $N^{2k+1}u = 0$. By the polarization hypothesis,

$$0 \ < \ -Q_\mathfrak{g}(u, N^{2k}\bar{u}) = (-1)^{k+1} Q_\mathfrak{g}(N^k u, N^k \bar{u})\,.$$

Lemma III.20 implies $0 \neq v = N^k u \in \mathfrak{g}^{-1,1}$ satisfies $Q_\mathfrak{g}(v, \bar{v}) > 0$. Therefore k is odd, and the length of the N–string is $2k + 1 = 4\ell + 3 \equiv 3 \bmod 4$. □

V.B The Hodge–Tate case

The polarized limiting mixed Hodge structure $(V, W_\bullet(N), F^\bullet)$ is *Hodge–Tate* if

$$I^{p,q} = 0 \quad \text{for all} \quad p \neq q\,. \tag{V.3}$$

We will now recall and prove

Proposition I.15. *The polarized limiting mixed Hodge structure $(V, W_\bullet(N), F^\bullet)$ is Hodge–Tate if and only if the induced limiting mixed Hodge structure $(\mathfrak{g}, W_\bullet(N)_\mathfrak{g}, F^\bullet_\mathfrak{g})$ is Hodge–Tate.*

Proposition I.15 is a consequence of the following lemma.

Lemma V.4. *Fix a real form $\mathfrak{g}_\mathbb{R}$ and let $\mathfrak{h} \subset \mathfrak{g}_\mathbb{C}$ be a Cartan subalgebra that is stable under complex conjugation. Let $L \in \mathfrak{h}$ be any grading element. Let V be a representation of $\mathfrak{g}_\mathbb{C}$, and let $\Lambda(V) \subset \Lambda_{\mathrm{wt}}$ denote the weights of V. Suppose that $\lambda(L) = \lambda(\bar{L})$ for all $\lambda \in \Lambda(V)$. (Equivalently, $\lambda(L) = \bar{\lambda}(L)$.) Then $\mu(L) = \mu(\bar{L})$ for the all weights $\mu \in \Lambda_{\mathrm{wt}}$ of $\mathfrak{g}_\mathbb{C}$.*

Proof of Lemma V.4. The weight lattice $\Lambda_{\mathrm{wt}} \subset \mathfrak{h}^*$ is spanned over $\mathbb{Q}$ by the weights of V; that is, $\Lambda_{\mathrm{wt}} = \mathrm{span}_\mathbb{Q}\, \Lambda(V)$. □

Proof of Proposition I.15. The nontrivial $I^{p,q}$ are precisely those with $(p, q) = (\lambda(L), \lambda(\bar{L}))$ for some $\lambda \in \Lambda(V)$. □

Proof of Proposition I.7. Follows directly from Proposition I.15 and Lemma III.20. □

We may now prove the following corollary to Proposition I.15 and Theorem I.17.

Corollary V.5. *If a limiting mixed Hodge structure $(V, W_\bullet(N), F^\bullet)$ polarizes a totally real (and necessarily closed) $G_\mathbb{R}$-orbit, then the limiting mixed Hodge structure is of Hodge–Tate type.*

Proof. If a limiting mixed Hodge structure $(V, W_\bullet(N), F^\bullet)$ polarizes the closed orbit, then Theorem I.17 asserts that the induced limiting mixed Hodge structure $(\mathfrak{g}, W_\bullet(N)_\mathfrak{g}, F^\bullet_\mathfrak{g})$ is as depicted in Figure III.3.b. If the closed orbit is totally real, then the fact that the N-strings in that figure are uninterrupted and centered on the line $p = -q$ implies that $(\mathfrak{g}, W_\bullet(N)_\mathfrak{g}, F^\bullet_\mathfrak{g})$ is Hodge–Tate. The corollary then follows from Proposition I.15. □

Together Corollary V.5 and Proposition I.15 yield the second half of Theorem I.16.

There are a number of conditions that a Hodge domain D must satisfy in order to admit a Hodge–Tate degeneration. Some of the conditions are Hodge-theoretic in nature (Lemma V.7), while others are representation theoretic (Lemma V.10 and Remark V.12).[14]

Given weight n, set

$$m = \lfloor n/2 \rfloor \,; \tag{V.6a}$$

that is, m is defined by

$$n \in \{2m, 2m+1\} \,. \tag{V.6b}$$

Lemma V.7. *Let D be a Hodge domain for polarized Hodge structures of weight n and with $\mathbf{h} = (h^{n,0}, h^{n-1,1}, \ldots, h^{1,n-1}, h^{0,n})$. If a point of D admits a Hodge–Tate degeneration, then the Hodge numbers satisfy*

$$h^{n,0} \le h^{n-1,1} \le \cdots \le h^{n-m,m} \,. \tag{V.8}$$

The converse to Lemma V.7 holds when D is a period domain (Theorem V.15), but fails for Hodge domains in general (Remark V.16).

Proof. If there is a Hodge–Tate degeneration of a point in D, then the limiting mixed Hodge structure looks like

$$\begin{array}{ccccccccccc}
H^0(n) & \to & H^0(n-1) & \to & H^0(n-2) & \to \cdots \to & H^0(2) & \to & H^0(1) & \to & H^0_0 \\
 & & H^0_2(n-1) & \to & H^0_2(n-2) & \to \cdots \to & H^0_2(2) & \to & H^0_2(1) & & \\
 & & & & H^0_4(n-2) & \to \cdots \to & H^0_4(2) & & & & \\
 & & & & & \vdots & & & & &
\end{array} \tag{V.9}$$

[14] With respect to the latter we present only an illustrative sketch. The general story will be discussed in a later work.

where H^0_{2k} is a Hodge–Tate structure of weight zero. The inequalities (V.8) are then simple consequences of this picture: the Deligne splitting must satisfy

$$i^{n,n} \leq i^{n-1,n-1} \leq \cdots \leq i^{n-m,n-m}.$$

Since $h^{p,q} = i^{p,n-q}$, this yields (V.8).

For example, when $n = 4$, the $I^{p,q}$ picture of the Hodge–Tate limiting mixed Hodge structure is

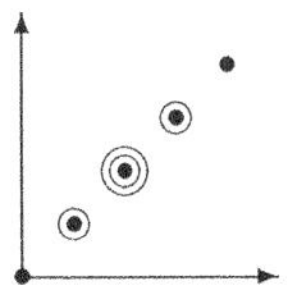

and $h^{4,0} = i^{4,4} \leq h^{3,1} = i^{3,3} \leq h^{2,2} = i^{2,2}$.

□

A second condition that D must satisfy to admit a Hodge–Tate degeneration is

Lemma V.10. *If a Hodge domain $D \subset \check{D} = G_{\mathbb{C}}/P$ admits a Hodge–Tate degeneration, then it is necessarily the case that $\mathfrak{p}$ is an even Jacobson–Morosov parabolic* (Remark II.40).

Example V.11 (Even Jacobson–Morosov parabolics in G_2). The exceptional simple Lie algebra $\mathfrak{g}_{\mathbb{C}} = \mathfrak{g}_2$ has three conjugacy classes of (proper) parabolic subalgebras. Each may be realized as Jacobson–Morosov parabolics, but only two are even; they are indexed by $\Sigma = \{\alpha_2\}$ (the simple root α_2 is long) and $\Sigma = \{\alpha_1, \alpha_2\}$ (indexing the Borel), cf. [9, Section 8.4]. In these two cases there exist Hodge domains $D \subset G_{\mathbb{C}}/P$, homogeneous with respect to the split real form of G_2, admitting Hodge–Tate degenerations, cf. [18, Section 6.1.3].

Proof. Suppose that $(V, W_\bullet(N), F^\bullet)$ is Hodge–Tate limiting mixed Hodge structure for the period domain D. Then the induced limiting mixed Hodge structure $(\mathfrak{g}, W_\bullet(N)_{\mathfrak{g}}, F^\bullet_{\mathfrak{g}})$ is also Hodge–Tate (Proposition I.15). Therefore the Lie algebra of the parabolic stabilizing $F^\bullet$ is

$$\mathfrak{p} = F^0_{\mathfrak{g}} \overset{\text{(II.32)}}{=} \bigoplus_{p\geq 0} I^{p,\bullet}_{\mathfrak{g}} \overset{\text{(V.3)}}{=} \bigoplus_{p\geq 0} I^{p,p}_{\mathfrak{g}} \overset{\text{(II.32)}}{=} W_0(N^+)_{\mathfrak{g}}$$

where N^+ is the nilpositive element of an $\mathfrak{sl}_2$–triple (II.25) containing N. It follows that $\mathfrak{p}$ is a Jacobson–Morosov parabolic. Moreover, since the neutral element Y acts on $I^{p,p}_{\mathfrak{g}}$ by the even scalar $2p$, it follows that $\mathfrak{p}$ is an even Jacobson–Morosov parabolic. □

Remark V.12 (Hodge–Tate degenerations in flag varieties $\check{D} = G_{\mathbb{C}}/B$). Lemma V.10 imposes constraints on the compact duals without reference to the real form $G_{\mathbb{R}}$. Once we pick a Jacobson–Morosov parabolic $P \subset G_{\mathbb{C}}$, there are constraints on the real forms $G_{\mathbb{R}}$ that may arise as Hodge groups with Hodge representation admitting a Hodge–Tate degeneration. To illustrate this we consider the case that

$G_{\mathbb{C}}$ is one of the classical groups and $P = B$ is the Borel. First a few general remarks.

The set $\mathcal{N} \subset \mathfrak{g}_{\mathbb{C}}$ of nilpotent endomorphisms decomposes into a finite union of $\mathrm{Ad}(G_{\mathbb{C}})$–orbits. Moreover, there is a unique open orbit $\mathcal{N}_{\mathrm{prin}}$, elements of which are the principal nilpotents. A Jacobson–Morosov parabolic $W_0(N)$ is even and a Borel subalgebra if and only if N is principal [9, Section 4.1]. In the case that $G_{\mathbb{C}}$ is one of the classical simple Lie groups, the principal nilpotents are nicely characterized by the decomposition of the standard representation under an $\mathfrak{sl}_2 \subset \mathfrak{g}_{\mathbb{C}}$. To be precise, given any nilpotent $N \in \mathcal{N}$, let $\mathfrak{sl}_2 \subset \mathfrak{g}_{\mathbb{C}}$ be the Lie subalgebra spanned by an $\mathfrak{sl}_2$–triple (II.25) containing N. Then:

(i) Let $V_{\mathbb{C}} = \mathbb{C}^n$ be the standard representation of any one of $\mathfrak{sl}_n\mathbb{C}$, $\mathfrak{sp}_{2m}\mathbb{C}$ with $n = 2m$, or $\mathfrak{so}_{2m+1}\mathbb{C}$ with $n = 2m+1$. Then the nilpotent N is principal if and only $V_{\mathbb{C}}$ is irreducible as an $\mathfrak{sl}_2$–module.

(ii) If $G_{\mathbb{C}} = \mathrm{SO}_{2m}\mathbb{C}$, then N is principal if and only if the standard representation $V_{\mathbb{C}} = \mathbb{C}^{2m}$ decomposes as $\mathbb{C} \oplus \mathbb{C}^{2m-1}$ (a trivial subrepresentation plus an irreducible subrepresentation).

The challenge to the real form $G_{\mathbb{R}}$ is that:

An open $G_{\mathbb{R}}$–orbit $D \subset G_{\mathbb{C}}/B$ can admit the structure of a Hodge domain with a Hodge–Tate degeneration only if $\mathfrak{g}_{\mathbb{R}} \cap \mathcal{N}_{\mathrm{prin}}$ is nonempty.

From the classification [9, Section 9.3] of $\mathrm{Ad}(G_{\mathbb{R}})$–orbits of nilpotent elements in $\mathfrak{g}_{\mathbb{R}}$ we deduce that $\mathfrak{g}_{\mathbb{R}} \cap \mathcal{N}_{\mathrm{prin}}$ is nonempty only for the following real forms (in the classical case):

$$\mathfrak{su}(m,m)\,,\ \mathfrak{su}(m,m+1)\,;\quad \mathfrak{sp}(n,\mathbb{R})\,;\quad \mathfrak{so}(m,m)\,,\ \mathfrak{so}(m+1,m)\,,\mathfrak{so}(m+2,m)\,.$$

The classification also determines the open $G_{\mathbb{R}}$–orbits $D \subset G_{\mathbb{C}}/B$:

(a) Suppose that $\mathfrak{g}_{\mathbb{R}} = \mathfrak{sp}(n,\mathbb{R}) = \mathrm{End}(\mathbb{R}^{2n}, Q)$. From the classification [9, Theorem 9.3.5] we may draw the following conclusions. The intersection $\mathfrak{g}_{\mathbb{R}} \cap \mathcal{N}_{\mathrm{prin}}$ consists of two $\mathrm{Ad}(G_{\mathbb{R}})$–orbits $\mathcal{N}^+_{\mathrm{prin},\mathbb{R}}$ and $\mathcal{N}^-_{\mathrm{prin},\mathbb{R}}$. Given $N \in \mathcal{N}^\pm_{\mathrm{prin},\mathbb{R}}$, there exists $v \in \mathbb{R}^{2n}$ such that $\{v, Nv, \ldots, N^{2n-1}v\}$ is a basis of $\mathbb{R}^{2n}$ and

$$\pm Q(N^a v, N^b v) = (-1)^a \delta^{a+b}_{2n-1}\,.$$

So we see that, if $N \in \mathcal{N}^+_{\mathrm{prin},\mathbb{R}}$ and

$$F^p = \mathrm{span}_{\mathbb{C}}\{N^a v \ :\ a \geq p\} \tag{V.13}$$

then $(\mathbb{C}^{2n}, W_\bullet(N), F^\bullet)$ is a limiting mixed Hodge structure for the period domain parameterizing weight $2n-1$, Q–polarized Hodge structures with Hodge numbers $\mathbf{h} = (1,1,\ldots,1)$. Note that

$$I^{p,p} = \mathrm{span}\{N^{2n-1-p}v\}\,.$$

(b) In the case that $\mathfrak{g}_{\mathbb{R}} = \mathfrak{so}(m+1,m) = \mathrm{End}(\mathbb{R}^{2m+1}, Q)$, the classification [9, Theorem 9.3.4] asserts that the intersection $\mathfrak{g}_{\mathbb{R}} \cap \mathcal{N}_{\mathrm{prin}}$ consists of one

$\mathrm{Ad}(G_\mathbb{R})$–orbit $\mathcal{N}_{\mathrm{prin},\mathbb{R}}$, and given an element N in that orbit, $\mathbb{R}^{2m+1}$ admits a basis $\{v, Nv, \ldots, N^{2m}v\}$ such that

$$(-1)^m Q(N^a v, N^b v) = (-1)^a \delta_{2m}^{a+b}\,.$$

Taking again the definition (V.13), we see that $(\mathbb{C}^{2n}, W_\bullet(N), F^\bullet)$ is a limiting mixed Hodge structure for the period domain parametrizing weight $2m$, $(-1)^m Q$–polarized Hodge structures with Hodge numbers $\mathbf{h} = (1, \ldots, 1)$.

(c) In the case that $\mathfrak{g}_\mathbb{R} = \mathfrak{so}(m,m) = \mathrm{End}(\mathbb{R}^{2m}, Q)$, the classification [9, Theorem 9.3.4] asserts that the intersection $\mathfrak{g}_\mathbb{R} \cap \mathcal{N}_{\mathrm{prin}}$ consists of two $\mathrm{Ad}(G_\mathbb{R})$–orbits $\mathcal{N}^+_{\mathrm{prin},\mathbb{R}}$ and $\mathcal{N}^-_{\mathrm{prin},\mathbb{R}}$. Given $N \in \mathcal{N}^\pm_{\mathrm{prin},\mathbb{R}}$ there exists a Q–orthogonal decomposition $\mathbb{R}^{2m} = \mathbb{R} \oplus \mathbb{R}^{2m-1}$ into $\mathfrak{sl}_2$–submodules admitting bases $\{w\}$ and $\{v, Nv, \ldots, N^{2m-2}\}$ such that

$$\pm Q(w,w) = 1 \quad \text{and} \quad \pm\, Q(N^a v, N^b v) = (-1)^{m+a} \delta_{2m-2}^{a+b}\,.$$

Let

$$\begin{aligned} F^p &= \mathrm{span}\{N^a v \;:\; a \geq p\}\,, \quad p \geq m\,, \\ F^p &= \mathrm{span}\{w\,,\, N^a v \;:\; a \geq p\}\,, \quad p \leq m-1\,. \end{aligned} \tag{V.14}$$

If m is even and $N \in \mathcal{N}^+_{\mathrm{prin},\mathbb{R}}$, then $(\mathbb{C}^{2m}, W_\bullet(N), F^\bullet)$ is a limiting mixed Hodge structure for the period domain parametrizing weight $2m-2$, Q–polarized Hodge structures with Hodge numbers $\mathbf{h} = (1, \ldots, 1, 2, 1 \ldots, 1)$. Note that

$$I^{p,p} = \begin{cases} \mathrm{span}\{N^{2m-2-p}v\}\,, & p \neq m-1\,, \\ \mathrm{span}\{w, N^{m-1}v\}\,, & p = m-1\,. \end{cases}$$

(d) In the case that $\mathfrak{g}_\mathbb{R} = \mathfrak{so}(m+2,m) = \mathrm{End}(\mathbb{R}^{2m}, Q)$, the classification [9, Theorem 9.3.4] asserts that the intersection $\mathfrak{g}_\mathbb{R} \cap \mathcal{N}_{\mathrm{prin}}$ consists of a single $\mathrm{Ad}(G_\mathbb{R})$–orbit $\mathcal{N}_{\mathrm{prin},\mathbb{R}}$. Given $N \in \mathcal{N}_{\mathrm{prin},\mathbb{R}}$ there exists a Q–orthogonal decomposition $\mathbb{R}^{2m} = \mathbb{R} \oplus \mathbb{R}^{2m-1}$ into $\mathfrak{sl}_2$–submodules admitting bases $\{w\}$ and $\{v, Nv, \ldots, N^{2m-2}\}$ such that

$$Q(w,w) = 1 \quad \text{and} \quad Q(N^a v, N^b v) = (-1)^{m+a} \delta_{2m}^{a+b}\,.$$

Taking (V.14) as above, we see that if m is even, then $(\mathbb{C}^{2m}, W_\bullet(N), F^\bullet)$ is a limiting mixed Hodge structure for the period domain parametrizing weight $2m-2$, Q–polarized Hodge structures with Hodge numbers $\mathbf{h} = (1, \ldots, 1, 2, 1 \ldots, 1)$.

V.C Hodge–Tate degenerations in period domains

In the event that D is a period domain, the converse to Lemma V.7 holds.

Theorem V.15. *Let D be the period domain for polarized Hodge structures of weight n and with Hodge numbers $\mathbf{h} = (h^{n,0}, h^{n-1,1}, \ldots, h^{1,n-1}, h^{0,n})$. If (V.8) holds, then a point of D admits a Hodge–Tate degeneration.*

Proof. This is an existence question, and we will construct a Hodge–Tate limiting mixed Hodge structure $(V, W_\bullet(N), F^\bullet)$. The construction will be given as direct sum of "atomic" Hodge–Tate limiting mixed Hodge structures with dimensions

$$\mathbf{i}_{k,d} = \left(i^{n,n}_{k,d}, \ldots, i^{0,0}_{k,d}\right) = \big(\underbrace{0,\ldots,0}_{k \text{ terms}}, d,\ldots,d, \underbrace{0,\ldots,0}_{k \text{ terms}}\big);$$

the latter are illustrated in Figure V.1. To make this precise, given $0 \le d \in \mathbb{Z}$ and $0 \le k \le m$, with m as in (V.6), define

$$H^0_{k,d}(s) = \mathrm{span}_{\mathbb{R}}\{e^s_1, \ldots, e^s_d\} \simeq \mathbb{R}^d, \quad k \le s \le n-k.$$

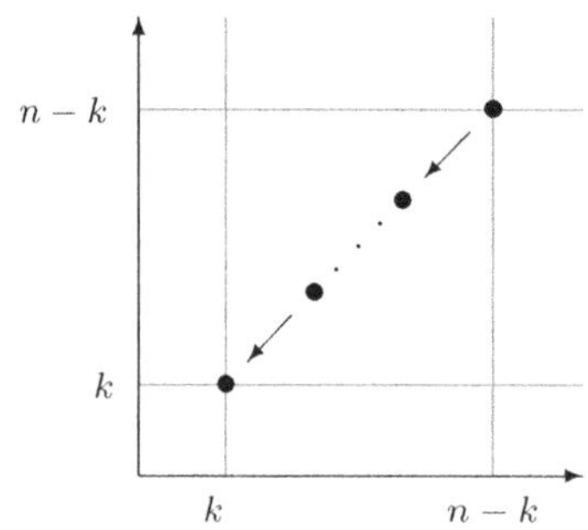

$$H^0_{k,d}(n-k) \to \cdots \to H^0_{k,d}(k)$$
$$\dim H^0_{k,d}(s) = d$$

Figure V.1: The atomic limiting mixed Hodge structures $(V_{k,d}, W_\bullet(N_{k,d}), F^\bullet_{k,d})$

Set

$$V_{k,d}(\mathbb{R}) = \bigoplus_{k \le s \le n-k} H^0_{k,d}(s),$$

and define a nilpotent $N_{k,d} \in \mathrm{End}(V_{k,d}, \mathbb{R})$ by

$$N_{k,d}(e^s_a) = \begin{cases} e^{s-1}_a, & s > k, \\ 0, & s = k. \end{cases}$$

Note that

$$N_{k,d}\left(H^0_{k,d}(s)\right) = \begin{cases} H^0_{k,d}(s-1), & s > k, \\ 0, & s = k. \end{cases}$$

Define a $(-1)^n$–symmetric $Q_{k,d}$ on $V_{k,d}$ by

$$(-1)^{n-k-s} Q(e^s_a, e^t_b) = \delta_{ab}\, \delta^{s+t}_n.$$

Then $N_{k,d} \in \mathrm{End}(V_{k,d}, Q_{k,d})$ is nilpotent with weight filtration

$$W_\ell(N_{k,d}) = \mathrm{span}\{e^s_a \ : \ 1 \le a \le d,\ s \le \ell\}.$$

Setting

$$F^p_{k,d} = \mathrm{span}\{e^s_a \ : \ 1 \le a \le d, s \ge p\}$$

defines an $\mathbb{R}$–split, polarized, Hodge–Tate limiting mixed Hodge structure $(V_{k,d}, W_\bullet(N_{k,d}), F^\bullet_{k,d})$, with dimensions $\mathbf{i}_{k,d}$.

We now define $(V, W_\bullet(N), F^\bullet)$ as follows. Set

$$d_k = \begin{cases} h^{n,0}, & k = 0, \\ h^{n-k,k} - h^{n-k+1,k-1}, & 1 \le k \le m, \end{cases}$$

and

$$V = \bigoplus_{0 \le k \le m} V_{k,d_k},$$
$$N = \bigoplus_{0 \le k \le m} N_{k,d_k},$$
$$F^p = \bigoplus_{0 \le k \le m} F^p_{k,d_k}.$$

Then $(V, W_\bullet(N), F^\bullet)$ is an $\mathbb{R}$–split, polarized, Hodge–Tate limiting mixed Hodge structure with dimensions

$$\mathbf{i} = \sum \mathbf{i}_{k,d_k} = \mathbf{h}.$$

Therefore $e^{zN} \cdot F^\bullet \in D$ for all $\mathrm{Im}(z) > 0$. □

Remark V.16. Theorem V.15 is false for for Hodge domains in general. That is, the inequalities (V.8) do not imply the existence of a Hodge–Tate degeneration. To see this suppose that (V, Q) and $(\tilde{V}, \tilde{Q})$ are two Hodge representations of G realizing $G_\mathbb{R}/R$ as a Hodge domain (Section II.D.4); denote the two (isomorphic) realizations by D and $\tilde{D}$.

(a) Any limiting mixed Hodge structure $(V, W_\bullet(N), F^\bullet)$ on (V, Q) determines a limiting mixed Hodge structure $(\tilde{V}, \tilde{W}_\bullet(N), \tilde{F}^\bullet)$ on $(\tilde{V}, \tilde{Q})$. Moreover, these two limiting mixed Hodge structures induce the same limiting mixed Hodge structure $(\mathfrak{g}, W_\bullet(N)_\mathfrak{g}, F^\bullet_\mathfrak{g})$.

(b) The reduced limit period mapping Φ_∞ sends $(F^\bullet, N)$ to the closed orbit if and only if it sends $(\tilde{F}^\bullet, N)$ to the closed orbit.

(c) By Proposition I.15 the limiting mixed Hodge structure $(V, W_\bullet(N), F^\bullet)$ is Hodge–Tate if and only if $(\tilde{V}, \tilde{W}_\bullet(N), \tilde{F}^\bullet)$ is Hodge–Tate.

(d) Suppose that (V.8) fails for $(\tilde{V}, \tilde{Q})$. Then $\tilde{D}$, as the Hodge domain for $(\tilde{V}, \tilde{Q})$, contains *no* point admitting a Hodge–Tate degeneration. Therefore, even if (V.8) holds for (V, Q), the Hodge domain D can not have a point admitting a Hodge–Tate degeneration. So to disprove the theorem if suffices to exhibit a pair (V, Q) and $(\tilde{V}, \tilde{Q})$ such that (V.8) holds for one but not the other.

Here is a counter–example.

Example V.17. Fix a symmetric nondegenerate bilinear form Q on $\mathbb{R}^5$ of signature $(1, 4)$. Set $G_\mathbb{R} = \mathrm{SO}(1, 4) = \mathrm{Aut}(\mathbb{R}^5, Q)$. Then $G_\mathbb{C} = \mathrm{SO}(5, \mathbb{C})$. Let the compact dual $\check{D} = G_\mathbb{C}/P = \mathrm{Gr}^Q(2, \mathbb{C}^5)$ be the variety of Q–isotropic 2–planes in $\mathbb{C}^5$, and let

D be the period domain parameterizing Q–polarized Hodge structures on $\mathbb{C}^5$ with Hodge numbers $\mathbf{h} = (2,1,2)$. Note that (V.8) fails. On the other hand the induced $Q_{\mathfrak{g}}$–polarized Hodge structure on $\mathfrak{g}_{\mathbb{R}}$ has Hodge numbers $\mathbf{h}_{\mathfrak{g}} = (1,2,4,2,1)$; in this case (V.8) holds.

V.D Non–Hodge–Tate degenerations in period domains

We recall and prove

Theorem I.8. *Let D be a period domain parameterizing weight n Hodge structures. If there exists a limiting mixed Hodge structure $(V, W_\bullet(N), F^\bullet)$ that maps to the closed $G_{\mathbb{R}}$–orbit in $\check{D}$, but is* not *of Hodge–Tate type, then $n = 2m$ is even and:*

(a) *For $k \neq 0$, $\mathrm{Gr}^{W_\bullet(N)}_{n+k,\mathrm{prim}}$ is of Hodge–Tate type.* (Thus k is even.)

(b) *For $k \neq 0$, $\mathrm{Gr}^{W_\bullet(N)}_{n+k,\mathrm{prim}} \neq 0$ implies $k \equiv 2$ (mod) 4.*

(c) $\mathrm{Gr}^{W_\bullet(N)}_{n,\mathrm{prim}} \neq 0$, *and the only nonzero $I^{p,q}_{\mathrm{prim}}$, with $p+q=n$, are*

$$I^{m+1,m-1}_{\mathrm{prim}} \quad \textit{and} \quad I^{m-1,m+1}_{\mathrm{prim}} .$$

The proof of Theorem I.8 proceeds in four steps.

Step 1: odd weight

For odd weight $n = 2m+1$, we have $G_{\mathbb{R}} = \mathrm{Sp}(2g, \mathbb{R})$. This real form is $\mathbb{R}$–split; whence the closed orbit $\mathcal{O}_{\mathrm{cl}}$ is totally real. It follows from Theorem I.16 that, if a degeneration to $\mathcal{O}_{\mathrm{cl}}$ exists, then the limiting mixed Hodge structure must be Hodge–Tate. So from this point on we assume that

$$n = 2m .$$

Step 2: preliminaries for even weight

By Theorem I.17 we are given the form of the induced limiting mixed Hodge structure $(\mathfrak{g}, W_\bullet(N)_{\mathfrak{g}}, F^\bullet_{\mathfrak{g}})$. The issue to is extract from the inclusion

$$\mathfrak{g} \subset \mathrm{End}(V, Q)$$

the form of original limiting mixed Hodge structure $(V, W_\bullet(N), F^\bullet)$. So, the question is: *given the Deligne splitting for $(\mathfrak{g}, W_\bullet(N)_{\mathfrak{g}}, F^\bullet_{\mathfrak{g}})$, what can we infer about the Deligne splitting for $(V, W_\bullet(N), F^\bullet)$?* An especially convenient way to do this is to use the relationship between the weights of $V_{\mathbb{C}}$ and the roots of $\mathfrak{g}_{\mathbb{C}} = \mathrm{End}(V_{\mathbb{C}}, Q)$; this will amount to comparing eigenvalues between the two spaces.

Define $L \in \mathfrak{g}_{\mathbb{C}}$ by

$$L|_{I^{p,q}_{\mathfrak{g}}} = q\,\mathbb{1} .$$

Then (III.8) implies the Deligne splitting is the $(L, \overline{L})$–eigenspace decomposition of $\mathfrak{g}_\mathbb{C}$,

$$(L, \overline{L})\big|_{I^{p,q}_\mathfrak{g}} = (q, p)\,\mathbb{1}\,.$$

Likewise, in analogy with the discussion of Sections II.C.2 and II.D.2, the Deligne splitting of $V_\mathbb{C}$ is an $(L, \overline{L})$–eigenspace decomposition. To be precise, $(L, \overline{L})$ acts on $I^{p+m,q+m}$ by the scalars (q, p). It will be convenient to shift the bigrading by

$$\tilde{I}^{p,q} = I^{p+m,q+m}\,, \quad \text{so that} \quad (L, \overline{L})\big|_{\tilde{I}^{p,q}} = (q, p)\,\mathbb{1}\,.$$

Note that, like $\oplus I^{p,q}_\mathfrak{g}$, the splitting $\oplus \tilde{I}^{p,q}$ is symmetric about the line $p + q = 0$.

Let $\mathfrak{h}$ be a Cartan subalgebra containing the grading elements L and $\overline{L}$. We may assume without loss of generality that $\mathfrak{h}$ is closed under conjugation. Let $\Lambda(V) \subset \mathfrak{h}^*$ denote the weights of the standard representation $V_\mathbb{C}$, and given $\lambda \in \Lambda(V)$, let $V^\lambda \subset V_\mathbb{C}$ denote the weight space. Then the $(L, \overline{L})$–eigenvalues of $V_\mathbb{C}$ are $\{(L(\lambda), \overline{L}(\lambda) \ : \ \lambda \in \Lambda(V)\}$; equivalently,

$$\tilde{I}^{p,q} = \bigoplus_{\substack{L(\lambda) = q \\ \overline{L}(\lambda) = p}} V^\lambda\,. \tag{V.18a}$$

Likewise, the $(L, \overline{L})$–eigenvalues of $\mathfrak{g}_\mathbb{C}$ are $\{(\alpha(L), \alpha(\overline{L})) \ : \ \alpha \in \Delta \cup \{0\}\}$; that is,

$$I^{p,q}_\mathfrak{g} = \bigoplus_{\substack{\alpha(L) = q \\ \alpha(\overline{L}) = p}} \mathfrak{g}^\alpha \ \Big(\oplus\ \mathfrak{h} \quad \text{if} \quad (p, q) = (0, 0)\Big)\,. \tag{V.18b}$$

The relationship between the roots of $\mathfrak{g}_\mathbb{C}$ and the weights $\Lambda(V)$ of $V_\mathbb{C}$ is

$$\Delta \cup \{0\} = \{\alpha = \lambda + \mu \ : \ \lambda \neq \mu \in \Lambda(V)\}\,. \tag{V.18c}$$

This yields relationships between the $(L, \overline{L})$–eigenvalues on $V_\mathbb{C}$ and $\mathfrak{g}_\mathbb{C}$. In particular, (V.18) implies the following:

$$\begin{array}{l} \text{Suppose } \tilde{I}^{p,q}, \tilde{I}^{r,s} \neq 0 \text{ and } (p, q) \neq (r, s). \\ \text{Then } I^{p+q,r+s}_\mathfrak{g} \neq 0. \end{array} \tag{V.19}$$

Remark V.20 (Properties of $\tilde{I}^{p,q}$). In the arguments that follow it will be helpful to keep in mind that both $\tilde{I}^{p,q}$ and $I^{p,q}_\mathfrak{g}$ are symmetric about the $p + q = 0$ and $p - q = 0$ lines: if $\tilde{I}^{p,q}$ is nonzero, then so are $\tilde{I}^{q,p}$, $\tilde{I}^{-q,-p}$ and $\tilde{I}^{-p,-q}$. (Similarly for $I^{p,q}_\mathfrak{g}$.) In fact they all have the same dimension. The symmetry about the line $p - q = 0$ is due to the fact that both the roots Δ and the weights $\Lambda(V)$ are closed under conjugation (because $\mathfrak{h}$, $\mathfrak{g}$ and V are defined over $\mathbb{R}$). The symmetry about the line $p + q = 0$ is due to the fact that the N–strings are uninterrupted and centered on the line $p + q = 0$. This also implies the following: suppose that $\tilde{I}^{p,q}$ is nonzero. Then $\tilde{I}^{k,-k}$ is nonzero if $p - q = 2k$, and $\tilde{I}^{k+1,-k}$ and $\tilde{I}^{k,-k-1}$ are nonzero if $p - q = 2k + 1$.

Step 3: Suppose there exists $\tilde{I}^{p,q} \neq 0$ with $p - q = 2k + 1$

Then $\tilde{I}^{k+1,-k}, \tilde{I}^{k,-k-1} \neq 0$ (Remark V.20). So (V.19) implies $I_{\mathfrak{g}}^{2b+1,-2b-1} = 0$, and (V.2b) forces $c = \pm 1$. So, suppose that $\tilde{I}^{\pm 1,0}$ and $\tilde{I}^{0,\pm 1}$ are all nonzero. (As discussed in Remark V.20, if any one of the four is nonzero, then all four are nonzero.) If $\tilde{I}^{r,s} \neq 0$, then (V.19) implies

$$I_{\mathfrak{g}}^{r\pm 1,s} \quad \text{and} \quad I_{\mathfrak{g}}^{r,s\pm 1} \quad \text{are nonzero.}$$

Given (V.2c), this forces $|r - s| \leq 1$. Whence the Deligne splitting must be of the form depicted in Figure V.2. But this implies that (V.8) holds. Whence Theorem V.15 implies the limiting mixed Hodge structure is Hodge–Tate, contradicting our hypothesis. To conclude:

If $\tilde{I}^{p,q} \neq 0$, then $p - q$ is even.

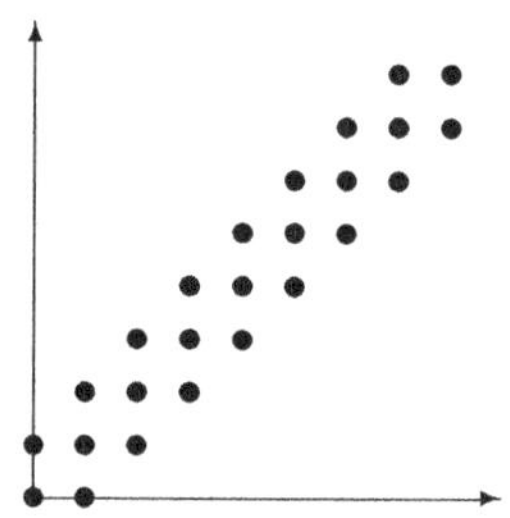

Figure V.2: The Deligne splitting $V_{\mathbb{C}} = \oplus I^{p,q}$.

Step 4: Suppose there exists $\tilde{I}^{p,q} \neq 0$ with $p - q = 2k > 0$

Then $\tilde{I}^{k,-k}$ and $\tilde{I}^{-k,k}$ are nonzero (Remark V.20). Since $N \neq 0$, there exists some nonzero $\tilde{I}^{r,s}$ with $r + s \neq 0$. Then (V.19) implies $I_{\mathfrak{g}}^{r-k,s+k}$ and $I_{\mathfrak{g}}^{r+k,s-k}$ are nonzero, and (V.2c) forces $|r - s \pm 2k| \leq 2$. By hypothesis $2k \geq 2$, so it must be the case that $k = 1$ and $r - s = 0$. Thus, the Deligne splitting $V_{\mathbb{C}} = \oplus I^{p,q}$ is as depicted in Figure V.3.a. (The induced Deligne splitting $\mathfrak{g}_{\mathbb{C}} = \oplus I_{\mathfrak{g}}^{p,q}$ is as in Figure V.3.b.) This establishes Theorem I.8(a).

The hypothesis that $(V, W_\bullet(N), F^\bullet)$ is not Hodge–Tate implies

$$\tilde{I}^{-1,1} = \tilde{I}^{-1,1}_{\text{prim}} \neq 0\,. \tag{V.21}$$

Moreover, if $\tilde{I}^{p,p}_{\text{prim}} \neq 0$, then (V.19) and (V.21) imply that $I^{p-1,p+1}_{\mathfrak{g},\text{prim}} \neq 0$. By Theorem V.1, this forces $p \geq 1$ to be odd. This establishes Theorem I.8(b). Finally, since $\tilde{I}^{-1,1} = I^{m-1,m+1}$, we see that (V.21) establishes Theorem I.8(c) and completes the proof.

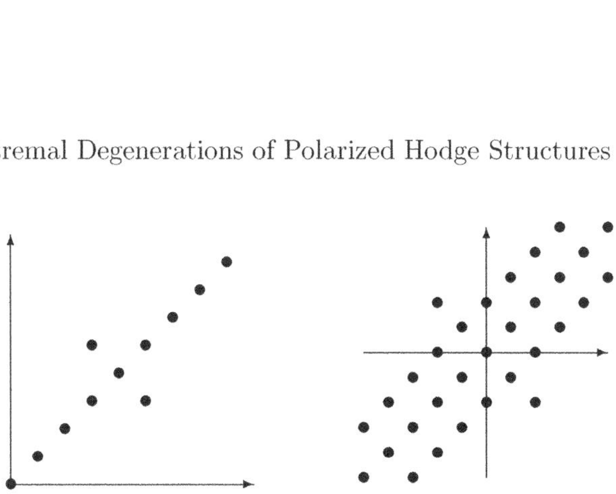

Figure V.3.a Figure V.3.b

Figure V.3: The Deligne splittings $V_{\mathbb{C}} = \oplus I^{p,q}$ and $\mathfrak{g}_{\mathbb{C}} = \oplus I_{\mathfrak{g}}^{p,q}$.

V.E All degenerations are induced from maximal Hodge–Tate degenerations

In a suitably interpreted sense all degenerations are induced from a (maximal) degeneration of Hodge–Tate type. Some care must be taken with this statement, as it is not necessarily the case that the underlying degeneration arises algebro–geometrically: this is a statement about the orbit structure and representation theory associated with the SL_2–orbit approximating an arbitrary degeneration, which may or may not arise algebro–geometrically.

Fix a Hodge domain $D \subset \check{D}$. Let $N \in \mathfrak{g}_{\mathbb{R}}$ be a nilpotent element and consider the corresponding boundary component $B(N) = \tilde{B}(N)/\exp(\mathbb{C}N)$. Given $F^\bullet \in \tilde{B}(N)$, let $(F^\bullet, W(N)_\bullet)$ denote the corresponding limiting mixed Hodge structure on $\mathfrak{g}$. Let $\mathfrak{g}_{\mathbb{C}} = \oplus I_{\mathfrak{g}}^{p,q}$ be the Deligne splitting. Without loss of generality, the limiting mixed Hodge structure is $\mathbb{R}$–split. The diagonal subalgebra

$$\mathfrak{s}_{\mathbb{C}} = \bigoplus_p I_{\mathfrak{g}}^{p,p}$$

is a conjugation stable subalgebra of $\mathfrak{g}_{\mathbb{C}}$ containing N. Let $\mathfrak{s}_{\mathbb{R}} = \mathfrak{s}_{\mathbb{C}} \cap \mathfrak{g}_{\mathbb{R}}$ denote the real form. Moreover, as the zero eigenspace for the grading element $L - \overline{L}$, the subalgebra $\mathfrak{s}_{\mathbb{R}}$ is necessarily a Levi subalgebra, and therefore reductive.

Lemma V.22. *The limiting mixed Hodge structure $(F_{\mathfrak{g}}^\bullet, W_\bullet(N)_{\mathfrak{g}})$ on $\mathfrak{g}_{\mathbb{R}}$ induces a sub-limiting mixed Hodge structure $(F_{\mathfrak{s}}^\bullet, W_\bullet(N)_{\mathfrak{s}})$ on $\mathfrak{s}_{\mathbb{R}}$ by*

$$F_{\mathfrak{s}}^\bullet = F^\bullet \cap \mathfrak{s}_{\mathbb{C}} = \bigoplus_{q \geq p} I_{\mathfrak{g}}^{q,q} \quad \text{and} \quad W_\bullet(N)_{\mathfrak{s}} = W_\bullet(N)_{\mathfrak{g}} \cap \mathfrak{s}_{\mathbb{R}} = \bigoplus_{q \leq p} I_{\mathfrak{g},\mathbb{R}}^{q,q} .$$

Proof. This follows directly from the definition of limiting mixed Hodge structures (Section II.F). □

Let $S_{\mathbb{C}} \subset G_{\mathbb{C}}$ be the connected Lie subgroup with Lie algebra $\mathfrak{s}_{\mathbb{C}}$, and set

$$\check{\mathcal{D}} = S_{\mathbb{C}} \cdot F^\bullet \quad \text{and} \quad \mathcal{D} = \check{\mathcal{D}} \cap D .$$

By [CKS], $F^\bullet_{\mathfrak{g}}(z) = e^{zN} F^\bullet_{\mathfrak{g}} \in D$ for all $\mathrm{Im}(z) > 0$; equivalently, the Hodge filtration $F^\bullet_{\mathfrak{g}}(z)$ defines a Hodge structure φ_z on $\mathfrak{g}_\mathbb{R}$ (Section II.C.1). Likewise, $F^\bullet_{\mathfrak{s}}(z) = e^{zN} F^\bullet_{\mathfrak{s}} \in \mathcal{D}$ defines a Hodge structure $\varphi_z|_{\mathfrak{s}}$ on $\mathfrak{s}_\mathbb{R}$ for all $\mathrm{Im}(z) > 0$. This implies

> *$\mathcal{D}$ carries the structure of a Hodge subdomain of D with compact dual $\check{\mathcal{D}}$.*

In particular, $\mathcal{D}$ is an open $S_\mathbb{R}$–orbit in $\check{\mathcal{D}}$. From Proposition I.7, and the fact that $(F^\bullet_{\mathfrak{s}}, W_\bullet(N)_{\mathfrak{s}})$ is Hodge–Tate, we see that

$$S_\mathbb{R} \cdot F^\bullet \text{ is the closed } S_\mathbb{R}\text{–orbit in } \check{\mathcal{D}}.$$

In this sense,

> *the nilpotent orbit $(F^\bullet_{\mathfrak{s}}, N)$ is a maximal degeneration of the Hodge structure on $\mathfrak{s}_\mathbb{R}$.*

So far we have viewed $\mathfrak{s}_\mathbb{R}$ as having sub–Hodge structures $\varphi_z|_{\mathfrak{s}}$ that are restrictions of Hodge structures φ_z on $\mathfrak{g}_\mathbb{R}$. In fact a stronger statement holds: the circle φ_z is contained in $S_\mathbb{R}$.

Lemma V.23. *The Hodge structure $(\mathfrak{g}, \varphi_z)$ is given by a Hodge representation of $S_\mathbb{R}$.*

In this sense,

> *the degeneration of Hodge structure on $\mathfrak{g}$ given by $(F^\bullet_{\mathfrak{g}}, N)$ is induced from a maximal degeneration of Hodge structure on $\mathfrak{s}_\mathbb{R}$.*

Proof. We need to show that the circle φ_z is contained in $S_\mathbb{C}$, cf. [14]. The corresponding grading element L_z (Sections II.C.3 and II.D.3) is equal to $\varphi'(1)/4\pi\mathbf{i}$, cf. [23, Section 2.3]. So $\varphi_z \subset S_\mathbb{R}$ if and only if L_z, a priori an element of $\mathfrak{g}_\mathbb{C}$, is an element of $\mathfrak{s}_\mathbb{C}$.

Decompose $\mathfrak{s}_\mathbb{C} = \mathfrak{z}_\mathbb{C} \oplus \mathfrak{s}^{\mathrm{ss}}_\mathbb{C}$ into its center $\mathfrak{z}_\mathbb{C}$ and semisimple factor $\mathfrak{s}^{\mathrm{ss}}_\mathbb{C} = [\mathfrak{s}_\mathbb{C}, \mathfrak{s}_\mathbb{C}]$. Since $\mathfrak{s}_\mathbb{R}$ is a sub–Hodge structure of $\mathfrak{g}_\mathbb{R}$, with respect to φ_z, it follows that $\mathfrak{s}^{\mathrm{ss}}_\mathbb{R}$ is also a sub–Hodge structure. Therefore, L_z determines a graded decomposition of $\mathfrak{s}^{\mathrm{ss}}_\mathbb{C}$. As discussed in Remark II.7, this graded decomposition is also induced by a grading element $L'_z \in \mathfrak{s}_\mathbb{C}$. It follows that $L_z - L'_z \in \mathfrak{g}_\mathbb{C}$ is contained in the centralizer of $\mathfrak{s}_\mathbb{C}$. It is here that the fact that $\mathfrak{s}_\mathbb{C}$ is a Levi subalgebra is key, for it is a well–known property of Levi subalgebras that the centralizer in $\mathfrak{g}_\mathbb{C}$ is equal to the center $\mathfrak{z}_\mathbb{C}$. Therefore, $L_z - L'_z \in \mathfrak{z}_\mathbb{C} \subset \mathfrak{s}_\mathbb{C}$. Whence $L_z \in \mathfrak{s}_\mathbb{C}$. □

Remark V.24. Since $\mathfrak{s}$ is reductive, $\mathfrak{g} = \mathfrak{s} \oplus \mathfrak{s}^\perp$ as a $\mathfrak{s}$–module. The idea here is that the essential structure/relationship is between N and the Levi subalgebra $\mathfrak{s}$; the remaining structure on $\mathfrak{g} = \mathfrak{s} \oplus \mathfrak{s}^\perp$, that is the structure on $\mathfrak{s}^\perp$, is induced from the $\mathfrak{s}$–module structure on $\mathfrak{s}^\perp$. This sort of idea does back to Bala and Carter's classification [4, 5] of nilpotent orbits $\mathcal{N} \subset \mathfrak{g}_\mathbb{C}$, where the idea is to look at minimal Levi subalgebras $\mathfrak{l}$ containing a fixed $N \in \mathcal{N}$, and to classify the pairs $(N, \mathfrak{l})$. (In fact, the idea goes back farther to Dynkin [11], who looked at minimal reductive subalgebras containing N, but this approach does not seem to work as well.)

References

[1] D. Abramovich and K. Karu. Weak semistable reduction in characteristic 0. *Invent. Math.*, 139(2):241–273, 2000.

[2] A. Ash, D. Mumford, M. Rapoport, and Y. Tai. *Smooth compactification of locally symmetric varieties*. Math. Sci. Press, Brookline, Mass., 1975. Lie Groups: History, Frontiers and Applications, Vol. IV.

[3] W. L. Baily, Jr. and A. Borel. Compactification of arithmetic quotients of bounded symmetric domains. *Ann. of Math. (2)*, 84:442–528, 1966.

[4] P. Bala and R. W. Carter. Classes of unipotent elements in simple algebraic groups. I. *Math. Proc. Cambridge Philos. Soc.*, 79(3):401–425, 1976.

[5] P. Bala and R. W. Carter. Classes of unipotent elements in simple algebraic groups. II. *Math. Proc. Cambridge Philos. Soc.*, 80(1):1–17, 1976.

[6] Andreas Čap and Jan Slovák. *Parabolic geometries. I*, volume 154 of *Mathematical Surveys and Monographs*. American Mathematical Society, Providence, RI, 2009. Background and general theory.

[7] James Carlson, Mark Green, and Phillip Griffiths. Variations of Hodge structure considered as an exterior differential system: old and new results. *SIGMA Symmetry Integrability Geom. Methods Appl.*, 5:Paper 087, 40, 2009.

[8] Eduardo Cattani, Aroldo Kaplan, and Wilfried Schmid. Degeneration of Hodge structures. *Ann. of Math. (2)*, 123(3):457–535, 1986.

[9] David H. Collingwood and William M. McGovern. *Nilpotent orbits in semisimple Lie algebras*. Van Nostrand Reinhold Mathematics Series. Van Nostrand Reinhold Co., New York, 1993.

[10] E. B. Dynkin. Semisimple subalgebras of semisimple Lie algebras. *Mat. Sbornik N. S.*, 30(72):349–462 (3 plates), 1952.

[11] E. B. Dynkin. Semisimple subalgebras of semisimple Lie algebras. *Amer. Math. Soc. Trans.*, 6:111–244, 1957. Translation of [10].

[12] Gregor Fels, Alan Huckleberry, and Joseph A. Wolf. *Cycle spaces of flag domains*, volume 245 of *Progress in Mathematics*. Birkhäuser Boston Inc., Boston, MA, 2006.

[13] M. Green, P. Griffiths, R. Laza, and C. Robles. Hodge theory and moduli of H-surfaces. In preparation.

[14] Mark Green, Phillip Griffiths, and Matt Kerr. *Mumford-Tate groups and domains: their geometry and arithmetic*, volume 183 of *Annals of Mathematics Studies*. Princeton University Press, Princeton, NJ, 2012.

[15] Mark Green, Phillip Griffiths, and Matt Kerr. *Hodge theory, complex geometry, and representation theory*, volume 118 of *CBMS Regional Conference Series in Mathematics*. Published for the Conference Board of the Mathematical Sciences, Washington, DC, 2013.

[16] Kazuya Kato and Sampei Usui. *Classifying spaces of degenerating polarized Hodge structures*, volume 169 of *Annals of Mathematics Studies*. Princeton University Press, Princeton, NJ, 2009.

[17] Matt Kerr and Gregory Pearlstein. Boundary components of Mumford-Tate domains. *Duke Math. J.*, 165(4):661–721, 2016.

[18] Matt Kerr and Gregory Pearlstein. Naive boundary strata and nilpotent orbits. *Ann. Inst. Fourier (Grenoble)*, 64(6):2659–2714, 2014.

[19] Matt Kerr and Colleen Robles. Hodge theory and real orbits in flag varieties. arXiv:1407.4507, 2014.

[20] Matt Kerr, Gregory Pearlstein, and Colleen Robles. Polarized relations on horizontal SL(2)'s. arXiv = 1705.03117.

[21] János Kollár. Moduli of varieties of general type. In *Handbook of moduli. Vol. II*, volume 25 of *Adv. Lect. Math. (ALM)*, pages 131–157. Int. Press, Somerville, MA, 2013.

[22] Bertram Kostant. The principal three-dimensional subgroup and the Betti numbers of a complex simple Lie group. *Amer. J. Math.*, 81:973–1032, 1959.

[23] C. Robles. Schubert varieties as variations of Hodge structure. *Selecta Math. (N.S.)*, 20(3):719–768, 2014.

[24] Colleen Robles. Classification of horizontal SL(2)s. *Compos. Math.*, 152(5):918–954, 2016.

[25] Ichirǫ Satake. On compactifications of the quotient spaces for arithmetically defined discontinuous groups. *Ann. of Math. (2)*, 72:555–580, 1960.

[26] Wilfried Schmid. Variation of Hodge structure: the singularities of the period mapping. *Invent. Math.*, 22:211–319, 1973.

[27] Joseph Steenbrink. Limits of Hodge structures. *Invent. Math.*, 31(3):229–257, 1975/76.

[28] Joseph A. Wolf. The action of a real semisimple group on a complex flag manifold. I. Orbit structure and holomorphic arc components. *Bull. Amer. Math. Soc.*, 75:1121–1237, 1969.

Hodge Theory and L^2-analysis
ALM 39, pp. 377–418

Deligne–Beilinson Cohomology of Affine Groups*

Richard Hain[†]

1 Introduction

The purpose of these notes is to develop the basic theory of Deligne–Beilinson cohomology of affine group schemes that are endowed with a mixed Hodge structure. These results are needed in the works [12] on the Hodge theory of modular groups and [13] on universal mixed elliptic motives. The results here generalize results in the unipotent case, which were established in [4]. One motivation for this work is to interpret Brown's computation [2] of the periods of iterated iterated integrals of Eisenstein series in terms of Deligne cohomology.

This work is built on several fundamental results proved by Zucker in the 1980s—most notably, his construction in [27] of a mixed Hodge complex for computing the mixed Hodge structure (MHS) on the cohomology of an affine curve with coefficients in a polarized variation of MHS, and his paper [25] with Joseph Steenbrink in which they define admissible variations of MHS over a curve. These foundational works were generalized from curves to higher dimensional varieties by Saito [24] and Kashiwara [20].

By an *affine group*, we mean an affine group scheme over a field of characteristic zero. The ring of functions on an affine group G will be denoted by $\mathcal{O}(G)$. Such a group is *algebraic* if $\mathcal{O}(G)$ is finitely generated. Every affine group over the field $\mathbb{F}$ is a proalgebraic $\mathbb{F}$-group; that is, it is an inverse limit of linear affine algebraic $\mathbb{F}$-groups, [26, p. 24].

The theory of cohomology of algebraic groups was developed by Hochschild [16] and Hochschild–Serre [17]. It generalizes to affine groups [18]. A modern

*To Steven Zucker on the occasion of his 65th birthday; Supported in part by grant DMS-1406420 from the National Science Foundation and by the Friends of the Institute for Advance Study.

[†]Department of Mathematics, Duke University, Durham, NC 27708-0320. E-mail: hain@math.duke.edu

exposition can be found in the early chapters of Jantzen's book [19], which we use as our basic reference.

Denote the category of representations of an affine $\mathbb{F}$-group G by $\mathsf{Rep}(G)$. We refer to objects of $\mathsf{Rep}(G)$ as G-modules. The most direct definition of the cohomology of an affine group G with coefficients in the G-module V is

$$H^\bullet(G,V) := \mathrm{Ext}^\bullet_{\mathsf{Rep}(G)}(\mathbb{F},V),$$

where the right hand side denotes the group of Yoneda extensions of the trivial G-module $\mathbb{F}$ by V in $\mathsf{Rep}(G)$.

Suppose now that $\mathbb{F}$ is a subfield of $\mathbb{R}$. A MHS on an affine $\mathbb{F}$-group G is an ind-MHS on its coordinate ring $\mathcal{O}(G)$ that is respected by its Hopf algebra operations. A Hodge representation of the affine group G with an $\mathbb{F}$-MHS is an $\mathbb{F}$-MHS V whose underlying vector space is a G-module with the property that the coaction

$$V \to V \otimes \mathcal{O}(G)$$

is a morphism of MHS. Denote the category of Hodge representations of G by $\mathsf{HRep}(G)$. It is an abelian category. The cohomology $H^\bullet(G,V)$ groups of G with coefficients in a Hodge representation have a natural ind-MHS. (Cf. Proposition 4.1.)

Suppose that G is an affine $\mathbb{F}$-group with a MHS. The Deligne–Beilinson cohomology of G with coefficients in the Hodge representation V, defined in Section 4, is isomorphic to the Yoneda ext group

$$H^\bullet_{\mathcal{D}}(G,V) := \mathrm{Ext}^\bullet_{\mathsf{HRep}(G)}(\mathbb{F},V),$$

where $\mathbb{F}$ denotes the Hodge structure $\mathbb{F}(0)$, and sits in an exact sequence

$$0 \to \mathrm{Ext}^1_{\mathsf{MHS}}(\mathbb{F}, H^{m-1}(G,V)) \to H^m_{\mathcal{D}}(G,V) \to \Gamma H^m(G,V) \to 0$$

of vector spaces, where $\Gamma = \mathrm{Hom}_{\mathsf{MHS}}(\mathbb{F},\ )$.

Affine groups with MHS and their Hodge representations arise as follows. Suppose that X is a smooth complex algebraic variety and that $\mathbb{H}$ is a polarized variation of $\mathbb{F}$-Hodge structures over X. Fix a base point $x \in X$. Denote the fiber of $\mathbb{H}$ over x by H_x. The Zariski closure of the image of the monodromy representation

$$\rho_x : \pi_1(X,x) \to \mathrm{Aut}(H_x)$$

is a reductive $\mathbb{F}$-group, which we denote by R_x. The completion of $\pi_1(X,x)$ relative to ρ_x is a proalgebraic (and thus an affine) $\mathbb{F}$-group $\mathcal{G}_x$. It has a natural MHS ([9]).

Denote by $\mathsf{MHS}(X,\mathbb{H})$ the category of admissible variations $\mathbb{V}$ of MHS over X with the property that the monodromy action of $\pi_1(X,x)$ on each weight graded quotient $\mathrm{Gr}^W_m V_x$ of its fiber over x factors through an action of R_x on $\mathrm{Gr}^W_m V_x$. Restricting to the fiber over the base point defines an equivalence of categories

$$\mathsf{MHS}(X,\mathbb{H}) \to \mathsf{HRep}(\mathcal{G}_x).$$

This statement generalizes the main result of [15] of the author and Zucker.[1] Combined with the result above, this result establishes that there is an isomorphism

$$H^\bullet_\mathcal{D}(\mathcal{G}_x, V) \cong \mathrm{Ext}^\bullet_{\mathsf{MHS}(X,\mathbb{H})}(\mathbb{F}, \mathbb{V})$$

that is compatible with all natural products.

In Section 6 we prove that, for all $\mathbb{V}$ in $\mathsf{MHS}(X, \mathbb{H})$, the natural homomorphism

$$H^\bullet(\mathcal{G}_x, V_x) \to H^\bullet(X, \mathbb{V})$$

is a morphism of MHS, where the right-hand group has Saito's MHS [23]. In Section 7, we lift this to a natural map

$$\theta : H^\bullet_\mathcal{D}(\mathcal{G}_x, V_x) \to H^\bullet_\mathcal{D}(X, \mathbb{V})$$

from the Deligne–Beilinson cohomology of $\mathcal{G}_x$ to the Deligne–Beilinson cohomology of X and show that it is an isomorphism in degrees 0, 1 and injective in degree 2.

Denote the category of all admissible variations of $\mathbb{F}$-MHS over X by $\mathsf{MHS}_\mathbb{F}(X)$. The following is a combination of Theorem 7.5 and its corollary.

Theorem 1. *If X is a quasi-projective manifold and $\mathbb{V}$ is an admissible variation of $\mathbb{F}$-MHS over X, there there is a homomorphism*

$$\mathrm{Ext}^\bullet_{\mathsf{MHS}(X)}(\mathbb{F}, \mathbb{V}) \to H^\bullet_\mathcal{D}(X, \mathbb{V})$$

that is compatible with products. It is an isomorphism in degrees ≤ 1 and injective in degree 2 for all X. When X is an affine curve, it is an isomorphism in all degrees.

When $\mathbb{V}$ is a polarized variation of $\mathbb{Z}$-Hodge structure over X of negative weight, the corresponding group of admissible normal functions is, by definition, the group of extensions of $\mathbb{Z}$ by $\mathbb{V}$ in the category of admissible variations of $\mathbb{Z}$-MHS over X. These are holomorphic sections of the corresponding bundle of intermediate jacobians $J(\mathbb{V})$ over X with good asymptotic properties. Each admissible normal function determines a class in $H^1(X, \mathbb{V})$. A special case of the previous result is the following modest generalization of a result [28] of Zucker.

Corollary 2. *The group of admissible normal function sections of $J(\mathbb{V})$, after tensoring with $\mathbb{Q}$, is isomorphic to $H^1_\mathcal{D}(X, \mathbb{V}_\mathbb{Q})$. Its image in $H^1(X, \mathbb{V})$ is the group $\Gamma H^1(X, \mathbb{V}_\mathbb{Q})$ of Hodge classes of type $(0, 0)$ in $H^1(X, \mathbb{V})$.*

The primary motivation for this work was to interpret Francis Brown's computation [2] of the periods of twice iterated integrals of Eisenstein series in terms of the cup product in Deligne cohomology. This is achieved in Sections 8 and 10. While not apparent in this paper, this interpretation uses Zucker's work on the Hodge theory of modular curves [27]. This topic is fully explained in [12].

Acknowledgments: I am grateful to Francis Brown for patiently explaining his period computations [2] and for his many helpful comments and corrections. The author gratefully acknowledges the support of the Friends of the Institute for Advanced Study which allowed him to spend the 2014–15 academic year at the Institute.

[1] A proof of the more general statement for affine curves is sketched in [12]. A detailed proof of the general case will be given in [14].

2 Conventions

Unless otherwise stated, all modules over a group G are *left* G-modules. If G is an affine group and V is a (left) G-module, then V is naturally a *right* comodule over $\mathcal{O}(G)$:

$$\nabla_V : V \to V \otimes \mathcal{O}(G).$$

The category of representations of an affine $\mathbb{F}$-group will be denoted by $\mathsf{Rep}(G)$. The category of representations of $G \otimes K$, where K is an extension field of $\mathbb{F}$, will be denoted by $\mathsf{Rep}_K(G)$.

We will use the algebraists' convention for path multiplication, which means that paths are composed in the "functional order". Suppose that X is a topological space. The composition $\gamma\mu$ of two paths $\gamma, \mu : [0,1] \to X$ is defined when $\mu(1) = \gamma(0)$. The path $\gamma\mu$ first traverses μ and then γ. We use this convention so that fundamental groups are compatible with Tannaka theory. The fundamental groupoid Π_X of X is the category whose objects are the points of X and where $\mathrm{Hom}(x,y)$ is the set of homotopy classes of paths from x to y. With our path multiplication convention, there is an equivalence of categories between local systems over X and *left* $\pi_1(X,x)$ modules, provided that X is path connected.

When $\mathbb{F}$ is a subfield of the real numbers, we will denote the category of $\mathbb{F}$ mixed Hodge structures by $\mathsf{MHS}_{\mathbb{F}}$ and the trivial Hodge structure $\mathbb{F}(0)$ by $\mathbb{F}$. For an $\mathbb{F}$-MHS A, set

$$\Gamma A := \mathrm{Hom}_{\mathsf{MHS}}(\mathbb{F}, A).$$

For a subring L of $\mathbb{C}$ that contains $\mathbb{F}$, A_L will denote the L-module that underlies A.

Unless otherwise stated, all varieties, with the exception of the affine groups, are defined over $\mathbb{C}$ and are of finite type.

3 Cohomology of affine groups

The cohomology theory of linear algebraic groups was initiated by Hochschild [16] and Hochschild–Serre [17]. This section is a quick review of the generalization of this theory to affine groups. It is mainly distilled from [19, Chapts. 3,4]. The purpose of this section is to present the theory in such a way that we will be able to check, in later sections, that various basic constructions in the theory are compatible with Hodge theory.

3.1 Preliminaries

Suppose that $\mathbb{F}$ is a field of characteristic zero, and that G is an affine group over $\mathbb{F}$. Denote the coordinate ring of G by $\mathcal{O} = \mathcal{O}(G)$. Denote the category of G-modules by $\mathsf{Rep}(G)$. Every G-module V is a direct limit of its finite dimensional submodules. This is an immediate consequence of the fact that if the image of $v \in V$ under the coaction

$$\Delta_V : V \to V \otimes \mathcal{O}$$

is $\sum_j v_j \otimes f_j$, then the smallest submodule of V that contains v is contained in the span of the v_j.

If V is a G-module, then the composite

$$V \xrightarrow{\Delta_V} V \otimes \mathcal{O} \xrightarrow{\mathrm{id}_V \otimes \epsilon} V \otimes \mathbb{F} \xrightarrow{\simeq} V$$

is the identity, where $\epsilon : \mathcal{O} \to F$ denotes the augmentation (i.e., evaluation at the identity e). It follows that Δ_V is an inclusion.

Regard $\mathcal{O}$ as a left G-module via right translation. On R-rational points, where R is an $\mathbb{F}$-algebra, this is given by

$$g : f \mapsto \{x \mapsto f(xg)\}, \quad f \in \mathcal{O} \text{ and } g, x \in G(R).$$

The corresponding right comodule is given by the coproduct $\Delta : \mathcal{O} \to \mathcal{O} \otimes \mathcal{O}$. A left G-module homomorphism $\phi : V \to \mathcal{O}$ is determined by the composite $r_\phi := \epsilon \circ \phi$

$$V \xrightarrow{\phi} \mathcal{O} \xrightarrow{\epsilon} \mathbb{F}$$

of ϕ with the augmentation. One recovers ϕ as the composite

$$V \xrightarrow{\Delta_V} V \otimes \mathcal{O} \xrightarrow{r_\phi \otimes \mathrm{id}_\mathcal{O}} \mathbb{F} \otimes \mathcal{O} \xrightarrow{\simeq} \mathcal{O}.$$

This implies easily that $\mathcal{O}$ is an injective object of $\mathsf{Rep}(G)$. Indeed, given a diagram

$$\begin{array}{ccccc} 0 & \longrightarrow & A & \xrightarrow{\psi} & B \\ & & & \searrow^{\phi_A} & \vdots\,\phi_B \\ & & & & \mathcal{O} \end{array}$$

of G-modules with exact top row, choose a lift $r_B : B \to \mathbb{F}$ of $r_A := r_{\phi_A} : A \to \mathbb{F}$. The lift ϕ_B is the composite $(r_B \otimes \mathrm{id}_\mathcal{O}) \circ \Delta_B : B \to \mathcal{O}$.

A consequence of this computation is that if V is a G-module, then the map $\Delta_V : V \hookrightarrow V \otimes \mathcal{O}$ is an embedding of V into an injective G-module. That is, $\mathsf{Rep}(G)$ has enough injectives. More generally, one has:

Lemma 3.1. *If M is an $\mathbb{F}$-vector space, then the right $\mathcal{O}$-comodule $M \otimes \mathcal{O}$ is an injective left G-module. For each G-module A, G-module maps $\phi : A \to M \otimes \mathcal{O}$ correspond bijectively with $\mathbb{F}$-linear maps $r : A \to M$. Two such maps correspond if and only if the diagram*

$$\begin{array}{ccc} A & \xrightarrow{\phi} & M \otimes \mathcal{O} \\ & \searrow^{r} & \downarrow \mathrm{id}_M \otimes \epsilon \\ & & M \end{array}$$

commutes.

Extensions of a G-module homomorphism $\phi_A : A \to M \otimes \mathcal{O}$ from a G-submodule A of a G-module B correspond to $\mathbb{F}$-linear extensions $r_B : B \to M$ of $(\mathrm{id}_M \otimes \epsilon) \circ \phi : A \to M$.

Remark 3.2. As is standard, right G-modules can be converted into left G-modules using inverses. This functor takes injectives to injectives. In particular, $\mathcal{O}(G)$ is injective when viewed as both a right and as a left G-module. In particular, it is injective with respect to the action defined by

$$(gf)(x) := f(g^{-1}x), \quad g, x \in G(R),\ f \in \mathcal{O}(G).$$

3.2 Cohomology

The cohomology of G with coefficients in the G-module V is defined to be the group

$$H^\bullet(G, V) = \operatorname{Ext}^\bullet_{\mathsf{Rep}(G)}(\mathbb{F}, V)$$

of Yoneda extensions. The fact that every G-module is a direct limit of its finitely generated submodules implies that if A and B are finite dimensional G-modules, then the natural map

$$\operatorname{Ext}^\bullet_{\mathsf{Rep}^{\mathsf{f}}(G)}(A, B) \to \operatorname{Ext}^\bullet_{\mathsf{Rep}(G)}(A, B) \tag{3.1}$$

is an isomorphism, where $\mathsf{Rep}^{\mathsf{f}}(G)$ denotes the category of finite dimensional G-modules. In particular,

$$H^\bullet(G, V) \cong \operatorname{Ext}^\bullet_{\mathsf{Rep}^{\mathsf{f}}(G)}(\mathbb{F}, V)$$

whenever V is a finite dimensional G-module.

Since $\mathsf{Rep}(G)$ has enough injectives,[2] $H^\bullet(G, V)$ can be defined by taking an injective resolution $V \to I^\bullet$ of V and then taking the homology of its invariants:

$$H^\bullet(G, V) = H^\bullet(\operatorname{Hom}_G(\mathbb{F}, I^\bullet)) = H^\bullet\big((I^\bullet)^G\big).$$

Remark 3.3. Note that an algebraic group G is reductive if and only if $H^j(G, V)$ vanishes for all G-modules V and all $j > 0$.

3.3 Standard cochains

Denote by $E_\bullet G$ the simplicial scheme whose scheme of n-simplices $E_n G$ is G^{n+1}. The group G acts on each $E_n G$ diagonally on the right. The face maps

$$d_j : E_n G \to E_{n-1} G,\ j \in \{0, \dots, n\},\ n > 0$$

are defined by $d_j(x_0, \dots, x_n) = (x_0, \dots, \widehat{x}_j, \dots, x_n)$. They are G-equivariant.

Dually, one has the cosimplicial object $\mathcal{O}^{\otimes \bullet}$ of $\mathsf{Rep}(G)$ whose module in degree n is

$$\mathcal{O}(E_n G) = \mathcal{O}(G^{n+1}) \cong \mathcal{O}^{\otimes(n+1)}.$$

The map $G \times G^n \to G^{n+1} = E_n G$ defined by

$$\big(g, (g_1, \dots, g_n)\big) \mapsto (g, gg_1, gg_1g_2, \dots, gg_1 \cdots g_n)$$

[2] Note that $\mathsf{Rep}^{\mathsf{f}}(G)$ does not have enough injectives when G is non-trivial.

is G-equivariant, where G acts on $G \times G^n$ by left multiplication on the first factor and trivially on G^n. The discussion above implies that each $\mathcal{O}(E_nG) \cong \mathcal{O}(G^n) \otimes \mathcal{O}(G)$ is an injective left G-module with respect to the action[3]

$$(gf)(g_0, \dots, g_n) = f(g^{-1}g_0, \dots, g^{-1}g_n), \quad f \in \mathcal{O}(G^{n+1}).$$

Denote the coface maps of $\mathcal{O}^{\otimes \bullet}$ by $d^j : \mathcal{O}^{\otimes n} \to \mathcal{O}^{\otimes(n+1)}$. Define a differential

$$d : \mathcal{O}^{\otimes n} \to \mathcal{O}^{\otimes(n+1)}$$

by $d = \sum_{0 \le j \le n} (-1)^j d^j$. The section $\sigma : E_nG \to E_{n+1}G$ defined by $\sigma(x_0, \dots, x_n) = (e, x_0, \dots, x_n)$ induces the G-module map $s : \mathcal{O}^{\otimes(n+1)} \to \mathcal{O}^{\otimes n}$ defined by

$$s(f_0 \otimes f_1 \otimes \cdots \otimes f_n) \mapsto \epsilon(f_0)\, f_1 \otimes \cdots \otimes f_n.$$

It satisfies $ds + sd = \mathrm{id} - \epsilon$. Consequently, the inclusion $V \to V \otimes \mathcal{O}(E_0G) = V \otimes \mathcal{O}$ of the constant V-valued functions gives an injective resolution $0 \to V \to V \otimes \mathcal{O}^{\otimes \bullet}$ of V as a left G-module.

Thus $H^\bullet(G, V)$ can be computed as the cohomology of the complex

$$\mathcal{C}^\bullet(G, V) := \mathrm{Hom}_G(\mathbb{F}, V \otimes \mathcal{O}^{\otimes \bullet}) = \mathrm{Mor}_G(E_\bullet G, V), \tag{3.2}$$

where G acts on the left of $E_\bullet G$ via $g : (g_0, \dots, g_n) \mapsto (gg_0, \dots, gg_n)$. Identifying E_nG with $G \times G^n$ as above, we see that $H^\bullet(G, V)$ can be computed using the standard bar complex whose cochains in degree n are

$$\mathcal{C}^n(G, V) = \mathrm{Mor}(G^n, V)$$

with the usual differential $\delta : \mathrm{Mor}(G^n, V) \to \mathrm{Mor}(G^{n+1}, V)$. Cf. Maclane [21, Ch. IV, §4] or Brown [3, Ch. I, §5].

Remark 3.4. The complex $\mathcal{C}^\bullet(G, V)$ is functorial in G and V, and exact in V. If $V^\bullet$ is a complex in $\mathsf{Rep}(G)$ that is bounded below, one can define $H^\bullet(G, V^\bullet)$ to be the hyper cohomology of the total complex associated to the double complex

$$\mathcal{C}^\bullet(G, V^\bullet).$$

Standard homological arguments imply that if $V_1^\bullet \to V_2^\bullet$ is a quasi-isomorphism, then $H^\bullet(G, V_1^\bullet) \to H^\bullet(G, V_2^\bullet)$ is an isomorphism.

Cohomology behaves well with respect to limits. Suppose that the affine group G is the inverse limit $\varprojlim_\alpha G_\alpha$ of algebraic groups G_α. Since

$$\mathcal{O}(G) = \varinjlim_\alpha \mathcal{O}(G_\alpha)$$

and since every G-module V can be written as a direct limit of G_α-modules V_α,

$$\left(V \otimes \mathcal{O}(G)^{\otimes \bullet}\right)^G = \varinjlim_\alpha \left(V \otimes \mathcal{O}(G_\alpha)^{\otimes \bullet}\right)^{G_\alpha}.$$

Since homology commutes with direct limits:

[3]This argument shows that $\mathcal{O}^{\otimes n}$ is an injective G module for all $n \ge 1$, where G acts diagonally. It implies that the tensor product $\mathcal{O}^{\otimes n} \otimes \mathcal{O}^{\otimes m}$ of two such G-modules is also injective.

Proposition 3.5. *If the affine group G is an inverse limit of algebraic groups G_α as above, then for all G-modules V, we can write $V = \varinjlim_\alpha V_\alpha$ where V_α is a G_α-module and there is a natural isomorphism*

$$\varinjlim H^\bullet(G_\alpha, V_\alpha) \xrightarrow{\simeq} H^\bullet(G, V).$$

3.4 The spectral sequence of a group extension

Suppose that N is a normal subgroup of the affine group G. The conjugation action of G on N induces an action of G on $H^\bullet(N,V)$. The usual argument shows that its restriction to N is trivial, so that $H^\bullet(N,V)$ is a G/N-module.

Taking a G-module V to its N-invariants V^N defines a functor $Q : \mathsf{Rep}(G) \to \mathsf{Rep}(G/N)$. It takes injectives to injectives. Denote the category of $\mathbb{F}$-vector spaces by $\mathsf{Vec}_\mathbb{F}$. Applying the Grothendieck spectral sequence to the diagram of functors

$$\mathsf{Rep}(G) \xrightarrow{Q} \mathsf{Rep}(G/N) \xrightarrow{\mathrm{Hom}_{G/N}(\mathbb{F},\ )} \mathsf{Vec}_\mathbb{F}, \qquad \mathsf{Rep}(G) \xrightarrow{\mathrm{Hom}_G(\mathbb{F},\ )} \mathsf{Vec}_\mathbb{F}$$

gives the spectral sequence for a group extension:

Proposition 3.6. *If N is a normal subgroup of the affine group G, then for all G-modules V, there is a spectral sequence satisfying*

$$E_2^{s,t} = H^s(G/N, H^t(N,V)) \implies H^{s+t}(G,V).$$

Corollary 3.7. *If the affine group G is an extension*

$$1 \to N \to G \to R \to 1$$

of a reductive group by an affine group N, then there is a natural isomorphism

$$H^\bullet(G,V) \cong H^\bullet(N,V)^R.$$

4 Mixed Hodge structures on affine group schemes

Throughout this section, $\mathbb{F}$ will be a subfield of $\mathbb{R}$. Denote the category of $\mathbb{F}$-mixed Hodge structures by $\mathsf{MHS}_\mathbb{F}$. It is a neutral tannakian category. Let $\omega^B : \mathsf{MHS}_\mathbb{F} \to \mathsf{Vec}_\mathbb{F}$ be the functor that takes a mixed Hodge structure to its underlying $\mathbb{F}$-vector space. It is a fiber functor. (The "Betti realization".) Set

$$\pi_1(\mathsf{MHS}_\mathbb{F}) := \mathrm{Aut}^\otimes \omega^B.$$

By Tannaka duality [8, Thm. 2.11], $\mathsf{Rep}(\pi_1(\mathsf{MHS}_\mathbb{F}))$ is equivalent to the category of ind-objects of $\mathsf{MHS}_\mathbb{F}$. The category $\mathsf{MHS}_\mathbb{F}$ is equivalent to $\mathsf{Rep}^\mathsf{f}(\pi_1(\mathsf{MHS}_\mathbb{F}))$ of finite dimensional $\pi_1(\mathsf{MHS}_\mathbb{F})$-modules.

An $\mathbb{F}$-mixed Hodge structure on an affine $\mathbb{F}$-group G is, by definition, an action of $\pi_1(\mathsf{MHS}_\mathbb{F})$ on G. Equivalently, the coordinate ring $\mathcal{O}(G)$ of G is a Hopf algebra in $\mathsf{Rep}(\pi_1(\mathsf{MHS}_\mathbb{F}))$, the category of ind-objects of $\mathsf{MHS}_\mathbb{F}$.

A *Hodge representation* of a group G with a mixed Hodge structure is an ind-object V of $\mathsf{MHS}_\mathbb{F}$ together with an action of G on V such that the coaction

$$\Delta_V : V \to V \otimes \mathcal{O}(G)$$

is a morphism in ind-$\mathsf{MHS}_\mathbb{F}$. Denote the category of Hodge representations of G by $\mathsf{HRep}(G)$.

If G has a mixed Hodge structure, then one can form the semi-direct product

$$\widehat{G} := \pi_1(\mathsf{MHS}_\mathbb{F}) \ltimes G,$$

which is also an affine $\mathbb{F}$-group. Note that $\mathsf{Rep}(\widehat{G}) = \mathsf{HRep}(G)$.

Proposition 4.1. *If G is an affine group with a mixed Hodge structure, and if V is a Hodge representation of G, then the cohomology group $H^\bullet(G,V)$ has a natural ind-MHS. It is compatible with products in the sense that if V_1 and V_2 are Hodge representations of G, then the product*

$$H^\bullet(G,V_1) \otimes H^\bullet(G,V_2) \to H^\bullet(G,V_1 \otimes V_2). \tag{4.1}$$

is a morphism of MHS.

Proof. Standard homological arguments (see Section 3.4) imply that $H^\bullet(G,V)$ is a $\widehat{G}/G = \pi_1(\mathsf{MHS}_\mathbb{F})$-module. So $H^\bullet(G,V)$ is an ind-object of $\mathsf{MHS}_\mathbb{F}$. They also imply that the product (4.1) is a $\pi_1(\mathsf{MHS}_\mathbb{F})$-module homomorphism. □

In particular, if V is in $\mathsf{HRep}(G)$, then V^G is a sub-MHS of V.

Proposition 4.2. *If G is an affine group with an $\mathbb{F}$-mixed Hodge structure and if V is a Hodge representation of G, then there is a natural exact sequence*

$$0 \to \mathrm{Ext}^1_{\mathsf{MHS}_\mathbb{F}}(\mathbb{F}, H^{j-1}(G,V)) \to \mathrm{Ext}^j_{\mathsf{HRep}(G)}(\mathbb{F}, V) \to \Gamma H^j(G,V) \to 0. \tag{4.2}$$

Proof. Since $\mathrm{Ext}^j_{\mathsf{MHS}_\mathbb{F}}(A,B)$ vanishes whenever $j > 1$, the result follows by applying Proposition 3.6 to the extension $1 \to G \to \widehat{G} \to \pi_1(\mathsf{MHS}_\mathbb{F}) \to 1$. □

4.1 Cochains

Here we dispose of several technical issues that arise in subsequent sections. Throughout this section, G is an affine group with an $\mathbb{F}$-MHS.

The first task is to give an alternative construction of the MHS on $H^\bullet(G,V)$ for each Hodge representation V of G. Note that $\mathsf{HRep}(G)$ has enough injectives as $\mathsf{HRep}(G) = \mathsf{Rep}(\widehat{G})$.

Definition 4.3 (cf. [4]). A *relatively injective G-module* is an object I of $\mathsf{HRep}(G)$ that is an injective G-module. A *relatively injective resolution* of an object V of $\mathsf{HRep}(G)$ is a resolution $V \to I^\bullet$ in $\mathsf{HRep}(G)$ which is an injective resolution of V in $\mathsf{Rep}(G)$. Equivalently, it is a resolution in $\mathsf{HRep}(G)$ in which each I^n is a relatively injective G-module.

Injective modules of the form $V \otimes \mathcal{O}(G)^{\otimes n}$, where V is in $\mathsf{HRep}(G)$, are relatively injective. The standard resolution $V \to V \otimes \mathcal{O}(G)^{\otimes \bullet}$ of V in $\mathsf{Rep}(G)$ is a relatively injective resolution of V in $\mathsf{HRep}(G)$.

Proposition 4.4. *If $V \to I^\bullet$ is a relatively injective resolution of V in $\mathsf{HRep}(G)$, then $(I^\bullet)^G$ is a complex in ind-$\mathsf{MHS}_\mathbb{F}$ and there is a natural isomorphism*

$$H^\bullet(G, V) \cong H^\bullet((I^\bullet)^G)$$

in ind-MHS_F.

Proof. Let $J^{s,t} = I^t \otimes \mathcal{O}(\widehat{G})^{\otimes(s+1)}$. Then $0 \to I^t \to J^{\bullet,t}$ is an injective resolution of I^t in $\mathsf{Rep}(\widehat{G})$, $V \to \operatorname{tot} J^{\bullet\bullet}$ is an injective resolution of V in $\mathsf{Rep}(\widehat{G})$, and $I^\bullet \to \operatorname{tot} J^{\bullet\bullet}$ is a quasi-isomorphism in $\mathsf{Rep}(\widehat{G})$.

Since $\mathcal{O}(\widehat{G}) \cong \mathcal{O}(G) \otimes \mathcal{O}(\pi_1(\mathsf{MHS}_\mathbb{F}))$ as a G-module, $\mathcal{O}(\widehat{G})$ is an injective G-module. Consequently, $V \to \operatorname{tot} J^{\bullet\bullet}$ is an injective resolution in $\mathsf{Rep}(G)$, which implies that

$$(I^\bullet)^G \to \operatorname{tot}(J^{\bullet\bullet})^G$$

is a quasi-isomorphism. It induces an isomorphism of MHS on cohomology. The right hand complex is a complex of $\widehat{G}/G \cong \pi_1(\mathsf{MHS})$-modules which computes the natural MHS on $H^\bullet(G, V)$. The result follows. □

The exactness properties of $\operatorname{Gr}^W_\bullet$ and $\operatorname{Gr}^\bullet_F$ imply that relatively injective modules have the following property:

Lemma 4.5. *Suppose that $0 \to A \to B$ is an exact sequence in $\mathsf{HRep}(G)$. If I is a relatively injective G-module, then the natural surjection*

$$\operatorname{Hom}_G(B, I) \to \operatorname{Hom}_G(A, I)$$

is strict with respect to the Hodge and weight filtrations $F^\bullet$ and $W_\bullet$. In particular the maps

$$W_0 \operatorname{Hom}_G(B_\mathbb{F}, I_\mathbb{F}) \to W_0 \operatorname{Hom}_G(A_\mathbb{F}, I_\mathbb{F}),\ W_0 \operatorname{Hom}_G(B_\mathbb{C}, I_\mathbb{C}) \to W_0 \operatorname{Hom}_G(A_\mathbb{C}, I_\mathbb{C})$$
$$\text{and } F^0 W_0 \operatorname{Hom}_G(B_\mathbb{C}, I_\mathbb{C}) \to F^0 W_0 \operatorname{Hom}_G(A_\mathbb{C}, I_\mathbb{C})$$

are surjective. □

4.2 Deligne–Beilinson cohomology of an affine group

It is convenient to have a complex which computes $\operatorname{Ext}^\bullet_{\mathsf{HRep}(G)}(\mathbb{F}, V)$ which depends only on G and its MHS. This is the complex of Deligne cochains defined below.

Definition 4.6. The *Deligne–Beilinson cohomology* $H^\bullet_{\mathcal{D}}(G,V)$ of an affine group G with an $\mathbb{F}$-mixed Hodge structure with coefficients in a Hodge representation V is defined to be the cohomology of the complex

$$\operatorname{cone}\big(F^0W_0\mathcal{C}^\bullet(G,V)_\mathbb{C} \oplus W_0\mathcal{C}^\bullet(G,V)_\mathbb{F} \to W_0\mathcal{C}^\bullet(G,V)_\mathbb{C}\big)[-1]. \tag{4.3}$$

As in the case of classical Deligne-Beilinson cohomology, it sits in an exact sequence

$$0 \to \operatorname{Ext}^1_{\mathsf{MHS}_\mathbb{F}}(\mathbb{F}, H^{j-1}(G,V)) \to H^j_{\mathcal{D}}(G,V) \to \Gamma H^j(G,V) \to 0. \tag{4.4}$$

Proposition 4.4 implies that the DB-cohomology of G can be computed with any relatively injective resolution of V.

The following result generalizes the main result of [4].

Theorem 4.7. *If G is an affine group with an $\mathbb{F}$-mixed Hodge structure, and if V is a Hodge representation of G, then there is a natural isomorphism*

$$\operatorname{Ext}^\bullet_{\mathsf{HRep}_\mathbb{F}(G)}(\mathbb{F},V) \xrightarrow{\simeq} H^\bullet_{\mathcal{D}}(G,V).$$

Sketch of Proof. This is a refinement of the standard proof that the Yoneda product in Ext groups corresponds to the cup product in cohomology constructed using injective resolutions. To prove the result, it suffices to construct a homomorphism

$$\operatorname{Ext}^\bullet_{\mathsf{HRep}_\mathbb{F}(G)}(\mathbb{F},V) \to H^\bullet_{\mathcal{D}}(G,V) \tag{4.5}$$

that induces a map from the exact sequence (4.2) to the exact sequence (4.4) and is an isomorphism on the kernels and cokernels.

Let

$$0 \to V \to E^0 \to E^1 \to \cdots \to E^{n-1} \to F \to 0$$

be a Yoneda n-extension in $\mathsf{HRep}_\mathbb{F}(G)$. Then one has the standard resolution

$$0 \to V \to I^0 \to I^1 \to I^2 \to \cdots$$

of V in $\mathsf{Rep}(G)$, where $I^n = V \otimes \mathcal{O}(G)^{\otimes(n+1)}$. This is a relatively injective resolution of V in $\mathsf{HRep}(G)$.

Standard homological arguments and Lemma 4.5 imply that we have the diagram:

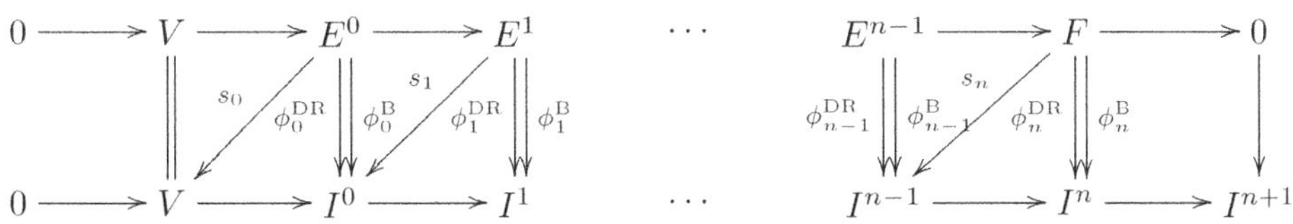

where $\phi^{DR}_\bullet$ and $\phi^B_\bullet$ are chain maps satisfying

$$\phi^{DR}_j \in F^0W_0\operatorname{Hom}_\mathbb{C}(E^j,I^j)^G \text{ and } \phi^B_j \in W_0\operatorname{Hom}_\mathbb{F}(E^j,I^j)^G$$

(convention $E^n = F$) and $s_\bullet$ is a chain homotopy from $\phi_\bullet^{\mathrm{DR}}$ to $\phi_\bullet^{\mathrm{B}}$ satisfying

$$s_j \in W_0 \operatorname{Hom}_{\mathbb{C}}(E^j, I^{j-1})^G, \quad s_0 = 0.$$

Since $\delta(s_n) = \phi_n^{\mathrm{DR}} - \phi_n^{\mathrm{B}}$, the triple $(\phi_n^{\mathrm{DR}}, \phi_n^{\mathrm{B}}, s_n)$ is a cocycle in the complex (4.3). The homotopy properties of injective resolutions imply that the cohomology class of this cocycle is independent of the choices of $\phi_\bullet^{\mathrm{DR}}$, $\phi_\bullet^{\mathrm{B}}$ and $s_\bullet$.[4] There is therefore a well defined map

$$\operatorname{Ext}^\bullet_{\mathsf{HRep}(G)}(\mathbb{F}, V) \to H^\bullet_{\mathcal{D}}(G, V).$$

It is easily seen to be a homomorphism with the property that it induces the identity on the kernels and cokernels of the exact sequences (4.2) and (4.4). □

4.3 Cochains

Suppose that $A^\bullet$ is a complex in ind-$\mathsf{MHS}_{\mathbb{F}}$. Then one can consider the complex

$$A^\bullet_{\mathcal{D}} := \operatorname{cone}\bigl(F^0 W_0 A^\bullet_{\mathbb{C}} \oplus W_0 A^\bullet_{\mathbb{F}} \to W_0 A^\bullet_{\mathbb{C}}\bigr)[-1].$$

Following Beilinson [1], we write an element of degree j in the form

$$\begin{bmatrix} & c & \\ w & & z \end{bmatrix}$$

where $w \in F^0 W_0 A^j_{\mathbb{C}}$, $z \in W_0 A^j_{\mathbb{F}}$ and $c \in W_0 A^{j-1}_{\mathbb{F}}$. With this notation, the differential is

$$\delta : \begin{bmatrix} & c & \\ w & & z \end{bmatrix} \mapsto \begin{bmatrix} & -dc + w - z & \\ dw & & dz \end{bmatrix}.$$

4.4 Products

Suppose that we have complexes $A^\bullet$, $B^\bullet$ and $C^\bullet$ in ind-$\mathsf{MHS}_{\mathbb{F}}$. Suppose that there is an associative product

$$\cup : A^\bullet \otimes B^\bullet \to C^\bullet$$

which is a morphism in ind-$\mathsf{MHS}_{\mathbb{F}}$ and commutes with the differentials. For each $t \in \mathbb{F}^\times$, define a product

$$\cup_t : A^\bullet_{\mathcal{D}} \otimes B^\bullet_{\mathcal{D}} \to C^\bullet_{\mathcal{D}}$$

by

$$\begin{bmatrix} & c' & \\ w' & & z' \end{bmatrix} \otimes \begin{bmatrix} & c'' & \\ w'' & & z'' \end{bmatrix} \mapsto \begin{bmatrix} & c_t & \\ w' \cup w'' & & z' \cup z'' \end{bmatrix}$$

where

$$c_t = (1-t)c_0 + tc_1$$

and

$$c_0 = c' \cup z'' + (-1)^{|w'|} w' \cup c'' \text{ and } c_1 = c' \cup w'' + (-1)^{|z'|} z' \cup c''.$$

Each $\cup_t$ commutes with the differentials and is chain homotopic to c_0, so that all $\cup_t$ induce the same map on cohomology.

[4] A detailed proof of the dual assertion in the unipotent case is given in [4, §8].

Proposition 4.8. *The natural isomorphism (4.5) is compatible with products.*

Proof. Suppose that

$$0 \to V_j \xrightarrow{\eta_j} E_j^0 \longrightarrow \dots \longrightarrow E_j^{n_j-1} \longrightarrow F \longrightarrow 0, \quad j = 1, 2 \qquad (\mathbf{e}_j)$$

are two Yoneda extensions in $\mathsf{HRep}(G)$ with $n_j > 0$. Let

$$0 \to V_j \to I_j^\bullet, \qquad j = 1, 2$$

be the standard resolution of V_j in $\mathsf{Rep}(G)$. These are relatively injective resolutions. Their tensor product

$$0 \to V_1 \otimes V_2 \to \operatorname{tot}(I_1^\bullet \otimes I_2^\bullet)$$

is a resolution of $V_1 \otimes V_2$ in $\mathsf{HRep}(G)$. Since $\mathcal{O}^{\otimes k}$ is an injective object of $\mathsf{Rep}(G)$ for all $k > 0$, each

$$I_1^j \otimes I_2^k \cong V_1 \otimes V_2 \otimes \mathcal{O}^{\otimes(j+k+2)}$$

is injective in $\mathsf{Rep}(G)$. It follows that $I^\bullet := \operatorname{tot}(I_1^\bullet \otimes I_2^\bullet)$ is a relatively injective resolution of $V_1 \otimes V_2$ in $\mathsf{HRep}(G)$ and that it can be used to compute $H_{\mathcal{D}}^\bullet(G, V_1 \otimes V_2)$.

For convenience we set $m = n_1$ and $n = n_2$. Let

$$(\phi_j^{\mathrm{DR}},\ \phi_j^{\mathrm{B}}, \sigma_j), \qquad j = 0, \dots, m$$

be two chain maps and a homotopy between them associated to the extension $(\mathbf{e}_1)$ as in the proof of Theorem 4.7. Let

$$(\psi_k^{\mathrm{DR}},\ \psi_k^{\mathrm{B}},\ \tau_k), \qquad k = 0, \dots, n$$

the corresponding chain maps and homotopy associated to $(\mathbf{e}_2)$. Note that $\sigma_0 = \tau_0 = 0$. The extensions $(\mathbf{e}_1)$ and $(\mathbf{e}_2)$ are represented by the Deligne cocycles

$$\begin{bmatrix} & \sigma_m & \\ \phi_m^{\mathrm{DR}} & & \phi_m^{\mathrm{B}} \end{bmatrix} \text{ and } \begin{bmatrix} & \tau_n & \\ \psi_n^{\mathrm{DR}} & & \psi_n^{\mathrm{B}} \end{bmatrix} \qquad (4.6)$$

respectively.

We need to show that the Deligne cohomology class that corresponds to their Yoneda product is a cup product of these cocycles. To this end, consider the diagram in Figure 1. The bottom and right hand edges form the Yoneda product of $(\mathbf{e}_1)$ and $(\mathbf{e}_2)$. We have to compute the corresponding chain maps $\Phi_\bullet^{\mathrm{DR}}$, $\Phi_\bullet^{\mathrm{B}}$ and the homotopy $s_\bullet$ for their Yoneda product.

For $? \in \{\mathrm{DR}, B\}$, set

$$\Phi_j^? = \begin{cases} \eta_1 \otimes \psi_j^? & j = 0, \dots, n-1, \\ \phi_k^? \otimes \psi_n^? & j = n+k,\ k = 0, \dots, m. \end{cases}$$

Both are G-invariant chain maps to $\operatorname{tot}(I_1^\bullet \otimes I_2^\bullet)$. Clearly $\Phi_\bullet^{\mathrm{DR}}$ preserves the Hodge and weight filtrations, while $\Phi_\bullet^{\mathrm{B}}$ is defined over $\mathbb{F}$ and preserves the weight filtration. A chain homotopy between them is defined by

$$s_j = \begin{cases} \eta_1 \otimes \tau_j & j = 0, \dots, n-1, \\ \sigma_k \otimes \psi_n^{\mathrm{B}} + (-1)^k \phi_k^{\mathrm{DR}} \otimes \tau_n & j = n+k,\ k = 0, \dots, m. \end{cases}$$

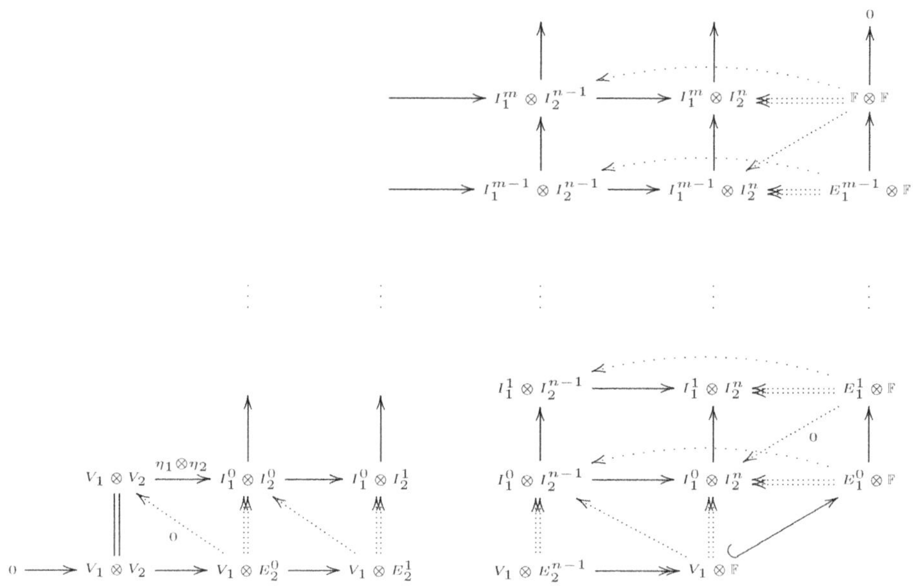

Figure 1

Note that the restriction of $s_n : E_1^0 \otimes \mathbb{F} \to I_0^1 \otimes I_2^{n-1}$ to $V_1 \otimes \mathbb{F}$ is $\eta_1 \otimes \tau_n$ as the restriction of ϕ_0^{DR} to V_1 is η_1 since $\sigma_0 = 0$.

The Yoneda product of $(\mathbf{e}_1)$ and $(\mathbf{e}_2)$ is thus represented by the Deligne cocycle

$$\begin{bmatrix} \phi_m^{\mathrm{DR}} \otimes \psi_n^{\mathrm{DR}} & \sigma_m \otimes \psi_n^{\mathrm{B}} + (-1)^m \phi_m^{\mathrm{DR}} \otimes \tau_n & \phi_m^B \otimes \psi_n^B \end{bmatrix}.$$

This is the $\cup_0$ product of the cocycles (4.6) that represent $(\mathbf{e}_1)$ and $(\mathbf{e}_2)$. □

4.5 Natural hypotheses

The MHSs on the affine groups G that arise in practice (e.g., in the next section) satisfy $W_{-1}\mathcal{O}(G) = 0$. In this case, $W_0\mathcal{O}(G)$ is a Hopf subalgebra. Set $\overline{G} = \operatorname{Spec} W_0\mathcal{O}(G)$. This is a quotient of G.

Lemma 4.9. *The kernel of the quotient homomorphism $G \to \overline{G}$ is prounipotent. Denote it by $W_{-1}G$. It has a natural weight filtration*

$$W_{-1}G \supseteq W_{-2}G \supseteq W_{-3}G \supseteq \cdots$$

which is a descending central series.

Proof. Denote the maximal ideal in $\mathcal{O}(G)$ of functions that vanish at the identity by $\mathfrak{m}_G$. The maximal ideal of the identity of $\overline{G}$ in $\mathcal{O}(\overline{G}) = W_0\mathcal{O}(G)$ is $W_0\mathfrak{m}_G$. The coordinate ring of the kernel $W_{-1}G$ of $G \to \overline{G}$ is thus $\mathcal{O}(G)/(W_0\mathfrak{m}_G)$.

Since $\mathcal{O}(G)$ is a Hopf algebra in the category of ind-objects of $\mathsf{MHS}_{\mathbb{F}}$, $\mathcal{O}(W_{-1}G)$ inherits a natural ind-MHS. In particular, it has a weight filtration $W_\bullet$ which is preserved by the product and coproduct. Using the unit and counit, we can write it as

$$\mathcal{O}(W_{-1}G) = F \oplus W_1\mathcal{O}(W_{-1}G),$$

so that it is connected. Since a finitely generated connected Hopf algebra is the ring of functions on a unipotent group, and since $\mathcal{O}(W_{-1}G)$ is a direct limit of finitely generated (necessarily connected) Hopf algebras, $W_{-1}G$ is prounipotent. □

5 Categories of variations of MHS

5.1 Relative completion

Here we recall Deligne's notion of the relative unipotent completion of a discrete group. Suppose that Γ is a discrete group, that $\mathbb{F}$ is a field of characteristic zero, that R is a reductive $\mathbb{F}$-group and that $\rho : \Gamma \to R(\mathbb{F})$ is a Zariski dense representation. The completion of G relative to ρ is an affine $\mathbb{F}$-group $\mathcal{G}$ that is an extension

$$1 \to \mathcal{U} \to \mathcal{G} \to R \to 1$$

of R by a prounipotent group. It can be defined as the tannakian fundamental group of the category $\mathcal{R}(\Gamma, \rho)$ of finite dimensional $F[\Gamma]$-modules V that admit a filtration

$$0 = V_0 \subset V_1 \subset \cdots \subset V_{N-1} \subset V_N = V$$

by Γ-submodules such that the action of Γ on each V_m/V_{m-1} factors through a representation of R via ρ. There is a canonical representation $\Gamma \to \mathcal{G}(\mathbb{F})$ whose composition with the quotient mapping $\mathcal{G}(\mathbb{F}) \to R(\mathbb{F})$ is ρ.

5.2 Hodge theory

In the remainder of this section, $\mathbb{F}$ will be a subfield of $\mathbb{R}$. Suppose that X is a smooth quasi-projective variety over $\mathbb{C}$ and that $\mathbb{H}$ is a semi-simple variation of $\mathbb{F}$-MHS over X. That is, it is a direct sum of polarized variations of $\mathbb{F}$-Hodge structure (PVHS) whose local monodromy operators are quasi-unipotent.

Fix a base point $x \in X$. Denote the fiber of the variation of mixed Hodge structure $\mathbb{V}$ over X over x by V_x. Let R_x be the closure of the monodromy representation $\pi_1(X, x) \to \mathrm{Aut}(H_x)$. A theorem of Deligne [6, 4.2.6] (see also [22, 7.2.5]) implies that R_x is a reductive $\mathbb{F}$-group. Let $\rho_x : \pi_1(X, x) \to R_x(\mathbb{F})$ be the corresponding representation. Denote the completion of $\pi_1(X, x)$ with respect to ρ_x by $\mathcal{G}_x$. This group has a natural $\mathbb{F}$-MHS, which was constructed in [9]. A more concrete description of it in the case where X is an affine curve is given in [12].[5] It is an extension

$$1 \to \mathcal{U}_x \to \mathcal{G}_x \to R_x \to 1$$

where $\mathcal{U}_x = W_{-1}\mathcal{G}_x$ is prounipotent.

[5] A more concrete description in the general case will be given in [14].

Denote by $\mathsf{MHS}(X, \mathbb{H})$ the category of admissible variations $\mathbb{V}$ of MHS over X whose weight graded quotients have the property that their monodromy factors through a representation of R_x via ρ_x. The fiber V_x of such a local system $\mathbb{V}$ is an object of the category $\mathcal{R}(\pi_1(X, x), \rho_x)$ and thus a representation of $\mathcal{G}_x$. Its monodromy representation factors through this homomorphism. That is, there is a homomorphism $\mathcal{G}_x \to \operatorname{Aut} V_x$ such that

$$\pi_1(X, x) \longrightarrow \mathcal{G}_x(\mathbb{F}) \longrightarrow \operatorname{Aut}(V_x)$$

is the monodromy representation.

5.3 Relative completion of path torsors

Suppose that X is a locally simply connected topological space, that R_x is a reductive $\mathbb{F}$-group and that $\rho : \pi_1(X, x) \to R_x(\mathbb{F})$ is a Zariski dense representation. Denote by $\mathcal{R}(X, \rho)$ the category of $\mathbb{F}$-local systems $\mathbb{V}$ over X whose monodromy representation $\pi_1(X, x) \to \operatorname{Aut}(V_x)$ is an object of the category $\mathcal{R}(\pi_1(X, x), \rho)$. This category is equivalent to $\mathcal{R}(\pi_1(X, x), \rho)$. For each $y \in X$ the functor

$$\omega_y : \mathcal{R}(X, \rho) \to \mathsf{Vec}_{\mathbb{F}}$$

that takes a local system $\mathbb{V}$ to the $\mathbb{Q}$-vector space that underlies its fiber V_y over y is a fiber functor.

Denote by $\Pi_X(x, y)$ the set of homotopy classes of paths in X from x to y. The completion of $\Pi_X(x, y)$ relative to ρ is defined to be the affine scheme

$$\mathcal{G}_{x,y} := \operatorname{Isom}^{\otimes}(\omega_x, \omega_y)$$

of tensor isomorphisms of ω_x and ω_y. When $x = y$, this is the completion $\mathcal{G}_x$ of $\pi_1(X, x)$ relative to ρ. There are natural functions $\Pi_X(x, y) \to \mathcal{G}_{x,y}(\mathbb{F})$, which are Zariski dense, and morphisms $\mathcal{G}_{y,z} \times \mathcal{G}_{x,y} \to \mathcal{G}_{x,z}$ such that the diagram

$$\begin{array}{ccc} \Pi_X(y, z) \times \Pi_X(x, y) & \longrightarrow & \Pi_X(x, z) \\ \downarrow & & \downarrow \\ \mathcal{G}_{y,z}(\mathbb{F}) \times \mathcal{G}_{x,y}(\mathbb{F}) & \longrightarrow & \mathcal{G}_{x,z}(\mathbb{F}) \end{array}$$

commutes.

5.4 Categories of variations of MHS

The following theorem generalizes the main result of [15], which is the unipotent case. This result will be proved in full generality in [14]. The proof of the case where X is an affine curve case was sketched in [12].

Theorem 5.1. *Suppose that X is a smooth quasi-projective variety and that $x \in X$. If $\mathbb{V}$ is an object of $\mathsf{MHS}(X, \mathbb{H})$, then the morphism*

$$V_x \to V_x \otimes \mathcal{O}(\mathcal{G}_x)$$

corresponding to the homomorphism $\mathcal{G}_x \to \operatorname{Aut} V_x$ *is a morphism of MHS. The corresponding functor*

$$\mathsf{MHS}(X, \mathbb{H}) \to \mathsf{HRep}(\mathcal{G}_x)$$

is an equivalence of categories.

The relevance of this result is that, combined with Theorem 4.7, it gives a computation of $\operatorname{Ext}^\bullet_{\mathsf{MHS}(X,\mathbb{H})}(\mathbb{F}, \mathbb{V})$.

Corollary 5.2. *If* $\mathbb{V}$ *is an object of* $\mathsf{MHS}(X, \mathbb{H})$, *there is a natural isomorphism*

$$H^\bullet_{\mathcal{D}}(\mathcal{G}_x, V_x) \cong \operatorname{Ext}^\bullet_{\mathsf{MHS}(X,\mathbb{H})}(\mathbb{F}, \mathbb{V})$$

which is compatible with products.

5.5 Tangential base points

Theorem 5.1 also holds when x is a tangential base point. Because the weight filtration is replaced by a relative weight filtration, which depends on the tangent vector, additional explanation is required.

Suppose that $X = \overline{X} - D$, where $\overline{X}$ is smooth and projective and D is a normal crossing divisor in $\overline{X}$. Suppose that $P \in D$. A tangent vector $\vec{v} \in T_P\overline{X}$ at P that is not tangent to any local component of D determines an element $\sigma_{\vec{v}} \in \pi_1(X, \vec{v})$. The class $\sigma_{\vec{v}}$ is the image of the positive generator of $\pi_1(\Delta^*)$ in $\pi_1(X, \vec{v})$ associated to any analytic arc $(\Delta, 0) \hookrightarrow (\overline{X}, P)$ that is tangent to $\vec{v}$.

The fiber $V_{\vec{v}}$ of an admissible variation of MHS $\mathbb{V}$ over X at $\vec{v}$ is, by definition, the limit MHS associated to $\vec{v}$. The underlying complex vector space is the fiber over P of Deligne's canonical extension of $\mathbb{V} \otimes \mathcal{O}_X$ to $\overline{X}$. Its rational structure is determined by $\vec{v}$. The local monodromy operator $h_{\vec{v}}$ is the image of $\sigma_{\vec{v}}$ in $\sigma_{\vec{v}}$ in $\operatorname{Aut} V_{\vec{v}}$. It is quasi-unipotent. The local monodromy logarithm N is defined by

$$N = \frac{1}{k} \log T_{\vec{v}}^k$$

where $k > 0$ and $T_{\vec{v}}^k$ is unipotent. The limit MHS $V_{\vec{v}}$ has two weight filtrations: $W_\bullet$, which is the restriction of the weight filtration of $\mathbb{V}$ to $V_{\vec{v}}$, and the relative weight filtration $M_\bullet$ associated to the local monodromy logarithm $N : (V_{\vec{v}}, W_\bullet) \to (V_{\vec{v}}, W_\bullet)$. The limit MHS is $(V_{\vec{v}}, M_\bullet, F^\bullet)$, where $F^\bullet$ is the limit Hodge filtration. The filtration $W_\bullet$ is a filtration of $V_{\vec{v}}$ by the mixed Hodge structures $(W_m V_{\vec{v}}, F^\bullet, M_\bullet)$.

The family $\{\mathcal{O}(\mathcal{G}_x)\}_{x \in X}$ over X underlies an ind-object of $\mathsf{MHS}(X, \mathbb{H})$, [12, 14]. In particular, there is a limit MHS on $\mathcal{O}(\mathcal{G}_{\vec{v}})$ with weight filtrations $W_\bullet$ and $M_\bullet$. These are respected by the Hopf algebra operations and thus pass to filtrations on $\mathcal{G}_{\vec{v}}$. There is a natural homomorphism

$$\rho_{\vec{v}} : \pi_1(X, \vec{v}) \to \mathcal{G}_{\vec{v}}(\mathbb{F}).$$

The universal monodromy logarithm $N_{\vec{v}} \in \mathfrak{g}_{\vec{v}}$ is the logarithm of the unipotent part of the Jordan decomposition of $\rho_{\vec{v}}(\sigma_{\vec{v}})$. It spans a copy of $\mathbb{Q}(1)$ in $\mathfrak{g}_{\vec{v}}$.

A Hodge representation of $\mathcal{G}_{\vec{v}}$ is a MHS $(V, F^\bullet, M_\bullet)$ where V is in $\mathsf{Rep}(\mathcal{G}_{\vec{v}})$. The coaction

$$V \to V \otimes \mathcal{O}(\mathcal{G}_{\vec{v}}) \tag{5.1}$$

is required to be a morphism of MHS with respect to the weight filtration $M_\bullet$.

In general, the relative weight filtration $M_\bullet$ of a filtration preserving nilpotent endomorphism of a filtered vector space $(V, W_\bullet)$ does not determine the filtration $W_\bullet$.[6] However, in the presence of a $\mathcal{G}_{\vec{v}}$-action, it does.

Lemma 5.3. *Each Hodge representation $(V, M_\bullet)$ of $\mathcal{G}_{\vec{v}}$ has a unique filtration $W_\bullet$ that is defined over $\mathbb{F}$, preserved by N and such that $M_\bullet$ is the relative weight filtration of $N \in W_0 \operatorname{End} V$. Moreover, each $W_m V$ is a sub-MHS of V and the monodromy representation $\mathcal{G}_{\vec{v}} \to \operatorname{Aut} V$ preserves $W_\bullet$.*

Sketch of proof. The coordinate rings $\{\mathcal{O}(R_x)\}$ of the reductive quotients of the $\mathcal{G}_x$ form an ind-PVHS over X. Denote its fiber over $\vec{v}$ by $\mathcal{O}(R_{\vec{v}})$ and the corresponding group by $R_{\vec{v}}$. It is the reductive quotient of $\mathcal{G}_{\vec{v}}$. To simplify notation, we set $\mathcal{G} = \mathcal{G}_{\vec{v}}$ and $R = R_{\vec{v}}$. Denote the kernel of the projection $\mathcal{G} \to R$ by $\mathcal{U}$. It is prounipotent.

We first prove the result in the case where V is an irreducible Hodge representation of R. Note that V is not necessarily irreducible as an R-module. Recall that its weight filtration is denoted by $M_\bullet$ and that

$$N : (V, M_\bullet) \to (V, M_\bullet)(-1)$$

is a morphism of MHS. We have to show that there is a unique $m \in \mathbb{Z}$ such that $M_\bullet$ is the weight filtration of $N : V \to V$ shifted by m.

Denote the Lie algebra of R by $\mathfrak{r}$. Since $\mathfrak{r}$ has weight 0, its relative weight filtration $M_\bullet$ with respect to N is the weight filtration of N and is centered about 0. Schmid's work [22] implies that there is a copy of ${}_2$ in $\mathfrak{r}_\mathbb{R}$ that contains N and which has a semi-simple element h_0 whose eigenspaces split the relative weight filtration of $\mathfrak{r}_\mathbb{R}$. Likewise, the relative weight filtration $M_\bullet$ of $\mathcal{O}(R)$ is centered about 0, is the weight filtration of N and is split by the eigenspaces of h_0.

The restriction of the monodromy action to ${}_2$ defines an ${}_2$ action on $V_\mathbb{R}$. Decompose V into its h_0 eigenspaces:

$$V_\mathbb{R} = \bigoplus V_r.$$

Since the coaction $V \to V \otimes \mathcal{O}(R)$ is a morphism of MHS (with respect to the weight filtration $M_\bullet$), it is strict with respect to $M_\bullet$. Since it is also ${}_2$-equivariant, it respects the grading by h_0 weight.

Say that an R-submodule A of $V_\mathbb{R}$ has weight m_A if

$$N^k : \operatorname{Gr}^M_{m_A+k} A \to \operatorname{Gr}^M_{m_A-k} A$$

---

[6]A simple example where $M_\bullet$ and N do not determine $W_\bullet$ is where V is the first homology $H^1(E)$ of an elliptic curve and N is the monodromy logarithm associated to degenerating to the nodal cubic. This satisfies $N \neq 0$ and $N^2 = 0$. The weight filtrations are

$$0 = W_0 V \subset W_1 V = V \text{ and } 0 = M_{-1} V \subset M_0 V = M_1 V \subset M_2 V = V,$$

where $M_0 V = \operatorname{im} N$. Apart from the original $W_\bullet$, we can take $W_\bullet$ to be $M_\bullet$. There is nothing special about this example as $M_\bullet$ is always the relative weight filtration of $N \in M_0 \operatorname{End}(V, M_\bullet)$.

is an isomorphism for all k. The weight m_A, if it exists, is unique. The discussion in the previous paragraph implies that every cyclic submodule A of $V_{\mathbb{R}}$ that is generated by an h_0 eigenvector

$$v \in (M_k A \setminus M_{k-1} A) \cap V_r$$

has weight $m_A = k - r$.

Let $\mathcal{X}$ be the set of R-submodules of $V_{\mathbb{R}}$ that have a weight. It is clear that if B is a non-zero submodule of $A \in \mathcal{X}$, then $B \in \mathcal{X}$ and $m_B = m_A$. This implies that if $A, B \in \mathcal{X}$ and $A \cap B \neq 0$, then $A + B \in \mathcal{X}$ and $m_{A+B} = m_A = m_B$. Since $\mathcal{X}$ contains all cyclic modules of the type discussed above, and since V is a simple Hodge representation of R, this implies that $V \in \mathcal{X}$. The weight filtration

$$0 = W_{m-1} V \subset W_m V = V,$$

where $m = m_V$, is defined over $\mathbb{F}$, preserved by N, and has the property that $M_\bullet$ is the relative weight filtration of $N \in W_0 \operatorname{End} V$. The monodromy representation $V \to V \otimes \mathcal{O}(\mathcal{G})$ respects $W_\bullet$. It is also a filtration by MHS.

We now consider the general case. Suppose that V is a $\mathcal{G}$-module. If $\mathcal{U}$ acts trivially, then we are in the semi-simple case above, and so have a weight filtration. If not, then $\mathcal{U}$ acts unipotently, and so $V^{\mathcal{U}} \neq 0$. This is an R-module, defined over $\mathbb{F}$, and thus has a weight filtration $W_\bullet$ by the semi-simple case above. Since $V^{\mathcal{U}}$ is defined over $\mathbb{F}$, the filtration $W_\bullet$ is also defined over $\mathbb{F}$.

Let m be the smallest integer such that $W_m V^{\mathcal{U}} \neq 0$. This is a sub-MHS of V. The quotient $V/W_m V^{\mathcal{U}}$ is a Hodge representation of $\mathcal{G}$ and $(V/W_m V^{\mathcal{U}})^{\mathcal{U}}$ is a Hodge representation of R. Let k be the smallest integer such that $W_k(V/W_m V^{\mathcal{U}})^{\mathcal{U}} \neq 0$. To complete the proof, it suffices to show that $m < k$. For if this holds, one can define $W_k V$ to be the inverse image of $W_k(V/W_m V^{\mathcal{U}})^{\mathcal{U}}$ in V. One can then continue inductively to define the $W_\bullet$ filtration of V.

To prove that $k > m$, observe that monodromy homomorphism $\mathcal{U} \times V \to V$ induces a homomorphism

$$H_1(\mathfrak{u}) \otimes W_k(V/W_m V^{\mathcal{U}})^{\mathcal{U}} \to W_m V^{\mathcal{U}}. \tag{5.2}$$

If the action is trivial, then $W_k(V/W_m V^{\mathcal{U}})^{\mathcal{U}} = \operatorname{Gr}_k^W V^{\mathcal{U}}$, which implies that $m < k$. If the action is non-trivial, there is a least ℓ such that the restriction of (5.2) to $W_\ell H_1(\mathfrak{u})$ is non-trivial. The induced map

$$\operatorname{Gr}_\ell^W \mathfrak{u} \otimes W_k(V/W_m V^{\mathcal{U}})^{\mathcal{U}} \to W_m V^{\mathcal{U}}$$

is R-invariant, and thus $_2$-invariant. It follows that $k + \ell = m$. Since $\ell < 0$, we have $k > m$. This completes the proof of the existence of a weight filtration $W_\bullet$ with the property that $M_\bullet$ is the relative weight filtration of $N \in W_0 \operatorname{End} V$ and where each $W_m V$ is a sub-MHS. The construction implies that each $W_m V$ is a $\mathcal{G}$-submodule of V and that the monodromy representation $\mathfrak{u} \times V \to V$ preserves $W_\bullet$. Together these imply that $\mathcal{G} \times V \to V$ does too. $\square$

Proof of Theorem 5.1 for tangential base points. First suppose that $\mathbb{V}$ is an object of $\mathsf{MHS}(X, \mathbb{H})$ with fiber $(V, F^\bullet, W_\bullet, M_\bullet)$ over $\vec{v}$. Denote by $\mathbb{O}$ the local system over X whose fiber over $x \in X$ is $\mathcal{O}(\mathcal{G}_x)$. The monodromy coactions

$$V_x \to V_x \otimes \mathcal{O}(\mathcal{G}_x)$$

give a flat section of the variation $\operatorname{Hom}_{\mathbb{F}}(\mathbb{V}, \mathbb{V} \otimes \mathbb{O})$. Theorem 5.1 implies that this section is in

$$\Gamma H^0(X, \operatorname{Hom}_{\mathbb{F}}(\mathbb{V}, \mathbb{V} \otimes \mathbb{O})).$$

The Theorem of the Fixed Part implies that

$$\Gamma H^0(X, \operatorname{Hom}_{\mathbb{F}}(\mathbb{V}, \mathbb{V} \otimes \mathbb{O})) = F^0 W_0 \operatorname{Hom}_{\mathcal{G}_b}(V_b, V_b \otimes \mathcal{O}(\mathcal{G}_b))$$

for all base points b of X, tangential or standard. This implies that the fiber $V_{\vec{v}}$ of $\mathbb{V}$ over $\vec{v}$ is in $\mathsf{HRep}(\mathcal{G}_{\vec{v}})$.

Now suppose that V is in $\mathsf{HRep}(\mathcal{G}_{\vec{v}})$. Let $\mathbb{V}$ be the corresponding $\mathbb{F}$ local system over X. We have to prove that $\mathbb{V}$ underlies an admissible variation of MHS whose fiber over $\vec{v}$ is V. Lemma 5.3 implies that V has a weight filtration $W_\bullet$ by $\mathcal{G}_{\vec{v}}$ submodules such that $M_\bullet$ is the relative weight filtration of $N \in W_0 \operatorname{End}(V)$. This implies that the local system $\mathbb{V}$ has a filtration $W_\bullet\mathbb{V}$ by local systems. Denote the fiber of $\mathbb{V}$ over $x \in X$ (possibly tangential) by V_x. We have to construct a MHS on each V_x.

To do this, we use the the local system $\mathbb{O}_{\vec{v}}$ over X whose fiber over x is $\mathcal{O}(\mathcal{G}_{x,\vec{v}})$. This is an ind-object of $\mathsf{MHS}(X; \mathbb{H})$. The key observation is that for all x, the image of the coaction

$$V_x \hookrightarrow V \otimes \mathcal{O}_{x,\vec{v}}$$

is a mixed Hodge structure. This is true because the image of V_x in $V \otimes \mathcal{O}_{\vec{v},x}$ is the kernel of

$$\Delta_V \otimes 1 - 1 \otimes \Delta : V \otimes \mathcal{O}_{\vec{v},x} \to V \otimes \mathcal{O}_{\vec{v},\vec{v}} \otimes \mathcal{O}_{\vec{v},x},$$

which has a MHS as both Δ_V and Δ are morphisms of MHS. From this it follows that $\mathbb{V}$ is isomorphic to the kernel of

$$\Delta_V \otimes 1 - 1 \otimes \Delta : V \otimes \mathbb{O}_{\vec{v}} \to V \otimes \mathbb{O} \otimes \mathbb{O}_{\vec{v}}$$

and is therefore an admissible variation of MHS. □

6 The natural homomorphism

A homomorphism $\Gamma \to G(\mathbb{F})$ from a discrete group Γ into the $\mathbb{F}$-points of an affine $\mathbb{F}$-group G induces a functor $\mathsf{Rep}(G) \to \mathsf{Rep}_{\mathbb{F}}(\Gamma)$ and therefore a homomorphism

$$H^\bullet(G, V) \to H^\bullet(\Gamma, V)$$

for each V in $\mathsf{Rep}(G)$. When Γ is the fundamental group of a (sufficiently nice) topological space X and $\mathbb{V}$ is a locally constant sheaf over X corresponding to the Γ-module V, there are natural homomorphisms

$$H^\bullet(G, V) \to H^\bullet(\Gamma, V) \to H^\bullet(X, \mathbb{V}).$$

Now suppose that $\mathcal{G}$ is the completion of Γ relative to $\rho : \Gamma \to R(\mathbb{F})$. The following result is standard. (Cf. [9].)

Proposition 6.1. *For all R-modules V, the natural homomorphism*

$$H^\bullet(\mathcal{G}, V) \to H^\bullet(\Gamma, V) \tag{6.1}$$

is an isomorphism in degrees ≤ 1 and injective in degree 2. If Γ is free, then (6.1) is an isomorphism in all degrees.

Applying this to the situation where X is a smooth complex algebraic variety, $\mathbb{H}$ is an $\mathbb{F}$-PVHS over X and $\mathcal{G}_x$ is the corresponding relative completion of $\pi_1(X, x)$, we see that for each $\mathbb{V}$ in $\mathsf{MHS}_{\mathbb{F}}(X, \mathbb{H})$, there is a natural homomorphism

$$H^\bullet(\mathcal{G}_x, V_x) \to H^\bullet(X, \mathbb{V}). \tag{6.2}$$

It is compatible with products.

Proposition 6.2. *For all objects $\mathbb{V}$ of $\mathsf{VMHS}(X, \mathbb{H})$, the natural homomorphism (6.2) is a morphism of MHS. If X is an affine curve, it is an isomorphism.*

Proof. The group $H^\bullet(X, \mathbb{V})$ can be computed by taking an injective resolution $V_x \hookrightarrow I^\bullet$ of the fiber V_x over x in $\mathsf{Rep}(\mathcal{G}_x)$, taking the corresponding resolution $\mathbb{V} \hookrightarrow \mathbb{I}^\bullet$ of $\mathbb{V}$, and then taking the cohomology of the total complex of the double complex

$$E^\bullet(X, \mathbb{I}^\bullet) := \bigoplus_{m \geq 0} E^\bullet(X, \mathbb{I}^m)$$

where $E^\bullet(X, \mathbb{I}^m)$ denotes the complex of smooth forms with coefficients in $\mathbb{I}^m$. The homomorphism (6.2) is induced by the inclusion $(I^\bullet)^{\mathcal{G}_x} \hookrightarrow E^0(X, \mathbb{I}^\bullet) \hookrightarrow E^\bullet(X, \mathbb{I}^\bullet)$. To prove the result, we need to realize this construction using mixed Hodge complexes.

Choose a relatively injective resolution $V_x \to I^\bullet$ of V_x in $\mathsf{HRep}(\mathcal{G}_x)$. Theorem 5.1 implies that this corresponds to a resolution $\mathbb{V} \to \mathbb{I}^\bullet$ in ind-$\mathsf{MHS}(X, \mathbb{H})$ with fiber $I^\bullet$ over x.

Write $X = \overline{X} - D$, where $\overline{X}$ is smooth and complete, and where D is a divisor with normal crossings. Denote by $K^\bullet(\overline{X}, D; \mathbb{A})$ Saito's mixed Hodge complex [23] that computes the MHS on $H^\bullet(X, \mathbb{A})$, where $\mathbb{A}$ is an admissible variation of MHS over X. When X is a curve, one can use Zucker's MHC [27] instead.

The inclusion $\mathbb{V} \hookrightarrow \mathbb{I}^\bullet$ induces a morphism of mixed Hodge complexes

$$K^\bullet(\overline{X}, D; \mathbb{V}) \to \operatorname{tot} K^\bullet(\overline{X}, D; \mathbb{I}^\bullet)$$

and thus an isomorphism of MHS on cohomology. It is a quasi-isomorphism as $\mathbb{V} \to \mathbb{I}^\bullet$ is a resolution. The theorem of the fixed part implies that $H^0(X, \mathbb{I}^\bullet)$ is a mixed Hodge complex and that its inclusion into $\operatorname{tot} K^\bullet(\overline{X}, D; \mathbb{I}^\bullet)$ is a morphism. This inclusion, coupled with the natural isomorphisms

$$H^\bullet(\mathcal{G}_x, V_x) \cong H^\bullet((I^\bullet)^{\mathcal{G}_x}) \cong H^\bullet(H^0(X, \mathbb{I}^\bullet))$$

and

$$H^\bullet(X,\mathbb{V}) \cong H^\bullet(K^\bullet(\overline{X},D;\mathbb{V})) \cong H^\bullet(\operatorname{tot} K^\bullet(\overline{X},D;\mathbb{I}^\bullet)),$$

induces the natural morphism (6.2). It is a morphism of MHS as it is induced by a morphism of mixed Hodge complexes.

The last assertion follows from Proposition 6.1 as every affine curve is a model of the classifying space of a free group. □

7 Comparison with Deligne–Beilinson cohomology

As in the previous section, X is the complement of a normal crossing divisor D in a smooth projective variety $\overline{X}$. Saito's mixed Hodge complex [23] with coefficients in an admissible variation of MHS $\mathbb{V}$ over X will be denoted by $K^\bullet(\overline{X},D;\mathbb{V})$.

Suppose that $(K^\bullet, W_\bullet)$ is a filtered complex, where $W_\bullet$ is increasing. Recall from [6, 1.3.3] that the filtered complex $\operatorname{Dec}_W K^\bullet$ is the same complex $K^\bullet$ endowed with the shifted filtration (*filtration decalée*) defined by

$$W_m \operatorname{Dec}_W K^j = \{k \in W_{m+j}K^j : dk \in W_{m+j+1}K^{j+1}\}.$$

The reason for shifting the filtration $W_\bullet$ is that if $K^\bullet$ is a mixed Hodge complex, then the inclusion $W_m \operatorname{Dec}_W K^\bullet \hookrightarrow K^\bullet$ induces isomorphisms

$$H^\bullet(W_m \operatorname{Dec}_W K^\bullet) \xrightarrow{\simeq} W_m H^\bullet(K^\bullet), \quad H^\bullet(F^p W_m \operatorname{Dec}_W K^\bullet) \xrightarrow{\simeq} F^p W_m H^\bullet(K^\bullet).$$

Definition 7.1. The Deligne–Beilinson (DB) cohomology $H^\bullet_{\mathcal{D}}(X,\mathbb{V})$ is defined to be the cohomology of the complex

$$\operatorname{cone}\big(F^0 W_0 \operatorname{Dec}_W K^\bullet_{\mathbb{C}}(\overline{X},D;\mathbb{V}) \oplus W_0 \operatorname{Dec}_W K^\bullet_{\mathbb{Q}}(\overline{X},D;\mathbb{V}) \to W_0 \operatorname{Dec}_W K^\bullet_{\mathbb{C}}(\overline{X},D;\mathbb{V})\big)[-1]. \quad (7.1)$$

Standard arguments imply that it is independent of the choice of the compactification $\overline{X}$ and that it can be expressed as the extension

$$0 \to \operatorname{Ext}^1_{\mathsf{MHS}_{\mathbb{F}}}(\mathbb{Q}, H^{j-1}(X,\mathbb{V})) \to H^j_{\mathcal{D}}(X,\mathbb{V}) \to \Gamma H^j(X,\mathbb{V}) \to 0. \quad (7.2)$$

7.1 Comparison of DB-cohomologies

Suppose that $\mathbb{H}$ is an $\mathbb{F}$-PVHS over X and that $\mathcal{G}_x$ is the corresponding relative completion of $\pi_1(X,x)$.

Theorem 7.2. *For all objects* $\mathbb{V}$ *of* $\mathsf{VMHS}(X,\mathbb{H})$, *there is a natural homomorphism*

$$H^\bullet_{\mathcal{D}}(\mathcal{G}_x, V_x) \to H^\bullet_{\mathcal{D}}(X,\mathbb{V})$$

which respects the products and whose degree j part fits into a commutative diagram

$$\begin{array}{ccccccccc} 0 & \longrightarrow & \mathrm{Ext}^1_{\mathsf{MHS}}(\mathbb{F}, H^{j-1}(\mathcal{G}_x, V_x)) & \longrightarrow & H^j_{\mathcal{D}}(\mathcal{G}_x, V_x) & \longrightarrow & \Gamma H^j(\mathcal{G}_x, V_x) & \longrightarrow & 0 \\ & & \downarrow & & \downarrow & & \downarrow & & \\ 0 & \longrightarrow & \mathrm{Ext}^1_{\mathsf{MHS}}(\mathbb{F}, H^{j-1}(X, \mathbb{V})) & \longrightarrow & H^j_{\mathcal{D}}(X, \mathbb{V}) & \longrightarrow & \Gamma H^j(X, \mathbb{V}) & \longrightarrow & 0 \end{array}$$

where the right and left hand vertical maps are induced by the natural homomorphism (6.2).

Proof. Choose a relatively injective resolution $V_x \hookrightarrow I^\bullet$ of V_x in $\mathsf{HRep}(\mathcal{G}_x)$. The corresponding resolution of $\mathbb{V}$ in $\mathsf{MHS}(X, \mathbb{H})$ will be denoted by $\mathbb{V} \hookrightarrow \mathbb{I}^\bullet$.

Define

$$\mathrm{Dec}_W \operatorname{tot} K^\bullet(\overline{X}, D; \mathbb{I}^\bullet) = \bigoplus_{n \geq 0} \mathrm{Dec}_W \, K^\bullet(\overline{X}, D; \mathbb{I}^n).$$

In other words, use only the degree in $K^\bullet$ to compute Dec_W, and ignore the degree in $\mathbb{I}^\bullet$. With this convention, the inclusion above induces an inclusion

$$(I^\bullet)^{\mathcal{G}_x} = H^0(X, \mathbb{I}^\bullet) \hookrightarrow \mathrm{Dec}_W \operatorname{tot} K^\bullet(\overline{X}, D; \mathbb{I}^\bullet)$$

that preserves the Hodge and weight filtrations. This inclusion induces a chain map from the complex (4.3) to the complex (7.1). This induces the morphism on Deligne cohomology that is compatible with products. □

Combining this with the second assertion of Proposition 6.2, we conclude:

Corollary 7.3. *If X is an affine curve, then the natural map $H^\bullet_{\mathcal{D}}(\mathcal{G}_x, V_x) \to H^\bullet_{\mathcal{D}}(X, \mathbb{V})$ is an isomorphism for all $\mathbb{V}$ in $\mathsf{MHS}(X, \mathbb{H})$. It is compatible with products.*

7.2 Extensions of variations of MHS

Denote the category of admissible variations of $\mathbb{F}$ mixed Hodge structure over X by $\mathsf{MHS}_{\mathbb{F}}(X)$. This is the "union" of the categories $\mathsf{MHS}_{\mathbb{F}}(X, \mathbb{H})$ over all semi-simple objects $\mathbb{H}$ of $\mathsf{MHS}_{\mathbb{F}}(X)$.

Lemma 7.4. *For all $\mathbb{V}$ in $\mathsf{MHS}_{\mathbb{F}}(X)$, the inclusions $\mathsf{MHS}(X, \mathbb{H}) \hookrightarrow \mathsf{MHS}_{\mathbb{F}}(X)$ induce an isomorphism*

$$\varinjlim_{\mathbb{H}} \mathrm{Ext}^\bullet_{\mathsf{MHS}(X,\mathbb{H})}(\mathbb{F}, \mathbb{V}) \to \mathrm{Ext}^\bullet_{\mathsf{MHS}_{\mathbb{F}}(X)}(\mathbb{F}, \mathbb{V}).$$

where $\mathbb{H}$ ranges over a set of representatives of the isomorphism classes of semi-simple objects of $\mathsf{MHS}_{\mathbb{F}}(X)$ that contain $\mathbb{V}$.

Proof. This follows from the fact that every finite diagram of objects in $\mathsf{MHS}_{\mathbb{F}}(X)$ lies in $\mathsf{MHS}(X, \mathbb{H})$ for some semi-simple object $\mathbb{H}$ of $\mathsf{MHS}_{\mathbb{F}}(X)$. □

Theorem 1 is obtained by combining the following two results.

Theorem 7.5. *If X is a quasi-projective manifold and $\mathbb{V}$ is an admissible variation of MHS over X, there there is a homomorphism*

$$\mathrm{Ext}^\bullet_{\mathsf{MHS}(X)}(\mathbb{F},\mathbb{V}) \to H^\bullet_{\mathcal{D}}(X,\mathbb{V})$$

which is compatible with products. It is an isomorphism in degrees ≤ 1 and injective in degree 2.

Proof. The first assertion follows directly from Lemma 7.4, Corollary 5.2 and Theorem 7.2. The last assertion follows by applying Proposition 6.2 to the map from the exact sequence (4.2) with $G = \mathcal{G}_x$ to the exact sequence (7.2). □

Combining this with Corollary 7.3 we conclude:

Corollary 7.6. *If X is an affine curve, then for all admissible variations of $\mathbb{F}$-MHS $\mathbb{V}$ over X, there is a natural isomorphism*

$$\mathrm{Ext}^\bullet_{\mathsf{MHS}_{\mathbb{F}}(X)}(\mathbb{F},\mathbb{V}) \cong H^\bullet_{\mathcal{D}}(X,\mathbb{V})$$

which is compatible with products.

8 Brown's computation of the cup products of Eisenstein series

In this section, we briefly explain how Brown's computations [2, §11] can be interpreted in terms of Deligne cohomology of the relative completion of $\mathrm{SL}_2(\mathbb{Z})$. This discussion is not self contained; the reader will need some familiarity with [2], [12] and [13].

Remark 8.1. The reader should be aware that, whereas we use *left* group cochains, Brown uses *right* cochains. He also multiplies paths in the topologist's order, the opposite of that used here. So, with the conventions in this paper, his fundamental group of a space is the opposite group of the fundamental group used here. His work coincides with what we have here once one makes the following observations:

Denote the opposite group of a group Γ by Γ^{op}. Every left Γ module V is naturally a right Γ^{op} module, denoted V^{op}. Suppose that V is a left Γ-module. The function that takes a left Γ-invariant function $f : \Gamma^{n+1} \to V$ to the right invariant function $F : (\Gamma^{\mathrm{op}})^{n+1} \to V^{\mathrm{op}}$ defined by

$$F(g_0,\dots,g_n) = f(g_0^{-1},\dots,g_n^{-1})$$

is an isomorphism and induces a chain isomorphism

$$C^\bullet(\Gamma,V) \to C^\bullet(\Gamma^{\mathrm{op}},V^{\mathrm{op}}).$$

8.1 Preliminaries

Let $\Gamma = \mathrm{SL}_2(\mathbb{Z})$. We identify this with the orbifold fundamental group $\pi_1(\mathcal{M}_{1,1}, \partial/\partial q)$ of the modular curve with base point the natural tangent vector at the cusp. Let $\mathcal{G}$ be the completion (with $\mathbb{F} = \mathbb{Q}$) of Γ with respect to the inclusion $\Gamma \hookrightarrow \mathrm{SL}_2(\mathbb{Q})$. It has a natural (limit) MHS with weight filtrations $W_\bullet$ (the "global" weight filtration) and $M_\bullet$ (the relative weight filtration). The defining representation of SL_2 will be denoted by H. The corresponding local system $\mathbb{H}$ over $\mathcal{M}_{1,1}$ underlies a polarized variation of Hodge structure of weight $+1$. Its fiber over $\partial/\partial q$ is H. It has a natural limit MHS with weight filtrations $W_\bullet$ and $M_\bullet$. It is a Hodge representation of $\mathcal{G}$.

Brown's computations are most naturally interpreted in the Deligne cohomology of $\mathcal{G}$. Theorem 7.2 implies that the natural map

$$H^\bullet_{\mathcal{D}}(\mathcal{G}, S^m H(r)) \to H^\bullet_{\mathcal{D}}(\mathcal{M}_{1,1}, S^m \mathbb{H}(r))$$

is an isomorphism. These groups vanish when m is odd.

For all $r \in \mathbb{Z}$, the homomorphism from $\mathcal{G}$ to the trivial group induces a natural isomorphism

$$H^j_{\mathcal{D}}(\mathcal{G}, S^m H(r)) \cong \mathrm{Ext}^j_{\mathsf{MHS}}(\mathbb{Q}, \mathbb{Q}(r)).$$

This group vanishes, except when $r = j = 0$ and when $j = 1$ and $r > 0$. When $m > 0$, standard results (Eichler-Shimura, Manin-Drinfeld) about the cohomology of modular curves imply that

$$H^1_{\mathcal{D}}(\mathcal{G}, S^m H(r)) \cong \begin{cases} \mathbb{Q} & m = 2n,\ r = 2n+1, \\ 0 & \text{otherwise} \end{cases}$$

and that

$$H^2_{\mathcal{D}}(\mathcal{G}, S^m \mathbb{H}(r)) \cong \mathrm{Ext}^1_{\mathsf{MHS}}(\mathbb{Q}, H^1(\mathcal{M}_{1,1}, S^m \mathbb{H}(r))).$$

This is non-zero only when $m = 2n > 0$.

We will describe the image of the cup product in *real* Deligne cohomology as in degree 2 it is more easily described:

$$\begin{aligned} H^2_{\mathcal{D}}(\mathcal{G}, S^{2n}\mathbb{H}_{\mathbb{R}}(r)) &\cong \mathrm{Ext}^1_{\mathsf{MHS}_{\mathbb{R}}}(\mathbb{R}, H^1(\mathcal{M}_{1,1}, S^{2n}\mathbb{H}(r))) \\ &= \mathrm{Ext}^1_{\mathsf{MHS}_{\mathbb{R}}}(\mathbb{R}, \mathbb{R}(r-2n-1)) \oplus \bigoplus_{f \in \mathfrak{B}_{2n+2}} \mathrm{Ext}^1_{\mathsf{MHS}_{\mathbb{R}}}(\mathbb{R}, V_f(r)) \\ &\cong \begin{cases} \mathbb{R} \oplus \bigoplus_{f \in \mathfrak{B}_{2n+2}} V_{f,\mathbb{R}} & n > 0,\ r \geq 2n+2, \\ 0 & \text{otherwise.} \end{cases} \end{aligned}$$

Here $\mathfrak{B}_{2n+2}$ denotes the set of normalized Hecke eigen cusp forms of weight $2n+2$ and V_f the 2-dimensional real Hodge structure corresponding to $f \in \mathfrak{B}_{2n+2}$.[7]

[7] The projection π_f is defined on cohomology with $\mathbb{F} = \mathbb{Q}_f$, the field generated by the Fourier coefficients of f.

The cup product is most easily described by giving its projections onto the components of the degree 2 cohomology corresponding to Hecke eigenforms. To this end, recall that, when $\mathbb{F} = \mathbb{Q}$ and $\mathbb{R}$, one has projections

$$\begin{aligned}\pi_{\text{eis}} : H^2_{\mathcal{D}}(\mathcal{G}, S^{2n}\mathbb{H}_{\mathbb{F}}(r)) &\to \operatorname{Ext}^1_{\mathsf{MHS}}(\mathbb{F}, \mathbb{F}(r-2n-1)) \\ \text{and } \pi_{\text{cusp}} : H^2_{\mathcal{D}}(\mathcal{G}, S^{2n}\mathbb{H}_{\mathbb{F}}(r)) &\to \operatorname{Ext}^1_{\mathsf{MHS}}(\mathbb{F}, H^1_{\text{cusp}}(\mathcal{M}_{1,1}, S^{2n}\mathbb{H}_{\mathbb{F}}(r))). \quad (8.1)\end{aligned}$$

onto the Eisenstein and cuspidal parts, respectively. When $\mathbb{F} = \mathbb{R}$, we can further decompose π_{cusp} into the projections

$$\pi_f : H^2_{\mathcal{D}}(\mathcal{G}, S^{2n}\mathbb{H}_{\mathbb{R}}(r)) \to \operatorname{Ext}^1_{\mathsf{MHS}}(\mathbb{R}, V_f(r)) \cong V_{f,\mathbb{R}}, \quad (8.2)$$

where $f \in \mathfrak{B}_{2n+2}$.

To describe the cup product, we also have to decompose the coefficients into their irreducible pieces. Suppose that $n = j + k$, with $j > k > 0$. The standard isomorphism

$$S^{2j}H \otimes S^{2k}H \cong \bigoplus_{r=0}^{k} S^{2n-2r}H(-r)$$

of $\mathrm{SL}(H)$-modules respects the mixed Hodge structures. Fix a normalization of the $\mathrm{SL}(H)$-invariant projection

$$p_r : S^{2j}H \otimes S^{2k}H \to S^{2n-2r}H(-r).$$

such as the one given in [2, §2.4]. When $r \leq 2\min\{j, k\}$, one can compose the cup product

$$H^1_{\mathcal{D}}(\mathcal{G}, S^{2j}H(2j+1)) \otimes H^1_{\mathcal{D}}(\mathcal{G}, S^{2k}H(2k+1)) \to H^2_{\mathcal{D}}(\mathcal{G}, S^{2j}H \otimes S^{2k}H(2n+2))$$

with the projection p_r and the projections (8.1) and (8.2) to obtain maps

$$\Phi^{\text{eis}}_r : H^1_{\mathcal{D}}(\mathcal{G}, S^{2j}H(2j+1)) \otimes H^1_{\mathcal{D}}(\mathcal{G}, S^{2k}H(2k+1)) \to \operatorname{Ext}^1_{\mathsf{MHS}}(\mathbb{R}, \mathbb{R}(2n-r+2)). \quad (8.3)$$

and

$$\Phi_{r,f} : H^1_{\mathcal{D}}(\mathcal{G}, S^{2j}H(2j+1)) \otimes H^1_{\mathcal{D}}(\mathcal{G}, S^{2k}H(2k+1)) \to V_{f,\mathbb{R}} \quad (8.4)$$

for each $f \in \mathfrak{B}_{2n-2r+2}$. Brown [2, §10, §11] shows that Φ^{eis}_r vanishes except in the extremal case where $r = 2\min\{j, k\}$.

Note that, since the generators of $H^1_{\mathcal{D}}(\mathcal{G}, S^{2n}\mathbb{H}(2n+1))$ are invariant under the de Rham involution $\overline{\mathcal{F}}_\infty$, so is the image of the cup product.

8.2 Modular forms and $\mathcal{O}(\mathcal{G})$

Denote the canonical extension of the flat bundle $\mathbb{H} \otimes \mathcal{O}_{\mathcal{M}_{1,1}}$ to $\overline{\mathcal{M}}_{1,1}$ by $\mathcal{H}$. Denote the upper half plane by $\mathfrak{h}$. Set $q = e^{2\pi i\tau}$. The pullback of $\mathcal{H}$ to the q-disk is a trivial holomorphic vector bundle. The trivializing sections are

$$\mathbf{a} \text{ and } \mathbf{w} = -2\pi i(\mathbf{b} - \tau\mathbf{a}),$$

where $\mathbf{a}$ and $\mathbf{b}$ is the standard basis of H, defined in [12, §9]. They can also be regarded as trivializing sections of $\mathbb{H}_{\mathbb{Q}}$ over $\mathfrak{h}$.

For a modular form $f : \mathfrak{h} \to \mathbb{C}$ of $\mathrm{SL}_2(\mathbb{Z})$ of weight $2n+2$, set

$$\omega_f = 2\pi i f(\tau)\mathbf{e}(\tau)d\tau.$$

where $\mathbf{e}(\tau) = \mathbf{w}^{2n} = (2\pi i)^{2n}(\mathbf{b} - \tau\mathbf{a})^{2n}$. This is a Γ-invariant 1-form on $\mathfrak{h}$ with coefficients in $\mathcal{H}$. When f is a normalized eigenform, it is section of $\Omega^1_{\overline{\mathcal{M}}_{1,1/\mathbb{Q}}}(\log P) \otimes F^{2n}S^{2n}\mathcal{H}$ defined over $\mathbb{Q}$, [11, §19]. The de Rham theorem for relative completion [9] implies that if f_j is a modular form of weight $2n_j + 2$, then the function

$$I_{[f_1|\cdots|f_m]} : \mathrm{SL}_2(\mathbb{Z}) \to S^{n_1}H \otimes S^{n_2}H \otimes \cdots \otimes S^{n_m}H$$

defined by the *regularized* (cf. [2]) iterated integral

$$I_{[f_1|\ldots|f_m]} : \gamma \mapsto \int_\gamma \omega_{f_1}\omega_{f_2}\ldots\omega_{f_m}$$

is an element of $F^{2n+m}\big(\mathcal{O}(\mathcal{G})\otimes S^{n_1}H\otimes S^{n_2}H\otimes\cdots\otimes S^{n_m}H\big)$ where $n = n_1+\cdots+n_m$. It is a 1-cochain satisfying

$$I_{[f_1|\cdots|f_m]} \in F^{2n+m}C^1(\mathcal{G}, S^{n_1}H \otimes S^{n_2}H \otimes \cdots \otimes S^{n_m}H).$$

When $m = 1$, it is a cocycle. Invariance under Γ and standard properties of iterated integrals imply that

$$I_{[f_1]} \cup I_{[f_2]} + \delta I_{[f_1|f_2]} = 0.$$

See [2, (5.2)].

Remark 8.2. Note that in [2], $\mathrm{SL}_2(\mathbb{Z})$ acts on the *right* of the vector space $V = \mathbb{Q}X \oplus \mathbb{Q}Y$ by the action

$$\begin{pmatrix} X \\ Y \end{pmatrix} \mapsto \begin{pmatrix} a & b \\ c & d \end{pmatrix}\begin{pmatrix} X \\ Y \end{pmatrix}.$$

This corresponds to the left action

$$\begin{pmatrix} \mathbf{a} & -\mathbf{b} \end{pmatrix} \mapsto \begin{pmatrix} \mathbf{a} & -\mathbf{b} \end{pmatrix}\begin{pmatrix} a & b \\ c & d \end{pmatrix}$$

of $\mathrm{SL}_2(\mathbb{Z})$ on $H = \mathbb{Q}\mathbf{a} \oplus \mathbb{Q}\mathbf{b}$ under the identification $X = \mathbf{b}$, $Y = \mathbf{a}$.

8.3 Interlude

For clarity, we make a few remarks about the Betti and de Rham structures of the coefficients $S^m H(r)$. Although we are not using the $\mathbb{Q}$-DR structure in the construction of DB-cohomology, we describe the $\mathbb{Q}$-DR structure of $S^m H(r)$ as that should help clarify the normalizations we choose in this section and the next as well as its role in applications, such as those in [13]. Relevant background can be found in [12].

Suppose that V is a Hodge structure with underlying $\mathbb{Q}$-Betti and $\mathbb{Q}$-de Rham spaces V^B and V^{DR}. One can identify the $\mathbb{Q}$-de Rham space of $V(r)$ with V^{DR} and its $\mathbb{Q}$-Betti space with

$$V(r)^B := (2\pi i)^r V^B.$$

We now explain how this works when $V = S^m H$.

As a Hodge structure of weight 1, H^B has $\mathbb{Q}$ basis $\mathbf{a}$ and $\mathbf{b}$ and H^{DR} has a $\mathbb{Q}$-basis $\mathbf{a}$ and $\mathbf{w} = -2\pi i\mathbf{b}$. The vector $\mathbf{w}$ spans $F^1 H^{\mathrm{DR}}$. The mth symmetric power $S^m H$ thus has $\mathbb{Q}$-Betti basis the monomials $\mathbf{a}^j\mathbf{b}^{m-j}$, and $\mathbb{Q}$-de Rham basis $(2\pi i)^{m-j}\mathbf{a}^j\mathbf{b}^{m-j}$. It decomposes as a sum

$$S^m H = \mathbb{Q}(-m) \oplus \mathbb{Q}(-m+1) \oplus \cdots \oplus \mathbb{Q}(1) \oplus \mathbb{Q}(0).$$

The $\mathbb{Q}$-Betti space of $S^m H(r)$ is $S^m H(r)^B = (2\pi i)^r S^m H^B$. For future reference, the $\mathbb{Q}$-de Rham and $\mathbb{Q}$-Betti bases of $S^m H(r)$ are given in Table 1.

Table 1: $\mathbb{Q}$-de Rham and $\mathbb{Q}$-Betti bases of $S^m H(r)$

	$\mathbb{Q}(r-m)$	$\dots$	$\mathbb{Q}(r+j-m)$	$\dots$	$\mathbb{Q}(r)$
DR	$\mathbf{w}^m$		$\mathbf{w}^{m-j}\mathbf{a}^j$		$\mathbf{a}^m$
B	$(2\pi i)^r\mathbf{b}^m$		$(2\pi i)^r\mathbf{b}^{m-j}\mathbf{a}^j$		$(2\pi i)^r\mathbf{a}^m$

Denote the *normalized* Eisenstein series of weight $2k$ by G_{2k}. (The definition is recalled below.) If we regard $\omega_{G_{2k}}$ as taking values in $S^{2k-2}H(2k-1)$, then

$$I_{[G_{2m_1}|\dots|G_{2m_r}]} \in F^0 W_0\big[\mathcal{O}(\mathcal{G}) \otimes S^{2m_1-2}H \otimes \cdots \otimes S^{2m_r-2}H(2m-r)\big],$$

where $m = m_1 + \cdots + m_r$.

8.4 The Eisenstein cocycle

The generator $\mathbf{G}_{2n+2}$ of $H^1_{\mathcal{D}}(\mathcal{G}, S^{2n}H(2n+1))$ corresponds to the normalized Eisenstein series

$$G_{2n}(q) = -\frac{B_{2n}}{4n} + \sum_{n=1}^{\infty} \sigma_{2n-1}(n)q^n$$

of weight $2n+2$, where B_{2n} denotes the $2n$th Bernoulli number. The $\mathbb{Q}$-DR 1-cocycle that corresponds to $\omega_{G_{2n+2}}$ is $I_{[G_{2n+2}]}$. Brown's computation [2, Lem. 7.1] implies that

$$Z_{2n} := \begin{bmatrix} & -\frac{(2n)!}{2}\zeta(2n+1)\mathbf{a}^{2n} & \\ I_{[G_{2n+2}]}(\mathbf{b},\mathbf{a}) & & (2\pi i)^{2n+1} e^0_{2n+2}(\mathbf{b},\mathbf{a}) \end{bmatrix}$$

is a 1-cocycle that represents $\mathbf{G}_{2n}$, where e^0_{2n+2} is the 1-cocycle defined in [2, §7.3]. It is the "rational part" of the normalized period polynomial of $G_{2n+2}/(2\pi i)^{2n+1}$.

Since $\Gamma \to \mathcal{G}(\mathbb{Q})$ is Zariski dense, the restriction map $\mathcal{C}^\bullet(\mathcal{G},V) \to \mathcal{C}^\bullet(\Gamma,V)$ on group cochains is injective. Since $I_{[G_{2n+2}]}$ and $\delta\mathbf{b}^{2n} \in \mathcal{C}^1(\mathcal{G}_\mathbb{Q}, S^{2n}H)$, e^0_{2n+2} is also in $\mathcal{C}^1(\mathcal{G}, S^{2n}H)$. It is rational as it takes rational values on Γ.

Remark 8.3. The class $\mathbf{G}_{2n+2}$ restricts to the base point $\partial/\partial q$ to give an element of

$$\operatorname{Ext}^1_{\mathsf{MHS}}(\mathbb{Q}, S^{2n}H(2n+1)) \cong \oplus_{r=1}^{2n+1} \operatorname{Ext}^1_{\mathsf{MHS}}(\mathbb{Q},\mathbb{Q}(r)).$$

This is the fiber of the corresponding VMHS over $\mathcal{M}_{1,1}$ at $\partial/\partial q$ and is the limit MHS of this variation with respect to the tangent vector $\partial/\partial q$. Brown's cocycle implies that this limit MHS is the coset of $-(2n)!\zeta(2n+1)\mathbf{a}^{2n}/2$ in

$$\operatorname{Ext}^1_{\mathsf{MHS}}(\mathbb{Q}, S^{2n}H(2n+1)) \cong S^{2n}H_\mathbb{C}/(2\pi i)^{2n+1}S^{2n}H_\mathbb{Q}.$$

It corresponds to the coset of $-(2n)!\zeta(2n+1)$ in $\operatorname{Ext}^1_{\mathsf{MHS}}(\mathbb{Q},\mathbb{Q}(2n+1))$.

8.5 The cup product of Eisenstein classes

In this section, we compute a formula for the external cup product $\mathbf{G}_{2j+2}\cup\mathbf{G}_{2k+2}$. This will allow us to use Brown's period computations [2] to compute the cup products of the classes $\mathbf{G}_{2m}$ in Deligne cohomology.

Suppose that $j \geq k > 0$. Set $n = j+k$. In the following discussion, the cup product of groups cochains is to be interpreted as the external cup product

$$\cup : \mathcal{C}^\bullet(\mathcal{G}, S^{2j}H(2j+1)) \otimes \mathcal{C}^\bullet(\mathcal{G}, S^{2k}H(2k+1)) \to \mathcal{C}^\bullet(\mathcal{G}, S^{2j}H \otimes S^{2k}H(2n+2)).$$

Set $A_{2m+2} = -\frac{(2m)!}{2}\zeta(2m+1)\mathbf{a}^{2m}$. Observe that

$$Z_{2j+2} \cup_{1/2} Z_{2k+2} = \begin{bmatrix} & (2\pi i)^{2k+1}A_{2j+2}\cup e^0_{2k+2} - (2\pi i)^{2j+1}e^0_{2j+2}\cup A_{2k+2} + C \\ I_{[G_{2j+2}]}\cup I_{[G_{2k+2}]} & (2\pi i)^{2n+2}e^0_{2j+2}\cup e^0_{2k+2} \end{bmatrix}$$

with $\delta C = 0$, where

$$C = \frac{1}{2}\big(A_{2j+2}\cup(\delta A_{2k+2}) - (\delta A_{2j+2})\cup A_{2k+2}\big).$$

For future reference, we note that this is *real*. Since the coboundary operator δ is strict with respect to $W_\bullet$, we can find $h \in W_0\mathcal{C}^1(\mathcal{G}_\mathbb{Q}, S^{2j}H_\mathbb{Q}\otimes S^{2k}H_\mathbb{Q})(2n+2)$ such that

$$e^0_{2j+2}\cup e^0_{2k+2} + \delta h = 0.$$

(Brown [2, §10.3] denotes a choice of h by $-e^0_{2j+2,2k+2}$.) Since

$$I_{[G_{2j+2}]}\cup I_{[G_{2k+2}]} + \delta I_{[G_{2j+2}|G_{2k+2}]} = 0$$

and since (after twisting)

$$I_{[G_{2j+2}|G_{2k+2}]} \in F^0W_0\mathcal{C}^1(\mathcal{G}, S^{2j+1}H\otimes S^{2k+1}H(2n+2)),$$

$$D := \begin{bmatrix} & 0 & \\ I_{[G_{2j+2}|G_{2k+2}]} & & (2\pi i)^{2n+2}h \end{bmatrix}$$

is a 1-cochain.

The cup product, as an element of $H^2_{\mathcal{D}}(\mathcal{M}_{1,1}, S^{2j}\mathbb{H} \otimes S^{2k}\mathbb{H}(2n+2))$, is represented by the cocycle

$$Z_{2j+2} \cup_{1/2} Z_{2k+2} + \delta D = \begin{bmatrix} & E & \\ 0 & & 0 \end{bmatrix}$$

where E is the 1-cocycle

$$I_{[G_{2j+2}|G_{2k+2}]} + (2\pi i)^{2k+1} A_{2j+2} \cup e^0_{2k+2} - (2\pi i)^{2j+1} e^0_{2j+2} \cup A_{2k+2} + C - (2\pi i)^{2n+2} h.$$

with values in $S^{2j}H \otimes S^{2k}H(2n+2)$.

Proposition 8.4. *With the notation above, the cup product* $\mathbf{G}_{2j+2} \cup \mathbf{G}_{2k+2}$ *in the* real *Deligne cohomology group*

$$H^2_{\mathcal{D}}(\mathcal{G}, S^{2j}H_{\mathbb{R}} \otimes S^{2k}H_{\mathbb{R}}(2n+2)) \cong \mathrm{Ext}^1_{\mathsf{MHS}}\left(\mathbb{R}, H^1(\mathcal{G}, S^{2j}H_{\mathbb{R}} \otimes S^{2k}H_{\mathbb{R}}(2n+2))\right)$$

is represented by the imaginary part of

$$\mathrm{Im}(I_{[G_{2j+2}|G_{2k+2}]} + (2\pi i)^{2k+1} A_{2j+2} \cup e^0_{2k+2} - (2\pi i)^{2j+1} e^0_{2j+2} \cup A_{2k+2}).$$

Proof. To compute the cup product in *real* Deligne cohomology, we can modify E by any element of the form

$$\delta \begin{bmatrix} & 0 & \\ 0 & & z \end{bmatrix}$$

where z is any cocycle in $W_0\mathcal{C}^1(\mathcal{G}_{\mathbb{R}}, S^{2j}H \otimes S^{2k}H(2n+2))$. Since C and h are real, this implies that

$$\mathbf{G}_{2j+2} \cup \mathbf{G}_{2k+2} \in H^2_{\mathcal{D}}(\mathcal{G}, S^{2j}H_{\mathbb{R}} \otimes S^{2k}H_{\mathbb{R}}(2n+2)) \cong H^1(\mathcal{G}, S^{2j}H_{\mathbb{R}} \otimes S^{2k}H_{\mathbb{R}}(2n+2))$$

is represented by the imaginary part

$$\mathrm{Im}(I_{[G_{2j+2}|G_{2k+2}]} + (2\pi i)^{2k+1} A_{2j+2} \cup e^0_{2k+2} - (2\pi i)^{2j+1} e^0_{2j+2} \cup A_{2k+2})$$

of E with respect to the isomorphisms

$$\begin{aligned} &\mathrm{Ext}^1_{\mathsf{MHS}}\left(\mathbb{R}, H^1(\mathcal{G}, S^{2j}H_{\mathbb{R}} \otimes S^{2k}H_{\mathbb{R}}(2n+2))\right) \\ &\cong H^1(\mathcal{G}, S^{2j}H_{\mathbb{C}} \otimes S^{2k}H_{\mathbb{C}})/(2\pi i)^{2n+2} H^1(\mathcal{G}, S^{2j}H_{\mathbb{R}} \otimes S^{2k}H_{\mathbb{R}}) \\ &\cong iH^1(\mathcal{G}, S^{2j}H_{\mathbb{R}} \otimes S^{2k}H_{\mathbb{R}}). \end{aligned}$$

□

8.6 The cuspidal projection of the cup product

In this section, we interpret Brown's period computations [2, §11] in terms of Deligne cohomology.

Recall that the *modular symbol* of a cusp form f of weight $2n+2$ is the homogeneous polynomial

$$\mathsf{r}_f := \int_0^\infty \omega_f \in S^{2n} H_{\mathbb{C}},$$

where the path of integration is the imaginary axis. When f has real Fourier coefficients (such as when $f \in \mathfrak{B}_{2n+2}$), this decomposes

$$\mathsf{r}_f = \mathsf{r}_f^+ + i\mathsf{r}_f^-$$

where $\mathsf{r}_f^\pm$ are real and where all terms of r_f^+ (resp. r_f^-) have even (resp. odd) degrees in $\mathbf{a}$ and $\mathbf{b}$. The polynomials $\mathsf{r}_f^\pm$ represent classes in $H^1_{\mathrm{cusp}}(\mathcal{M}_{1,1}, S^{2n} H_{\mathbb{R}})$.

Theorem 8.5 (Brown). *Suppose that $j \geq k > 0$ and that $0 \leq r \leq 2k$. Set $n = j+k$. If f is a normalized Hecke eigen cusp form of weight $2n-2r+2$, then*

$$\Phi_{r,f}(\mathbf{G}_{2j+2} \otimes \mathbf{G}_{2k+2}) = (2\pi i)^r \Lambda(f, 2n-r+2)\mathsf{r}_f^\epsilon$$

where $\Lambda(f,s) = (2\pi)^{-s}\Gamma(s)L(f,s)$ is the completed L-function associated to f and $\epsilon \in \{\pm\}$ is the sign of $(-1)^{r+1}$.

Proof. This follows directly from Equations (11.2) and (11.4) and Corollary 11.2 in [2] and Proposition 8.4. □

8.7 An external cup product

We can also use Brown's cocycle $\mathbf{G}_{2n+2}$ to compute the cup product

$$\mathrm{Ext}^1_{\mathsf{MHS}}(\mathbb{Q}, \mathbb{Q}(2m-1)) \otimes H^1_{\mathcal{D}}(\mathcal{G}, S^{2n} H(2n+1)) \to H^2_{\mathcal{D}}(\mathcal{G}, S^{2n} H(2n+2m))$$

of the class of $\zeta(2m-1)$ with the class G_{2n+1} of the Eisenstein series, where we are identifying $H^1_{\mathcal{D}}(\mathcal{G}, \mathbb{Q}(2m-1))$ with $\mathrm{Ext}^1_{\mathsf{MHS}}(\mathbb{Q}, \mathbb{Q}(2m-1))$ via the homomorphism from $\mathcal{G}$ to the trivial group.

One has the projection

$$\pi_{\mathrm{cusp}} : H^2_{\mathcal{D}}(\mathcal{G}, S^{2n} H_{\mathbb{Q}}(2n+2m)) \to \mathrm{Ext}^1_{\mathsf{MHS}}(\mathbb{Q}, H^1_{\mathrm{cusp}}(\mathcal{M}_{1,1}, S^{2n}\mathbb{H}(2n+2m)))$$

onto its cuspidal part. One also has the Eisenstein projection

$$\mathrm{id} - \pi_{\mathrm{cusp}} : H^2_{\mathcal{D}}(\mathcal{G}, S^{2n} H_{\mathbb{Q}}(2n+2m)) \to \mathrm{Ext}^1_{\mathsf{MHS}}(\mathbb{Q}, \mathbb{Q}(2m-1)).$$

Denote the class of $\zeta(2m-1)$ in $\mathrm{Ext}^1_{\mathsf{MHS}}(\mathbb{Q}, \mathbb{Q}(2m-1))$ by $\mathbf{Z}_{2m-1}$. It is represented by the Deligne cocycle

$$\begin{bmatrix} & \zeta(2m-1) & \\ 0 & & 0 \end{bmatrix}.$$

Proposition 8.6. *The cuspidal projection of*

$$\mathbf{Z}_{2m-1} \cup \mathbf{G}_{2n+2} \in H^2_{\mathcal{D}}(\mathcal{G}, S^{2n}H(2n+2m))$$

is trivial. Its Eisenstein projection is

$$\mathbf{Z}_{2m-1} \in \operatorname{Ext}^1_{\mathsf{MHS}}(\mathbb{Q}, \mathbb{Q}(2m-1)).$$

Proof. The cup product is represented by

$$\begin{bmatrix} & \zeta(2m-1) & \\ 0 & & 0 \end{bmatrix} \cup_0 \begin{bmatrix} & -\frac{(2n)!}{2}\zeta(2n+1)\mathbf{a}^{2n} & \\ I_{[G_{2n+2}]}(\mathbf{b},\mathbf{a}) & & (2\pi i)^{2n+1}e^0_{2n+2}(\mathbf{b},\mathbf{a}) \end{bmatrix}$$

$$= \begin{bmatrix} & (2\pi i)^{2n+1}\zeta(2m-1)e^0_{2n+2}(\mathbf{b},\mathbf{a}) & \\ 0 & & 0 \end{bmatrix}.$$

This has trivial projection under π_{cusp}. The top entry lies in $\mathbb{C}e^0_{2n+2}(\mathbf{b},\mathbf{a})$ and is well-defined mod $(2\pi i)^{2m+2n}e^0_{2n+2}(\mathbf{b},\mathbf{a})$. It therefore corresponds to $\zeta(2m-1)$ mod $(2\pi i)^{2m-1}\mathbb{Q}$ in $\operatorname{Ext}^1_{\mathsf{MHS}}(\mathbb{Q}, \mathbb{Q}(2m-1))$. □

9 Lie algebra aspects

One can also define the Deligne–Beilinson cohomology of a Lie algebra with MHS. This was done in the unipotent case in [4] where it was related to extensions of unipotent variations of MHS. Here we briefly consider the general case and its relation to the DB-cohomology of affine groups.

9.1 Generalities

Suppose that $\mathfrak{h}$ is a Lie algebra over the field $\mathbb{F}$ of characteristic zero. Denote its enveloping algebra by $U\mathfrak{h}$. The cohomology of $\mathfrak{h}$ with coefficients in an $\mathfrak{h}$-module V is defined by

$$H^\bullet(\mathfrak{h}, V) := \operatorname{Ext}^\bullet_{U\mathfrak{h}}(\mathbb{F}, V).$$

It can be computed using the standard projective resolution $U\mathfrak{h} \otimes \Lambda^\bullet\mathfrak{h} \to \mathbb{F}$, which computes $H^\bullet(\mathfrak{h}, V)$ from the Chevalley–Eilenberg cochains

$$\mathcal{C}^\bullet(\mathfrak{h}, V) = \operatorname{Hom}_{\mathbb{F}}(\Lambda^\bullet\mathfrak{h}, V) \cong \operatorname{Hom}_{U\mathfrak{h}}(U\mathfrak{h} \otimes \Lambda^\bullet\mathfrak{h}, V)$$

with its standard differential. (See [5] for details.)

If $\mathfrak{h}$ is a topological Lie algebra (such as a pronilpotent Lie algebra) which is topologized as the limit of a projective system of Lie algebras $\mathfrak{h}_\alpha$, we define the continuous cohomology $H^\bullet_{\mathrm{cts}}(\mathfrak{h}, V)$ of $\mathfrak{h}$ by

$$H^\bullet_{\mathrm{cts}}(\mathfrak{h}, V) = \varinjlim_\alpha H^\bullet(\mathfrak{h}_\alpha, V).$$

It can be computed using the complex

$$\mathcal{C}^\bullet_{\mathrm{cts}}(\mathfrak{h}, V) := \varinjlim_\alpha \mathcal{C}^\bullet(\mathfrak{h}_\alpha, V)$$

of continuous cochains.

9.2 Hodge theory

Suppose that $\mathfrak{h}$ is a (finite dimensional) Lie algebra in $\mathsf{MHS}_{\mathbb{F}}$. A Hodge representation of $\mathfrak{h}$ is a finite dimensional $\mathfrak{h}$-module V endowed with an $\mathbb{F}$-MHS for which the action $\mathfrak{h}\otimes V \to V$ is a morphism of MHS. Denote the category of Hodge representations of $\mathfrak{h}$ by $\mathsf{HRep}(\mathfrak{h})$. When V is a Hodge representation of $\mathfrak{h}$, the Chevalley–Eilenberg complex $\mathcal{C}^\bullet(\mathfrak{h},V)$ is a complex in $\mathsf{MHS}_{\mathbb{F}}$, which implies that $H^\bullet(\mathfrak{h},V)$ has a MHS.

More generally, suppose that $\mathfrak{h}$ is a topological Lie algebra which is topologized as the inverse limit of a projective system $\{\mathfrak{h}_\alpha\}$ of Lie algebras in $\mathsf{MHS}_{\mathbb{F}}$. We will refer to such a Lie algebra as a Lie algebra in pro-MHS_F. Define $\mathsf{HRep}(\mathfrak{h})$ by

$$\mathsf{HRep}(\mathfrak{h}) = \varinjlim_{\alpha} \mathsf{HRep}(\mathfrak{h}_\alpha).$$

For each V in $\mathsf{HRep}(\mathfrak{h})$, the continuous cochain complex $\mathcal{C}^\bullet_{\text{cts}}(\mathfrak{h},V)$ is a complex in ind-$\mathsf{MHS}_{\mathbb{F}}$.

Definition 9.1. The Deligne–Beilinson cohomology $H^\bullet_{\mathcal{D}}(\mathfrak{h},V)$ of a Lie algebra $\mathfrak{h}$ in pro-$\mathsf{MHS}_{\mathbb{F}}$ and coefficients in an object V of $\mathsf{HRep}(\mathfrak{h})$ is defined to be the cohomology of the complex

$$\operatorname{cone}\big(F^0W_0\mathcal{C}^\bullet_{\text{cts}}(\mathfrak{h},V)_{\mathbb{C}} \oplus W_0\mathcal{C}^\bullet_{\text{cts}}(\mathfrak{h},V)_{\mathbb{F}} \to W_0\mathcal{C}^\bullet_{\text{cts}}(\mathfrak{h},V)_{\mathbb{C}}\big)[-1].$$

Standard arguments imply that, for all $j\geq 0$, there is a short exact sequence

$$0\to \operatorname{Ext}^1_{\mathsf{MHS}}\left(\mathbb{F},H^{j-1}(\mathfrak{h},V)\right) \to H^j_{\mathcal{D}}(\mathfrak{h},V)\to \operatorname{Hom}_{\mathsf{MHS}}\left(\mathbb{F},H^j(\mathfrak{h},V)\right)\to 0.$$

9.3 Relation to group cohomology

Suppose that G is an affine (and therefore proalgebraic) $\mathbb{F}$-group. Its Lie algebra $\mathfrak{g}$ is the inverse limit of a finite dimensional Lie algebras. The functor $\mathsf{Rep}(G)\to \mathsf{Rep}^{\text{cts}}(\mathfrak{g})$ induces a map

$$H^\bullet(G,V)\to H^\bullet_{\text{cts}}(\mathfrak{g},V) \tag{9.1}$$

for all V in $\mathsf{Rep}(G)$. This is an isomorphism when G is prounipotent. This map can be realized on the chain level by choosing an injective resolution $V\hookrightarrow I^\bullet$ of V in $\mathsf{Rep}(G)$. Since the complex $I^\bullet$ is a resolution of V in $\mathsf{Rep}(\mathfrak{g})$, the induced chain map

$$\mathcal{C}^\bullet(\operatorname{Hom}^{\text{cts}}_{\mathbb{F}}(\Lambda^\bullet\mathfrak{g},V))\hookrightarrow \operatorname{tot}\operatorname{Hom}^{\text{cts}}_{\mathbb{F}}(\Lambda^\bullet\mathfrak{g},I^\bullet)$$

is a quasi-isomorphism. So the right hand complex computes $H^\bullet_{\text{cts}}(\mathfrak{g},V)$. The inclusion

$$(I^\bullet)^G\hookrightarrow \operatorname{tot}\operatorname{Hom}^{\text{cts}}_{\mathbb{F}}(\Lambda^\bullet\mathfrak{g},I^\bullet)$$

is a chain map which induces (9.1).

Denote the maximal ideal of $\mathcal{O}(G)$ of functions that vanish at the identity by $\mathfrak{m}_e$. The Lie algebra $\mathfrak{g}$ of G is

$$\mathfrak{g} = \operatorname{Hom}_{\mathbb{F}}(\mathfrak{m}_e/\mathfrak{m}_e^2,\mathbb{F}).$$

Its bracket is dual to the skew symmetrization $\mathfrak{m}_e/\mathfrak{m}_e^2 \to (\mathfrak{m}_e/\mathfrak{m}_e^2)^{\otimes 2}$ of the reduced diagonal $\mathfrak{m}_e \to \mathfrak{m}_e \otimes \mathfrak{m}_e$. This implies that if G has an $\mathbb{F}$-MHS, then its Lie algebra is a Lie algebra in the category pro-$\mathsf{MHS}_\mathbb{F}$ and that the restriction functor $\mathsf{Rep}(G) \to \mathsf{Rep}^{\mathrm{cts}}(\mathfrak{g})$ takes Hodge representations to Hodge representations.

Proposition 9.2. *If G is an affine group with an $\mathbb{F}$-MHS, then for all Hodge representations V of G, the natural map (9.1) is a morphism of MHS and there is a natural map*

$$H^\bullet_\mathcal{D}(G,V) \to H^\bullet_\mathcal{D}(\mathfrak{g},V)$$

which is an isomorphism when G is prounipotent.

Proof. Let $V \to I^\bullet$ be the standard injective resolution of V in $\mathsf{Rep}(G)$. The linear map $(I^\bullet)^G = V \otimes \mathcal{O}(G)^{\otimes n} \to \mathrm{Hom}_\mathbb{F}(\Lambda^n \mathfrak{g}, V)$ defined by

$$v \otimes f_1 \otimes \cdots \otimes f_n \mapsto v \otimes (df_1 \wedge \cdots \wedge df_n)$$

is a chain map that induces the natural map $H^\bullet(G,V) \to H^\bullet(\mathfrak{g},V)$. It is a morphism of MHS, which implies that the natural map is a morphism of MHS. It also induces a map of Deligne cochains, and therefore a map of Deligne cohomology. □

When G is not prounipotent, the homomorphism (9.1) is generally not an isomorphism. To consider the difference, we adopt the natural hypotheses of Section 4.5. Namely, $G = W_0 G$ and that $R := \mathrm{Gr}^W_0 G$ is reductive. In this case, the restriction map $H^\bullet(G,V) \to H^\bullet(U,V)$ is an inclusion of MHS which induces an isomorphism of MHS

$$H^\bullet(G,V) \xrightarrow{\simeq} H^\bullet(U,V)^R.$$

We can thus express $H^\bullet_\mathcal{D}(G,V)$ as an extension

$$0 \to \mathrm{Ext}^1_{\mathsf{MHS}}(\mathbb{F}, H^{j-1}(U,V)^R) \to H^j_\mathcal{D}(G,V) \to \Gamma H^j(U,V)^R \to 0.$$

This sequence can also be deduced from the Hochschild-Serre spectral sequence of the extension

$$1 \to U \to \pi_1(\mathsf{MHS}) \ltimes G \to \pi_1(\mathsf{MHS}) \ltimes R \to 1.$$

It seems that the group R does not (in general) act on $H^j_\mathcal{D}(U,V)$, so it does not make sense to write $H^\bullet_\mathcal{D}(G,V) = H^\bullet_\mathcal{D}(U,V)^R$.

10 The Eisenstein projection of the cup product

Here we give an alternative approach to Brown's computation [2, Thm. 10.1] of the Eisenstein projection of $\mathbf{G}_{2n+2m} \cup \mathbf{G}_{2m}$. We do this by restricting to the punctured q-disk, where the variations of MHS that correspond to the $\mathbf{G}_{2k}$ become unipotent. This allows us to use the Lie algebra version of DB-cohomology explained in the previous section. While not essential, it does simplify the computation.

The inclusion $\Delta^* \to \mathcal{M}_{1,1}$ of the punctured q-disk into $\mathcal{M}_{1,1}$ induces a homomorphism

$$\pi_1(\Delta^*, \partial/\partial q) \to \pi_1(\mathcal{M}_{1,1}^{\text{an}}, \partial/\partial q) \cong \mathrm{SL}(H_{\mathbb{Z}})$$

whose image is the unipotent element T whose logarithm is the nilpotent endomorphism $N = -\mathbf{a}\partial/\partial\mathbf{b}$. The unipotent completion of $\Gamma_\infty := \pi_1(\Delta^*, \partial/\partial q)$ is a unipotent group isomorphic to $\mathbb{G}_a$ whose Lie algebra $\mathfrak{g}_\infty$ is isomorphic to $\mathbb{Q}(1)$ in MHS. The inclusion $\Gamma_\infty \hookrightarrow \Gamma$ induces a homomorphism $\mathcal{G}_\infty \hookrightarrow \mathcal{G}$ which is a morphism of MHS.

Remark 10.1. Hodge representations of $\mathcal{G}_\infty$ correspond to admissible variations of MHS on the punctured tangent space $T'_{e_o}\overline{\mathcal{M}}_{1,1} := T_{e_o}\overline{\mathcal{M}}_{1,1} - \{0\}$ at the cusp e_o. These are nilpotent orbits of MHS.

10.1 DB-cohomology of $\mathcal{G}_\infty$

The Deligne cohomology of $\mathcal{G}_\infty$ is particularly simple, which makes it well suited to computations. If V is a $\mathcal{G}_\infty$ module, then

$$H^0(\mathcal{G}_\infty, V) = V^N, \quad H^1(\mathcal{G}^\infty, V) = (V/NV)(-1),$$

and all other cohomology groups vanish. If V is an $\mathrm{SL}(H)$-module, viewed as a $\mathcal{G}_\infty$-module via the homomorphism, $\mathcal{G}_\infty \to \mathcal{G} \to \mathrm{SL}(H)$, then V/NV is isomorphic to the space of highest weight vectors in V.

If V is a Hodge representation of $\mathcal{G}_\infty$, then there is a natural isomorphism

$$H^2_{\mathcal{D}}(\mathcal{G}_\infty, V(r)) \cong \mathrm{Ext}^1_{\mathsf{MHS}}\big(\mathbb{Q}, (V/NV)(r-1)\big) \cong V^{\mathrm{DR}}_{\mathbb{C}}/\big((2\pi i)^{r-1}V^B_{\mathbb{Q}} + NV^{\mathrm{DR}}_{\mathbb{C}}\big).$$

10.2 Lie algebra cochains

Denote the coordinate in $T_{e_o}\overline{\mathcal{M}}_{1,1}$ with respect to the basis vector $\partial/\partial q$ by q_o. The coordinate ring of $\mathcal{G}^{\mathrm{DR}}_\infty$ is the polynomial algebra over $\mathbb{Q}$ generated by the iterated integral $[dq_o/q_o]$. Since

$$\overbrace{[dq_o/q_o|\cdots|dq_o/q_o]}^{r} = [dq_o/q_o]^r/r!,$$

the cotangent space of $\mathcal{G}^\infty$ is spanned by the coset of $[dq_o/q_o]$ mod iterated integrals of dq_o/q_o of length ≥ 2. Let λ_{DR} be the linear function on the tangent space $\mathfrak{g}^{\mathrm{DR}}_\infty$ of $\mathcal{G}^{\mathrm{DR}}_\infty$ at 1 that takes the value 1 on $[dq_o/q_o]$. Let $\lambda_B = \lambda_{\mathrm{DR}}/2\pi i$. It is a linear functional on $\mathfrak{g}^B_\infty$.

Suppose that V is a Hodge representation of $\mathfrak{g}_\infty$. Then

$$\mathcal{C}^\bullet(\mathfrak{g}^{\mathrm{DR}}_\infty, V^{\mathrm{DR}}) = \Lambda^\bullet(d\lambda_{\mathrm{DR}}) \otimes V^{\mathrm{DR}}$$

with the differential $\delta : V^{\mathrm{DR}} \to V^{\mathrm{DR}} \otimes d\lambda^{\mathrm{DR}}$ defined by $v \mapsto Nv\, d\lambda_{\mathrm{DR}}$. The Betti complex $\mathcal{C}^\bullet(\mathfrak{g}^B_\infty, V^B)$ is defined similarly. It has differential $\delta : V^B \to V^B \otimes d\lambda^B$ defined by $v \mapsto N^B v\, d\lambda_B$, where $N_B = N/2\pi i$. When $V = S^m H$, $N = \mathbf{a}\partial/\partial\mathbf{w}$ and $N_B = -\mathbf{a}\partial/\partial\mathbf{b}$.

10.3 Computation of the Eisenstein projection

Denote by $\mathbf{G}_{2k}^{\infty}$ the image of $\mathbf{G}_{2k}$ under the restriction map

$$H^1_{\mathcal{D}}\big(\mathcal{G}, S^{2k-2}H(2k-1)\big) \to H^1_{\mathcal{D}}\big(\mathcal{G}_\infty, S^{2k-2}H(2k-1)\big).$$

Proposition 10.2. *If $n > 0$ and $m > 1$, then*

$$\mathbf{G}_{2n+2m}^{\infty} \cup \mathbf{G}_{2m}^{\infty} \in H^2_{\mathcal{D}}\big(\mathcal{G}_\infty, S^{2n+2m-2}H \otimes S^{2m-2}H(2n+4m-2)\big)$$

is the coset of

$$-\zeta(2m-1)\frac{(2m-2)!}{2}\frac{B_{2n+2m}}{4n+4m}\mathbf{w}^{2n+2m-2} \otimes \mathbf{a}^{2m-2}$$

in

$$\begin{aligned} &\operatorname{Ext}^1_{\mathsf{MHS}}\Big(\mathbb{Q}, H^1\big(\mathcal{G}^\infty, S^{2n+2m-2}H \otimes S^{2m-2}H(2n+4m-2)\big)\Big) \\ &\cong \operatorname{Ext}^1_{\mathsf{MHS}}\big(\mathbb{Q}, [S^{2n+2m-2}H \otimes S^{2m-2}H(2n+4m-3)]/\operatorname{im} N\big) \\ &\cong \bigoplus_{r=0}^{2m-2} \operatorname{Ext}^1_{\mathsf{MHS}}(\mathbb{Q}, \mathbb{Q}(r+1)). \end{aligned}$$

Its projection to

$$\operatorname{Ext}^1_{\mathsf{MHS}}\Big(\mathbb{Q}, H^1\big(\mathcal{G}^\infty, S^{2n+4m-4-2r}H(2n+4m-2-r)\big)\Big) \cong \operatorname{Ext}^1_{\mathsf{MHS}}(\mathbb{Q}, \mathbb{Q}(r+1))$$

vanishes when $0 \le r < 2m-2$.

To compute the projection of $\mathbf{G}_{2n+2m}^{\infty} \cup \mathbf{G}_{2m}^{\infty}$ to $H^2_{\mathcal{D}}(\mathcal{G}_\infty, S^{2n}H(2n+2m))$, we need to specify an $\mathrm{SL}(H)$-invariant projection $S^{2n+2m-2}H \otimes S^{2m-2}H \to S^{2n}H$. To this end, define

$$\partial : S^a H(a) \otimes S^b H(b) \to S^{a-1}H \otimes S^{b-1}H(a+b-1)$$

to be the operator

$$\partial := \partial/\partial\mathbf{w} \otimes \partial/\partial\mathbf{a} - \partial/\partial\mathbf{a} \otimes \partial/\partial\mathbf{w} = (2\pi i)^{-1}(\partial/\partial\mathbf{a} \otimes \partial/\partial\mathbf{b} - \partial/\partial\mathbf{b} \otimes \partial/\partial\mathbf{a}).$$

It is $\mathrm{SL}(H)$ equivariant and preserves the $\mathbb{Q}$-DR and $\mathbb{Q}$-Betti structures.

Identify $S^{2n}H$ with $S^{2n}H \otimes 1$. Then ∂^{2m-2} is an $\mathrm{SL}(H)$-invariant operator

$$\partial^{2m-2} : S^{2n+2m-2}H \otimes S^{2m-2}H(2n+4m-2) \to S^{2n}H(2n+2m).$$

Since

$$\partial^{2m-2}(\mathbf{w}^{2n+2m-2} \otimes \mathbf{a}^{2m-2}) = (2m-2)!\frac{(2n+2m-2)!}{(2n)!}\mathbf{w}^{2n} \otimes 1,$$

and since the Eisenstein projection (8.1) is just the restriction mapping

$$\begin{aligned} &H^2_{\mathcal{D}}(\mathcal{G}, S^{2n}H(r)) \to H^2_{\mathcal{D}}(\mathcal{G}_\infty, S^{2n}H(r)) \\ &\qquad \cong H^2_{\mathcal{D}}(\mathcal{G}_\infty, S^{2n}H(r)) \cong \operatorname{Ext}^1_{\mathsf{MHS}}(\mathbb{Q}, \mathbb{Q}(2n+r)), \end{aligned}$$

we have:

Corollary 10.3. *If $n > 0$ and $m > 1$, then*

$$\partial^{2m-2}(\mathbf{G}_{2n+2m} \cup \mathbf{G}_{2m}) \in H^2_{\mathcal{D}}(\mathcal{G}_\infty, S^{2n}H(2n+2m))$$

is the coset of

$$-\zeta(2m-1)\frac{[(2m-2)!]^2}{2}\frac{(2n+2m-2)!}{(2n)!}\frac{B_{2n+2m}}{4n+4m}\mathbf{w}^{2n}$$

in

$$\begin{aligned}&\operatorname{Ext}^1_{\mathsf{MHS}}\Big(\mathbb{Q}, H^1(\mathcal{G}^\infty, S^{2n}H(2n+2m))\Big)\\ &\cong \operatorname{Ext}^1_{\mathsf{MHS}}\Big(\mathbb{Q}, [S^{2n}H(2n+2m-1)]/\operatorname{im} N\Big) \cong \operatorname{Ext}^1_{\mathsf{MHS}}(\mathbb{Q}, \mathbb{Q}(2m-1)).\end{aligned}$$ □

Proof of Proposition 10.2. The restriction of the cocycle $\mathbf{G}_{2k}$ to $\mathcal{G}^\infty$ is

$$\begin{bmatrix} & -\frac{(2k-2)!}{2}\zeta(2k-1)\mathbf{a}^{2k-2} & \\ -\frac{B_{2k}}{4k}\mathbf{w}^{2k-2}[dq_o/q_o] & & -\frac{B_{2k}}{4k}(2\pi i\mathbf{b})^{2k-2}[dq_o/q_o]\end{bmatrix}.$$

Consequently, the corresponding 1-cocycle on $\mathfrak{g}_\infty$ is

$$Z^\infty_{2k} := \begin{bmatrix} & -\frac{(2k-2)!}{2}\zeta(2k-1)\mathbf{a}^{2k-2} & \\ -\frac{B_{2k}}{4k}\mathbf{w}^{2k-2}d\lambda_{\mathrm{DR}} & & -2\pi i\frac{B_{2k}}{4k}(2\pi i\mathbf{b})^{2k-2}d\lambda_B\end{bmatrix}.$$

The class of $\mathbf{G}_{2n+2m} \cup \mathbf{G}_{2m}$ is represented by

$$Z^\infty_{2n+2m} \cup_0 Z^\infty_{2m} = \begin{bmatrix} & C & \\ 0 & & 0\end{bmatrix}$$

where

$$\begin{aligned}C = \bigg(&\frac{(2n+2m-2)!}{2}\frac{B_{2m}}{4m}\zeta(2n+2m-1)\,\mathbf{a}^{2n+2m-2}\otimes\mathbf{w}^{2m-2}\\ &-\frac{(2m-2)!}{2}\frac{B_{2n+2m}}{4n+4m}\zeta(2m-1)\,\mathbf{w}^{2n+2m-2}\otimes\mathbf{a}^{2m-2}\bigg)\,d\lambda_{\mathrm{DR}}.\end{aligned}$$

Since the set of (H) weights of highest weight vectors in $S^{2n+2m-2}H \otimes S^{2m-2}$ is $\{2n, 2n+2, \dots, 2n+2m-2\}$, and since $\mathbf{a}^{2n+2m-2}\otimes\mathbf{w}^{2m-2}$ has weight $-2n$, it is in the image of N. So the class of C in $H^1(\mathcal{G}_\infty, S^{2n+2m-2}H \otimes S^{2m-2})$ is also represented by

$$C' = -\frac{(2m-2)!}{2}\frac{B_{2n+2m}}{4n+4m}\zeta(2m-1)\,\mathbf{w}^{2n+2m-2}\otimes\mathbf{a}^{2m-2}\,d\lambda_{\mathrm{DR}}.$$

Since this vector has (H)-weight $2n$, its projection to $S^{2n+4m-4-2r}H$ is trivial, except possibly when $r = 2m-2$. □

10.4 Comparison with $\mathbf{Z}_{2m-1} \cup \mathbf{G}_{2n+2}^{\infty}$

The classes $\mathbf{G}_{2n+2m}^{\infty} \cup \mathbf{G}_{2m}^{\infty}$ and $\mathbf{Z}_{2m-1} \cup \mathbf{G}_{2n+2}^{\infty}$ both live in $H_{\mathcal{D}}^2(\mathcal{G}_{\infty}, S^{2n}H(2n+2m))$. In order to compare them, we will use the $\mathrm{SL}(H)$-invariant embedding

$$D^{\ell} : S^k H \to S^{k+\ell}H \otimes S^{\ell}H(\ell)$$

defined by

$$f(\mathbf{a}, \mathbf{w}) \mapsto (\mathbf{w} \otimes \mathbf{a} - \mathbf{a} \otimes \mathbf{w})^{\ell} f(\mathbf{a}, \mathbf{w}) \otimes 1,$$

which preserves the $\mathbb{Q}$-DR and $\mathbb{Q}$-Betti structures. The case of interest to us is when $\ell = 2m-2$ and $k = 2n$.

Lemma 10.4. *For all $k, \ell \geq 0$, we have*

$$D^{\ell}(\mathbf{w}^k \otimes 1) \equiv \frac{k+\ell+1}{k+1} \mathbf{w}^{k+\ell} \otimes \mathbf{a}^{\ell} \bmod \operatorname{im} N.$$

Proof. Identify $S^{k+\ell}H$ with polynomials in x of degree $\leq k+\ell$ and $S^{\ell}H$ with polynomials in y of degree $\leq \ell$, where $x^{k+\ell}$ corresponds to the highest weight vector of $S^{k+\ell}H$ and y^{ℓ} corresponds to the highest weight vector of $S^{\ell}H$. With these conventions, D corresponds to multiplication by $x-y$, the vector $D^{\ell}\mathbf{w}^k$ corresponds to $(x-y)^{\ell}x^k$, and the nilpotent endomorphism N of $S^{k+\ell}H \otimes S^{\ell}H$ corresponds to the derivation $\partial/\partial x + \partial/\partial y$. Since

$$\left(\frac{\partial}{\partial x} + \frac{\partial}{\partial y}\right) \sum_{i=1}^{j} \frac{(-1)^i}{i} \binom{k+\ell}{i-1} x^{k+\ell-i+1} y^i = -x^{k+\ell} + (-1)^j \binom{k+\ell}{j} x^{k+\ell-j} y^j,$$

$$x^{k+\ell} \equiv (-1)^j \binom{k+\ell}{j} x^{k+\ell-j} y^j \bmod \operatorname{im} N.$$

So

$$(x-y)^{\ell} x^k = \sum_{j=0}^{\ell} (-1)^j \binom{\ell}{j} x^{k+\ell-j} y^j \equiv x^{k+\ell} \sum_{j=0}^{\ell} \binom{\ell}{j} \Big/ \binom{k+\ell}{j} \bmod \operatorname{im} N.$$

The result now follows from the identity

$$\sum_{j=0}^{\ell} \binom{\ell}{j} \Big/ \binom{k+\ell}{j} = \frac{k+\ell+1}{k+1},$$

which can be proved by showing that the both sides satisfy the recursion

$$(k+\ell)F(k,\ell) = kF(k-1,\ell) + \ell F(k,\ell-1)$$

and verifying equality when either k or ℓ is zero. It can also be proved using the binomial transformation. □

Theorem 10.5. *If $n > 0$ and $m > 1$, then*

$$\mathbf{G}^{\infty}_{2n+2m} \cup \mathbf{G}^{\infty}_{2m} = \frac{(2m-2)!}{2} \frac{\binom{2n+2}{2}}{\binom{2n+2m}{2}} \frac{B_{2n+2m}}{B_{2n+2}} D^{2m-2}\big(\mathbf{Z}_{2m-1} \cup \mathbf{G}_{2n+2}\big)$$

in $H^2_{\mathcal{D}}\big(\mathcal{G}, S^{2n+2m-2}H \otimes S^{2m-2}(2n+2m)\big) \cong \operatorname{Ext}^1_{\mathsf{MHS}}(\mathbb{Q}, \mathbb{Q}(2m-1))$.

Proof. Proposition 8.6 implies that $\mathbf{Z}_{2m-1} \cup \mathbf{G}^{\infty}_{2n+2}$ is represented by the coset of

$$-\zeta(2m-1)\,\frac{B_{2n+2}}{4n+4}\,\mathbf{w}^{2n} \otimes 1$$

in

$$\operatorname{Ext}^1_{\mathsf{MHS}}\big(\mathbb{Q}, H^1(\mathcal{G}_\infty, S^{2n}H(2n+2m))\big) \cong \operatorname{Ext}^1_{\mathsf{MHS}}\big(\mathbb{Q}, S^{2n}H(2n+2m-1)/\operatorname{im} N\big).$$

Proposition 10.2 and Lemma 10.4 imply that $\mathbf{G}^{\infty}_{2n+2m} \cup \mathbf{G}^{\infty}_{2m}$ is represented by

$$\frac{(2m-2)!}{2} \frac{\binom{2n+2}{2}}{\binom{2n+2m}{2}} \frac{B_{2n+2m}}{B_{2n+2}} \left[-\zeta(2m-1)\,\frac{B_{2n+2}}{4n+4}\, D^{2m-2}(\mathbf{w}^{2n} \otimes 1)\right]. \qquad \square$$

A The orbifold case

For applications to (say) modular curves, one needs to know that Theorem 5.1 holds for orbifolds that are globally the quotient of a smooth variety by a finite group. Here we sketch a proof that this is indeed the case.

Suppose that X is the orbifold quotient of a smooth connected variety Y by a finite group Γ, which acts on Y on the right. We write $X = Y//\Gamma$. Note that we are not assuming that $\Gamma \to \operatorname{Aut} Y$ is injective.

To work on X is to work with Γ-equivariant objects on Y. For example, a local system (resp. admissible VMHS) over X is (by definition) a local system (resp. admissible VMHS) over Y with a right Γ action that commutes with the projection to X.

The fundamental group $\pi_1(X, y_o)$, where $y_o \in Y$ is a "geometric point" of X, is the group whose elements are pairs (α, γ), where $\alpha : [0,1] \to Y$ is a homotopy class of paths from y_o to $y_o\gamma$. Multiplication is

$$(\alpha, \gamma) \circ (\beta, \mu) = (\alpha|_\mu \circ \beta, \gamma\mu),$$

where $\circ$ denotes path composition (in the functional order) and $\alpha|_\mu$ denotes the translate of α by μ.[8] The projection $(\alpha, \gamma) \mapsto \gamma$ defines a homomorphism $\pi_1(X, y) \to \Gamma$. There is an exact sequence

$$1 \to \pi_1(Y, y_o) \to \pi_1(X, y_o) \to \Gamma \to 1.$$

[8] This is equivalent to the usual definition of the orbifold fundamental group of X. The description given here is the same as the one given in [10], except that it is modified for the change in the path multiplication convention.

If $\mathbb{V}$ is a local system over X, then one has a monodromy representation

$$\pi_1(X, y_o) \to \operatorname{Aut} V_o$$

where V_o denotes the fiber over y_o. If $\mathbb{H}$ is a PVHS over X, then the coordinate ring of the Zariski closure R_o of the monodromy representation $\pi_1(X, y_o) \to \operatorname{Aut} H_o$ has a Hodge structure of weight 0.

The construction of the MHS on $\mathcal{O}(\mathcal{G}_X)$ is similar to that in the non-orbifold case. First recall that the C^∞ de Rham complex of X with coefficients in a (real or complex) local system $\mathbb{V}$ over X is defined to be the complex of Γ-invariant forms on Y with values in $\mathbb{V}$:

$$E^\bullet(X, \mathbb{V}) := E^\bullet(Y, \mathbb{V})^\Gamma.$$

One uses the construction from [9] using the complex

$$E^\bullet_{\mathrm{fin}}(X, \mathbb{O}_o) := E^\bullet_{\mathrm{fin}}(Y, \mathbb{O}_o)^\Gamma$$

of R_o-finite vectors,[9] where $\mathbb{O}_o$ is the local system over X whose fiber over y_o is $\mathcal{O}(R_o)$ and whose monodromy representation factors through right multiplication by R_o.

To show that the orbifold version of Theorem 5.1 follows from the standard version, Theorem 5.1, we use a standard trick. Fix a simply connected smooth variety W on which Γ acts fixed point freely.[10] Fix $w_o \in W$ and set $z_o = (y_o, w_o)$. The projection $Y \times W \to Y$ induces an isomorphism on fundamental groups. Set

$$Z = (Y \times W)/\Gamma.$$

This is a smooth variety. The projection $Y \times W \to Y$ induces an orbifold map $(Z, z_o) \to (X, x_o)$ and an isomorphism $\pi_1(Z, z_o) \to \pi_1(X, y_o)$. So the relative completion of $\pi_1(Z, z_o)$ is isomorphic to the relative completion of $\pi_1(X, y_o)$. Their MHSs are also isomorphic as the projection

$$E^\bullet_{\mathrm{fin}}(Z, \mathbb{O}_o) = E^\bullet_{\mathrm{fin}}(Y \times W, \mathbb{O}_o)^\Gamma \to E^\bullet_{\mathrm{fin}}(Y, \mathbb{O}_o)^\Gamma = E^\bullet_{\mathrm{fin}}(X, \mathbb{O}_o)$$

induced by the Γ-invariant inclusion $Y \times w_o \hookrightarrow Y \times W$ is a morphism of mixed Hodge complexes which induces an isomorphism on homology in degrees < 2 and an injection in degree 2.

Theorem 5.1 implies that $\mathsf{HRep}(\mathcal{G}_X)$ is equivalent to $\mathsf{MHS}(Z, \mathbb{H})$. Variations of MHS over Z are equivalent to Γ-invariant variations over $Y \times W$. Since W is simply connected, variations of MHS over Z are constant when pulled back to any W-slice $y \times W \to Z$ and are therefore determined by their pullback to the Y-slice $Y \times w_o$. In this way, Γ-invariant variations of MHS over $Y \times W$ correspond to Γ-invariant variations over $Y \cong Y \times w_o$.

[9] These are the elements of $E^\bullet(Y, \mathbb{O}_o)^\Gamma$ that lie in a finite dimensional R_o-submodule.

[10] For example, one can take W to be $(\mathbb{C}^N)^\Gamma - \Delta$, where Δ is the fat diagonal and $N > 1$. This is $N - 1$ connected. The group Γ acts by permuting the coordinates.

References

[1] A. Beilinson: *Higher regulators and values of L-functions*, Current problems in mathematics, Vol. 24, 181–238, Itogi Nauki i Tekhniki, Akad. Nauk SSSR, Vsesoyuz. Inst. Nauchn. i Tekhn. Inform., Moscow, 1984.

[2] F. Brown: *Multiple modular values for* $\mathrm{SL}_2(\mathbb{Z})$, [arXiv:1407.5167]

[3] K. Brown: *Cohomology of groups*, Graduate Texts in Mathematics, 87, Springer-Verlag, New York-Berlin, 1982.

[4] J. Carlson, R. Hain: *Extensions of variations of mixed Hodge structure*, Théorie de Hodge, Luminy, Juin, 1987, Asterisque no. 179-180, 39–65.

[5] H. Cartan, S. Eilenberg: *Homological algebra*, Princeton University Press, 1956.

[6] P. Deligne: *Théorie de Hodge, II*, Inst. Hautes Études Sci. Publ. Math. No. 40 (1971), 5–57.

[7] P. Deligne, A. Goncharov: *Groupes fondamentaux motiviques de Tate mixte*, Ann. Sci. École Norm. Sup. (4) 38 (2005), 1–56.

[8] P. Deligne, J. Miline: *Tannakian categories*, in: Deligne, Pierre; Milne, James S.; Ogus, Arthur; Shih, Kuang-yen *Hodge cycles, motives, and Shimura varieties*, Lecture Notes in Mathematics, 900. Springer-Verlag, Berlin-New York, 1982.

[9] R. Hain: *Hodge-de Rham theory of relative Malcev completion*, Ann. Sci. École Norm. Sup., t. 31 (1998), 47–92.

[10] R. Hain: *Lectures on moduli spaces of elliptic curves*, in Transformation groups and moduli spaces of curves, 95–166, Adv. Lect. Math. (ALM), 16, Int. Press, 2011. [arXiv:0812.1803]

[11] R. Hain: *Notes on the Universal Elliptic KZB Equation*, [arXiv:1309.0580]

[12] R. Hain: *The Hodge-de Rham Theory of Modular Groups*, in Recent Advances in Hodge Theory, M. Kerr and G. Pearlstein editors, Cambridge University Press, 2016 [arXiv:1403.6443]

[13] R. Hain, M. Matsumoto: *Universal mixed elliptic motives*, submitted 2015, [arxiv:1512.03975]

[14] R. Hain, G. Pearlstein: *Tannakian fundamental groups of categories of variations of mixed Hodge structure*, in preparation.

[15] R. Hain, S. Zucker: *Unipotent variations of mixed Hodge structure*, Invent. Math. 88 (1987), 83–124.

[16] G. Hochschild: *Cohomology of algebraic linear groups*, Illinois J. Math. 5 (1961), 492–519.

[17] G. Hochschild, J.-P. Serre: *Cohomology of Lie algebras*, Ann. of Math. 57 (1953), 591–603.

[18] G. Hochschild, G. Mostow: *Pro-affine algebraic groups*, Amer. J. Math. 91 (1969), 1127–1140.

[19] J. C. Jantzen, *Representations of algebraic groups*, Pure and Applied Mathematics Vol.131, Academic Press, 1987.

[20] M. Kashiwara: *A study of variation of mixed Hodge structure*, Publ. Res. Inst. Math. Sci. 22 (1986), no. 5, 991–1024.

[21] S. Maclane: *Homology*, Die Grundlehren der mathematischen Wissenschaften, Bd. 114. Springer-Verlag, Berlin-Göttingen-Heidelberg, 1963.

[22] W. Schmid: *Variation of Hodge structure: the singularities of the period mapping*, Invent. Math. 22 (1973), 211–319.

[23] M. Saito: *Mixed Hodge modules and admissible variations*, C. R. Acad. Sci. Paris, t. 309, 1989, Série I, pp. 351–356.

[24] M. Saito: *Mixed Hodge modules*, Publ. Res. Inst. Math. Sci. 26 (1990), 221–333.

[25] J. Steenbrink, S. Zucker: *Variation of mixed Hodge structure, I*, Invent. Math. 80 (1985), 489–542.

[26] W. Waterhouse: *Introduction to affine group schemes*, Graduate Texts in Mathematics 66, Springer-Verlag, 1979.

[27] S. Zucker: *Hodge theory with degenerating coefficients, L_2 cohomology in the Poincaré metric*, Ann. of Math. (2) 109 (1979), 415–476.

[28] S. Zucker: *Intermediate Jacobians and normal functions*, in Topics in transcendental algebraic geometry (Princeton, N.J., 1981/1982), 259–267, Ann. of Math. Stud., 106, Princeton University Press, 1984.

Hodge Theory and L^2-analysis
ALM 39, pp. 419–449

Singularities in Arbitrary Characteristic via Jet Schemes

Shihoko Ishii[*], Ana J. Reguera[†]

Abstract

This paper summarizes recent results concerning singularities with respect to the Mather–Jacobian log discrepancies over an algebraically closed field of arbitrary characteristic. The basic point is that the inversion of adjunction with respect to Mather–Jacobian discrepancies holds under arbitrary characteristic. Using this fact, we will reduce several geometric properties of the singularities to jet scheme problems and try to avoid discussions that are distinctive to characteristic 0.

1 Introduction

Canonical and log-canonical singularities play important roles in birational geometry over the base field of characteristic 0 and are recognized as "good singularities", which we admit on a minimal model. These singularities can be well described by jet schemes when they are locally complete intersections.

On the other hand, [3] and [14] independently introduced singularities that can be described by jet schemes in general. These are based on Mather–Jacobian (MJ) discrepancies. We define MJ-canonical (MJ-log-canonical) singularities based on MJ discrepancy in a similar way to canonical (log-canonical) singularities. These MJ-canonical and MJ-log-canonical singularities have the following good properties: rationality (MJ-canonical), stability under small deformations (see [8]), and MJ-multiplier ideals (see [7]). These are all for a characteristic 0 case.

Now, for a positive characteristic case, one can also define canonical and log-canonical singularities as well as MJ-canonical and MJ-log-canonical singularities.

*Graduate School of Mathematical Science, University of Tokyo, 3-8-1 Komaba, Meguro, 153-8914 Tokyo, Japan. Department of Mathematics, Tokyo Woman's Christian University, 2-6-1 Zenpukuji, Suginami, 167-8585 Tokyo, Japan.

†Dep. de Álgebra, Geometría y Topología, Universidad de Valladolid, Paseo Belén 7, 47011 Valladolid, Spain. E-mail: areguera@agt.uva.es

However, canonical and log-canonical singularities of positive characteristic are difficult to treat, because the following results are unavailable:

1. resolution of singularities;
2. Bertini's second theorem (generic smoothness); and
3. the Kodaira vanishing theorem,

all of which are used in discussions of singularities over characteristic 0.

For example, even the seemingly simple statement that "a general hyperplane section of a quasi-projective variety has at worst canonical singularities if the variety has such singularities" is not yet proved for positive characteristic case. This statement has been proved in the characteristic 0 case by resolution of the singularities and Bertini's second theorem. If the variety is of dimension 3 over a base field of positive characteristic, then a resolution of the singularities exists [2], but this problem is not yet proved (although a result has been found under certain conditions [11]).

On the contrary, the MJ-version of this problem has the possibility to be solved, because MJ-singularities are well described by the jet schemes, which act as an alternative to Bertini's second theorem. As evidence, in Corollary 4.11, we prove this statement for three-dimensional MJ-canonical quasi-projective variety, thereby showing the possibility of MJ discrepancy for discussion of positive characteristics.

One of the aims of this paper is to summarize the results of MJ-singularities in the positive characteristic case and to clarify those which have been proved and those which have not yet been proved. Since such a review does not presently exist elsewhere, we think that it would be useful for us to summarize them here. The basic theorems showing that MJ-singularities are well described by jet schemes are Theorems 3.18 and 3.23. These theorems are proved in the same line as their proof in the characteristic 0-case, by carefully avoiding the use of resolutions of the singularities.

The other aim of this paper is to show that jet schemes are useful in the study of singularities over positive characteristic base fields (Corollary 4.11).

This paper is structured as follows: in Section 2, we give characteristic-free discussions of MJ discrepancies. In Section 3, we give preliminary information about jet schemes, define the codimension of a cylinder, and describe minimal log MJ discrepancies using jet schemes in the arbitrary-characteristic case. Note that these have been already established in the characteristic 0 case. In Section 4, we show some properties of MJ-canonical and MJ-log-canonical singularities of positive characteristic that have been proved and list some open problems for the positive-characteristic case.

Acknowledgements The authors express their hearty thanks to Masayuki Hirokado for providing them information on varieties over the base field of positive characteristic. We also thank the referees for constructible suggestions to improve the paper.

2 Mather–Jacobian discrepancy

Throughout this paper, a variety refers to a reduced pure-dimensional scheme of finite type over an algebraically closed field k of arbitrary characteristic, unless otherwise stated.

Definition 2.1. Let X be a variety of dimension d. Note that the projection

$$\pi\colon \mathbb{P}_X(\wedge^n \Omega_X) \to X$$

is an isomorphism over the smooth locus $X_{reg} \subseteq X$. In particular, we have a section $\sigma\colon X_{reg} \to \mathbb{P}_X(\wedge^n \Omega_X)$.

The closure of the image of the section σ is called the *Nash blow-up* of X, and is denoted by $\widehat{X}$:

$$\begin{array}{ccc} & \mathbb{P}_X(\wedge^n \Omega_X) & \supseteq \widehat{X} := \overline{\sigma(X_{reg})} \\ {}^{\sigma}\nearrow & \downarrow{\scriptstyle \pi} & \\ X_{reg} \hookrightarrow & X. & \end{array}$$

Definition 2.2. Let X be a variety. We call a morphism $\varphi : Y \to X$ a *partial resolution*, if φ is proper birational and Y is normal. We also sometimes call Y a partial resolution. A prime divisor over X is a prime divisor that appears in a partial resolution of X. A prime divisor E over X is called an *exceptional prime divisor* if a partial resolution on which E appears is not isomorphic at the generic point of E.

Definition 2.3. Let E be a prime divisor over X, then the Mather discrepancy $\widehat{k}_E \in \mathbb{Z}_{\geq 0}$ and the Jacobian discrepancy j_E at E are defined as follows:

Let $\varphi : Y \to X$ be a partial resolution of X such that E appears on Y, let $\widehat{X} \to X$ be the Nash blow-up and let $Y \times_X \widehat{X}$ be the fiber product. In the fiber product, the union of the irreducible components each of which dominates an irreducible component of X is called the *main part*.

Replacing Y by the normalization of the main part of the fiber product, we may assume that $\varphi : Y \to X$ factors through the Nash blow-up $\widehat{X} \to X$.

By the universality of the Nash blow-up, the image Im of the following homomorphism is invertible:

$$\varphi^*(\wedge^d \Omega_X) \to \omega_Y,$$

where $d = \dim X$. Restricting the above homomorphism to the smooth locus Y_{sm}, we can describe

$$Im|_{Y_{sm}} = \mathcal{I}\omega_{Y_{sm}}$$

with some invertible ideal sheaf $\mathcal{I}$, as $\omega_{Y_{sm}}$ is invertible. As Y is normal, the generic point η of E is in Y_{sm}. Let us define

$$\widehat{k}_E = \mathrm{val}_E(\mathcal{I})$$

and call it the *Mather discrepancy* of X at E. We note that if $\mathcal{I} = \mathcal{O}_Y(-\widehat{K}_{Y/X})$ is expressed with a Cartier divisor $\widehat{K}_{Y/X}$ on Y_{sm}, then $\widehat{k}_E$ is the coefficient of $\widehat{K}_{Y/X}$ at E.

For a coherent ideal sheaf $\mathfrak{a} \subset \mathcal{O}_X$, let us define

$$\mathrm{val}_E(\mathfrak{a}) = \min\{\mathrm{val}_E(f) \mid f \in \mathfrak{a}\}.$$

Note that if $\varphi : Y \to X$ factors go through the blow-up of X by the ideal $\mathfrak{a}$, then $\mathrm{val}_E(\mathfrak{a})$ is the coefficient at E of the divisor Z defined by $\mathcal{O}_Y(-Z) = \mathfrak{a}\mathcal{O}_Y$. In particular, when $\mathfrak{a}$ is the Jacobian ideal $\mathcal{J}_X$ of X, we define

$$j_E = \mathrm{val}_E(\mathcal{J}_X)$$

and call it the *Jacobian discrepancy* of X at E.

Then, the difference $\widehat{k}_E - j_E$ is called the *Mather–Jacobian (MJ) discrepancy* of X at E.

Definition 2.4. Let X be a variety over k, let $\mathfrak{a} \subset \mathcal{O}_X$ be a non-zero coherent ideal sheaf, and let n be a non-negative real number. A real number

$$a_{MJ}(E; X, \mathfrak{a}^n) = \widehat{k}_E - j_E - n \cdot \mathrm{val}_E(\mathfrak{a}) + 1$$

is called the *MJ-log discrepancy* of the pair $(X, \mathfrak{a}^n)$ at E.

A pair $(X, \mathfrak{a}^n)$ is called *MJ-canonical* (resp. *MJ-log-canonical*) at a (not necessarily closed) point $x \in X$, if

$$a_{MJ}(E; X, \mathfrak{a}^n) \geq 1 \quad (\text{resp. } \geq 0)$$

for every exceptional prime divisor E over X whose center contains x.

A pair $(X, \mathfrak{a}^n)$ is called *MJ-klt* at a (not necessarily closed) point $x \in X$, if

$$a_{MJ}(E; X, \mathfrak{a}^n) > 0$$

for every prime divisor E over X whose center on X contains x. Here, klt means "Kawamata log-terminal" which appears in Minimal Model Program and plays an important role (see, for example, [17]). Our "MJ-klt" is MJ-version of the usual klt.

We call a pair $(X, \mathfrak{a}^n)$ *MJ-canonical* (resp. *MJ-log-canonical, MJ-klt*) if it is MJ-canonical (resp. MJ-log-canonical, MJ-klt) at every point of X.

Here, MJ-klt is the MJ-version of the usual klt, i.e., Kawamata log terminal (see, for example, [17]).

Remark 2.5. 1. Note that the definitions of MJ-canonical and MJ-log-canonical require the condition only for "exceptional" prime divisors, while the definition of MJ-klt requires the condition for all prime divisors over X. However, regarding the MJ-log-canonical case, if $(X, \mathfrak{a}^n)$ is MJ-log-canonical at a point $x \in X$, then $a_{MJ}(E; X, \mathfrak{a}^n) \geq 0$ holds for every prime divisor over X whose center contains x.

2. If X is normal and locally a complete intersection, the image of the canonical map $\wedge^d \Omega_X \to \omega_X$ is $\mathcal{J}_X \omega_X$ (see, for example, [6, Remark 9.6]). In this case $\widehat{k_E} - j_E = k_E$ for every prime divisor E over X. Therefore, the MJ-canonical and MJ-log-canonical cases are equivalent to the usual canonical and log-canonical cases, respectively.

Definition 2.6. Let X be a variety over k. The Mather–Jacobian minimal log discrepancies (MJ-mld) of the pair $(X, \mathfrak{a}^n)$ at a proper closed subset $W \subset X$ and at a point $\eta \in X$ are defined as follows:

$$\begin{aligned} \mathrm{mld}_{\mathrm{MJ}}(W; X, \mathfrak{a}^n) &= \inf\{a_{MJ}(E; X, \mathfrak{a}^n) \mid E \text{ is a prime exceptional divisor} \\ &\qquad \text{over } X \text{ with the center in } W\}, \\ \mathrm{mld}_{\mathrm{MJ}}(\eta; X, \mathfrak{a}^n) &= \inf\{a_{MJ}(E; X, \mathfrak{a}^n) \mid E \text{ is a prime exceptional divisor} \\ &\qquad \text{over } X \text{ with the center in } \overline{\{\eta\}}\}, \end{aligned}$$

when $\dim X \geq 2$. When $\dim X = 1$ and the right-hand side is ≥ 0, then we define $\mathrm{mld}_{\mathrm{MJ}}$ by the right-hand side. Otherwise, we define $\mathrm{mld}_{\mathrm{MJ}} = -\infty$.

Remark 2.7. (i) We strictly distinguish "center in Z" from "center Z."

(ii) For a point $x \in X$, the pair $(X, \mathfrak{a}^n)$ is MJ-canonical (resp. MJ-log-canonical) if and only if $\mathrm{mld}_{\mathrm{MJ}}(\eta; X, \mathfrak{a}^n) \geq 1$ (resp. $\mathrm{mld}_{\mathrm{MJ}}(\eta; X, \mathfrak{a}^n) \geq 0$) for every $\eta \in X$ such that $x \in \overline{\{\eta\}}$.

(iii) When $\dim X \geq 2$, we can prove that $\mathrm{mld}_{\mathrm{MJ}}(W; X, \mathfrak{a}^n) = -\infty$, if $\mathrm{mld}_{\mathrm{MJ}}(W; X, \mathfrak{a}^n) < 0$.

(iv) If X is normal and of a complete intersection, then the MJ-log discrepancy coincides with the usual log discrepancy. Therefore, in this case, the MJ-canonical and MJ-log-canonical cases coincide with the usual canonical and log-canonical cases, respectively. In particular, if X is non-singular

$$\mathrm{mld}_{\mathrm{MJ}}(W; X, \mathfrak{a}^n) = \mathrm{mld}(W; X, \mathfrak{a}^n),$$

where the right-hand side is the usual minimal log discrepancy.

3 The arc space and jet schemes of a variety

Throughout this section, X is always a d-dimensional variety over an algebraically closed field k of an arbitrary characteristic. In this section, we prove the inversion of adjunction of minimal MJ-log discrepancies by means of discussions of arc spaces and jet schemes. This theorem has been independently proved in [3] and [14] based on the concept of [6], when the base field is of characteristic 0. Here, we will present a characteristic-free proof for this theorem.

The basic tool is the arc space. So first we introduce arcs and jets of a variety.

Definition 3.1. Let X be a scheme of finite type over k and a $K \supset k$ a field extension. For $m \in \mathbb{Z}_{\geq 0}$ a k-morphism $\mathrm{Spec} K[t]/(t^{m+1}) \to X$ is called an *m-jet* of X and k-morphism $\mathrm{Spec} K[[t]] \to X$ is called an *arc* of X. The unique point of $\mathrm{Spec} K[t]/(t^{m+1})$ and the closed point of $\mathrm{Spec} K[[t]]$ are both denoted by 0. and the image of 0 by the jet or the arc is called the *center* of the jet or the arc.

Theorem 3.2. *Let X be a scheme of finite type over k, Sch/k the category of k-schemes and Set the category of sets. A contravariant functor $F_m^X : Sch/k \to Set$ is defined as follows:*

$$F_m^X(Z) = \mathrm{Hom}_k(Z \times_{\mathrm{Spec} k} \mathrm{Spec} k[t]/(t^{m+1}), X).$$

Then F_m^X is representable by a scheme X_m of finite type over k, therefore, there is a bijection as follows:

$$\mathrm{Hom}_k(Z, X_m) \simeq \mathrm{Hom}_k(Z \times_{\mathrm{Spec} k} \mathrm{Spec} k[t]/(t^{m+1}), X).$$

This scheme X_m is called the space of m-jets or the jet scheme of X.

There exists the projective limit

$$X_\infty := \varprojlim_m X_m$$

and it is called the space of arcs or the arc space of X. There is also a bijection for a k-algebra A as follows:

$$\mathrm{Hom}_k(\mathrm{Spec} A, X_\infty) \simeq \mathrm{Hom}_k(\mathrm{Spec} A[[t]], X).$$

Definition 3.3. For a variety X over k, let X_m ($m \in \mathbb{N}$) and X_∞ be the m-jet scheme and the arc space of X. Denote the canonical truncation morphisms by $\psi_m : X_\infty \to X_m$ and $\pi_m : X_m \to X$. In particular we denote the extremal morphism $\psi_0 = \pi_\infty : X_\infty \to X$ by π. We also denote the canonical truncation morphism $X_{m'} \to X_m$ ($m' > m$) by $\psi_{m',m}$. To specify the space X, we sometimes write $\psi^X_{m',m}$.

Definition 3.4. The pull-back $\psi_m^{-1}(S) \subset X_\infty$ of a constructible set $S \subset X_m$ ($m \in \mathbb{Z}_{\geq 0}$) is called a *cylinder*.

A subset $C \subset X_\infty$ is called a *thin set* if there is a closed subset $Z \subset X$ with dimension less than the dimension of X such that $C \subset Z_\infty$.

A subset $C \subset \psi_m^{-1}(S)$ is called an *irreducible component* of the cylinder $\psi_m^{-1}(S)$ if it is a maximal irreducible closed subset of the cylinder. Here, we note that a maximal irreducible closed subset of a cylinder exists because the set consisting irreducible closed subsets in the cylinder is an inductive set with respect to the inclusion order (the existence of a maximal irreducible component follows from Zorn's lemma, which we always assume). If char $k = 0$ or $\dim X \leq 3$, then every cylinder in X_∞ has only a finite number of irreducible components (a resolution of singularities [2] is used to prove this result).

Definition 3.5. For an arc $\gamma \in X_\infty$, the order of an ideal $\mathfrak{a} \subset \mathcal{O}_X$ measured by γ is defined as follows: let $\gamma^* : \mathcal{O}_{X,\gamma(0)} \to k[[t]]$ be the corresponding ring homomorphism of γ. Then, we define

$$\mathrm{ord}_\gamma(\mathfrak{a}) = \sup\{r \in \mathbb{Z}_{\geq 0} \mid \gamma^*(\mathfrak{a}) \subset (t^r)\}.$$

We define the subsets "contact loci" in the arc space as follows:

$$\mathrm{Cont}^m(\mathfrak{a}) = \{\gamma \in X_\infty \mid \mathrm{ord}_\gamma(\mathfrak{a}) = m\}.$$

In a similar manner, we define

$$\mathrm{Cont}^{\geq m}(\mathfrak{a}) = \{\gamma \in X_\infty \mid \mathrm{ord}_\gamma(\mathfrak{a}) \geq m\}.$$

By this definition, we can see that

$$\mathrm{Cont}^{\geq m}(\mathfrak{a}) = \psi_{m-1}^{-1}(Z(\mathfrak{a})_{m-1}).$$

Here, $Z(\mathfrak{a})$ is the closed subscheme defined by the ideal $\mathfrak{a}$ in X. Therefore, the contact loci are cylinders in X_∞.

Definition 3.6. Let E be a prime divisor over X and $\varphi : Y \to X$ be a partial resolution of X on which E appears. Let $p \in E$ be the generic point. We define $C_X(\mathrm{val}_E) = \overline{\varphi_\infty((\pi_\infty^Y)^{-1}(p))}$, also denoted by N_E in the literature on the Nash problem. Furthermore, for $q \in \mathbb{N}$, let $p_{q-1} \in (\pi_{q-1}^Y)^{-1}((E \cap Y_{reg})_{q-1})$ be the generic point. Then, we define

$$C_X(q \cdot \mathrm{val}_E) = \overline{\varphi_\infty(\psi_{q-1}^Y)^{-1}(p_{q-1})}$$

and call it the *maximal divisorial set* corresponding to the divisorial valuation $q \cdot \mathrm{val}_E$. This definition is the same as that in [4] and [13] in case char $k = 0$.

An irreducible closed subset $C \subset X_\infty$ is called a *divisorial set*, if there exist $q \in \mathbb{N}$ and a prime divisor E over X such that the divisorial valuation $q \cdot \mathrm{val}_E$ is given by ord_α, where α is the generic point of C.

The generic point $\alpha \in C$ gives a divisorial valuation $q \cdot \mathrm{val}_E$ by ord_α. A maximal divisorial set $C_X(q \cdot \mathrm{val}_E)$ is the maximal set among all divisorial sets corresponding to the valuation $q \cdot \mathrm{val}_E$.

Proposition 3.7 ([5, Lemma 4.1], [20, before lemma 3.2] or [21, 3.4],[6, Proposition 4.1]). *For $e \in \mathbb{Z}_{\geq 0}$, the following subsets in X_∞ and in X_m are defined by using the Jacobian ideal $\mathcal{J}_X$:*

$$X_\infty^e := \{\gamma \in X_\infty \mid ord_\gamma(\mathcal{J}_X) = e\} \quad \textit{and} \quad X_{m,\infty}^e := \psi_m(X_\infty^e).$$

Then, there exists $c \in \mathbb{N} \setminus \{0\}$ such that for an integer $m \geq ce$ the canonical map $X_{m+1,\infty}^e \to X_{m,\infty}^e$ is a piecewise trivial fibration with fibers isomorphic to $\mathbb{A}^d$, where $d = \dim X$. Therefore, we also know that $X_{m+1,\infty}^{\leq e} \to X_{m,\infty}^{\leq e}$ is a piecewise trivial fibration with fibers isomorphic to $\mathbb{A}^d$, where we define

$$X_\infty^{\leq e} := \{\gamma \in X_\infty \mid ord_\gamma(\mathcal{J}_X) \leq e\} \quad \textit{and} \quad X_{m,\infty}^{\leq e} := \psi_m(X_\infty^{\leq e}).$$

This proposition is stated in [5] under the condition that char $k = 0$; however in [20] and [6], its proof was also confirmed to apply to perfect fields in the positive-characteristic case.

Remark 3.8. Using Proposition 3.7, *stable points* of the space of arcs were defined as follows in [21]: let γ be a point of X_∞ (i.e., γ is a prime ideal of $\mathcal{O}_{X_\infty}$) and let $Z(\gamma)$ be the set of zeros of γ on X_∞; then, γ is a stable point if there exists $m_0 \in \mathbb{N}$ and $G \in \mathcal{O}_{X_\infty}$, $G \notin \gamma$, $G \in \mathcal{O}_{X_{m_0}}$, such that, for $m \geq m_0$, the map $\psi_{m+1,m} : X_{m+1} \to X_m$ induces a trivial fibration $\overline{\psi_{m+1}(Z(\gamma))} \cap (X_{m+1})_G \to \overline{\psi_m(Z(\gamma))} \cap (X_m)_G$ with fiber $\mathbb{A}^d$, where $(X_m)_G$ is the open subset $X_m \setminus Z(G)$ of X_m.

The following is an easy consequence of the previous proposition when *char* $k = 0$, since in that case a cylinder has only finite number of irreducible components ([4, Proposition 3.6]), and therefore, an irreducible component of a cylinder is regarded generically as a cylinder. Herein, we give a proof that does not use the finiteness of the irreducible components of a cylinder; thus, this proof also works for a positive characteristic case.

Lemma 3.9. *For an irreducible component C of a cylinder in X_∞ such that $C \not\subset \mathrm{Sing}(X)_\infty$, its generic point γ is a stable point of X_∞. In particular, there exists e such that*

$$C_m^{\leq e} := \psi_m(C) \cap X_{m,\infty}^{\leq e}$$

is a nonempty open subset of $\psi_m(C)$ and the codimension of $C_m^{\leq e}$ inside $X_{m,\infty}^{\leq e}$ stabilizes for $m \gg e$.

Proof. This lemma is proved in the same way as in the case of characteristic 0. Let C be an irreducible component of a cylinder Γ in X_∞ such that $C \not\subset \mathrm{Sing}(X)_\infty$. Let $\gamma \in C$ be the generic point. Then by $\gamma \notin (\mathrm{Sing}X)_\infty$, we have $\mathrm{ord}_\gamma(\mathcal{J}_X) = e > 0$. Then, $C_m^{\leq e}$ is a nonempty open subset of $\psi_m(C)$. From this and Proposition 3.5, the result follows. □

Here, we show a result about a stable point, for the readers who are interested in this direction.

Proposition 3.10. *If γ is a stable point of X_∞, then there exists an irreducible cylinder $C \subset X_\infty$ whose generic point is γ.*

Proof. We may assume that X is affine. Let γ be a stable point of X_∞. Then, by [20, Theorem 4.1] (see also the finiteness property of the stable points in [21, 3.10]), there exist $G \in \mathcal{O}_{X_\infty}$, $G \notin \gamma$ and $L_1, \ldots, L_r \in \mathcal{O}_{X_\infty}$ such that

$$\gamma\left(\mathcal{O}_{(X_\infty)_{\mathrm{red}}}\right)_G = (L_1, \ldots, L_r)\ \left(\mathcal{O}_{(X_\infty)_{\mathrm{red}}}\right)_G. \tag{1}$$

Since $\mathcal{O}_{X_\infty} = \lim_{\to} \mathcal{O}_{X_n}$, there exists $n_0 \in \mathbb{N}$ such that $G, L_1, \ldots, L_r \in \mathcal{O}_{X_{n_0}}$. Let us consider the constructible subset

$$S := Z(L_1, \ldots, L_r) \cap (G \neq 0)\ \subset\ X_{n_0}.$$

Then $C = \psi_{n_0}^{-1}(S)$ is an irreducible cylinder whose generic point is γ (see (1) and the definition of cylinder in [6, beginning of section 5]). □

Definition 3.11. For a variety X, let C be an irreducible component of a cylinder in X_∞ such that $C \not\subset (\mathrm{Sing}X)_\infty$, then we define the codimension of C in X_∞ as follows:

$$\mathrm{codim}(C, X_\infty) := \mathrm{codim}(C_m^{\leq e}, X_{m,\infty}^{\leq e})$$

for $m \gg e$, where $e = \mathrm{ord}_\gamma(\mathcal{J}_X)$ for a generic point $\gamma \in C$.

Let $\Gamma \subset X_\infty$ be a cylinder not contained in $\mathrm{Sing}(X)_\infty$, then we define the codimension of the cylinder Γ in X_∞ as the minimal value of the codimensions of the irreducible components not contained in $\mathrm{Sing}(X)_\infty$.

The following lemma was proved in [5, Lemma 3.4] and in [6, Theorem 6.2, Lemma 6.3]. The former one was stated under the condition of characteristic 0, while the latter ones were stated under an arbitrary characteristic. Note that the statement of Lemma 6.2 in [6] assumed the properness of the morphism f, but the properness was not actually used in the proof.

Lemma 3.12 ([5], [6]). *Let $f : Y \to X$ be a birational morphism from a smooth variety Y and $f_m : Y_m \to X_m$ the induced morphism. Let $\gamma \in Y_\infty$ be any arc such that $\tau := \mathrm{ord}_\gamma(\widehat{K}_{Y/X}) < \infty$. Then for $m \gg 0$, letting $\gamma_m = \psi^Y_m(\gamma)$, we have*

$$f_m^{-1}(f_m(\gamma_m)) \simeq \mathbb{A}^\tau.$$

Moreover, for every $\gamma'_m \in f_m^{-1}(f_m(\gamma_m))$, we have $\psi^Y_{m,m-\tau}(\gamma_m) = \psi^Y_{m,m-\tau}(\gamma'_m)$, where $\psi^Y_{m,m-\tau} : Y_m \to Y_{m-\tau}$ is the canonical truncation morphism.

Theorem 3.13. *Let E be a prime divisor over X and $g : \tilde{X} \to X$ be a partial resolution of X such that E appears on $\tilde{X}$. Then $C_X(q \cdot \mathrm{val}_E)$ is a cylinder of X_∞ of codimension*

$$\mathrm{codim}(C_X(q \cdot \mathrm{val}_E), X_\infty) = q \cdot (\mathrm{val}_E(\widehat{K}_{Y/X}) + 1).$$

Proof. Let $f : Y \to X$ be the restriction of g on an open subset $Y \subset \tilde{X}$ such that Y and $E \cap Y$ are smooth. Then, by using Lemma 3.12, we can prove the required equality in the same way as [4, Theorem 3.9]. Note that Theorem 3.9 in [4] was stated under the condition that $f : Y \to X$ is a resolution of the singularities of X, but the resolution is not actually needed for the proof as the condition in Lemma 3.12 does not require the properness of f. □

The following is a modified version of Lemma 3.12 for a blow-up f.

Lemma 3.14. *Let $B \subset X$ be an irreducible reduced closed subset of dimension s with the defining ideal I_B in X and let $\gamma \in X_\infty$ be an arc that is not contained in $\mathrm{Sing}(X)_\infty \cup B_\infty$ and the center is a smooth closed point of B with $e := \mathrm{ord}_\gamma(I_B) > 0$. Let $f : Y \to X$ be the blow-up with the center B and $\gamma' \in Y_\infty$ be the lifting of γ. Let $f_m : Y_{m,\infty} \to X_{m,\infty}$ be the morphism induced by f. Then for $m \gg 0$, letting $\gamma'_m = \psi^Y_m(\gamma')$, we have*

$$\dim f_m^{-1}(f_m(\gamma'_m)) \geq (d - s - 1)e.$$

Moreover, for every $\gamma''_m \in f_m^{-1}(f_m(\gamma'_m))$ we have $\psi^Y_{m,m-e}(\gamma''_m) = \psi^Y_{m,m-e}(\gamma'_m)$.

Proof. Since the problem is local, we may assume that X is an affine scheme embedded in $A = \mathbb{A}^N_k$, B is defined by the ideal $(x_1, \ldots, x_{N-s})$ in $A = \mathrm{Spec}\, k[x_1, \ldots, x_N]$ and the center of γ is the origin $0 = (0, \ldots, 0) \in B \subset X \subset A$. Let $\gamma_m = \psi^X_m(\gamma)$, then by definition $f_m(\gamma'_m) = \gamma_m$. Then, we can express the ring homomorphism $\gamma^*_m : k[x_1, \ldots, x_N] \to k[t]/(t^{m+1})$ corresponding to γ_m as follows:

$$\gamma^*_m(x_i) = \sum_{j=e}^{m} a_i^{(j)} t^j \quad (i = 1, \ldots, N - s),$$

$$\gamma^*_m(x_i) = \sum_{j=1}^{m} a_i^{(j)} t^j \quad (i = N - s + 1, \ldots, N),$$

where $a_i^{(j)} \in k$. Here, by this assumption, we may assume that $a_1^{(e)} \neq 0$, without loss of generality.

Let $f' : A' \to A$ be the blow-up with the center B. Then A' is covered by $(N - s)$ affine spaces

$$U(i) = \operatorname{Spec} k[x_i, \frac{x_1}{x_i}, \stackrel{i}{\vee} \cdots, \frac{x_{N-s}}{x_i}, x_{N-s+1}, \ldots, x_N] \quad (i = 1, \ldots, N - s).$$

By this assumption, we may assume that $\gamma'_m \in U(1)_m$. We denote the restriction $f'_m|_{U(1)_m} : U(1)_m \to A_m$ of $f'_m : A'_m \to A_m$ by the same symbol f'_m. Let $y_i = \frac{x_i}{x_1}$ for $i = 2, \ldots, N - s$, then $x_1, y_2, \ldots, y_{N-s}, x_{N-s+1}, \ldots, x_N$ form a coordinate system of $U(1)$. We can express $\gamma'_m \in U(1)_m$ as follows:

$$\gamma'_m{}^*(x_1) = \sum_{j=e}^{m} a_1^{(j)} t^j,$$

$$\gamma'_m{}^*(y_i) = \sum_{j=0}^{m} b_i^{(j)} t^j \quad (i = 2, \ldots, N - s),$$

$$\gamma'_m{}^*(x_i) = \sum_{j=1}^{m} a_i^{(j)} t^j \quad (i = N - s + 1, \ldots, N),$$

where $b_i^{(j)}$'s satisfy $\gamma'_m{}^*(y_i)\gamma'_m{}^*(x_1) = \gamma_m{}^*(x_i) = \sum_{j=e}^{m} a_i^{(j)} t^j$ for $i = 2, \ldots, N - s$.

Then, we can see that a jet $\alpha \in (f'_m)^{-1}(\gamma_m)$ is expressed by

$$\alpha^*(x_1) = \sum_{j=e}^{m} a_1^{(j)} t^j,$$

$$\alpha^*(y_i) = \sum_{j=0}^{m-e} b_i^{(j)} t^j + \sum_{j=m-e+1}^{m} y_i^{(j)} t^j \quad (i = 2, \ldots, N - s),$$

$$\alpha^*(x_i) = \sum_{j=1}^{m} a_i^{(j)} t^j \quad (i = N - s + 1, \ldots, N),$$

where $y_i^{(j)}$ $(i = 2, \ldots, N - s, j = m - e + 1, \ldots, m)$ can be an arbitrary element in k. Therefore, we have that every $\alpha \in (f'_m)^{-1}(\gamma_m)$ is mapped to the same element

$\psi^{A'}_{m,m-e}(\gamma'_m)$ by the truncation morphism $\psi^{A'}_{m,m-e}$, in particular, which yields the second statement of the lemma.

Let $\gamma'_{m-e} = \psi^{A'}_{m,m-e}(\gamma'_m)$, then any element $\beta \in (\psi^{A'}_{m,m-e})^{-1}(\gamma'_{m-e})$ is expressed as

$$\beta^*(x_1) = \sum_{j=e}^{m-e} a_1^{(j)} t^j + \sum_{j=m-e+1}^{m} x_1^{(j)} t^j,$$

$$\beta^*(y_i) = \sum_{j=0}^{m-e} b_i^{(j)} t^j + \sum_{j=m-e+1}^{m} y_i^{(j)} t^j \quad (i = 2, \ldots, N-s),$$

$$\beta^*(x_i) = \sum_{j=1}^{m-e} a_i^{(j)} t^j + \sum_{j=m-e+1}^{m} x_i^{(j)} t^j \quad (i = N-s+1, \ldots, N),$$

where $y_i^{(j)}$ and $x_i^{(j)}$ are arbitrary elements of k. Here, comparing the expressions of α^* and β^* above, we obtain that

$$(f'_m)^{-1}(\gamma_m) = (f'_m)^{-1}(\gamma_m) \cap (\psi^{A'}_{m,m-e})^{-1}(\gamma'_{m-e}) \subset (\psi^{A'}_{m,m-e})^{-1}(\gamma'_{m-e})$$

is defined by $(s+1)e$ equations:

$$x_1^{(j)} = a_1^{(j)} \quad (j = m-e+1, \ldots, m),$$

$$x_i^{(j)} = a_i^{(j)} \quad (i = N-s+1, \ldots, N, \ \ j = m-e+1, \ldots, m)$$

in $(\psi^{A'}_{m,m-e})^{-1}(\gamma'_{m-e})$.

Now, remember that Y is the strict transform of X in A', $\mathcal{J}_Y$ be the Jacobian ideal of Y and let $r := \mathrm{ord}_{\gamma'}(\mathcal{J}_Y)$. Take m such that it is sufficiently large with respect to r and e. Let $\rho_{m,m-1} : Y^{\leq r}_{m,\infty} \to Y^{\leq r}_{m-1,\infty}$ be the restriction of the truncation morphism $\psi^Y_{m,m-1} : Y_m \to Y_{m-1}$. Then, $\rho_{m,m-1}$ is a piecewise trivial morphism with the fiber $\mathbb{A}^d$. Therefore $\rho_{m,m-e} : Y^{\leq r}_{m,\infty} \to Y^{\leq r}_{m-e,\infty}$ has exactly de-dimensional fibers. Here we note that $\rho^{-1}_{m,m-e}(\gamma'_{m-e}) \subset (\psi^{A'}_{m,m-e})^{-1}(\gamma'_{m-e})$. By this and the above observation, we obtain that

$$(f'_m)^{-1}(\gamma_m) \cap \rho^{-1}_{m,m-e}(\gamma'_{m-e}) \subset \rho^{-1}_{m,m-e}(\gamma'_{m-e})$$

is defined by at most $(s+1)e$ equations in de-dimensional variety $\rho^{-1}_{m,m-e}(\gamma'_{m-e})$. This implies that $f_m : Y_m \to X_m$ has a fiber $f_m^{-1}(\gamma_m)$ of dimension $\geq (d-s-1)e$. □

Remark 3.15. In the previous lemma, the center of the lifting $\gamma' \in Y_\infty$ may be a singular point of X, while the center of $\gamma \in Y_\infty$ is non-singular in Lemma 3.12. This, the situation of Lemma 3.14 is different from that of Lemma 3.12.

As a corollary, we obtain the following. But the result has been proved in [21, Proposition 3.7 (vii)] using the concept of stable points in the space of arcs [21, Definition 3.6] (see Lemma 3.6).

This corollary was also proved in [4, Proposition 2.12] for the characteristic 0 case by using resolutions of singularities.

Corollary 3.16. *Every irreducible component C of a cylinder such that $C \not\subset \mathrm{Sing}(X)_\infty$ gives a divisorial valuation. In particular, the closure of an irreducible component C of finite intersection of contact loci* $\mathrm{Cont}^{m_i}(\mathfrak{a}_i)$ $(i = 1, \ldots, s)$ *is a maximal divisorial set.*

Proof. Let α be the generic point of C. First, we prove that there exists a prime divisor E over X such that the lifting $\tilde{\alpha}$ of α onto Y on which E appears has the center E. Let $B \subset X$ be the center of α. If $\dim B = d - 1$, then we have the required conclusion. Therefore, we assume that $\dim B = s < d - 1$. By replacing X by a small affine neighborhood, we may assume that X is embedded in the affine space $A = \mathbb{A}_k^N$. Let $f^{(1)} : A^{(1)} \to A$ be the blow-up with the center B. Then, $f^{(1)}$ is isomorphic outside B and the image of α is not contained in B, therefore α is lifted onto the proper transform $Y^{(1)}$ of X on $A^{(1)}$ by the properness criteria applied to $f^{(1)}$. Then, the lifting $\alpha^{(1)}$ of α on $Y^{(1)}$ gives an irreducible component $C^{(1)}$ of a cylinder on $Y^{(1)}$. Indeed, if C is an irreducible component of the cylinder $\Gamma = (\psi_m^X)^{-1}(S)$ for a constructible subset $S \subset X_m$, then $C^{(1)}$ is an irreducible component of the cylinder $(\psi_m^{Y^{(1)}})^{-1}\left((f_m^{(1)})^{-1}(S)\right)$.

Now, take m to be sufficiently large and a general arc $\gamma \in C$ whose center is a closed point in B. Let $e = \mathrm{ord}_\gamma I_B$. Then, we obtain that the morphism $f^{(1)} : C_m^{(1)} \to C_m$ has relative dimension $\geq (d - s - 1)e \geq 1$. Therefore, $\dim C_m^{(1)} > \dim C_m$, which yields $\mathrm{codim}(C^{(1)}, Y_\infty^{(1)}) < \mathrm{codim}(C, X_\infty)$. If the center of the generic point of $C^{(1)}$ is not of dimension $d - 1$, we blow-up at the center and obtain the irreducible component whose codimension is less than the previous one. In this way, we obtain the successive blow-ups:

$$A^{(n)} \to \cdots \to A^{(2)} \to A^{(1)} \to A$$

with the sequence of irreducible components

$$C^{(n)}, \ldots, C^{(2)}, C^{(1)}, C$$

of cylinders in $Y^{(n)}, \ldots, Y^{(2)}, Y^{(1)}, X$ such that

$$\mathrm{codim} C^{(n)} < \cdots < \mathrm{codim} C^{(2)} < \mathrm{codim} C^{(1)} < \mathrm{codim} C.$$

Because the codimension is finite, this procedure should terminate, *i.e.*, there exists a number n such that the dimension of the center of the generic point of $C^{(n)}$ has dimension $d - 1$. This is the conclusion to our first claim.

Then, we prove that the lifting $\alpha^{(n)}$ of α gives a divisorial valuation $q \cdot \mathrm{val}_E$. Replacing $Y^{(n)}$ by its normalization if necessary, we may assume that $Y^{(n)}$ is normal and the center of the lifting $\alpha^{(n)}$ of α on $Y^{(n)}$ is a prime divisor, say E. The arc $\alpha^{(n)}$ gives a ring homomorphism of local rings $\alpha^{(n)*} : \mathcal{O}_{Y^{(n)}, \alpha^{(n)}(0)} \to K[[t]]$, where $\alpha^{(n)}(0)$ is the center of $\alpha^{(n)}$ on $Y^{(n)}$ and K is the residue field of $\alpha^{(n)} \in Y_\infty^{(n)}$. Here, we note that the center $\alpha^{(n)}(0)$ is the generic point of E. Then, we can see that $\alpha^{(n)*}$ factors through

$$\beta : \widehat{\mathcal{O}}_{Y^{(n)}, \alpha^{(n)}(0)} = k(E)[[\tau]] \to K[[t]],$$

where $k(E)$ is the rational function field of E and τ is the generator of the maximal ideal of the local ring $\mathcal{O}_{Y^{(n)},\alpha^{(n)}(0)}$. If we denote $\beta(\tau) = t^q$, then it implies that $\mathrm{ord}_{\alpha^{(n)}} = q \cdot \mathrm{val}_E$.

Finally we assume that C is an irreducible component of the intersection of $\mathrm{Cont}^{m_i}(\mathfrak{a}_i)$'s. The generic point α of C gives a divisorial valuation $q \cdot \mathrm{val}_E$ for a prime divisor over X by the above discussions. By definition of $\mathrm{Cont}^{m_i}(\mathfrak{a}_i)$, it contains also the generic point of the maximal divisorial set $C(q \cdot \mathrm{val}_E)$ for every i. This complete the proof of the last assertion of the corollary. □

For the interpretation of MJ-minimal log discrepancies with center at a (not necessarily closed) point in terms of jet schemes, we need the following definition:

Definition 3.17. Let X be a variety and $\eta \in X$ a (not necessarily closed) point. For a cylinder $S \subset X_\infty$ we define the codimension of $S \cap \pi^{-1}(\eta)$ as follows:

$$\mathrm{codim}(S \cap \pi^{-1}(\eta), X_\infty)$$
$$:= \inf \left\{ \mathrm{codim} C \mid \begin{array}{l} C \text{ is an irreducible component of } S \cap \pi^{-1}(\overline{\{\eta\}}) \\ \text{dominating } \overline{\{\eta\}} \text{ and not contained in } (\mathrm{Sing} X)_\infty \end{array} \right\}.$$

By Corollary 3.16, we obtain the following interpretation of the MJ minimal log discrepancy by the arc space. This is proved [14] for the characteristic zero case.

Theorem 3.18. *Let $(X, \mathfrak{a})$ be a pair consisting of an arbitrary variety X and a non-zero coherent ideal sheaf $\mathfrak{a} \subset \mathcal{O}_X$. Let $n \geq 0$ be a real number, let W be a proper closed subset of X and let I_W be the (reduced) ideal of W. Then,*

$$\mathrm{mld}_{\mathrm{MJ}}(W; X, \mathfrak{a}^n)$$
$$= \inf_{m_i \in \mathbb{N}} \{\mathrm{codim}(\mathrm{Cont}^{m_1}(\mathfrak{a}) \cap \mathrm{Cont}^{m_2}(\mathcal{J}_X) \cap \mathrm{Cont}^{\geq 1}(I_W)) - m_1 n - m_2\}, \quad (2)$$

$$\mathrm{mld}_{\mathrm{MJ}}(W; X, \mathfrak{a}^n)$$
$$= \inf_{m_i \in \mathbb{N}} \{\mathrm{codim}(\mathrm{Cont}^{\geq m_1}(\mathfrak{a}) \cap \mathrm{Cont}^{\geq m_2}(\mathcal{J}_X) \cap \mathrm{Cont}^{\geq 1}(I_W)) - m_1 n - m_2\}. \quad (3)$$

For a (not necessarily closed) point $\eta \in X$, we also have

$$\mathrm{mld}_{\mathrm{MJ}}(\eta; X, \mathfrak{a}^n)$$
$$= \inf_{m_i \in \mathbb{N}} \{\mathrm{codim}(\mathrm{Cont}^{m_1}(\mathfrak{a}) \cap \mathrm{Cont}^{m_2}(\mathcal{J}_X) \cap \pi^{-1}(\eta)) - m_1 n - m_2\}, \quad (4)$$

$$\mathrm{mld}_{\mathrm{MJ}}(\eta; X, \mathfrak{a}^n)$$
$$= \inf_{m_i \in \mathbb{N}} \{\mathrm{codim}(\mathrm{Cont}^{\geq m_1}(\mathfrak{a}) \cap \mathrm{Cont}^{\geq m_2}(\mathcal{J}_X) \cap \pi^{-1}(\eta)) - m_1 n - m_2\}. \quad (5)$$

Proof. For char $k = 0$, we proved these in [14, Proposition 3.8 and Remark 3.8], using [4, Theorem 3.9]. The proof for positive characteristic case follows just in the same way by using Theorem 3.13 and Corollary 3.16. However, for the reader's convenience, we write down the proof here. For the proof of the equality (2), we will show first the following inequality:

$$\begin{aligned}&\mathrm{mld}_{\mathrm{MJ}}(\eta; X, \mathfrak{a}^n)\\ &\geq \inf_{m_i\in\mathbb{N}}\{\mathrm{codim}(\mathrm{Cont}^{m_1}(\mathfrak{a})\cap \mathrm{Cont}^{m_2}(\mathcal{J}_X)\cap \pi^{-1}(\eta)) - m_1 n - m_2\}.\end{aligned}$$

Let E be any prime divisor over X with the center in W. Let $m_1 = \mathrm{val}_E(\mathfrak{a})$, $m_2 = \mathrm{val}_E(\mathcal{J}_X)$ and $v = \mathrm{val}_E$. Then, there is a non-empty open subset C of the maximal divisorial set $C_X(v)$ such that $C \subset \mathrm{Cont}^{m_1}(\mathfrak{a}) \cap \mathrm{Cont}^{m_2}(\mathcal{J}_X) \cap \mathrm{Cont}^{\geq 1}(I_W)$. Hence,

$$\begin{aligned}\widehat{k}_E - \mathrm{val}_E(\mathcal{J}_X) - n\cdot \mathrm{val}_E(\mathfrak{a}) + 1 &= \mathrm{codim}(C_X(v)) - m_1 n - m_2\\ &\geq \mathrm{codim}(\mathrm{Cont}^{m_1}(\mathfrak{a})\cap \mathrm{Cont}^{m_2}(\mathcal{J}_X)\cap \mathrm{Cont}^{\geq 1}(I_W)) - m_1 n - m_2,\end{aligned}$$

which yields the required inequality unless $\dim X = 1$ and $\mathrm{mld}_{\mathrm{MJ}}(X, \mathfrak{a}^n) = -\infty$.

When $\dim X = 1$ and $\mathrm{mld}_{\mathrm{MJ}}(X, \mathfrak{a}^n) = -\infty$, there is a prime divisor E over X with the center in W such that $\widehat{k}_E - \mathrm{val}_E(\mathcal{J}_X) - n\cdot \mathrm{val}_E(\mathfrak{a}) + 1 < 0$. Let $m_1 = \mathrm{val}_E(\mathfrak{a})$ and $m_2 = \mathrm{val}_E(\mathcal{J}_X)$, then $\mathrm{codim} C_X(\mathrm{val}_E) - m_1 n - m_2 < 0$. Here, for every $q \in \mathbb{N}$, by Proposition 3.13,

$$\mathrm{codim} C_X(q\cdot \mathrm{val}_E) - qm_1 n - qm_2 = q(\mathrm{codim} C_X(\mathrm{val}_E) - m_1 n - m_2) < 0.$$

As a non-empty open subset of $C_X(q\cdot \mathrm{val}_E)$ is contained in

$$\mathrm{Cont}^{qm_1}(\mathfrak{a})\cap \mathrm{Cont}^{qm_2}(\mathcal{J}_X)\cap \mathrm{Cont}^{\geq 1}(I_W),$$

we have

$$\begin{aligned}&\mathrm{codim}(\mathrm{Cont}^{qm_1}(\mathfrak{a})\cap \mathrm{Cont}^{qm_2}(\mathcal{J}_X)\cap \mathrm{Cont}^{\geq 1}(I_W)) - qm_1 n - qm_2\\ &\qquad\leq \mathrm{codim}(C_X(q\cdot \mathrm{val}_E)) - qm_1 n - qm_2\\ &\qquad= q(\mathrm{codim}(C_X(\mathrm{val}_E)) - m_1 n - m_2) < 0.\end{aligned}$$

Here, if $q \to \infty$, then, we have

$$\mathrm{codim}(\mathrm{Cont}^{qm_1}(\mathfrak{a})\cap \mathrm{Cont}^{qm_2}(\mathcal{J}_X)\cap \mathrm{Cont}^{\geq 1}(I_W)) - qm_1 n - qm_2 \to -\infty,$$

which implies that the right-hand side of (2) in the theorem is $-\infty$.

For the proof of the converse inequality,

$$\begin{aligned}&\mathrm{mld}_{\mathrm{MJ}}(\eta; X, \mathfrak{a}^n)\\ &\leq \inf_{m_i\in\mathbb{N}}\{\mathrm{codim}(\mathrm{Cont}^{m_1}(\mathfrak{a})\cap \mathrm{Cont}^{m_2}(\mathcal{J}_X)\cap \pi^{-1}(\eta)) - m_1 n - m_2\},\end{aligned}$$

we may assume that $\widehat{k}_E - \mathrm{val}_E(\mathcal{J}_X) - n\cdot \mathrm{val}_E(\mathfrak{a}) + 1 \geq 0$ for every prime divisor E over X with the center in W. Indeed if there is a prime divisor E with the center in W and $\widehat{k}_E - \mathrm{val}_E(\mathcal{J}_X) - n\cdot \mathrm{val}_E(\mathfrak{a}) + 1 < 0$, then $\mathrm{mld}_{\mathrm{MJ}}(W; X, \mathfrak{a}^n) = -\infty$ by Remark 2.7, (iii), and therefore, the required inequality is trivial.

For $m_i \in \mathbb{N}$, let $C \subset \mathrm{Cont}^{m_1}(\mathfrak{a})\cap \mathrm{Cont}^{m_2}(\mathcal{J}_X)\cap \mathrm{Cont}^{\geq 1}(I_W)$ be an irreducible component that gives the codimension of the cylinder. Then, the closure $\overline{C}$ is $C_X(v)$ for some divisorial valuation v by Corollary 3.16. Let $v = q\cdot \mathrm{val}_E$. Then

$m_1 = v(\mathfrak{a})$, $m_2 = v(\mathcal{J}_X)$ and E is a prime divisor over X with the center in W and

$$\operatorname{codim}(\operatorname{Cont}^{m_1}(\mathfrak{a}) \cap \operatorname{Cont}^{m_2}(\mathcal{J}_X) \cap \operatorname{Cont}^{\geq 1}(I_W)) - m_1 n - m_2$$
$$= \operatorname{codim}(C_X(v)) - m_1 n - m_2$$
$$= q(\widehat{k}_E + 1) - qn \cdot \operatorname{val}_E(\mathfrak{a}) - q \cdot \operatorname{val}_E(\mathcal{J}_X) \geq \widehat{k}_E + 1 - n \cdot \operatorname{val}_E(\mathfrak{a}),$$

which yields the required inequality.

For the proof of (3) of the theorem, let

$$a_m = \operatorname{codim}(\operatorname{Cont}^{m_1}(\mathfrak{a}) \cap \operatorname{Cont}^{m_2}(\mathcal{J}_X) \cap \operatorname{Cont}^{\geq 1}(I_W)) - m_1 n - m_2,$$

$$b_m = \operatorname{codim}(\operatorname{Cont}^{\geq m_1}(\mathfrak{a}) \cap \operatorname{Cont}^{\geq m_2}(\mathcal{J}_X) \cap \operatorname{Cont}^{\geq 1}(I_W)) - m_1 n - m_2.$$

As $\operatorname{Cont}^m(\mathfrak{a}) \subset \operatorname{Cont}^{\geq m}(\mathfrak{a})$, we have $a_m \geq b_m$. Therefore, it follows

$$\inf_m \{a_m\} \geq \inf_m \{b_m\}.$$

Next, we prove the converse inequality. For every $m \in \mathbb{N}$, let $C_X(v)$ be the irreducible component of $\operatorname{codim}(\operatorname{Cont}^{\geq m_1}(\mathfrak{a}) \cap \operatorname{Cont}^{\geq m_2}(\mathcal{J}_X) \cap \operatorname{Cont}^{\geq 1}(I_W))$ that gives the codimension. Then, for $m_1' := v(\mathfrak{a}) \geq m_1$ and $m_2' := v(\mathcal{J}_X) \geq m_2$ we have

$$\operatorname{codim}(\operatorname{Cont}^{\geq m_1}(\mathfrak{a}) \cap \operatorname{Cont}^{\geq m_2}(\mathcal{J}_X) \cap \operatorname{Cont}^{\geq 1}(I_W))$$
$$= \operatorname{codim}(\operatorname{Cont}^{m_1'}(\mathfrak{a}) \cap \operatorname{Cont}^{m_2'}(\mathcal{J}_X) \cap \operatorname{Cont}^{\geq 1}(I_W)).$$

Hence, $b_m \geq a_{m'}$, which yields $\inf_m\{b_m\} \geq \inf_m\{a_m\}$.

The equalities (4) and (5) follow in the same way, indeed one has only to be careful to replace "center in W" by "center $\overline{\{\eta\}}$." □

Remark 3.19. Our formula can be easily extended for the combination of ideals $\mathfrak{a}_1, \mathfrak{a}_2, \ldots, \mathfrak{a}_r$ instead of one ideal $\mathfrak{a}$. I.e., we have

$$\operatorname{mld}_{\mathrm{MJ}}(W; X, \mathfrak{a}_1^{e_1} \mathfrak{a}_2^{e_2} \cdots \mathfrak{a}_r^{e_r}) =$$
$$\inf_{m_i \in \mathbb{N}} \left\{ \operatorname{codim} \left(\left(\bigcap_{i=1}^{r} \operatorname{Cont}^{m_i}(\mathfrak{a}_i) \right) \cap \operatorname{Cont}^{m_{r+1}}(\mathcal{J}_X) \cap \operatorname{Cont}^{\geq 1}(I_W) \right) \right.$$
$$\left. - \sum_{i=1}^{r} m_i e_i - m_{r+1} \right\},$$

where e_i's are positive real numbers. Here, any of $\operatorname{Cont}^{m_i}(\mathfrak{a}_i)$'s can be replaced by $\operatorname{Cont}^{\geq m_i}(\mathfrak{a}_i)$. For simplicity of the notation and the proofs, we keep formulating the forthcoming formulas for one ideal only. But note that the formulas in this section are also valid under this combination form.

The description of log-canonical threshold on a non-singular variety by jet schemes in positive characteristic case is obtained by Zhu [22].

Here, we will show the inversion of adjunction Formula for MJ-minimal log discrepancies for the base field of arbitrary characteristic. The proof basically follows the idea in the case of characteristic 0. For the proof, we prepare some lemmas. The first one was proved in [6] for an arbitrary characteristic.

Lemma 3.20 ([6, Lemma 8.4]). *Let A be a non-singular variety and $M = H_1 \cap \cdots \cap H_c$ a codimension c complete intersection in A. If C is an irreducible locally closed cylinder in A_∞ such that*

$$C \subset \bigcap_{i=1}^{c} \mathrm{Cont}^{\geq d_i}(H_i),$$

and if there is an arc $\gamma \in C \cap M_\infty$ with $\mathrm{ord}_\gamma(\mathcal{J}_M) = e$, then

$$\mathrm{codim}(C \cap M_\infty, M_\infty) \leq \mathrm{codim}(C, A_\infty) + e - \sum_{i=1}^{c} d_i.$$

3.21. Consider $\mathbb{G}_m = \mathbb{A}^1 \setminus \{0\} = \mathrm{Spec}k[s, s^{-1}]$ as a multiplicative group scheme. For $m \in \mathbb{N} \cup \{\infty\}$, the morphism $k[[t]] \to k[s, s^{-1}][[t]]$ defined by $t \mapsto s \cdot t$ gives an action

$$\phi : \mathbb{G}_m \times_{\mathrm{Spec}k} \mathrm{Spec}k[[t]] \to \mathrm{Spec}k[[t]]$$

of $\mathbb{G}_m$ on $\mathrm{Spec}k[[t]]$. Therefore, it gives an action

$$\Phi : \mathbb{G}_m \times_{\mathrm{Spec}k} X_\infty \to X_\infty$$

of $\mathbb{G}_m$ on X_∞. As ϕ is extended to a morphism: $\overline{\phi} : \mathbb{A}^1 \times_{\mathrm{Spec}k} k[[t]] \to k[[t]]$, we obtain the extension

$$\overline{\Phi} : \mathbb{A}^1 \times_{\mathrm{Spec}k} X_\infty \to X_\infty$$

of Φ.

In a similar way, we have an action of $\mathbb{G}_m$ on X_r for $r \in \mathbb{N}$ and note that the truncation morphisms are $\mathbb{G}_m$-equivariant.

The following lemma is proved in Lemma 8.3 in [6] for characteristic 0 case by using a resolution of singularities. To give a characteristic free proof, we have only to check that the proof in [6] works under the normalized blow-up instead of a resolution. Let $\overline{\Phi} : \mathbb{A}^1 \times X_\infty \to X_\infty$ be the morphism given as above. Note that if an arc γ has the center $x \in X$, then $\Phi(0, \gamma)$ is the constant arc over x.

Lemma 3.22 ([6]). *Let C be a non-empty cylinder on X_∞ for a variety X. If $\overline{\Phi}(\mathbb{A}^1 \times C) \subset C$, then $C \not\subset (\mathrm{Sing}X)_\infty$.*

Proof. Take an arc $\gamma \in C$ and denote $C = (\psi_m^X)^{-1}(S)$ for some constructible set $S \subset X_m$. Let $x \in X$ be the center of γ, then by $\mathbb{A}^1$-invariance of C, the trivial m-jet $\mathbf{0}_m^x$ at x belongs to S. Let $f : X' \to X$ be the normalized blow-up at the point x and take a non-singular point x' of X' inside $f^{-1}(x)$. Let C' be the cylinder $(\psi_m^{X'})^{-1}(\mathbf{0}_m^{x'})$, then $f_\infty(C') \subset C$. As C' is not a thin set in X'_∞, C is not thin in X_∞ either, which gives that $C \not\subset (\mathrm{Sing}X)_\infty$. □

The following is the inversion of adjunction formula for MJ-minimal log discrepancies which was proved for a characteristic 0 base field in [3] and [14] independently. We give here a proof for the arbitrary characteristic case. We prepared necessary lemmas for the proof in arbitrary characteristic case that were used in the proof in characteristic 0 case. So the outline of the proof can be the same as that of the characteristic 0 case (presented in [3] and [14] which are based on [6]). Here, we remind the reader that if $\dim X = 1$ and there is a prime divisor E over X such that $a_{MJ}(E; X, \mathfrak{a}^n) < 0$, then we define $\mathrm{mld_{MJ}}(x; X, \mathfrak{a}^n) = -\infty$ (see, Definition 2.6).

Theorem 3.23 (inversion of adjunction). *Let X be a variety over an algebraically closed field k of an arbitrary characteristic and A a smooth variety containing X as a closed subscheme of codimension c. Let $\widetilde{\mathfrak{a}} \subset \mathcal{O}_A$ be a coherent ideal sheaf such that its image $\mathfrak{a} := \widetilde{\mathfrak{a}}\mathcal{O}_X \subset \mathcal{O}_X$ is non-zero. Denote the ideal of X in A by I_X. Then, for a proper closed subset W of X, we have*

$$\mathrm{mld_{MJ}}(W; X, \mathfrak{a}^n) = \mathrm{mld}(W; A, \widetilde{\mathfrak{a}}^n I_X^c). \tag{6}$$

For a point $\eta \in X$, we have

$$\mathrm{mld_{MJ}}(\eta; X, \mathfrak{a}^n) = \mathrm{mld}(\eta; A, \widetilde{\mathfrak{a}}^n I_X^c). \tag{7}$$

Proof. First, we prove the inequality $\geq$ in (6). We assume the contrary and will induce a contradiction. By the assumption, there exist $e, m \in \mathbb{N}$ and an irreducible component $C \subset \mathrm{Cont}^{\geq m}(\mathfrak{a}) \cap \mathrm{Cont}^e(\mathcal{J}_X) \cap \mathrm{Cont}^{\geq 1}(I_W)$ such that $\mathrm{codim} C - mn < \mathrm{mld}(W; A, \widetilde{\mathfrak{a}}^n I_X^c)$. Then, for a sufficiently large $s \in \mathbb{N}$, we obtain

$$(s+1)d - \dim \psi_m(C) - mn = \mathrm{codim} C - mn < \mathrm{mld}(W; A, \widetilde{\mathfrak{a}}^n I_X^c), \tag{8}$$

where I_W is the defining ideal of W in X. As $\psi_s(C) \subset \mathrm{Cont}^{\geq m}(\mathfrak{a})_s \cap \mathrm{Cont}^{\geq 1}(I_W)_s = \mathrm{Cont}^{\geq m}(\widetilde{\mathfrak{a}})_s \cap \mathrm{Cont}^{\geq 1}(\widetilde{I}_W)_s \cap X_s$, where $\widetilde{I}_W$ is the defining ideal of W in $\mathcal{O}_A$, we have

$$C \subset (\psi_s^A)^{-1}(\psi_s(C)) \subset \mathrm{Cont}^{\geq m}(\widetilde{\mathfrak{a}}) \cap \mathrm{Cont}^{\geq 1}(\widetilde{I}_W) \cap \mathrm{Cont}^{\geq s+1}(I_X) =: S,$$

where $\psi_s^A : A_\infty \to A_s$ is the truncation morphism.

Now we obtain

$$\begin{aligned}
&(d+c)(s+1) - \dim \psi_s(C) \\
&\quad = \mathrm{codim}(\psi_s(C), A_s) = \mathrm{codim}((\psi_s^A)^{-1}(\psi_s(C), A_\infty) \\
&\quad \geq \mathrm{codim}(S, A_\infty) \geq \mathrm{mld}(W; A, \widetilde{\mathfrak{a}}^n I_X^c) + mn + c(s+1),
\end{aligned}$$

which is a contradiction to (8).

For the converse inequality in (6), we have only to show the following claim

Claim 3.24. For a prime divisor F over A with center in W, there is a prime divisor E over X with center in W and an integer $q \geq 1$ such that

$$q \cdot a_{MJ}(E; X, \mathfrak{a}^n) \leq a_{MJ}(F, A, \tilde{\mathfrak{a}}^n I_X^c). \tag{9}$$

We can prove that this induces the inequality

$$\mathrm{mld}_{\mathrm{MJ}}(W; X, \mathfrak{a}^n) \le \mathrm{mld}(W; A, \tilde{\mathfrak{a}}^n I_X^c)$$

as follows: If $\mathrm{mld}_{\mathrm{MJ}}(W; X, \mathfrak{a}^n) = -\infty$, then the required inequality is trivial. If $\mathrm{mld}_{\mathrm{MJ}}(W; X, \mathfrak{a}^n) \ge 0$ then $a_{MJ}(E; X, \mathfrak{a}^n) \ge 0$ in (9), which yields $a_{MJ}(F, A, \tilde{\mathfrak{a}}^n I_X^c) \ge \frac{1}{q} a_{MJ}(F, A, \tilde{\mathfrak{a}}^n I_X^c) \ge a_{MJ}(E; X, \mathfrak{a}^n) \ge \mathrm{mld}_{\mathrm{MJ}}(W; X, \mathfrak{a}^n)$ for every prime divisor F over A with center in W.

In the following, we prove Claim 3.24. Consider the maximal divisorial set

$$V = C_A(\mathrm{val}_F) \subset A_\infty;$$

then we have

$$\mathrm{codim}(V, A_\infty) = k_F + 1.$$

The intersection $V \cap X_\infty \subset X_\infty$ is a non-empty cylinder in X_∞ and is not contained in $(\mathrm{Sing} X)_\infty$ by Lemma 3.22. Let C be an irreducible component of $V \cap X_\infty$ that is not contained in the arc space of the singular locus of X, and the generic point $\gamma \in C$ gives the minimal value $\mathrm{ord}_\gamma(\mathcal{J}_X) = e$ among the points in $V \cap X_\infty$. We can also assume that C has the minimal codimension among the components with $\mathrm{ord}_\gamma(\mathcal{J}_X) = e$. Then, by Corollary 3.16, there exists a prime divisor E over X with the center in W, such that the generic point γ of C gives the divisorial valuation $q \cdot \mathrm{val}_E$ for some $q \in \mathbb{N}$. Therefore, $C \subset C_X(q \cdot \mathrm{val}_E)$ and we have the following inequalities:

$$q \cdot (\widehat{k}_E + 1) = \mathrm{codim}(C_X(q \cdot \mathrm{val}_E), X_\infty) \le \mathrm{codim}(C, X_\infty). \tag{10}$$

Now, by an appropriate choice of c generators of I_X, we can take a complete intersection scheme M of dimension d containing X such that $\mathrm{ord}_\gamma(\mathcal{J}_M) = e$. Then, by [6, Corollary 9.2], we have

$$\mathcal{J}_M \cdot \mathcal{O}_X \subset ((I_M : I_X) + I_X)/I_X.$$

It follows that γ belongs to the open cylinder

$$V_0 := V \cap \mathrm{Cont}^{\le e}(\mathcal{J}_M) \cap \mathrm{Cont}^{\le e}(I_M : I_X).$$

Here, we note that

$$V_0 \cap X_\infty = V_0 \cap M_\infty.$$

Indeed, by the definition of V_0 an arc $\alpha \in V_0 \cap M_\infty$ has finite order along $(I_M : I_X)$ which is a defining ideal of the union X' of the components of M other than X. This means $\alpha \notin X'_\infty$.

Here, for every $\beta \in M_\infty$, we note that $\mathrm{ord}_\beta(\mathcal{J}_X) \le \mathrm{ord}_\beta(\mathcal{J}_M)$; therefore, by the definition of C, we obtain

$$\mathrm{codim}(C, X_\infty) = \mathrm{codim}(V_0 \cap X_\infty, X_\infty) = \mathrm{codim}(V_0 \cap M_\infty, M_\infty). \tag{11}$$

Apply Lemma 3.20 to V_0 and we obtain

$$\mathrm{codim}(V_0 \cap M_\infty, M_\infty) \le \mathrm{codim}(V_0, A_\infty) + e - \sum_{i=1}^{c} d_i,$$

where d_i's satisfy $V_0 \subset \bigcap_{i=1}^{c} \mathrm{Cont}^{\geq d_i}(H_i)$ for $M = H_1 \cap \cdots \cap H_c$. Then, as V_0 is an open subset of $V = C_A(\mathrm{val}_F)$, the term $\sum d_i$ in the above inequality satisfies $\sum d_i \geq c \cdot \mathrm{val}_F(I_X)$ and the equality $\mathrm{codim}(V_0, A_\infty) = \mathrm{codim}(V, A_\infty)$ holds. On the other hand, $e = \mathrm{ord}_\gamma(\mathcal{J}_X) = q \cdot \mathrm{val}_E(\mathcal{J}_X) = q \cdot j_E$. Therefore, we have the following inequality

$$\mathrm{codim}(V_0 \cap M_\infty, M_\infty) \leq \mathrm{codim}(V_0, A_\infty) + q \cdot \mathrm{val}_E(\mathcal{J}_X) - c \cdot \mathrm{val}_F(I_X). \tag{12}$$

Now combining (10), (11), and (12), we obtain

$$q \cdot (\widehat{k}_E - j_E + 1) \leq \mathrm{codim}(V, A_\infty) - c \cdot \mathrm{val}_F(I_X).$$

Note that for any proper coherent ideal sheaf $\mathfrak{b} \subset \mathcal{O}_M$ not vanishing on any component of X, we have $\mathrm{val}_C(\mathfrak{b}|_X) \geq \mathrm{val}_V(\mathfrak{b})$ by the inclusion $C \subset V$. In particular, this implies that

$$q \cdot \mathrm{val}_E(\mathfrak{b}|_X) \geq \mathrm{val}_F(\mathfrak{b}),$$

which yields the inequality in Claim 3.24.

The proof of the equality (7) follows in the same way. Indeed, we have only to be careful to replace "with the center in W" by "with the center $\overline{\{\eta\}}$" in the proof above. □

Remark 3.25. The formula in the theorem can be easily extended for the combination of ideals $\mathfrak{a}_1, \mathfrak{a}_2, \ldots, \mathfrak{a}_r$ instead of one ideal $\mathfrak{a}$. I.e., we have

$$\mathrm{mld}_{\mathrm{MJ}}(W; X, \mathfrak{a}_1^{e_1}\mathfrak{a}_2^{e_2} \cdots \mathfrak{a}_r^{e_r}) = \mathrm{mld}(W; A, \tilde{\mathfrak{a}}_1^{e_1}\tilde{\mathfrak{a}}_2^{e_2} \cdots \tilde{\mathfrak{a}}_r^{e_r} I_X^c),$$

where e_i's are non-negative real numbers and $\tilde{\mathfrak{a}}_1, \tilde{\mathfrak{a}}_2, \ldots, \tilde{\mathfrak{a}}_r$ are coherent ideal sheaves of $\mathcal{O}_A$ such that the extensions $\mathfrak{a}_i = \tilde{\mathfrak{a}}_i\mathcal{O}_X$'s are not zero.

Corollary 3.26. *Let $\eta \in X$ be a point of a variety X of dimension d. For each $m \in \mathbb{N}$ we denote $r_m = \dim \pi_m^{-1}(x)$ for a general closed point $x \in \overline{\{\eta\}}$. Then, we have the equality:*

$$\begin{aligned}\mathrm{mld}_{\mathrm{MJ}}(\eta; X) := \mathrm{mld}_{\mathrm{MJ}}(x; X, \mathcal{O}_X) &= \inf_m \left\{(m+1)d - \dim \overline{\pi_m^{-1}(\eta)}\right\} \\ &= \inf_m \left\{(m+1)d - (\dim \overline{\{\eta\}} + r_m)\right\}.\end{aligned} \tag{13}$$

Proof. Since the statement is local, we can assume that X is a closed subvariety of codimension c of a non-singular variety A. By Theorem 3.23, we have

$$\mathrm{mld}_{\mathrm{MJ}}(\eta; X) = \mathrm{mld}(\eta; A, I_X^c)$$

and by Theorem 3.18, we have

$$\mathrm{mld}(\eta; A, I_X^c) = \inf_{m \geq 0} \left\{\mathrm{codim}(\mathrm{Cont}^{m+1}(I_X) \cap (\pi^A)^{-1}(\eta), A_\infty) - (m+1)c\right\},$$

where, for our convenience later, we shift m to $m+1$ on the right-hand side. By Definition 3.17, $\mathrm{codim}(\mathrm{Cont}^{m+1}(I_X) \cap (\pi^A)^{-1}(\eta), A_\infty)$ is the minimal codimension of irreducible components $C \subset (\psi_m^A)^{-1}(X_m) \cap \pi^{-1}(\overline{\{\eta\}})$ in A_∞ that dominate $\overline{\{\eta\}}$. Therefore,

$$\operatorname{codim}(\operatorname{Cont}^{m+1}(I_X) \cap (\pi^A)^{-1}(\eta), A_\infty) = \operatorname{codim}\left(\overline{(\pi_m^X)^{-1}(\eta)}, A_m\right).$$

As $\operatorname{codim}\left(\overline{(\pi_m^X)^{-1}(\eta)}, A_m\right) = (m+1)(d+c) - \dim \overline{(\pi_m^X)^{-1}(\eta)}$, we have the first equality in the corollary. For the second equality, we need to know that

$$\dim \overline{\pi_m^{-1}(\eta)} = \dim \overline{\{\eta\}} + r_m. \qquad \square$$

Corollary 3.27. *Let X be a variety of dimension d and $\mathfrak{a}$ a coherent ideal sheaf of $\mathcal{O}_X$. Let $V \subset W$ be two irreducible proper closed subsets of X and η_V and η_W are the generic points of V and W, respectively. Then,*

(i) We have the following inequality:

$$\operatorname{mld}_{\mathrm{MJ}}(\eta_V; X, \mathfrak{a}) \leq \operatorname{codim}(V, X),$$

where the equality holds if and only if η_V is a regular point and $\mathfrak{a} = \mathcal{O}_X$ around η_V.

In particular, if $x \in X$ is a closed point, then

$$\operatorname{mld}_{\mathrm{MJ}}(x; X, \mathfrak{a}) \leq d$$

and the equality holds if and only if X is smooth at x and $\mathfrak{a} = \mathcal{O}_X$ around x. (This statement for the usual mld *is Shokurov's conjecture and is not yet proved.)*

(ii)
$$\operatorname{mld}_{\mathrm{MJ}}(\eta_V; X, \mathfrak{a}) \leq \operatorname{mld}_{\mathrm{MJ}}(\eta_W; X, \mathfrak{a}) + \operatorname{codim}(V, W),$$
where the equality holds for very general V in W; i.e., η_V is in the complement of a countable number of closed subsets in W. (If k is an uncountable field, then this subset is non-empty.)

(iii) If char $k = 0$, *then the equality in (ii) holds for general V in W.*

Proof. First note that if $\mathfrak{a} \neq \mathcal{O}_X$ around η_V, then

$$\operatorname{mld}_{\mathrm{MJ}}(\eta_V; X, \mathfrak{a}) < \operatorname{mld}_{\mathrm{MJ}}(\eta_V; X, \mathcal{O}_X)$$

by the definition of MJ-log discrepancy. Here, by Corollary 3.26, we have

$$\operatorname{mld}_{\mathrm{MJ}}(\eta_V; X, \mathcal{O}_X) \leq (m+1)d - r_m - \dim V \tag{14}$$

for every $m \geq 0$. Therefore, in particular for $m = 0$, we obtain

$$\operatorname{mld}_{\mathrm{MJ}}(\eta_V; X, \mathcal{O}_X) \leq \operatorname{codim}(V, X).$$

Now assume the equality in (i): $\mathrm{mld}_{\mathrm{MJ}}(\eta_V; X, \mathcal{O}_X) = d - \dim V$. then, by the first comment in this proof, we have $\mathfrak{a} = \mathcal{O}_X$. Consider the inequality (14) for $m = 1$, we obtain

$$d - \dim V \leq (d - \dim V) + d - r_1. \tag{15}$$

Here, we note that r_1 is the dimension of the tangent space of X at a general closed point $x \in V$. Then, the inequality (15) gives $r_1 = d$, which means that general points of V are non-singular in X.

For the proof of (ii), let

$$s_V(m,n) := \mathrm{codim}\left(\mathrm{Cont}^{\geq m}(I_X) \cap \mathrm{Cont}^{\geq n}(\mathfrak{a}) \cap (\psi^A)^{-1}(\eta_V), A_\infty\right) - mc - n,$$

$$s_W(m,n) := \mathrm{codim}\left(\mathrm{Cont}^{\geq m}(I_X) \cap \mathrm{Cont}^{\geq n}(\mathfrak{a}) \cap (\psi^A)^{-1}(\eta_W), A_\infty\right) - mc - n.$$

Then, by Theorem 3.18 and Theorem 3.23, we have

$$\mathrm{mld}_{\mathrm{MJ}}(\eta_V; X, \mathfrak{a}) = \inf_{m,n} s_V(m,n) \text{ and } \mathrm{mld}_{\mathrm{MJ}}(\eta_W; X, \mathfrak{a}) = \inf_{m,n} s_W(m,n).$$

Remember the action of $\mathbb{G}_m$ on A_r (3.21). Then, for each m, n, by an appropriate $r = r(m,n) \in \mathbb{N}$ and a $\mathbb{G}_m$-invariant closed subset

$$S_{m,n} = \psi_r((\mathrm{Cont}^{\geq m}(I_X) \cap \mathrm{Cont}^{\geq n}(\mathfrak{a})) \subset A_r,$$

we can express

$$\begin{aligned} s_V(m,n) - s_W(m,n) &= \dim S_{m,n} \cap \pi_r^{-1}(\eta_W) - \dim S_{m,n} \cap \pi_r^{-1}(\eta_V) \\ &= (\dim W + \delta_W) - (\dim V + \delta_V), \end{aligned}$$

where δ_V and δ_W are the dimensions of general fibers of $\pi_r|_{S_{m,n}\cap\pi_r^{-1}(V)}$ and $\pi_r|_{S_{m,n}\cap\pi_r^{-1}(W)}$, respectively. As $S_{m,n}$ is $\mathbb{G}_m$-invariant, the restricted morphism $\pi_r|_{S_{m,n}\setminus\sigma(A)} : S_{m,n} \setminus \sigma(A) \to A$ factors through the projective morphism $\pi'_r : (S_{m,n} \setminus \sigma(A))/\mathbb{G}_m \to A$. Here, $\sigma(A) \subset A_r$ is the subset consisting of the trivial r-jets on A. Therefore, dimension of fibers of π'_r and also $\pi_r|_{S_{m,n}}$ are upper-semicontinuous, which implies the inequality $\delta_W \leq \delta_V$. This yields

$$s_V(m,n) - s_W(m,n) \leq \mathrm{codim}(V, W)$$

for every m, n, which yields the inequality in (ii).

For a fixed m, n, there exists a closed subset $F_{m,n} \subset W$ such that for every point $\eta_V \in W$ which is not contained in $F_{m,n}$ satisfies $\delta_V = \delta_W$. Then, for these V we obtain

$$s_V(m,n) - s_W(m,n) = \mathrm{codim}(V, W).$$

Therefore, if η_V is not contained in $F = \bigcup_{m,n} F_{m,n}$, then, we obtain the equality

$$\mathrm{mld}_{\mathrm{MJ}}(\eta_V; X, \mathfrak{a}) = \mathrm{mld}_{\mathrm{MJ}}(\eta_W; X, \mathfrak{a}) + \mathrm{codim}(V, W).$$

For the proof of (iii), assume that char $k = 0$. Let $f : A' \to A$ be an embedded log resolution of $(X, \mathfrak{a})$ with at least one exceptional divisor with the center $\overline{\{\eta_W\}}$. Then, there is an exceptional prime divisor $E \subset A'$ computing the $\mathrm{mld}_{\mathrm{MJ}}(\eta_W; X, \mathfrak{a})$. If $\eta_V \in W$ is not in the union of the lower dimensional centers of the exceptional divisors of f, then, the divisor obtained by the blow-up with the center $f^{-1}(V) \cap E$ computes

$$\mathrm{mld}_{\mathrm{MJ}}(\eta_V; X, \mathfrak{a}) = \mathrm{mld}_{\mathrm{MJ}}(\eta_W; X, \mathfrak{a}) + \mathrm{codim}(V, W). \qquad \square$$

Remark 3.28. As is seen in the proof of (iii), one can see that for a positive characteristic case if (iii) in Corollary 3.27 does not hold, it shows a counter example of the existence of resolution of singularities.

4 MJ-canonical and MJ-log-canonical singularities

We say that X has MJ-canonical (resp. MJ-log-canonical) singularities if the pair $(X, \mathcal{O}_X)$ has MJ-canonical (resp. MJ-log-canonical) singularities. We denote $\mathrm{mld}_{\mathrm{MJ}}(x; X, \mathcal{O}_X)$ by $\mathrm{mld}_{\mathrm{MJ}}(x; X)$. In this section, we will study the nature of MJ-canonical singularities and MJ-log-canonical singularities.

Lemma 4.1 ([8, Proposition 3.3]). *Let $x \in X$ be a closed point. If X is MJ-canonical at x, then the embedding dimension satisfies*

$$\mathrm{emb}(X, x) \leq 2d - 1.$$

If X is MJ-log-canonical at x, then the embedding dimension satisfies

$$\mathrm{emb}(X, x) \leq 2d.$$

Proof. This is proved in [8, Proposition 3.3] in case char $k = 0$. However, the main point of the proof is the formula in Corollary 3.26 and we have seen in the proof of the corollary that the formula holds true for an arbitrary characteristic. Therefore, the same proof works for the lemma in the positive characteristic case too. $\square$

Definition 4.2. We say that a variety X has a pseudo rational singularity at $x \in X$ if

(1) X is normal around x;

(2) For every partial resolution $f : Y \to X$, which means a proper birational morphism with normal Y, the equality

$$f_*\omega_Y = \omega_X$$

holds around x;

(3) X is Cohen–Macaulay around x.

Note that if char $k = 0$ or if $\dim X = 2$, this definition is equivalent to the following:

(1) X is normal around x;

(2') For every resolution $f : Y \to X$ the vanishing

$$R^i f_* \mathcal{O}_Y = 0$$

holds for $i > 0$ around x.

The singularity (X, x) satisfying (1) and (2') is called a *rational singularity*.

Proposition 4.3. *Let a variety X has at worst MJ-canonical singularities. Then X has normal hypersurface singularities in codimension 2.*

If char $k = 0$, *then X is normal; furthermore, the singularities on X are rational.*

Proof. The second statement is proved in [3, Theorem 7.7] and [7, Corollary 3.7] independently. Regarding the first statement, since the problem is local, we may assume that X is a closed subvariety in the affine space $A = \mathbb{A}^N$. Then, as X is MJ-canonical, it follows that $\mathrm{mld}_{\mathrm{MJ}}(\eta; X) = \mathrm{mld}(\eta, A, I_X^c) \geq 0$, where I_X is the defining ideal of X in A and $c = \mathrm{codim}(X, A)$.

The first statement is proved by blow-up A at an irreducible component of the singular locus of X and checking the discrepancy of (A, I_X^c) as in the proof of Proposition 4.7. But, here, we present a proof using jet schemes discussions.

Let $\eta \in X$ be the generic point of an irreducible closed subset of codimension one. Then, as X is MJ-canonical, it follows that $\mathrm{mld}_{\mathrm{MJ}}(\eta; X) \geq 1$. On the other hand, we have $\mathrm{mld}_{\mathrm{MJ}}(\eta; X) \leq d - \dim \overline{\{\eta\}} = 1$ by Corollary 3.27. Then, we obtain that the equality in (i) in Corollary 3.27 holds, which yields that X is regular at η. Now, $\dim \mathrm{Sing}(X) \leq d - 2$. Let $\zeta \in X$ be the generic point of an irreducible component of $\mathrm{Sing}(X)$ of codimension 2. Then, by inequality (14), it follows

$$1 \leq \mathrm{mld}_{\mathrm{MJ}}(\zeta, X) \leq (m+1)d - r_m - \dim \overline{\{\zeta\}},$$

where r_m is the dimension of a general fiber of π_m. Considering the case $m = 1$, we obtain

$$1 \leq \mathrm{mld}_{\mathrm{MJ}}(\zeta, X) \leq 2d - r_1 - (d-2) \leq 1. \tag{16}$$

Here, the last inequality is showed as follows: note that r_1 is the dimension of Zariski tangent space of X at a general point; therefore, $r_1 \geq d+1$, as the point is a singular point. Therefore, all inequalities in (16) become equalities, in particular $r_1 = d+1$, which means that a general point is a hypersurface singularity. Hence, X is a Gorenstein variety in codimension 2 and satisfies R_1, which yields that X is normal in codimension 2 by Serre's criteria. □

Corollary 4.4. *A two-dimensional singularity (X, x) is MJ-canonical if and only if it is a rational double point in arbitrary characteristic.*

Proof. In case char $k = 0$, this statement is proved in [8]. The following is a characteristic free proof. As a rational double point of dimension two is a normal hypersurface singularity, the canonicity in the usual sense is equivalent to MJ-canonicity. Therefore, a rational double point of dimension two is MJ-canonical.

Conversely, if (X, x) is MJ-canonical, then, by Proposition 4.3, it is a normal hypersurface singularity. Therefore it is canonical in the usual sense, which yields that it is rational double. □

Corollary 4.5. *If an MJ-canonical variety X is locally a complete intersection, then X is normal and has pseudo rational singularities. In particular, for* char $k = 0$*, X has rational singularities.*

Proof. As X is locally a complete intersection, it is Gorenstein. By Proposition 4.3, X satisfies R_1, therefore by Serre's criteria, it is normal. Since X is locally a complete intersection, we have that $\widehat{k}_E - j_E = k_E$ (Remark 2.5) for every prime divisor E over X. On the other hand, when we take a partial resolution $f : Y \to X$, for every prime divisor E on Y, we obtain $\mathrm{val}_E(K_Y - f^*K_X) = k_E = \widehat{k}_E - j_E \geq 0$ by the assumption that X is MJ-canonical. Therefore, on Y, we have the inequality $K_Y \geq f^*K_X$ which yields

$$f_*\omega_Y \supset f_*f^*\omega_X = \omega_X. \tag{17}$$

Now, we obtain $f_*\omega_Y = \omega_X$, since the opposite inclusion of (17) is trivial. □

Definition 4.6. Let $x \in X$ be a closed point of a variety X. We say that x is a normal crossing double point in X if $\widehat{\mathcal{O}_{X,x}} = k[[x_1, \ldots, x_N]]/(x_1 \cdot x_2)$.

Proposition 4.7. *Let a variety X have at worst MJ-log-canonical singularities. Then a general point of the singular locus of codimension one is normal crossing double.*

Proof. Since the problem is local, we may assume that X is a closed subvariety in a smooth affine variety A. Then, as X is MJ-log-canonical, it follows that $\mathrm{mld}_{\mathrm{MJ}}(\eta; X) = \mathrm{mld}(\eta, A, I_X^c) \geq 0$, for every (not necessarily closed) point $\eta \in X$, where I_X is the defining ideal of X in A and $c = \mathrm{codim}(X, A)$. It seems to be well known that if (A, I_X^c) is log-canonical, then X is a hypersurface with at worst normal crossing double singularities in codimension 1 when the base field is of characteristic 0. We will write the proof that works for an arbitrary characteristic case.

Let $S \subset X$ be an irreducible component of codimension 1 in the singular locus and let $\eta \in S$ be the generic point. Take A as a minimal dimensional smooth ambient space around η. Let $f : A_1 \to A$ be the blow-up of A with the center S and let E_1 be the exceptional divisor dominating S. Then, we obtain

$$a(E_1; A, I_X^c) = k_{E_1} - c \cdot \mathrm{val}_{E_1}(I_X) + 1 \geq 0.$$

Here, we note that $k_{E_1} = c$. By the minimality of $\dim A$, every element of I_X has multiplicity ≥ 2 at η, which yields $\mathrm{val}_E(I_X) \geq 2$. Then, we obtain

$$0 \leq k_{E_1} - c \cdot \mathrm{val}_{E_1}(I_X) + 1 \leq c - 2c + 1 = -c + 1 \leq 0.$$

Therefore all equalities should hold, in particular, $a(E_1; A, I_X^c) = 0$, $c = 1$ and $\mathrm{val}_{E_1}(I_X) = 2$. These mean that X has hypersurface double points in codimension 1.

If a general point of S is not a normal crossing double point, then E_1 and the proper transform X_1 contact with the order ≥ 2 along a closed subset S_1 dominating S. Let $A_2 \to A_1$ be the blow-up with the center S_1 and E_2 the exceptional divisor dominating S_1. Let $E_1' \subset A_2$ and $X_2 \subset A_2$ be the proper transforms of E_1 and X_1, respectively.

Then, three divisors E_2, E_1' and X_2 in A_2 still have an intersection at a closed subset S_2 of dimension $d-1$. Now, blow up $A_3 \to A_2$ with the center S_2 and let E_3 be the exceptional divisor dominating S_2. Then, the log discrepancy at E_3 is

$$a(E_3; A, I_X) = k_{E_3} - \mathrm{val}_{E_3} I_X + 1 \leq 3 - 5 + 1 = -1,$$

a contradiction. Therefore, E_1 and X_1 have the reduced intersection at S_1 over general points of S, which implies that X is normal crossing double at a general point of S. $\square$

Theorem 4.8. *Let k be an algebraically closed field. Let $p \in X$ be a closed point of an arbitrary variety X over k. A pair (X, B) consisting of X and an effective $\mathbb{R}$-Cartier divisor B on X satisfy*

$$\dim X - 1 \leq \mathrm{mld}_{\mathrm{MJ}}(p; X, B) \tag{18}$$

if and only if either

(i) *$\dim X \geq 2$, $B = 0$ and (X, p) is a compound Du Val singularity,*

(ii) *$B = 0$, and $(X, p) \subset (\mathbb{A}^3, 0)$ is given by:*

$$\begin{aligned}
&xy = 0,\\
&z^2 + xy^2 = 0 \textit{ if } \mathrm{char}\, k \neq 2, \textit{ or}\\
&z^2 + xy^2 + yzg(x,y) = 0 \textit{ if } \mathrm{char}\, k = 2,\\
&\qquad \textit{where } \mathit{mult} g \geq 1 \textit{ and either } g = 0 \textit{ or } g(x,0) \neq 0.
\end{aligned}$$

(iii) *(X, p) is non-singular and $0 \leq \mathrm{mult}_p B \leq 1$.*

In cases (i) *and* (ii), *we have $\mathrm{mld}_{\mathrm{MJ}}(p; X, \mathcal{J}_X) = \dim X - 1$ and in case* (iii) *we have $\mathrm{mld}_{\mathrm{MJ}}(p; X, B) = \mathrm{mld}(p; X, B) = \dim X - \mathrm{mult}_p B$ and the minimal log discrepancy is computed by the exceptional divisor of the first blow-up at p.*

Proof. As in [15, Theorem 4.1], it suffices to show that equality in (18) holds if and only if (i) or (ii) are satisfied. Germs of varieties (X, p), where p is a closed point, satisfying $\mathrm{mld}_{\mathrm{MJ}}(p, X) = \dim X - 1$ are called *top singularities* in [15].

Recall the invariant $\tau(X, p)$ introduced in [10]. In the case of a germ of an hypersurface (X, p) in $(\mathbb{A}_k^{d+1}, 0)$ defined by $f \in k[x_1, \ldots, x_{d+1}]$, τ is the smallest possible dimension of a linear subspace V_0 of $V = kx_1 + kx_2 + \cdots + kx_{d+1}$ such that the initial term $\mathrm{in} f$ of f lies in the subalgebra $k[V_0]$ of $k[x_1, \ldots, x_{d+1}]$. It is known to be an invariant of (X,p).

The following *results*, proved in [15] if char $k = 0$, are based on equality (13), hence, remain true if char $k > 0$:

(R0) [15, Lemma 3.5] Let X be a d-dimensional variety and let $X' \subset X$ be a $(d-c)$-dimensional subvariety, which is defined as the zero locus of c elements of $\mathcal{O}_X$. Let p be a closed point in X'. If (X',p) is a top singularity, then (X,p) is a top singularity.

(R1) [15, Lemma 3.6] If X has a top singularity at p, then X has a hypersurface singularity of multiplicity 2 at p.

(R2) [15, Lemma 3.20] Let $(X,p) \subset (\mathbb{A}_k^{d+1}, 0)$ be a germ of a hypersurface singularity of multiplicity 2 at p and $\tau(X,p) = 1$. Hence, the equation of $(X,p) \subset (\mathbb{A}^{d+1}, 0)$ is expressed as

$$f = x_1^2 - g(x_2, \ldots, x_{d+1}) + x_1 g'(x_1, \ldots, x_{d+1}) = 0.$$

Here, we note that $x_1 g' = x_1^2 - g(x_2, \ldots, x_{d+1}) \in x_1(x_1, \ldots, x_{d+1})^2$. Then, if (X,p) is a top singularity, we have $d \geq 2$ and mult $g = 3$.

(R3) [15, Proposition 3.23] In the conditions of (R2), suppose also that the initial form of g has only one factor. Hence, the equation of $(X,p) \subset (\mathbb{A}^{d+1}, 0)$ gives

$$\begin{aligned} &x_1^2 + x_2^3 + g_3(x_3, \ldots, x_{d+1})\, x_2 + g_4(x_3, \ldots, x_{d+1}) \\ &\quad \in x_1(x_3, \ldots, x_{d+1})^2 + x_1 x_2 (x_3, \ldots, x_{d+1}) + (x_1 x_2^2) + x_2^2 (x_3, \ldots, x_{d+1})^2 \end{aligned}$$

where mult $g_i \geq i$, for $i = 3, 4$. Then, if (X,p) is a top singularity, we have either mult $g_3 = 3$ or $4 \leq$ mult $g_4 \leq 5$.

On the other hand, in pages 256–268 of [18], a characterization of pseudo rational double points over any field is given.

Restricting to an algebraically closed field k, the following is proved: Let us assume the *hypothesis* H1:

H1) $X \subset \mathbb{A}_k^3$ be a surface of multiplicity 2. Then $\tau \leq 3$ and we have:

Case I : $\tau = 3$ if and only if $(X,p) \subset (\mathbb{A}^3, 0)$ has A_1-singularity.

Case II (II a) in [18]): $\tau = 2$ is equivalent to $(X,p) \subset (\mathbb{A}^3, 0)$ having A_n-singularity ($n \geq 2$) if we assume that (X,p) is pseudo rational.

If $\tau = 1$, the equation of X gives $z^2 - G(x,y) \in zM^2$ with mult $G \geq 3$, where M is the maximal ideal at the origin. Besides, the hypothesis X pseudo rational implies that

H2) mult $G = 3$.

Let $\overline{G}$ be the initial form of G. Then one of the following cases occurs:

Case III (III c in [18]): $\tau = 1$ and $\overline{G}$ has 3 factors. This is equivalent to X being a D_4-singularity

Case IV : $\tau = 1$ and $\overline{G}$ is the product of a linear factor and the square of another factor. This is equivalent to $(X, p) \subset (\mathbb{A}^3, 0)$ having a D_n-singularity ($n \geq 5$) if we assume that (X, p) is pseudo rational.

Case V : $\tau = 1$ and $\overline{G}$ has only 1 factor. Then, either (X, p) is an E_6-singularity or it can be expressed as

$$z^2 + y^3 + \rho x^3 y + \sigma x^5 \in (zxy, zy^2, x^2y^2, x^3z).$$

Again (X, p) pseudo rational implies:

H3) either ρ is a unit or σ is a unit.

Finally, we have

Case V c : ρ being a unit is equivalent to (X, p) being an E_7-singularity.

Case V d : ρ not a unit and σ a unit, is equivalent to (X, p) being an E_8-singularity.

Note that rational double points are top singularities (apply the proof for char $k = 0$ given in [15], example 3.12). From this, applying (R0), it follows that (i) in the theorem implies equality in (16). The fact that (ii) implies equality in (16) can also be checked. In fact, for the last case in (ii), note that the surface given by $z^2 + xy^2 + yzg(x, y) = 0$ (char $k = 2$), where mult$g \geq 1$ and either $g = 0$ or $g(x, 0) = 0$, can be desingularized by the blow-up of (y, z); if E is the exceptional curve appearing then $\widehat{k}_E = 1$ and $\mathrm{ord}_E(\mathcal{J}_X) = 1$.

Finally, let us prove that if (X, p) is a top singularity, then either (i) or (ii) hold. In fact, let (X', p) be the surface obtained by a general cut of (X, p) with $d - 2$ hyperplanes. Since (R1) holds for (X, p), we have that (H1) holds for (X', p). Analogously, since (R2) holds for (X, p), it follows that for (X', p) we have that if $\tau = 1$ then (H2) holds. Finally, (R3) for (X, p) implies that if (X', p) satisfies the hypothesis of Case V, either (X', p) is an E_6-singularity or (H3) holds.

We thus conclude that if we assume that (X', p) is pseudo rational in Cases II and IV, then (X, p) being a top singularity would imply that (X', p) is a rational double point; hence, (X, p) is a compound Du Val singularity.

Finally, for cases II and IV without the hypothesis of pseudo rational, we obtain

1. If (X', p) is in case II and not pseudo rational, then it can be expressed as $xy = 0$.

2. If (X', p) is in case IV and not pseudo rational, then it can be expressed as:

$$z^2 + xy^2 = 0 \text{ if char } k \neq 2, \text{ or}$$
$$z^2 + xy^2 + yzg(x, y) = 0 \text{ if char } k = 2,$$
where mult$g \geq 1$ and either $g = 0$ or $g(x, 0) \neq 0$.

Thus, we conclude the result. □

Remark 4.9. For a closed point $x \in X$ in a variety X. Define

$$s_m(x) := (m+1)d - \dim \pi_m^{-1}(x).$$

We know that $\mathrm{mld}_{\mathrm{MJ}}(x; X) = \inf_{m\in\mathbb{N}} s_m(x)$ by the formula (13) in Corollary 3.26. The proof of the previous theorem shows that $s_m \geq d-1$ $(m \leq 5)$ yields that $\mathrm{mld}_{\mathrm{MJ}}(x; X) \geq d-1$.

Proposition 4.10. *Let X be a variety over an algebraically closed field k. Assume that X has at worst MJ-canonical singularities. Then, X has at worst cDV singularities in codimension 2.*

Proof. In Proposition 4.3, it is proved that X has normal in codimension 2. Let $\eta \in X$ be the generic point of an irreducible component of the singular locus of codimension 2. As X is MJ-canonical, it follows $\mathrm{mld}_{\mathrm{MJ}}(\eta; X) \geq 1$; therefore, by the Corollary 3.26

$$1 \leq \mathrm{mld}_{\mathrm{MJ}}(\eta; X) = \inf_m \left\{(m+1)d - (\dim \overline{\{\eta\}} + r_m)\right\},$$

where $r_m = \dim \pi_m^{-1}(x)$ $(m \in \mathbb{N})$ for a general closed point $x \in \overline{\{\eta\}}$. Considering the cases for $m = 1, \ldots, 5$, we obtain

$$s_m(x) := (m+1)d - r_m \geq d-1$$

for a general point $x \in \overline{\{\eta\}}$. Therefore, by Remark 4.9, we obtain that $\mathrm{mld}_{\mathrm{MJ}}(x; X) = d-1$ at a general point $x \in \overline{\{\eta\}}$. □

Corollary 4.11. *Let X be an MJ-canonical quasi-projective variety of dimension 3 over an algebraically closed field k. Then, a general hyperplane section H of X has at worst Du Val singularities.*

Proof. Let X be a locally closed subvariety of $\mathbb{P}^N$. As the linear system of hyperplane in $\mathbb{P}^N$ is very ample, we can apply the original Bertini's theorem (see, for example, [9, Theorem 8.18, II]) that works for arbitrary characteristic. Then, it follows that a general hyperplane section H is non-singular away from the singular locus of X. Therefore, if $\dim \mathrm{Sing} X = 0$, then H is non-singular. So we assume that $\dim \mathrm{Sing} X = 1$. By Proposition 4.10, a general hyperplane section intersects the singular locus at finite number of cDV points. We have only to show that the intersection is general at each point $x \in H \cap \mathrm{Sing} X$. Let $|\mathcal{O}_{\mathbb{P}^N}(1)|$ be the complete linear system of hyperplanes on $\mathbb{P}^N$. For a point $x \in \mathrm{Sing} X$, we define a subset $D_x \subset |\mathcal{O}_{\mathbb{P}^N}(1)|$ as

$$D_x = \{\mathcal{H} \in |\mathcal{O}_{\mathbb{P}^N}(1)| \mid X \subset \mathcal{H} \text{ or } (\mathcal{H} \cap X, x) \text{ not rational double}\}.$$

As $\mathcal{O}_{\mathbb{P}^N}(1)$ is very ample, the canonical k-linear map

$$\varphi_x : \Gamma(\mathbb{P}^N, \mathcal{O}_{\mathbb{P}^N}(1)) \to \mathcal{O}_{\mathbb{P}^N}(1) \otimes \mathcal{O}_X/\mathfrak{m}_x^2 \simeq \mathcal{O}_X/\mathfrak{m}_x^2$$

is surjective. Let $\widetilde{D}_x \subset \Gamma(\mathbb{P}^N, \mathcal{O}_{\mathbb{P}^N}(1))$ be the subset corresponding to D_x for a cDV point $x \in \mathrm{Sing}X$, then $\widetilde{D}_x$ is the pull-back by φ_x of the proper closed subset of $k^4 = \mathfrak{m}_x/\mathfrak{m}_x^2 \subset \mathcal{O}_X/\mathfrak{m}_x^2$. Then,

$$\dim \widetilde{D}_x \leq N + 1 - 5 + 3 = N - 1;$$

therefore

$$\dim D_x \leq N - 2.$$

On the other hand, if $x \in \mathrm{Sing}X$ is not a cDV point, then

$$\dim D_x \leq N - 1.$$

Let $D \subset \mathrm{Sing}X \times |\mathcal{O}_{\mathbb{P}^N}(1)|$ be the set $\{\langle x, \mathcal{H}\rangle \mid \mathcal{H} \in D_x\}$. Then, as $\mathrm{Sing}X$ is of 1-dimensional and non cDV singularities are isolated, we have

$$\dim D \leq N - 1.$$

Hence, the image $p(D) \subset |\mathcal{O}_{\mathbb{P}^N}(1)|$ of D by the projection

$$p : \mathrm{Sing}X \times |\mathcal{O}_{\mathbb{P}^N}(1)| \to |\mathcal{O}_{\mathbb{P}^N}(1)|$$

has dimension $N - 1 \leq N = \dim |\mathcal{O}_{\mathbb{P}^N}(1)|$, which yields that general elements of $|\mathcal{O}_{\mathbb{P}^N}(1)|$ is not in $p(D)$. This completes the proof. □

The usual canonical version of this statement is proved in [11] under certain conditions.

Here, we list open problems for the positive characteristic case, which are all proved for characteristic 0 (see [7] for (1) and see [8] for (2), (3), (4)):

Open problems for positive characteristic case:

1. Is an MJ-canonical singularity normal? Cohen–Macaulay?
2. Is the map $X \to \mathbb{Z}$, $x \mapsto \mathrm{mld}_{\mathrm{MJ}}(x, X)$ lower semi-continuous?
3. Is an MJ-canonicity (MJ-log-canonicity) open condition?
4. Is a small deformation of MJ-canonical singularity again MJ-canonical?

Here, we note that if there exist resolutions of singularities, then we would have the affirmative answer to (2), (3), and (4). On the other hand, without resolutions we can prove these if the following natural conjecture holds (this will be discussed in a forthcoming paper by one of the authors):

Conjecture 4.12. *Let $0 \leq \delta \leq d$, there is a number $N_{\delta,d}$ depending only on δ and d such that if*

$$s_m(x) \geq \delta \geq 0, \textit{ for all } m \leq N_{\delta,d}$$

then $\mathrm{mld}_{\mathrm{MJ}}(x, X) \geq \delta$.

We observe that this holds true for $\delta = d - 1$. Indeed, we can take $N_{d-1,d} = 5$ as is seen in the proof of Remark 4.9.

This conjecture is equivalent to the following conjecture (this will be showed also in the forthcoming paper):

Conjecture 4.13. *There exists a number N_d depending only on d such that*

$$\min\{s_m(x) \mid m \leq N_d\} = \mathrm{mld_{MJ}}(x; X), \text{ if } \mathrm{mld_{MJ}}(x; X) \geq 0,$$

and

$$s_m(x) < 0, \text{ for some } m \leq N_d, \text{ if } \mathrm{mld_{MJ}}(x; X) = -\infty.$$

Remark 4.14. We stated the results in this section under the condition that k is algebraically closed. However, we can weaken this condition such that k is perfect in all results except for Theorem 4.8 and Corollary 4.11.

References

[1] M. Artin, *Coverings of the rational double points in characteristic p,* Complex Analytic and Algebraic Geometry, Iwanami and Cambridge UP, (1977) 12–22.

[2] V. Cossart and O. Piltant, *Resolution of singularities of threefolds in positive characteristic. II Artin-Schereier and purely inseparable coverings*, J. Algebra **321** (2009), no.1, 1836–1976.

[3] T. De Fernex and R. Docampo, *Jacobian discrepancies and rational singularities*, J. Eur. Math. Soc. **16** (2014), 165–199.

[4] T. De Fernex, L. Ein, and S. Ishii, *Divisorial valuations via arcs*, Publ. Res. Inst. Math. Sci. **44** (2008), no. 2, 425–448.

[5] J. Denef and F. Loeser, *Germs of arcs on singular algebraic varieties and motivic integration*, Invent. Math. **135** (1999), 201–232.

[6] L. Ein and M. Mustaţă, *Jet schemes and singularities*, Proc. Symp. Pure Math. **80** (2009), no. 2, 505–546.

[7] L. Ein, S. Ishii, and M. Mustaţă, *Multiplier ideals via Mather discrepancy,* to appear in Adv. Studies in Pure Math.

[8] L. Ein and S. Ishii, *Singularities with respect to Mather–Jacobian discrepancies*, MSRI Publications **67** (2015), 125–168.

[9] R. Hartshorne, *Algebraic Geometry*, Graduate Texts in Math. 52, Springer-Verlag, 1977.

[10] H. Hironaka, *Resolution of singularities of an algebraic variety over a field of characteristic zero: I and II*, Annals of Maths. **79** (1964), nos.1 and 2, 109–326.

[11] M. Hirokado, M. Ito, and N. Saito, *Three dimensional canonical singularities in codimension two in positive characteristic*, J. Algebra **373** (2013), 207–222.

[12] S. Ishii, *Jet schemes, arc spaces and the Nash problem,* C.R.Math. Rep. Acad. Canada **29** (2007), 1–21.

[13] S. Ishii, *Maximal divisorial sets in arc spaces,* Adv. St. Pure Math. **50** (2008), 237–249.

[14] S. Ishii, *Mather discrepancy and the arc spaces*, Ann. de l'Institut Fourier **63** (2013), 89–111.

[15] S. Ishii and Ana Reguera, *Singularities with the highest Mather minimal log discrepancy,* Math. Zeitschrift. **275** (2013), nos. 3 and 4, 1255–1274.

[16] E. R. Kolchin, *Differential algebra and algebraic groups*, Pure and Applied Mathematics, **54**, Academic Press, New York-London, 1973.

[17] J. Kollár, *Singularities of the Minimal Model Program,* Cambridge Tracts in Math., **200**, Cambridge UP. 2013.

[18] J. Lipman, *Rational singularities with applications to algebraic surfaces and unique factorization*, Publ. Math. I.H.E.S. **36** (1969), 195–279.

[19] J. Nicaise and J. Sebag, *Le théorème d'irréductibilité de Kolchin*, C. R. Math. Acad. Sci. Paris **341** (2005), 103–106.

[20] A. J. Reguera, *A curve selection lemma in spaces of arcs and the image of the Nash map*, Compositio Math. **142** (2006), 119–130.

[21] A. J. Reguera, *Towards the singular locus of the space of arcs*, Amer. J. Math. **131** (2009), no. 2, 313–350.

[22] Z. Zhu, *Log canonical thresholds in positive characteristic*, preprint, arXiv:1308.5445.

Hodge Theory and L^2-analysis
ALM 39, pp. 451–473

Extended Period Domains, Algebraic Groups, and Higher Albanese Manifolds*

Kazuya Kato[†], Chikara Nakayama[‡], Sampei Usui[§]

Abstract

For a linear algebraic group G over $\mathbf{Q}$, we consider the period domains D classifying G-mixed Hodge structures, and construct the extended period domains D_Σ. We give an interpretation of higher Albanese manifolds by Hain and Zucker by using the above D for some G, and extend them via D_Σ.

2000 Mathematics Subject Classification: Primary 14C30; Secondary 14D07, 32G20.

1 Introduction

For a linear algebraic group G over $\mathbf{Q}$, we consider the period domains D for G-mixed Hodge structures. We construct the extended period domains D_Σ, the space of nilpotent orbits. In this paper, we give an interpretation of higher Albanese manifolds by Hain and Zucker by using the above D for some G, and extend them via D_Σ.

In Section 2, we review a work on higher Albanese manifolds by Hain and Zucker in [11]. In Section 3, we define D by modifying the definition of Shimura variety over $\mathbf{C}$ by Deligne [4]. In Section 4, we introduce the extended period domain D_Σ and state the main results 4.3.1, 4.3.3. In Section 5, we explain the relation of this D_Σ with the theory for the usual period domains ([14]), and as examples, the Mumford–Tate domains, and mixed Shimura varieties. In the case

*Dedicated to Professor Steven Zucker on his 65th birthday.

†Department of Mathematics, University of Chicago, Chicago, Illinois, 60637, USA. E-mail: kkato@math.uchicago.edu

‡Department of Economics, Hitotsubashi University, 2-1 Naka, Kunitachi, Tokyo 186-8601, Japan. E-mail: c.nakayama@r.hit-u.ac.jp

§Graduate School of Science, Osaka University, Toyonaka, Osaka, 560-0043, Japan. E-mail: usui@math.sci.osaka-u.ac.jp

where D is the pure Mumford–Tate domain (cf. Green–Griffiths–Kerr's book [8]), D_Σ essentially coincides with the one by Kerr–Pearlstein ([16]). In Section 6, we study higher Albanese manifolds by Hain and Zucker by using the present D for some G, and extend them via D_Σ.

We omit the details of constructions and proofs of the general theory in Section 4 in this paper, which are to be published elsewhere.

For the p-adic variant of this paper, Koshikawa and the first author are preparing [13].

2 Relation of the work [11] and the present article

In this section, we briefly review the work of Hain and Zucker on unipotent variations of mixed Hodge structure [11] with which we compare our present article.

2.1 The main theorem of [11]

Let X be a connected smooth algebraic variety over $\mathbf{C}$ and $b \in X$, and let $\mathbf{F}$ be a real field, the field of rational numbers, or the ring of integers. Then, [11] (1.6) asserts that there is an equivalence of categories:

$$\begin{pmatrix}\text{good unipotent variations of mixed}\\ \text{Hodge structure on } X \text{ with}\\ \text{unipotency} \le r\text{, defined over } \mathbf{F}\end{pmatrix} \overset{\sim}{\to} \begin{pmatrix}\text{mixed Hodge theoretic}\\ \text{representations of } \mathbf{C}\pi_1(X,b)/J^{r+1},\\ \text{defined over } \mathbf{F}\end{pmatrix}$$

where J is the augmentation ideal, i.e., the kernel of $\varepsilon : \mathbf{C}\pi_1(X,b) \to \mathbf{C}$, $\gamma \mapsto 1$ $(\gamma \in \pi_1(X,b))$.

An outline of the proof is as follows.

2.1.1. The functor from the left-hand-side to the right-hand-side is given by taking the monodromy representation on the fiber over the base point b.

2.1.2. The correspondence from the right-hand-side to the left-hand-side is given by using higher Albanese manifold of X.

2.1.3. The rigidity of variations of mixed Hodge structure is shown under "good" condition ([11], (1.5)) at the boundary of X. This rigidity ensures that 2.1.2 yields the inverse functor of 2.1.1.

2.2 Iterated integration theory of Chen [3]

We review the result of Chen in [3].

2.2.1. Let I be the interval $[0,1]$. A loop on X with base point b is a C^∞ map $\gamma : I \to X$ with $\gamma(0) = \gamma(1) = b$. Let $PX = P_bX$ be the loop space on (X,b) which is the topological space consisting of all loops on X with base point b endowed with compact-open topology.

A local parameter system of PX is a pair (U,ϕ) of an open set U of $\mathbf{R}^n$ and a map $\phi : U \to PX$ such that $\varphi : I \times U \to X$ with $\varphi(t,u) := \phi(u)(t)$ is a C^∞

map. Here n is a non-negative integer. For an open set V of $\mathbf{R}^m$ and a C^∞ map $f : V \to U$, $(V, \phi \circ f)$ is also a local parameter system. For a non-negative integer k, a k-differential form on PX is a collection $\omega = (\omega_\phi)_\phi$ of a k-differential form ω_ϕ on U for a local parameter system (U, ϕ) such that, for $f : V \to U$ as above, $f^*\omega_\phi = \omega_{\phi\circ f}$.

2.2.2. Let r be a positive integer. Let $\Delta_r = \{(t_1, \ldots, t_r) \in \mathbf{R}^r \,|\, 0 \le t_1 \le \cdots \le t_r \le 1\}$ be the r-simplex. Let $\pi_j : X^r \to X$ be the j-th projection, $1 \le j \le r$. Let $\pi : \Delta_r \times PX \to PX$ be the projection and let $\varphi : \Delta_r \times PX \to X^r$ be the map defined by $\varphi(t_1, \ldots, t_r, \gamma) := (\gamma(t_1), \ldots, \gamma(t_r))$. Let $A^k(X)$ and $A^k(PX)$ be k-th forms on X and on PX, respectively. For positive integers $p_1, \ldots, p_r$, put $q = \sum_{j=1}^r (p_j - 1)$. The iterated integral

$$I : A^{p_1}(X) \times \cdots \times A^{p_r}(X) \to A^q(PX), \quad (\omega_1, \ldots, \omega_r) \mapsto \int \omega_1 \cdots \omega_r,$$

along $PX \xleftarrow{\;\pi\;} \Delta_r \times PX \xrightarrow{\;\varphi\;} X^r$, is defined as follows.

It is enough to define it on each local parameter system $\phi : U \to PX$ with $U \subset \mathbf{R}^n$. Let ι_j be the contraction of a differential form with $\frac{\partial}{\partial t_j}$ and set $\alpha_j := \iota_j \varphi^* \pi_j^* \omega_j$, $1 \le j \le r$. Then α_j is a $(p_j - 1)$-th form on $\Delta_r \times U$. Write

$$\alpha_1 \wedge \cdots \wedge \alpha_r =: \sum_{1 \le i_1 \le \cdots \le i_q \le n} \alpha_{i_1 \cdots i_q}(t_1, \ldots, t_r, u_1, \ldots, u_n) du_{i_1} \wedge \cdots \wedge du_{i_q}.$$

Define

$$\int \omega_1 \cdots \omega_r := \sum_{1 \le i_1 \le \cdots \le i_q \le n} \Big(\int_{\Delta_r} \alpha_{i_1 \cdots i_q} dt_1 \cdots dt_r\Big) du_{i_1} \wedge \cdots \wedge du_{i_q}.$$

2.2.3. The bar complex $B^\bullet(X)$ is the subcomplex of the de Rham complex $A^\bullet(PX)$ on PX, which is determined by the de Rham complex $A^\bullet(X)$ of X as follows.

$B^q(X)$ is the subspace of $A^q(PX)$ generated by the images of

$$I : A^{p_1}(X) \times \cdots \times A^{p_r}(X) \to A^q(PX), \quad (\omega_1, \ldots, \omega_r) \mapsto \int \omega_1 \cdots \omega_r,$$

for all positive integers r and $p_1, \ldots, p_r$ such that $q = \sum_{j=1}^r (p_j - 1)$.

For $1 \le j \le r$, let $\nu_j := \sum_{k=1}^j (p_k - 1)$, and let $\nu_0 := 0$. The exterior differential $d : B^q(X) \to B^{q+1}(X)$ is described as

$$\begin{aligned} d \int \omega_1 \cdots \omega_r = &\sum_{j=1}^r (-1)^{\nu_{j-1}+1} \int \omega_1 \cdots \omega_{j-1} d\omega_j \omega_{j+1} \cdots \omega_r \\ &+ \sum_{j=1}^{r-1} (-1)^{\nu_j+1} \int \omega_1 \cdots \omega_{j-1} (\omega_j \wedge \omega_{j+1}) \omega_{j+2} \cdots \omega_r. \end{aligned}$$

Length filtration $(L_r B^\bullet(X))_r$ is an increasing filtration by subcomplexes of $B^\bullet(X)$, where $L_r B^\bullet(X)$ is generated by $\int \omega_1 \cdots \omega_j$ for differential forms $\omega_1, \dots, \omega_j$ on X of positive degree over $j \le r$.

Denote by $B^\bullet(X)_{\mathbf{C}}$ etc. the $\mathbf{C}$-valued iterated integrals etc. Hodge filtration $(F^p B^\bullet(X)_{\mathbf{C}})_p$ is a decreasing filtration by subcomplexes of $B^\bullet(X)_{\mathbf{C}}$, where $F^p B^\bullet(X)_{\mathbf{C}}$ is generated by $\int \omega_1 \cdots \omega_r$ for differential forms $\omega_1 \in F^{p_1} A^\bullet(X)_{\mathbf{C}}, \dots, \omega_r \in F^{p_r} A^\bullet(X)_{\mathbf{C}}$ of positive degree such that $\sum_j p_j \ge p$.

2.2.4. An element of $B^0(X)$ is a function on the loop space PX:

$$B^0(X) \times PX \to \mathbf{R}, \quad (\int \omega_1 \cdots \omega_r, \gamma) \mapsto \int_\gamma \omega_1 \cdots \omega_r.$$

In this case, every ω_j is a 1-form on X and, by writing $\gamma^* \omega_j =: f_j(t)dt$, it is described as

$$\int_\gamma \omega_1 \cdots \omega_r = \int_{0 \le t_1 \le \cdots \le t_r \le 1} f_1(t_1) \cdots f_r(t_r) dt_1 \cdots dt_r.$$

This induces

$$H^0(B^\bullet(X)) \times \pi_1(X, b) \to \mathbf{R}$$

and

$$H^0(B^\bullet(X)) \to \mathrm{Hom}\,(\mathbf{Z}\pi_1(X, b), \mathbf{R})$$

and also

$$L_r H^0(B^\bullet(X)) \to \mathrm{Hom}\,(\mathbf{Z}\pi_1(X, b)/J^{r+1}, \mathbf{R}).$$

The theorem of Chen in [3] asserts that the last homomorphism is an isomorphism.

The filtrations on $\mathbf{C}\pi_1(X, b)/J^{r+1}$, which are induced from the length filtration and the Hodge filtration on the iterated integrals $B^\bullet(X)$, form a mixed Hodge structure called the r-th *canonical variation of mixed Hodge structure* as b varies over X ([11], (4.21)).

2.3 Higher Albanese manifolds in [11]

Put $G := \pi_1(X, b)$. Let R be a commutative ring with unity. Let $\varepsilon : RG \to R$ be the augmentation map. Let $\Delta : RG \to RG \otimes RG$ be the coproduct $\Delta(g) := g \otimes g$. Let $RG^\wedge := \varprojlim_r RG/J^{r+1}$ be the J-adic completion and $\hat{J}$ the closed ideal of $RG^\wedge$ generated by J. Define

$$\hat{G}_R := \{h \in RG^\wedge \mid \varepsilon(h) = 1,\ \Delta(h) = h \hat{\otimes} h\} \subset 1 + \hat{J},$$

$$\mathfrak{g}_R := \{h \in \hat{J} \mid \Delta(h) = 1 \hat{\otimes} h + h \hat{\otimes} 1\}.$$

Let $\hat{G}_{r,R} := \hat{G}_R/(\hat{G}_R \cap (1 + \hat{J}^{r+1}))$ and $\mathfrak{g}_{r,R} := \mathfrak{g}_R/(\mathfrak{g}_R \cap \hat{J}^{r+1})$ ([11], (2.13)). Let F be the Hodge filtration on $\mathfrak{g}_{r,\mathbf{C}}$ induced by the one on $\mathbf{C}G/J^{r+1}$. Let $F^0\hat{G}_{r,\mathbf{C}}$ be the corresponding subgroup of $\hat{G}_{r,\mathbf{C}}$. The higher Albanese manifold in [11] (5.15) is defined by

$$\mathrm{Alb}^r(X) := \hat{G}_{r,\mathbf{Z}} \setminus \hat{G}_{r,\mathbf{C}}/F^0\hat{G}_{r,\mathbf{C}}.$$

2.4 Inverse correspondence

We explain 2.1.2. An inverse correspondence is constructed as follows ([11], (5.21)). Given a mixed Hodge theoretic representation V, i.e., a ring homomorphism $\mathbf{C}\pi_1(X,b)/J^{r+1} \to \mathrm{End}(V)$ of mixed Hodge structures, we have a map $\mathrm{Alb}^r(X) \to D(V)$, where $D(V)$ is the classifying space of Hodge filtrations on V. Composing the higher Albanese map $X \to \mathrm{Alb}^r(X)$ with the above map and pulling back the universal variation of mixed Hodge structure on the classifying space, we get a variation of mixed Hodge structure on X.

2.5 The aim of the present paper

In this paper, we have the following two contributions 2.5.1 and 2.5.2 to the work of Hain and Zucker.

2.5.1. We give a description of the functor represented by the higher Albanese manifold in terms of tensor functors (see Section 6.1). Here we give a rough sketch of it (the precise and more general statement is given in Theorem 6.1.10).

Let Γ_r be the image of $\pi_1(X,b) \to \mathbf{C}\pi_1(X,b)/J^{r+1}$. By [19] p.85, p.474, cf. also [11] (2.17) (iii), if we define the subgroups Γ^r of $\pi_1(X,b)$ by $\Gamma^0 := \pi_1(X,b)$ and $\Gamma^{i+1} := [\pi_1(X,b), \Gamma^i]$ for $i \geq 0$, then Γ_r is the quotient group of $\pi_1(X,b)/\Gamma^r$ such that the kernel of $\pi_1(X,b)/\Gamma^r \to \Gamma_r$ consists of all elements of $\pi_1(X,b)/\Gamma^r$ of finite orders. Let $\mathcal{C}_{X,\Gamma_r}$ be the category of variations of **Q**-MHS $\mathcal{H}$ on X satisfying the following conditions.

(i) For any $w \in \mathbf{Z}$, $\mathrm{gr}^W_w\mathcal{H}$ is a constant polarizable Hodge structure.

(ii) $\mathcal{H}$ is good at infinity in the sense of [11] (1.5).

(iii) The monodromy action of $\pi_1(X,b)$ on $\mathcal{H}_{\mathbf{Q}}(b)$ (which is unipotent under (i)) factors through Γ_r.

Then our result is roughly that for a complex analytic space S, there is a functorial bijection between the set $\mathrm{Mor}(S, \mathrm{Alb}^r(X))$ of morphisms and the set of isomorphism classes of exact tensor functors $\mathcal{C}_{X,\Gamma_r} \to \mathrm{MHS}(S)$, where $\mathrm{MHS}(S)$ denotes the category of analytic families of **Q**-MHS parametrized by S, which sends h_X to h_S for any **Q**-MHS h (more precisely, see 6.1.9 (i)). Here objects of $\mathrm{MHS}(S)$ need not satisfy Griffiths transversality, though objects of $\mathcal{C}_{X,\Gamma_r}$ should satisfy it. h_X (resp. h_S) denotes the constant variation (resp. family) of **Q**-MHS on X (resp. S) associated to h.

2.5.2. We construct toroidal partial compactifications of $\mathrm{Alb}^r(X)$, and describe the functors represented by them generalizing 2.5.1 to its log version. See Section 6.2 for details.

2.5.3. We will deduce these results 2.5.1 and 2.5.2 from the work of Hain and Zucker and from our general theory of (extended) period domains for mixed Hodge structures associated to algebraic groups, which we develop in Section 3 and Section 4.

3 The period domain D

Let G be a linear algebraic group over $\mathbf{Q}$. Let G_u be the unipotent radical of G. Let $\mathrm{Rep}(G)$ be the category of finite-dimensional linear representations of G over $\mathbf{Q}$.

3.1 G-mixed Hodge structures and period domain D

We define and consider the notion of *G-mixed Hodge structure* (*G-MHS* for short).

3.1.1. As in [4], let $S_{\mathbf{C}/\mathbf{R}}$ be the Weil restriction of the multiplicative group $\mathbf{G}_m$ from $\mathbf{C}$ to $\mathbf{R}$. It represents the functor $R \mapsto (\mathbf{C} \otimes_{\mathbf{R}} R)^\times$ for commutative rings R over $\mathbf{R}$. We have $S_{\mathbf{C}/\mathbf{R}}(\mathbf{R}) = \mathbf{C}^\times$, and $S_{\mathbf{C}/\mathbf{R}}$ is understood as $\mathbf{C}^\times$ regarded as an algebraic group over $\mathbf{R}$.

Let $w : \mathbf{G}_{m,\mathbf{R}} \to S_{\mathbf{C}/\mathbf{R}}$ be the homomorphism induced from the natural maps $R^\times \to (\mathbf{C} \otimes_{\mathbf{R}} R)^\times$ for commutative rings R over $\mathbf{R}$.

3.1.2. A linear representation of $S_{\mathbf{C}/\mathbf{R}}$ over $\mathbf{R}$ is equivalent to a finite-dimensional $\mathbf{R}$-vector space V endowed with a decomposition

$$V_{\mathbf{C}} := \mathbf{C} \otimes_{\mathbf{R}} V = \bigoplus_{p,q \in \mathbf{Z}} V_{\mathbf{C}}^{p,q}$$

such that for any p, q, $V_{\mathbf{C}}^{q,p}$ coincides with the complex conjugate of $V_{\mathbf{C}}^{p,q}$ (that is, the image of $V_{\mathbf{C}}^{p,q}$ under $\mathbf{C} \otimes_{\mathbf{R}} V \to \mathbf{C} \otimes_{\mathbf{R}} V$; $z \otimes v \mapsto \bar{z} \otimes v$). For a linear representation V of $S_{\mathbf{C}/\mathbf{R}}$, the corresponding decomposition is defined by

$$V_{\mathbf{C}}^{p,q} = \{v \in V_{\mathbf{C}} \mid [z]v = z^p \bar{z}^q v \text{ for } z \in \mathbf{C}^\times\}.$$

Here $[z]$ denotes z regarded as an element of $S_{\mathbf{C}/\mathbf{R}}(\mathbf{R})$.

3.1.3. Let $h_0 : S_{\mathbf{C}/\mathbf{R}} \to (G/G_u)_{\mathbf{R}}$ be a homomorphism. Assume that the composite $\mathbf{G}_{m,\mathbf{R}} \overset{w}{\to} S_{\mathbf{C}/\mathbf{R}} \to (G/G_u)_{\mathbf{R}}$ is $\mathbf{Q}$-rational and central. Assume also that for one (and hence any) lifting $\mathbf{G}_{m,\mathbf{R}} \to G_{\mathbf{R}}$ of this composite, the adjoint action of $\mathbf{G}_{m,\mathbf{R}}$ on $\mathrm{Lie}(G_u)_{\mathbf{R}} = \mathbf{R} \otimes_{\mathbf{Q}} \mathrm{Lie}(G_u)$ is of weight ≤ -1.

Then, for any $V \in \mathrm{Rep}(G)$, the action of $\mathbf{G}_m$ on V via a lifting $\mathbf{G}_m \to G$ of the above $\mathbf{G}_m \to G/G_u$ defines a rational increasing filtration W on V called the *weight filtration*, which is independent of the lifting.

In the above situation, a *G-mixed Hodge structure* (*G-MHS*, for short) is defined as an exact $\otimes$-functor from $\mathrm{Rep}(G)$ to the category of $\mathbf{Q}$-MHS keeping the underlying vector spaces with the weight filtrations.

3.1.4. Let H be a G-MHS. By 3.1.2 and Tannaka duality ([7]), the Hodge decompositions of gr^W of $H(V)$ for $V \in \mathrm{Rep}(G)$ give a homomorphism $S_{\mathbf{C}/\mathbf{R}} \to (G/G_u)_{\mathbf{R}}$ such that the composite $\mathbf{G}_{m,\mathbf{R}} \overset{w}{\to} S_{\mathbf{C}/\mathbf{R}} \to (G/G_u)_{\mathbf{R}}$ is $\mathbf{Q}$-rational and central. We call this homomorphism the *homomorphism associated with* H.

3.1.5. We define the *period domain* D *associated to* G *and* h_0 as the set of all isomorphism classes of G-MHS whose associated homomorphism $S_{\mathbf{C}/\mathbf{R}} \to (G/G_u)_{\mathbf{R}}$ is $(G/G_u)(\mathbf{R})$-conjugate to h_0. This D is also called the *period domain of type* (G, h_0).

3.2 Complex analytic structure of D

We first give a real analytic understanding of D, and then consider the complex analytic structure of it.

Fix a homomorphism $h_0 : S_{\mathbf{C}/\mathbf{R}} \to (G/G_u)_{\mathbf{R}}$ as in 3.1.3.

3.2.1. Let $h : S_{\mathbf{C}/\mathbf{R}} \to G_{\mathbf{R}}$ be a homomorphism such that the composite $\mathbf{G}_{m,\mathbf{R}} \xrightarrow{w} S_{\mathbf{C}/\mathbf{R}} \xrightarrow{h} G_{\mathbf{R}} \to (G/G_u)_{\mathbf{R}}$ coincides with $h_0 \circ w$. We define an $\mathbf{R}$-subspace $L(h)$ of $\mathrm{Lie}(G)_{\mathbf{R}} = \mathbf{R} \otimes_{\mathbf{Q}} \mathrm{Lie}(G)$ as the set of all $\delta \in \mathrm{Lie}(G)_{\mathbf{R}}$ such that the (p,q)-Hodge component of δ with respect to the adjoint action of $S_{\mathbf{C}/\mathbf{R}}$ via h (3.1.2) is 0 unless $p < 0$ and $q < 0$.

3.2.2. For $\delta \in L(h)$, we obtain a G-MHS $H(h,\delta)$ as follows. For a linear representation V of G over $\mathbf{Q}$, $H(h,\delta)(V)$ is (V,W,F), where W is the weight filtration on V (3.1.3) and F is the Hodge filtration on $V_{\mathbf{C}} = \mathbf{C} \otimes_{\mathbf{Q}} V$ defined in the following way. Let $V_{\mathbf{C}} = \bigoplus_{p,q} V_{\mathbf{C}}^{p,q}$ be the Hodge decomposition defined by the action of $S_{\mathbf{C}/\mathbf{R}}$ via h (3.1.2). Let

$$F^p := \exp(i\delta)\left(\bigoplus_{p' \geq p,\, q \in \mathbf{Z}} V_{\mathbf{C}}^{p',q}\right).$$

Proposition 3.2.3. *The above construction $(h,\delta) \mapsto H(h,\delta)$ gives a bijection from the set of all (h,δ) as above onto the set of all isomorphism classes of G-MHS.*

Proof. By [2] and by Tannaka duality (cf. [7]). □

3.2.4. Consider the action of the subgroup $G(\mathbf{R})G_u(\mathbf{C})$ of $G(\mathbf{C})$ on D defined by changing Hodge filtrations.

Proposition 3.2.5. *The action of $G(\mathbf{R})G_u(\mathbf{C})$ on D is transitive.*

Proof. This follows from the definition of D in 3.1.5 and Proposition 3.2.3. □

3.2.6. Let $\mathcal{C}$ be the category of triples (V,W,F), where V is a finite-dimensional $\mathbf{Q}$-vector space, W is an increasing filtration on V (called the weight filtration), and F is a decreasing filtration on $V_{\mathbf{C}}$ (called Hodge filtration).

Let Y be the set of all isomorphism classes of exact $\otimes$-functors from $\mathrm{Rep}(G)$ to the category $\mathcal{C}$ preserving the underlying vector spaces and the weight filtrations.

Then $G(\mathbf{C})$ acts on Y by changing the Hodge filtration. We have $D \subset Y$ and D is stable in Y under the action of $G(\mathbf{R})G_u(\mathbf{C})$.

Let

$$\check{D} := G(\mathbf{C})D \subset Y.$$

Since the action of $G(\mathbf{C})$ on $\check{D}$ is transitive and the isotropy group of each point of $\check{D}$ is an algebraic subgroup of $G(\mathbf{C})$, $\check{D}$ has a natural structure of a complex analytic manifold as a $G(\mathbf{C})$-homogeneous space.

Proposition 3.2.7. *D is open in $\check{D}$.*

Proof. Let $G_r = G/G_u$. By considering the Hodge decomposition of $\mathrm{Lie}(G_r)_{\mathbf{C}}$, we can see the equality $\mathrm{Lie}(G_r)_{\mathbf{C}} = \mathrm{Lie}(G_r)_{\mathbf{R}} + F^0\mathrm{Lie}(G_r)_{\mathbf{C}}$. Since $\mathrm{Lie}(G_r)_{\mathbf{R}} \cap F^0\mathrm{Lie}(G_r)_{\mathbf{C}} = 0$ in $\mathrm{Lie}(G_r)_{\mathbf{C}}$, we have

$$\dim_{\mathbf{R}} \mathrm{Lie}(G_r)_{\mathbf{C}}/F^0\mathrm{Lie}(G_r)_{\mathbf{C}} = \dim_{\mathbf{R}} \mathrm{Lie}(G_r)_{\mathbf{R}},$$

and hence the proposition follows. □

Corollary 3.2.8. *D is a complex analytic manifold.*

3.3 Polarizability

For a linear algebraic group G, let G' be the commutator algebraic subgroup.

3.3.1. Let $h_0 : S_{\mathbf{C}/\mathbf{R}} \to (G/G_u)_{\mathbf{R}}$ be as in 3.1.3. Let C be the image of $i \in \mathbf{C}^\times = S_{\mathbf{C}/\mathbf{R}}(\mathbf{R})$ by h_0 in $(G/G_u)(\mathbf{R})$. We say that h_0 is **R**-*polarizable* if $\{a \in (G/G_u)'(\mathbf{R}) \mid Ca = aC\}$ is a maximal compact subgroup of $(G/G_u)'(\mathbf{R})$.

3.3.2. A relationship with the usual **R**-polarizability is as follows ([5], 2.11). Let h_0 be as in 3.1.3. Let H be a G-MHS such that the associated $S_{\mathbf{C}/\mathbf{R}} \to (G/G_u)_{\mathbf{R}}$ is **R**-polarizable. Let $V \in \mathrm{Rep}(G)$. Then for each $w \in \mathbf{Z}$, there is an **R**-bilinear form on $\mathrm{gr}^W_w(V)_{\mathbf{R}}$ which is stable under $(G/G_u)'$ and which polarizes $\mathrm{gr}^W_w H(V)$.

3.3.3. We will often consider a subgroup Γ of $G(\mathbf{Q})$ satisfying the following condition.

There is a faithful $V \in \mathrm{Rep}(G)$ and a **Z**-lattice L in V such that L is stable under the action of Γ.

Proposition 3.3.4. *Let $h_0 : S_{\mathbf{C}/\mathbf{R}} \to (G/G_u)_{\mathbf{R}}$ be as in* 3.1.3. *Assume that $h_0 : S_{\mathbf{C}/\mathbf{R}} \to (G/G_u)_{\mathbf{R}}$ is* **R**-*polarizable* (3.3.1). *Let Γ be a subgroup of $G(\mathbf{Q})$ satisfying the condition in* 3.3.3.

Then the following holds.

(1) *The action of Γ on D is proper, and the quotient space $\Gamma \backslash D$ is Hausdorff.*

(2) *If Γ is torsion-free and if $\gamma p = p$ with $\gamma \in \Gamma$ and some $p \in D$, then $\gamma = 1$.*

(3) *If Γ is torsion-free, then the projection $D \to \Gamma \backslash D$ is a local homeomorphism.*

Proof. (1) By the assumption of **R**-polarizability, the action of Γ on D is proper. By [14] Part III 4.2.4.1, the quotient space $\Gamma \backslash D$ is Hausdorff.

(2) By the condition in 3.3.3, Γ is discrete. (2) follows then from **R**-polarizability and torsion-freeness of Γ.

(3) Since Γ is discrete and D is Hausdorff, (2) implies (3) by [14] Part III 4.2.4.2. □

4 Space of nilpotent orbits D_Σ

We define the extended period domain $D_\Sigma \supset D$ as the space of nilpotent orbits, and state the main results. We fix G and h_0 as in 3.1.3. Assume that h_0 is **R**-polarizable.

4.1 Definition of D_Σ

4.1.1. A *nilpotent cone* is a subset σ of $\mathrm{Lie}(G)_{\mathbf{R}}$ satisfying the following (i)–(iii).

(i) $\sigma = \mathbf{R}_{\geq 0}N_1 + \cdots + \mathbf{R}_{\geq 0}N_n$ for some $N_1, \ldots, N_n \in \mathrm{Lie}(G)_{\mathbf{R}}$.

(ii) For any $V \in \mathrm{Rep}(G)$, the image of σ under the induced map $\mathrm{Lie}(G)_{\mathbf{R}} \to \mathrm{End}_{\mathbf{R}}(V)$ consists of nilpotent operators.

(iii) $[N, N'] = 0$ for any $N, N' \in \sigma$.

4.1.2. Let $F \in \check{D}$ and let σ be a nilpotent cone. We say that the pair (σ, F) *generates a nilpotent orbit* if the following (i)–(iii) are satisfied.

(i) There is a faithful $V \in \mathrm{Rep}(G)$ such that the action of σ on $V_{\mathbf{R}}$ is admissible with respect to W, i.e., there exist a family $(M(\tau, W))_\tau$ of finite increasing filtrations $M(\tau, W)$ on V given for each face τ of σ which satisfy the compatibility conditions (1)–(4) in [14] Part III 1.2.2.

(ii) $NF^p \subset F^{p-1}$ for any $N \in \sigma$ and $p \in \mathbf{Z}$.

(iii) Let $N_1, \ldots, N_n$ be as in (i) in 4.1.1. Then $\exp(\sum_{j=1}^n z_jN_j)F \in D$ if $z_j \in \mathbf{C}$ and $\mathrm{Im}(z_j) \gg 0$ $(1 \leq j \leq n)$.

A *nilpotent orbit* is a pair (σ, Z) of a nilpotent cone σ and an $\exp(\sigma_{\mathbf{C}})$-orbit Z in $\check{D}$ satisfying that for any $F \in Z$, (σ, F) generates a nilpotent orbit. Here $\sigma_{\mathbf{C}}$ denotes the $\mathbf{C}$-linear span of σ in $\mathrm{Lie}(G)_{\mathbf{C}}$.

4.1.3. A *weak fan* Σ *in* $\mathrm{Lie}(G)$ is a nonempty set of sharp rational nilpotent cones satisfying the conditions that it is closed under taking faces and that any $\sigma, \sigma' \in \Sigma$ coincide if they have a common interior point and if there is an $F \in \check{D}$ such that both (σ, F) and (σ', F) generate nilpotent orbits.

For a weak fan Σ, let D_Σ be the set of all nilpotent orbits (σ, Z) such that $\sigma \in \Sigma$. Then D is naturally embedded in D_Σ by $F \mapsto (\{0\}, F)$.

Let Γ be a subgroup of $G(\mathbf{Q})$ satisfying the condition in 3.3.3. We say that Σ and Γ are *strongly compatible* if Σ is stable under the adjoint action of Γ and if any $\sigma \in \Sigma$ is generated by elements whose exp in $G(\mathbf{R})$ belong to Γ. If this is the case, Γ naturally acts on D_Σ.

4.2 Log mixed Hodge structures

4.2.1. We work in the category $\mathcal{B}(\log)$ of locally ringed spaces over $\mathbf{C}$ with fs log structures satisfying a certain condition, which contains the category of fs log analytic spaces over $\mathbf{C}$ ([14], Part III, 1.1). For an object $S = (S, \mathcal{O}_S, M)$ of $\mathcal{B}(\log)$, there exists the associated ringed space $S^{\log} = (S^{\log}, \mathcal{O}_S^{\log})$ and a proper surjective morphism $S^{\log} \to S$ of ringed spaces ([14], Part III). We denote by $\mathrm{LMH}(S)$ the category of log $\mathbf{Q}$-mixed Hodge structures over S ([14], Part III, 1.3).

Let Γ be a subgroup of $G(\mathbf{Q})$ satisfying the condition in 3.3.3. A *G-LMH over S with a Γ-level structure* is a pair (H, μ) of an exact $\otimes$-functor $H : \mathrm{Rep}(G) \to \mathrm{LMH}(S)$ and a global section μ of the quotient sheaf $\Gamma \setminus \mathcal{I}$, where $\mathcal{I}$ is the following sheaf on $S^{\log}$. For an open set U of $S^{\log}$, $\mathcal{I}(U)$ is the set of all isomorphisms $H_{\mathbf{Q}}|_U \xrightarrow{\cong} \mathrm{id}$ of $\otimes$-functors from $\mathrm{Rep}(G)$ to the category of local systems of $\mathbf{Q}$-modules over U preserving the weight filtrations.

4.2.2. Let (G, h_0) be as in 3.1.3, let Γ be a subgroup of $G(\mathbf{Q})$ satisfying the condition in 3.3.3 and let Σ be a weak fan in $\mathrm{Lie}(G)$ which is strongly compatible with Γ. A G-LMH over S with a Γ-level structure (H, μ) is said to be *of type* (h_0, Σ) if the following (i) and (ii) are satisfied for any $s \in S$ and any $t \in s^{\log}$. Take a $\otimes$-isomorphism $\tilde{\mu}_t : H_{\mathbf{Q},t} \cong \mathrm{id}$ which belongs to μ_t.

(i) There is a $\sigma \in \Sigma$ such that the logarithm of the action of the local monodromy cone $\mathrm{Hom}((M_S/\mathcal{O}_S^\times)_s, \mathbf{N}) \subset \pi_1(s^{\log})$ on $H_{\mathbf{Q},t}$ is contained, via $\tilde{\mu}_t$, in $\sigma \subset \mathrm{Lie}(G)_{\mathbf{R}}$.

(ii) Let $\sigma \in \Sigma$ be the smallest cone satisfying (i). Let $a : \mathcal{O}_{S,t}^{\log} \to \mathbf{C}$ be a ring homomorphism which induces the evaluation $\mathcal{O}_{S,s} \to \mathbf{C}$ at s and consider the element $F : V \mapsto \tilde{\mu}_t(a(H(V)))$ of Y (3.2.6). Then this element belongs to $\check{D}$ and (σ, F) generates a nilpotent orbit (4.1.2).

If (H, μ) is of type (h_0, Σ), we have a map $S \to \Gamma \backslash D_\Sigma$, called the *period map* associated to (H, μ), which sends $s \in S$ to the class of the nilpotent orbit $(\sigma, Z) \in D_\Sigma$ obtained in the above (ii).

4.2.3. Let (G, h_0, Γ, Σ) be as in 4.2.2. We endow $\Gamma \backslash D_\Sigma$ with a topology, a sheaf of rings $\mathcal{O}$ over $\mathbf{C}$ and a log structure M defined as follows. The topology is the strongest topology for which the period map $S \to \Gamma \backslash D_\Sigma$ is continuous for any (S, H, μ), where S is an object of $\mathcal{B}(\log)$, H is a G-LMH on S, and μ is a Γ-level structure which is of type (h_0, Σ). For an open set U of $\Gamma \backslash D_\Sigma$, $\mathcal{O}(U)$ (resp. $M(U)$) is the set of all $\mathbf{C}$-valued functions f on U such that for any (S, H, μ) as above with the period map $\varphi : S \to \Gamma \backslash D_\Sigma$, the pullback of f on $U' := \varphi^{-1}(U)$ belongs to the image of $\mathcal{O}_{U'}$ (resp. $M_{U'}$) in the sheaf of $\mathbf{C}$-valued functions on U'.

These structures of $\Gamma \backslash D_\Sigma$ are defined also by defining spaces E_σ $(\sigma \in \Sigma)$ in a similar way as [14] Part III. We get the same structures when we use only $S = E_\sigma$ for $\sigma \in \Sigma$ and the universal objects (H, μ) over E_σ in the above definitions of the structures.

4.2.4. Let S be an object of $\mathcal{B}(\log)$. Let S° be the underlying locally ringed space over $\mathbf{C}$ of S with the trivial log structure.

By an MHS over S, we mean an LMH over S°.

Let (G, h_0) be as in 3.1.3 and let Γ be a subgroup of $G(\mathbf{Q})$ satisfying the condition in 3.3.3. By a *G-MHS over S with Γ-level structure*, we mean a G-LMH over S° with Γ-level structure. By a *G-MHS over S with Γ-level structure of type* h_0, we mean a G-LMH over S° with Γ-level structure of type (h_0, Σ) where Σ is the fan consisting of the one cone $\{0\}$.

4.3 Main results

We state main results for moduli of G-log mixed Hodge structures in our general theory.

Theorem 4.3.1. *Let (G, h_0, Γ, Σ) be as in 4.2.2. Assume that h_0 is $\mathbf{R}$-polarizable. Then*

(1) *$\Gamma \backslash D_\Sigma$ is Hausdorff.*

(2) *When Γ is neat, $\Gamma \backslash D_\Sigma$ is a log manifold (*[14], Part III, 1.1.5*). In particular, $\Gamma \backslash D_\Sigma$ belongs to $\mathcal{B}(\log)$.*

Here we say that Γ is *neat* if there is a faithful $V \in \mathrm{Rep}(G)$ such that for any $\gamma \in \Gamma$, the subgroup of $\mathbf{C}^\times$ generated by all eigenvalues of $\gamma : V_{\mathbf{C}} \to V_{\mathbf{C}}$ is torsion-free.

4.3.2. The outline of the proof is as follows. As in [14], we can define various spaces $D_{\mathrm{SL}(2)}$, D_{BS}, E_σ etc., and have the theory of CKS map. Then, as in [14], by using the CKS map, good properties of $\Gamma \setminus D_\Sigma$ are deduced from those of the space of SL(2)-orbits $D_{\mathrm{SL}(2)}$, which reduce to the $\mathbf{R}$-polarizable version of [14]. We remark that what were shown in [14] by using $\mathbf{Q}$-polarizations still hold under $\mathbf{R}$-polarizations (3.3.2) .

Theorem 4.3.3. *Let (G, h_0, Γ, Σ) be as in Theorem 4.3.1. When Γ is neat, $\Gamma \setminus D_\Sigma$ represents the contravariant functor from $\mathcal{B}(\log)$ to (Set):*
$S \mapsto \{$isom. *class of G-LMH over S with a Γ-level structure of type (h_0, Σ)*$\}$.

The proof of 4.3.3 is similar to the proof of [14] Part III 2.6.6.

Concerning extensions of period maps to the boundary, we have:

Theorem 4.3.4. *Let (G, h_0) be as in 3.1.3. Assume that h_0 is $\mathbf{R}$-polarizable. Let S be a connected, log smooth, fs log analytic space, and let U be the open subspace of S consisting of all points of S at which the log structure of S is trivial. Let Γ be a subgroup of $G(\mathbf{Q})$ as in 3.3.3. Assume that Γ is neat.*

Let (H, μ) be a G-MHS over U with a Γ-level structure of type h_0 (4.2.4). Let $\varphi : U \to \Gamma \setminus D$ be the associated period map. Assume that (H, μ) extends to a G-LMH over S with a Γ-level structure (4.2.1). Then:

(1) *For any point $s \in S$, there exist an open neighborhood V of s, a log modification V' of V ([15], 3.6.12), a subgroup Γ' of Γ, and a fan Σ in* Lie(G) *which is strongly compatible with Γ' such that the period map $\varphi|_{U \cap V}$ lifts to a morphism $U \cap V \to \Gamma' \setminus D$ which extends uniquely to a morphism $V' \to \Gamma' \setminus D_\Sigma$ of log manifolds. Furthermore, we can take a commutative group Γ'.*

$$\begin{array}{ccccc} U & \supset & U \cap V & \subset & V' \\ {\scriptstyle\varphi}\downarrow & & \downarrow & & \downarrow \\ \Gamma \setminus D & \longleftarrow & \Gamma' \setminus D & \subset & \Gamma' \setminus D_\Sigma. \end{array}$$

(2) *Assume $S \smallsetminus U$ is a smooth divisor. Then we can take $V = V' = S$ and $\Gamma' = \Gamma$. That is, we have a commutative diagram*

$$\begin{array}{ccc} U & \subset & S \\ {\scriptstyle\varphi}\downarrow & & \downarrow \\ \Gamma \setminus D & \subset & \Gamma \setminus D_\Sigma. \end{array}$$

(3) *Assume that Γ is commutative. Then we can take $\Gamma' = \Gamma$.*

(4) *Assume that Γ is commutative and that the following condition* (i) *is satisfied.*

(i) *There is a finite family* $(S_j)_{1\leq j\leq n}$ *of connected locally closed analytic subspaces of* S *such that* $S = \bigcup_{j=1}^n S_j$ *as a set and such that, for each* j, *the inverse image of the sheaf* $M_S/\mathcal{O}_S^\times$ *on* S_j *is locally constant.*

Then we can take $\Gamma' = \Gamma$ *and* $V = S$.

Note that in (1), we can take a fan Σ (we do not need a weak fan).

This is the G-mixed Hodge version of [14] Part III Theorem 7.5.1 in mixed case and of [15] Theorem 4.3.1 in pure case. The proof goes exactly in the same way as in the pure case treated in [15].

5 Basic examples

We discuss basic examples of D to which our theory can be applied so that we can give D_Σ for these D.

5.1 Usual period domains

We explain that the classical Griffiths domains [9] and their mixed Hodge generalization in [20] are essentially regarded as special cases of the period domains of this paper. In this case, our partial compactifications essentially coincide with those in [14] Part III.

Let $\Lambda = (H_0, W, (\langle\ ,\ \rangle_w)_w, (h^{p,q})_{p,q})$ be as usual as in [14] Part III. Let G be the subgroup of $\mathrm{Aut}(H_{0,\mathbf{Q}}, W)$ consisting of elements which induce *similitudes for* $\langle\ ,\ \rangle_w$ for each w. That is, $G := \{g \in \mathrm{Aut}(H_{0,\mathbf{Q}}, W) \mid$ for any w, there is a $t_w \in \mathbf{G}_m$ such that $\langle gx, gy\rangle_w = t_w\langle x, y\rangle_w$ for any $x, y \in \mathrm{gr}^W_w\}$. Let $G_1 := \mathrm{Aut}(H_{0,\mathbf{Q}}, W, (\langle\ ,\ \rangle_w)_w) \subset G$.

Let $D(\Lambda)$ be the period domain of [20]. Then $D(\Lambda)$ is identified with an open and closed part of D in this paper as follows.

Assume that $D(\Lambda)$ is not empty and fix an $\mathbf{r} \in D(\Lambda)$. Then the Hodge decomposition of $\mathrm{gr}^W\mathbf{r}$ induces $h_0 : S_{\mathbf{C}/\mathbf{R}} \to (G/G_u)_\mathbf{R}$. (We have $\langle [z]x, [z]y\rangle_w = |z|^{2w}\langle x, y\rangle_w$ for $z \in \mathbf{C}^\times$ (see 3.1.2 for $[z]$).) Consider the associated period domain D (3.1.3). Then D is a finite disjoint union of $G_1(\mathbf{R})G_u(\mathbf{C})$-orbits which are open and closed in D. Let $\mathcal{D}$ be the $G_1(\mathbf{R})G_u(\mathbf{C})$-orbit in D consisting of points whose associated homomorphisms $S_{\mathbf{C}/\mathbf{R}} \to (G/G_u)_\mathbf{R}$ are $(G_1/G_u)(\mathbf{R})$-conjugate to h_0. Then the map $H \mapsto H(H_{0,\mathbf{Q}})$ gives a $G_1(\mathbf{R})G_u(\mathbf{C})$-equivariant isomorphism $\mathcal{D} \xrightarrow{\cong} D(\Lambda)$.

5.2 Mixed Mumford–Tate domains

5.2.1. Let H be a $\mathbf{Q}$-MHS whose gr^W are $\mathbf{R}$-polarizable.

The *Mumford–Tate group* G *of* H is the Tannaka group (cf. [18]) of the Tannaka category generated by H (cf. [1]). Explicitly, it is the smallest $\mathbf{Q}$-subgroup G of $\mathrm{Aut}(H_\mathbf{Q})$ such that $G_\mathbf{R}$ contains the image of the homomorphism $h : S_{\mathbf{C}/\mathbf{R}} \to \mathrm{Aut}(H_\mathbf{R})$ and such that $\mathrm{Lie}(G)_\mathbf{R}$ contains δ. Here h and δ are determined by the canonical splitting of H ([2], [14], Part II, 1.2). In the case where H is pure,

G is the smallest $\mathbf{Q}$-subgroup of $\mathrm{Aut}(H_{\mathbf{Q}})$ such that $G_{\mathbf{R}}$ contains the image of $S_{\mathbf{C}/\mathbf{R}} \to \mathrm{Aut}(H_{\mathbf{R}})$.

The following proposition is well-known in the pure case.

Proposition 5.2.2. *The two definitions of G in 5.2.1 coincide.*

Proof. Let G be the group explicitly defined in the latter part of 5.2.1.

Let J be the Tannaka group defined in the beginning of 5.2.1. By Tannaka duality, the theory of (h, δ) of MHS gives a homomorphism $S_{\mathbf{C}/\mathbf{R}} \to J_{\mathbf{R}}$ and $\delta \in \mathrm{Lie}(J)_{\mathbf{R}}$. The homomorphism $J \to \mathrm{Aut}(H_{\mathbf{Q}})$ is injective (otherwise, if K denotes the kernel, representations of J/K would form a smaller Tannaka category). Hence we have $G \subset J$. We will use

Claim. *For linear representations V_1 and V_2 of J over $\mathbf{Q}$, we have* $\mathrm{Hom}_J(V_1, V_2) = \mathrm{Hom}_G(V_1, V_2)$.

This is because

$$\mathrm{Hom}_J(V_1, V_2) \subset \mathrm{Hom}_G(V_1, V_2) \subset \mathrm{Hom}_{\mathrm{MHS}}(V_1, V_2) = \mathrm{Hom}_J(V_1, V_2).$$

By the pure case, we have $G/G_u = J/J_u$. (For this, a point is that G_u coincides with the kernel G_1 of $G \to \mathrm{Aut}(\mathrm{gr}^W)$. We have $G_1 \subset G_u$. It is sufficient to prove that G/G_1 is reductive. This is seen from the polarizability of gr^W.)

Assume $G \neq J$. Then by $G/G_u = J/J_u$, we have $G_u \neq J_u$. Hence the map $G_u \to J_u/[J_u, J_u]$ is not surjective. Since the image of this map is stable under the adjoint action of $G/G_u = J/J_u$, the image is a normal subgroup of $J/[J_u, J_u]$. Let Q be the quotient of $J/[J_u, J_u]$ by this image. Let Q_1 be the quotient of $J_u/[J_u, J_u]$ by the image of G_u. Then Q is a semi-direct product of Q_1 and G/G_u. We consider the following representations V_1 and V_2 of Q over $\mathbf{Q}$. Let $V_1 = \mathbf{Q}$ with the trivial action of Q. Let $V_2 = \mathbf{Q} \oplus Q_1$ on which G/G_u acts by the trivial action on $\mathbf{Q}$ and by the adjoint action on Q_1, and $v \in Q_1$ acts by sending $(1, 0)$ to $(1, v)$ and trivially on Q_1. The $\mathbf{Q}$-linear map $V_1 \to V_2$ which sends 1 to $(1, 0)$ is a G-homomorphism but not a J-homomorphism. This contradicts the Claim. □

5.2.3. The *Mumford–Tate domain associated to H* is defined as the period domain D associated to G and $h_0 : S_{\mathbf{C}/\mathbf{R}} \to (G/G_u)_{\mathbf{R}}$ which is defined by $\mathrm{gr}^W H$.

In the pure case, our $\Gamma \setminus D_\Sigma$ is essentially the same as the one by Kerr–Pearlstein ([16]).

5.3 Mixed Shimura varieties

See [18] for the generality of mixed Shimura varieties. This is the case where the universal object satisfies Griffiths transversality. $\mathrm{gr}^W_w \mathrm{Lie}(G)$ should be 0 unless $w = 0, -1, -2$. The (p, q)-Hodge component of $\mathrm{gr}^W_w \mathrm{Lie}(G)$ for $w = 0$ (resp. $w = -1$, resp. $w = -2$) should be 0 unless (p, q) is $(1, -1)$, $(0, 0)$, and $(-1, 1)$ (resp. $(0, -1)$ and $(-1, 0)$, resp. $(-1, -1)$). (If this condition is satisfied by one point of D, it is satisfied by all points of D.)

For example, the universal abelian variety over a Shimura variety of PEL (polarizations, endomorphisms, and level structures) type is a mixed Shimura variety. Toroidal compactifications of these universal abelian varieties are expressed as $\Gamma \setminus D_\Sigma$.

6 Higher Albanese manifolds and their toroidal partial compactifications

6.1 Understanding higher Albanese manifolds by D

6.1.1. Let X be a connected smooth algebraic variety over $\mathbf{C}$. Fix $b \in X$. Let Γ be a quotient group of $\pi_1(X, b)$ and assume that Γ is a torsion-free nilpotent group.

Let $\mathcal{G} = \mathcal{G}_\Gamma$ be the unipotent algebraic group over $\mathbf{Q}$ whose Lie algebra is defined as follows. Let I be the augmentation ideal $\mathrm{Ker}(\mathbf{Q}[\Gamma] \to \mathbf{Q})$ of $\mathbf{Q}[\Gamma]$. Then $\mathrm{Lie}(\mathcal{G})$ is the $\mathbf{Q}$-subspace of $\mathbf{Q}[\Gamma]^\wedge := \varprojlim_n \mathbf{Q}[\Gamma]/I^n$ generated by all $\log(\gamma)$ $(\gamma \in \Gamma)$. The Lie product of $\mathrm{Lie}(\mathcal{G})$ is defined by $[x, y] = xy - yx$. We have $\Gamma \subset \mathcal{G}(\mathbf{Q})$.

We have

$$\mathrm{Lie}(\mathcal{G}) = \{h \in \mathbf{Q}[\Gamma]^\wedge \mid \Delta(h) = h \otimes 1 + 1 \otimes h\},$$

$$\mathcal{G}(R) = \{g \in (R[\Gamma]^\wedge)^\times \mid \Delta(g) = g \otimes g\}$$

for any commutative ring R over $\mathbf{Q}$, where $\Delta : R[\Gamma]^\wedge \to R[\Gamma \times \Gamma]^\wedge$ is the ring homomorphism induced by the ring homomorphism $R[\Gamma] \to R[\Gamma \times \Gamma]$; $\gamma \mapsto \gamma \otimes \gamma$ $(\gamma \in \Gamma)$.

6.1.2. For $r \geq 0$, let Γ_r be the torsion-free nilpotent quotient group of $\pi_1(X, b)$ defined in 2.5.1. Then for a given Γ as in 6.1.1, there is an $r \geq 1$ such that Γ is a quotient of Γ_r. We define the weight filtration on $\mathrm{Lie}(\mathcal{G}_\Gamma)$ (resp. the Hodge filtration on $\mathrm{Lie}(\mathcal{G}_\Gamma)_\mathbf{C}$) as the image of that of $\mathrm{Lie}(\mathcal{G}_{\Gamma_r})$ (resp. $\mathrm{Lie}(\mathcal{G}_{\Gamma_r})_\mathbf{C}$) (2.2.4, 2.3). This gives a structure of an MHS on $\mathrm{Lie}(\mathcal{G}_\Gamma)$ which is independent of the choice of r.

Note that $\mathcal{G}_{\Gamma_r}$ is written as $\hat{G}_r$ in 2.3.

6.1.3. The higher Albanese manifold $A_{X,\Gamma}$ of X for Γ is as follows. Let $F^0\mathcal{G}(\mathbf{C})$ be the algebraic subgroup of $\mathcal{G}(\mathbf{C})$ over $\mathbf{C}$ corresponding to the Lie subalgebra $F^0\mathrm{Lie}(\mathcal{G})_\mathbf{C}$ of $\mathrm{Lie}(\mathcal{G})_\mathbf{C}$. Define

$$A_{X,\Gamma} := \Gamma \setminus \mathcal{G}(\mathbf{C})/F^0\mathcal{G}(\mathbf{C}).$$

Let Γ_r be as in 2.5.1. For $\Gamma = \Gamma_r$, $A_{X,\Gamma}$ coincides with $\mathrm{Alb}^r(X)$ in 2.3.

In the case where Γ is $H_1(X, \mathbf{Z})/(\text{torsion})$ regarded as a quotient group of $\pi_1(X, b)$, $A_{X,\Gamma}$ coincides with the Albanese variety $\Gamma \setminus H_1(X, \mathbf{C})/F^0H_1(X, \mathbf{C})$ of X.

We will give an understanding of $A_{X,\Gamma}$ by using D of this paper in Theorem 6.1.6.

We will describe the functor represented by $A_{X,\Gamma}$ in Theorem 6.1.10.

6.1.4. Take a $\mathbf{Q}$-MHS V_0 with polarizable gr^W having the $\mathbf{Q}$-MHS $\mathrm{Lie}(\mathcal{G})$ as a direct summand (6.1.2). Let Q be the Mumford–Tate group associated to the V_0 (5.2.1). The action of Q on $\mathrm{Lie}(\mathcal{G})$ induces an action of Q on $\mathcal{G}$. By using this action, define the semidirect product G of Q and $\mathcal{G}$ with an exact sequence $1 \to \mathcal{G} \to G \to Q \to 1$. We have $\mathcal{G} \subset G_u$. We have $h_0 : S_{\mathbf{C}/\mathbf{R}} \to (Q/Q_u)_\mathbf{R} = (G/G_u)_\mathbf{R}$ given by the Hodge decomposition of $\mathrm{gr}^W V_0$.

Then (G,Γ) satisfies the condition in 3.3.3, and Γ is a neat subgroup of $G(\mathbf{Q})$.

Let D_G (resp. D_Q) be the period domain D for G (resp. Q) and h_0 (3.1.5). We have a canonical map $\Gamma\setminus D_G \to D_Q$ induced by the canonical homomorphism $G \to Q$.

6.1.5. We define $b_Q \in D_Q$ and $b_G \in D_G$ as follows. Let $b_Q \in D_Q$ be the isomorphism class of the evident functor $\mathrm{Rep}(Q) \to \mathbf{Q}$-MHS (cf. 5.2). Since $Q \subset G$ is a semidirect summand, we have the restriction functor $\mathrm{Rep}(G) \to \mathrm{Rep}(Q)$. Let $b_G \in D_G$ be the isomorphism class of the composite functor $\mathrm{Rep}(G) \to \mathrm{Rep}(Q)$ and $b_Q : \mathrm{Rep}(Q) \to \mathbf{Q}$-MHS. Then we see that the map $D_G \to D_Q$, induced by the canonical homomorphism $G \to Q$, sends b_G to b_Q. Let $\mathcal{D}$ be the fiber of the map $D_G \to D_Q$ over b_Q.

The following theorem is a generalization of [11] (5.10) into the present context of tensor functors.

Theorem 6.1.6. *The map $\mathcal{G}(\mathbf{C}) \to D_G$; $g \mapsto gb_G$ induces isomorphisms:*

(1) $\mathcal{G}(\mathbf{C})/F^0\mathcal{G}(\mathbf{C}) \cong \mathcal{D}$.

(2) $A_{X,\Gamma} \cong \Gamma\setminus\mathcal{D}$.

Proof. We prove (1), from which (2) follows. Define

$$F^0(G(\mathbf{R})G_u(\mathbf{C})) := \{g \in G(\mathbf{R})G_u(\mathbf{C}) \mid gb_G = b_G\},$$

$$F^0(Q(\mathbf{R})Q_u(\mathbf{C})) := \{g \in Q(\mathbf{R})Q_u(\mathbf{C}) \mid gb_Q = b_Q\}.$$

Then we have a commutative diagram of exact sequences

$$\begin{array}{ccccccccc}
1 & \longrightarrow & \mathcal{G}(\mathbf{C}) & \longrightarrow & G(\mathbf{R})G_u(\mathbf{C}) & \longrightarrow & Q(\mathbf{R})Q_u(\mathbf{C}) & \longrightarrow & 1 \\
 & & \cup & & \cup & & \cup & & \\
1 & \longrightarrow & F^0(\mathcal{G}(\mathbf{C})) & \longrightarrow & F^0(G(\mathbf{R})G_u(\mathbf{C})) & \longrightarrow & F^0(Q(\mathbf{R})Q_u(\mathbf{C})) & \longrightarrow & 1.
\end{array}$$

Here the surjectivity of $F^0(G(\mathbf{R})G_u(\mathbf{C})) \to F^0(Q(\mathbf{R})Q_u(\mathbf{C}))$ follows from $F^0(Q(\mathbf{R})Q_u(\mathbf{C})) \subset F^0(G(\mathbf{R})G_u(\mathbf{C}))$ which is induced from $Q(\mathbf{R})Q_u(\mathbf{C}) \subset G(\mathbf{R})G_u(\mathbf{C})$.

Combining this with $(G(\mathbf{R})G_u(\mathbf{C}))/F^0 \xrightarrow{\sim} D_G$; $g \mapsto gb_G$ and $(Q(\mathbf{R})Q_u(\mathbf{C}))/F^0 \xrightarrow{\sim} D_Q$; $g \mapsto gb_Q$, we get $\mathcal{G}(\mathbf{C})/F^0 \xrightarrow{\sim} \mathcal{D}$; $g \mapsto gb_G$. □

6.1.7. Let $\mathcal{C}_{X,\Gamma}$ be the category of variations of $\mathbf{Q}$-MHS $\mathcal{H}$ on X satisfying the following conditions.

(i) For any $w \in \mathbf{Z}$, $\mathrm{gr}^W_w\mathcal{H}$ is a constant polarizable Hodge structure.

(ii) $\mathcal{H}$ is good at infinity in the sense of [11] (1.5).

(iii) The monodromy action of $\pi_1(X,b)$ on $\mathcal{H}_{\mathbf{Q}}(b)$ (which is unipotent under (i)) factors through Γ.

Let $\mathcal{C}'_{X,\Gamma}$ be the category of $\mathbf{Q}$-MHS H with polarizable gr^W endowed with an action of the Lie algebra $\mathrm{Lie}(\mathcal{G})$ on $H_{\mathbf{Q}}$ such that $\mathrm{Lie}(\mathcal{G}) \otimes H \to H$ is a homomorphism of MHS.

6.1.8. Let $\mathcal{C}_{X,r}$ (resp. $\mathcal{C}'_{X,r}$) be the left (resp. right) category in the equivalence of categories at the beginning of 2.1. Then we have

$$\mathcal{C}_{X,\Gamma_r} \supset \mathcal{C}_{X,r}, \quad \mathcal{C}'_{X,\Gamma_r} \supset \mathcal{C}'_{X,r},$$

$$\bigcup_{\Gamma} \mathcal{C}_{X,\Gamma} = \bigcup_{r} \mathcal{C}_{X,\Gamma_r} = \bigcup_{r} \mathcal{C}_{X,r}, \quad \bigcup_{\Gamma} \mathcal{C}'_{X,\Gamma} = \bigcup_{r} \mathcal{C}'_{X,\Gamma_r} = \bigcup_{r} \mathcal{C}'_{X,r}.$$

The equivalences $\mathcal{C}_{X,r} \simeq \mathcal{C}'_{X,r}$ in 2.1 for all r induce an equivalence $\bigcup_r \mathcal{C}_{X,r} \simeq \bigcup_r \mathcal{C}'_{X,r}$, and it induces an equivalence

$$\mathcal{C}_{X,\Gamma} \simeq \mathcal{C}'_{X,\Gamma}$$

between the full subcategories.

6.1.9. Define a contravariant functor

$$\mathcal{F}_\Gamma : \mathcal{B}(\log) \to (\mathrm{Set})$$

as follows.

$\mathcal{F}_\Gamma(S)$ is the set of isomorphism classes of pairs (H, μ), where H is an exact $\otimes$-functor $\mathcal{C}_{X,\Gamma} \to \mathrm{MHS}(S)$ and μ is a Γ-level structure, satisfying the following condition (i). Here a Γ-*level structure* means a global section of the sheaf $\Gamma \setminus \mathcal{I}$, where $\mathcal{I}$ is the sheaf of functorial $\otimes$-isomorphisms $H(\mathcal{H})_{\mathbf{Q}} \overset{\cong}{\to} \mathcal{H}(b)_{\mathbf{Q}}$ of $\mathbf{Q}$-local systems preserving weight filtrations.

(i) For any $\mathbf{Q}$-MHS h, we have a functorial $\otimes$-isomorphism $H(h_X) \cong h_S$ such that the induced isomorphism of local systems $H(h_X)_{\mathbf{Q}} \cong h_{\mathbf{Q}} = h_X(b)_{\mathbf{Q}}$ belongs to μ. Here h_X (resp. h_S) denotes the constant variation (resp. family) of $\mathbf{Q}$-MHS over X (resp. S) associated to h.

Theorem 6.1.10. *The higher Albanese manifold* $A_{X,\Gamma}$ *represents* $\mathcal{F}_\Gamma$.

Proof. For $S \in \mathcal{B}(\log)$, we show $\mathrm{Mor}(S, A_{X,\Gamma}) \simeq \mathcal{F}_\Gamma(S)$.

The map from the right-hand-side to the left-hand-side is as follows. For an element $\mathcal{C}_{X,\Gamma} \to \mathrm{MHS}(S)$ of $\mathcal{F}_\Gamma(S)$, consider the composition

$$\mathrm{Rep}(G) \subset \mathcal{C}'_{X,\Gamma} \simeq \mathcal{C}_{X,\Gamma} \to \mathrm{MHS}(S).$$

Here $\subset$ is given by the induced action of $\mathrm{Lie}(\mathcal{G})$. By the non-log version of the general theorem 4.3.4, this yields a morphism $S \to \Gamma \setminus D_G$ whose image is sent to b_Q under $\Gamma \setminus D_G \to D_Q$. Thus we get an element $S \to \Gamma \setminus \mathcal{D} = A_{X,\Gamma}$ (6.1.6) of $\mathrm{Mor}(S, A_{X,\Gamma})$.

As for the map from the left-hand-side to the right-hand-side, which is inverse to the above map, we give two constructions.

The first construction is as follows. Assume that we are given a morphism $S \to A_{X,\Gamma}$. Similarly as in 2.4, for an object V of $\mathcal{C}'_{X,\Gamma}$, we have a Lie algebra homomorphism $\mathrm{Lie}(\mathcal{G}) \to \mathrm{End}(V)$ which is a homomorphism of MHS, and it induces a morphism from $A_{X,\Gamma}$ to the classifying space $\Gamma \setminus D(V)$ for V. Pulling back the universal variation of MHS on $\Gamma \setminus D(V)$ by the composition $S \to A_{X,\Gamma} \to \Gamma \setminus D(V)$, we get an object of MHS(S). This gives a desired pair (H, μ) of a functor $H : \mathcal{C}_{X,\Gamma} \simeq \mathcal{C}'_{X,\Gamma} \to \mathrm{MHS}(S)$ and a Γ-level structure μ.

The second construction is as follows. Assume we are given a morphism $S \to A_{X,\Gamma}$ and an object V of $\mathcal{C}'_{X,\Gamma}$. Let Q be the Mumford–Tate group of the $\mathbf{Q}$-MHS $V_0 := \mathrm{Lie}(\mathcal{G}) \oplus V$, and define G as in 6.1.4. Then we have $S \to A_{X,\Gamma} \simeq \Gamma \setminus \mathcal{D} \hookrightarrow \Gamma \setminus D_G$. By the non-log version of Theorem 4.3.3, the object V of $\mathrm{Rep}(G)$ gives an object of MHS(S). □

6.1.11. The *higher Albanese map* $\varphi : X \to A_{X,\Gamma}$ corresponds in 6.1.10 to the evident functor $H : \mathcal{C}_{X,\Gamma} \to \mathrm{MHS}(X)$.

6.2 Toroidal partial compactifications

6.2.1. Let G be as in 6.1.4. Let Σ be a weak fan in $\mathrm{Lie}(G)$ such that $\sigma \subset \mathrm{Lie}(\mathcal{G})_{\mathbf{R}}$ for any $\sigma \in \Sigma$. Assume that Σ and Γ in 6.1.1 are strongly compatible (4.1.3). Then, we have a canonical morphism $\Gamma \setminus D_{G,\Sigma} \to D_Q$, extending the morphism $\Gamma \setminus D_G \to D_Q$, induced by the homomorphism $G \to Q$.

6.2.2. Define the toroidal partial compactification $A_{X,\Gamma,\Sigma}$ of $A_{X,\Gamma}$ as the subspace of $\Gamma \setminus D_{G,\Sigma}$ which is defined to be the inverse image of b_Q. We can endow $A_{X,\Gamma,\Sigma}$ with a structure of a log manifold such that for any object S of $\mathcal{B}(\log)$, $\mathrm{Mor}(S, A_{X,\Gamma,\Sigma})$ coincides with the set of all morphisms $S \to \Gamma \setminus D_{G,\Sigma}$ whose images in D_Q are b_Q (6.1.5).

This $A_{X,\Gamma,\Sigma}$ is independent of the choice of V_0 in 6.1.4 which is used in the definitions of Q and G.

6.2.3. Define a contravariant functor

$$\mathcal{F}_{\Gamma,\Sigma} : \mathcal{B}(\log) \to (\mathrm{Set})$$

as follows.

$\mathcal{F}_{\Gamma,\Sigma}(S)$ is the set of isomorphism classes of pairs (H, μ) where H is an exact $\otimes$-functor $\mathcal{C}_{X,\Gamma} \to \mathrm{LMH}(S)$ and μ is a Γ-level structure satisfying the condition (i) in 6.1.9 and also the following condition (ii).

(ii) The following (ii-1) and (ii-2) are satisfied for any $s \in S$ and any $t \in s^{\log}$. Let $\tilde{\mu}_t : H(\mathcal{H})_{\mathbf{Q},t} \cong \mathcal{H}(b)_{\mathbf{Q}}$ be a functorial $\otimes$-isomorphism which belongs to μ_t.

(ii-1) There is a $\sigma \in \Sigma$ such that the logarithm of the action of the local monodromy cone $\mathrm{Hom}\,((M_S/\mathcal{O}_S^\times)_s, \mathbf{N}) \subset \pi_1(s^{\log})$ on $H_{\mathbf{Q},t}$ is contained, via $\tilde{\mu}_t$, in $\sigma \subset \mathrm{Lie}(\mathcal{G})_{\mathbf{R}}$.

(ii-2) Let $\sigma \in \Sigma$ be the smallest cone which satisfies (ii-1) and let $a : \mathcal{O}_{S,t}^{\log} \to \mathbf{C}$ be a ring homomorphism which induces the evaluation $\mathcal{O}_{S,s} \to \mathbf{C}$ at s. Then, for each $\mathcal{H} \in \mathcal{C}_{X,\Gamma}$, $(\sigma, \tilde{\mu}_t(a(H(\mathcal{H}))))$ generates a nilpotent orbit in the sense of [14] Part III 2.2.2.

Theorem 6.2.4. *The functor $\mathcal{F}_{\Gamma,\Sigma}$ is represented by $A_{X,\Gamma,\Sigma}$.*

Proof. This follows from Theorem 4.3.3. The proof is similar to the one of Theorem 6.1.10. We replace MHS(S) there by LMH(S) and, in the latter half of the proof, we use the second construction of the inverse map. □

6.2.5. Let Ξ be the set of all rational nilpotent cones in $\mathrm{Lie}(\mathcal{G})_{\mathbf{R}}$ of rank ≤ 1. Then Ξ is a fan and is strongly compatible with Γ.

Theorem 6.2.6. *Let $\overline{X}$ be a smooth algebraic variety over $\mathbf{C}$ which contains X as a dense open subset such that the complement $\overline{X} \smallsetminus X$ is a smooth divisor. Endow $\overline{X}$ with the log structure associated to this divisor.*

Then the higher Albanese map $\varphi : X \to A_{X,\Gamma}$ extends uniquely to a morphism $\overline{\varphi} : \overline{X} \to A_{X,\Gamma,\Xi}$ of log manifolds giving a commutative diagram

$$\begin{array}{ccc} X & \subset & \overline{X} \\ {\scriptstyle\varphi}\downarrow & & \downarrow{\scriptstyle\overline{\varphi}} \\ A_{X,\Gamma} & \subset & A_{X,\Gamma,\Xi}. \end{array}$$

Proof. Since an object of $\mathcal{C}_{X,\Gamma}$ is good at infinity, it extends to an LMH over $\overline{X}$. Hence this theorem follows from (2) of the general theorem 4.3.4 by using Theorem 6.2.4. □

6.3 Example

6.3.1. We consider

$$X := \mathbf{P}^1(\mathbf{C}) \smallsetminus \{0, 1, \infty\} \subset \overline{X} := \mathbf{P}^1(\mathbf{C}).$$

We will consider the toroidal partial compactification of the second higher Albanese manifold $\mathrm{Alb}^2(X)$ (2.3) and the extended higher Albanese map from $\overline{X}$ to it (6.2.6). The description of the degeneration at the boundary of $\overline{X}$ becomes simpler if we take the base point b of the theory of Hain–Zucker in the boundary outside X. For this, we can use the idea of tangential base point of Deligne ([6], Section 15 "Points base à l'infini") and its variant described in 6.3.6.

As in [12],

$$\mathrm{Alb}^2(X) \cong \begin{pmatrix} 1 & \mathbf{Z} & \mathbf{Z} \\ 0 & 1 & \mathbf{Z} \\ 0 & 0 & 1 \end{pmatrix} \Big\backslash \begin{pmatrix} 1 & \mathbf{C} & \mathbf{C} \\ 0 & 1 & \mathbf{C} \\ 0 & 0 & 1 \end{pmatrix}.$$

The right-hand-side is actually a period domain $G_{u,\mathbf{Z}} \backslash D(\Lambda)$ of classical type (see 6.3.2 below) and the toroidal partial compactification of $\mathrm{Alb}^2(X)$ with respect to the fan Ξ (6.2.5) is isomorphic to the toroidal partial compactification $G_{u,\mathbf{Z}} \backslash D(\Lambda)_{\Xi}$ of this classical period domain considered in [14] Part III. We first consider this period domain of the classical type in 6.3.2–6.3.5.

6.3.2. Let $\Lambda = (H_0, W, (\langle\ ,\ \rangle_w)_{w \in \mathbf{Z}}, (h^{p,q})_{p,q \in \mathbf{Z}})$ be as follows. H_0 is a free $\mathbf{Z}$-module of rank 3 with basis $(e_j)_{1 \leq j \leq 3}$, W is the increasing filtration on $H_{0,\mathbf{Q}}$ defined as

$$W_{-5} = 0 \subset W_{-4} = W_{-3} = \mathbf{Q}e_1 \subset W_{-2} = W_{-1} = \mathbf{Q}e_1 + \mathbf{Q}e_2 \subset W_0 = H_{0,\mathbf{Q}},$$

$\langle\ ,\ \rangle_w : \mathrm{gr}^W_w(H_{0,\mathbf{Q}}) \times \mathrm{gr}^W_w(H_{0,\mathbf{Q}}) \to \mathbf{Q}$ are the $\mathbf{Q}$-bilinear forms characterized by $\langle e_3, e_3 \rangle_0 = \langle e_2, e_2 \rangle_{-2} = \langle e_1, e_1 \rangle_{-4} = 1$, and $h^{0,0} = h^{-1,-1} = h^{-2,-2} = 1$, $h^{p,q} = 0$

for the other (p, q). For $R = \mathbf{Z}, \mathbf{C}$, let $G_{u,R}$ be the group of automorphisms of the R-module $H_{0,R}$ which preserve W and induce the identity map on gr^W. Then $G_{u,R}$ is identified with the group of unipotent upper triangular $(3,3)$-matrices with entries in R.

The period domain $D(\Lambda)$ is isomorphic to $G_{u,\mathbf{C}}$ where the matrix

$$\begin{pmatrix} 1 & \beta & \lambda \\ 0 & 1 & \alpha \\ 0 & 0 & 1 \end{pmatrix}$$

corresponds to the following decreasing filtration $F = F(\alpha, \beta, \lambda)$ on $H_{0,\mathbf{C}}$: $F^1 = 0$, F^0 is generated by $e_3 + \alpha e_2 + \lambda e_1$, F^{-1} is generated by F^0 and $e_2 + \beta e_1$, and $F^{-2} = H_{0,\mathbf{C}}$.

The natural action of $G_{u,\mathbf{C}}$ on $D(\Lambda)$ is identified with the natural action of $G_{u,\mathbf{C}}$ on itself from the left.

6.3.3. Let Ξ be the set of all cones of the form $\mathbf{R}_{\geq 0}N$ where N is a $\mathbf{Q}$-linear map $H_{0,\mathbf{Q}} \to H_{0,\mathbf{Q}}$ such that $NW_w \subset W_{w-1}$ for all $w \in \mathbf{Z}$. We consider the extended period domain $D(\Lambda)_\Xi$ ([14], Part III).

For $N : H_{0,\mathbf{Q}} \to H_{0,\mathbf{Q}}$ defined by $Ne_3 = ae_2 + ce_1$, $Ne_2 = be_1$, $Ne_1 = 0$ $(a, b, c \in \mathbf{Q})$, $(N, F(\alpha, \beta, \lambda))$ generates a nilpotent orbit if and only if it satisfies the Griffiths transversality $NF^p \subset F^{p-1}$ $(p \in \mathbf{Z})$, and hence if and only if $c = a\beta - b\alpha$. Hence for $\sigma = \mathbf{R}_{\geq 0}N$, $D_\sigma \neq D$ if and only if either $a \neq 0$ or $b \neq 0$.

6.3.4. By [14] Part III, the quotient $G_{u,\mathbf{Z}} \setminus D(\Lambda)_\Xi$ has a structure of a log manifold.

Let $N : H_{0,\mathbf{Q}} \to H_{0,\mathbf{Q}}$ be the $\mathbf{Q}$-linear map defined by $Ne_3 = e_2$, $Ne_2 = Ne_1 = 0$ and let $\sigma = \mathbf{R}_{\geq 0}N$.

We describe the local structure of $G_{u,\mathbf{Z}} \setminus D(\Lambda)_\Xi$ at the point corresponding to a σ-nilpotent orbit. Let p be the image of an element of $D(\Lambda)_\sigma \smallsetminus D(\Lambda)$ in $G_{u,\mathbf{Z}} \setminus D(\Lambda)_\Xi$. Then for some $\lambda_0 \in \mathbf{C}$, p is the class of the σ-nilpotent orbit generated by $(N, F(0, 0, \lambda_0))$. Let Y be the log manifold $\{(q, \beta, \lambda) \in \mathbf{C}^3 \mid \beta = 0 \text{ if } q = 0\}$ with the strong topology ([15], Section 3.1), with the structure sheaf of rings which is the inverse image of the sheaf of holomorphic functions on $\mathbf{C}^3$, and with the log structure generated by q. Then there is an open neighborhood U of $(0, 0, \lambda_0)$ in $\mathbf{C}^3$ and an open immersion

$$Y \cap U \overset{\subset}{\to} G_{u,\mathbf{Z}} \setminus D(\Lambda)_\Xi$$

of log manifolds which sends $(q, \beta, \lambda) \in Y \cap U$ with $q \neq 0$ to the class of $F(\alpha, \beta, \lambda) = \exp(\alpha N)F(0, \beta, \lambda)$, where $\alpha \in \mathbf{C}$ is such that $q = e^{2\pi i \alpha}$, and which sends $(0, 0, \lambda_0)$ to p.

6.3.5. We can show that for any $p \in G_{u,\mathbf{Z}} \setminus D(\Lambda)_\Xi$ which does not belong to $G_{u,\mathbf{Z}} \setminus D(\Lambda)$, there are an open neighborhood U of $(0, 0, 0)$ in $\mathbf{C}^3$ and an open immersion $Y \cap U \to G_{u,\mathbf{Z}} \setminus D(\Lambda)_\Xi$ of log manifolds which sends $(0, 0, 0)$ to p.

6.3.6. We describe how to formulate a base point in the boundary in the theory of Hain–Zucker. The following is a variant of tangential base point of Deligne and matches log Hodge theory well.

Let $\overline{X}$ be a connected smooth algebraic variety over $\mathbf{C}$, let D be a divisor on $\overline{X}$ with normal crossings, and let $X := \overline{X} \smallsetminus D$. Endow $\overline{X}$ with the log structure associated to D. We formulate a base point in $\overline{X}$ outside X as follows.

In our definition, a base point in the boundary of $\overline{X}$ is a pair $b = (y, a)$ where y is a point of $\overline{X}^{\log}$ which does not belong to X, and a is a specialization $\mathcal{O}^{\log}_{\overline{X},y} \to \mathbf{C}$. That is, y is a pair (x, h) where x is a point of $\overline{X}$ which does not belong to X, h is a homomorphism $M^{\mathrm{gp}}_{\overline{X},x} \to \mathbf{S}^1 := \{z \in \mathbf{C}^\times \mid |z| = 1\}$ whose restriction to the subgroup $\mathcal{O}^\times_{\overline{X},x}$ of $M^{\mathrm{gp}}_{\overline{X},x}$ coincides with $f \mapsto f(x)/|f(x)|$, and a is a ring homomorphism $\mathcal{O}^{\log}_{\overline{X},y} \to \mathbf{C}$ whose restriction to the subring $\mathcal{O}_{\overline{X},x}$ of $\mathcal{O}^{\log}_{\overline{X},y}$ coincides with $f \mapsto f(x)$.

A path between y and a point b' of X induces an isomorphism

$$\pi_1(\overline{X}^{\log}, y) \cong \pi_1(X, b').$$

We can use $\pi_1(\overline{X}^{\log}, y)$ in place of $\pi_1(X, b')$ $(b' \in X)$ in the theory of Hain–Zucker.

Let Γ be a nilpotent torsion-free quotient group of $\pi_1(\overline{X}^{\log}, y)$. Then we have the unipotent algebraic group $\mathcal{G}$ over $\mathbf{Q}$ by using $\pi_1(\overline{X}^{\log}, y)$ by the method of 6.1.1. We obtain a structure of a $\mathbf{Q}$-MHS on Lie($\mathcal{G}$) as follows. For $b' \in X$, let $\mathcal{G}(b')$ be the unipotent group $\mathcal{G}$ in 6.1.1 obtained by using the base point b'. Then when $b' \in X$ moves, the MHS Lie($\mathcal{G}(b')$) forms an object $\mathcal{H}$ of $\mathcal{C}_{X,\Gamma}$. Since $\mathcal{H}$ is good at infinity, it extends uniquely to a $\mathbf{Q}$-LMH $\overline{\mathcal{H}}$ on $\overline{X}$. By the specialization of $\overline{\mathcal{H}}$ by a at y, we obtain a structure of $\mathbf{Q}$-MHS on Lie($\mathcal{G}$).

The main theorem of Hain–Zucker [11] introduced in Section 2.1 and results in Sections 6.1 and 6.2 remain true when we use the base point in the boundary, and are deduced from the work [11] and by the arguments in Sections 6.1 and 6.2.

6.3.7. A tangential base point of Deligne in the boundary of $\overline{X}$ ([6], Section 15) gives a base point b in the boundary in our sense (6.3.6). We explain this in the case where X is a curve. In this case, a tangential base point in the boundary is a non-zero element v of the tangent space $T_x(\overline{X}) = \mathrm{Hom}_{\mathbf{C}}(m_x/m_x^2, \mathbf{C})$ with $x \in \overline{X} \smallsetminus X$ and m_x being the maxmal ideal of $\mathcal{O}_{\overline{X},x}$. We have the corresponding base point $b = (y, a)$, $y = (x, h)$ in the boundary in our sense as follows.

$h : M^{\mathrm{gp}}_{\overline{X},x} \to \mathbf{S}^1$ is the unique group homomorphism which sends any element f of $\mathcal{O}^\times_{\overline{X},x}$ to $f(x)/|f(x)|$ and any prime element t of $\mathcal{O}_{\overline{X},x}$ to $v(t)/|v(t)|$. a is the unique ring homomorphism $\mathcal{O}^{\log}_{\overline{X},y} \to \mathbf{C}$ satisfying the following (i) and (ii).

(i) $a(f) = f(x)$ for any $f \in \mathcal{O}_{\overline{X},x}$.

(ii) Let t be a prime element of $\mathcal{O}_{\overline{X},x}$ such that $h(t) = 1$ (that is, $v(t) \in \mathbf{R}_{>0}$). Let $f \in \mathcal{O}^{\log}_{\overline{X},y}$ be the branch of $\log(t)$ such that the imaginary part of $f(x')$ converges to 0 if $x' \in X$ converges to y in $\overline{X}^{\log}$. Then $a(f) = \log(v(t)) \in \mathbf{R}$.

6.3.8. Now let $X = \mathbf{P}^1(\mathbf{C}) \smallsetminus \{0, 1, \infty\}$, $\overline{X} = \mathbf{P}^1(\mathbf{C})$. We take the base point b in the boundary of $\overline{X}$ corresponding to the tangent vector v at $0 \in \overline{X}$ which sends the class of the coordinate function x of $\mathbf{C} \subset \mathbf{P}^1(\mathbf{C})$ in m_0/m_0^2 to 1. That is,

$b = (y, a)$, $y = (0, h) \in \overline{X}^{\log}$ where h sends the coordinate function x to 1 and a sends the branch of $\log(x)$ which has real value on $\mathbf{R}_{>0}$ to 0.

The group $\pi_1(\overline{X}^{\log}, y)$ is a free group of rank 2 generated by elements γ_0 and γ_1. Here for $\alpha = 0, 1$, γ_α is the class of the following loop $[0, 1] \to \overline{X}^{\log}$ which we denote also by γ_α. Let x be the coordinate function of $\mathbf{C} \subset \mathbf{P}^1(\mathbf{C})$ as above. Then $\gamma_0(t) = (0, h)$ where h sends x to $e^{2\pi i t}$. $\gamma_1(0) = \gamma_1(1) = (0, h)$ where h sends x to 1. For $0 < t < 1/3$, $\gamma_1(t) = 3t \in X$. For $1/3 \leq t \leq 2/3$, $\gamma_1(t) = (1, h)$ where h sends $1 - x$ to $e^{2\pi i(3t-1)}$. For $2/3 < t < 1$, $\gamma_1(t) = 3(1-t) \in X$.

6.3.9. Let Γ be the quotient group $\pi_1(\overline{X}^{\log}, y)/[\pi_1, [\pi_1, \pi_1]]$ of $\pi_1 = \pi_1(\overline{X}^{\log}, y)$. We consider $A_{X,\Gamma}$ (6.1.3) which is the second higher Albanese manifold of X by using the above base point b in the boundary (6.3.8).

For $\alpha = 0, 1$, let $N_\alpha = \log(\gamma_\alpha) \in \mathrm{Lie}(\mathcal{G})$. Then $\mathrm{Lie}(\mathcal{G})$ is three-dimensional over $\mathbf{Q}$ with basis N_0, N_1, $[N_1, N_0]$ (cf. 6.1.1).

The mixed Hodge structure on $\mathrm{Lie}(\mathcal{G})$ is as follows. The weight filtration is given by

$$W_{-5} = 0 \subset W_{-4} = W_{-3} = \mathbf{Q} \cdot [N_1, N_0] \subset W_{-2} = \mathrm{Lie}(\mathcal{G}).$$

N_0 and N_1 are of Hodge type $(-1, -1)$, and $[N_1, N_0]$ is of Hodge type $(-2, -2)$.

We have $F^0\mathcal{G}(\mathbf{C}) = \{1\}$ and hence

$$A_{X,\Gamma} = \Gamma \setminus \mathcal{G}(\mathbf{C}).$$

The following 6.3.10 and 6.3.11 are seen from [6] (cf. also [12]).

6.3.10. Consider the following $\mathbf{Q}$-MHS V and the Lie action of $\mathrm{Lie}(\mathcal{G})$ on V. $V = H_{0,\mathbf{Q}}$ with the Hodge filtration $F(0, 0, 0)$ on $V_{\mathbf{C}}$ (6.3.2). The action of $\mathrm{Lie}(\mathcal{G})$ is as follows. $N_0 e_3 = e_2$, $N_0 e_j = 0$ for $j = 1, 2$; $N_1 e_2 = e_1$, $N_1 e_j = 0$ for $j = 1, 3$. Then the action $\mathrm{Lie}(\mathcal{G}) \otimes V \to V$ is a homomorphism of MHS.

This induces an isomorphism $A_{X,\Gamma} \xrightarrow{\cong} G_{u,\mathbf{Z}} \setminus D(\Lambda)$ of complex analytic manifolds. It extends to an isomorphism $A_{X,\Gamma,\Xi} \cong G_{u,\mathbf{Z}} \setminus D(\Lambda)_\Xi$ of log manifolds.

The composition $X \to G_{u,\mathbf{Z}} \setminus D(\Lambda)$ of the higher Albanese map $X \to A_{X,\Gamma}$ and the above isomorphism sends $x \in X$ to the class of

$$F((2\pi i)^{-1} \log(x), (2\pi i)^{-1} l_1(x), (2\pi i)^{-2} l_2(x)),$$

where $l_1(x) = -\log(1 - x)$ and $l_2(x)$ is the dilog function.

6.3.11. There is another isomorphism $A_{X,\Gamma} \cong G_{u,\mathbf{Z}} \setminus D(\Lambda)$ which may be more popular. Consider the $\mathbf{Q}$-MHS V as in 6.3.10 and consider the Lie action of $\mathrm{Lie}(\mathcal{G})$ on V such that $N_0 e_2 = e_1$, $N_0 e_j = 0$ for $j = 1, 3$; $N_1 e_3 = e_2$, $N_1 e_j = 0$ for $j = 1, 2$. Then the action $\mathrm{Lie}(\mathcal{G}) \otimes V \to V$ is a homomorphism of MHS. This induces an isomorphism $A_{X,\Gamma} \xrightarrow{\cong} G_{u,\mathbf{Z}} \setminus D(\Lambda)$ of complex analytic manifolds and an isomorphism $A_{X,\Gamma,\Xi} \cong G_{u,\mathbf{Z}} \setminus D(\Lambda)_\Xi$ of log manifolds.

In this case, the composition $X \to G_{u,\mathbf{Z}} \setminus D(\Lambda)$ of the higher Albanese map $X \to A_{X,\Gamma}$ and the above isomorphism sends $x \in X$ to the class of

$$\begin{pmatrix} 1 & -(2\pi i)^{-1}\log(x) & (2\pi i)^{-2}l_2(x) \\ 0 & 1 & -(2\pi i)^{-1}l_1(x) \\ 0 & 0 & 1 \end{pmatrix}^{-1} \in G_{u,\mathbf{C}} \cong D(\Lambda).$$

The pullback on X of the universal object on $G_{u,\mathbf{Z}} \setminus D(\Lambda)$ under this composite map is the so-called *dilog sheaf* on X.

6.3.12. Consider the extended higher Albanese map $\overline{X} \to A_{X,\Gamma,\Xi}$ (6.2.6). Let $\overline{X} \to G_{u,\mathbf{Z}} \setminus D(\Lambda)_{\Xi}$ be the composite of this extended map and the isomorphism $A_{X,\Gamma,\Xi} \cong G_{u,\mathbf{Z}} \setminus D(\Lambda)_{\Xi}$ in 6.3.10. Then the image of $0 \in \overline{X}$ under this composite map is the class of the nilpotent orbit generated by $(N, F(0,0,0))$ with N as in 6.3.4 (i.e., $N = N_0$ in 6.3.10). Let $Y \cap U \to G_{u,\mathbf{Z}} \setminus D(\Lambda)_{\Xi}$ be the open immersion given in 6.3.4 with $\lambda_0 = 0$. Then if $x \in X$ is near to $0 \in \overline{X}$, the image of x in $G_{u,\mathbf{Z}} \setminus D(\Lambda)_{\Xi}$ is the image of

$$(x, (2\pi i)^{-1}l_1(x), (2\pi i)^{-2}l_2(x)) \in Y \cap U.$$

The last element converges to $(0,0,0)$ when x converges to $0 \in \overline{X}$.

Acknowledgments. The authors thank Teruhisa Koshikawa for helpful discussions. The book [17] of Toshitake Kohno was helpful for the authors. The authors thank to the referee for valuable comments. K. Kato was partially supported by NFS grants DMS 1001729 and DMS 1303421. C. Nakayama was partially supported by JSPS Grants-in-Aid for Scientific Research (C) 22540011, (B) 23340008, and (C) 16K05093. S. Usui was partially supported by JSPS Grants-in-Aid for Scientific Research (B) 23340008 and (C) 17K05200.

References

[1] Y. André, Mumford-Tate groups of mixed Hodge structures and the theorem of the fixed part, Comp. Math. **82-1** (1992), 1–24.

[2] E. Cattani, A. Kaplan and W. Schmid, Degeneration of Hodge structures, Ann. of Math. **123** (1986), 457–535.

[3] K.-T. Chen, Extension of C^∞ function algebra by integrals and Malcev compeletion of π_1, Advances in Math. **23** (1977), 181–210.

[4] P. Deligne, Travaux de Shimura, Séminaire Bourbaki (1970/71), Exp. 389, Lecture Notes in Math. **244**, Springer, (1971), 123–165.

[5] P. Deligne, La conjecture de Weil pour les surfaces K3, Invent. Math. **15** (1972), 206–226.

[6] P. Deligne, Groupe fondamental de la droite projective moins trois points, in Galois group over **Q**, MSRI Publications **16** (1989), 79–297.

[7] P. Deligne and J. S. Milne, Tannakian categories, in Hodge cycles, Motives, and Shimura varieties, Lecture Notes in Math. **900**, Springer, (1982), 101–228.

[8] M. Green, P. Griffiths and M. Kerr, Mumford-Tate Groups and Domains: Their Geometry and Arithmetic, Ann. Math. Studies **183**, Princeton University Press (2012).

[9] P. Griffiths, Periods of integrals on algebraic manifolds. I. Construction and properties of modular varieties, Amer. J. Math. **90** (1968), 568–626.

[10] R. M. Hain, The geometry of mixed Hodge structures on the fundamental group, Proc. Symp. Pure Math. **46** (1987), 247–282.

[11] R. M. Hain and S. Zucker, Unipotent variation of mixed Hodge structure, Invent. Math. **88** (1987), 83–124.

[12] R. M. Hain and S. Zucker, A guide to unipotent variations of mixed Hodge structures, Lecture Notes in Math. **1246**, Springer, (2006), 92–106.

[13] K. Kato and T. Koshikawa, Arithmetic models of Griffiths period domains, in preparation.

[14] K. Kato, C. Nakayama and S. Usui, Classifying spaces of degenerating mixed Hodge structures, I., Adv. Stud. Pure Math. **54** (2009), 187–222, II., Kyoto J. Math. **51** (2011), 149–261, III., J. Algebraic Geometry **22** (2013), 671–772, IV., Kyoto J. Math., to appear.

[15] K. Kato and S. Usui, Classifying spaces of degenerating polarized Hodge structures, Ann. Math. Studies **169**, Princeton University Press (2009).

[16] M. Kerr and G. Pearlstein, Boundary components of Mumford-Tate domains, Duke Math. J. **165-4** (2016), 661–721.

[17] T. Kohno, Geometry of iterated integrals, Springer contemporary mathematics series **14** (2012), 295 pages, Maruzen Publ. Co., Ltd., (in Japanese).

[18] J. S. Milne, Canonical models of (mixed) Shimura varieties and automorphic vector bundles, in Automorphic forms, Shimura varieties, and L-functions **1** (1990), 283–414.

[19] D. Passman, The algebraic structure of group rings, New York, John Wiley (1977).

[20] S. Usui, Variation of mixed Hodge structure arising from family of logarithmic deformations. II. Classifying space, Duke Math. J. **51-4** (1984), 851–875.

Hodge Theory and L^2-analysis
ALM 39, pp. 475–488

Motivic and Automorphic Aspects of the Reductive Borel-Serre Compactification*

Arvind N. Nair†

Abstract

This article is a survey of work on two aspects of the cohomology of the reductive Borel-Serre compactification of a locally symmetric space. On the one hand, when the locally symmetric space has an algebraic structure the cohomology is like that of a singular algebraic variety. On the other hand, the cohomology can be computed analytically using spaces of automorphic forms. We discuss these results and an application to the properties of Chern classes of automorphic vector bundles.

The reductive Borel-Serre compactification M^{rbs} of a noncompact locally symmetric space M was introduced in [Z2] and has since played a fundamental role in work on cohomology of arithmetic groups and locally symmetric spaces. This paper is a survey of two aspects of its cohomology, motivic (when M has an algebraic structure) and automorphic, and their interaction.

It is a well-known idea that when M is a locally symmetric variety M^{rbs} behaves like a partial resolution of the singularities of the minimal (or Satake-Baily-Borel) compactification M^{bb}. Although M^{rbs} is not an algebraic variety, its cohomology looks motivic: It carries a mixed Hodge structure [Z4], underlies a Voevodsky motive [AZ], and is part of a mixed realization [N2]. In §1 we outline the approach of [N2] to the mixed realization of $\mathrm{H}^*(M^{rbs})$ based on Morel's weight truncations in categories of mixed sheaves constructed by Saito and on a comparison between weight truncations and the weighted complexes of [GHM]. A special case of the comparison puts a mixed realization on the cohomology of M^{rbs}. We also discuss motivic results of Ayoub and Zucker [AZ] and Vaish [V].

The cohomology of M^{rbs} is also automorphic, in a precise sense: There are C^∞ de Rham models for its cohomology and homology in terms of relative Lie algebra cohomology of spaces of functions on the group (by specializing the main result of

*For Steven Zucker.

†School of Mathematics, Tata Institute of Fundamental Research, Homi Bhabha Road, Colaba, Mumbai 400005, India. E-mail: arvind@math.tifr.res.in

[N1]). For homology we can further reduce to a space of automorphic forms [F1]. In §2 we review these results and describe some consequences drawn in [N3]. The results in §2 work for any locally symmetric space, algebraic or not.

In §3 we describe a result of [N3] which exploits both the motivic and automorphic results on M^{rbs} to show that the Chern classes of the natural topological extensions to M^{rbs} of automorphic vector bundles on a locally symmetric variety M behave as they would if M^{rbs} were an algebraic variety over k and automorphic vector bundles extended algebraically to M^{rbs}.

Much of the interest in M and its compactifications comes from the hope of relating the mixed motives appearing in the cohomology of M to the special values of L-functions (see e.g. [H]). It follows from the result described in §3 that the summand of cohomology with trivial Hecke eigenvalues consists of mixed Tate realizations. In [N4] we treat this summand using automorphic methods from [F2] (see Remark 4); the RBS compactification plays an important role in this.

It should be obvious how much Steve Zucker's work (both on locally symmetric spaces and in Hodge theory) has influenced the matters discussed here and it is an honour to contribute to this volume for him. My own interest in Steve's "favorite space" M^{rbs} was rekindled by a discussion with him at Banff in May 2008 where he told me about [Z4]. I realized soon after that Morel's work [Mo] gives a simple and natural approach to such questions. The main result of [N2] was worked out during a stay at the IAS Princeton in the Fall of 2008 and this later led to [NV, N3, V].

I thank the referee and Vaibhav Vaish for helpful comments.

1 Mixed realizations and M^{rbs}

After recalling the categories of mixed sheaves of relevance to us from Saito's [S1, S2, S3] and Morel's method of weight truncations from [Mo], we sketch how the cohomology M^{rbs} carries a natural mixed realization, following [N2]. We also mention the motivic results of Ayoub-Zucker [AZ] and some further developments due to Vaish [V].

1.1 Some categories of mixed sheaves

Let k be a number field in $\mathbb{C}$ and Var/k the category of varieties (i.e. separated schemes of finite type) and morphisms defined over k. The algebraic closure of k in $\mathbb{C}$ is denoted $\bar{\mathbb{Q}}$.

The cohomology of an algebraic variety over k carries extra structures: The singular cohomology $\mathrm{H}^*(X(\mathbb{C}), \mathbb{Q})$ with its increasing weight filtration W and decreasing Hodge filtration F on $\mathrm{H}^*(X(\mathbb{C}), \mathbb{C})$ is a mixed Hodge structure. If X is smooth the algebraic de Rham cohomology $\mathbb{H}^*(X, \Omega^\bullet_{X/k})$ with its weight and Hodge filtrations is a bifiltered k-vector space. For each prime l the l-adic etale cohomology $\mathrm{H}^*(X \times_k \bar{\mathbb{Q}}, \mathbb{Q}_l)$ is a continuous l-adic $\mathrm{Gal}(\bar{\mathbb{Q}}/k)$-representation with a Galois-stable filtration by Frobenius weights. There are comparison isomorphisms $\mathrm{H}^*(X(\mathbb{C}), \mathbb{Q}) \otimes \mathbb{Q}_l \cong \mathrm{H}^*(X \times_k \bar{\mathbb{Q}}, \mathbb{Q}_l)$ and $\mathrm{H}^*(X(\mathbb{C}), \mathbb{C}) \otimes \mathbb{C} \cong \mathbb{H}^*(X, \Omega^\bullet_{X/k}) \otimes_k \mathbb{C}$ relating filtrations.

This leads to the notion (due to Deligne and Jannsen [J]) of a *mixed realization over k*. This is a collection $(H_B, H_{dR}, (H_l)_l, I, (I_l)_l)$ where the Betti part $H_B = (H_{\mathbb{Q}}, (H_{\mathbb{Q}}, W), (H_{\mathbb{C}}, F))$ is a $\mathbb{Q}$-mixed Hodge structure, the de Rham part $H_{dR} = (H_k, W, F)$ is a bifiltered k-vector space, and for each prime l we have $\mathbb{Q}_l$-vector space H_l with a continuous representation of $\mathrm{Gal}(\bar{\mathbb{Q}}/k)$ and a Galois-stable filtration W. The map $I : H_{\mathbb{Q}} \otimes_{\mathbb{Q}} \mathbb{C} \to H_k \otimes_k \mathbb{C}$ is an isomorphism identifying the W and F filtrations, and for each l, $I_l : H_{\mathbb{Q}} \otimes_{\mathbb{Q}} \mathbb{Q}_l \to H_l$ is an isomorphism identifying the W filtrations. Mixed realizations over k form an abelian category $\mathscr{MR}_k$ and $X \mapsto \mathrm{H}^*(X) = (\mathrm{H}^*_B(X), \mathrm{H}^*_{dR}(X/k), (\mathrm{H}^*_l(X))_l)$ with $\mathrm{H}^*_B(X) = (\mathrm{H}^*(X(\mathbb{C}), \mathbb{Q}), W, F)$, $\mathrm{H}^*_{dR}(X/k) = (\mathbb{H}^*(X, \Omega^\bullet_{X/k}), W, F)$, and $\mathrm{H}^*_l(X) = (\mathrm{H}^*(X \times_k \bar{\mathbb{Q}}, \mathbb{Q}_l), W)$ defines a functor from smooth varieties to $\mathscr{MR}_k$. (We will drop the data $I, (I_l)_l$ from the notation when there is no ambiguity.)

Relative versions of mixed Hodge structures and mixed realizations have been defined by M. Saito [S1, S2, S3]. The category $\mathrm{MHM}(X)$ of algebraic mixed Hodge modules ([S1], especially §3) on a complex algebraic variety X is an abelian category which for $X = \mathrm{Spec}(\mathbb{C})$ is the category of graded-polarizable rational mixed Hodge structures and which for X smooth contains admissible variations of mixed Hodge structure on X. The objects of $\mathrm{MHM}(X)$ are triples $((M, W, F), (K, W), \alpha)$ where (M, W, F) is a bifiltered regular holonomic D_X-module, (K, W) is a filtered algebraically constructible perverse $\mathbb{Q}_X$-sheaf, and $\alpha : \mathrm{DR}(M, W) \to (K, W) \otimes \mathbb{C}$ is a filtered quasiisomorphism. These are subject to complicated inductively defined conditions. (The theory rests, in the end, on [Z1] over curves.) Remembering only the underlying perverse sheaf K defines a faithful exact functor *rat* to the category $\mathrm{P}(X)$ of algebraically constructible perverse sheaves on X. This derives to a functor $rat : D^b\mathrm{MHM}(X) \longrightarrow D^b_c(\mathbb{Q}_X)$ to the derived category of complexes of $\mathbb{Q}_X$-sheaves with algebraically constructible cohomology sheaves (because $D^b(\mathrm{P}(X))$ is equivalent to $D^b_c(\mathbb{Q}_X)$ by [B]).

For a k-variety X, the category $\mathscr{M}(X)$ of mixed realizations over X is an abelian category defined in [S2, 1.8] and [S3, 1.1] by enriching mixed Hodge modules on $X(\mathbb{C})$: An object in $\mathscr{M}(X)$ consists of $((M_k, W, F), (K, W), (K_l, W)_l, \alpha, (\alpha_l)_l)$ where (M_k, F) is a regular holonomic D_X-module with finite increasing filtration W, (K, W) is a filtered perverse $\mathbb{Q}_{X(\mathbb{C})}$-sheaf, and for each l, (K_l, W) is a filtered perverse l-adic sheaf on X, and α and α_l are comparison isomorphisms. We refer to [S3, 1.1] and [S2, 1.8] for more details. For $X = \mathrm{Spec}(k)$ we get mixed realizations in $\mathscr{MR}_k$ with graded-polarizable Betti part. Remembering $((M_k, W, F) \otimes \mathbb{C}, (K, W), \alpha)$ gives a faithful exact functor $\mathscr{M}(X) \to \mathrm{MHM}(X(\mathbb{C}))$ and remembering only K gives a faithful exact functor *For* to perverse sheaves. The induced functor

$$For : D^b\mathscr{M}(X) \longrightarrow D^b_c(\mathbb{Q}_{X(\mathbb{C})})$$

on derived categories factors through *rat*.

Both $\mathrm{MHM}(\cdot)$ and $\mathscr{M}(\cdot)$ are examples of theories of mixed sheaves in the sense of [S2]. It then follows that there is a functorial formalism: For $f : X \to Y$ in Var/k there are functors $f_*, f^*, f_!, f^!$ between the derived categories $D^b\mathrm{MHM}(X(\mathbb{C}))$ and $D^b\mathrm{MHM}(Y(\mathbb{C}))$ (resp. between $D^b\mathscr{M}(X)$ and $D^b\mathscr{M}(Y)$) compatible under *rat* (resp. under *For*) with the usual functors between $D^b_c(\mathbb{Q}_{X(\mathbb{C})})$ and $D^b_c(\mathbb{Q}_{Y(\mathbb{C})})$, and these satisfy all the usual identities. (See [S1] for MHM(·) and [S3, Theorem 1.2] and [S2] for $\mathscr{M}(\cdot)$.) Thus for $K \in D^b\mathscr{M}(X)$ and $a_X : X \to \mathrm{Spec}(k)$ the structure morphism, the object

$$\mathrm{H}^i(a_{X*}K) \in \mathscr{M}(\mathrm{Spec}(k)) \tag{1}$$

gives a mixed realization with rational vector space underlying the Betti part given by $\mathbb{H}^i(X, \mathit{For}(K))$.

There is a unit object $\mathbb{Q}^{\mathscr{M}}$ in $\mathscr{M}(\mathrm{Spec}(k))$ with $\mathit{For}(\mathbb{Q}^{\mathscr{M}}) = \mathbb{Q}$. For $X \in \mathit{Var}/k$ with structure morphism $a_X : X \to \mathrm{Spec}(k)$ the object $\mathbb{Q}_X^{\mathscr{M}} = a_X^*\mathbb{Q}^{\mathscr{M}}$ ("the constant sheaf on X") in $D^b\mathscr{M}(X)$ has $\mathit{For}(\mathbb{Q}_X^{\mathscr{M}}) = \mathbb{Q}_{X(\mathbb{C})}$. Taking $\mathrm{H}^i(a_{X*}\mathbb{Q}_X^{\mathscr{M}})$ gives a mixed realization

$$\mathrm{H}^i(X) := (\mathrm{H}^i_B(X), \mathrm{H}^i_{dR}(X/k), (\mathrm{H}^i_l(X))_l)$$

with Betti part $\mathrm{H}^i(X(\mathbb{C}), \mathbb{Q})$ and this defines a functor $\mathit{Var}/k \to \mathscr{MR}_k$ extending the earlier functor from smooth varieties. (Note that the underlying k-vector space of $\mathrm{H}^*_{dR}(X/k)$ will not usually be $\mathbb{H}^*(X, \Omega^\bullet_{X/k})$ if X is singular.) Homology becomes a mixed realization by duality and intersection cohomology becomes a mixed realization using $IC_X = (j_{!*}\mathbb{Q}_U^{\mathscr{M}}[\dim X])[-\dim X]$ in (1) where $j : U \hookrightarrow X$ is an open dense smooth subset. We denote it $\mathrm{IH}^*(X) = (\mathrm{IH}^*_B(X), \mathrm{IH}^*_{dR}(X), (\mathrm{IH}^*_l(X))_l)$.

The Tate object is $\mathbb{Q}^{\mathscr{M}}(1) = \mathrm{H}_2(\mathbb{P}^1) = (\mathbb{Q}_B(1), \mathbb{Q}_{dR}(1), (\mathbb{Q}_l(1))_l)$, where $\mathbb{Q}_B(1)$ is the usual Tate Hodge structure of weight -2 and $\mathbb{Q}_l(1)$ is one-dimensional with Galois action by the l-adic cyclotomic character $\chi_l : \mathrm{Gal}(\bar{\mathbb{Q}}/k) \to \mathbb{Z}_l^*$. For $n \geq 0$ set $\mathbb{Q}^{\mathscr{M}}(n) = \mathbb{Q}^{\mathscr{M}}(1)^{\otimes n}$ and $\mathbb{Q}^{\mathscr{M}}(-n) = \mathbb{Q}^{\mathscr{M}}(-1)^{\otimes n}$. For an object $K \in \mathscr{M}(X)$ we let $K(n) = K \otimes \mathbb{Q}_X^{\mathscr{M}}(n)$ where $\mathbb{Q}_X^{\mathscr{M}}(n) = a_X^*\mathbb{Q}^{\mathscr{M}}(n)$.

1.2 Truncation by weights

A beautiful idea of S. Morel [Mo] is to define truncation functors using t-structures coming from the theory of weights. Morel's construction was originally for mixed l-adic complexes on schemes over finite fields, but works mutatis mutandis in any reasonable category of sheaves in which one has weight filtrations with the correct properties, in particular in any theory of mixed sheaves in the sense of Saito [S2]. We work with $\mathscr{M}(\cdot)$.

For $a \in \mathbb{Z}$, Morel defines a t-structure $({}^wD^{\leqslant a}, {}^wD^{>a})$ on $D^b\mathscr{M}(X)$ as follows: ${}^wD^{\leqslant a}$ (resp. ${}^wD^{>a}$) is the full subcategory of complexes K such that $\mathrm{H}^i(K) \in \mathscr{M}(X)$ has weights $\leq a$ for all i (resp. $> a$ for all i). ([1]) This defines a t-structure with heart ${}^wD^{\leqslant a} \cap {}^wD^{>a} = \{0\}$ and the corresponding truncation functors are denoted $w_{\leqslant a}, w_{>a}$. These t-structures have many nice properties, in particular the subcategories ${}^wD^{\leqslant a}, {}^wD^{>a}$ are stable under translation and triangulated and the truncation functors commute with translation.

If $X = \bigsqcup_{S \in \mathscr{S}} S$ is a stratification of X by equidimensional varieties over k, and $\underline{d} : \mathscr{S} \to \mathbb{Z} \cup \{\pm\infty\}$ is a function on the set of strata, then the standard machinery of [BBD] allows us to glue the t-structures $({}^wD^{\leqslant \underline{d}(S)}, {}^wD^{>\underline{d}(S)})$ on $D^b\mathscr{M}(S)$ for

[1] Contrast this with the property "K has weights $\leq a$" (resp. "K has weights $\geq a$") of [BBD, S1], which holds if and only if $\mathrm{H}^i(K)$ is of weights $\leq i+a$ for all i (resp. of weights $\geq i+a$ for all i).

$S \in \mathscr{S}$ to get a t-structure $({}^wD^{\leqslant \underline{d}}, {}^wD^{>\underline{d}})$ on $D^b\mathscr{M}(X)$. If $i_S : S \hookrightarrow X$ is the locally closed immersions of a stratum $S \in \mathscr{S}$ then a complex K belongs to ${}^wD^{\leqslant \underline{d}}$ (resp. ${}^wD^{>\underline{d}}$) if $i_S^*K \in {}^wD^{\leqslant \underline{d}(S)}$ (resp. $i_S^!K \in {}^wD^{>\underline{d}(S)}$) for all $S \in \mathscr{S}$. The associated truncation functors are denoted $w_{\leqslant \underline{d}}, w_{>\underline{d}}$.

There are two canonical functions $\mathscr{S} \to \mathbb{Z} \cup \{\pm\infty\}$ associated with any equidimensional stratified variety $X = \bigsqcup_{S\in\mathscr{S}} S$: The constant function $\underline{d} \equiv n = \dim X$ and the function $\underline{dim}$ defined by $\underline{dim}(S) := \dim S$. For the former, Morel [Mo] proved the following remarkable formula: If $j : U \hookrightarrow X$ is the inclusion of an open dense smooth subset then

$$w_{\leqslant n} j_* \mathbb{Q}_U^{\mathscr{M}}[n] = j_{!*}\mathbb{Q}_U^{\mathscr{M}}[n]. \tag{2}$$

So one recovers the intersection complex $IC_X = w_{\leqslant n} j_* \mathbb{Q}_U^{\mathscr{M}}$ as a weight truncation.

In [NV] V. Vaish and I found that the glued t-structure given by the function $\underline{dim}$ defined by $\underline{dim}(S_i) = \dim S_i$ has particularly nice properties. (The interest in $\underline{dim}$ came from Theorem 2 below, so we are inverting the chronology here.) In particular, the following was proven:

Theorem 1. *([NV, Proposition 4.1.2]) Suppose that U is smooth of dimension n, $j : U \hookrightarrow X$ is an open immersion as a Zariski-dense subset of an irreducible variety, and $\pi : Y \to X$ is a proper morphism from Y smooth such that $\pi|_{\pi^{-1}(U)}$ is an isomorphism. Assume X is given a stratification such that $j_!\mathbb{Q}_U^{\mathscr{M}}[n], j_*\mathbb{Q}_U^{\mathscr{M}}[n]$, and $\pi_*\mathbb{Q}_Y^{\mathscr{M}}[n]$ are constructible. Then*

$$w_{\leqslant \underline{dim}} j_* \mathbb{Q}_U^{\mathscr{M}}[n] = w_{\leqslant \underline{dim}} j_{!*} \mathbb{Q}_U^{\mathscr{M}}[n] = w_{\leqslant \underline{dim}} \pi_* \mathbb{Q}_Y^{\mathscr{M}}[n]$$

in $D^b\mathscr{M}(X)$.

Informally, this shows that the object $EC_X := w_{\leqslant \underline{dim}} j_* \mathbb{Q}_U^{\mathscr{M}}$ can be constructed "from the inside" using $j : U \hookrightarrow X$ or "from the outside" using $\pi : Y \to X$. This defines a "cohomology theory" $X \mapsto EH^*(X) := \mathbb{H}^*(X, w_{\leqslant \underline{dim}} j_* \mathbb{Q}_U^{\mathscr{M}})$ lying somewhere between the usual cohomology and intersection cohomology and enjoying a number of pleasant formal properties established in [NV]. Among these properties are ring structure and functoriality for certain morphisms. Whether there is more topological content to the theory seems to be an interesting question.

1.3 Locally symmetric varieties

Let M be a locally symmetric variety and $n = \dim_{\mathbb{C}} M$. We work over the field of definition k of M, a number field in $\mathbb{C}$, which can be made quite explicit if M is a Shimura variety. We will not distinguish notationally between the varieties over k and their complex points.

Consider the diagram

$$\begin{array}{ccc} & & M^{rbs} \\ & & \downarrow p \\ M & \stackrel{j}{\hookrightarrow} & M^{bb} \end{array}$$

where M^{bb} is the minimal (or minimal Satake or Baily-Borel) compactification. The main technical result of [N2] (Theorem 4.3.1 of loc. cit.) compared various weight truncations $w_{\leqslant d}j_*\mathbb{Q}_M^{\mathscr{M}}$ with the pushforward of the weighted complexes of Goresky-Harder-MacPherson [GHM]. This was done to prove that certain spectral sequences appearing in work of Franke [F1] on Borel's conjecture are spectral sequences of mixed realizations (the main result of [N2]). For the special case of the $\underline{dim}$ truncation we get:

Theorem 2. *There is a natural isomorphism* $For(w_{\leqslant \underline{dim}}j_*\mathbb{Q}_M^{\mathscr{M}}) = p_*\mathbb{Q}_{M^{rbs}}$.

Another special case of [N2] is the constant function $n = \dim M$, when one gets

$$For(w_{\leqslant n}j_*\mathbb{Q}_M^{\mathscr{M}}) = p_*WC_{M^{rbs}}(\mathbb{Q}_M)$$

where $WC_{M^{rbs}}(\mathbb{Q}_M)$ is the (upper or lower) middle weighted cohomology complex of [GHM]. By the main result of [GHM] $p_*WC_{M^{rbs}}(\mathbb{Q}_M) = (j_{!*}\mathbb{Q}_M[n])[-n]$, so this reduces to Morel's formula (2) (or could be deduced from it).

The proof of Theorem 2 used several ingredients that had been around for some time. The main technical problem is that one is comparing the functorial image (under *For*) of a complex defined using a t-structure in $D^b\mathscr{M}(M^{bb})$ to a complex in $D^b_c(\mathbb{Q}_{M^{bb}})$ where there is no similar t-structure available. The key point is to use the local Hecke operator introduced by Looijenga in his proof of Zucker's conjecture and a splitting property in the derived category for its action on each of the objects. For the weighted complexes the splitting property was proved in [GHM]. For weight truncations $w_{\leqslant \underline{dim}}j_*\mathbb{Q}_M$ the splitting property is established in [N1] using results of Looijenga and Rapoport [LR] and Jordan decomposition for endomorphisms in the derived category.

Corollary 3. *The cohomology of the RBS compactification is part of a mixed realization, i.e. there is a mixed realization*

$$\mathrm{H}^*(M^{rbs}) = (\mathrm{H}^*_B(M^{rbs}), \mathrm{H}^*_{dR}(M^{rbs}/k), (\mathrm{H}^*_l(M^{rbs}))_l)$$

in $\mathscr{MR}_k$ *with Betti part a mixed Hodge structure on* $\mathrm{H}^*(M^{rbs}, \mathbb{Q})$.

Now we combine Theorems 1 and 2. Choose a smooth toroidal compactification $\pi : M^{\Sigma} \to M^{bb}$ and consider the following diagram:

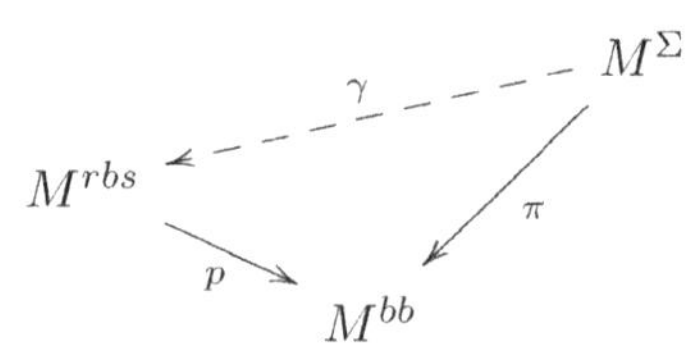

(3)

Theorem 1 gives identities

$$w_{\leqslant \underline{dim}}j_*\mathbb{Q}_M = w_{\leqslant \underline{dim}}j_{!*}\mathbb{Q}_M = w_{\leqslant \underline{dim}}\pi_*\mathbb{Q}_{M^{\Sigma}}. \tag{4}$$

The consequences are summarized in the following commutative diagram in $\mathscr{M}(\mathrm{Spec}(k))$:

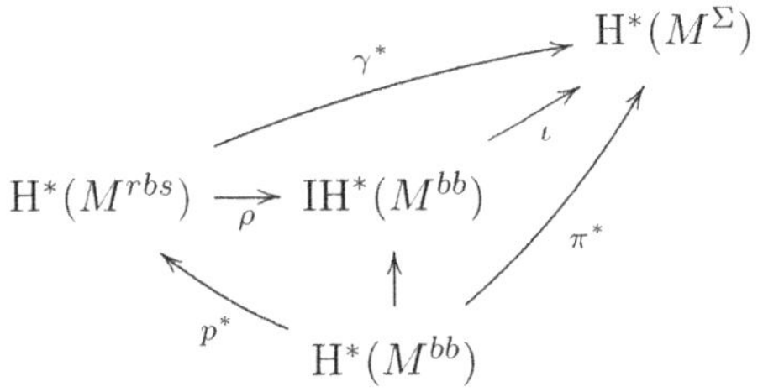

(5)

The maps ρ and γ^* are given by (4) and $\iota : \mathrm{IH}^*(M^{bb}) \hookrightarrow \mathrm{H}^*(M^{\Sigma})$ is given by any homomorphism $j_{!*}\mathbb{Q}^{\mathscr{M}}_{M_K}[n] \to \pi_*\mathbb{Q}^{\mathscr{M}}_{M^{\Sigma}_K}[n]$ coming from the decomposition theorem ([S2, 6.10]). (That $\pi^* = \gamma^* \circ p^*$ recovers the result of [GT] in rational cohomology. Thus, although no γ extending the identity exists in (3), cohomologically it looks like it does.) The morphisms other than ι are canonical and Hecke-equivariant in the appropriate sense (e.g. for π^*, γ^* one keeps in mind that a Hecke operator goes from $\mathrm{H}^*(M^{\Sigma}) \to \mathrm{H}^*(M^{\Sigma'})$ for some Σ').

We also have the following consequence of the formalism:

$$\mathrm{Gr}^W_i \mathrm{H}^i(M^{rbs}) = \mathrm{im}[\mathrm{H}^i(M^{rbs}) \to \mathrm{IH}^i(M^{bb})] = \mathrm{im}[\mathrm{H}^i(M^{rbs}) \to \mathrm{H}^i(M^{\Sigma})]$$

which is analogous to standard formulas in mixed Hodge theory.

1.4 Motivic results

The categories above depend on realizations. A more geometric construction is Ayoub's triangulated category of (etale) motivic sheaves, which we denote here by $\mathrm{DM}(X)$; then $\mathrm{DM}(\mathrm{Spec}(k)) = \mathrm{DM}_k$ is Voevodsky's triangulated category of motives over k. (See [A] for a survey of motivic sheaves; $\mathrm{DM}(X)$ is the category $\mathbf{DA}^{\mathrm{et}}(X;\mathbb{Q})$ of loc. cit.) For a morphism $f : X \to Y$ in Var/k there are functors $f_*, f^*, f_!, f^!$ between $\mathrm{DM}(X)$ and $\mathrm{DM}(Y)$ satisfying the usual adjunctions.

The result of Ayoub and Zucker [AZ] gives an object $\mathbb{E}_{M^{bb}}$ in $\mathrm{DM}(M^{bb})$ which, on pushforward to $\mathrm{Spec}(k)$, gives an object of DM_k. Using the Betti realization constructed by Ayoub, this is shown to realize to $R\Gamma(M^{rbs}, \mathbb{Q}) = \bigoplus_k \mathrm{H}^k(M^{rbs}, \mathbb{Q})$ $[-k]$. This shows that M^{rbs} is motivic in the sense of Voevodsky. Unfortunately, although Huber has constructed a functor $\mathrm{DM}_k \to \mathscr{MR}_k$, this is not quite enough to show that $\mathrm{H}^*(M^{rbs})$ is a mixed realization (i.e. Corollary 3) because some compatibilities are unknown (see [N3, 2.5] or [AZ, 4.4, 4.9] for a discussion).

In [V] Vaish generalizes the approach to M^{rbs} using Morel's truncation functor $w_{\leqslant \underline{dim}}$ to categories of motivic sheaves. He defines the truncation $w_{\leqslant \underline{dim}}$ (denoted $w_{\leq \mathrm{Id}}$ in loc. cit.) and develops its formalism in a very general motivic sheaf context, including in positive characteristic. He also shows that $w_{\leqslant \underline{dim}}$ (which is defined on a suitable subcategory of $\mathrm{DM}(X)$) agrees with the relative Artin motive functor ω^0 defined in [AZ] in characteristic zero (using abstract representability theorems). In particular, this gives another approach to $\mathbb{E}_{M^{bb}}$ and clarifies the relation of ω^0 with weights. In addition to the formulas $\mathbb{E}_{M^{bb}} = w_{\leqslant \underline{dim}} j_* \mathbf{1}_M = w_{\leqslant \underline{dim}} \pi_* \mathbf{1}_M$ in (4) which are in [AZ] (where $\mathbf{1}_M \in \mathrm{DM}(M)$ is the unit object over M, the analogue of the constant sheaf), he is able to prove the relation with the intersection complex in (4) in the motivic context, as we now explain. The notion of *motivic intersection complex* has been defined by Wildeshaus [W1]. The object IM_X in $\mathrm{DM}(X)$, when it exists, is unique up to isomorphism (but not up to unique isomorphism). It exists if $X = M^{bb}$ [W2].

Theorem 4. *(Vaish [V, Theorem 4.1.2]) If the motivic intersection complex IM_X exists, then $w_{\leqslant \underline{dim}} IM_X = \mathbb{E}_X$. In particular, this holds for $X = M^{bb}$.*

This gives a diagram like (5) in DM_k. (The outer triangle is in [AZ].)

Remark 1. Taken together, the results of [N2] (i.e. Theorem 2) and [V] give a different approach to the main theorem of [AZ]. (The results of [NV] are not required for this, they merely served as the model for the development in [V].) Curiously, while the local Hecke operator plays a key role in our approach (in the proof of Theorem 2), it is does not appear in [AZ]. On the other hand, [AZ] uses explicit information about toroidal compactifications [AMRT], which our approach does not (although some use of them is hidden in our use of [LR]). The approach of [N2, V] is simpler in that much of the burden of proof is shifted to a cohomological setting.

Remark 2. The category $\mathscr{M}_{go}(X)$ of objects of geometric origin is the smallest full subcategory of $\mathscr{M}(X)$ closed under subquotients and containing the objects $\mathrm{H}^i(f_* \mathbb{Q}_Y^{\mathscr{M}})(n)$ for any projective morphism $f : Y \to X$ and all $i, n \in \mathbb{Z}$. There is a functor $\mathrm{DM}(X) \longrightarrow D^b\mathscr{M}(X)$ and it seems to be expected that this gives an equivalence of the heart of the (conjectured) perverse t-structure on $\mathrm{DM}(X)$ with $\mathscr{M}_{go}(X)$.

2 Automorphic forms and M^{rbs}

The cohomology of the RBS compactification has an analytic interpretation in terms of automorphic forms, thanks to results of [N1] and [F1]. We review this and some of its consequences. The notation is as follows: G is a semisimple group over $\mathbb{Q}$, $K \subset G(\mathbb{R})$ is a maximal compact subgroup, $\Gamma \subset G(\mathbb{Q})$ is an arithmetic subgroup, $\mathfrak{g}$ is the Lie algebra of $G(\mathbb{R})$ and $U(\mathfrak{g})$ it universal enveloping algebra. The locally symmetric space $M = \Gamma\backslash G(\mathbb{R})/K$ is not assumed to have a complex structure in this section.

2.1 The RBS compactification and automorphic forms

The de Rham theorem allows us to compute the cohomology of M as relative Lie algebra cohomology:

$$\mathrm{H}^*(M, \mathbb{C}) = \mathrm{H}^*(\mathfrak{g}, K, C^\infty(\Gamma\backslash G(\mathbb{R}))).$$

Here $C^\infty(\Gamma\backslash G(\mathbb{R}))$ is the space of smooth and K-finite functions. We will define certain $(\mathfrak{g}, K)$-submodules of $C^\infty(\Gamma\backslash G(\mathbb{R}))$ in terms of growth on Siegel sets and use this to compute $\mathrm{H}^*(M^{rbs}, \mathbb{C})$.

Fix a minimal $\mathbb{Q}$-parabolic P_0 and a Levi subgroup $M_0 \subset P_0$, and let N_0 be the unipotent radical and $A_0 \subset M_0$ the maximal $\mathbb{Q}$-split central torus in M_0. Let Δ_0 be the set of simple positive roots of A_0 in P_0 and ρ_0 the half-sum of positive roots. Recall that a Siegel set in $G(\mathbb{R}) = P_0(\mathbb{R})K$ has the form $\mathfrak{S} = \omega \cdot A_0(t) \cdot K$ where $\omega \subset (N_0M_0)(\mathbb{R})$ is a relatively compact subset and $A_0(t) = \{a \in A_0(\mathbb{R})^0 : a^\alpha > t \text{ for all } \alpha \in \Delta_0\}$. For ω large enough and t small enough there are finitely many elements $g_1, \ldots, g_r \in G(\mathbb{Q})$ such that $\bigsqcup_{i=1}^r g_i\mathfrak{S}$ forms a coarse fundamental domain for Γ, i.e. $\bigsqcup_{i=1}^r g_i\mathfrak{S} \to \Gamma\backslash G(\mathbb{R})$ is surjective with finite fibres and bijective onto the complement of a compact set. Fix a norm $\|\cdot\|$ on $\mathfrak{a}_0 = \mathrm{Lie}\, A_0(\mathbb{R})$.

Let $B = B(\Gamma\backslash G(\mathbb{R}))$ be the subspace of $f \in C^\infty(\Gamma\backslash G(\mathbb{R}))$ for which there exists $N \in \mathbb{N}$ such that for all $1 \leq i \leq r$ and $D \in U(\mathfrak{g})$, there exists $C > 0$ such that

$$|Df(g_ipak)| \leq C(1 + \|\log(a)\|)^N \quad \text{for } pak \in \mathfrak{S}.$$

Informally, we say that f is "uniformly bounded up to logarithmic terms".

Let $R = R(\Gamma\backslash G(\mathbb{R}))$ be the subspace of $C^\infty(\Gamma\backslash G(\mathbb{R}))$ of functions f such that for $D \in U(\mathfrak{g})$ and $1 \leq i \leq r$, and for all $N \in \mathbb{N}$, there exists $C > 0$ such that

$$|Df(g_ipak)| \leq C(1 + \|\log(a)\|)^{-N} \cdot a^{2\rho_0} \quad \text{for } pak \in \mathfrak{S}.$$

If $f \in B, g \in R$ then fg is L^1 and this defines a $(\mathfrak{g}, K)$-invariant pairing of B and R.

Let $\mathfrak{J}$ be the ideal of the centre of $U(\mathfrak{g})$ killing the trivial representation. Then

$$\mathfrak{A}_R = \mathfrak{A}_R(\Gamma\backslash G(\mathbb{R})) = \{f \in R : \mathfrak{J}^m f = 0 \text{ for some } m\}$$

is a $(\mathfrak{g}, K)$-module of automorphic forms. Specializing the main result of [N1] to the cases $\lambda = 0, \lambda = 2\rho_0$ gives:

Theorem 5. *([N1]) There are natural isomorphisms*

$$\mathrm{H}^*(M^{rbs}, \mathbb{C}) = \mathrm{H}^*(\mathfrak{g}, K, B)$$

and

$$\mathrm{H}_{\dim_{\mathbb{R}} M - *}(M^{rbs}, \mathbb{C}) = \mathrm{H}^*(\mathfrak{g}, K, R) = \mathrm{H}^*(\mathfrak{g}, K, \mathfrak{A}_R).$$

We remark that replacing R by the submodule of automorphic forms $\mathfrak{A}_R$ relies on the difficult results of Franke [F1]. The map induced by $B \subset R$ is the fundamental class map $\mathrm{H}^*(M^{rbs}, \mathbb{C}) \to \mathrm{H}_{\dim_{\mathbb{R}} M - *}(M^{rbs}, \mathbb{C})$.

The constant functions belong to B. The $(\mathfrak{g}, K)$-cohomology of the trivial representation is the cohomology of the compact dual symmetric space $\hat{X}$ of $X = G/K$, so this gives a map

$$\theta : \mathrm{H}^*(\hat{X}, \mathbb{C}) \to \mathrm{H}^*(M^{rbs}, \mathbb{C}). \tag{6}$$

In the compact case this is injective and is an isomorphism onto the Hecke invariants by Matsushima's formula, see [N3, 1.2]. Using Theorem 5 and structural results of Franke on the space of automorphic forms, we prove in [N3, §1] that:

Theorem 6. *([N3]) The inclusion of the constant functions $\mathbb{C} \subset B$ induces an isomorphism of* $\mathrm{H}^*(\hat{X}, \mathbb{C})$ *onto the algebra of Hecke-invariants and Hecke-coinvariants in* $\mathrm{H}^*(M^{rbs}, \mathbb{C})$.

Note that the algebra of invariants splits off as a direct summand. It is essential here to first work with $\mathfrak{A}_R$ and compute the invariants in $\mathrm{H}_*(M^{rbs}, \mathbb{C})$ and then use duality, because B cannot be replaced by a module of automorphic forms to compute cohomology.

Remark 3. Another C^∞ de Rham model for $\mathrm{H}^*(M^{rbs}, \mathbb{C})$ using L^p cohomology (for large p) was found in [Z3]. It is not clear to me if Theorem 6 can be proven using this model.

2.2 Functoriality of RBS cohomology: A question

Suppose that $\varphi : H \to G$ is a homomorphism of semisimple $\mathbb{Q}$-groups with finite kernel. Let $\Gamma \subset G(\mathbb{Q})$ be arithmetic and $\Gamma_H := \varphi^{-1}(\Gamma)$. Let $X_H \subset X$ be the embedding of symmetric spaces induced by φ. For $M_H = \Gamma_H \backslash X_H$ and $M = \Gamma \backslash X$ there is a proper map of locally symmetric spaces

$$\varphi : M_H \longrightarrow M.$$

It is easy to give examples to show that this does *not* extend continuously to RBS compactifications in general. (Morally the reason is clear: Two non-Γ-conjugate parabolics in G might have preimages in H which are Γ_H-conjugate. Then the corresponding stratum of M_H^{rbs} "does not know where to go" in M^{rbs}.) Nevertheless, it is a consequence of Theorem 5 that there is a natural restriction map

$$\mathrm{H}^*(M^{rbs}, \mathbb{C}) \longrightarrow \mathrm{H}^*(M_H^{rbs}, \mathbb{C}) \tag{7}$$

because functions in $B(\Gamma \backslash G(\mathbb{R}))$ pull back to functions in $B(\Gamma_H \backslash H(\mathbb{R}))$. In the case where both M_H and M are locally symmetric varieties, there is an extension $M_H^{bb} \to M^{bb}$, and this can be used (by the properties of EC_X alluded to at the end of 1.2) to show that (7) is $\mathbb{Q}$-rational and even a homomorphism of mixed realizations.

For general M and M_H, the following topological explanation for (7), inspired by [GT], suggests itself: Embed M_H in the $(M_H \times M)^{rbs} = M_H^{rbs} \times M^{rbs}$ by $x \mapsto (x, \varphi(x))$ and let M'_H be its closure. There are maps

$$M_H^{rbs} \xleftarrow{p_1} M'_H \xrightarrow{p_2} M^{rbs}$$

induced by the two projections. *Are the fibres of* p_1 *contractible?* If so, the resulting pullback map $\mathrm{H}^*(M^{rbs}, \mathbb{Z}) \to \mathrm{H}^*(M_H^{rbs}, \mathbb{Z})$ should give (7).

3 An application: Chern classes

In [N3] we combined the results surveyed in §1 and §2 with results in the literature due to several authors to prove that the Chern classes of automorphic vector bundles on M^{rbs} have motivic properties in mixed realizations. The result is best formulated in the setting of Shimura varieties, so we will only give an approximate statement, referring to [N3] for details.

Recall the map defined in (6) above

$$\theta : \mathrm{H}^*(\hat{X}, \mathbb{C}) \to \mathrm{H}^*(M^{rbs}, \mathbb{C})$$

for a general locally symmetric space using the C^∞ de Rham model of Theorem 5 (or the L^p cohomology model of [Z3]). In general this map does not relate the two rational structures (presumably irrational quantities like $\zeta(k)/\pi^k$ for odd k are involved when $G = SL(n)$ thanks to Borel's theorem).

When M is a locally symmetric variety, however, θ is Betti rational. This is because θ has another description in terms of Chern classes of automorphic vector bundles. Recall that $\hat{X} = G(\mathbb{C})/Q$ where Q is a parabolic with Levi subgroup the complexification $K_{\mathbb{C}}$ of K. A finite-dimensional complex representation $\rho : K \to GL(V)$ gives an algebraic vector bundle $\hat{\mathscr{V}}$ on $\hat{X}$. Restricting by the Borel embedding $X \subset \hat{X}$ and dividing by Γ gives an vector bundle $\mathscr{V}$ on M which has a canonical algebraic structure. The underlying topological vector bundle of $\mathscr{V}$ extends naturally to a topological vector bundle $\mathscr{V}^{rbs}$ on M^{rbs} (see [GT, 9.2] or [Z3, 1.9]). Then θ satisfies

$$\theta(c_i(\hat{\mathscr{V}})) = (-1)^i c_i(\mathscr{V}^{rbs}).$$

In the compact case this is essentially Hirzebruch proportionality; in general it follows from results of Zucker [Z3], see [N3, 3.6]. Since the rational cohomology of $\hat{X}$ is generated by Chern classes of homogeneous bundles, θ is rational.

Since $\mathrm{H}^*(M^{rbs})$ has a mixed realization (Corollary 3), it is natural to ask if θ is a homomorphism of mixed realizations. This amounts to asking if the *topologically defined* classes $c_i(\mathscr{V}^{rbs})$ have "motivic" properties. The following is an informal version of the main result of [N3]; we refer to that paper for precise statements:

Theorem 7. *The Chern classes of the topological extensions to M^{rbs} of automorphic vector bundles on M have the properties one would expect if M^{rbs} were an algebraic variety over k and automorphic vector bundles extended algebraically over M^{rbs} respecting fields of definition.*

Thus for example, the Chern class $c_i(\mathscr{V}^{rbs})$ belongs to the correct step of the Hodge filtration on $\mathrm{H}^{2i}(M^{rbs}, \mathbb{C})$, and if $\mathscr{V}$ is defined over L/k the action of $\mathrm{Gal}(\bar{\mathbb{Q}}/L)$ on $c_i(\mathscr{V}^{rbs})$, considered in etale cohomology via the comparison isomorphism, is via the ith power of the cyclotomic character χ_l.

The proof of this theorem combines Theorem 6 with results about automorphic vector bundles on Shimura varieties and their canonical extensions to toroidal compactifications due to Mumford and Harris, and also results due to Goresky-Pardon and Zucker. As a corollary, for the correct de Rham k-structure on $\hat{X}$ (see [N3]), we have:

Corollary 8. *The map $\theta : \mathrm{H}^*(\hat{X}) \to \mathrm{H}^*(M^{rbs})$ given by $\theta(c_k(\hat{\mathscr{V}})) = (-1)^k c_k(\mathscr{V})$ is a ring isomorphism of mixed realizations onto the Hecke-invariants in $\mathrm{H}^*(M^{rbs})$.*

This can be seen as a generalization of the classical Hirzebruch-Mumford proportionality theorem [Mu] to M^{rbs} in mixed realizations. The construction of a motivic θ (i.e in a context like 1.4) seems like an interesting problem.

Remark 4. The corollary can be used to show that the direct summand $\mathrm{H}^*(M)_{\mathcal{I}}$ of generalized Hecke-invariants of the cohomology of $\mathrm{H}^*(M)$, i.e. the summand on which the Hecke operators act with the same eigenvalues as in the trivial Hecke-module, is mixed Tate in $\mathscr{MR}_k$ (i.e. the graded pieces for the weight filtration are sums of Tate objects). The extensions in this summand can further be analyzed using methods of automorphic forms from [F2], see [N4].

Remark 5. A remarkable fact is that the analogue of the corollary is *not* true for M^{bb} due to the appearance of nontrivial mixed Tate extensions in $\mathrm{H}^*(M^{bb})$ (see the remarks in [N3, 0.4,4.3] and [N4] and [L]).

References

[AMRT] Avner Ash, David Mumford, Michael Rapoport, and Yung-Sheng Tai, *Smooth Compactification of Locally Symmetric Varieties*, Math Sci Press, Brookline, MA, 1975.

[A] J. Ayoub, A guide to (étale) motivic sheaves, to appear in the proceedings of ICM 2014, available at `http://user.math.uzh.ch/ayoub/`

[AZ] J. Ayoub and S. Zucker, Relative Artin motives and the reductive Borel-Serre compactification of a locally symmetric variety, *Invent. Math.* 188 (2012), 277-427.

[B] A. A. Beilinson, On the derived category of perverse sheaves, *K-Theory, Arithmetic and Geometry*, Lecture Notes in Math. 1289 (1987), 27–41.

[BBD] A. Beilinson, J. Bernstein, and P. Deligne, Faisceaux pervers, *Astérisque* 100 (1982).

[F1] J. Franke, Harmonic analysis in weighted L^2 spaces, *Ann. Sci. ÉNS* t. 31 (1998), 181–279.

[F2] J. Franke, A topological model for some summand of the Eisenstein cohomology of congruence subgroups, *Eisenstein Series*, Prog. in Math. 258 (2008), 27–85.

[GHM] M. Goresky, G. Harder, and R. MacPherson, Weighted cohomology, *Invent. Math.* 116 (1994), 139–213.

[GT] M. Goresky and Y.-S. Tai, Toroidal and reductive Borel-Serre compactifications of locally symmetric spaces, *Amer. J. Math.* 121 (1999) 1095–1151.

[H] G. Harder, *Eisensteinkohomologie und die Konstruktion gemischter Motive*, Lecture Notes in Math. 1562, Springer (1993).

[J] U. Jannsen, *Mixed Motives and Algebraic K-theory*, Lecture Notes in Math. 1400, Springer-Verlag (1990).

[LR] E. Looijenga and M. Rapoport, Weights in the local cohomology of a Baily-Borel compactification, *Complex Geometry and Lie Theory*, Proc. Symp. Pure Math. 53 (1991), 223–260.

[L] E. Looijenga, Goresky-Pardon Chern classes and extensions which may be mixed Tate, preprint (2015).

[Mo] S. Morel, Complexes pondérés sur les compactifications de Baily-Borel: Le cas des variétés de Siegel, *J. Amer. Math. Soc.* 21 (2008), 23–61.

[Mu] D. Mumford, Hirzebruch's proportionality theorem in the noncompact case, *Invent. Math.* 42 (1977), 239–272.

[N1] A. N. Nair, Weighted cohomology of arithmetic groups, *Ann. of Math.* 150 (1999), 1–31.

[N2] A. N. Nair, Mixed motives in Shimura varieties and automorphic forms, preprint (2013), available at the author's homepage.

[NV] A. N. Nair and V. Vaish, Weightless cohomology of algebraic varieties, *J. Algebra* 424 (2015), 147–189.

[N3] A. N. Nair, Chern classes of automorphic vector bundles and the reductive Borel-Serre compactification, preprint (2014), available at the author's homepage.

[N4] A. N. Nair, Mixed Tate motives in A_g, in preparation.

[S1] M. Saito, Mixed Hodge modules, *Publ. Math. RIMS* 26 (1990), 221–333.

[S2] M. Saito, On the formalism of mixed sheaves, `arXiv:math/0611597v1` (2006).

[S3] M. Saito, Arithmetic mixed sheaves, *Invent. Math.* 144 (2001), 533–569.

[V] V. Vaish, Motivic weightless complex and the relative Artin motive, *Advances in Math.* 292 (2016), 316–373.

[W1] J. Wildeshaus, Motivic intersection complex, Regulators, *Contemp. Math.* 571 (2012), 255–276.

[W2] J. Wildeshaus, Intermediate extension of chow motives of abelian type, `arXiv.math:1211.5327v2` (2012).

[Z1] S. Zucker, Hodge theory with degenerating coefficients: L^2 cohomology in the Poincaré metric, *Ann. of Math.* 109 (1979), 415–476.

[Z2] S. Zucker, L^2 cohomology of warped products and arithmetic groups, *Invent. Math.* 70 (1982), 169–218.

[Z3] S. Zucker, On the reductive Borel-Serre compactification: L^p cohomology of arithmetic groups (for large p), *Amer. J. Math.* 123 (2001), 951–984.

[Z4] S. Zucker, On the reductive Borel-Serre compactification III: Mixed Hodge structures, *Asian. J. Math.* 8 (2004), 881–912.

Hodge Theory and L^2-analysis
ALM 39, pp. 489–516

An Update of Extension Theorems by the L^2 Estimates for $\bar{\partial}$*

Takeo Ohsawa[†]

Abstract

Recent activity around extension of holomorphic functions is reviewed. Highlighted materials are the solution of Suita's conjecture by Błocki, its generalization by Guan and Zhou including its striking application to the variation of the Bergman kernel, Berndtsson-Lempert's surprising proof of a sharp L^2 extension theorem, and a concise variant of Błocki-Guan-Zhou's proof. Some remarks on the classical theories of Grauert and Hörmander will be given in a related context.

2010 Mathematics Subject Classification: Primary 32A36; Secondary 32T27.

Introduction

In function theory of several complex variables, there are basically two types of extension problems. One of these is analytic continuation from the boundary. Hartogs [H] initiated this research in the study of the series

$$\sum_{k=0}^{\infty} a_k(z)w^k, \quad z \in D, \quad w \in \mathbb{C},$$

where D is a domain in $\mathbb{C}^n$ and $a_k(z)$ are holomorphic on D. The other is on the extension from analytic subsets. This question has a long history as an interpolation problem since Newton and Lagrange. Oka [O] and Cartan [C] studied it systematically and founded a basic theory of several complex variables.

From the viewpoint that holomorphic functions are nothing but the weak solutions of the Cauchy-Riemann equation, these two extension problems can be

*Dedicated to Steven M. Zucker on his 65th birthday.

[†]Graduate School of Mathematics Nagoya University 464-8602 Chikusaku Furocho Nagoya Japan. E-mail: ohsawa@math.nagoya-u.ac.jp

formulated in a context of partial differential equations (=PDEs), so that they become boundary value problems (=BVPs) for the $\bar{\partial}$-equation say $\bar{\partial}\alpha = v$. Here $\bar{\partial}$ denotes the complex exterior derivative of type (0,1).

The purpose of the present survey article is to outline the recent development of Oka-Cartan theory in this PDE-context. For that, let us first recall that a penetrating general method was established by Andreotti and Vesentini [A-V-1,2] and by Hörmander [Hm], independently. An advantage of this so called L^2 method is that one can obtain quantitatively definite results on holomorphic functions by exploring the L^2 estimates for the $\bar{\partial}$-operator. Andreotti and Vesentini analyzed the $\bar{\partial}$-equation on complete Hermitian manifolds by extending the method of Kodaira [K] and Nakano [N]. The result was later applied to locally symmetric varieties (cf. [M-Z], [C-V] and [Oh-4]). On the other hand, Hörmander's analysis is more focused on smoothly bounded domains as a development of the method initiated by Garabedian and Spencer [G-S], Morrey [Mo] and Kohn [Kn]. By this L^2 method, Hörmander solved a question posed by Bergman on the boundary behavior of the Bergman kernel affirmatively. Namely, he proved that the Bergman kernel $K_D(z,w)$ of a pseudoconvex domain $D \subset \mathbb{C}^n$ behaves near a strictly pseudoconvex boundary point $x \in \partial D$ in such a way that

$$\lim_{z \to x} K_D(z,z) \mathrm{dist}(z, \partial D)^{n+1} = \frac{\pi^n}{n!},$$

where $\mathrm{dist}(z, \partial D) = \inf_{w \in \partial D} \|z - w\|$, if the $2n-1$ dimensional unit sphere touches ∂D at x to the order at least 3. The L^2 method was later applied to the extension problem from analytic sets to solve another question of Bergman whether or not

$$\liminf_{z \to x} K_D(z,z)\|z - x\|^2 > 0$$

holds if D is a bounded pseudoconvex domain and ∂D is C^2-smooth around x. An affirmative answer is included in [Oh-T] whose main result is the following.

Theorem 0.1. *There exists a constant $C > 0$ such that, for any pseudoconvex domain $D \subset \mathbb{C}^n$ satisfying $\sup_D |z_n| \le 1$, for any plurisubharmonic function φ on D and for any holomorphic function f on $D' = D \cap \{z_n = 0\}$, one can find a holomorphic function $\tilde{f}$ on D extending f such that*

$$\int_D |\tilde{f}|^2 e^{-\varphi} \le C \int_{D'} |f|^2 e^{-\varphi},$$

where the integrals are with respect to the Lebesgue measures.

Theorem 0.1 was recently improved by Błocki [Bl-1] so that one can take π as the above constant C, and further by Guan and Zhou [G-Z-1,2] so that one has a gain in the estimate of $\tilde{f}$.

For any complex manifold M, we shall denote by $\mathcal{O}(M)$ the set of holomorphic functions on M and by $PSH(M)$ the set of plurisubharmonic functions on M. A refined variant of Theorem 0.1 established in [G-Z-1,2] as an optimized version of [Oh-5] is then stated as follows.

Theorem 0.2. *For any pseudoconvex domain $D \subset \mathbb{C}^n$, for any $\varphi, \psi \in PSH(D)$ satisfying $\sup_D (\psi + \log |z_n|^2) \leq 0$ and for any $f \in \mathcal{O}(D')$, one can find a holomorphic extension $\tilde{f}$ of f to D such that*

$$\int_D |\tilde{f}|^2 e^{-\varphi+\psi} \leq \pi \int_{D'} |f|^2 e^{-\varphi}. \tag{0.1}$$

We shall report on the recent activity around Theorem 0.1 and Theorem 0.2, starting from Błocki's groundbreaking work [Bl-1] which settled a long conjectured inequality on the Bergman kernel, the Suita conjecture, as well as from a generalized and refined solution by Guan and Zhou [G-Z-1,2] (a striking application of Theorem 0.2 (with $\psi = 0$) is in [G-Z-2]). After recalling some prerequisites, i.e. a part of Andreotti-Vesentini-Hörmander's theory, we shall review the new L^2 estimate for $\bar{\partial}$ by Błocki (cf. Theorem 4.1). At this point, we shall not proceed to prove Theorem 0.2. Before that, we shall give a review on the relation between the L^2 extension and a variational property of the Bergman kernels following Berndtsson [B-2] and [G-Z-2]. We shall then state a variant of Theorem 0.2 also formulated in [G-Z-2] and outline its surprising proof by Berndtsson and Lempert [B-L].

The point of [G-Z-2] is that, for any pseudoconvex family $\{D_\zeta\}$ of domains parametrized by the unit disc $\mathbb{D} = \{\zeta \in \mathbb{C}; |\zeta| < 1\}$, the inequality (0.1) can be regarded as a sub-mean value property so that it implies in particular the plurisubharmonicity of $\log K_{D_\zeta}(z, z)$ in (ζ, z). (It was really surprising as well as [Bl-1], at least to the author.)

The point of [B-L] is that the L^2 extension can be deduced from a theorem of Berndtsson [B-2] on the semipositivity of the direct image of the relative canonical sheaf twisted (= multiplied) by any semipositive line bundle (see Theorem 5.1 and the remark after that). It was even more unexpected than [G-Z-2] because this semipositivity of the direct image is a generalization of the plurisubharmonicity of $\log K_{D_\zeta}(z, z)$. Nevertheless, it is also natural and regitimate, because although Berndtsson's argument is based on an L^2 estimate for $\bar{\partial}$, his proof does not rely on the L^2 extension theorem itself. We recall also that, for any proper holomorphic map with smooth fibers between Kähler manifolds, Berndtsson's result can be verified by a direct comuputation in the non-twisted case by using the expression of the fiber metric in question in terms of the primitive forms (cf. [F]). In the context of algebraic geometry, this was on the way of a natural development of the Hodge theory (cf. [Gri-1, 3]). The new analytic approach of Berndtsson was inspired by a pioneering work of Maitani and Yamaguchi [M-Y] which arose from a totally different background. This connection seems to suggest the width of the scope of several complex variables as was anticipated by Oka and Cartan.

After an overview of these remarkable achievements, a not so surprising but quite concise proof of Theorem 0.2 will also be presented after [Oh-6]. Finally, we would like to add a few remarks and questions related to the L^2 $\bar{\partial}$-cohomology into which this new attractive context might bring some new insight.

The author is very grateful to the referee who read the manuscript carefully and gave valuable comments.

1 Extension problem as a BVP

Let M be a connected and paracompact complex manifold of dimension n and let $E \to M$ be a holomorphic vector bundle. Given any closed subset $A \subset M$, we consider a C^∞ section $s : A \to E$ such that s extends to a C^∞ section of E over M, say $\hat{s}$, such that $\bar{\partial}\hat{s} = 0$ holds at every point of A. We want to extend such s holomorphically to M as a section of E. For that, we try to solve the equation

$$\bar{\partial}\alpha = \bar{\partial}\hat{s}$$

with the boundary condition

$$\alpha|_A = 0.$$

If such α exists, then $\hat{s} - \alpha$ will be a desired holomorphic extension of s.

This BVP is easily transferred to a problem of solving the $\bar{\partial}$-equation with an integrability condition. Namely, for any locally integrable function $\varphi : M \to [-\infty, \infty)$ such that $e^{-\varphi}$ is not integrable around any point of A, it follows from a basic property of holomorphic functions that the extension problem is solvable if one can find a weak solution to $\bar{\partial}\alpha = \bar{\partial}\hat{s}$ such that $\alpha e^{-\varphi/2}$ is locally square integrable.

To let this reformulation work, one needs a situation where (M, E, φ) satisfies a certain curvature assumption. The point is that, given a Hermitian metric (= a positive (1,1)-form) ω on M and a fiber metric h of E, the question is to solve the equation $\bar{\partial}\alpha = v$ for a given E-valued differential form v with conditions $\bar{\partial}v = 0$ and

$$\int_M e^{-\varphi}|v|^2\omega^n < \infty$$

in such a way that a solution α satisfies

$$\int_M e^{-\varphi}|\alpha|^2\omega^n \leq C\int_M e^{-\varphi}|v|^2\omega^n$$

for some $C = C(\varphi, \omega, h) > 0$. A guiding principle is that such a solution exists under a reasonable curvature condition. In the following two sections, we shall briefly recall this strategy by restricting ourselves to the case where M is a pseudoconvex manifold, ω is a Kähler metric, (E, h) is Nakano semipositive, and A is a complex submanifold.

2 Nakano's inequality

Let (M, ω) be a Kähler manifold of dimension n and let (E, h) be a Hermitian holomorphic vector bundle over M. We shall denote by $C^{p,q}(M, E)$ the space of E-valued C^∞ (p, q)-forms on M. Recall that an E-valued (p, q)-form is a section of the bundle

$$\bigwedge^p T_M^* \otimes \bigwedge^q \overline{T_M^*} \otimes E \longrightarrow M,$$

where T_M denotes the holomorphic tangent bundle of M, T_M^* the dual of T_M, and $\overline{T_M^*}$ its conjugate. By $C_0^{p,q}(M, E)$ we denote the pre-Hilbert space of compactly supported elements of $C^{p,q}(M,E)$ whose inner product is with respect to ω and h. Denoting by $\langle u,v\rangle$ the pointwise inner product of u and v, the inner product of u and v in $C_0^{p,q}(M,E)$ is defined as

$$(u,v)=\int_M \langle u,v\rangle \frac{\omega^n}{n!}.$$

We put $\|u\|^2(=\|u\|_h^2)=(u,u)$. We define a map $\Lambda : C^{p,q}(M,E) \longrightarrow C^{p-1,q-1}(M,E)$ as the pointwise adjoint of the exterior multiplication by ω, i.e. by $\langle \omega\wedge u, v\rangle = \langle u, \Lambda v\rangle$.

Let $L_{(2)}^{p,q}(M,E)$ be the completion of $C_0^{p,q}(M,E)$, i.e. the Hilbert space of E-valued L^2 (p,q)-forms on M with respect to ω and h. By taking the closure of the graph of $\bar\partial : C_0^{p,q}(M,E) \to C_0^{p,q+1}(M,E)$ in $L_{(2)}^{p,q}(M,E)\times L_{(2)}^{p,q+1}(M,E)$, a closed operator

$$\bar\partial : L_{(2)}^{p,q}(M,E) \to L_{(2)}^{p,q+1}(M,E)$$

is defined.

We put

$$H^{p,q}(M,E)=\frac{\operatorname{Ker}(\bar\partial : C^{p,q}(M,E)\to C^{p,q+1}(M,E))}{\operatorname{Im}(\bar\partial : C^{p,q-1}(M,E)\to C^{p,q}(M,E))},$$

$$H_0^{p,q}(M,E)=\frac{\operatorname{Ker}(\bar\partial : C_0^{p,q}(M,E)\to C_0^{p,q+1}(M,E))}{\operatorname{Im}(\bar\partial : C_0^{p,q-1}(M,E)\to C_0^{p,q}(M,E))},$$

and

$$H_{(2)}^{p,q}(M,E)=\frac{\operatorname{Ker}(\bar\partial : L_{(2)}^{p,q}(M,E)\to L_{(2)}^{p,q+1}(M,E))}{\operatorname{Im}(\bar\partial : L_{(2)}^{p,q-1}(M,E)\to L_{(2)}^{p,q}(M,E))}.$$

$H_{(2)}^{p,q}(M,E)$ is called the E-valued L^2 $\bar\partial$-cohomology group of M of type (p,q). Solvability of the $\bar\partial$-equation with L^2 norm estimates is interpreted as the vanishing of the L^2 $\bar\partial$ cohomology groups. This vanishing is a surjectivity assertion for $\bar\partial$, which is equivalent to the injectivity of the adjoint of $\bar\partial$ in some circumstance. In this context, an L^2 estimate naturally arises.

The adjoint of $\bar\partial$ as well as the formal adjoint of $\bar\partial$ will be denoted by $\bar\partial_h^*$. We put

$$\Box_h = \bar\partial\bar\partial_h^* + \bar\partial_h^*\bar\partial.$$

Let $D_h=\bar\partial+\partial_h$ be the Chern connection of (E,h). We put

$$\overline{\Box}_h = \partial_h\partial_h^* + \partial_h^*\partial_h,$$

where ∂_h^* denotes the formal adjoint of ∂_h (which actually does not depend on h). Let D_h^* be the formal adjoint of D_h. Then

$$D_hD_h^* + D_h^*D_h = \Box_h + \overline{\Box}_h \tag{2.1}$$

and

$$i(D_h^2\Lambda - \Lambda D_h^2) = \Box_h - \overline{\Box}_h \tag{2.2}$$

hold since $d\omega = 0$. Let us put $D_h^2 = \Theta_h (\in C^{1,1}(M, E^* \otimes E))$ for simplicity. Then, by integration by parts one deduces from (2.2) that

$$(i\Theta_h \Lambda u, u) \leq \|\bar{\partial} u\|^2 + \|\bar{\partial}_h^* u\|^2 \tag{2.3}$$

holds for any $u \in C_0^{n,q}(M, E)$ (cf. [N]). If $q = 1$, $\langle i\Theta_h \Lambda u, u\rangle$ is naturally identified with a Hermitian form on $E \otimes T_M$ up to the isomorphism $T_M \cong \overline{T_M^*}$ induced by ω.

3 Curvature and extension

Let the notation be as above. For any point $x \in M$, let $\gamma_h(x)$ be the least eigenvalue of $\langle i\Theta_h \Lambda u, u\rangle$ $(q = 1)$. γ_h is then a continuous real-valued function on M. (E, h) and the curvature form Θ_h are said to be Nakano positive (resp. Nakano semipositive) if γ_h is everywhere positive (resp. semipositive) on M. In other words, (E, h) is Nakano (semi-)positive if, for any point $x_0 \in M$ there is a neighborhood U with local coordinate $z = (z_1, \ldots, z_n)$ around x_0 and a coordinate $w = (w_1, \ldots, w_r)$ on the fibers of $E|_U$ coming from a holomorphic trivialization such that

1) over U the fiber metric is $\sum_{\mu,\nu} h_{\mu\bar{\nu}}(z) w_\mu \overline{w_\nu}$,
2) the matrix $(h_{\mu\bar{\nu}}(x_0))$ is the unit matrix,
3) the total derivative $dh_{\mu\bar{\nu}}(x_0) = 0$,

and

4) the Hermitian form $\sum (\partial^2 h_{\mu\bar{\nu}} / \partial z_\alpha \partial \overline{z_\beta})(x_0) \gamma_{\mu\alpha} \overline{\gamma_{\nu\beta}}$ is positive (semi-)definite.

Nakano will not be referred to if the rank of E is one.

Let us recall that one can deduce the following from (2.3).

Theorem 3.1 (cf. [Dm] and [Oh-1,2]). *Let (M, ω) be a Kähler manifold of dimension n and let (E, h) be a Hermitian holomorphic vector bundle over M. Assume that M admits a complete Kähler metric and (E, h) is Nakano positive. Then, for any locally square integrable E-valued (n, q)-form v on M satisfying $\bar{\partial} v = 0$ and*

$$((i\Theta_h \Lambda)^{-1} v, v) \left(= \int_M \langle (i\Theta_h \Lambda)^{-1} v, v\rangle \omega^n \right) < \infty,$$

one can find $\alpha \in L_{(2)}^{n,q-1}(M, E)$ satisfying $\bar{\partial}\alpha = v$ and $\|\alpha\|^2 \leq ((i\Theta_h \Lambda)^{-1} v, v)$.

If M is a Stein manifold, it admits a C^∞ strictly plurisubharmonic exhaustion function, say Φ. Then M admits a complete Kähler metric $i\partial\bar{\partial} e^\Phi$. Moreover, for any Hermitian holomorphic vector bundle (E, h), the curvature form of $he^{-\lambda \circ \Phi}$ becomes Nakano positive for some convex increasing function λ, since

$$\Theta_{he^{-\lambda \circ \Phi}} = \Theta_h + i\partial\bar{\partial}(\lambda \circ \Phi) Id_E.$$

Hence Theorem 3.1 implies the following.

Corollary 1. *For any holomorphic vector bundle E over a Stein manifold M, $H^{p,q}(M,E) = 0$ holds for any $q \in \mathbb{N}$.*

Before proceding to the sharp L^2 extension theorems, let us recall how one can apply Theorem 3.1 to prove an extension theorem without any (or with a loose) growth condition. The point is that any C^∞ form becomes square integrable with respect to a modified fiber metric of the form $he^{-\Phi}$ for some exhaustion function Φ on M.

Let $A \subset M$ be a (regularly embedded) closed complex submanifold. We consider a class of continuous functions $\varphi : M \to [-\infty, 0)$ such that

1) $\varphi^{-1}(-\infty) = A$,
2) $e^{-\varphi}$ is not locally integrable around any point of A,
3) for any open set $U \subset M$ and for any C^∞ function v on U satisfying $v|_{U\cap A} = 0$, $|v|^2 e^{-\varphi}$ is locally integrable on U,

and

4) $\varphi|_{M\setminus A}$ is C^∞.

Let us denote by Π_A the set of such functions φ. Clearly $\Pi_A \neq \varnothing$ for any A. We shall state an extension theorem on a pseudoconvex manifold in terms of the curvature of the modified fiber metric $he^{-\varphi}$. Here M is called pseudoconvex if it admits a C^∞ plurisubharmonic exhaustion function. For simplicity, an extendability criterion is stated with respect to the bundle $\omega_M \otimes E$ instead of E. Here ω_M denotes the canonical bundle of M.

Theorem 3.2. *Let (M, ω) be a pseudoconvex Kähler manifold of dimension n, let (E, h) be a Hermitian vector bundle over M, and let $A \subset M$ be a closed complex submanifold. Assume that there exists $\varphi \in \Pi_A$ and a positive continuous function c on M such that $\gamma_{he^{-\varphi}} \geq c$ holds on $M \setminus A$. Then every holomorphic section of $\omega_M \otimes E$ over A is holomorphically extendable to M as a section of $\omega_M \otimes E$.*

Proof. Let $s : A \to \omega_M \otimes E$ be any holomorphic section and let $\hat{s} : M \to \omega_M \otimes E$ be a C^∞ section such that $\bar\partial \hat{s} = 0$ holds at every point of A. Let $\Phi : M \to [0, \infty)$ be a C^∞ plurisubharmonic exhaustion function such that

$$\int_M c^{-1} e^{-\Phi-\varphi} |\bar\partial \hat{s}|^2 \omega^n < \infty.$$

Here the length of $\bar\partial \hat{s}$ is measured as an E-valued $(n,1)$-form with respect to ω and h. Since $M \setminus A$ admits a complete Kähler metric, that $\gamma_{he^{-\varphi}} \geq c$ holds on $M \setminus A$ implies that one can find a locally L^2 section α of $K_M \otimes E|_{M\setminus A}$ such that $\bar\partial \alpha = \bar\partial \hat{s}|_{M\setminus A}$ and

$$\int_{M\setminus A} e^{-\Phi-\varphi} |\alpha|^2 \omega^n \leq \int_{M\setminus A} c^{-1} e^{-\Phi-\varphi} |\bar\partial \hat{s}|^2 \omega^n$$

is satisfied. Here Φ is replaced if necessary, by $\lambda \circ \Phi$ for a sufficiently rapidly increasing convex function λ. Clearly $\hat{s} - \alpha$ is holomorphically extendable to M as a section of $\omega_M \otimes E$ which extends s. □

Corollary 2. *For any holomorphic vector bundle E over a Stein manifold M and for any closed complex submanifold $A \subset M$, every holomorphic global section of $E|_A$ is holomorphically extendable to that of E.*

A step of refining Theorem 3.2 towards Theorem 0.1 was made in [Oh-1] when M is a complete Kähler domain with C^1-smooth boundary and A is a hyperplane section. The method employed in [Oh-1] was improved in [Oh-T] and [Oh-5], and became optimal in [Bl-1] and [G-Z-1,2]. There was a question on the Bergman kernel which gave a strong impetus to this development. We shall review it in the next section.

4 Suita conjecture and Błocki's theorem

Let D be a pseudoconvex domain in $\mathbb{C}^n$, let $A^2(D)$ be the Hilbert space of L^2 holomorphic functions on D, and let $K_D(z,w)$ be the Bergman kernel function on D. Recall that $K_D(z,w)$ is the reproducing kernel of the space $A^2(D)$, i.e. a function on $D \times D$ defined by the properties:

1) $K_D(\cdot, w) \in A^2(D)$ for any $w \in D$,
2) $K_D(z,w) = \overline{K_D(w,z)}$,

and

3) $f(z) = \int_D K_D(z,w)\overline{f(w)}\omega_w^n/n!$ for any $f \in A^2(D)$,

where

$$\omega_w = \frac{i}{2}\sum_{j=1}^{n} dw_j \wedge d\overline{w_j}.$$

Recall also that $K_D(z,w)$ has a series representation

$$K_D(z,w) = \sum f_j(z)\overline{f_j(w)},$$

where $\{f_j\}$ is a complete orthonormal system of $A^2(D)$.

In 1967, Carleson [Cl] observed a relation between $K_D(z) := K_D(z,z)$ and a function $c_{\beta,D} : D \to [0,\infty)$ defined by $c_{\beta,D} = e^{\gamma_D}$, where

$$\gamma_D(z) = \varlimsup_{w \to z} \sup\{u(w) - \log\|z-w\|; u \in PSH_z^-(D)\}$$

and $PSH_z^-(D)$ denotes the set of negative plurisubharmonic functions u on D satisfying $u(\cdot) - \log\|\cdot - z\| \in L^\infty_{loc}(D)$. In the case $n = 1$, Carleson proved that $K_D(z) \equiv 0$ holds if and only if $c_{\beta,D}(z) \equiv 0$. In [S-O], Sario and Oikawa raised a question to compare K_D and $c_{\beta,D}$. Suita [Su] answered it by showing that, if $D = \{z \in \mathbb{C}; r < |z| < 1\}$ one has $\pi K_D \geq c_{\beta,D}^2$, where the inequality is strict if and only if $0 < r < 1$. It became therefore natural to ask for a generalized variant of the inequality. Suita conjectured that it holds for any Riemann surface and, with respect to the strictness, it holds if and only if the surface is not conformally equivalent to a subdomain of the unit disc whose complement is contained in the preimage of $-\infty$ for some nonconstant subharmonic function.

That the inequality $\pi K_D \geq c_{\beta,D}^2$ holds for any domain $D \subset \mathbb{C}$ was first proved by Błocki [Bl-1] by improving the method of [G-Z-Z] to find appropriate weights and "twists" by solving an ordinary differential equation (=ODE). In his proof, the following L^2 estimate inspired by B.Y. Chen's work [Ch] is crucial.

Theorem 4.1. *Assume that v is a $\bar{\partial}$-closed $(0,1)$-form with L^2_{loc} coefficients on a pseudoconvex domain $D \subset \mathbb{C}^n$. Let φ be plurisubharmonic on D and let ψ be in $W^{1,2}_{loc}$ on D, locally bounded from above, such that $|\bar{\partial}\psi|^2_{i\partial\bar{\partial}\varphi} \leq 1$ on D and $|\bar{\partial}\psi|^2_{i\partial\bar{\partial}\varphi} \leq \delta$ on* supp$v$. *Then there exists $u \in L^2_{loc}(D)$ satisfying $\bar{\partial}u = v$ and*

$$\int_D |u|^2(1 - |\bar{\partial}\psi|^2_{i\partial\bar{\partial}\varphi})e^{2\psi-\varphi} \leq \frac{1+\sqrt{\delta}}{1-\sqrt{\delta}} \int_D |v|^2_{i\partial\bar{\partial}\varphi} e^{2\psi-\varphi}. \tag{4.1}$$

Here $|v|_{i\partial\bar{\partial}\varphi}$ denotes the length of v with respect to $i\partial\bar{\partial}\varphi$ when φ is strictly plurisubharmonic. For the general case, see [Bl-1].

The estimate (4.1) works after it is combined with the solution of a system of ODE to prove a sharp L^2 extension theorem (i.e. Theorem 0.1 for $C = \pi$) that implies the Suita-Błocki inequality (=SBI) $\pi K_D \geq c_{\beta,D}^2$. Later, Guan and Zhou [G-Z-1,2] proved its generalized variants as Theorem 0.2 based on the estimate in [Oh-5] and [Oh-T]. In the crucial step of [G-Z-1,2], the same system of ODE arises naturally. However, we shall not go into the detail of these arguments here, because we prefer to present a simpler proof of Theorem 0.2 later.

Instead of Błocki's first proof of SBI, let us review his another approach in [Bl-2] to Suita conjecture. It is an accurate estimate for K_D implying SBI;

Theorem 4.2. *For any pseudoconvex domain $D \subset \mathbb{C}^n$, for any $z \in D$ and for any $a \geq 0$, one has*

$$K_D(z) \geq \frac{1}{e^{2na}\mathrm{volume}(\{w; G_{D,z}(w) < -a\})},$$

where

$$G_{D,z}(w) = \sup\{u(w); u \in PSH_z^-(D)\}.$$

Proof. Let $G = G_{D,w}$ and

$$\chi(t) = \begin{cases} 0 & t \geq -a \\ \int_a^{-t} s^{-1}e^{-ns}ds & t < -a. \end{cases}$$

Then, solving the $\bar{\partial}$-equation $\bar{\partial}\alpha = \bar{\partial}(\chi \circ G)$ with $\alpha \in L^2(D, e^{-2nG})$ in such a way that

$$\int_D |\alpha|^2 \leq C\mathrm{volume}(\{G < -a\})$$

where C is an absolute constant, one has a holomorphic function $f := \chi \circ G - \alpha$ satisfying

$$f(w) = \chi(-\infty) = \int_{na}^{\infty} s^{-1}e^{-s}ds =: \mathcal{E}(na).$$

Since

$$\|f\| \leq \|\chi \circ G\| + \|\alpha\| \leq (\chi(-\infty) + \sqrt{C})\sqrt{\text{volume}(\{G < -a\})},$$

the estimate

$$K_D(w) \geq \frac{|f(w)|^2}{\|f\|^2} \geq \frac{c(n,a)}{\text{volume}(\{G < -a\})} \tag{4.2}$$

holds, where

$$c(n,a) = \frac{\mathcal{E}(na)^2}{(\mathcal{E}(na) + \sqrt{C})^2}.$$

For any $m \in \mathbb{N}$, let $\tilde{D} = D^m \subset \mathbb{C}^{nm}$ and $\tilde{w} = (w, \ldots, w) \in \tilde{D}$. Then $K_{\tilde{D}}(\tilde{w}) = (K_D(w))^m$. Since $G_{\tilde{D},\tilde{w}}(z^1, \ldots, z^m) = \max_{1 \leq j \leq m} G(z^j)$, one has

$$\text{volume}(\{G_{\tilde{D},\tilde{w}} < -a\}) = \text{volume}(\{G < -a\})^m.$$

Hence, by (4.2) applied for $\tilde{D}$,

$$(K_D(w))^m \geq \frac{c(nm,a)}{(\text{volume}(\{G < -a\}))^m}.$$

The desired inequality follows from the fact that

$$\lim_{m \to \infty} c(nm,a)^{1/m} = e^{-2na}. \qquad \square$$

It has to be noted that the above method is a natural consequence of a series of studies since Bergman conjectured in 1934 that the Bergman kernel of a bounded pseudoconvex domain D in $\mathbb{C}^2$ with C^2-smooth boundary grows near the boundary at least as $dist(z, \partial D)^{-2}$ (cf. [Hm], [P], [Oh-3], [Bl-P] and [Hb]). A remarkable point is that it was Błocki's this alternate approach to SBI which suggested a completely new proof of a sharp L^2 extension theorem by Berndtsson and Lempert [B-L]. Let us review this topic together with an application of Theorem 0.2 in [G-Z-2] to the variation of the Bergman kernels. These two results seem to suggest a new perspective in complex analysis.

5 Applying L^2 extension to variation of Bergman kernels

Let $D \subset \mathbb{C}^{n+1}$ be a pseudoconvex domain and let $D_\zeta = \{z \in \mathbb{C}^n; (z,\zeta) \in D\}$, $\zeta \in \mathbb{C}$. For any $\varphi \in PSH(D)$, let

$$A^2_\varphi(D_\zeta) = \{f \in L^2_{loc}(D_\zeta); \int_{D_\zeta} e^{-\varphi}|f|^2 < \infty, \bar{\partial} f = 0\},$$

let $K_{D_\zeta,\varphi}(z,w)$ be the reproducing kernel of $A^2_\varphi(D_\zeta)$, and let $K_{D_\zeta,\varphi}(z) = K_{D_\zeta,\varphi}(z,z)$. In [B-2] the following was proved.

Theorem 5.1. $\log K_{D_\zeta,\varphi}(z) \in PSH(D)$.

Outline of the proof: It is not difficult to see that it suffices to prove the assertion when $D = \{(z,\zeta) \in \Omega \times \mathbb{D}; \psi(z,\zeta) < 0\}$ for some pseudoconvex domain $\Omega \subset \mathbb{C}^n$ and a C^∞ plurisubharmonic function ψ on a neighborhood of $\overline{\Omega \times \mathbb{D}}$ in $\mathbb{C}^{n+1}$, and when φ is a C^∞ strictly plurisubharmonic function on a neighborhood of $\overline{\Omega \times \mathbb{D}}$. Indeed, the plurisubharmonicity of $\log K_{D_\zeta,\varphi}(z)$ will then follow from that of $\log K_{\Omega,\varphi+t\psi}$ by letting $t \to \infty$ and then by applying a standard limiting argument. To prove this simplified variant, a curvature formula of Griffiths (cf. [Gri-2]) is applied to the pair of the trivial bundle $E = L^2(\Omega) \times \mathbb{D}$ equipped with fiber metrics $\int_\Omega e^{-\varphi(\cdot,\zeta)}|f|^2$ and its subbundle $S = A^2(\Omega) \times \mathbb{D}$ with the induced metric say h. The curvature formula for the exact sequence

$$0 \to S \hookrightarrow E \to E/S \to 0$$

yields a formula for Θ_h, which turns out to be Nakano semipositive when it is estimated from below by using (2.3). Since (S,h) is Nakano semipositive and $\dim \mathbb{D} = 1$, its dual bundle (S^*,h^*) is Nakano seminegative. Therefore, for any holomorphic section $\xi : \mathbb{D} \to S^*$, $\log\|\xi\|$ is subharmonic. Recall here that subbundles of negative bundles are (more) negative. If $\xi(\zeta)(f) = f(z)$ for all $\zeta \in \mathbb{D}$, one has $\|\xi\|^2 = K_{\Omega,\varphi}(z)$, hence

$$\frac{\partial^2 \log K_{\Omega,\varphi}(z)}{\partial\zeta\partial\bar\zeta} \geq 0,$$

so that

$$\frac{\partial^2 \log K_{D_\zeta,\varphi}(z)}{\partial\zeta\partial\bar\zeta} \geq 0$$

as mentioned above. Applying this result to the families $\{D_\zeta + \zeta v\}$ for $v \in \mathbb{C}^n$, the desired conclusion is obtained. □

We recall that Theorem 5.1 was first proved by Maitani and Yamaguchi [M-Y] for the special case $\varphi = 0$ when $n = 1$. For the proof they exploited the formula

$$\frac{\partial^2 \gamma_{D_\zeta}(z)}{\partial z\partial\bar z} = \pi K_{D_\zeta}(z)$$

due to Suita [Su]. Before [M-Y], Yamaguchi [Y] had noticed that $\gamma_{D_\zeta} \in PSH(D)$ follows from Suita's formula and applied it to answer a question of Nishino on pseudoconvex families of open Riemann surfaces, whether or not the fibers are parabolic if they are so outside a set of logarithmic capacity 0. Nishino raised this question in the study of classifying entire functions of several variables (cf. [Ni]).

In [B-2], Theorem 5.1 is extended for a proper and smooth Kähler morphism. Namely, given a Kähler manifold M, a Nakano semipositive vector bundle $(E,h) \to$

M and a proper and surjective holomorphic map $\pi : M \to S$ which is smooth, i.e. everywhere of maximal rank, the direct image $\pi_*(\omega_{M/S} \otimes E)$, where $\omega_{M/S} = \omega_M \otimes \pi^*\omega_S^{-1}$, equipped with the natural L^2 metric is Nakano semipositive. This result had been known by T. Fujita [F] for the trivial (E, h) as a consequence of Griffiths's theory of the variation of Hodge structures (cf. [Gri-1,3]). As for the applications of the semipositivity of $\pi_*(\omega_{M/S} \otimes E)$, see [E-V] and [L]. Combining the method of [B-2] with a torsion freeness theorem for the higher direct images of $\omega_{M/S} \otimes E$ (cf. [T]), the result was further extended to the semipositivity of $R^q\pi_*(\omega_{M/S} \otimes E)$ in [M-T]. Explicit formulas involving the Kodaira-Spencer class are presented in [B-3] and [L-Y].

In [G-Z-2], Guan and Zhou showed that Theorem 5.1 is a corollary of Theorem 0.2. Their argument is completely straightforward as follows.

Proof of Theorem 5.1 by Theorem 0.2. To make notation simpler, we set

$$k(z, \zeta) = K_{D_\zeta, \varphi}(z), \quad z \in D_\zeta.$$

It suffices to show that, for any fixed $z_0 \in D_0$, the average of $\log k(z_0, \zeta)$ over $|\zeta| < r$ is not less than $\log k(z_0, 0)$. For that, let us put

$$f(z) = \frac{K_{D_0,\varphi}(z, z_0)}{\sqrt{k(z_0, 0)}}, \quad z \in D_0.$$

By Theorem 0.2, for any $0 < r \leq 1$ there exists a holomorphic function $f_r(z, \zeta)$ on $\{(z, \zeta) \in D; |\zeta| < r\}$ such that $f_r|_{D_0} = f$ and $\|f_r\|^2 \leq \pi r^2$. Clearly

$$k(z, \zeta) \geq \frac{|f_r(z, \zeta)|^2}{\|f_r(\cdot, \zeta)\|^2} \tag{5.1}$$

if $|\zeta| < r$. Rewriting the inequality $\|f_r\|^2 \leq \pi r^2$ as

$$\frac{1}{\pi r^2} \int_{|\zeta|<r} \|f_r(\cdot, \zeta)\|^2 \leq 1$$

and deducing from it

$$\exp \frac{1}{\pi r^2} \int_{|\zeta|<r} (2 \log |f_r(z_0, \zeta)| - \log k(z_0, \zeta)) \leq 1$$

by combining the convexity of e^x and (5.1), we arrive at the desired sub-mean value property

$$\log k(z_0, 0) = |f_r(z_0, 0)|^2 \leq \frac{1}{\pi r^2} \int_{|\zeta|<r} \log k(z_0, \zeta)$$

from the subharmonicity of $\log |f_r(z_0, \zeta)|$. $\square$

In [Ca], J. Cao (after a suggestion by H. Tsuji) applied the above argument of Guan and Zhou to establish a generalization of a theorem of Berndtsson and Păun in [B-P] for projective morphisms, a positivity result of the curvature current of twisted pluricanonical bundles, to Kähler morphisms. It includes the following as a special case.

Theorem 5.2. *Let X be a connected Kähler manifold, let $\pi : X \to Y$ be a proper and smooth holomorphic map, let $L \to X$ be a semipositive line bundle and let $X_y = \pi^{-1}(y)$. Then*

i) the restriction maps

$$R^0\pi_*(\omega_{X/Y} \otimes L) \to H^{n,0}(X_y, L) \qquad (n = \dim X_y)$$

are surjective

and

ii) for any holomorphic section ξ of $Hom(R^0\pi_*(\omega_{X/Y} \otimes L), \mathcal{O}_Y)$, the norm $\|\xi\|_y$ of the induced maps $\xi_y : H^{n,0}_{(2)}(X_y, L) \to \mathbb{C}$ satisfies $\log \|\xi_y\| \in PSH(Y)$.

It must be noted that the assertion i) was first obtained in [B-2] by refining the method of applyiing the Siu $\partial\bar{\partial}$-Bochner formula in [B-1]. In [Ca], i) is a part of a generalization of Theorem 0.2 to a sharp L^2 extension theorem on pseudoconvex Kähler manifolds, whose proof is a natural refinement (with the optimization required for the proof of ii)) of an extension theorem of L.Yi [Yi] on compact Kähler manifolds.

In the next section, we shall review the idea of [B-L] to deduce a sharp L^2 extension theorem from the subharmonicity theorem of the above type.

6 Convexity implies a sharp extension

Let us present an optimized variant of Theorem 0.1 following [B-L]. It will include Theorem 0.2 with a slight restriction that $\varphi - \psi$ is plurisubharmonic.

Let D be a pseudoconvex domain in $\mathbb{C}^n$ and let V be a closed complex submanifold of codimension k in D. Let $\varphi \in PSH(D)$ and

$$A^2_\varphi(V) = \{f \in \mathcal{O}(V); \int_V |f|^2 e^{-\varphi} < \infty\},$$

where the integration is with respect to the volume element induced by the euclidean metric. Let $d_V(z)$ be the distance from z to V and let G be a negative plurisubharmonic function on D which satisfies

$$G(z) \leq \log d_V(z)^2 + C$$

and

$$G(z) \geq \log d_V(z)^2 + B(z)$$

as z goes to V. Here C is some constant and B is a continuous function on D.

Theorem 6.1. *In the above situation, every* $f \in A^2_\varphi(D)$ *has an extension* $\tilde{f} \in \mathcal{O}(D)$ *such that*

$$\int_D |\tilde{f}|^2 e^{-\varphi} \leq \sigma_k \int_V |f|^2 e^{-\varphi - kB},$$

where σ_k *is the volume of the unit ball in* $\mathbb{C}^k$.

Outline of the proof: We may assume that V extends to a neighborhood of $\bar{D}$, that D is a bounded strongly pseudoconvex domain and that B is continuous on $\bar{D}$. We may also assume that f has some extension $\hat{f}$ in $A^2_\varphi(D)$. Let $\rho : A^2_\varphi(D) \to \mathcal{O}(V)$ be the restriction map and let $I_V = \mathrm{Ker}\rho$. In order to prove the assertion, we need to estimate the projection of $\hat{f}$ to the orthogonal complement of I_V, say $\tilde{f}$. For that, the equality

$$\|\tilde{f}\| = \sup |(\xi, f)| / \|\xi\| \tag{6.1}$$

is applied. Here the supremum is taken over all $\xi \neq 0$ in the dual of $A^2_\varphi(D)$ that vanish on I_V.

For any $g \in C^\infty_0(V)$, we define an element ξ_g of $A^2_\varphi(D)^*$ by

$$(\xi_g, f) = \sigma_k \int_V f\bar{g} e^{-\varphi - kB}.$$

Clearly, $f|_V = 0$ if $(\xi_g, f) = 0$ for all $g \in C^\infty_0(V)$. Let $\tilde{g}$ denote the orthogonal projection of g to $A^2_\varphi(V)$ with respect to the norm $\|\cdot\|_V$ defined by

$$\|f\|^2_V = \sigma_k \int_V |f|^2 e^{\varphi - kB}.$$

Then, by (6.1) and Cauchy-Schwarz inequality

$$\|\tilde{f}\| \leq \sup_g \|f\|_V \|\tilde{g}\|_V / \|\xi_g\|.$$

Thus, it suffices to prove that

$$\|\tilde{g}\|_V \leq \|\xi_g\|. \tag{6.2}$$

To prove (6.2), we consider a family

$$A^2_{t,p} := \{h \in \mathcal{O}; \|h\|^2_{t,p} := \int_D |h|^2 e^{-\varphi - p\psi(t,\cdot)} < \infty\}$$

for $p > 0$ and $t \leq 0$, where we put

$$\psi(\zeta, z) := \max(G(z) - \mathrm{Re}\zeta, 0).$$

Since $\psi(0, z) = 0$,

$$\|\xi_g\| = \|\xi_g\|_0 = \lim_{p \to \infty} \|\xi_g\|_0.$$

Here $\|\xi_g\|_t$ are defined for $t \leq 0$ as the norm of ξ_g in the dual of $A^2_{t,p}$. Then, $\log \|\xi\|_{\mathrm{Re}\zeta}$ is subharmonic in ζ as was noted in the proof of Theorem 5.1, so that $\log \|\xi_g\|_t^2 + kt$ is convex in t. Since $\|\xi_g\|_t^2 e^{kt}$ is bounded by our setting, it follows that $\|\xi_g\|_t^2 e^{kt}$ is an increasing function. Therefore

$$\|\xi_g\|_0^2 \geq \lim_{t\to-\infty} \|\xi_g\|_t^2 e^{kt} \quad \text{for any } p > 0. \tag{6.3}$$

On the other hand, it is straightforward to see that

$$\lim_{p\to\infty} \lim_{t\to-\infty} \|\xi_g\|_t^2 e^{kt} \geq \|\tilde{g}\|_V^2, \tag{6.4}$$

although it requires some computation.

Combining (6.3) with (6.4) we obtain (6.2). □

The above argument is a generalization of an alternate proof of Suita conjecture proposed by Lempert (cf. [Bl-3]). Let us just have a glimpse of Lempert's idea in an oversimplified situation that $D = \mathbb{D}$.

Illustration of Lempert's idea: It suffices to show that $\pi k_{\mathbb{D}}(0) \geq 1$, or equivalently that $\log(\pi k_{\mathbb{D}}(0)) \geq 0$. For that, we put

$$\mathbb{D}_\zeta = \{z; |z| < |\zeta|\} \quad \text{for} \quad |\zeta| < 1.$$

Then, by Theorem 5.1, $\log k_{\mathbb{D}_\zeta}(0)$ is subharmonic in $\log \zeta$. Since $\log k_{\mathbb{D}_\zeta}(0)$ depends only on $\log |\zeta|$, it is convex in $\log |\zeta|$. Therefore, the function

$$-\log \pi k_{\mathbb{D}}(0) + \log \pi k_{\mathbb{D}_\zeta}(0) + \log |\zeta|^2$$

turns out to be increasing in $\log |\zeta|$, since it is convex and bounded. Hence

$$\begin{aligned}
&\lim_{\zeta\to 0}(-\log \pi k_{\mathbb{D}}(0) + \log \pi k_{\mathbb{D}_\zeta}(0) + \log |\zeta|^2) \\
&\leq \lim_{\zeta\to 1}(-\log \pi k_{\mathbb{D}}(0) + \log \pi k_{\mathbb{D}_\zeta}(0) + \log |\zeta|^2) = 0.
\end{aligned}$$

Admitting that Suita conjecture is infinitesimally true, i.e. that

$$\lim_{\zeta\to 0} \pi k_{\mathbb{D}_\zeta}(0)|\zeta|^2 = 1,$$

we obtain the desired conclusion. □

The author believes that the reader will accept the above illustration not just as a mockery but as a prototype of the aforementioned whole argument of the proof of a sharp L^2 extension theorem. In the next section, we shall present a proof of Theorem 0.2 following [Oh-6].

7 A proof of Theorem 0.2

It suffices to show that, given any relatively compact pseudoconvex subdomain $\Omega \subset D$, one can find $f^\Omega \in \mathcal{O}(\Omega)$ such that

$$f^\Omega|_{\Omega\cap D'} = f|_{\Omega\cap D'} \tag{7.1}$$

and

$$\int_{\Omega} e^{-\varphi+\psi}|f^{\Omega}|^2 \leq \pi \int_{D'} e^{-\varphi}|f|^2. \tag{7.2}$$

We may assume that φ and ψ are C^{∞}. To find f^{Ω}, it suffices to solve a set of $\bar{\partial}$-equations

$$\bar{\partial} u_{\epsilon} = \tilde{f}\bar{\partial}\chi_{\epsilon}(\psi/2 + \log|z_n|) \wedge dz_1 \wedge \cdots \wedge dz_n \tag{7.3}$$

on $\Omega \setminus D'$ for $0 < \epsilon < 1$, where $\tilde{f}$ is any fixed holomorphic extension of f to D and

$$\chi_{\epsilon}(-t) = \begin{cases} 0 & \text{if} \quad t > \log\epsilon, \\ \log(-t) - \log(-\log\epsilon) & \text{if} \quad \log\epsilon \geq t \geq e\log\epsilon, \\ 1 & \text{if} \quad t < e\log\epsilon. \end{cases}$$

Clearly, if one can find solutions $u_{\epsilon} = f_{\epsilon} dz_1 \wedge \cdots \wedge dz_n$ such that u_{ϵ} are continuously extendible to Ω by letting $u_{\epsilon} = 0$ on $D' \cap \Omega$ and

$$\liminf_{\epsilon \to 0} \int_{\Omega} e^{-\varphi+\psi}|f_{\epsilon}|^2 \leq \pi \int_{D'} e^{-\varphi}|f|^2,$$

then a subsequence of $\tilde{f}\chi_{\epsilon}(\psi/2 + \log|z_n|) - f_{\epsilon}$ will converge to a desired extension f^{Ω}.

To let this strategy work, we first establish a general criterion.

Theorem 7.1. *Let X be a Stein manifold of dimension n, let (E, h) be a Hermitian holomorphic vector bundle over X and let (B, b) be a Hermitian holomorphic line bundle over X. Assume that there exists a holomorphic section σ of B with $S := \sigma^{-1}(0) \neq \varnothing$, $\Theta_h|_{S \setminus \mathrm{Sing} S}$ is Nakano semipositive and there exist positive C^{∞} functions η and c on $X \setminus S$ satisfying*

$$\eta\Theta_h - (\partial\bar{\partial}\eta + c\partial\eta\bar{\partial}\eta) \otimes Id_E \geq 0 \quad on \ X \setminus S \tag{7.4}$$

and

$$\eta(z) + \log|\sigma(z)|^2 \quad \textit{is bounded on a neighborhood of } S. \tag{7.5}$$

Then, for any positive continuous function Q on X satisfying

$$|\sigma|^2(\eta + c^{-1}) \leq Q, \tag{7.6}$$

the following holds.

For any holomorphic section f of $\omega_S \otimes E$ over S, where $\omega_S = (\omega_X \otimes B)|_S$, there exists a holomorphic section $\tilde{f}$ of $\omega_X \otimes B \otimes E$ over X satisfying

$$\tilde{f}|_S = f \wedge d\sigma \tag{7.7}$$

and

$$\|Q^{-\frac{1}{2}}\tilde{f}\|^2 \leq 2\pi\|f\|^2. \tag{7.8}$$

Proof. In view of the Steinness of X and because of the required properties (7.7) and (7.8) on $\tilde{f}$, we may assume in advance that S is nonsingular and σ has only simple zeros. Moreover, it suffices to find for each bounded open set D of X, a holomorphic extension of $f \wedge d\sigma$ to D, say f_D such that $\|Q^{-\frac{1}{2}} f_D\|^2 \leq 2\pi \|f\|^2$.

In this situation, let $\hat{f}$ be any holomorphic extension of $f \wedge d\sigma$ to X, let (η, c) be as in the assumption, and let D be any bounded pseudoconvex open set.

We shall solve a set of $\bar{\partial}$-equations

$$\bar{\partial} u_\epsilon = \hat{f} \bar{\partial} \chi_\epsilon(-\log|\sigma|) \quad \text{on } D \setminus S, \tag{7.9}$$

where $0 < \epsilon < \frac{1}{e}$ and

$$\chi_\epsilon(t) = \begin{cases} 0 & \text{if } t < -\log\epsilon, \\ \log t - \log(-\log\epsilon) & \text{if } -\log\epsilon \leq t \leq -e\log\epsilon, \\ 1 & \text{if } t > -e\log\epsilon. \end{cases}$$

with side conditions of u_ϵ such that u_ϵ are extendable to D continuously in such a way that $u_\epsilon|_{S \cap D} = 0$ and

$$\liminf_{\epsilon \to 0} \|Q^{-\frac{1}{2}} u_\epsilon\|^2 \leq 2\pi \|f\|^2. \tag{7.10}$$

Then a subsequence of $\hat{f}\chi_\epsilon(\log|\sigma|) - u_\epsilon$ will converge to a desired extension. In order to find u_ϵ, we fix a Kähler metric on $D \setminus S$ and apply a general formula

$$(i(\eta\Theta_h - \partial\bar{\partial}\eta - c\partial\eta \wedge \bar{\partial}\eta)\Lambda u, u) \leq \|\sqrt{\eta}\bar{\partial} u\|^2 + \|\sqrt{\eta + c^{-1}}\bar{\partial}^* u\|^2 \tag{7.11}$$

for $u \in C_0^{n,q}(D \setminus S, E)$ (cf. [Oh-T]), by modifying the given η and c to η_ϵ and c_ϵ as follows.

We first put

$$\hat{\eta}_\epsilon(z) = \begin{cases} \eta(z) \quad \text{if } \epsilon > |\sigma(z)|, \\ \min\left\{\eta(z), -\log|\sigma(z)|^2 + \log|\sigma(z)| \log\left(\frac{\log|\sigma(z)|}{\log\epsilon}\right)\right\} \quad \text{if } \epsilon^e \leq |\sigma(z)| \leq \epsilon, \\ -e\log\epsilon \quad \text{if } |\sigma(z)| < \epsilon^e. \end{cases}$$

Since $\Theta_h|_S \geq 0$, $\eta + \log|\sigma|^2$ is bounded near S and X is a Stein manifold, by approximating $\hat{\eta}_\epsilon$ one can find positive $\mathcal{C}^\infty$ functions η_ϵ, c_ϵ and d_ϵ on X such that

$$\eta_\epsilon \Theta_{h \otimes b|\sigma|^{-2}} - \partial\bar{\partial}\eta_\epsilon - c_\epsilon \partial\eta_\epsilon \bar{\partial}\eta_\epsilon \geq 0, \tag{7.12}$$

$$d_\epsilon \to 0 \qquad \text{locally uniformly as } \epsilon \to 0, \tag{7.13}$$

$$\eta_\epsilon \Theta_{h \otimes b|\sigma|^{-2}} - \partial\bar{\partial}\eta_\epsilon - c_\epsilon \partial\eta_\epsilon \bar{\partial}\eta_\epsilon \geq (1 - d_\epsilon)\partial\log|\sigma|\bar{\partial}\log|\sigma|/(-\log|\sigma|) \tag{7.14}$$

if $\epsilon^e < |\sigma(z)| < \epsilon$, and

$$(1 - d_\epsilon)(\eta_\epsilon + c_\epsilon^{-1})|\sigma|^2 \leq Q. \tag{7.15}$$

By (7.12)—(7.15), one has solutions u_ϵ to (7.9) on $D \setminus S$ which are extendable holomorphically along $D \cap S$, in such a way that their restrictions to $D \cap S$ are zero and

$$(1 - d_\epsilon)^{-1} \|(\eta_\epsilon + c_\epsilon^{-1})^{-1/2} u_\epsilon\|^2_{h \otimes b|\sigma|^{-2}} \leq 2\pi \|f\|^2. \tag{7.16}$$

Combining (7.16) with (7.15) one has

$$\liminf_{\epsilon\to 0} \|Q^{-\frac{1}{2}}u_\epsilon\|^2_{h\otimes b} \le 2\pi\|f\|^2,$$

which is the end of the proof. □

To apply Theorem 7.1 for the proof of Theorem 0.2, it suffices to show that one can find η and c for $X = D$ and $\sigma = z_n$ with respect to $(B, b) = (\mathbb{C}\times D, e^{\psi})$ and $(E, h) = (\mathbb{C}\times D, e^{-\varphi-\phi})$, in such a way that (7.6) holds for $Q = e^{-\phi}$. Here φ and ψ are assumed to be C^∞ and plurisubharmonic on D and ϕ is required to satisfy $\phi|_{D'} = 0$. For that, let

$$s(t) = \frac{t}{1-e^{-t}} - 1,\ u(t) = -\log(1-e^{-t}),\ v(t) = \frac{1-e^{-t}}{e^t - 1 - t} \tag{7.17}$$

and put

$$\eta(z) = s\left(\log\frac{1}{|z_n|^2}\right),\ \phi(z) = u\left(\log\frac{1}{|z_n|^2}\right),\ c(z) = v\left(\log\frac{1}{|z_n|^2}\right). \tag{7.18}$$

Then

$$\eta\Theta_h - \partial\bar\partial\eta - c\partial\eta\wedge\bar\partial\eta \ge (su'' - s'' - vs'^2)|_{t=\log\frac{1}{|z_n|^2}} \frac{dz_n\wedge d\overline{z_n}}{|z_n|^2}.$$

Note that

$$s(t) = e^{u(t)}\int_0^t e^{-u(x)}dx$$

so that

$$\frac{s'}{s} = (\log s)' = u' + \frac{1}{s}.$$

Therefore

$$s' - su' = 1. \tag{7.19}$$

Hence one has

$$\frac{s'^2}{su'' - s''} = -\frac{s'}{u'} = \frac{e^t - 1 - t}{1-e^{-t}} = \frac{1}{v}. \tag{7.20}$$

Thus, by the above choice of η, c and h, one has

$$\eta\Theta_h - \partial\bar\partial\eta - c\partial\eta\wedge\bar\partial\eta \ge 0.$$

On the other hand,

$$\eta + c^{-1} = \left(\frac{t}{1-e^{-t}} - 1 + \frac{e^t-1-t}{1-e^{-t}}\right)\Bigg|_{t=\log\frac{1}{|z_n|^2}}$$

and

$$\frac{t}{1-e^{-t}} - 1 + \frac{e^t-1-t}{1-e^{-t}} = e^t(1-e^{-t}) = e^t e^{-u}.$$

Hence (7.6) is satisfied for $Q = e^{-\phi}$. □

Similarly as above, one can prove the following.

Theorem 7.2 (cf. [G-Z] and [Oh-5]). *Let M be a connected complex manifold of dimension n containing a dense Stein open set Ω whose complement is locally pluripolar. Let w be a holomorphic function on M such that every irreducible component of $W := w^{-1}(0)$ intersects with Ω. Let W_0 be the set of regular points of W. Then, for any $\varphi, \psi \in PSH(M)$ such that*

$$\sup_M (\psi + \log |w|^2) \leq 0$$

and for any holomorphic $(n-1)$-form f on W_0 satisfying

$$\left| \int_{W_0} e^{-\varphi} f \wedge \bar{f} \right| < \infty,$$

there exists a holomorphic n-form F on M such that $F = f \wedge dw$ holds at every point of W_0 and

$$\left| \int_M e^{-\varphi+\psi} F \wedge \bar{F} \right| \leq 2\pi \left| \int_{W_0} e^{-\varphi} f \wedge \bar{f} \right| .$$

We recall that a subset of M is said to be locally pluripolar if it is locally contained in the preimage of $-\infty$ by a nonconstant plurisubharmonic function, and that L^2 holomorphic functions extend holomorphically across locally pluripolar sets (cf. [S]).

It is obvious that Theorem 7.1 is generalized to the holomorphic n forms whose coefficients are in any Nakano semipositive bundle, say E. The semipositivity of the curvature is required everywhere on M, although it is suspected that the same extendability statement will follow if the curvature form of E is semipositive *along* W, provided that we do not care about the estimate for the extension. In fact, in case M is pseudoconvex and W is compact, it follows from Grauert's direct image theorem that holomorphic sections of $K_M \otimes E$ over W are holomorphically extendible to M. Since this observation was recently generalized to some extent in [Oh-7] in a context of the Hartogs type extension theorem, we would like to review it in the next section following the presentation of [Oh-7].

8 Grauert's isomorphism in the L^2 setting

In [Oh-0] the following was proved:

Theorem 8.1. *Let M be a pseudoconvex manifold of dimension n and let $B \to M$ be a holomorphic line bundle. Assume that B admits a fiber metric whose curvature form is positive outside some compact subset of M. Then*

$$\dim H^{n,q}(M, B) < \infty \quad for \ \ q \geq 1.$$

This is an extension of Grauert's theorem in [G] asserting that the q-th cohomology groups of strongly pseudoconvex domains with coefficients in coherent analytic sheaves are finite dimensional for all $q \geq 1$. A significance of the finite-dimensionality lies in that it implies the existence of holomorphic sections of bundles with prescribed singularities at infinity. For a generalization of Theorem 8.1 to the bundles of higher rank, see [N-R]. In what follows, a pseudoconvex manifold M will be referred to by specifying a plurisubharmonic exhaustion function say φ. We set $M_c := \{x \in M; \varphi(x) < c\}$.

For the proof of Theorem 8.1 and its generalization in [N-R], the method of Hörmander [Hm] is applied. The main idea is to approximate $H^{n,q}(M, B)$ by $H^{n,q}_{(2)}(M, B)$ with respect to a fixed Hermitian metric ω and a fixed fiber metric of B say h such that $\omega = i\Theta_h$ holds on $M \setminus M_c$. More precisely, $H^{n,q}_{(2)}(M, B)$ are considered with respect to a family of fiber metrics. Namely, letting $H^{n,q}_{(2),\lambda}(M, B)$ be the L^2 $\bar{\partial}$-cohomology groups with respect to $(\omega, he^{-\lambda(\varphi)})$, Theorem 8.1 is a corollary of the following more intricate assertion.

Theorem 8.2. *Let (M, φ) be a pseudoconvex manifold of dimension n, let ω be a Hermitian metric on M, and let B be a holomorphic line bundle over M with fiber metric h whose curvature form Θ_h satisfies $\omega = i\Theta_h$ on $M \setminus M_c$ for some $c \in \mathbb{R}$. Then, for any nonconstant C^∞ convex increasing function λ, one can find a positive number $\mu(\lambda)$ such that for all $\mu \geq \mu(\lambda)$ the following are true.*

1) $\dim H^{n,q}_{(2),\mu\lambda}(M, B) < \infty, \quad q \geq 1.$

2) The natural restriction homomorphisms

$$H^{n,q}_{(2),\mu\lambda}(M, B) \longrightarrow H^{n,q}_{(2)}(M_c, B), \quad q \geq 1$$

with respect to the restrictions of h and ω are bijective.

3) The natural inclusion homomorphisms

$$\alpha^q_\mu : H^{n,q}_{(2),\mu\lambda}(M, B) \to H^{n,q}(M, B), \quad q \geq 1$$

are bijective.

Theorem 8.2, which is essentially what we have proved in [Oh-0], is a higher dimensional analogue of Mittag-Leffler's theorem and Runge's approximation theorem in function theory of one variable.

For the reader's convenience, let us mention Hörmander's original result in [Hm] in a slightly modified but essentially equivalent form:

Theorem 8.3. *Let M be a complex manifold of dimension n and let $\phi : M \to \mathbb{R}$ be an exhaustion function of class C^2. Suppose that the Levi form of ϕ has at least $n - q + 1$ positive eigenvalues everywhere on $M \setminus M_c$. Then, for any holomorphic vector bundle $E \to M$, the following hold.*

1) For any fiber metric h on E, there exist a Hermitian metric ω on M, a convex increasing function λ and $\mu_0 > 0$ such that

$$\dim H^{0,p}_{(2)}(M, E) < \infty$$

for all $p \geq q$ with respect to ω and $he^{-\mu\lambda(\phi)}$ for any $\mu > 0$, and

$$H^{0,p}_{(2)}(M, E) \cong H^{0,p}(M, E)$$

holds with respect to ω and $he^{-\mu\lambda(\phi)}$ if $\mu \geq \mu_0$ and $p \geq q$.

2) $H^{0,p}(M, E) \cong H^{0,p}(M_c, E) \cong H^{0,p}_{(2)}(M_c, E)$ for all $p \geq q$.

3) The image of

$$H^{0,q-1}(M, E) \to H^{0,q-1}(M_c, E)$$

is dense.

Roughly speaking, the stability of weighted L^2 cohomology groups for sufficiently large μ as above can be regarded as the vanishing of cohomology groups along the divisor at infinity, under some curvature conditions. We shall make this similarly more explicit by establishing an extension theorem on compact complex manifolds. From now on, we assume that M is compact, W is an effective divisor on M, and (E, h) is a Hermitian holomorphic vector bundle over M. Let $\mathcal{I}_W$ be the ideal sheaf of W, and put $\mathcal{O}_W = \mathcal{O}_M/\mathcal{I}_W$, where $\mathcal{O}_M$ denotes the structure sheaf of M. By $(|W|, \mathcal{O}_W)$ we denote the complex space whose underlying space is the support $|W|$ of W and structure sheaf is $\mathcal{O}_W$. By K_M and $[W]$ we denote the canonical line bundle of M and the line bundle associated to W, respectively. Then our remark here on Theorem 8.3 is the following.

Theorem 8.4. *Let M, E and W be as above. Suppose that $[W]$ is semipositive and $E|_{|W|}$ is Nakano positive. Then there exists $\mu_0 \in \mathbb{N}$ such that*
1) The restriction map

$$H^{0,0}(M, K_M \otimes E \otimes [W]^\mu) \to H^0(|W|, \mathcal{O}_W(K_M \otimes E \otimes [W]^\mu))$$

is surjective if $\mu \geq \mu_0$.
and
2) $H^{n,q}(M, E \otimes [W]^\mu) \cong H^{n,q}(M \setminus |W|, E)$ if $q \geq 1$ and $\mu \geq \mu_0$.

For the reader's convenience, the proof of Theorem 8.2 will be given at first.

9 Proof of an L^2 isomorphism theorem

Let (M, φ), (B, h), ω and Θ_h be as in the assumption of Theorem 8.2, and let λ_0 be any C^∞ convex increasing function such that

$$\omega_\epsilon := \omega + \epsilon i\partial\bar{\partial}(\lambda_0 \circ \varphi) \tag{9.1}$$

is a complete metric on M for any $\epsilon > 0$. (One may take $\lambda_0(t) = e^t$ for instance.) Let λ be any nonconstant C^∞ convex increasing function and let $L^{n,q}_{\lambda,\epsilon}(M, B)(= L^{n,q}_{(2),\lambda,\epsilon}(M, B))$ denote the Hilbert space of square integrable B-valued (n, q)-forms on M with respect to $(\omega_\epsilon, he^{-(\lambda+\epsilon\lambda_0)\circ\varphi})$.

As before, by $\bar{\partial} : L^{n,q}_{\lambda,\epsilon}(M,B) \to L^{n,q+1}_{\lambda,\epsilon}(M,B)$ we denote the closed extension of $\bar{\partial}$ whose domain of definition is

$$\mathrm{Dom}\bar{\partial} \cap L^{n,q}_{\lambda,\epsilon}(M,B) := \{u \in L^{n,q}_{\lambda,\epsilon}(M,B); \bar{\partial}u \in L^{n,q+1}_{\lambda,\epsilon}(M,B)\}.$$

We put

$$H^{n,q}_{(2),\lambda,\epsilon}(M,B) = \frac{\mathrm{Ker}(\bar{\partial} : L^{n,q}_{\lambda,\epsilon}(M,B) \to L^{n,q+1}_{\lambda,\epsilon}(M,B))}{\mathrm{Im}(\bar{\partial} : L^{n,q-1}_{\lambda,\epsilon}(M,B) \to L^{n,q}_{\lambda,\epsilon}(M,B))}.$$

Let $L^{n,q}_{(2)}(M_c,B)$ and $H^{n,q}_{(2)}(M_c,B)$ be defined similarly with respect to the restrictions of ω and h. Let $\bar{\partial}^*(=\bar{\partial}^*_{\lambda,\epsilon}) : L^{n,q+1}_{\lambda,\epsilon}(M,B) \to L^{n,q}_{\lambda,\epsilon}(M,B)$ denote the adjoint of $\bar{\partial}$ and put

$$\mathcal{H}^{n,q}_{\lambda,\epsilon} = \mathrm{Ker}\bar{\partial} \cap \mathrm{Ker}\bar{\partial}^* \cap L^{n,q}_{\lambda,\epsilon}(M,B).$$

Similarly we define

$$\bar{\partial} : L^{n,q}_{(2)}(M_c,B) \to L^{n,q+1}_{(2)}(M_c,B),$$

$$\bar{\partial}^*(=\bar{\partial}^*_c) : L^{n,q+1}_{(2)}(M_c,B) \to L^{n,q}_{(2)}(M_c,B)$$

and

$$\mathcal{H}^{n,q}_c = \mathrm{Ker}\bar{\partial} \cap \mathrm{Ker}\bar{\partial}^* \cap L^{n,q}_{(2)}(M_c,B).$$

The proof of Theorem 8.2 is based on the following.

Proposition 9.1. *Under the above situation, there exists a constant $C > 0$ which does not depend on the choices of λ_0, λ and ϵ such that*

$$\|u\|^2 \le C\{\|\bar{\partial}u\|^2 + \|\bar{\partial}^*u\|^2 + \int_{M_{c+1}} |u|^2 e^{-(\lambda+\epsilon\lambda_0)\circ\varphi}\omega_\epsilon^n\} \tag{9.2}$$

holds for any $u \in L^{n,q}_{\lambda,\epsilon}(M,B) \cap \mathrm{Dom}\bar{\partial} \cap \mathrm{Dom}\bar{\partial}^*$ with $q \ge 1$. Here $\|\cdot\|$ denotes the L^2 norm with respect to $(\omega_\epsilon, he^{-(\lambda+\epsilon\lambda_0)(\varphi)})$ and $|u|$ the length of u with respect to ω_ϵ and h.

Corollary 3. $\dim \mathcal{H}^{n,q}_{\lambda,\epsilon} < \infty$ *and* $H^{n,q}_{(2),\lambda,\epsilon}(M,B) \cong \mathcal{H}^{n,q}_{\lambda,\epsilon}$ *hold if* $q \ge 1$.

Corollary 4. *There exists μ_0 such that for any $\mu' \ge \mu \ge \mu_0$ and $\epsilon > 0$ the natural inclusion homomorphisms*

$$H^{n,q}_{(2),\mu\lambda,\epsilon}(M,B) \to H^{n,q}_{(2),\mu'\lambda,\epsilon}(M,B), \quad q \ge 1$$

are injective.

Proof of Proposition 9.1. By the assumption on the curvature form of h, for any compactly supported C^∞ B-valued (n,q)-form u such that $\mathrm{supp}u \subset M \setminus M_c$ and $q \ge 1$,

$$\|u\|^2 \le \|\bar{\partial}u\|^2 + \|\bar{\partial}^*u\|^2$$

holds with respect to ω_ϵ and $he^{-(\lambda+\epsilon\lambda_0)\circ\varphi}$, by Nakano's identity on Kähler manifolds. Hence, for any u in $L^{n,q}_{\lambda,\epsilon}(M,B) \cap \mathrm{Dom}\bar{\partial} \cap \mathrm{Dom}\bar{\partial}^*$ with $q \ge 1$, by multiplying a C^∞ cut-off function χ with supp$\chi \subset M \setminus M_c$ and supp$(1-\chi) \subset M_{c+1}$ to u and approximating χu by C^∞ forms, one has (9.2) by the completeness of ω_ϵ. C depends only on the gradient of χ. □

Proof of Corollary 3. From (9.2) and Rellich's lemma, it follows that, for any sequence u_μ ($\mu = 1, 2, \ldots$) in $L^{n,q}_{\lambda,\epsilon}(M,B)$ ($q \geq 1$) such that $u_\mu \perp \mathcal{H}^{n,q}_{\lambda,\epsilon}$, $\|u_\mu\| = 1$, $\|\bar\partial u_\mu\| \to 0$ and $\|\bar\partial^* u_\mu\| \to 0$, one can choose a strongly convergent subsequence, once we fix λ and ϵ. Therefore, there has to exist a constant $C_0 = C_0(\lambda, \epsilon)$ such that

$$\|u\|^2 \leq C_0\{\|\bar\partial u\|^2 + \|\bar\partial^* u\|^2\}$$

holds for any $u \in (L^{n,q}_{\lambda,\epsilon}(M,B) \ominus \mathcal{H}^{n,q}_{\lambda,\epsilon}) \cap \mathrm{Dom}\bar\partial \cap \mathrm{Dom}\bar\partial^*$ with $q \geq 1$, where $\ominus$ denotes the orthocomplement. Indeed, if there were no such C_0, one would have a sequence $u_\mu \in (L^{n,q}_{\lambda,\epsilon}(M,B) \ominus \mathcal{H}^{n,q}_{\lambda,\epsilon}) \cap \mathrm{Dom}\bar\partial \cap \mathrm{Dom}\bar\partial^*$, $q \geq 1$, such that $\|u_\mu\| = 1$, $\|\bar\partial u_\mu\| \to 0$ and $\|\bar\partial^* u_\mu\| \to 0$. Then the limit, say u, of a subsequence of u_μ would satisfy $\|u\| = 1$, $\bar\partial u = \bar\partial^* u = 0$ and $u \perp \mathcal{H}^{n,q}_{\lambda,\epsilon}$, which is an absurdity. Hence the assertion follows by a standard argument using Hahn-Banach's theorem. □

Proof of Corollary 4. Let λ_0 be as in (9.1). We may assume in advance that suppλ_0 = supp$\lambda = (-\infty, c]$ because for any convex increasing function λ_1 one can find a convex increasing function λ_2 such that supp$\lambda_2 = (-\infty, c]$ and $|\lambda_1 - \lambda_2|$ is bounded. Then, similarly as in the proof of Corollary 3, one can find $\mu_0 = \mu_0(\lambda) > 0$ and a constant C_1 such that, for any $\epsilon > 0$ and $\mu \geq \mu_0$,

$$\|u\|^2 \leq C_1\{\|\bar\partial u\|^2 + \|\bar\partial^* u\|^2\}$$

holds for any $u \in L^{n,q}_{\mu\lambda,\epsilon}(M,B) \cap \mathrm{Dom}\bar\partial \cap \mathrm{Dom}\bar\partial^*$ satisfying

$$\int_{M_c} \langle u, v \rangle \omega^n = 0$$

for all $v \in \mathcal{H}^{n,q}_c$, where $\langle \cdot, \cdot \rangle$ denotes the pointwise inner product with respect to ω and h. Hence, by Hahn-Banach again, we obtain the desired conclusion. □

Proof of Theorem 8.2. Since the argument is essentially the repetition of the above, we shall only give a sketch. Let μ_0 be as above. Then, similarly as above, the natural homomorphisms

$$\beta^q_{\mu,\epsilon} : H^{n,q}_{(2),\mu\lambda,\epsilon}(M,B) \to H^{n,q}_{(2)}(M_c,B), \ q \geq 1$$

are injective for all $\mu \geq \mu_0$, $\dim H^{n,q}_{(2)}(M_c,B) < \infty$ for $q \geq 1$, and

$$\overline{(\cup_{\mu=1}^\infty \mathrm{Im}\beta^q_{\mu,\epsilon})} = H^{n,q}_{(2)}(M_c,B)$$

for all $q \geq 0$.

Therefore, there exists $\mu_1 \geq \mu_0$ such that $\beta^q_{\mu,\epsilon}$ are bijective if $q \geq 1, \mu \geq \mu_1$ and $\epsilon > 0$. Since C_1 is independent of μ and ϵ, we are allowed to let $\epsilon \to 0$ and obtain the bijectivity of $\beta^q_{\mu,0}$ for $\mu \geq \mu_1$ as well. Similarly, it is easy to see that α^q_μ are bijective if $\mu \geq \mu_1$. Hence it suffices to put $\mu(\lambda) = \mu_1$. □

10 Extension and isomorphism on compact manifolds

Proof of Theorem 8.4. Let M, E and W be as in the assumption, and let s be a canonical section of $[W]$. Since $[W]$ is semipositive, there exists a fiber metric of $[W]$ for which the function $-\log|s|$ is plurisubharmonic on $M \setminus |W|$. Here $|s|$ denotes the length of s. Since E is Nakano positive along $|W|$, there exists a fiber metric h_E of E and a neighborhood $U \supset |D|$ such that (E, h_E) is Nakano positive on U. In this situation, one can find a Hermitian metric ω_M on M such that the estimate $\|u\|^2 \leq \|\bar\partial u\|^2 + \|\bar\partial^* u\|^2$ holds for any compactly supported C^∞ E-valued (n,q)-form u on $U \setminus |W|$, in virtue of Nakano's identity. Hence, similarly as in the proof of Corollary 4, one can find a complete Hermitian metric ω_ϵ on $M \setminus |W|$ and $\mu_0 \in \mathbb{N}$ such that the homomorphisms

$$H^{n,1}_{(2),-\mu\log|s|,\epsilon}(M \setminus |W|, E) \to H^{n,1}_{(2),-\mu'\log|s|,\epsilon}(M \setminus |W|, E),\ \mu' \geq \mu \geq \mu_0 - 1$$

are injective. Given any $f \in H^0(|W|, K_M \otimes [W]^\mu \otimes E)$, let $\tilde f$ be any extension of f to M as a C^∞ section of $K_M \otimes [W]^\mu \otimes E$ such that $\bar\partial \tilde f|_{|W|} = 0$. Then

$$\frac{\bar\partial \tilde f}{s^\mu} \in L^{n,1}_{-(\mu-1)\log|s|,\epsilon}(M \setminus |W|, E) \cap \mathrm{Ker}\bar\partial.$$

Since $\bar\partial(\tilde f/s^\mu) = \bar\partial \tilde f/s^\mu$ and $\tilde f/s^\mu \in L^{n,0}_{-\mu\log|s|,\epsilon}(M \setminus |W|, E)$, the above injectivity implies that there exists $g \in L^{n,0}_{-(\mu-1)\log|s|,\epsilon}(M \setminus |W|, E)$ such that $\bar\partial g = \bar\partial \tilde f/s^\mu$ holds on $M \setminus |W|$. Then $\tilde f - s^\mu g$ is the desired extension of f in $H^{0,0}(M, K_M \otimes [W]^\mu \otimes E)$, which is the end of the proof of 1). The proof of 2) is similar as in Theorem 8.2. □

Corollary of Theorem 8.4. *Let M be a compact complex manifold. If there exist an effective divisor W on M and a Hermitian holomorphic vector bundle $E \to M$ such that $[W]$ is semipositive on M and E is Nakano positive on some neighborhood of $|W|$. Then M is a Moishezon manifold.*

In particular, by a theorem of Chow and Kodaira [C-K], M is projective algebraic if moreover $\dim M = 2$. If there exists a compact Riemann surface S of genus $g \geq 2$ and a holomorphic embedding $p : S \to M$ such that $[p(S)] \geq 0$, one may take K_M or $[p(S)]$ as E. In other words, *a compact complex surface containing a semipositively embedded irreducible smooth curve of genus ≥ 2 is algebraic.*

In view of this phenomenon, we would like to ask whether or not the following is true.

Conjecture. Let M be a compact complex manifold and let W be an effective divisor on M such that $[W]$ is semipositive. Then one can find a constant $C = C(W)$ such that, for any Hermitian holomorphic vector bundle $(E, h) \to M$ such that $\gamma_h|_{|W|} > 0$ and for any $\mu \in \mathbb{N}$ such that $\mu > \max\{-C \inf_M \gamma_h, 0\}$, the restriction map

$$H^{0,0}(M, K_M \otimes E \otimes [W]^\mu) \to H^0(|W|, \mathcal{O}_W(K_M \otimes E \otimes [W]^\mu))$$

is surjective.

References

[A-V-1] Andreotti, A. and Vesentini, E., *Sopra un teorema di Kodaira*, Ann. Scuola Norm. Sup. Pisa **15** (1961), 283–309.

[A-V-2] —, *Carleman estimates for the Laplace-Beltrami equation on complex manifolds*, Inst. Hautes Études Sci. Publ. Math. **25** (1965), 81–130. Erratum: Inst. Hautes Études Sci. Publ. Math. **27** (1965), 153–155.

[B-1] Berndtsson, B., *The extension theorem of Ohsawa-Takegoshi and the theorem of Donnelly-Fefferman*, Ann. Inst. Fourier **46** (1996), 1083–1094.

[B-2] —, *Curvature of vector bundles associated to holomorphic fibrations*, Ann. of Math. **169** (2009), 531–560.

[B-3] —, *Strict and nonstrict positivity of direct image bundles*, Math. Z. **269** (2011), 1201–1218.

[B-L] Berndtsson, B. and Lempert, L., *A proof of Ohsawa-Takegoshi theorem with sharp estimates*, to appear in Journal of Math. Soc. Japan.

[B-P] Bermdtsson, B. and Păun, M., emphBergman kernels and the pseudoeffectivity of relative canonical bundles, Duke Math. J. **145** (2008), 341–378.

[Bl-1] Błocki, Z., *Suita conjecture and the Ohsawa-Takegoshi extension theorem*, Invent. Math. **193** (2013), 149–158.

[Bl-2] —, *A lower bound for the Bergman kernel and the Bourgain-Milman inequality*, Geometric Aspects of Functional Analysis, Israel Seminar (GAFA) 2011–2013, eds. B. Klartag, E. Milman, Lecture Notes in Mathematics 2116, Springer, 2014, pp. 53–63.

[Bl-3] —, *Bergman kernel and pluripotential theory*, to appear in Proceedings of the Conference in honor of Duong Phong, Contemporary Mathematics, American Mathematical Society.

[Bl-P] Błocki, Z. and Pflug, P., *Hyperconvexity and Bergman completeness*, Nagoya Math. J. **151** (1998), 221–225.

[C-V] Calabi, E. and Vesentini, E., *On compact, locally symmetric Kähler manifolds*, Ann. of Math. **71** (1960), 472–507.

[Ca] Cao, J., *Ohsawa-Takegoshi extension theorem for compact Kähler manifolds and applications*, arXiv:1404.6937v1 [math.AG] 28 Apr 2014.

[Cl] Carleson, L., *Selected problems on exceptional sets*, Van Nostrand Mathematical Studies, No. 13 D. Van Nostrand Co., Inc., Princeton, N.J.-Toronto, Ont.-London 1967.

[C] Cartan, H., *Idéaux et modules de fonctions analytiques de variables complexes*, Bull. Soc. Math. France **78** (1950), 28–64.

[Ch] Chen, B.-Y., *A simple proof of the Ohsawa-Takegoshi extension theorem*, arXiv:1105.2430v1 [math.CV] 12 May 2011.

[C-K] Chow, W.-L. and Kodaira, K., *On analytic surfaces with two independent meromorphic functions*, Proc. Nat. Acad. Sci. U. S. A. **38** (1952), 319–325.

[Dm] Demailly, J.-P., *Estimations L^2 pour l'opérateur $\bar{\partial}$ d'un fibré vectoriel holomorphe semi-positif au-dessus d'une variété kählérienne complète*, Ann. Sci. École Norm. Sup. **15** (1982), 457–511.

[E-V] Esnault, H. and Viehweg, E., *Lectures on vanishing theorems*, DMV Seminar **20**, Birkhäuser (1992).

[F] Fujita, T., *On Kähler fiber spaces over curves*, L. Math. Soc. Japan **30** (1978), 779–794.

[G-S] Garabedian, P. R. and Spencer, D. C., *Complex boundary value problems*, Trans. Amer. Math. Soc. **73** (1952), 223–242.

[G] Grauert, H.,*On Levi's problem and the embedding of real-analytic manifolds*, Ann. Math. **68** (1958), 460–472.

[Gri-1] Griffiths, P.A., *Periods of integrals on algebraic manifolds. I and II*, Amer. J. Math. **90** (1968), 568–626 and 805–865.

[Gri-2] —, *Hermitian differential geometry, Chern classes, and positive vector bundles*, Global Analysis (Papers in Honor of K. Kodaira) Univ. Tokyo Press, Tokyo, 1969, pp. 185–251.

[Gri-3] —, *Periods of integrals on algebraic manifolds. III. Some global differential-geometric properties of the period mapping*, Inst. Hautes Études Sci. Publ. Math. **38** (1970), 125–180.

[G-Z-1] Guan, Q.-A. and Zhou, X.-Y., *Optimal constant problem in the L^2 extension theorem*, C. R. Math. Acad. Sci. Paris **350** (2012), 753–756.

[G-Z-2] —, *A solution of an L^2 extension problem with optimal estimate and applications*, Ann. of Math. **181** (2015), 1139–1208.

[G-Z-Z] Guan, Q., Zhou, X.-Y. and Zhu, L., *On the Ohsawa-Takegoshi L^2 extension theorem and the Bochner-Kodaira identity with non-smooth twist factor*, J. Math. Pures Appl. **97** (2012), 579–601.

[H] Hartogs, F., *Zur Theorie der analytischen Funktionen mehrerer unabhängiger Veränderlichen insbesondere über die Darstellung derselben durch Reihen, welche nach Potenzen einer Veränderlichen fortschreiten*, Math. Ann. **62** (1906), 1–88.

[Hb] Herbort, G., *The Bergman metric on hyperconvex domains*, Math. Z. **232** (1999), 183–196.

[Hm] Hörmander, L., *L^2 estimates and existence theorems for the $\bar\partial$ operator*, Acta Math. **113** (1965), 89–152.

[K] Kodaira, K., *On a differential-geometric method in the theory of analytic stacks*, Proceedings of the National Academy of Sciences, U.S.A. **39** (1953), 1268–1273.

[Kn] Kohn, J. J., *Harmonic integrals on strongly pseudo-convex manifolds. I*, Ann. of Math. **78** (1963), 112–148.

[L] Lazarsfeld, R., *Positivity in Algebraic Geometry, I, II*, Springer-Verlag (2004).

[L-Y] Liu, K.-F. and Yang, X.-K., *Curvatures of direct image sheaves of vector bundles and applications*, J. Diff. G. **98** (2014), 117–145.

[M-Y] Maitani, F. and Yamaguchi, H., *Variation of Bergman metrics on Riemann surfaces*, Math. Ann. **330** (2004), 477–489.

[M-Z] Mok, N. and Zhong, J.-Q., *Compactifying complete K̈hler-Einstein manifolds of finite topological type and bounded curvature*, Ann. of Math. **129** (1989), 427–470.

[Mo] Morrey, C. B., *The analytic embedding of abstract real-analytic manifolds*, Ann. of Math. **68** (1958), 159–201.

[M-T] Mourougane, C. and Takayama, S., *Hodge metrics and the curvature of higher direct images*, Ann. Sci. Éc. Norm. Supér. **41** (2008), 905–924.

[N] Nakano, S., *On complex analytic vector bundles*, J. Math. Soc. Japan **7** (1955), 1–12.

[N-R] Nakano, S. and Rhai, T. S., *Vector bundle version of Ohsawa's finiteness theorems*, Math. Jap. **24** (1980), 657–664.

[Ni] Nishino, T., *Sur une propriété des familles de fonctions analytiques de deux variables complexes*, J. Math. Kyoto Univ. **4** (1965), 255–282.

[Oh-0] Ohsawa, T., *Finiteness theorems on weakly 1-complete manifolds*, Publ. RIMS, Kyoto Univ. **15** (1979), 853–870 (master thesis).

[Oh-1] —, *On complete Kähler domains with C^1-boundary*, Publ. Res. Inst. Math. Sci. **16** (1980), 929–940.

[Oh-2] —, *Vanishing theorems on complete Kähler manifolds*, Publ. Res. Inst. Math. Sci. **20** (1984), 21–38.

[Oh-3] —, *Boundary behavior of the Bergman kernel function on pseudoconvex domains*, Publ. Res. Inst. Math. Sci. **23** (1984), 897–902.

[Oh-4] —, *On the rigidity of noncompact quotients of bounded symmetric domains*, Publ. Res. Inst. Math. Sci. **23** (1987), 881–894.

[Oh-5] —, *On the extension of L^2 holomorphic functions. III. Negligible weights*, Math. Z. **219** (1995), 215–225.

[Oh-6] —, *On the extension of L^2 holomorphic functions VIII — a remark on a theorem of Guan and Zhou*, preprint.

[Oh-7] —, *A remark on Hörmander's isomorphism*, Complex Analysis and Geometry, Springer Proc. in Math. and Stat. edit. Kang-Tae Kim et al. pp. 273–280.

[Oh-T] Ohsawa, T. and Takegoshi, K., *On the extension of L^2 holomorphic functions*, Math. Z. **195** (1987), 197–204.

[O] Oka, K., *Sur quelques notions arithmétiques*, Bull. Soc. Math. France **78** (1950), 1–27.

[P] Pflug, P., *Quadratintegrable holomorphe Funktionen und die Serre-Vermutung*, Math. Ann. **216** (1975), 285–288.

[S-O] Sario, L. and Oikawa, K., *Capacity functions*, Die Grundlehren der mathematischen Wissenschaften, Band **149** Springer-Verlag New York Inc., New York 1969.

[S] Siciak, J., *On removable singularities of L^2 holomorphic functions of several variables,* Prace matematyczno-fizyczne, edited by Skwarczynski, M and Wasilewski, W., WSI Radom 1982, pp. 73–81.

[Su] Suita, N.,*Capacities and kernels on Riemann surfaces*, Arch. Rational Mech. Anal. **46** (1972), 212–217.

[T] Takegoshi, K., *Higer direct images of canonical sheaves tensorized with semi-positive vector bundles by proper Kähler morphisms*, Math Ann. **303** (1995), 389–416.

[Y] Yamaguchi, H., *Sur le mouvement des constantes de Robin*, J. Math. Kyoto Univ. **15** (1975), 53–71.

[Yi] Yi, L., *An Ohsawa-Takegoshi theorem on compact Kähler manifolds*, Science China Math. **57** (2014), 9–30.

Hodge Theory and L^2-analysis
ALM 39, pp. 517–553

A Young Person's Guide to Mixed Hodge Modules*

Morihiko Saito†

Abstract

We give a rather informal introduction to the theory of mixed Hodge modules for young mathematicians.

Introduction

The theory of mixed Hodge modules ([Sa3], [Sa7]) was originally constructed as a Hodge-theoretic analogue of the theory of ℓ-adic mixed perverse sheaves ([De2], [De7], [BBD]) and also as an extension of Deligne's mixed Hodge theory ([De3], [De6]) including the theory of degenerations of variations of pure or mixed Hodge structures (see [Sch], [St], [CK], [CKS1], [StZ], [Ka5], etc.) The main point is the "stability" by the direct images and the pull-backs under morphisms of complex algebraic varieties and also by the dual and the nearby and vanishing cycle functors. Here "stability" means more precisely the "stability theorems" saying that there are *canonically defined functors* between the derived categories. These stability theorems become quite useful by combining them with the fundamental theorem of mixed Hodge modules asserting that any admissible variations of mixed Hodge structure on smooth complex algebraic varieties in the sense of [StZ], [Ka5] are mixed Hodge modules [Sa7]; in particular, mixed Hodge modules on a point are naturally identified with graded-polarizable mixed $\mathbb{Q}$-Hodge structures in the sense of Deligne [De3].

Technically the *strictness* of the Hodge filtration F on the underlying complexes of $\mathcal{D}$-modules is very important in the theory of mixed Hodge modules. Here $\mathcal{D}$-modules are indispensable for the generalization of Deligne's theory in the absolute

*Dedicated to Steven Zucker.

†RIMS Kyoto University, Kyoto 606-8502 Japan.

case ([De3], [De6]) to the *relative* case. For instance, this includes the assertion that any morphism of mixed Hodge modules is *bistrict* for the Hodge and weight filtrations F, W, generalizing the case of mixed Hodge structures by Deligne [De3]. $\mathcal{D}$-modules are also essential for the construction of a relative version of $\mathrm{Dec}\, W$ in the proof of the stability theorem of mixed Hodge modules by the direct images under projective morphisms extending Deligne's argument in the absolute case, where we have *bistrict* complexes of $\mathcal{D}$-modules with filtrations $(F, \mathrm{Dec}\, W)$ under the direct images by projective morphisms, see [Sa7, Proposition 2.15].

In order to understand the theory of mixed Hodge modules, the general theory of $\mathcal{D}$-modules does not seem to be absolutely indispensable. In fact, $\mathcal{D}$-modules appearing in the theory of mixed Hodge modules are rather special ones, and we have to deal with them always as *F-filtered $\mathcal{D}$-modules*, where a slightly different kind of argument is usually required. For instance, although the *regularity* of $\mathcal{D}$-modules is quite useful for the construction of the direct images by affine open immersions, this can be reduced essentially to the normal crossing case by using Beilinson's functor together with the stability theorem under the direct images by projective morphisms (see [Sa29]), and in the latter case, a generalization of the Deligne canonical extensions [De1] to the case of $\mathcal{D}$-modules with normal crossing singular supports is sufficient. Also the pull-backs of mixed Hodge modules under closed immersions are constructed by using nearby and vanishing cycle functors, which is entirely different from the usual construction of pull-backs of $\mathcal{D}$-modules, see [Sa7, Section 4.4].

Note finally that we need only Zucker's Hodge theory in the curve case [Zu] for the proof of the stability theorem of mixed Hodge modules under the direct images by proper morphisms, and classical Hodge theory is not used, see 2.5.

I thank the referee for useful comments. The author is partially supported by Kakenhi 15K04816.

In Section 1 we explain the main properties of pure Hodge modules. In Section 2 we give an inductive definition of pure Hodge modules, and explain an outline of proofs of Theorems 1.3 and 1.4. In Section 3 we explain a simplified definition of mixed Hodge modules following [Sa29]. In Appendices we explain some basics of hypersheaves, $\mathcal{D}$-modules, and compatible filtrations.

Conventions 1. We assume that algebraic varieties in this paper are always defined over $\mathbb{C}$ and are reduced (but not necessarily irreducible). More precisely, a variety means a separated reduced scheme of finite type over $\mathbb{C}$, but we consider only its *closed points*, that is, its *$\mathbb{C}$-valued points*. So it is close to a variety in the sense of Serre (except that reducible varieties are allowed here). We also assume that a variety is always quasi-projective, or more generally, globally embeddable into a smooth variety (where morphisms of varieties are assumed quasi-projective) in order to simplify some arguments. The reader may assume that all the varieties in this paper are reduced quasi-projective complex algebraic varieties.

2. We use *analytic* sheaves on complex algebraic varieties; in particular, any $\mathcal{D}$-modules are analytic $\mathcal{D}$-modules. (These are suitable for calculations using local coordinates.) For the underlying filtered $\mathcal{D}$-module (M, F) of a mixed Hodge module on X, one can pass to the corresponding *algebraic* filtered $\mathcal{D}$-module by applying GAGA to each F_pM after taking an extension of the mixed Hodge module over a compactification of X.

3. In this paper, perverse sheaves (which do not seem appropriate at least for book titles, see [Di], [KS]) are mainly called hypersheaves by an analogy with hypercohomology versus cohomology (although the word "sheaf" may not be of Greek origin). The abelian category of hypersheaves on X are denoted by $\mathbf{HS}(X, A)$ where A is a subfield of $\mathbb{C}$. Hypersheaves are not sheaves in the usual sense, but they behave like sheaves in some sense; for instance, they can be defined locally provided that gluing data satisfying some compatibility condition are also given.

1 Main properties of pure Hodge modules

In this section we explain the main properties of pure Hodge modules.

1.1. Filtered $\mathcal{D}$-modules with $\mathbb{Q}$-structure. A pure Hodge module $\mathcal{M}$ on a smooth complex algebraic variety X of dimension d_X is basically a coherent left $\mathcal{D}_X$-module M endowed with the Hodge filtration F such that (M, F) is a filtered left $\mathcal{D}_X$-module with F_pM coherent over $\mathcal{O}_X$. The filtration F on $\mathcal{D}_X$ is by the order of differential operators, and we have the filtered $\mathcal{D}$-module condition

$$(F_p\mathcal{D}_X)\, F_qM \subset F_{p+q}M \qquad (p \in \mathbb{N},\, q \in \mathbb{Z}), \tag{1.1.1}$$

(which is equivalent to *Griffiths transversality* in the case of variations of Hodge structure), and the equality holds in (1.1.1) for $q \gg 0$. Moreover M has a $\mathbb{Q}$-*structure* given by an isomorphism

$$\alpha_{\mathcal{M}} : \mathrm{DR}_X(M) \cong \mathbb{C} \otimes_{\mathbb{Q}} K \quad \text{in } \mathbf{HS}(X, \mathbb{C}), \tag{1.1.2}$$

with $K \in \mathbf{HS}(X, \mathbb{Q})$ (see A.1). Here $\mathrm{DR}_X(M)$ is the de Rham complex of the $\mathcal{D}_X$-module M viewed as a quasi-coherent $\mathcal{O}_X$-module with an integrable connection:

$$\mathrm{DR}_X(M) := \big[M \to \Omega_X^1 \otimes_{\mathcal{O}_X} M \to \cdots \to \Omega_X^{d_X} \otimes_{\mathcal{O}_X} M\big], \tag{1.1.3}$$

with last term put at degree 0. In (1.1.2) we *assume*

$$\mathrm{DR}_X(M) \in \mathbf{HS}(X, \mathbb{C}), \tag{1.1.4}$$

and moreover (1.1.4) holds with M replaced by any subquotients of M as coherent $\mathcal{D}$-modules. (More precisely, these properties follow from the condition that M is *regular holonomic*, see (B.5.2). We effectively assume the last condition in the definition of Hodge modules, see Remark (ii) after Theorem 1.3.) In this paper we also assume

$$M, K \text{ are quasi-unipotent (see B.6 below).} \tag{1.1.5}$$

We will denote by $\mathrm{MF}(X, \mathbb{Q})$ the category of

$$\mathcal{M} = \big((M, F), K, \alpha_{\mathcal{M}}\big)$$

satisfying the above conditions. (Sometimes $\alpha_{\mathcal{M}}$ will be omitted to simplify the notation.) We also use the notation

$$\mathrm{rat}(\mathcal{M}) := K. \tag{1.1.6}$$

The category $\mathrm{MH}(X,w)$ of pure Hodge modules of weight w on X will be defined as a full subcategory of $\mathrm{MF}(X,\mathbb{Q})$.

1.2. Strict support decomposition. The first condition for pure Hodge modules is the *decomposition by strict support*

$$\mathcal{M} = \bigoplus\nolimits_{Z\subset X} \mathcal{M}_Z \quad \text{in } \mathrm{MF}(X,\mathbb{Q}), \tag{1.2.1}$$

where Z runs over irreducible closed subvarieties of X, and

$$\mathcal{M}_Z = \big((M_Z,F),K_Z,\alpha_{\mathcal{M}_Z}\big)$$

has *strict support* Z (that is, its support is Z, and it has no nontrivial sub nor quotient object supported on a proper subvariety of Z). More precisely, we assume that the last condition is satisfied for both M_Z and K_Z. Note that the condition for K_Z is equivalent to that K_Z is an intersection complex with local system coefficients, see [BBD]. We then get

$$\mathrm{Hom}(\mathcal{M}_Z,\mathcal{M}_{Z'}) = 0 \quad \text{if } Z \neq Z'. \tag{1.2.2}$$

In fact, this holds with $\mathcal{M}_Z,\mathcal{M}_{Z'}$ replaced by $M_Z,M_{Z'}$ or by $K_Z,K_{Z'}$.

In 2.2, we will define the full subcategory

$$\mathrm{MH}_Z(X,w) \subset \mathrm{MF}(X,\mathbb{Q})$$

consisting of *pure Hodge modules of weight w with strict support Z* by increasing induction on $\dim Z$, and put

$$\mathrm{MH}(X,w) := \bigoplus\nolimits_{Z\subset X} \mathrm{MH}_Z(X,w) \subset \mathrm{MF}(X,\mathbb{Q}), \tag{1.2.3}$$

where the direct sum over closed irreducible subvarieties Z of X is justified by (1.2.1) and (1.2.2).

This full subcategory $\mathrm{MH}_Z(X,w) \subset \mathrm{MF}(X,\mathbb{Q})$ can be defined effectively by the following *fundamental theorem of pure Hodge modules*, which may be viewed as a *working definition* of $\mathrm{MH}_Z(X,w)$.

Theorem 1.3 ([Sa7, Theorem 3.21]). *For any closed irreducible subvariety $Z \subset X$, the restriction to sufficiently small open subvarieties of Z induces an equivalence of categories*

$$\mathrm{MH}_Z(X,w) \xrightarrow{\sim} \mathrm{VHS}_{\mathrm{gen}}(Z,w-\dim Z)^p, \tag{1.3.1}$$

where the right-hand side is the category of polarizable variations of pure Hodge structure of weight $w-\dim Z$ defined on smooth dense open subvarieties U of Z. (More precisely, we take the inductive limit over $U \subset Z$.) Moreover, (1.3.1) *induces a one-to-one correspondence between polarizations of* $\mathcal{M} \in \mathrm{MH}_Z(X, w)$ *(see* (2.2.2)*) and those of the corresponding generic variation of Hodge structure.*

Remarks. (i) The equivalence of categories (1.3.1) means that any pure Hodge module with strict support Z is generically a polarizable variation of pure Hodge structure, and conversely any polarizable variation of pure Hodge structure defined on a smooth dense Zariski-open subset $U \subset Z$ can be extended uniquely to a pure Hodge module with strict support Z.

(ii) By using the stability theorem of pure Hodge modules under the direct images by projective morphisms (see Theorem 1.4 below), the proof of Theorem 1.3 can be reduced to the case where $Z = X$ and a variation of Hodge structure is defined on the complement U of a divisor with normal crossings. In this case we use the original definition of pure Hodge modules in [Sa3] given by induction on the dimension of the support of M and using the nearby and vanishing cycle functors.

In the case $D := X \setminus U$ is a divisor with normal crossings, the above extension of a variation of Hodge structure on U to a Hodge module on X is rather easy to describe by using Deligne's canonical extension [De1]. In fact, we have the following explicit formula (see [Sa7, (3.10.12)]):

$$F_pM = \textstyle\sum_{i\geqslant 0} F_i\mathcal{D}_X\big(j_*F_{p-i}\mathcal{L} \cap \widehat{\mathcal{L}}^{>-1}\big) \subset \widehat{\mathcal{L}}^{>-1}(*D), \tag{1.3.2}$$

with $j : U \hookrightarrow X$ is the inclusion. Here $\mathcal{L}$ is the locally free $\mathcal{O}_U$-module underlying the variation of Hodge structure with F the Hodge filtration, $\widehat{\mathcal{L}}^{>-1}$ is the Deligne extension of $\mathcal{L}$ over X such that the eigenvalues of the residues of the connection are contained in $(-1, 0]$, and the last term of (1.3.2) is the Deligne meromorphic extension, see [De1]. (Note that a decreasing filtration F is identified with an increasing filtration by setting $F_p := F^{-p}$.) Taking the union over $p \in \mathbb{Z}$, we have

$$M = \mathcal{D}_X\, \widehat{\mathcal{L}}^{>-1} \subset \widehat{\mathcal{L}}^{>-1}(*D). \tag{1.3.3}$$

The underlying $\mathbb{Q}$-local system L of a polarizable variation of Hodge structure on U is canonically extended over X as an intersection complex (see [BBD]) where L must be shifted by $\dim Z$. This can be done without assuming that $D = X \setminus U$ is a divisor with normal crossings on a smooth variety.

The second main theorem in the theory of pure Hodge modules is the stability theorem of pure Hodge modules under the direct images by projective morphisms.

Theorem 1.4 ([Sa3, Theorem 5.3.1]). *Let* $f : X \to Y$ *be a projective morphism of smooth complex algebraic varieties, and* $\mathcal{M} = ((M, F), K, \alpha_{\mathcal{M}}) \in \mathrm{MH}_Z(X, w)$. *Let* ℓ *be the first Chern class of an* f*-ample line bundle. Then the direct image* $f_*^{\mathcal{D}}(M, F)$ *as a filtered* $\mathcal{D}$*-module (see* B.3*) is strict, and we have*

$$\mathcal{H}^i f_*\mathcal{M} := (\mathcal{H}^i f_*^{\mathcal{D}}(M, F), {}^{\mathrm{m}}\mathcal{H}^i f_*K, {}^{\mathrm{m}}\mathcal{H}^i f_*\alpha_{\mathcal{M}}) \in \mathrm{MH}(Y, w+i) \quad (i \in \mathbb{Z}), \tag{1.4.1}$$

together with the isomorphisms

$$\ell^i : \mathcal{H}^{-i} f_*\mathcal{M} \xrightarrow{\sim} \mathcal{H}^i f_*\mathcal{M}(i) \quad (i > 0), \tag{1.4.2}$$

where (i) *denotes the Tate twist shifting the filtration* F *by* i*, see a remark after* (2.1.10).

Moreover, if $S : K \otimes K \to \mathbb{D}_X(-w)$ *is a polarization of* $\mathcal{M}$ *(see* (2.2.2)*), then a polarization of the* ℓ*-primitive part*

$$^P\mathcal{H}^{-i}f_*\mathcal{M} := \operatorname{Ker} \ell^{i+1} \subset \mathcal{H}^{-i}f_*\mathcal{M} \quad (i \geqslant 0)$$

is given by the restriction to the ℓ*-primitive part of the induced pairing*

$$(-1)^{i(i-1)/2}\, {}^{\mathrm{m}}\mathcal{H}f_*S \circ (id \otimes \ell^i) : {}^{\mathrm{m}}\mathcal{H}^{-i}f_*K \otimes {}^{\mathrm{m}}\mathcal{H}^{-i}f_*K \to \mathbb{D}_Y(i-w). \tag{1.4.3}$$

Remarks. (i) The action of ℓ can be defined by using C^∞ forms and the Dolbeault resolution to get a filtered complex which is filtered quasi-isomorphic to the relative de Rham complex

$$\mathrm{DR}_{X\times Y/Y}\big((i_f)_*^{\mathcal{D}}(M,F)\big),$$

which is used in the definition of the direct image in B.3. Note that the first Chern class is represented by a closed 2-form of type $(1,1)$ on X, see also [Sa3, Lemma 5.3.2]. (It is also possible to define the action of ℓ by using the restriction to a sufficiently general hyperplane section of (M,F).)

(ii) We use Deligne's sign convention of a polarization S of a Hodge structure $((H_{\mathbb{C}},F),H_{\mathbb{Q}})$, see [De3, Definition 2.1.15]; that is,

$$S(v, C\bar{v}) > 0 \quad (v \in H_{\mathbb{C}} \setminus \{0\}), \tag{1.4.4}$$

where C is the Weil operator defined by i^{p-q} on $H_{\mathbb{C}}^{p,q}$, and the Tate twist $(2\pi i)^w$ is omitted to simplify the notation. (This sign convention seems to be theoretically natural if one considers the action of the Weil restriction of $\mathbf{G_m}$, see [De2, Section 2.1].)

Recall that the usual sign convention is

$$S(Cv, \bar{v}) > 0 \quad (v \in H_{\mathbb{C}} \setminus \{0\}), \tag{1.4.5}$$

and the difference is given by the multiplication by

$$(-1)^w, \tag{1.4.6}$$

where w is the weight of the Hodge structure.

If we use the usual sign convention (1.4.5) instead of (1.4.4), then the difference (1.4.6) implies a considerable change of the formula (1.4.3) of Theorem 1.4. In fact, we have to change the sign for each direct factor with strict support Z *depending on* $d_Z := \dim Z$, since the difference of sign (1.4.6) depends on the *pointwise weight*

$$w - i - d_Z \tag{1.4.7}$$

of the generic variations of Hodge structure of each direct factor with strict support Z:

$$(\mathcal{H}^{-i}f_*\mathcal{M})_Z \subset \mathcal{H}^{-i}f_*\mathcal{M}.$$

(iii) For a polarization S of a generic variation of Hodge structure of a pure Hodge module with strict support Z, the associated polarization of the Hodge modules is defined by

$$(-1)^{d_Z(d_Z-1)/2}\, S : K|_U \otimes K|_U \to \mathbb{D}_U(-w), \tag{1.4.8}$$

on an open subvariety $U \subset Z$ where the variation of Hodge structure is defined, see [Sa3, Proposition 5.2.16] for the constant coefficient case. This can be extended to a pairing of K by using the theory of intersection complexes [BBD].

For a smooth projective variety X, it is well-known that

$$\int_X (-1)^{j(j-1)/2}\, i^{p-q}\, v \wedge \bar{v} \wedge \omega^{d_X-j} > 0, \tag{1.4.9}$$

for a nonzero element v in the primitive cohomology ${}^P H^j(X,\mathbb{C})$ of type (p,q), where ω is a Kähler form. One may think that (1.4.9) contradicts (1.4.8) combined with (1.4.3) in Theorem 1.4 for $f : X \to pt$. In fact, by setting $\xi(k) = k(k-1)/2$ for $k \in \mathbb{N}$, the difference between the sign $(-1)^{\xi(j)}$ in (1.4.9) and the product of $(-1)^{\xi(d_X)}$ in (1.4.8) with $Z = X$ and $(-1)^{\xi(d_X-j)}$ in (1.4.3) with $i = d_X - j$ does not coincide with the difference between the two sign conventions (1.4.4) and (1.4.5) which is equal to $(-1)^j$ by (1.4.6) for $w = j$, since

$$\xi(d_X) - \xi(d_X - j) - \xi(j) - j = jd_X \mod 2.$$

However, the remaining sign $(-1)^{jd_X}$ just comes from the isomorphism

$$\mathbb{R}\Gamma(X,\mathbb{C}_X)[d_X]\otimes_{\mathbb{C}}\mathbb{R}\Gamma(X,\mathbb{C}_X)[d_X] \cong \big(\mathbb{R}\Gamma(X,\mathbb{C}_X)\otimes_{\mathbb{C}}\mathbb{R}\Gamma(X,\mathbb{C}_X)\big)[2d_X]. \tag{1.4.10}$$

In fact, if we set $\mathcal{K}^\bullet := \mathbb{R}\Gamma(X,\mathbb{C}_X)$, then $\mathcal{K}^\bullet[d_X]$ is identified with $\mathbb{C}_X[d_X]\otimes_{\mathbb{C}}\mathcal{K}^\bullet$, and the above isomorphism is given by

$$\mathbb{C}[d_X]\otimes_{\mathbb{C}}\mathcal{K}^\bullet\otimes_{\mathbb{C}}\mathbb{C}[d_X]\otimes_{\mathbb{C}}\mathcal{K}^\bullet \cong \mathbb{C}[d_X]\otimes_{\mathbb{C}}\mathbb{C}[d_X]\otimes_{\mathbb{C}}\mathcal{K}^\bullet\otimes_{\mathbb{C}}\mathcal{K}^\bullet,$$

where the middle two components are exchanged, but the remaining ones are unchanged (see [De4] for the sign about single complexes associated with n-ple complexes). A similar argument is used in the proof of the anti-commutativity of the last diagram in [Sa3, Section 5.3.10] where $d_X = 1$.

2 Outline of proofs of Theorems 1.3 and 1.4

In this section we give an inductive definition of pure Hodge modules (see [Sa2], [Sa3]), and explain an outline of proofs of Theorems 1.3 and 1.4.

2.1. Admissibility condition along $g = 0$. Let X be a smooth algebraic variety, and Z be an irreducible closed subvariety of X. Let

$$\mathcal{M} = ((M,F),K,\alpha_{\mathcal{M}}) \in \mathrm{MF}(X,\mathbb{Q}),$$

with strict support Z, see 1.2. Let g be a function on X, that is, $g \in \Gamma(X, \mathcal{O}_X)$. Let $i_g : X \hookrightarrow X \times \mathbb{C}$ be the graph embedding by g. Set

$$(\widetilde{M}, F) := (i_g)_*^{\mathcal{D}}(M, F),$$

see B.3 for $(i_g)_*^{\mathcal{D}}$. (Note that the filtration F is *shifted by* 1, which is the codimension of the embedding.) We have the filtration V on $\widetilde{M}$ (see B.6).

Definition. We say that (M, F) is *admissible along* $g = 0$ (or *g-admissible* for short) in this paper if the following two conditions are satisfied:

$$t(F_p V^\alpha \widetilde{M}) = F_p V^{\alpha+1} \widetilde{M} \qquad (\forall\, \alpha > 0), \tag{2.1.1}$$

$$\partial_t(F_p \mathrm{Gr}_V^\alpha \widetilde{M}) = F_{p+1} \mathrm{Gr}_V^{\alpha-1} \widetilde{M} \qquad (\forall\, \alpha < 1,\, p \in \mathbb{Z}), \tag{2.1.2}$$

see [Sa3, 3.2.1]. (These properties were first found in the one-dimensional case, see [Sa1].)

In the case $Z \subset g^{-1}(0)$, (M, F) is g-admissible if and only if the following condition is satisfied:

$$g\, F_p M \subset F_{p-1} M \qquad (\forall\, p \in \mathbb{Z}), \tag{2.1.3}$$

see [Sa3, Lemma 3.2.6].

In the case $Z \not\subset g^{-1}(0)$, let $j : X \times \mathbb{C}^* \hookrightarrow X \times \mathbb{C}$ be the natural inclusion. We have the isomorphisms

$$F_p \widetilde{M} = \textstyle\sum_{i \geqslant 0} \partial_t^i (j_* F_{p-i} \widetilde{M} \cap V^{>0} M) \qquad (p \in \mathbb{Z}), \tag{2.1.4}$$

if (M, F) is g-admissible and moreover the following condition holds:

$$\partial_t : \mathrm{Gr}_V^1(\widetilde{M}, F) \to \mathrm{Gr}_V^0(\widetilde{M}, F[-1]) \text{ is strictly surjective}, \tag{2.1.5}$$

see [Sa3, Remark 3.2.3]. This is closely related to (1.3.2). (Forgetting F, condition (2.1.5) is equivalent to that M has no nontrivial quotient supported in $g^{-1}(0)$, see (B.6.6). The strictness of F follows from the properties of Hodge modules as is seen in (2.3.5).)

Assume (M, F) is g-admissible. We have the *nearby and vanishing cycle functors* ψ_g, φ_g defined by

$$\psi_g(M, F) := \textstyle\bigoplus_{\lambda \in \mathbb{C}_1^*} \psi_{g,\lambda}(M, F), \quad \varphi_g(M, F) := \textstyle\bigoplus_{\lambda \in \mathbb{C}_1^*} \varphi_{g,\lambda}(M, F), \tag{2.1.6}$$

$$\begin{aligned} \psi_{g,\mathbf{e}(-\alpha)}(M, F) &:= \mathrm{Gr}_V^\alpha(\widetilde{M}, F) \quad (\alpha \in (0, 1]), \\ \varphi_{g,1}(M, F) &:= \mathrm{Gr}_V^0(\widetilde{M}, F[-1]), \end{aligned} \tag{2.1.7}$$

where $\mathbb{C}_1^* := \{\lambda \in \mathbb{C}^* \mid |\lambda| = 1\}$, $\mathbf{e}(-\alpha) := \exp(-2\pi i \alpha)$, and $\psi_{g,\lambda} = \varphi_{g,\lambda}$ $(\lambda \neq 1)$ as in (A.2.8). We have $\varphi_{g,1}(M, F) = (M, F)$ by [Sa3, Lemma 3.2.6] if $\operatorname{supp} M \subset g^{-1}(0)$ and $gF_pM \subset F_{p-1}M$ $(p \in \mathbb{Z})$. Note that F is shifted by 1 when the direct image $(i_g)_*^{\mathcal{D}}$ is taken. This is the reason for which F is shifted for φ in (2.1.7), and not for ψ (for left $\mathcal{D}$-modules).

Combining these with (B.6.7) below, we get

$$\begin{aligned}\psi_g\mathcal{M} &:= (\psi_g(M,F), {}^{\mathrm{m}}\psi_g K, {}^{\mathrm{m}}\psi_g\alpha_{\mathcal{M}}),\\ \varphi_g\mathcal{M} &:= (\varphi_g(M,F), {}^{\mathrm{m}}\varphi_g K, {}^{\mathrm{m}}\varphi_g\alpha_{\mathcal{M}}) \quad \text{in } \ \mathrm{MF}(X,\mathbb{Q}).\end{aligned} \tag{2.1.8}$$

We have similarly $\psi_{g,1}\mathcal{M}$, $\varphi_{g,1}\mathcal{M}$ together with the morphisms

$$\begin{aligned}&\mathrm{can}: \psi_{g,1}\mathcal{M} \to \varphi_{g,1}\mathcal{M}, \quad \mathrm{Var}: \varphi_{g,1}\mathcal{M} \to \psi_{g,1}\mathcal{M}(-1),\\ &N: \psi_g\mathcal{M} \to \psi_g\mathcal{M}(-1), \quad N: \varphi_{g,1}\mathcal{M} \to \varphi_{g,1}\mathcal{M}(-1),\end{aligned} \tag{2.1.9}$$

such that the restrictions of can, Var, N to the $\mathcal{D}$-module part $\mathrm{Gr}_V^1\mathcal{M}$, $\mathrm{Gr}_V^0\mathcal{M}$, $\mathrm{Gr}_V^\alpha\mathcal{M}$ ($\alpha \in [0,1]$) are respectively given by $-\partial_t$, t, $-(\partial_t t - \alpha)$, and we have

$$\mathrm{Var}\circ\mathrm{can} = N \ \text{ on } \varphi_{g,1}\mathcal{M}, \quad \mathrm{can}\circ\mathrm{Var} = N \ \text{ on } \varphi_{g,1}\mathcal{M}. \tag{2.1.10}$$

Here the Tate twist (k) for $k \in \mathbb{Z}$ in general is essentially the shift of the filtration F by $[k]$. For the $\mathbb{Q}$-coefficient part, it is defined by the tensor product of $\mathbb{Q}(k) := (2\pi i)^k\mathbb{Q} \subset \mathbb{C}$ over $\mathbb{Q}$, see [De3, Definition 2.1.13]. (Similarly with $\mathbb{Q}$ replaced by any subfield $A \subset \mathbb{C}$)

By [Sa3, Lemma 5.1.4] we see that the strict support decomposition (1.2.1) holds if for any $g \in \Gamma(U, \mathcal{O}_U)$ with U an open subvariety of X, $(M,F)|_U$ is g-admissible and moreover

$$\varphi_{g,1}\mathcal{M}|_U = \mathrm{Im}\,\mathrm{can} \oplus \mathrm{Ker}\,\mathrm{Var} \quad \text{in} \quad \mathrm{MF}(U,\mathbb{Q}), \tag{2.1.11}$$

2.2. Inductive definition of pure Hodge modules. For a smooth complex algebraic variety X and an irreducible closed subvariety $Z \subset X$, we define the full subcategory $\mathrm{MH}_Z(X,w) \subset \mathrm{MF}(X,\mathbb{Q})$ by increasing induction on $d_Z := \dim Z$ as follows (see [Sa2], [Sa3]):

Case 1. If Z is a point $x \in X$, then we have an equivalent of categories

$$\begin{aligned}&(i_x)_*: \mathrm{HS}(w)^p \xrightarrow{\sim} \mathrm{MH}_{\{x\}}(X,w)\\ \text{with} \quad &(i_x)_*\big((H_{\mathbb{C}},F),H_{\mathbb{Q}}\big) = \big((i_x)_*^{\mathcal{D}}(H_{\mathbb{C}},F),(i_x)_*H_{\mathbb{Q}}\big),\end{aligned} \tag{2.2.1}$$

where $i_x : \{x\} \hookrightarrow X$ denotes the canonical inclusion, and $\mathrm{HS}(w)^p$ denotes the category of polarizable $\mathbb{Q}$-Hodge structures of weight w (see [De3]). The latter is naturally identified with a full subcategory of $\mathrm{MF}(\{x\},\mathbb{Q})$ (by setting $F_p = F^{-p}$ as usual).

Case 2. If $d_Z > 0$, then $\mathcal{M} = ((M,F),K) \in \mathrm{MF}(X,\mathbb{Q})$ with strict support Z belongs to $\mathrm{MH}_Z(X,w)$ if there is a perfect pairing (see A.3)

$$S: K \otimes_{\mathbb{Q}} K \to \mathbb{D}_X(-w) = \mathbb{Q}_X(d_X - w)[2d_X], \tag{2.2.2}$$

which is called a *polarization* of $\mathcal{M}$, and the following two conditions are satisfied:

(i) The pairing S is compatible with the Hodge filtration F in the following sense:

There is an isomorphism of filtered $\mathcal{D}$-modules

$$\mathbb{D}(M,F)=(M,F)(w),$$

which corresponds (by using (B.4.6) below) to an isomorphism defined over $\mathbb{Q}$:

$$\mathbb{D}(K)=K(w),$$

and the latter is identified with the perfect pairing S via (A.3.1).

(ii) For any Zariski-open subset $U\subset X$ and $g\in\Gamma(U,\mathcal{O}_U)$, the restriction of (M,F) to U is g-admissible, and moreover, in the case $Z\cap U\not\subset g^{-1}(0)$, we have

$$\mathrm{Gr}_k^W\psi_g\mathcal{M}|_U,\ \mathrm{Gr}_k^W\varphi_{g,1}\mathcal{M}|_U\in \mathrm{MH}_{<d_Z}(U,w), \tag{2.2.3}$$

$${}^{\mathrm{m}}\psi_g S\circ(id\otimes N^i)\text{ gives a polarization of }{}^P\mathrm{Gr}^W_{w-1+i}\psi_g\mathcal{M}|_U\quad(i\geqslant 0). \tag{2.2.4}$$

The last term of (2.2.3) is the direct sum of $\mathrm{MH}_{Z'}(U,w)$ with Z' running over closed irreducible subvarieties of U with $d_{Z'}<d_Z$. The weight filtrations W on $\psi_g\mathcal{M}|_U$, $\varphi_{g,1}\mathcal{M}|_U$ are the *monodromy filtration* associated with the action of $N:=(2\pi i)^{-1}\log T_u$ (see [De7]) which are shifted by $w-1$ and w respectively. This means that W on $\psi_g\mathcal{M}|_U$ *with filtration F forgotten* is uniquely determined by the following conditions:

$$\begin{aligned} &N(W_i\psi_g\mathcal{M}|_U)\subset(W_{i-2}\psi_g\mathcal{M}|_U)(-1)\quad(i\in\mathbb{Z}),\\ &N^i:\mathrm{Gr}^W_{w-1+i}\psi_g\mathcal{M}|_U\xrightarrow{\sim}(\mathrm{Gr}^W_{w-1-i}\psi_g\mathcal{M}|_U)(-i)\quad(i\in\mathbb{N}).\end{aligned} \tag{2.2.5}$$

Here the Tate twists may be neglected since F is forgotten in (2.2.5). (However, the last isomorphism of (2.2.5) is strictly compatible with F by (2.3.3) below if $\psi_g\mathcal{M}|_U$ belongs to $\mathrm{MHW}(U)$ in 2.3.) A similar assertion holds for $\varphi_{g,1}\mathcal{M}|_U$ with $w-1$ replaced by w.

In (2.2.4), the primitive part ${}^P\mathrm{Gr}^W_{w-1+i}\psi_g\mathcal{M}|_U$ is defined by

$${}^P\mathrm{Gr}^W_{w-1+i}\psi_g\mathcal{M}|_U:=\mathrm{Ker}\,N^{i+1}\subset\mathrm{Gr}^W_{w-1+i}\psi_g\mathcal{M}|_U, \tag{2.2.6}$$

by using the induced filtration F on the kernel. For $\psi_g S$, see (A.3.2). Note that the condition for a polarization S is also by induction on d_Z. For each direct factor of ${}^P\mathrm{Gr}^W_{w-1+i}\psi_g\mathcal{M}|_U$ with 0-dimensional strict support, we assume that $\psi_g S\circ(id\otimes N^i)$ induces a polarization of $\mathbb{Q}$-Hodge structure in the sense of [De3] where the place of the Weil operator is different from the usual one as is noted in Remark (ii) after Theorem 1.4.

2.3. Some properties of pure Hodge modules. Let $\mathrm{MH}(X,w)$ be as in (1.2.3). Let

$$\mathrm{MHW}(X)$$

be the category of *weakly mixed Hodge modules* consisting of $(\mathcal{M},W)$ with $\mathcal{M}\in\mathrm{MF}(X,\mathbb{Q})$ and W a finite increasing filtration of $\mathcal{M}$, which satisfy

$$\mathrm{Gr}^W_w\mathcal{M}\in\mathrm{MH}(X,w)\quad(\forall\,w\in\mathbb{Z}). \tag{2.3.1}$$

We have by definition the stability of pure Hodge modules by the nearby and vanishing cycle functors:

$$\psi_g\mathcal{M}|_U,\, \varphi_{g,1}\mathcal{M}|_U \in \mathrm{MHW}(U). \tag{2.3.2}$$

It is easy to show the following (see [Sa3, Proposition 5.1.14]):

MHW(X) and MH(X, w) ($w \in \mathbb{Z}$) are abelian categories such that any morphisms are strictly compatible with (F, W) and F respectively. (2.3.3)

This assertion is proved by using

$$\mathrm{Hom}(\mathcal{M}, \mathcal{M}') = 0 \text{ if } \mathcal{M} \in \mathrm{MH}(X, w),\ \mathcal{M}' \in \mathrm{MH}(X, w') \text{ with } w > w'. \tag{2.3.4}$$

This is reduced to [De3] by using the assertion that any $\mathcal{M} \in \mathrm{MH}_Z(X, w)$ is generically a variation of Hodge structure of weight $w - d_Z$. (The latter is an easy part of Theorem 1.3.)

These assertions hold without assuming polarizability (see [Sa3, Section 5.1]), and imply

$$\mathrm{can} : \psi_{g,1}\mathcal{M} \to \varphi_{g,1}\mathcal{M} \text{ is strictly surjective for } (F, W), \tag{2.3.5}$$

This assures the strict surjectivity in (2.1.5). It also gives a reason for which condition (2.2.4) is imposed only for ψ.

We prove Theorems 1.3 and 1.4 by induction on $\dim Z$ using the following rather technical key theorem:

Theorem 2.4. *Let $f : X \to Y$ be as in Theorem 1.4. Let $g \in \Gamma(Y, \mathcal{O}_Y)$. Put $h := fg$. Let $\mathcal{M} = ((M, F), K) \in \mathrm{MF}(X, \mathbb{Q})$ with strict support $Z \not\subset h^{-1}(0)$. Let $S : K \otimes K \to \mathbb{D}_X(-w)$ be a perfect pairing compatible with the filtration F as in condition (ii) in Case 2 of 2.2. Assume that (M, F) is h-admissible, we have*

$$\mathrm{Gr}_i^W \psi_h\mathcal{M},\, \mathrm{Gr}_i^W \varphi_{h,1}\mathcal{M} \in \mathrm{MH}(X, i) \quad (\forall\, i \in \mathbb{Z}), \tag{2.4.1}$$

with W as in (2.2.5), and the conclusions of Theorem 1.4 are satisfied for the N-primitive part

$$^{P_N}\mathrm{Gr}_{w-1+i}^W \psi_h\mathcal{M} \quad \textit{with polarization} \quad \psi_h S \circ (id \otimes N^i) \quad (i \geqslant 0). \tag{2.4.2}$$

Then

(i) The filtered direct image $f_*^{\mathcal{D}}(M, F)$ is strict on a sufficiently small neighborhood of $g^{-1}(0)$ (in the classical topology), and the $\mathcal{H}^i f_*^{\mathcal{D}}(M, F)$ are g-admissible.

(ii) The shifted direct image filtration $f_*^{\mathcal{D}}W[j]$ induces the monodromy filtration shifted by $w + j - 1$ on

$$\psi_g\mathcal{H}^j f_*^{\mathcal{D}}\mathcal{M} = \mathcal{H}^j f_*^{\mathcal{D}}\psi_h\mathcal{M} \quad (\forall\, j \in \mathbb{Z}), \tag{2.4.3}$$

which is denoted by W so that

$$\mathrm{Gr}_i^W \psi_g\mathcal{H}^j f_*^{\mathcal{D}}\mathcal{M} \in \mathrm{MH}(Y, i) \quad (\forall\, i \in \mathbb{Z}). \tag{2.4.4}$$

(iii) *We have isomorphisms on a sufficiently small neighborhood of* $g^{-1}(0)$*:*

$$\ell^j : \mathcal{H}^{-j} f_*^{\mathcal{D}} \mathcal{M} \xrightarrow{\sim} (\mathcal{H}^j f_*^{\mathcal{D}} \mathcal{M})(j) \quad (\forall j \geqslant 0). \tag{2.4.5}$$

(iv) *On the bi-primitive part* ${}^{P_\ell P_N}\mathrm{Gr}^W_{w-1-j+i}\psi_g \mathcal{H}^{-j} f_*^{\mathcal{D}} \mathcal{M}$ *defined by*

$$\operatorname{Ker} \ell^{j+1} \cap \operatorname{Ker} N^{i+1} \subset \mathrm{Gr}^W_{w-1-j+i}\psi_g \mathcal{H}^{-j} f_*^{\mathcal{D}} \mathcal{M}, \tag{2.4.6}$$

we have a polarization of Hodge module given by the induced pairing

$$(-1)^{j(j-1)/2}\mathrm{Gr}^W \psi_g \mathcal{H} f_* S \circ (id \otimes N^i \ell^j) \quad (\forall i, j \geqslant 0). \tag{2.4.7}$$

We first explain how the above theorem is used in the proofs of Theorems 1.3 and 1.4.

2.5. Outline of proofs of Theorems 1.3 and 1.4. We show the assertions by increasing induction on the dimension of the strict support Z. The order of the induction is rather complicated as is explained below:

Assume Theorems 1.3 and 1.4 are proved for $\dim Z < d$. Then Theorem 1.4 for $\dim Z = d$ with $f(Z) \neq pt$ follows from Theorem 2.4 where the decomposition by strict support follows from (2.1.11) (which is satisfied by using [Sa3, Corollary 4.2.4]). Using this, we can reduce the proof of Theorem 1.3 to the normal crossing case where the singular locus of M is a divisor with normal crossings. Here the filtration F can be defined by (1.3.2), and we have to show that the conditions for Hodge modules are satisfied for any locally defined functions g. This can be further reduced by using Theorem 2.4 (but not Theorem 1.4) to the case where the union of $g^{-1}(0)$ and the singular locus of M is a divisor with normal crossings. Here we can calculate explicitly the nearby and vanishing cycle functors together with the induced pairing (although these are rather complicated), see [Sa7] for details.

Now we have to prove Theorem 1.4 in the case $\dim Z = d$ and $f(Z) = pt$. Here we may assume that $X = \mathbb{P}^n$, $Y = pt$. Let $\pi : \widetilde{X} \to X$ be the blow-up along the intersection Z of two sufficiently general hyperplanes of $\mathbb{P}^n$. By Theorem 1.3 for $\dim Z = d$, there is

$$\widetilde{\mathcal{M}} = ((\widetilde{M}, F), \widetilde{K}) \in \mathrm{MH}(\widetilde{X}, w) \quad \text{with} \quad \widetilde{\mathcal{M}}|_{\widetilde{X}\setminus\pi^{-1}(Z)} = \mathcal{M}|_{X\setminus Z}.$$

We have a polarization $\widetilde{S}$ of $\widetilde{\mathcal{M}}$ extending the restriction of a polarization S of $\mathcal{M}$ to the complement of $Z \subset X$.

By Theorem 1.4 for $f(Z) \neq 0$, we see that $\mathcal{M}$ is a direct factor of $\mathcal{H}^0\pi_*\mathcal{M}$. Moreover its complement is isomorphic to $\mathcal{M}_Z(-1)$ since Z is sufficiently general, where $\mathcal{M}_Z$ is the noncharacteristic restriction of $\mathcal{M}$ to Z; more precisely,

$$\mathcal{M}_Z = ((M_Z, F), K_Z[-2]),$$

with (M_Z, F) and K_Z respectively the noncharacteristic restrictions of (M, F) and K to Z.

So we have the direct sum decomposition

$$\mathcal{H}^0\pi_*\widetilde{\mathcal{M}} = \mathcal{M} \oplus \mathcal{M}_Z(-1), \tag{2.5.1}$$

with $\mathcal{H}^j\pi_*\widetilde{\mathcal{M}} = 0$ for $j \neq 0$. Here $\pi_*\widetilde{S}$ is compatible with this decomposition, and its restriction to $\mathcal{M}$ coincides with S (by using (A.3.1) together with the remark after (1.2.2)). We then get the direct sum decompositions

$$H^j(\widetilde{X}, \widetilde{\mathcal{M}}) = H^j(X, \mathcal{M}) \oplus H^{j-2}(Z, \mathcal{M}_Z)(-1) \qquad (j \in \mathbb{Z}). \tag{2.5.2}$$

where $H^j(X, \mathcal{M}) := H^j(a_X)_*\mathcal{M}$ for the structure morphism $a_X : X \to pt$, and similarly for $H^j(\widetilde{X}, \widetilde{\mathcal{M}})$, etc. The above argument implies that these direct sum decompositions are compatible with the induced pairing by $\widetilde{S}$, and moreover its restriction to the first factors coincides with the induced pairing by S.

We have the Lefschetz pencil

$$p : \widetilde{X} \to \mathbb{P}^1,$$

and Theorem 1.4 for $f(Z) \neq pt$ can be applied to this. So the proof of Theorem 1.4 for $\dim Z = d$ and $f(Z) = pt$ is reduced to the case $X = \mathbb{P}^1$, where we can apply Zucker's result [Zu]. (Note that Zucker gave an "algebraic description" of the Hodge filtration F using holomorphic differential forms, see [Zu, Corollary 6.15 and Proposition 9.1].) However, we have to make some more calculations about polarizations on primitive classes related to the Leray spectral sequence, etc. (which are not quite trivial, see [Sa3, Sections 5.3.8—11] for details).

The Lefschetz pencil is also used in an essential way for the proof of the Weil conjecture and the hard Lefschetz theorem in [De7]. The reduction to the curve case is by analogy with the ℓ-adic case in some sense.

Remarks. (i) For the calculation of the nearby and vanishing cycle functors in the normal crossing case, we use the so-called "combinatorial description" of mixed Hodge modules with normal crossing singular loci. Here the *compatibility* of the $d_X + 1$ filtrations $F, V_{(i)}$ $(i \in [1, d_X])$ is quite essential, where $V_{(i)}$ is the V-filtration along $x_i = 0$, and the x_i are local coordinates compatible with the singular locus of a mixed Hodge module. Note, however, that this description *never* gives an equivalence of categories (consider, for instance, the case of variations of mixed Hodge structure having no singular loci; in fact, this "Hodge-theoretic combinatorial description" gives only the information of the fiber at the fixed point). Nevertheless it is quite useful when it is combined with the Verdier-type extension theorem [Ve2] inductively, see also [Sa7, Proposition 3.13], etc.

It seems rather easy to predict Hodge-theoretic combinatorial formulas for the nearby and vanishing cycle functors together with the induced pairing in the normal crossing case. These are implicitly related with Beilinson's construction of nearby cycles [Bei2], see [Sa8] for the mixed Hodge module case. It seems more difficult to prove that these formulas actually hold, see for instance [Sa4]. (Note that the argument in the Appendix of [Sa7] was simplified by the writer. The original argument used the reduction to the 2-dimensional case, and was much more complicated.)

(ii) The results of Cattani, Kaplan, Schmid ([CK], [CKS1], [CKS2], [CKS3]) are used in an essential way for the above "Hodge-theoretic combinatorial description". For instance, the *descent lemma* in [CKS1], [CKS3] is crucial to the "combinatorial description" of the pure Hodge module corresponding to the intersection complex. (This lemma is called "the vanishing cycle theorem" in [KK4], which does not seem to be contained in [KK2].)

(iii) It is still an open problem whether the Hodge structure obtained by the L^2 method in [CKS3], [KK4] has an "algebraic description" using holomorphic differential forms, and in the algebraic case, whether it coincides with the Hodge structure obtained by the theory of mixed Hodge modules. (It has been expected that the detailed versions of [KK3], [KK5] would give a positive answer to these problems, and some people thought that [KK2], [KK4] were written for this purpose. As for [BrSY], it seems rather difficult to apply it to filtered L^2-sheaf complexes.) We have to assume that polarizable variations of Hodge structure are geometric ones in [Sa9], and also in [PS] for the analytic case.

(iv) In the curve case, the answer to the above first problem was already given in [Zu, Corollary 6.15 and Proposition 9.1], and the second problem is then easy to solve, see [Sa3, Section 5.3.10].

(v) It does not seem easy to generalize the results in [CKS3], [KK4] to the case of a "tubular neighborhood" of a subvariety in a smooth complex algebraic variety, since we would have to take a *complete* metric to get a Hodge structure by applying a standard method.

2.6. Outline of proof of Theorem 2.4. We have to study the weight spectral sequence

$$E_1^{-k,j+k} = \mathcal{H}^j f_*\mathrm{Gr}_k^W \psi_h\mathcal{M} \Longrightarrow \mathcal{H}^j f_*\psi_h\mathcal{M}, \tag{2.6.1}$$

and similarly with ψ_h replaced by $\varphi_{h,1}$. This spectral sequence is defined in $\mathrm{MF}(Y,\mathbb{Q})$ if the differentials d_r $(r \geqslant 1)$ are strictly compatible with the filtration F inductively, see [Sa3, 1.3.6]. The last assertion for $r=1$ follows from (2.3.3), since d_1 preserves the Hodge filtration F and also the weight filtration W (the latter is shifted depending on the degree j of the cohomology sheaf $\mathcal{H}^j f_*\psi_h\mathcal{M}$). For $r \geqslant 2$, we can show $d_r = 0$ inductively by using (2.3.4). In particular, the E_2-degeneration of the spectral sequence follows. The above argument also implies the strictness of the filtration F on the direct image $f_*^{\mathcal{D}}\psi_h\mathcal{M}$ by using the theory of compatible filtrations (*loc. cit.*). A similar assertion holds for $\varphi_{g,1}\mathcal{M}$. These assertions imply the assertion (i) of Theorem 2.4 by using the completion by the V-filtration, see [Sa3, Section 3.3] for details.

To show the remaining assertions, we have to show that the "bi-symmetry" for the actions of ℓ, N on the E_1-term is preserved on the E_2-terms, and moreover the "bi-primitive part" of the E_2-term is *represented* by the bi-primitive part of the E_1-term. Using the strict support decomposition together with the easy part of Theorem 1.3, we can reduce the assertions to the case where the spectral sequence is defined in the category of Hodge structures. Theorem 2.4 then follows from the theory of bi-graded Hodge structures of Lefschetz type as in the proof of [Sa3, Proposition 4.2.2] (see also [Sa18] where a slightly better explanation is given).

Remarks. (i) Signs were not determined in [Sa3, Proposition 4.2.2], since this was a very subtle issue at that time (see for instance [De3, Section 2.2.5] where a problem of sign for Chern classes was raised.) In the case of the nearby cycles of the constant sheaf $\mathbb{Q}_X$ in the normal crossing case (that is, in the case of Steenbrink [St]), the primitive part

$$^P\mathrm{Gr}^W_{d_X-1+i}\psi_{h,1}(\mathbb{Q}_X[d_X-1])$$

is the direct sum of the constant sheaves supported on intersections of $i+1$ irreducible components of $h^{-1}(0)$. In particular, its support is pure dimensional, and the signs should depend only on this dimension. Then there would be no problem for using [Sa3, Proposition 4.2.2] in this case. In the general case, however, the situation is much more complicated. In fact, there may be direct factors $(^P\mathrm{Gr}^W_k\psi_h\mathcal{M})_Z$ of the primitive part $^P\mathrm{Gr}^W_k\psi_h\mathcal{M}$ which have strict supports Z of *various dimensions*, and surject to the *same* closed subvariety of Y (for instance, if $g^{-1}(0)=\{0\}$). In this case, we have to determine exactly the sign for each direct factor so that the *positivity* becomes compatible among the direct images of direct factors with various strict support dimensions.

(ii) Precise signs are written in [Sa3, Lemma 5.3.6] following Deligne's sign system [De8] (see also [GN]). Note that the conclusion of Lemma 5.3.6 follows from the *proof* of [Sa3, Proposition 4.2.2], since the hypothesis of the lemma is stronger than that of the proposition and the conclusion is essentially the same (that is, the E_2-term is bi-symmetric for ℓ, N and its bi-primitive part is represented by the bi-primitive part of the E_1-term). A pairing on the E_1-term of the Steenbrink weight spectral sequence [St] satisfying Deligne's sign system [De8] is constructed in [GN] although its relation with the pairing induced by the nearby cycle functor does not seem to be clear (for instance, an isomorphism like (1.4.10) does not seem to be used there).

(iii) An argument showing the decomposition (2.1.11) is noted in [Sa3, Lemma 5.2.5] which can replace [Sa3, Corollary 4.2.4] if one is quite sure that the signs in the lemma really hold in the case one is considering. In fact, the assertion is very sensitive to the signs: if the signs are modified, then the role of H and H' can be reversed, and we may get a decomposition of H instead of H'. (There is a misprint in the last line of the lemma: S' should be H'.) If one is not completely sure whether the signs of the lemma really hold in the case under consideration, then it is still possible to use [Sa3, Corollary 4.2.4] at least in the constant sheaf case with normal crossing singularities as is explained in Remark (i) above.

3 Mixed Hodge modules

In this section we explain a simplified definition of mixed Hodge modules following [Sa29].

3.1. Admissibility condition for weakly mixed Hodge modules. Let X be a smooth complex algebraic variety, and $g \in \Gamma(X, \mathcal{O}_X)$. For a weakly mixed Hodge module

$$(\mathcal{M}, W) = ((M; F, W), (K, W)) \in \mathrm{MHW}(X),$$

(see (2.3.1)), set

$$(\widetilde{M}; F, W) = (i_g)_*^{\mathcal{D}}(M; F, W),$$

where W is not shifted under the direct image. We have the filtration V on $\widetilde{M}$ as in B.6. We define the filtration L on the nearby and vanishing cycle functors by

$$L_k\psi_g\mathcal{M} := \psi_g W_{k+1}\mathcal{M}, \quad L_k\varphi_{g,1}\mathcal{M} = \varphi_{g,1}W_k\mathcal{M} \quad (\forall\, k \in \mathbb{Z}). \tag{3.1.1}$$

Here the filtration F can be neglected when the filtration L is defined.

We say that $\mathcal{M}$ is *admissible* along $g = 0$ (or *g-admissible* for short) if the following two conditions are satisfied:

$$\text{Three filtrations } F, W, V \text{ on } \widetilde{M} \text{ are compatible filtrations (see C.2).} \tag{3.1.2}$$

$$\text{There is the relative monodromy filtration } W \text{ on } (\psi_g\mathcal{M}, L),\ (\varphi_{g,1}\mathcal{M}, L). \tag{3.1.3}$$

The last condition means that there is a unique filtration W on $\psi_g\mathcal{M}$ satisfying the following two conditions:

$$\begin{aligned} N(W_i\psi_g\mathcal{M}) &\subset W_{i-2}\psi_g\mathcal{M}(-1) \quad (\forall\, i \in \mathbb{Z}),\\ N^i : \mathrm{Gr}^W_{k+i}\mathrm{Gr}^L_k\psi_g\mathcal{M} &\xrightarrow{\sim} \mathrm{Gr}^W_{k-i}\mathrm{Gr}^L_k\psi_g\mathcal{M}(-i) \quad (\forall\, i \in \mathbb{N},\ k \in \mathbb{Z}), \end{aligned} \tag{3.1.4}$$

and similarly for $\varphi_{g,1}\mathcal{M}$, see [De7], [StZ]. (Here the filtration F can be forgotten. However, the last isomorphism is compatible with F by (2.3.3) if $(\psi_f\mathcal{M}, W)$, $(\varphi_{f,1}\mathcal{M}, W) \in \mathrm{MHW}(X)$.)

Remark. Let X be a smooth complex variety, and $\overline{X}$ be a smooth compactification such that $D := \overline{X} \setminus X$ is a divisor with simple normal crossings. Let

$$(\mathcal{M}, W) = \big((M; , F, W), (H, W)\big)$$

be a variation of mixed Hodge structure on X such that $\mathrm{Gr}^W_k\mathcal{M}$ are polarizable pure Hodge structures of weight k for any $k \in \mathbb{Z}$, where $(M; F, W)$ is the underlying bi-filtered $\mathcal{O}_X$-module, and (H, W) is the underlying filtered $\mathbb{Q}$-local system.

Assume the local monodromies of H are all *unipotent*. Let $\overline{M}$ be the *canonical* Deligne extension of M over $\overline{X}$ (that is, the residues of the logarithmic connections are all *nilpotent*, see [De1]). The filtrations F, W are naturally extended on $\overline{M}$ by taking the intersection with $\overline{M}$ of the open direct images of F, W under the inclusion $j : X \hookrightarrow \overline{X}$. Let D_i be the irreducible components of D ($i \in [1, r]$). Let T_i be the local monodromy around a general point of D_i, which is defined on the fiber H_{x_0} at a base point $x_0 \in X$, and is unipotent by hypothesis. (This is well-defined up to a conjugate compatible with W.) Set $N_i := (2\pi i)^{-1}\log T_i$. We denote by L the filtration W on H_{x_0}.

Under the above notation and assumption, $(\mathcal{M}, W)$ is an *admissible variation of mixed Hodge structure* in the sense of [StZ], [Ka5] if and only if the following two conditions are satisfied:

$$\mathrm{Gr}_F^p \mathrm{Gr}_k^W \overline{M} \text{ are locally free } \mathcal{O}_{\overline{X}}\text{-modules for any } p, k \in \mathbb{Z}. \tag{3.1.5}$$

$$\text{There is the relative monodromy filtration on } (H_{x_0}, L) \text{ for each } N_i. \tag{3.1.6}$$

(The last condition means that (3.1.4) holds with $\psi_g \mathcal{M}$ replaced by H_{x_0}.) In fact, this equivalence follows from [Ka5, Theorem 4.5.2].

In the non-unipotent local monodromy case, let $\rho : X' \to X$ be a generically finite morphism of complex algebraic varieties such that $\rho^*\mathcal{M}$ has unipotent local monodromies around the divisor at infinity of a compactification of X' (by replacing X with a non-empty open subvariety if necessary). Then $\mathcal{M}$ is an admissible variation if and only if $\rho^*\mathcal{M}$ is. (In fact, may assume that ρ is finite étale by shrinking X. Then we can take the direct image.)

3.2. Well-definedness of open direct images. Let D be a locally principal divisor on a smooth complex algebraic variety X. Set $X' := X \setminus D$ with $j : X' \hookrightarrow X$ the inclusion. We say that the open direct images $j_!, j_*$ are well-defined for $\mathcal{M}' \in \mathrm{MHW}(X')$, if there are

$$\begin{gathered} \mathcal{M}'_!,\ \mathcal{M}'_* \in \mathrm{MHW}(X), \quad \text{satisfying} \\ \mathrm{rat}(\mathcal{M}'_!) = j_! K', \quad \mathrm{rat}(\mathcal{M}'_*) = \mathbb{R}j_* K' \quad \text{with} \quad K' := \mathrm{rat}(\mathcal{M}'), \end{gathered} \tag{3.2.1}$$

(see (1.1.6) for rat), and moreover the following condition is satisfied:

$$\mathcal{M}'_!,\ \mathcal{M}'_* \text{ are } g\text{-admissible for any } g \in \Gamma(U, \mathcal{O}_U) \text{ with } g^{-1}(0)_{\mathrm{red}} = D_{\mathrm{red}} \cap U. \tag{A3}$$

Here U is any open subvariety of X. If the above condition is satisfied, we then define

$$j_! \mathcal{M}' := \mathcal{M}'_!, \quad j_* \mathcal{M}' := \mathcal{M}'_*. \tag{3.2.2}$$

If $\mathcal{M}' = j^{-1}\mathcal{M}$ with $\mathcal{M} \in \mathrm{MHW}(X)$ and $\mathcal{M}$ is g-admissible, then we have the canonical morphisms (see [Sa7, Proposition 2.11])

$$j_! j^{-1} \mathcal{M} \to \mathcal{M}, \quad \mathcal{M} \to j_* j^{-1} \mathcal{M}. \tag{3.2.3}$$

3.3. Definition of mixed Hodge modules. Let X be a smooth complex algebraic variety. The category of mixed Hodge modules $\mathrm{MHM}(X)$ is the abelian full subcategory of $\mathrm{MHW}(X)$ in 2.3 defined by increasing induction on the dimension d of the support as follows:

For $\mathcal{M} \in \mathrm{MHW}(X)$ with $\mathrm{supp}\, \mathcal{M} = Z$, we have $\mathcal{M} \in \mathrm{MHM}(X)$ if and only if, for any $x \in X$, there is a Zariski-open neighborhood U_x of x in X together with $g_x \in \Gamma(U_x, \mathcal{O}_{U_x})$ such that

$$\dim Z \cap U_x \cap g_x^{-1}(0) < \dim Z,$$

$Z'_x := Z \cap U_x \setminus g_x^{-1}(0)$ is smooth, and moreover the following two conditions are satisfied:

$$\mathcal{M}|_{Z'_x} \text{ is an admissible variation of mixed Hodge structure.} \tag{3.3.1}$$

$$\mathcal{M}|_{U_x} \text{ is } g_x\text{-admissible, and } \varphi_{g_x,1}\mathcal{M}|_{U_x} \in \mathrm{MHM}(U_x). \tag{3.3.2}$$

More precisely, (3.3.1) means that $\mathcal{M}|_{U'_x}$ is isomorphic to the direct image of an admissible variation of mixed Hodge structure on Z'_x by the closed embedding

$$i_{Z'_x} : Z'_z \hookrightarrow U'_x := U_x \setminus g_x^{-1}(0).$$

If $Z = \{x\}$ for $x \in X$, then we set

$$\mathrm{MHM}_{\{x\}}(X) := \mathrm{MHW}_{\{x\}}(X) = \mathrm{MHS}(\mathbb{Q}), \tag{3.3.3}$$

where the first and second categories are respectively full subcategories of $\mathrm{MHM}(X)$ and $\mathrm{MHW}(X)$ consisting of objects supported on x, and the last one is the category of graded-polarizable mixed $\mathbb{Q}$-Hodge structures [De3]. (Here the direct image by $\{x\} \hookrightarrow X$ is used.)

This definition is justified by the following (see [Sa29, Theorem 1]).

Theorem 3.4. *Conditions* (3.3.1) *and* (3.3.2) *are independent of the choice of* U_x, g_x. *More precisely, if they are satisfied for some* U_x, g_x *for each* $x \in Z$, *then* (3.3.2) *is satisfied for any* U_x, g_x, *and* (3.3.1) *is satisfied in case* $\mathrm{rat}(\mathcal{M}')$ *is a local system up to a shift of complex, see* (1.1.6) *for* rat.

We have moreover the following (see [Sa29, Theorem 2]).

Theorem 3.5. *The categories* $\mathrm{MHM}(X)$ *for smooth complex algebraic varieties* X *are stable by the canonically defined cohomological functors* $\mathcal{H}^j f_*$, $\mathcal{H}^j f_!$, $\mathcal{H}^j f^*$, $\mathcal{H}^j f^!$, ψ_g, $\varphi_{g,1}$, $\boxtimes$, $\mathbb{D}$, *where* f *is a morphism of smooth complex algebraic varieties and* $g \in \Gamma(X, \mathcal{O}_X)$. *Moreover these functors are compatible with the corresponding functors of the underlying* $\mathbb{Q}$*-complexes via the forgetful functor* rat *in* (1.1.6).

The proofs of these theorems use Beilinson's maximal extension together with the stability by subquotients systematically. The well-definedness of open direct images in 3.3 is reduced to the normal crossing case, see [Sa29]. Combining Theorem 3.5 with the construction in [Sa7], we can get the following (see [Sa29, Corollary 1]).

Theorem 3.6. *There are canonically defined functors* f_*, $f_!$, f^*, $f^!$, ψ_g, $\varphi_{g,1}$, $\boxtimes$, $\mathbb{D}$, $\otimes$, $\mathcal{H}om$ *between the bounded derived categories* $D^b\mathrm{MHM}(X)$ *for smooth complex algebraic varieties* X *so that we have the canonical isomorphisms* $H^j f_* = \mathcal{H}^j f_*$, *etc., where* f *is a morphism of smooth complex algebraic varieties,* $g \in \Gamma(X, \mathcal{O}_X)$, H^j *is the usual cohomology functor of the derived categories, and* $\mathcal{H}^j f_*$, *etc. are as in Theorem* 3.5. *Moreover the above functors between the* $D^b\mathrm{MHM}(X)$ *are compatible with the corresponding functors of the underlying* $\mathbb{Q}$*-complexes via the forgetful functor* rat.

3.7. Some notes on references about applications. Since there is no more space to explain about applications of mixed Hodge modules, we indicate some references here. These are not intended to be complete.

For applications to algebraic cycles, see [BRS], [BFNP], [MüS], [NS], [RS1], [RS2], [Sa6], [Sa12], [Sa13], [Sa15], [Sa20], [Sa21], [Sa22], [Sa25], [Sa26], [Sa27], [SS2], etc. Some of them are related to normal functions. For the latter, see also [Sa8], [Sa17], [Sa30], [SS1], [Schn], etc.

Related to mixed Hodge structures on cohomologies of algebraic varieties, see [BDS], [DS1], [DS2], [DS3], [DS5], [DS6], [DSW], [DuS], [OS], [PS], [Sa19], [Sa28], [SZ], etc.

About Bernstein-Sato polynomials, Steenbrink spectra, and multiplier ideals, see [Bu], [BMS], [BS1], [BS2], [BSY], [DMS], [DMST], [DS4], [DS7], [MP], [Sa10], [Sa14], [Sa16], [Sa23], [Sa24], [Sa31], [Sa32], etc.

For direct images of dualizing sheaves and vanishing theorems, see [FFT], [Sa11], [Su], etc. Concerning Hirzebruch characteristic classes, see [BrScY], [MS], [MSS1], [MSS2], [MSS3], [MSS4], etc.

Appendix A Hypersheaves

In this appendix we review some basics of hypersheaves, see Convention 3.

A.1. A-complexes. Let X be a complex algebraic variety or a complex analytic space, and A be a subfield of $\mathbb{C}$. We denote by $D_c^b(X,A)$ the derived category of bounded complexes of A-modules with constructible cohomology sheaves, see [Ve1], etc. In the algebraic case, we use the classical topology for the sheaf complexes although we assume that *stratifications are algebraic.*

The category of hypersheaves $\mathbf{HS}(X,A)$ is the *full subcategory* of $D_c^b(X,A)$ consisting of objects K satisfying the condition:

$$\dim\operatorname{supp}\mathcal{H}^iK \leqslant -i, \quad \dim\operatorname{supp}\mathcal{H}^i\mathbb{D}(K) \leqslant -i \quad (\forall\, i\in\mathbb{Z}). \tag{A.1.1}$$

Here $\mathcal{H}^iK$ is the ith cohomology sheaf of K in the usual sense, and $\mathbb{D}(K)$ is the dual of K. The latter can be defined by

$$\mathbb{D}(K) := \mathbb{R}\mathcal{H}om_A(K,\mathbb{D}_X), \tag{A.1.2}$$

with $\mathbb{D}_X$ the dualizing sheaf in $D_c^b(X,A)$. In the *smooth* case (by taking an embedding into smooth varieties), it can be defined by

$$\mathbb{D}_X := A_X(d_X)[2d_X], \tag{A.1.3}$$

with $d_X := \dim X$.

By [BBD], $\mathbf{HS}(X,A)$ is an *abelian category*, and there are canonical cohomological functors

$${}^{\mathrm{m}}\mathcal{H}^i : D_c^b(X,A) \to \mathbf{HS}(X,A) \quad (i\in\mathbb{Z}), \tag{A.1.4}$$

where the superscript ${}^{\mathrm{m}}$ means the "middle perversity".

A.2. Nearby and vanishing cycles. Let g be a holomorphic function on an analytic space X. Let $\Delta \subset \mathbb{C}$ be a sufficiently small open disk with center 0, and $\pi : \widetilde{\Delta^*} \to \Delta^*$ be a universal covering of the punctured disk Δ^*. Let $\pi' : \widetilde{\Delta^*} \to \mathbb{C}$ be its composition with the inclusion $\Delta^* \hookrightarrow \mathbb{C}$. Let X_∞ be the base change of X by π'. We denote by $\widetilde{j} : X_\infty \to X$ the base change of π' by g. Set $X_0 := f^{-1}(0)$ with $i_0 : X_0 \hookrightarrow X$ the canonical inclusion. The nearby and vanishing cycle functors

$$\psi_g, \varphi_g : D_c^b(X, A) \to D_c^b(X_0, A)$$

are defined as in [De5] by

$$\psi_g K := i_0^* \mathbb{R}\widetilde{j}_* \widetilde{j}^* K, \ \ \varphi_g K := C(i_0^* K \to \psi_g K) \quad \text{for} \ \ K \in D_c^b(X, A), \tag{A.2.1}$$

where we take a flasque resolution of K to define $\mathbb{R}\widetilde{j}_*\widetilde{j}^*K$ and also the mapping cone. By definition we have a distinguished triangle

$$i^*K \to \psi_g K \to \varphi_g K \xrightarrow{+1}. \tag{A.2.2}$$

The action of the monodromy T is defined by γ^* with γ the automorphism of $\widetilde{\Delta^*}$ over Δ^* defined by $z \mapsto z+1$. Here $\widetilde{\Delta^*}$ is identified with $\{z \in \mathbb{C} \mid \operatorname{Im} z > r\}$ for some $r > 0$ and π is given by $z \mapsto t := \exp(2\pi i z)$. (This is compatible with the usual definition of the monodromy of a local system L on Δ^*. In fact, $(\gamma^*\sigma)(z_0) = \sigma(z_0 + 1)$ for $\sigma \in \Gamma(\widetilde{\Delta^*}, \pi^*L)$ with z_0 a base point of $\widetilde{\Delta^*}$, and the monodromy is given by the composition of canonical isomorphisms: $L_{\pi(z_0)} = (\pi^*L)_{z_0} = \Gamma(\widetilde{\Delta^*}, \pi^*L) = (\pi^*L)_{z_0+1} = L_{\pi(z_0)}$.) There is a nonzero minimal polynomial for T locally on X, and this implies the Jordan decomposition $T = T_sT_u$ (with T_s, T_u polynomials in T locally on X).

Assume $K \in \mathbf{HS}(X, A)$. Set

$$^{\mathrm{m}}\psi_g K := \psi_g K[-1], \ ^{\mathrm{m}}\varphi_g K := \varphi_g K[-1]. \tag{A.2.3}$$

Then

$$^{\mathrm{m}}\psi_g K, \ ^{\mathrm{m}}\varphi_g K \in \mathbf{HS}(X_0, A). \tag{A.2.4}$$

This follows for instance from [Ka3], [Ma3] by using the Riemann-Hilbert correspondence.

In the case $A = \mathbb{C}$, this implies the decompositions in the abelian category $\mathbf{HS}(X_0, A)$:

$$^{\mathrm{m}}\psi_g K = \textstyle\bigoplus_{\lambda \in \mathbb{C}^*} {}^{\mathrm{m}}\psi_{g,\lambda} K, \quad ^{\mathrm{m}}\varphi_g K = \textstyle\bigoplus_{\lambda \in \mathbb{C}^*} {}^{\mathrm{m}}\varphi_{g,\lambda} K, \tag{A.2.5}$$

(which are locally finite direct sum decompositions), where

$$^{\mathrm{m}}\psi_{g,\lambda} K := \operatorname{Ker}(T_s - \lambda) \subset {}^{\mathrm{m}}\psi_g K, \quad ^{\mathrm{m}}\varphi_{g,\lambda} K := \operatorname{Ker}(T_s - \lambda) \subset {}^{\mathrm{m}}\varphi_g K. \tag{A.2.6}$$

In the case $A \subset \mathbb{C}$, we have only the decompositions

$$^{\mathrm{m}}\psi_g K = {}^{\mathrm{m}}\psi_{g,1} K \oplus {}^{\mathrm{m}}\psi_{g,\neq 1} K, \quad ^{\mathrm{m}}\varphi_g K = {}^{\mathrm{m}}\varphi_{g,1} K \oplus {}^{\mathrm{m}}\varphi_{g,\neq 1} K, \tag{A.2.7}$$

which are compatible with the above decompositions after taking the scalar extension by $A \hookrightarrow \mathbb{C}$.

By (A.2.2) we have the canonical isomorphisms

$$ {}^{\mathrm{m}}\psi_{g,\neq 1}K \xrightarrow{\sim} {}^{\mathrm{m}}\varphi_{g,\neq 1}K, \quad {}^{\mathrm{m}}\psi_{g,\lambda}K \xrightarrow{\sim} {}^{\mathrm{m}}\varphi_{g,\lambda}K \quad (\lambda \neq 1,\, A = \mathbb{C}), \tag{A.2.8} $$

since the action of T on i^*K is trivial.

If $K = A_X$ and X is a smooth algebraic variety (or a complex manifold with X_0 compact), then the nearby cycle functor $\psi_g A_X$ is also defined by

$$ \psi_g A_X = \mathbb{R}\rho_* A_{X_c}, \tag{A.2.9} $$

where $X_c := f^{-1}(c) \subset X$ with $c \in \mathbb{C}^*$ sufficiently near 0, and $\rho : X_c \to X_0$ is an appropriate contraction morphism. The latter is constructed by using an embedded resolution of $X_0 \subset X$.

A.3. Compatibility with the dual functor $\mathbb{D}$. We say that a pairing

$$ K \otimes_A K' \to \mathbb{D}_X(k) $$

is a *perfect pairing* (with $k \in \mathbb{Z}$) if its corresponding morphism

$$ K \to \mathbb{D}(K')(k) = \mathbb{R}\mathcal{H}om_A(K', \mathbb{D}_X)(k) $$

is an isomorphism in $D^b_c(X, A)$. Here $\mathbb{D}_X$ is as in (A.1.2), and the above correspondence comes from the isomorphism

$$ \mathrm{Hom}(K \otimes K', \mathbb{D}_X(k)) = \mathrm{Hom}(K, \mathbb{R}\mathcal{H}om_A(K', \mathbb{D}_X(k))). \tag{A.3.1} $$

The Tate twist (k) is defined as in [De2, Definition 2.1.13], see also a remark after (2.1.10). This can be neglected if $i = \sqrt{-1}$ is chosen. However, this twist is quite useful in order to keep track of "weight". In fact, if $K = K'$ and it has pure weight w, then $\mathbb{D}K$ should have weight $-w$, and the above k must be equal to $-w$, since the Tate twist (k) changes the weight by $-2k$ (*loc. cit.*).

Assume there is a perfect pairing

$$ S : K \otimes_A K' \to \mathbb{D}_X(-w) \quad \text{for } K, K' \in \mathbf{HS}(X, A). $$

It induces a canonical pairing

$$ \psi_g S : \psi_g K \otimes_A \psi_g K' \to \psi_g \mathbb{D}_X(-w). \tag{A.3.2} $$

Assume X is *smooth* (by taking an embedding into a smooth variety), and X_0 is also *smooth* (by replacing K with its direct image under the graph embedding by g). Then we have

$$ \psi_g \mathbb{D}_X = A_{X_0}(d_X - w)[2d_X] = \mathbb{D}_{X_0}(1 - w)[2], \tag{A.3.3} $$

and (A.3.2) indues a canonical perfect pairing

$$ {}^{\mathrm{m}}\psi_g S : {}^{\mathrm{m}}\psi_g K \otimes_A {}^{\mathrm{m}}\psi_g K' \to \mathbb{D}_{X_0}(1 - w). \tag{A.3.4} $$

Here some sign appears, and this is closely related to the sign in (1.4.8).

The above construction is compatible with the monodromy T, that is,

$$^{\mathrm{m}}\psi_g S = {}^{\mathrm{m}}\psi_g S \circ (T \otimes T).$$

Since T^e is unipotent for some $e \in \mathbb{Z}_{>0}$, this implies

$$^{\mathrm{m}}\psi_g S \circ (N \otimes id) = -{}^{\mathrm{m}}\psi_g S \circ (id \otimes N), \quad {}^{\mathrm{m}}\psi_g S = {}^{\mathrm{m}}\psi_g S \circ (T_s \otimes T_s), \tag{A.3.5}$$

where $T = T_s T_u$ is the Jordan decomposition, and $N := (2\pi i)^{-1} \log T_u$.

We then get the induced perfect pairings

$$\begin{aligned} {}^{\mathrm{m}}\psi_{g,1} S &: {}^{\mathrm{m}}\psi_{g,1} K \otimes_A {}^{\mathrm{m}}\psi_{g,1} K' \to \mathbb{D}_{X_0}(1-w), \\ {}^{\mathrm{m}}\psi_{g,\neq 1} S &: {}^{\mathrm{m}}\psi_{g,\neq 1} K \otimes_A {}^{\mathrm{m}}\psi_{g,\neq 1} K' \to \mathbb{D}_{X_0}(1-w), \\ {}^{\mathrm{m}}\psi_{g,\lambda} S &: {}^{\mathrm{m}}\psi_{g,\lambda} K \otimes_A {}^{\mathrm{m}}\psi_{g,\lambda^{-1}} K' \to \mathbb{D}_{X_0}(1-w) \quad (A = \mathbb{C}). \end{aligned} \tag{A.3.6}$$

For the vanishing cycle functor φ_g, we have the induced perfect pairing

$$^{\mathrm{m}}\varphi_{g,1} S : {}^{\mathrm{m}}\varphi_{g,1} K \otimes_A {}^{\mathrm{m}}\varphi_{g,1} K' \to \mathbb{D}_{X_0}(-w), \tag{A.3.7}$$

satisfying

$$^{\mathrm{m}}\varphi_{g,1} S \circ (\mathrm{can} \otimes id) = {}^{\mathrm{m}}\psi_{g,1} S \circ (id \otimes \mathrm{Var}), \tag{A.3.8}$$

where the morphisms

$$\mathrm{can} : {}^{\mathrm{m}}\psi_{g,1} K \to {}^{\mathrm{m}}\varphi_{t,1} K, \quad \mathrm{Var} : {}^{\mathrm{m}}\varphi_{g,1} K' \to {}^{\mathrm{m}}\psi_{t,1} K'(-1), \tag{A.3.9}$$

are constructed in [Sa3, Section 5.2.1]. (These correspond respectively to the morphisms $-\mathrm{Gr}_V \partial_t$, $\mathrm{Gr}_V t$ in (B.6.9) below if $K = \mathrm{DR}_X(M)$, $K' = \mathrm{DR}_X(M')$ with $A = \mathbb{C}$.) We have

$$\mathrm{Var} \circ \mathrm{can} = N, \quad \mathrm{can} \circ \mathrm{Var} = N. \tag{A.3.10}$$

Note that the target of (A.3.7) is different from that of (A.3.6) by the Tate twist, and ${}^{\mathrm{m}}\varphi_{g,\neq 1} S$ is given by ${}^{\mathrm{m}}\psi_{g,\neq 1} S$ together with the isomorphism (A.2.8). The construction of (A.3.7) is not quite trivial, see [Sa3, Sections 5.2.1 and 5.2.3 and Lemma 5.2.4]. For instance, we used there the isomorphism

$$i^! K' = [i^* K' \to \psi_{g,1} K' \xrightarrow{-N} \psi_{g,1} K'(-1)], \tag{A.3.11}$$

where $i : X_0 \hookrightarrow X$ is the inclusion. If we replace this with

$$i^! K' = [i^* K' \to \psi_{g,1} K' \xrightarrow{id-T} \psi_{g,1} K'], \tag{A.3.12}$$

then (A.3.8) would hold with Var, N respectively replaced by var, $T - id$, where the Tate twist should be omitted (since $T - id$ is not compatible with the weight structure).

Appendix B $\mathcal{D}$-modules

In this appendix we review some basics of $\mathcal{D}$-modules.

B.1. Holonomic $\mathcal{D}$-modules. Let X be a complex manifold of dimension d_X, and M be a coherent left $\mathcal{D}_X$-module. This means that M has locally a finite presentation

$$\bigoplus^p \mathcal{D}_U \to \bigoplus^q \mathcal{D}_U \to M|_U \to 0,$$

over sufficiently small open subsets $U \subset X$. (This is equivalent to the condition that M is quasi-coherent over $\mathcal{O}_X$ and is locally finitely generated over $\mathcal{D}_X$.)

A filtration F on M is called a *good filtration* if (M, F) satisfies

$$(F_p\mathcal{D}_X)(F_qM) \subset F_{p+q}M \quad (p \in \mathbb{N},\, q \in \mathbb{Z}), \tag{B.1.1}$$

and $\mathrm{Gr}^F_\bullet M$ is a *coherent* $\mathrm{Gr}^F_\bullet\mathcal{D}_X$-module. (The last condition is equivalent to the conditions that each F_pM is coherent over $\mathcal{O}_X$ and the equality holds for $q \gg 0$ in (B.1.1).) Here F on $\mathcal{D}_X$ is by the order of differential operators, that is, we have for local coordinates $(x_1, \dots, x_{d_X})$

$$F_p\mathcal{D}_X = \textstyle\sum_{|\nu| \leqslant p} \mathcal{O}_X \partial_{x_i}^{\nu_i}.$$

The *characteristic variety* $\mathrm{Ch}(M) \subset T^*X$ of a coherent left $\mathcal{D}_X$-module M is defined to be the support of the $\mathrm{Gr}^F_\bullet\mathcal{D}_X$-module $\mathrm{Gr}^F_\bullet M$ in the cotangent bundle T^*X. (The latter can be defined to be the union of the analytic subspaces of T^*X defined by the ideal of $\mathrm{Gr}^F_\bullet\mathcal{D}_X$ annihilating g_i with g_i local generators of $\mathrm{Gr}^F_\bullet M$. Here $\mathrm{Gr}^F_\bullet\mathcal{D}_X$ is identified with the sheaf of holomorphic functions on T^*X which are polynomials on fibers of $T^*X \to X$.) This is independent of a choice of a good filtration F.

By the involutivity of the characteristic varieties (see [SKK], [Ma2], [Ga]), it is known that

$$\dim \mathrm{Ch}(M) \geqslant \dim X. \tag{B.1.2}$$

(See also [Bo] for the algebraic $\mathcal{D}$-module case.)

A coherent left $\mathcal{D}_X$-module M is called *holonomic* if

$$\dim \mathrm{Ch}(M) = \dim X. \tag{B.1.3}$$

We will denote by $M_{\mathrm{hol}}(\mathcal{D}_X)$ the abelian category of holonomic $\mathcal{D}_X$-modules.

B.2. Left and right $\mathcal{D}$-modules. We have the transformation between filtered left and right $\mathcal{D}_X$-modules on a complex manifold X of dimension d_x which associates the following to a filtered left $\mathcal{D}_X$-module (M, F):

$$(\Omega_X^{d_X}, F) \otimes_{\mathcal{O}_X} (M, F), \tag{B.2.1}$$

where the filtration F on $\Omega_X^{d_X}$ is defined by the condition

$$\mathrm{Gr}^F_p \Omega_X^{d_X} = 0 \quad (p \neq -d_X). \tag{B.2.2}$$

So the filtration is shifted by $-d_X$. Here it is better to distinguish $\Omega_X^{d_X}$ and the dualizing sheaf ω_X, since the Hodge filtration F on ω_X is usually defined by

$$\mathrm{Gr}_p^F \omega_X = 0 \quad (p \neq 0). \tag{B.2.3}$$

By choosing local coordinates $x_1, \dots, x_{d_X}$, the sheaf $\Omega_X^{d_X}$ is trivialized by $dx_1 \wedge \cdots \wedge dx_{d_X}$ locally on X, and forgetting F, the transformation is given by the anti-involution * of $\mathcal{D}_X$ defined by the conditions (see for instance [Ma1]):

$$\partial_{x_i}^* = -\partial_{x_i}, \quad g^* = g \ (g \in \mathcal{O}_X), \quad (PQ)^* = Q^*P^* \ (P, Q \in \mathcal{D}_X). \tag{B.2.4}$$

For a right $\mathcal{D}_X$-module N, the left $\mathcal{D}_X$-module corresponding to it is denoted often by

$$N \otimes_{\mathcal{O}_X} (\Omega_X^{d_X})^\vee. \tag{B.2.5}$$

Here $L^\vee$ denotes the dual of a locally free sheaf L in general, that is, $L^\vee := \mathcal{H}om_{\mathcal{O}_X}(L, \mathcal{O}_X)$.

B.3. Direct images. For a closed embedding $i : X \to Y$ of smooth complex algebraic varieties, the direct image of a filtered *right* $\mathcal{D}$-module (M, F) is defined by

$$i_*^{\mathcal{D}}(M, F) = (M, F) \otimes_{\mathcal{D}_X} (\mathcal{D}_{X \hookrightarrow Y}, F), \tag{B.3.1}$$

where the sheaf-theoretic direct image is omitted to simplify the notation, and

$$(\mathcal{D}_{X \hookrightarrow Y}, F) := \mathcal{O}_X \otimes_{\mathcal{O}_Y} (\mathcal{D}_Y, F). \tag{B.3.2}$$

For a filtered *left* $\mathcal{D}$-module (M, F), the $\mathcal{D}$-module is twisted by $\omega_{X/Y}$ and the filtration F is shifted by $r : = \mathrm{codim}_X Y$ because of the transformation between filtered left and right $\mathcal{D}$-modules in (B.2). If X is locally defined by $y_1 = \cdots = y_r$ with $y_1, \dots, y_m$ local coordinates of Y, then, setting $\partial_{y_i} := \partial/\partial y_i$, the direct image is *locally* defined by

$$i_*^{\mathcal{D}}(M, F) = (M, F[r]) \otimes_{\mathbb{C}} (\mathbb{C}[\partial_{y_1}, \dots, \partial_{y_r}], F). \tag{B.3.3}$$

For a smooth projection $p : Z := X \times Y \to Y$ with X, Y smooth, the direct image of a filtered *left* $\mathcal{D}_Z$-module (M, F) is defined by the sheaf-theoretic direct image of the relative de Rham complex $\mathrm{DR}_{Z/Y}(M, F)$, that is,

$$p_*^{\mathcal{D}}(M, F) := \mathbb{R}p_*\mathrm{DR}_{Z/Y}(M, F), \tag{B.3.4}$$

where $\mathrm{DR}_{Z/Y}(M, F)$ is the filtered complex defined by

$$(M, F) \to \Omega_{Z/Y}^1 \otimes_{\mathcal{O}_Z} (M, F[-1]) \to \cdots \to \Omega_{Z/Y}^{d_X} \otimes_{\mathcal{O}_Z} (M, F[-d_X]), \tag{B.3.5}$$

with the last term put at degree 0. The differential of this complex is defined as in the absolute case (see (1.2.3)), and is locally given as the Koszul complex associated with the action of $\partial/\partial x_i$ on M if $x_1, \dots, x_{d_X}$ are local coordinates of X. (Note, however, that this does not work for *smooth morphisms* which are not necessarily *smooth projections*, since there is no canonical lift of vector fields on Y to Z.)

In general, the direct image of a filtered right $\mathcal{D}_X$-module (M,F) by a morphism of smooth varieties $f: X \to Y$ is defined by

$$f_*^{\mathcal{D}}(M,F) = p_*^{\mathcal{D}} \circ (i_f)_*^{\mathcal{D}}(M,F), \tag{B.3.6}$$

where $i_f : X \to X \times Y$ is the graph embedding, and $p : X \times Y \to Y$ is the second projection so that $f = p \circ i_f$.

Remarks. (i) We can verify that the direct image in (B.3.6) is naturally isomorphic to the complex of the induced $\mathcal{D}_Y$-module associated with the (sheaf-theoretic) direct image of the filtered differential complex $\mathrm{DR}_X(M,F)$. This means the compatibility between the direct images of filtered differential complexes and filtered $\mathcal{D}$-modules.

(ii) It seems simpler to use the above construction of direct images instead of the induced $\mathcal{D}$-module construction as in [Sa3, 3.3.6] for the definition of direct images of V-filtrations.

(iii) The direct image for a morphism of singular varieties is rather complicated. For the direct image of mixed Hodge modules, we may assume that the morphism is projective by using a Beilinson-type resolution (see [Beil, Section 3] and the proof of [Sa7, Theorem 4.3]), and the cohomological direct image is actually enough. So it is reduced to the case of a morphism of smooth varieties.

B.4. Dual functor. For a holonomic $\mathcal{D}_X$-module M on a complex manifold X of dimension n, its dual $\mathbb{D}(M)$ is defined so that

$$(\Omega_X^{d_X})^\vee \otimes_{\mathcal{O}_X} \mathbb{D}(M) = \mathcal{E}xt^{d_X}_{\mathcal{D}_X}(M, \mathcal{D}_X). \tag{B.4.1}$$

and $\mathbb{D}$ is called the dual functor. This is a contravariant functor. It is well-known that

$$\mathcal{E}xt^p_{\mathcal{D}_X}(M, \mathcal{D}_X) = 0 \quad (p \neq \dim X), \tag{B.4.2}$$

and

$$\mathbb{D}^2 = id. \tag{B.4.3}$$

We say that a filtered holonomic $\mathcal{D}_X$-module (M,F) is *Cohen-Macaulay* if $\mathrm{Gr}^F_\bullet M$ is a Cohen-Macaulay $\mathrm{Gr}^F_\bullet$-module. In this case, we have

$$\mathcal{E}xt^i_{\mathrm{Gr}^F_\bullet \mathcal{D}_X}(\mathrm{Gr}^F_\bullet M, \mathrm{Gr}^F_\bullet \mathcal{D}_X) = 0 \qquad (i \neq d_X), \tag{B.4.4}$$

and the dual filtered $\mathcal{D}_X$-module $\mathbb{D}(M,F)$ can be defined so that

$$(\Omega_X^{d_X}, F) \otimes_{\mathcal{O}_X} \mathbb{D}(M,F) = \mathbb{R}\mathcal{H}om_{\mathcal{D}_X}\big((M,F),(\mathcal{D}_X, F[d_X])\big)[d_X]. \tag{B.4.5}$$

This means that the last filtered complex is filtered quasi-isomorphic to a filtered $\mathcal{D}$-module.

It is known that the dual functor $\mathbb{D}$ commutes with the de Rham functor DR_X, that is, for a regular holonomic $\mathcal{D}_X$-module M, there is a canonical isomorphism (see for instance [Sa3, Proposition 2.4.12]):

$$\mathbb{D}(\mathrm{DR}_X(M)) = \mathrm{DR}_X(\mathbb{D}(M)). \tag{B.4.6}$$

B.5. Regular holonomic $\mathcal{D}$-modules. Let Z be a closed analytic subset Z of a complex manifold X. Let $\mathcal{I}_Z \subset \mathcal{O}_X$ be the ideal of Z. For a bounded complex of $\mathcal{D}_X$-modules $M^\bullet$, set

$$\mathcal{H}^i_{[Z]}M^\bullet := \varinjlim_k \mathcal{E}xt^i_{\mathcal{O}_X}(\mathcal{O}_X/\mathcal{I}_Z^k, M^\bullet) \bigl(= \varinjlim_k \mathcal{E}xt^i_{\mathcal{D}_X}(\mathcal{D}_X/\mathcal{D}_X\mathcal{I}_Z^k, M^\bullet)\bigr),$$
$$\mathcal{H}^i_{[X|Z]}M^\bullet := \varinjlim_k \mathcal{E}xt^i_{\mathcal{O}_X}(\mathcal{I}_Z^k, M^\bullet) \bigl(= \varinjlim_k \mathcal{E}xt^i_{\mathcal{D}_X}(\mathcal{D}_X\mathcal{I}_Z^k, M^\bullet)\bigr),$$

so that we have a long exact sequence of $\mathcal{D}_X$-modules

$$\to \mathcal{H}^i_{[Z]}M^\bullet \to \mathcal{H}^i M^\bullet \to \mathcal{H}^i_{[X|Z]}M^\bullet \to \mathcal{H}^{i+1}_{[Z]}M^\bullet \to \tag{B.5.1}$$

see [Ka1] (and also [Gr] for the algebraic case). Note that $\mathcal{H}^0_{[Z]}M$ for a holonomic $\mathcal{D}_X$-module M is the largest holonomic $\mathcal{D}_X$-submodule supported in Z.

It is known that a holonomic $\mathcal{D}_X$-module M with support Z is *regular holonomic* if and only if there is a closed analytic subset $Z' \subset Z$ together with a proper morphism $\pi : \widetilde{Z} \to Z$ such that $\dim Z' < \dim Z$, $Z \setminus Z'$ is smooth and equi-dimensional, $\widetilde{Z}$ is smooth, π induces an isomorphism over $Z \setminus Z'$, $\pi^{-1}(Z')$ is a divisor with normal crossings on $\widetilde{Z}$, and moreover, by setting $\pi_X := j_Z \circ \pi$ with $j_Z : Z \hookrightarrow X$ the canonical inclusion, the following two conditions are satisfied:

(i) $\mathcal{H}^0_{[Z']}M$ is a regular holonomic $\mathcal{D}_X$-module.

(ii) $\mathcal{H}^0_{[X|Z']}M = \mathcal{H}^0(\pi_X)^{\mathcal{D}}_*\widetilde{M}$ with $\widetilde{M}$ the Deligne meromorphic extension of $M|_{Z\setminus Z'}$ on $\widetilde{Z}$.

We may assume that Z' is the union of $\mathrm{Sing}\, Z$ and the lower dimensional irreducible components of Z, and π is given by the desingularization of the union of the maximal dimensional irreducible components of Z. Note that in the case of algebraic $\mathcal{D}$-modules, we have $\mathcal{H}^0_{[X|Z']}M = j_*j^*M$ with $j : X \setminus Z' \hookrightarrow X$ the canonical inclusion.

The above criterion is by induction on the dimension of the support by using [De1], [Ka1]. In fact, it is known that regular holonomic $\mathcal{D}$-modules are stable by subquotients and extensions in the category of holonomic $\mathcal{D}$-modules and also by the direct images under proper morphisms, and contain Deligne meromorphic extensions in the normal crossing case.

Let $M_{rh}(\mathcal{D}_X) \subset M_{\mathrm{hol}}(\mathcal{D}_X)$ be the full category of regular holonomic $\mathcal{D}_X$-modules on a complex manifold X. Let $D^b_{rh}(\mathcal{D}_X) \subset D^b(\mathcal{D}_X)$ be the full subcategory of bounded complexes with regular holonomic cohomologies. We have the equivalence of categories, that is, the Riemann-Hilbert correspondence:

$$\mathrm{DR}_X : D^b_{rh}(\mathcal{D}_X) \xrightarrow{\sim} D^b_c(X,\mathbb{C}) \ \bigl(\text{inducing } \mathrm{DR}_X : M_{rh}(\mathcal{D}_X) \xrightarrow{\sim} \mathbf{HS}(X,\mathbb{C})\bigr), \tag{B.5.2}$$

see [Ka2], [Ka4], [KK1], [Me1], [Me2], [Me4] (and also [Bo], [HT], etc. for the algebraic case).

Remarks. (i) There are many ways to define the full subcategory $M_{rh}(\mathcal{D}_X) \subset M_{\mathrm{hol}}(\mathcal{D}_X)$. Their equivalences may follow by using the above argument in certain cases.

(ii) There is a nontrivial point about the commutativity of some diagram used in certain proofs of (B.5.2), and this is studied in [Sa5, Section 4].

(iii) Some people say that (B.5.2) is essentially proved in [KK1] where the full faithfulness of the de Rham functor is essentially shown (compare the assertion (ii) of Theorem 5.4.1 written in [KK1, p. 825] with [Me3, Proposition 3.3]). The essential surjectivity is not difficult to show by using the full faithfulness. In [Ka2], [Ka4], a quasi-inverse is explicitly constructed.

(iv) In the algebraic case, the argument is more complicated, since we have the regularity *at infinity*, see [Bo], [De1], [HT], etc. (This is essential to get a *canonical algebraic structure* on the vector bundle associated with a local system on a smooth complex variety by using the Deligne extension, see [De1].) The Riemann-Hilbert correspondence is used in an essential way in representation theory, see [BB], [BK]. (This point does not seem to be sufficiently clarified in the last one.)

B.6. V-filtration. For a complex manifold X, set $Y := X \times \mathbb{C}$ with t the coordinate of $\mathbb{C}$. We have the filtration V on $\mathcal{D}_Y$ indexed by $\mathbb{Z}$ and such that

(i) $V^0\mathcal{D}_Y \subset \mathcal{D}_Y$ is the subring generated by $\mathcal{O}_Y$, ∂_{y_i}, and $t\partial_t$,

(ii) $V^j\mathcal{D}_Y = t^j\, V^0\mathcal{D}_Y$, $V^{-j}\mathcal{D}_Y = \sum_{0\leqslant k\leqslant j} \partial_t^k\, V^0\mathcal{D}_Y \quad (j \in \mathbb{Z}_{>0})$,

where the y_i are local coordinates of Y, and $\partial_{y_i} := \partial/\partial y_i$.

Let M be a regular holonomic $\mathcal{D}_Y$-module. We say that M is *quasi-unipotent* if so is $K := \mathrm{DR}_Y(M)$, that is, if there is a stratification $\{S\}$ of Y such that the restrictions of the cohomology sheaves $\mathcal{H}^i K$ to each stratum S are $\mathbb{C}$-local systems having quasi-unipotent local monodromies around $\overline{S} \setminus S$.

For a quasi-unipotent regular holonomic left $\mathcal{D}_Y$-module M, there is a unique exhaustive filtration V of Kashiwara [Ka3] and Malgrange [Ma3] indexed discretely by $\mathbb{Q}$ and satisfying the following three conditions:

(iii) $V^\alpha M$ ($\forall\, \alpha \in \mathbb{Q}$) are locally finitely generated $V^0\mathcal{D}_Y$-submodules,

(iv) $t\,V^\alpha M \subset V^{\alpha+1}M$ (with equality if $\alpha > 0$), $\partial_t\, V^\alpha M \subset V^{\alpha-1}M$ for any $\alpha \in \mathbb{Q}$,

(v) $\partial_t t - \alpha$ is locally nilpotent on $\mathrm{Gr}_V^\alpha M \quad (\forall\, \alpha \in \mathbb{Q})$.

Here we say that V is *indexed discretely by* $\mathbb{Q}$ if there is a positive integer m satisfying

$$V^\alpha M = V^{j/m}M \quad \text{if} \quad (j-1)/m < \alpha \leqslant j/m \quad \text{with} \quad j \in \mathbb{Z}. \tag{B.6.1}$$

The existence of V follows from that of b-functions in [Ka1] where the holonomicity is actually sufficient (see [Sa14, Proposition 1.9]).

From condition (v) we can easily deduce the isomorphisms

$$t : \mathrm{Gr}_V^\alpha M \xrightarrow{\sim} \mathrm{Gr}_V^{\alpha+1} M, \quad \partial_t : \mathrm{Gr}_V^{\alpha+1} M \xrightarrow{\sim} \mathrm{Gr}_V^\alpha M \qquad (\forall\, \alpha \neq 0). \tag{B.6.2}$$

It is also easy to show the following:

$$V^\alpha M = 0 \ \ (\forall\, \alpha > 0) \quad \text{if} \quad \operatorname{supp} M \subset X \times \{0\} \subset Y, \tag{B.6.3}$$

$$t : V^\alpha M \xrightarrow{\sim} V^{\alpha+1} M \qquad (\forall\, \alpha > 0), \tag{B.6.4}$$

$$M \mapsto V^\alpha M \ (\text{or } \mathrm{Gr}_V^\alpha M) \text{ are exact functors } (\forall\, \alpha \in \mathbb{Q}), \tag{B.6.5}$$

see [Sa3, Lemma 3.1.3, Lemma 3.1.4, Corollary 3.1.5], where right $\mathcal{D}$-modules are used so that the action of $t\partial_t$ there corresponds to that of $-\partial_t t$ in this paper by (B.2.4), and an increasing filtration $V_\bullet$ is used there so that $V_\alpha = V^{-\alpha}$.

Set $Z := g^{-1}(0) \subset X$. Let M'_Z be the largest holonomic $\mathcal{D}_X$-submodule of M supported in Z, and similarly for M''_Z with submodule replaced with quotient module. Then we have the following canonical isomorphisms of $\mathcal{D}_X$-modules (see [Sa3, Proposition 3.1.8]):

$$\begin{aligned} M'_Z &= \mathrm{Ker}\big(t : \mathrm{Gr}_V^0 \widetilde{M} \to \mathrm{Gr}_V^1 \widetilde{M}\big), \\ M''_Z &= \mathrm{Coker}\big(\partial_t : \mathrm{Gr}_V^1 \widetilde{M} \to \mathrm{Gr}_V^0 \widetilde{M}\big). \end{aligned} \tag{B.6.6}$$

Set $K := \mathrm{DR}_Y(M) \in \mathbf{HS}(Y, \mathbb{C})$. In the notation of A.2, there are canonical isomorphisms

$$\begin{aligned} \mathrm{DR}_X(\mathrm{Gr}_V^\alpha M) &\xrightarrow{\sim} {}^{\mathrm{m}}\psi_{t,\mathbf{e}(-\alpha)} K \quad (\alpha \in (0,1]), \\ \mathrm{DR}_X(\mathrm{Gr}_V^0 M) &\xrightarrow{\sim} {}^{\mathrm{m}}\varphi_{t,1} K, \end{aligned} \tag{B.6.7}$$

such that $\exp(-2\pi i(\partial_t t - \alpha))$ on the left-hand side corresponds to the monodromy T on the right-hand side, where $\mathbf{e}(-\alpha) := \exp(-2\pi i \alpha)$. In particular, $-(\partial_t t - \alpha)$ corresponds to $N := (2\pi i)^{-1} \log T_u$ with $T = T_s T_u$ the Jordan decomposition. Moreover the morphisms

$$-\mathrm{Gr}_V \partial_t : \mathrm{Gr}_V^1 M \to \mathrm{Gr}_V^0 M, \quad \mathrm{Gr}_V t : \mathrm{Gr}_V^0 M \to \mathrm{Gr}_V^1 M \tag{B.6.8}$$

respectively correspond to the morphisms can and Var in (A.3.9) with $K' = K$ and $f = t$. (The sign before $\mathrm{Gr}_V \partial_t$ in (B.6.8) comes from the transformation between left and right $\mathcal{D}$-modules as in (B.2.4).)

In the case $M = (i_g)_*^{\mathcal{D}} \mathcal{O}_X$ with $i_g : X \hookrightarrow X \times \mathbb{C}$ the graph embedding by a holomorphic function g on X, the proof of (B.6.7) is given in [Ma3], and this can be extended to the general regular holonomic case (see also the proof of [Sa3, Proposition 3.4.12]). There are canonical morphisms inducing the isomorphisms of (B.6.7) by using logarithmic functions, and these canonical morphisms are quite important for the proof of the stability theorem of Hodge modules by direct images.

Appendix C Compatible filtrations

In this appendix we review some basics of compatible filtrations.

C.1. Compatible subobjects. Let $\mathcal{A}$ be an abelian category, and $n \in \mathbb{Z}_{>0}$. We say that subobjects B_i $(i \in [1,n])$ of $A \in \mathcal{A}$ are *compatible subobjects* if there is a *short exact n-ple complex* K in $\mathcal{A}$ such that

(i) $K^p = 0$ if $|p_i| > 1$ for some $i \in [1,n]$, where $p = (p_1, \ldots, p_n) \in \mathbb{Z}^n$.

(ii) $K^{p-\mathbf{1}_i} \to K^p \to K^{p+\mathbf{1}_i}$ is exact for any $p \in \mathbb{Z}^n$, $i \in [1,n]$.

(iii) $K^0 = A$, $K^{-\mathbf{1}_i} = B_i$ for any $i \in [1,n]$.

Here $\mathbf{1}_i = ((\mathbf{1}_i)_1, \ldots, (\mathbf{1}_i)_n)$ with $(\mathbf{1}_i)_j = \delta_{i,j}$ Note that conditions (i) and (ii) respectively correspond to "short" and "exact". In the case $n = 2$, K is the diagram of the *nine lemma*.

C.2. Compatible filtrations. We say that n filtrations $F_{(i)}$ $(i \in [1,n])$ of $A \in \mathcal{A}$ form *compatible filtrations* if

$$F_{(1)}^{\nu_1}A, \ \ldots \ , F_{(n)}^{\nu_n}A$$

are compatible subobjects of A for any $\nu = (\nu_1, \ldots, \nu_n) \in \mathbb{Z}^n$.

If $n = 2$, then any 2 filtrations $F_{(1)}$, $F_{(2)}$ are compatible filtrations. However, this does not necessarily hold for $n > 2$.

We can show that if the $F_{(i)}$ $(i \in [1,n])$ form compatible filtrations of A, then their restrictions to $F^\nu A := F_{(1)}^{\nu_1} \cdots F_{(n)}^{\nu_n} A$ also form compatible filtrations (see the proof of [Sa3, Corollary 1.2.13]). Using the short exact complex K with

$$K^0 = F^\nu A, \quad K^{-\mathbf{1}_i} = F^{\nu+\mathbf{1}_i} A \ \ (i \in [1,n]),$$

we can show that

$$\mathrm{Gr}_{F_{(1)}}^{\nu_1} \cdots \mathrm{Gr}_{F_{(n)}}^{\nu_n} A \text{ does not depend on the order of } \{1, \ldots, n\}. \tag{C.2.1}$$

In fact, $\mathrm{Gr}_{F_{(i)}}^{\nu_i}$ corresponds to restricting K to the subcomplex defined by $p_i = 1$, and

$$\mathrm{Gr}_{F_{(1)}}^{\nu_1} \cdots \mathrm{Gr}_{F_{(n)}}^{\nu_n} A = K^{1,\ldots,1}. \tag{C.2.2}$$

Note that (C.2.1) is not completely trivial even in the case $n = 2$, where Zassenhaus lemma is usually used, and we can replace it by the diagram of the nine lemma as is explained above.

C.3. Strict complexes. Let $A^\bullet$ be a complex in $\mathcal{A}$ with n filtrations $F_{(i)}$ $(i \in [1,n])$. We say that $(A^\bullet, F_{(i)}\ (i \in [1,n]))$ is *strict* if for any $j \in \mathbb{Z}$, $\nu = (\nu_1, \ldots, \nu_n) \in \mathbb{Z}^n$, there is a short exact n-ple complex K as in (C.1.1) such that

$$H^j\big(\textstyle\bigcap_{i\in I} F_{(i)}^{\nu_i} A^\bullet\big) = K^{-\mathbf{1}_I} \quad (\forall I \subset \{1, \ldots, n\}), \tag{C.3.1}$$

where $\mathbf{1}_I = ((\mathbf{1}_i)_I, \ldots, (\mathbf{1}_I)_n)$ with $(\mathbf{1}_I)_j = 1$ if $j \in I$, and 0 otherwise (and I can be empty in (C.3.1)).

We can show (see [Sa3, Corollary 1.2.13]) that if $\big(A^\bullet, F_{(i)}\,(i\in[1,n])\big)$ is strict, then

$$H^j,\ \mathrm{Gr}^{\nu_1}_{F_{(1)}},\ \dots,\ \mathrm{Gr}^{\nu_n}_{F_{(n)}}\ \text{on}\ \mathcal{A}^\bullet\ \text{commute with each other}\ (\forall\, j\in\mathbb{Z},\ \nu\in\mathbb{Z}^n). \tag{C.3.2}$$

$$\text{The induced filtrations}\ F_{(i)}\ (i\in[1,n])\ \text{on}\ H^jA^\bullet\ \text{form compatible filtrations.} \tag{C.3.3}$$

Note that $\big(A^\bullet, F_{(i)}\,(i\in[1,2])\big)$ is strict if $(A^\bullet, F_{(2)})$, $(\mathrm{Gr}^p_{F_{(2)}}A^\bullet, F_{(1)})$ $(\forall\, p\in\mathbb{Z})$ are strict, and $F^p_{(2)}A^\bullet=0$ for $p\gg 0$, where we assume that the filtered inductive limit in $\mathcal{A}$ is an exact functor, see [Sa3, Theorem 1.2.9].

References

[BRS] Barbieri-Viale, L., Rosenschon, A. and Saito, M., Deligne's conjecture on 1-motives, Ann. Math. (2) 158 (2003), 593–633.

[Bei1] Beilinson, A., On the derived category of perverse sheaves, Lect. Notes in Math. 1289, Springer, Berlin (1987), 27–41.

[Bei2] Beilinson, A., How to glue perverse sheaves, Lect. Notes in Math. 1289, Springer, Berlin (1987), 42–51.

[BB] Beilinson, A. and Bernstein, J., Localisation de $\mathfrak{g}$-modules, C. R. Acad. Sci. Paris, Sér. I Math. 292 (1981), 15–18.

[BBD] Beilinson, A., Bernstein, J. and Deligne, P., Faisceaux pervers, Astérisque 100, Soc. Math. France, Paris, 1982.

[Bo] Borel, A. et al., Algebraic $\mathcal{D}$-Modules, Perspectives in Math. 2, Academic Press, Boston, 1987.

[BrSY] Bressler, P., Saito, M. and Youssin, B., Filtered perverse complexes, Math. Res. Lett. 5 (1998), 119–136.

[BrScY] Brasselet, J.-P., Schürmann, J. and Yokura, S., Hirzebruch classes and motivic Chern classes of singular spaces, Journal of Topology and Analysis 2 (2010), 1–55.

[BFNP] Brosnan, P., Fang, H., Nie, Z. and Pearlstein, G., Singularities of admissible normal functions (with an appendix by N. Fakhruddin), Inv. math. 177 (2009), 599–629.

[BK] Brylinski, J.-L. and Kashiwara, M., Kazhdan-Lusztig conjecture and holonomic systems, Inv. Math. 64 (1981), 387–410.

[Bu] Budur, N., On Hodge spectrum and multiplier ideals, Math. Ann. 327 (2003), 257–270.

[BDS] Budur, N., Dimca, A. and Saito, M., First Milnor cohomology of hyperplane arrangements, Contemp. Math. 538, Amer. Math. Soc., Providence, RI, 2011, pp. 279–292.

[BMS] Budur, N., Mustaţă, M. and Saito, M., Bernstein-Sato polynomials of arbitrary varieties, Compos. Math. 142 (2006), 779–797.

[BS1] Budur, N. and Saito, M., Multiplier ideals, V-filtration, and spectrum, J. Alg. Geom. 14 (2005), 269–282.

[BS2] Budur, N. and Saito, M., Jumping coefficients and spectrum of a hyperplane arrangement, Math. Ann. 347 (2010), 545–579.

[BSY] Budur, N., Saito, M. and Yuzvinsky, S., On the local zeta functions and the b-functions of certain hyperplane arrangements (with Appendix by W. Veys), J. London Math. Soc. (2) 84 (2011), 631–648.

[CK] Cattani, E. and Kaplan, A., Polarized mixed Hodge structures and the local monodromy of a variation of Hodge structure, Inv. Math. 67 (1982), 101–115.

[CKS1] Cattani, E., Kaplan, A. and Schmid, W., L^2 and intersection cohomologies for a polarized variation of Hodge structure (unpublished announcement preprint) 1985.

[CKS2] Cattani, E., Kaplan, A. and Schmid, W., Degeneration of Hodge structures, Ann. Math. (2) 123 (1986), 457–535.

[CKS3] Cattani, E., Kaplan, A. and Schmid, W., L^2 and intersection cohomologies for a polarized variation of Hodge structure, Inv. Math. 87 (1987), 217–252.

[De1] Deligne, P., Equations différentielles à points singuliers réguliers, Lect. Notes in Math. 163, Springer, Berlin, 1970.

[De2] Deligne, P., Théorie de Hodge I, Actes du Congrès International des Mathématiciens (Nice, 1970), Tome 1, Gauthier-Villars, Paris, 1971, pp. 425–430.

[De3] Deligne, P., Théorie de Hodge II, Publ. Math. IHES 40 (1971), 5–57.

[De4] Deligne, P., Cohomologie à supports propres, SGA4 XVII, Lect. Notes in Math. 305, Springer, Berlin (1973), 250–461.

[De5] Deligne, P., Le formalisme des cycles évanescents, in SGA7 XIII, Lect. Notes in Math., 340, Springer, Berlin (1973), 82–115.

[De6] Deligne, P., Théorie de Hodge III, Publ. Math. IHES 44 (1974), 5–77.

[De7] Deligne, P., La conjecture de Weil, Publ. Math. IHES 52 (1980), 137–252.

[De8] Deligne, P., Positivité: signe (manuscrit, 16-2-84 et 6-11-85).

[Di] Dimca, A., Sheaves in topology, Universitext Springer, Berlin, 2004.

[DMS] Dimca, A., Maisonobe, Ph. and Saito, M., Spectrum and multiplier ideals of arbitrary subvarieties, Ann. Inst. Fourier 61 (2011), 1633–1653.

[DMST] Dimca, A., Maisonobe, Ph., Saito, M. and Torrelli, T., Multiplier ideals, V-filtrations and transversal sections, Math. Ann. 336 (2006), 901–924.

[DS1] Dimca, A. and Saito, M., Some consequences of perversity of vanishing cycles, Ann. Inst. Fourier 54 (2004), 1769–1792.

[DS2] Dimca, A. and Saito, M., A generalization of Griffiths theorem on rational integrals, Duke Math. J. 135 (2006), 303–326.

[DS3] Dimca, A. and Saito, M., Vanishing cycle sheaves of one-parameter smoothings and quasi-semistable degenerations, J. Alg. Geom. 21 (2012), 247–271.

[DS4] Dimca, A. and Saito, M., Koszul complexes and spectra of projective hypersurfaces with isolated singularities (arXiv:1212.1081).

[DS5] Dimca, A. and Saito, M., Weight filtration of the limit mixed Hodge structure at infinity for tame polynomials, Math. Z. 275 (2013), 293–306.

[DS6] Dimca, A. and Saito, M., Number of Jordan blocks of the maximal size for local monodromies, Compos. Math. 150 (2014), 344–368.

[DS7] Dimca, A. and Saito, M., Generalization of theorems of Griffiths and Steenbrink to hypersurfaces with ordinary double points (arXiv:1403.4563).

[DSW] Dimca, A., Saito, M. and Wotzlaw, L., A generalization of the Griffiths theorem on rational integrals II, Michigan Math. J. 58 (2009), 603–625.

[DuS] Durfee, A.H. and Saito, M., Mixed Hodge structures on the intersection cohomology of links, Compos. Math. 76 (1990), 49–67.

[FFT] Fujino, O., Fujisawa, T. and Saito, M., Some remarks on the semipositivity theorems, Publ. RIMS, Kyoto Univ. 50 (2014), 85–112.

[Ga] Gabber, O., The integrability of the characteristic variety, Amer. J. Math. 103 (1981), 445–468.

[GN] Guillén, F. and Navarro Aznar, V., Sur le théorème local des cycles invariants, Duke Math. J. 61 (1990), 133–155.

[Gr] Grothendieck, A., Local Cohomology, Lect. Notes in Math. 41, Springer, Berlin (1967).

[HT] Hotta, R. and Tanisaki, T., $\mathcal{D}$-Modules, Perverse Sheaves, and Representation Theory (translated from a Japanese edition by K. Takeuchi who also added some chapters on analytic $\mathcal{D}$-modules, perverse sheaves, etc.), Birkhäuser, Boston, 2008.

[Ka1] Kashiwara, M., On the holonomic systems of linear differential equations II, Inv. Math. 49 (1978), 121–135.

[Ka2] Kashiwara, M. Faisceaux constructibles et systèmes holonômes d'équations aux dérivées partielles linéaires à points singuliers réguliers, Séminaire Goulaouic-Schwartz, 1979-1980, Exp. 19 (7 pp.) Ecole Polytech., Palaiseau, 1980.

[Ka3] Kashiwara, M., Vanishing cycle sheaves and holonomic systems of differential equations, Lect. Notes in Math. 1016, Springer, Berlin (1983), 134–142.

[Ka4] Kashiwara, M., The Riemann-Hilbert problem for holonomic systems, Publ. RIMS, Kyoto Univ. 20 (1984), 319–365.

[Ka5] Kashiwara, M., A study of variation of mixed Hodge structure, Publ. RIMS, Kyoto Univ. 22 (1986), 991–1024.

[KK1] Kashiwara, M. and Kawai, T., On holonomic systems of microdifferential equations III. Systems with regular singularities, Publ. RIMS, Kyoto Univ. 17 (1981), 813–979.

[KK2] Kashiwara, M. and Kawai, T., The Poincaré lemma for a variation of polarized Hodge structure, Proc. Japan Acad. Ser. A Math. Sci. 61 (1985), 164–167.

[KK3] Kashiwara, M. and Kawai, T., Hodge structure and holonomic systems, Proc. Japan Acad. Ser. A Math. Sci. 62 (1986), 1–4.

[KK4] Kashiwara, M. and Kawai, T., The Poincaré lemma for variations of polarized Hodge structure, Publ. RIMS, Kyoto Univ. 23 (1987), 345–407.

[KK5] Kashiwara, M. and Kawai, T., A particular partition of unity: an auxiliary tool in Hodge theory, in Theta functions - Bowdoin 1987, Proc. Sympos. Pure Math., 49, Part 1, AMS, Providence, RI, 1989, pp. 19–26.

[KS] Kashiwara, M. and Schapira, P., Sheaves on manifolds, Springer, Berlin, 1990.

[Ma1] Malgrange, B., Le polynôme de Bernstein d'une singularité isolée, Lect. Notes in Math., 459, Springer, Berlin, 1975, pp. 98–119.

[Ma2] Malgrange, B., L'involutivité des caractéristiques des systèmes différentiels et microdifférentiels, Séminaire Bourbaki, Exp. 522, Lect. Notes in Math., 710, Springer, Berlin, 1979, pp. 277–289.

[Ma3] Malgrange, B., Polynôme de Bernstein-Sato et cohomologie évanescente, Astérisque 101–102 (1983), 243–267.

[MS] Maxim, L. and Schürmann, J., Characteristic classes of mixed Hodge modules and applications (arXiv:1307.5152).

[MSS1] Maxim, L., Saito, M. and Schürmann, J., Symmetric products of mixed Hodge modules, J. Math. Pures Appl. (9) 96 (2011), 462–483.

[MSS2] Maxim, L., Saito, M. and Schürmann, J., Hirzebruch-Milnor classes of complete intersections, Adv. Math. 241 (2013), 220–245.

[MSS3] Maxim, L., Saito, M. and Schürmann, J., Hirzebruch-Milnor classes and Steenbrink spectra of certain projective hypersurfaces (arXiv:1312.0392), to appear in the Proceedings of the Arbeitstagung 2013.

[MSS4] Maxim, L., Saito, M. and Schürmann, J., Spectral Hirzebruch-Milnor classes of singular hypersurfaces (arXiv:1606.02218).

[Me1] Mebkhout, Z., Cohomologie locale des espaces analytiques complexes. Thèse de Doctorat d'Etat, Université de Paris VII, 126 pages (Fév. 1979).

[Me2] Mebkhout, Z., Sur le problème de Hilbert-Riemann, in Complex Analysis, in Microlocal Calculus and Relativistic Quantum Theory (Proceedings of the International Colloquium, Centre Physique, Les Houches, 1979), Lect. Notes in Physics, 126, Springer, Berlin, 1980, pp. 90–110.

[Me3] Mebkhout, Z., Théorèmes de bidualité locale pour les $\mathcal{D}_X$-modules holonômes, Ark. Mat. 20 (1982), 111–124.

[Me4] Mebkhout, Z., Une autre équivalence de catégories, Compos. Math. 51 (1984), 63–88.

[MüS] Müller-Stach, S. and Saito, M., Relative Chow-Künneth decompositions for morphisms of threefolds, J. reine angew. Math. 666 (2012), 141–161.

[MP] Mustaţă, M. and Popa, P., Hodge ideals (arXiv:1605.08088).

[NS] Nagel, J. and Saito, M., Relative Chow-Künneth decompositions for conic bundles and Prym varieties, Int. Math. Res. Not. 2009, 16, 2978–3001.

[OS] Otwinowska, A. and Saito, M., Monodromy of a family of hypersurfaces containing a given subvariety, Ann. Sci. Ecole Norm. Sup. (4) 38 (2005), 365–386.

[PS] Peters, C. and Saito, M., Lowest weights in cohomology of variations of Hodge structure, Nagoya Math. J. 206 (2012), 1–24.

[RS1] Rosenschon, A. and Saito, M., Cycle map for strictly decomposable cycles, Amer. J. Math. 125 (2003), 773–790.

[RS2] Rosenschon, A. and Saito, M., Nonvanishing of external products for higher Chow groups, J. Alg. Geom. 13 (2004), 441–459.

[Sa1] Saito, M., Hodge filtrations on Gauss-Manin systems II, Proc. Japan Acad. Ser. A Math. Sci. 59 (1983), 37–40.

[Sa2] Saito, M., Hodge structure via filtered D-modules, in Differential systems and singularities (Luminy, 1983), Astérisque 130 (1985), 342–351.

[Sa3] Saito, M., Modules de Hodge polarisables, Publ. RIMS, Kyoto Univ. 24 (1988), 849–995.

[Sa4] Saito, M., Duality for vanishing cycle functors, Publ. RIMS, Kyoto Univ. 25 (1989), 889–921.

[Sa5] Saito, M., Induced $\mathcal{D}$-modules and differential complexes, Bull. Soc. Math. France 117 (1989), 361–387.

[Sa6] Saito, M., Introduction to mixed Hodge modules, in Actes du Colloque de Théorie de Hodge (Luminy, 1987), Astérisque 179–180 (1989), 145–162.

[Sa7] Saito, M., Mixed Hodge modules, Publ. RIMS, Kyoto Univ. 26 (1990), 221–333.

[Sa8] Saito, M., Extension of mixed Hodge modules, Compos. Math. 74 (1990), 209–234.

[Sa9] Saito, M., Decomposition theorem for proper Kähler morphisms, Tohoku Math. J. (2) 42 (1990), 127–147.

[Sa10] Saito, M., On Steenbrink's conjecture, Math. Ann. 289 (1991), 703–716.

[Sa11] Saito, M., On Kollár's conjecture, in Several complex variables and complex geometry, Proc. Sympos. Pure Math., 52, Part 2, AMS Providence, RI, 1991, pp. 509–517.

[Sa12] Saito, M., Hodge conjecture and mixed motives I, in Complex geometry and Lie theory, Proc. Sympos. Pure Math., 53, AMS, Providence, RI, 1991, pp. 283–303.

[Sa13] Saito, M., Hodge conjecture and mixed motives II, in Algebraic geometry, Lect. Notes in Math., 1479, Springer, Berlin, 1991, pp. 196–215.

[Sa14] Saito, M., On b-function, spectrum and rational singularity, Math. Ann. 295 (1993), 51–74.

[Sa15] Saito, M., Some remarks on the Hodge type conjecture, in Motives, Proc. Sympos. Pure Math., 55, Part 1, AMS Providence, RI, 1994, pp. 85–100.

[Sa16] Saito, M., On microlocal b-function, Bull. Soc. Math. France 122 (1994), 163–184.

[Sa17] Saito, M., Admissible normal functions, J. Alg. Geom. 5 (1996), 235–276.

[Sa18] Saito, M., Monodromy filtration and positivity (arXiv:math/0006162).

[Sa19] Saito, M., Mixed Hodge complexes on algebraic varieties, Math. Ann. 316 (2000), 283–331.

[Sa20] Saito, M., Arithmetic mixed sheaves, Inv. Math. 144 (2001), 533–569.

[Sa21] Saito, M., Refined cycle maps, in Algebraic geometry 2000, Azumino (Hotaka), Adv. Stud. Pure Math., 36, Math. Soc. Japan, Tokyo, 2002, pp. 115–143.

[Sa22] Saito, M., Filtrations on Chow groups and transcendence degree, Publ. RIMS, Kyoto Univ. 40 (2004), 933–948.

[Sa23] Saito, M., Bernstein-Sato polynomials of hyperplane arrangements (arXiv:math/0602527), to appear in Selecta Math.

[Sa24] Saito, M., Multiplier ideals, b-function, and spectrum of a hypersurface singularity, Compos. Math. 143 (2007), 1050–1068.

[Sa25] Saito, M., Direct image of logarithmic complexes and infinitesimal invariants of cycles, in Algebraic cycles and motives, Vol. 2, London Math. Soc. Lecture Note Series, 344, Cambridge University Press, Cambridge, 2007, pp. 304–318.

[Sa26] Saito, M., Generalized Thomas hyperplane sections and relations between vanishing cycles (arXiv:0806.1461).

[Sa27] Saito, M., Hodge-type conjecture for higher Chow groups, Pure Appl. Math. Q. 5, Special Issue: In honor of Friedrich Hirzebruch. Part 2 (2009), 947–976.

[Sa28] Saito, M., On the Hodge filtration of Hodge modules, Moscow Math. J. 9 (2009), 161–191,

[Sa29] Saito, M., On the definition of mixed Hodge modules (arXiv:1307.2140).

[Sa30] Saito, M., Normal functions and spread of zero locus, in Recent Advances in Hodge Theory, London Math. Soc. Lecture Note Series, 427, Cambridge University Press, 2016, pp. 264–274.

[Sa31] Saito, M., $\mathcal{D}$-modules generated by rational powers of holomorphic functions (arXiv:1507.01877).

[Sa32] Saito, M., Bernstein-Sato polynomials and graded Milnor algebras for projective hypersurfaces with weighted homogeneous isolated singularities (arXiv:1609.04801).

[SS1] Saito, M. and Schnell, C., A variant of Néron models over curves, Manuscripta Math. 134 (2011), 359–375.

[SS2] Saito, M. and Schnell, C., Fields of definition of Hodge loci, in Recent Advances in Hodge Theory, London Math. Soc. Lecture Note Series, 427, Cambridge University Press, 2016, pp. 275–291.

[SZ] Saito, M. and Zucker, S., The kernel spectral sequence of vanishing cycles, Duke Math. J. 61 (1990), 329–339.

[SKK] Sato, M., Kawai, T. and Kashiwara, M., Microfunctions and pseudo-differential equations, Lect. Notes in Math., 287, Springer, Berlin, 1973, pp. 265–529.

[Sch] Schmid, W., Variation of Hodge structure: The Singularities of the period mapping, Inv. Math. 2 (1973), 211–319.

[Schn] Schnell, C., Complex analytic Néron models for arbitrary families of intermediate Jacobians, Inv. Math. 188 (2012), 1–81.

[St] Steenbrink, J.H.M., Limits of Hodge structures, Inv. Math. 31 (1976), 229–257.

[StZ] Steenbrink, J.H.M. and Zucker, S., Variation of mixed Hodge structure I, Inv. Math. 80 (1985), 489–542.

[Su] Suh, J., Kawamata-Viehweg type vanishing theorems with coefficients; arithmetic and birational applications (preprint).

[Zu] Zucker, S., Hodge theory with degenerating coefficients: L_2 cohomology in the Poincaré metric, Ann. Math. (2) 109 (1979), 415–476.

[Ve1] Verdier, J.-L., Des catégories dérivées des catégories abéliennes, Astérisque 239 (1996).

[Ve2] Verdier, J.-L., Extension of a perverse sheaf over a closed subspace, Asterisque 130 (1985), 210–217.

Hodge Theory and L^2-analysis
ALM 39, pp. 555–581

Perverse Sheaves and the Reductive Borel-Serre Compactification*

Leslie Saper†

Abstract

We briefly introduce the theory of perverse sheaves with special attention to the topological situation where strata can have odd dimension. This is part of a project to use perverse sheaves on the topological reductive Borel-Serre compactification of a Hermitian locally symmetric space as a tool to study perverse sheaves on the Baily-Borel compactification, a projective algebraic variety. We sketch why the decomposition theorem holds for the natural map between the reductive Borel-Serre and the Baily-Borel compactifications. We demonstrate how to calculate extensions of simple perverse sheaves on the reductive Borel-Serre compactification and illustrate with the example of $\mathrm{Sp}(4, \mathbf{R})$.

1 Introduction

This paper is a report on work in progress to better understand the category of perverse sheaves on the Baily-Borel compactification X^* of a locally symmetric variety by using perverse sheaves on the reductive Borel-Serre compactification $\widehat{X}$.

1.1 Definition

Perverse sheaves were introduced in [5] by Beilinson, Bernstein, and Deligne following Goresky and MacPherson's introduction of intersection homology [19]. (For a full account of the exciting discoveries and interactions around that time, see Kleiman's excellent history [26]). We begin by briefly recalling the definition (and to avoid confusion, we point out, as in [5], that a perverse sheaf is not actually a sheaf nor is it perverse).

*Dedicated to Steve Zucker, with great respect and admiration.

†Department of Mathematics, Duke University, Durham, NC 27708. E-mail: saper@math.duke.edu

Let Y be a stratified pseudomanifold and fix an integer-valued function on the set of strata of Y called the *perversity*;[1] if all the strata of Y have even dimension, it is most useful to take for p the *middle perversity*, $p(S) = -(1/2)\dim S$. Let $\mathbf{D}^b_c(Y)$ be the constructible bounded derived category of Y; objects may be represented by complexes of sheaves on Y (up to quasi-isomorphism) whose cohomology lives in a bounded range of degrees and is locally constant along each stratum. By definition, a perverse sheaf $\mathcal{P}$ on Y (constructible with respect to the given stratification) is an object $\mathcal{S}$ whose local cohomology $H(i_S^*\mathcal{S})$ along each stratum $i_S\colon S \hookrightarrow Y$ lives in degrees $\leq p(S)$ and whose local cohomology supported on S, $H(i_S^!\mathcal{S})$, lives in degrees $\geq p(S)$. The full subcategory of perverse sheaves on Y is an abelian category $\mathbf{P}(Y)$ which is Artinian and Noetherian. For every connected stratum S and irreducible local system $\mathbf{E}$ on S there is a simple perverse sheaf $\mathcal{P}_S(\mathbf{E}) = \mathcal{P}_{p,S}(\mathbf{E})$ and all simple objects are obtained in this fashion. In fact $\mathcal{P}_S(\mathbf{E})$ is the intersection cohomology sheaf of $\overline{S}$ with coefficients in $\mathbf{E}$ shifted by $-p(S)$.

When Y is a complex algebraic variety, we will always consider the middle perversity. In addition, one usually does not fix the (algebraic) stratification in the definitions of $\mathbf{D}^b_c(Y)$ and $\mathbf{P}(Y)$.

1.2 Two original applications

Perverse sheaves have played a critical role in the topology of algebraic varieties and representation theory from their beginning; we mention just two examples from the early days. For a fuller survey of applications, see [14].

The first example is the Kazdhan-Lusztig conjecture. Let $Y = G/B$ be the flag variety associated to a simply connected semisimple complex algebraic group G. It is stratified by its B-orbits S_w which are indexed by w in the Weyl group W of G; each closure $\overline{S_w}$ is a Schubert variety. The correspondence $w \leftrightarrow S_w$ is order preserving in the sense that $y \leq w$ if and only if $S_y \subset \overline{S_w}$. For $y \leq w \in W$, Kazdhan and Lusztig [24] gave a conjectural formula for the multiplicity of the irreducible $\mathfrak{g}$-module L_y with highest weight $y\rho - \rho$ in the composition series of the Verma module M_w with highest weight $w\rho - \rho$. The formula involved a combinatorially defined polynomial $P_{y,w}(q)$ which appeared to be related to the failure of Poincaré duality on $\overline{S_w}$; later they showed [25] that the coefficients in $P_{y,w}(q)$ were $\dim H^k(i_{S_y}^*\mathcal{P}_{S_w}(\mathbf{C}))$, the local Betti numbers along S_y of the simple perverse sheaf associated to S_w.

The conjecture was resolved independently by Beilinson and Bernstein [6] and by Brylinski and Kashiwara [11] by transferring the problem from representations to perverse sheaves. Specifically one shows that a certain category of $\mathfrak{g}$-modules (including all L_y and M_w) is equivalent to the category of regular holonomic $\mathscr{D}_Y$-modules and that by the Riemann-Hilbert correspondence this category is equivalent to $\mathbf{P}(Y)$.

[1] The origin of the term "perversity" goes back to the creation of intersection homology by Goresky and MacPherson [19] where chains were constrained so that the dimension of their intersection with a singular stratum S was at most $p(S) - p(S_0) - 1$ more than that allowed by transversality (S_0 being the open dense stratum).

The second original example of the importance of perverse sheaves is the decomposition theorem. This deep result about the topology of proper algebraic maps was first proved by Beilinson, Bernstein, Deligne, and Gabber [5, *Théorème* 6.2.5] and later extended by Saito [39, 40]. It says that if $X \to Y$ is a proper morphism of algebraic varieties, then the direct image of a simple perverse sheaf of Hodge type on X decomposes into the direct sum of shifted simple perverse sheaves, likewise of Hodge type, supported on subvarieties of Y. Here by a simple perverse sheaf of Hodge type we mean one that corresponds to a polarizable Hodge module in the sense of Saito; by [39, *Lemmes* 5.1.10, 5.2.12; 40, Theorem 0.2]; such simple perverse sheaves are precisely those of the form $\mathcal{P}_S(\mathbf{E})$ where $\mathbf{E}$ underlies a real polarizable variation of Hodge structure. As a special case, if $\widetilde{Y} \to Y$ is a resolution of singularities, the theorem implies that the ordinary cohomology $H^{\cdot}(\widetilde{Y};\mathbf{C})$ is the direct sum of shifted middle perversity intersection cohomology groups of subvarieties of Y with various local systems as coefficients. Furthermore one summand will be $IH^{\cdot}(Y;\mathbf{C})$, the intersection cohomology of Y itself.

The proof in [5] is for a simple perverse sheaf of geometric origin and proceeds by first dealing with the theorem over the algebraic closure of a finite field and then lifting the result to $\mathbf{C}$. The version of the theorem stated above was proved by Saito [39, 40] using purely characteristic 0 methods. It uses mixed Hodge modules, which are regular holonomic $\mathscr{D}$-modules equipped with various filtrations and satisfying certain conditions; they correspond to the mixed perverse sheaves over finite fields considered in [5]. A pure Hodge module corresponds under the Riemann-Hilbert correspondence to a simple perverse sheaf of Hodge type.

More recently, other proofs and generalizations of the decomposition theorem have been given. A proof of the constant coefficient case using classical Hodge theory was given by de Cataldo and Migliorini [13]; for more details and a discussion of the different approaches to the decomposition theorem see also their survey [14]. Kashiwara has conjectured [23] that the decomposition theorem should hold for *any* simple perverse sheaf. This conjecture has been proven analytically using polarizable pure twistor $\mathscr{D}$-modules by work of Sabbah [38] and Mochizuki [33, 34]. At about the same time, an arithmetic proof was given independently by Böckle and Khare [7] and by Gaitsgory [17]; they proved de Jong's conjecture [15] which by Drinfeld [16] implies Kashiwara's conjecture. Note that Kashiwara actually conjectures the theorem should hold for a simple holonomic $\mathscr{D}_X$-module with possibly irregular singularities; this has been settled by Mochizuki [35].

1.3 Locally symmetric varieties

Our main interest here is the category of perverse sheaves on a locally symmetric variety. Let $X = \Gamma\backslash D = \Gamma\backslash G(\mathbf{R})/A_G(\mathbf{R})K$ be an arithmetic quotient of a symmetric space of noncompact type; here G is a reductive algebraic group defined over $\mathbf{Q}$, K a maximal compact subgroup of $G(\mathbf{R})$, A_G is the maximal $\mathbf{Q}$-split torus in the center of G, and Γ is an arithmetic subgroup. Our main focus will be when D is Hermitian symmetric and unless otherwise noted we shall assume that in this introduction, however some non-Hermitian symmetric spaces will arise as well.

A natural choice of compactification in the Hermitian case is the Baily-Borel compactification X^* [3] (topologically it is one of Satake's compactifications). The Baily-Borel compactification is a projective algebraic variety defined over a number field; it is commonly called a locally symmetric variety. Locally symmetric varieties and Shimura varieties, their adelic variants, play an important role in number theory, automorphic forms, and the Langlands's program.

The Baily-Borel compactification X^* has a natural stratification $\coprod_Q F_Q$, where Q ranges over all Γ-conjugacy classes of saturated parabolic $\mathbf{Q}$-subgroups of G; when G is almost $\mathbf{Q}$-simple the saturated condition simply means Q is maximal parabolic $\mathbf{Q}$-subgroup. Each F_Q is again a Hermitian locally symmetric space associated to a reductive group $L_{Q,h}$, a quotient of Q; its closure $\overline{F_Q}$ in X^* is a subvariety and is the Baily-Borel compactification F_Q^* of F_Q. We fix this stratification when considering perverse sheaves on X^*.

On the locally symmetric space X one usually considers local systems $\mathbf{E}$ that arise from a representation E of G as opposed to merely a representation of Γ. We say that the simple perverse sheaf $\mathcal{P}_X(\mathbf{E})$ on X^* is *reductively constructible* if the coefficient system $\mathbf{E}$ arises from a representation E of G as above; similarly $\mathcal{P}_{F_Q}(\mathbf{E})$ is reductively constructible if $\mathbf{E}$ arises from a representation of $L_{Q,h}$.

Zucker's conjecture [46] gives an analytic realization of a simple reductively constructible perverse sheaf $\mathcal{P}_X(\mathbf{E})$. The choice of an admissible inner product [32] on E induces a metric on $\mathbf{E}$ which we assume fixed. The conjecture then states that there is a natural isomorphism $\mathcal{P}_X(\mathbf{E}) \cong \mathscr{L}_2(X^*;\mathbf{E})[-p(X)]$, where $\mathscr{L}_2(X^*;\mathbf{E})$ is the sheafification of the presheaf that associates to $U \subset X^*$ the complex of $\mathbf{E}$-valued differential forms on $U \cap X$ which, together with their exterior derivative, are L^2 with respect to the natural locally symmetric metric on X and the above metric on $\mathbf{E}$. The conjecture was independently settled at about the same time by Looijenga [29] and Saper and Stern [44] using quite different methods. A few years later an important new proof was given by Looijenga and Rapoport [30].

The conjecture thus shows that the reductively constructible simple perverse sheaves $\mathcal{P}_{F_Q}(\mathbf{E})$ are related to representations $L^2(\Gamma_{L_{Q,h}} \backslash L_{Q,h}(\mathbf{R}))$ (see [10]) and hence to automorphic forms. The category of (not necessarily simple) perverse sheaves on X^* is thus clearly worth further study.

1.4 The reductive Borel-Serre compactification

In general X^* is very singular so it may be profitable to study its perverse sheaves by using a resolution.

A natural choice for an algebraic resolution, though non-unique, is one of the smooth toroidal compactifications $\tilde{\pi}\colon \widetilde{X}_\Sigma \to X^*$ [1]. Let $\mathcal{P}_X(\mathbf{E})$ be a reductively constructible simple perverse sheaf on X^*. Let $U = \tilde{\pi}^{-1}(X) \subset \widetilde{X}_\Sigma$; via $\tilde{\pi}$ we can view $\mathbf{E}$ also as a local system on U. The decomposition theorem applies[2] to show

[2]We only need the version proved by Saito here by the following argument. In the Hermitian case, Zucker [48] has shown $\mathbf{E}$ underlies a complex polarizable variation of Hodge structure; it only a real polarizable variation of Hodge structure if the representation E is real. However $\mathbf{E} \otimes_{\mathbf{R}} \mathbf{C}$ does underlie a real polarizable variation of Hodge structure so Saito's decomposition theorem applies to it. Now $\mathbf{E} \otimes_{\mathbf{R}} \mathbf{C} \cong \mathbf{E} \oplus \overline{\mathbf{E}}$ and one can prove that the decomposition theorem holds for a direct sum if and only if it holds for each summand.

that $\tilde{\pi}_* \mathcal{P}_U(\mathbf{E})$ is the direct sum of (shifted) simple perverse sheaves on X^*, one of which is $\mathcal{P}_X(\mathbf{E})$.

However this argument does not easily extend to treat simple perverse sheaves supported on smaller closed strata of X^*. For example $\tilde{\pi}^{-1}(F_Q)$ is rarely smooth nor irreducible.

To get around this we look instead at the reductive Borel-Serre compactification $\widehat{X}$. This was introduced by Zucker in the same paper [46] where he made his conjecture about the L^2-cohomology of X. (It is actually defined for any arithmetic locally symmetric space, not necessarily Hermitian.) It is non-algebraic (in fact it may have odd dimensional strata) but it is canonically associated to X and its singularities are easy to describe. Zucker [47] showed that there is a continuous quotient map $\pi\colon \widehat{X} \to X^*$ extending the identity on X so that $\widehat{X}$ may be viewed as a partial resolution of singularities. Goresky and Tai [22] show that up to homotopy any $\tilde{\pi}$ factors through $\widehat{X}$. So in this sense it is a minimal partial resolution.

The reductive Borel-Serre compactification $\widehat{X}$ has a natural stratification $\coprod_P X_P$, where P ranges over all Γ-conjugacy classes of parabolic $\mathbf{Q}$-subgroups of G. Each X_P is again an arithmetic locally symmetric space associated to a reductive group, namely L_P, the Levi quotient of P; its closure $\overline{X_P}$ in $\widehat{X}$ is the reductive Borel-Serre compactification $\widehat{X}_P$ of X_P. We fix this stratification when considering perverse sheaves on $\widehat{X}$. Since $\widehat{X}$ may have odd dimensional strata, there are two middle perversities, p_- and p_+ (see (3.1)), that we will consider, both depending only on the dimension of the stratum. A simple perverse sheaf $\mathcal{P}_{X_P}(\mathbf{E})$ on $\widehat{X}$ is *reductively constructible* if $\mathbf{E}$ is induced from a representation E of L_P.

Despite $\widehat{X}$ not being algebraic, it "wants" to be algebraic. For example, Zucker showed [49] that its cohomology carries a mixed Hodge structure such that

$$H^{\cdot}(X^*) \xrightarrow{\pi^*} H^{\cdot}(\widehat{X}) \xrightarrow{i_X^*} H^{\cdot}(X)$$

are morphisms of mixed Hodge structures. (Here the outer two cohomology groups carry Deligne's canonical mixed Hodge structure.) Furthermore, Ayoub and Zucker [2] construct a motive corresponding to $\widehat{X}$.

As another example of the algebraic-like nature of $\widehat{X}$ is the conjecture of Rapoport [37] and of Goresky and MacPherson [21]. This conjecture was proved by Saper and Stern [37, Appendix] when the $\mathbf{Q}$-rank of G is 1, and by Saper [41] in general. It states that for a local system $\mathbf{E}$ associated to a representation of G,

$$\pi_* \mathcal{P}_{X_G}(\mathbf{E}) = \mathcal{P}_{F_G}(\mathbf{E}); \tag{1.1}$$

Here the perversity on the left (for $\widehat{X}$) can be either of the two middle perversities—one obtains the same pushforward.

Unlike in the situation for $\tilde{\pi}\colon \widetilde{X}_\Sigma \to X^*$, the methods of [41] can be used to generalize (1.1) to all reductively constructible simple perverse sheaves $\mathcal{P}_{X_P}(\mathbf{E})$ on $\widehat{X}$. Specifically (see [41, §21]) for every parabolic $\mathbf{Q}$-subgroup P, there is a saturated parabolic $\mathbf{Q}$-subgroup $P^\dagger$ so that $\pi(X_P) = F_{P^\dagger}$. In fact, $\pi|_{X_P}\colon X_P \to F_{P^\dagger}$ is a flat bundle which becomes trivial over a finite cover of $F_{P^\dagger}$; the fiber is $X_{P,\ell}$, a locally symmetric space *not* usually of Hermitian type. We have the following extension of the decomposition theorem despite $\widehat{X}$ not being algebraic:

Theorem (Decomposition theorem for $\pi\colon \widehat{X} \to X^*$). *Let* $\mathbf{E}$ *be an irreducible local system on a stratum* X_P *of* $\widehat{X}$ *which is induced from an algebraic representation of* L_P. *Let* p *be a middle perversity. Then*

$$\pi_* \mathcal{P}_{p,X_P}(\mathbf{E}) = \bigoplus_i \mathcal{P}_{F_{P^\dagger}}(\mathbf{H}^i(\widehat{X}_{P,\ell}; \mathcal{P}_{p,X_{P,\ell}}(\mathbf{E})))[-i]. \tag{1.2}$$

The theorem makes it clear that π_* is not injective on objects since the map $P \mapsto P^\dagger$ is not injective.

Here is a sketch of the proof. It suffices to pass to a finite cover so one can arrange that $X_P = X_{P,\ell} \times F_{P^\dagger}$ whence $\widehat{X}_P = \widehat{X}_{P,\ell} \times \widehat{F}_{P^\dagger}$, with $\pi|_{\widehat{X}_P}$ being projection on the second factor. Now apply the Künneth formula of Cohen, Goresky, and Ji [12] (which one checks is applicable for middle perversities) and apply Rapoport's conjecture to the second factor.

1.5 Summary of this paper

The theorem above shows that it is reasonable to study the category of reductively constructible perverse sheaves on the Baily-Borel compactification X^* by studying the category of perverse sheaves on the reductive Borel-Serre compactification $\widehat{X}$. By the above version of the decomposition theorem, we understand π_* on simple objects; the goal of this very modest paper is to begin to understand extensions better in this context. After recalling the basics of t-structures in §2, we carefully define middle perversity perverse sheaves on a stratified pseudomanifold in §3, playing special attention to the issues that arise due to odd dimensional strata. We also note certain non-trivial extensions that arise due to odd codimension strata. In §4 we indicate how one may calculate extensions between two simple perverse sheaves. After discussing needed background on the reductive Borel-Serre compactification in §5 and the link cohomology of simple perverse sheaves in §6, we conclude in §7 by doing the exercise of calculating all extensions between simple perverse sheaves for $\widehat{\mathscr{A}_2}$, the reductive Borel-Serre compactification of the moduli space of principally polarized abelian surfaces.

1.6 A potential application: perverse cohomology

One reason to better understand extensions is to be able to calculate the *perverse cohomology* of an object $\mathcal{S}$ in $\mathbf{D}^b_{rc}(\widehat{X})$ (or $\pi_*\mathcal{S}$ in $\mathbf{D}^b_{rc}(X^*)$), the reductively constructible bounded derived category of sheaves. Recall that the definition of a perverse sheaf started with sheaves as the basic objects, proceeded to the derived category, and within it found the abelian category $\mathbf{P}(Y)$ of perverse sheaves. Thus a perverse sheaf is represented as a complex of ordinary sheaves. However it has become clear that it is also useful to view perverse sheaves themselves as the basic objects and in some cases the roles of the two types of objects can be reversed. Beilinson [4] shows that if Y is an algebraic variety and we consider all algebraic stratifications (not a fixed one), then the bounded derived category of $\mathbf{P}(Y)$ is equivalent to $\mathbf{D}_c^b(Y)$ within which one finds the category of ordinary constructible sheaves. Thus in this setting, a constructible sheaf, or in fact any object $\mathcal{S}$ of $\mathbf{D}_c^b(Y)$, can be represented as a complex of perverse sheaves $\cdots \to \mathcal{P}^{i-1} \to \mathcal{P}^i \to \mathcal{P}^{i+1} \to \cdots$. The cohomology of this complex $H^i(\mathcal{P}^\cdot)$ is a perverse sheaf called the perverse cohomology ${}^pH^i(\mathcal{S})$.

Even when Beilinson's theorem does not apply, as it does not for $\widehat{X}$ since we have fixed a stratification, the perverse cohomology can be defined as ${}^pH^i(\mathcal{S}) = {}^p\tau_{\leq 0}{}^p\tau_{\geq 0}\mathcal{S}[i]$. Furthermore, just like for the ordinary cohomology sheaves of a complex of sheaves, there is a spectral sequence with $E_2^{ij} = H^j(Y; {}^pH^i(\mathcal{S}))$ that abuts to $H^{i+j}(Y;\mathcal{S})$. Many important invariants of X are realized as the cohomology of complexes of sheaves on $\widehat{X}$, for example, the cohomology of the arithmetic group is $H^\cdot(\Gamma; E) = H^\cdot(\widehat{X}; i_{X*}\mathbf{E})$. Thus calculating ${}^pH^i(\mathcal{S})$, or at least having bounds on the degrees in which $H^\cdot(\widehat{X}; {}^pH^i(\mathcal{S}))$ can be nonzero, is important.

One potential approach to calculating, or at least approximating, the perverse cohomology of $\mathcal{S}$ is through its *micro-support*. In [41, §7], we defined the micro-support $\mathrm{SS}(\mathcal{S})$ and proved a vanishing theorem [41, §10] for $H^\cdot(\widehat{X};\mathcal{S})$ based on $\mathrm{SS}(\mathcal{S})$. As an application, we were able to prove [43], for example, that in the Hermitian case, $H^i(\Gamma;\mathbf{E}) = 0$ for $i < (1/2)\dim X$ provided E had regular highest weight; this result was independently proved by Li and Schwermer [28] by different methods.

We also determined [41, §17] the micro-support of a simple perverse sheaf $\mathcal{P}_{X_P}(\mathbf{E})$ provided the $\mathbf{Q}$-root system of L_P did not have a factor of type D_n, F_4, or E_n (a restriction that should be removable). The result shows that if a perverse sheaf is semi-simple and satisfies a certain conjugate self-contragredient assumption on the coefficients, the micro-support determines the perverse sheaf. In fact we will show elsewhere that under certain conditions, even for a not necessarily semi-simple perverse sheaf, the micro-support determines the simple constituents and thus potentially the micro-support can be used as a tool to calculate the perverse cohomology.

1.7 Acknowledgments

I would like to thank Amnon Neeman for helpful discussions regarding extensions. I would also like to thank Lizhen Ji, Steve Zucker, and an expert anonymous referee for thoughtful comments concerning this paper. Needless to say, any failings of this paper are solely due to myself.

It was a great honor to be invited to speak at the birthday conference for Steve Zucker and to contribute to this volume. Steve has always been a source of inspiration and encouragement to me. When I was a graduate student and he was a visitor at the Institute, I remember him sitting down with me in the common room and carefully showing me the derivation of the Poincaré punctured disk metric. In later encounters he always took a deep interest in my work which I appreciated. I also enjoyed long evenings with him spent listening to classical music as well as many delicious meals (particularly Maryland crabs whacked open with a wooden mallet!).

Notation. Morphisms in a category $\mathbf{C}$ will be denoted $\mathrm{Mor}_{\mathbf{C}}(A,B)$; when $\mathbf{C}$ is a category of representations of a group G, we simply write $\mathrm{Mor}_G(A,B)$.

If A is an object of some derived category of sheaves on a space Y and $V \subset Y$ is an open subset, we will sometimes abuse notation and also denote by A the inverse image of A to the derived category of sheaves on V, that is $A|_V$.

If $S \subset Y$, we let $i_S \colon S \hookrightarrow Y$ denote the inclusion; we use the same notation to denote the inclusion into any subset of Y containing S. If V is an open subset of Y containing S for which S is closed in V, we let $j_S \colon U = V \setminus S \hookrightarrow V$. We thus have inclusions of complementary open and closed subsets

$$U \xhookrightarrow{j_S} V \xhookleftarrow{i_S} S. \tag{1.3}$$

Often S is a stratum of a stratification of Y and V is an open union of strata for which S is a minimal stratum.

If S, $S' \subset Y$, we write $S \preccurlyeq S'$ if $S \subseteq \overline{S'}$ and $S \prec S'$ if $S \subsetneq \overline{S'}$.

2 t-structures

We briefly recall the theory of t-structures on a triangulated category $\mathbf{D}$ following [5, §1.3].

A *t-structure* on $\mathbf{D}$ consists of two full subcategories $\mathbf{D}^{\leq 0}$ and $\mathbf{D}^{\geq 0}$, closed under isomorphism, satisfying the following conditions. First set $\mathbf{D}^{\leq k} = \mathbf{D}^{\leq 0}[-k]$ and $\mathbf{D}^{\geq k} = \mathbf{D}^{\geq 0}[-k]$. We require

$$\mathrm{Mor}_{\mathbf{D}}(A,B) = 0 \text{ for } A \in \mathbf{D}^{\leq 0},\ B \in \mathbf{D}^{\geq 1}. \tag{2.1}$$

Secondly we require that

$$\mathbf{D}^{\leq 0} \subseteq \mathbf{D}^{\leq 1} \qquad \text{and} \qquad \mathbf{D}^{\geq 0} \supseteq \mathbf{D}^{\geq 1}. \tag{2.2}$$

And finally we assume that for any object $X \in \mathbf{D}$, there exists a distinguished triangle

$$A \to X \to B \xrightarrow{[1]} \qquad (A \in \mathbf{D}^{\leq 0},\ B \in \mathbf{D}^{\geq 1}). \tag{2.3}$$

In fact the distinguished triangle above is unique up to unique isomorphism and one can use it to define truncation functors $\tau_{\leq 0}X = A$ and $\tau_{\geq 1}X = B$. By shifting we obtain truncation functors $\tau_{\leq k} \colon \mathbf{D} \to \mathbf{D}^{\leq k}$ and $\tau_{\geq k} \colon \mathbf{D} \to \mathbf{D}^{\geq k}$ for all $k \in \mathbf{Z}$. There are natural morphisms $\tau_{\leq k} \to \mathrm{id}$ and $\mathrm{id} \to \tau_{\geq k}$, which induce adjoint relations

$$\mathrm{Mor}_{\mathbf{D}}(A,B) \cong \mathrm{Mor}_{\mathbf{D}^{\leq k}}(A, \tau_{\leq k}B) \qquad (A \in \mathbf{D}^{\leq k},\ B \in \mathbf{D}), \tag{2.4}$$

$$\mathrm{Mor}_{\mathbf{D}}(A,B) \cong \mathrm{Mor}_{\mathbf{D}^{\geq k}}(\tau_{\geq k}A, B) \qquad (A \in \mathbf{D},\ B \in \mathbf{D}^{\geq k}). \tag{2.5}$$

The *heart* of a t-structure is $\mathbf{C} = \mathbf{D}^{\leq 0} \cap \mathbf{D}^{\geq 0}$. It is a full abelian subcategory of $\mathbf{D}$. Furthermore, a short exact sequence $0 \to C' \to C \to C'' \to 0$ in $\mathbf{C}$ corresponds to a distinguished triangle $C' \to C \to C'' \xrightarrow{[1]}$ in $\mathbf{D}$ with all objects belonging to $\mathbf{C}$ and vice-versa. Thus $\mathrm{Ext}^1_{\mathbf{C}}(C'', C') = \mathrm{Mor}_{\mathbf{D}}(C'', C'[1])$.

The functor $H^0 \colon \mathbf{D} \to \mathbf{C}$ given by $H^0(A) = \tau_{\geq 0}\tau_{\leq 0}A = \tau_{\leq 0}\tau_{\geq 0}A$ is a cohomological functor. We set $H^k(A) = H^0(A[k])$

Lemma 2.1. *If* $A \in \mathbf{D}^{\leq k}$ *and* $B \in \mathbf{D}^{\geq k}$, *then* $\mathrm{Mor}_{\mathbf{D}}(A, B) \cong \mathrm{Mor}_{\mathbf{C}}(H^k(A), H^k(B))$.

Proof.

$$\begin{aligned} \mathrm{Mor}_{\mathbf{D}}(A, B) &= \mathrm{Mor}_{\mathbf{D}}(\tau_{\leq k} A, B) && (A = \tau_{\leq k} A \text{ since } A \in \mathbf{D}^{\leq k} A) \\ &= \mathrm{Mor}_{\mathbf{D}}(\tau_{\geq k} \tau_{\leq k} A, B) && (\text{by (2.5) since } B \in \mathbf{D}^{\geq k}) \\ &= \mathrm{Mor}_{\mathbf{D}}(H^k(A)[-k], B). \end{aligned}$$

Similarly one may replace B here by $H^k(B)[-k]$ whence the lemma since $\mathbf{C}$ is full. □

In this paper, $\mathbf{D}$ will be the constructible bounded derived category of sheaves on a stratified pseudomanifold Y. The *standard t-structure* has $\mathbf{D}^{\leq 0}$ consisting of objects $\mathcal{S}$ satisfying

$$H^k(\mathcal{S}) = 0 \text{ for } k > 0, \tag{2.6}$$

and $\mathbf{D}^{\leq 0}$ consisting of objects $\mathcal{S}$ satisfying

$$H^k(\mathcal{S}) = 0 \text{ for } k < 0. \tag{2.7}$$

In this case, the truncation functors are the usual truncations of a complex, and objects in the heart may be represented by constructible sheaves viewed as complexes with only one nonzero term in degree 0.

In §3 we will define the perverse t-structure whose heart is the category of perverse sheaves. The subcategories, truncation functors, and cohomology functors for this t-structure will always be distinguished by a left superscript p, as in ${}^p\mathbf{D}^{\leq 0}$.

3 Perverse sheaves

3.1 Definition

Let Y be a stratified topological pseudomanifold of dimension n with stratification $\mathcal{X} = \{S\}$. We will assume that all strata of Y are connected which implies that the frontier condition holds: the closure of any stratum is a union of strata. We also assume that there are finitely many strata. The strata of Y are partially ordered by $S \preccurlyeq T$ if and only if $S \subseteq \overline{T}$.

We now briefly recall the definition of the category of perverse sheaves on Y following [5, §2.1].

Fix a $\mathbf{Z}$-valued function p on the set of strata, the *perversity* (in the sense of [5]); we assume the perversity satisfies $p(S) \geq p(T)$ when $S \preccurlyeq T$. Our main interest is when p is one of the two *middle perversities*:

$$p_-(S) = -\left\lfloor \frac{\dim S + 1}{2} \right\rfloor, \qquad p_+(S) = -\left\lfloor \frac{\dim S}{2} \right\rfloor. \tag{3.1}$$

If Y has only even dimensional strata, both p_- and p_+ are equal to the self-dual perversity $p_{1/2}(S) = -(1/2)\dim S$. (A non-middle perversity is introduced in §6.4 for technical reasons.)

Let $\mathbf{D}^b_c(Y)$ denote the bounded derived category of $\mathbf{C}$-sheaves on Y, constructible with respect to $\mathcal{X}$. (For an algebraic variety one usually does not fix a stratification, but since our main focus here is $Y = \widehat{X}$ it seems appropriate.) An object in $\mathbf{D}^b_c(Y)$ may be represented by a complex of sheaves $\mathcal{S}$ on Y whose local cohomology sheaves $H(i_S^*\mathcal{S})$ along a stratum S are locally constant and finitely generated; such a locally constant sheaf may be equivalently viewed as a $\pi_1(S)$-module, once we have picked as base point in S (omitted in the notation). Since Y is a stratified pseudomanifold, it can be shown that $H(i_S^!\mathcal{S})$ is likewise locally constant and finitely generated [9, V,§3].

Define the full subcategory ${}^p\mathbf{D}^{\leq 0}_c(Y)$ of $\mathbf{D}^b_c(Y)$ to consist of objects $\mathcal{S}$ which satisfy

$$H^k(i_S^*\mathcal{S}) = 0 \text{ for } k > p(S), \text{ for all strata } S; \tag{3.2}$$

Likewise define ${}^p\mathbf{D}^{\geq 0}_c(Y)$ to consist of objects $\mathcal{S}$ which satisfy

$$H^k(i_S^!\mathcal{S}) = 0 \text{ for } k < p(S), \text{ for all strata } S. \tag{3.3}$$

We refer to (3.2) and (3.3) as the *perverse vanishing* condition and the *perverse covanishing* condition respectively. As usual, we let ${}^p\mathbf{D}^{\leq k}_c(Y) = {}^p\mathbf{D}^{\leq 0}_c(Y)[-k]$ and ${}^p\mathbf{D}^{\geq k}_c(Y) = {}^p\mathbf{D}^{\geq 0}_c(Y)[-k]$.

The pair $({}^p\mathbf{D}^{\leq 0}_c(Y), {}^p\mathbf{D}^{\geq 0}_c(Y))$ forms a t-structure; its heart, ${}^p\mathbf{D}^{\leq 0}_c(Y) \cap {}^p\mathbf{D}^{\geq 0}_c(Y)$, is the category of *p-perverse sheaves* on Y, denoted $\mathbf{P}(Y) = \mathbf{P}_p(Y)$. For $p = p_-$ (resp. p_+) we simply write $\mathbf{P}_-(Y)$ (resp. $\mathbf{P}_+(Y)$).

For $\mathcal{S} \in \mathbf{D}^b_c(Y)$, $D_S(i_S^!\mathcal{S}) = i_S^* D_Y(\mathcal{S})$ where D_Y and D_S denote the respective Verdier duality involutions. Set

$$p^*(S) = -p(S) - \dim S,$$

the *dual* perversity. Then D_Y sends p-perverse sheaves to p^*-perverse sheaves. From (3.1) we see that $p_-(S) + p_+(S) = -\dim S$ and thus $p_-^* = p_+$ and $p_+^* = p_-$. In particular, D_Y interchanges $\mathbf{P}_-(Y)$ and $\mathbf{P}_+(Y)$.

Warning. In §5.2, after introducing the reductive Borel-Serre compactification, we will impose a condition we call *reductively constructible* on our objects which is more appropriate to use on $\widehat{X}$. All of the material of the current section as well as §4 hold without change under the assumption of reductively constructible (with a minor exception noted later).

3.2 Simple perverse sheaves

For every stratum T of Y and every irreducible local system $\mathbf{E}$ on T (corresponding to an irreducible representation E of $\pi_1(T)$) there is a simple object

$$\mathcal{P}_T(\mathbf{E}) = \mathcal{P}_{p,T}(\mathbf{E}) = \tau_{\leq p(S_1)-1} j_{S_1 *} \cdots \tau_{\leq p(S_N)-1} j_{S_N *}(i_{T*}\mathbf{E}[-p(T)]) \tag{3.4}$$

of $\mathbf{P}(Y)$ and these are all the simple objects. Here $S_1, \ldots, S_N$ is an enumeration of the strata $S \prec T$ such that if $S_i \prec S_j$ then $i < j$.

$\mathcal{P}_T(\mathbf{E})$ is supported on $\overline{T}$ and satisfies the usual perverse vanishing and covanishing condition on T:

$$H^k(i_T^*\mathcal{P}_T(\mathbf{E})) = 0 \text{ for } k > p(T) \text{ and } H^k(i_T^!\mathcal{P}_T(\mathbf{E})) = 0 \text{ for } k < p(T).$$

In fact, $i_T^*\mathcal{P}_T(\mathbf{E}) = i_T^!\mathcal{P}_T(\mathbf{E}) = \mathbf{E}[-p(T)]$. However $\mathcal{P}_T(\mathbf{E})$ satisfies stronger conditions on strata smaller than T:

$$H^k(i_S^*\mathcal{P}_T(\mathbf{E})) = 0 \text{ for } k \geq p(S), \text{ for all strata } S \prec T \tag{3.5}$$

and

$$H^k(i_S^!\mathcal{P}_T(\mathbf{E})) = 0 \text{ for } k \leq p(S), \text{ for all strata } S \prec T. \tag{3.6}$$

We call (3.5) and (3.6) the *intersection cohomology vanishing* condition and the *intersection cohomology covanishing* condition respectively.

Let S be a stratum of Y, V_S an open union of strata containing S as a minimal stratum, and $j_S \colon U_S = V_S \setminus S \hookrightarrow V_S$. Define the *link cohomology functor* $\kappa_S \colon \mathbf{D}_c^b(U_S) \to \mathbf{D}_c^b(S)$ by

$$\kappa_S A = i_S^* j_{S*} A.$$

The local cohomology of $\kappa_S A$ is the cohomology of A pulled back to the topological link of the stratum S. If A is defined on a subset larger than U_S, for example Y, we write $\kappa_S A$ for $\kappa_S(A|_{U_S})$. Thus it fits into a distinguished triangle for $A \in \mathbf{D}_c^b(Y)$:

$$i_S^! A \to i_S^* A \to \kappa_S A \xrightarrow{[1]} .$$

Now the values of the cohomology groups appearing in (3.5) and (3.6) in the other degrees can be easily calculated:

$$H^k(i_S^*\mathcal{P}_T(\mathbf{E})) \cong H^k(\kappa_S\mathcal{P}_T(\mathbf{E})) \text{ for } k < p(S), \text{ for all strata } S \prec T \tag{3.7}$$

and

$$H^k(i_S^!\mathcal{P}_T(\mathbf{E})) = H^{k-1}(\kappa_S\mathcal{P}_T(\mathbf{E})) \text{ for } k > p(S), \text{ for all strata } S \prec T. \tag{3.8}$$

All degrees of the $\mathcal{P}_T(\mathbf{E})$-cohomology of the link of S appear, split between $i_S^*\mathcal{P}_T(\mathbf{E})$ and (with a shift) $i_S^!\mathcal{P}_T(\mathbf{E})$.

3.3 Comparison with intersection cohomology

Assume p is one of the middle perversities p_- or p_+. Define

$$\bar{p}_T(S) = p(S) - p(T) - 1 \qquad (S \preccurlyeq T). \tag{3.9}$$

Then by pulling out the shift in (3.4) we see that

$$\mathcal{P}_{p,T}(\mathbf{E}) = \mathcal{I}_{\bar{p}_T}\mathcal{C}(\overline{T}, \mathbf{E})[-p(T)]$$

where for $\bar{p} \colon \mathcal{X}(\overline{T}) \to \mathbf{Z}$ we set

$$\mathcal{I}_{\bar{p}}\mathcal{C}(\overline{T}, \mathbf{E}) = \tau_{\leq \bar{p}(S_1)} j_{S_1*} \cdots \tau_{\leq \bar{p}(S_N)} j_{S_N*} i_{T*}\mathbf{E}. \tag{3.10}$$

When $\bar{p}(S)$ depends only on $k = \operatorname{codim}_T S$, as is the case for $\bar{p}_T(S)$ from (3.9), the object $\mathcal{I}_{\bar{p}}\mathcal{C}(\overline{T}, \mathbf{E})$ is essentially Deligne's sheaf for intersection cohomology as in [9, V].

In fact $\bar{p}_T(S)$ agrees with one of the two "classical" middle perversities of Goresky and MacPherson [20]:

$$m(k) = \left\lfloor \frac{k-2}{2} \right\rfloor, \qquad n(k) = \left\lfloor \frac{k-1}{2} \right\rfloor.$$

Specifically for $p = p_-$,

$$\bar{p}_T(S) = \begin{cases} n(\operatorname{codim}_T S) & \text{if } \dim T \text{ is odd,} \\ m(\operatorname{codim}_T S) & \text{if } \dim T \text{ is even,} \end{cases}$$

and for $p = p_+$,

$$\bar{p}_T(S) = \begin{cases} m(\operatorname{codim}_T S) & \text{if } \dim T \text{ is odd,} \\ n(\operatorname{codim}_T S) & \text{if } \dim T \text{ is even.} \end{cases}$$

Thus $\mathcal{P}_{p,T}(\mathbf{E}) = \mathcal{I}_{\bar{p}_T}\mathcal{C}(\overline{T}, \mathbf{E})[-p(T)]$ is either $\mathcal{I}_m\mathcal{C}(\overline{T}, \mathbf{E})[-p(T)]$ or $\mathcal{I}_n\mathcal{C}(\overline{T}, \mathbf{E})[-p(T)]$ depending on p and the parity of $\dim T$.

3.4 A non-trivial extension of perverse sheaves

Let p be a middle perversity and let $\mathbf{E}$ be an irreducible local system on a stratum T of Y. While $\mathcal{P}_{p,T}(\mathbf{E}) = \mathcal{I}_{\bar{p}_T}\mathcal{C}(\overline{T}, \mathbf{E})[-p(T)]$ is a simple perverse sheaf in $\mathbf{P}_p(Y)$, in fact the object

$$\widetilde{\mathcal{P}}_{p,T}(\mathbf{E}) \equiv \mathcal{P}_{p^*,T}(\mathbf{E})[p^*(T) - p(T)] = \mathcal{I}_{\overline{p^*}_T}\mathcal{C}(\overline{T}, \mathbf{E})[-p(T)]$$

is also a perverse sheaf in $\mathbf{P}_p(Y)$, though not necessarily simple.

To see this, note that the p-perversity condition on the stratum T is satisfied due to the shift. To deal with the other strata, note first that for a middle perversity

$$p^*(S) - p(S) \in \begin{cases} \{0, 1\} & \text{if } p = p_-, \\ \{0, -1\} & \text{if } p = p_+, \end{cases}$$

for any stratum S; here the value 0 occurs if and only if $\dim S$ is even. Consequently

$$-1 \leq (p^*(S) - p(S)) - (p^*(T) - p(T)) \leq 1 \tag{3.11}$$

for all strata S and T. Now for $S \prec T$, the intersection cohomology vanishing condition (3.5) with respect to p^* implies that

$$H^k(i_S^* \mathcal{P}_{p^*,T}(\mathbf{E})[p^*(T) - p(T)]) = 0 \text{ for } k \geq p^*(S) - (p^*(T) - p(T)).$$

This implies the p-perverse vanishing condition (3.2) for $k > p(S)$ since

$$k \geq p(S) + 1 \geq p(S) + (p^*(S) - p(S)) - (p^*(T) - p(T)) = p^*(S) - (p^*(T) - p(T))$$

by (3.11). The p-perverse covanishing condition (3.3) follows similarly.

The natural morphism $\mathcal{I}_m\mathcal{C}(\overline{T},\mathbf{E})[-p(T)] \to \mathcal{I}_n\mathcal{C}(\overline{T},\mathbf{E})[-p(T)]$ thus induces short exact sequences

$$0 \to \mathcal{P}_{p,T}(\mathbf{E}) \to \widetilde{\mathcal{P}}_{p,T}(\mathbf{E}) \to \mathcal{P} \to 0 \qquad (\text{if } \bar{p}_T = m)$$

or

$$0 \to \mathcal{P} \to \widetilde{\mathcal{P}}_{p,T}(\mathbf{E}) \to \mathcal{P}_{p,T}(\mathbf{E}) \to 0 \qquad (\text{if } \bar{p}_T = n)$$

in $\mathbf{P}_p(Y)$ and hence extensions in $\mathbf{P}_p(Y)$.

If $\mathcal{P}_{p,T}(\mathbf{E}) \neq \widetilde{\mathcal{P}}_{p,T}(\mathbf{E})$, the resulting extensions are not trivial. For if $\widetilde{\mathcal{P}}_{p,T}(\mathbf{E}) = \mathcal{P} \oplus \mathcal{P}_{p,T}(\mathbf{E})$ in $\mathbf{P}_p(Y)$, then applying i_S^* or $i_S^!$ and shifting by $p(T) - p^*(T)$, we see that $\mathcal{P}[p(T)-p^*(T)]$ and $\widetilde{\mathcal{P}}_{p^*,T}(\mathbf{E}) = \mathcal{P}_{p,T}(\mathbf{E})[p(T)-p^*(T)]$ satisfy the p^*-perverse conditions (since $\mathcal{P}_{p^*,T}(\mathbf{E}) = \widetilde{\mathcal{P}}_{p,T}(\mathbf{E})[p(T) - p^*(T)]$ does). This implies that $\mathcal{P}_{p^*,T}(\mathbf{E}) = \mathcal{P}[p(T) - p^*(T)] \oplus \widetilde{\mathcal{P}}_{p^*,T}(\mathbf{E})$ in $\mathbf{P}_{p^*}(Y)$, contradicting the fact that $\mathcal{P}_{p^*,T}(\mathbf{E})$ is simple in $\mathbf{P}_{p^*}(Y)$.

Remarks. (i) The inequality $\mathcal{P}_{p,T}(\mathbf{E}) \neq \widetilde{\mathcal{P}}_{p,T}(\mathbf{E})$ will occur precisely when there is a odd codimension stratum $S \prec T$ with nonvanishing middle degree link cohomology

$$H^k(\kappa_S \mathcal{P}_{p,T}(\mathbf{E})) \neq 0 \qquad (k = (\operatorname{codim}_T S - 1)/2 + p(T)).$$

(ii) If $Y = \widehat{X}$, the reductive Borel-Serre compactification of a locally symmetric variety, and $\mathbf{E}$ is induced from an algebraic representation of L_T, then the decomposition theorem for $\pi\colon \widehat{X} \to X^*$ (see §1.4) implies that this extension always becomes trivial when pushed down to the Baily-Borel compactification. In fact $\pi_*\mathcal{P}_{p,X}(\mathbf{E}) = \pi_*\widetilde{\mathcal{P}}_{p,X}(\mathbf{E})$ by the solution to Rapoport's conjecture [41] since it applies to both middle perversities and they agree on X^*; more generally the theorem shows that $\pi_*\mathcal{P}_{p,X_{P}}(\mathbf{E})$ and $\pi_*\widetilde{\mathcal{P}}_{p,X_{P}}(\mathbf{E})$ differ only in the coefficient system, but in the reductively constructible setting, the category of coefficient systems on a stratum is semisimple.

(iii) When $Y = \widehat{X}$ (with X not necessarily Hermitian), the inequality $\mathcal{P}_{p,T}(\mathbf{E}) \neq \widetilde{\mathcal{P}}_{p,T}(\mathbf{E})$ has an intimate relation with the presence of infinite dimensional local L^2-cohomology on $\widehat{X}$.

4 Extensions

Fix a perversity p (not necessarily middle). In this section we indicate how one may compute the extensions $\operatorname{Ext}^1_{\mathbf{P}(Y)}(\mathcal{P}_T(\mathbf{E}), \mathcal{P}_{T'}(\mathbf{E}'))$ between two simple perverse sheaves corresponding to strata T, $T' \in \mathcal{X}(Y)$. Note that by the theory of t-structures,

$$\operatorname{Ext}^1_{\mathbf{P}(Y)}(\mathcal{P}_T(\mathbf{E}), \mathcal{P}_{T'}(\mathbf{E}')) = \operatorname{Mor}_{\mathbf{D}^b_c(Y)}(\mathcal{P}_T(\mathbf{E}), \mathcal{P}_{T'}(\mathbf{E}')[1]).$$

Lemma 4.1. *Let V be an open union of strata with S a minimal stratum of V; set $U = V \setminus S$. For A, $B \in \mathbf{D}_c^b(V)$ and $k \in \mathbf{Z}$, assume* $\operatorname{Mor}_{\mathbf{D}_c^b(U)}(j_S^* A, j_S^* B[k-1]) = \operatorname{Mor}_{\mathbf{D}_c^b(U)}(j_S^* A, j_S^* B[k]) = 0$. *Then* $\operatorname{Mor}_{\mathbf{D}_c^b(V)}(A, B[k]) \cong \operatorname{Mor}_{\mathbf{D}_c^b(S)}(i_S^* A, i_S^! B[k])$.

Proof. The distinguished triangle $i_{S*} i_S^! B \to B \to j_{S*} j_S^* B \xrightarrow{[1]}$ yields a long exact sequence

$$\operatorname{Mor}_{\mathbf{D}_c^b(V)}(A, j_{S*} j_S^* B[k-1]) \longrightarrow \operatorname{Mor}_{\mathbf{D}_c^b(V)}(A, i_{S*} i_S^! B[k]) \longrightarrow \\ \operatorname{Mor}_{\mathbf{D}_c^b(V)}(A, B[k]) \longrightarrow \operatorname{Mor}_{\mathbf{D}_c^b(V)}(A, j_{S*} j_S^* B[k]).$$

The outer two groups are zero by hypothesis, thus

$$\operatorname{Mor}_{\mathbf{D}_c^b(V)}(A, B[k]) \cong \operatorname{Mor}_{\mathbf{D}_c^b(V)}(A, i_{S*} i_S^! B[k]) \cong \operatorname{Mor}_{\mathbf{D}_c^b(S)}(i_S^* A, i_S^! B[k]). \qquad \square$$

Lemma 4.2. *If all maximal strata of $\overline{T} \cap \overline{T'}$ are strictly smaller than T and T', then* $\operatorname{Mor}_{\mathbf{D}_c^b(Y)}(\mathcal{P}_T(\mathbf{E}), \mathcal{P}_{T'}(\mathbf{E}')[k]) = 0$ *for $k \le 1$.*

Proof. Let V be an open union of strata containing $Y \setminus \overline{T} \cap \overline{T'}$. We will prove that

$$\operatorname{Mor}_{\mathbf{D}_c^b(V)}(\mathcal{P}_T(\mathbf{E}), \mathcal{P}_{T'}(\mathbf{E}')[k]) = 0, \qquad k \le 1,$$

by induction on the number of strata in V. When $V = Y$ this proves the lemma.

If $V = Y \setminus \overline{T} \cap \overline{T'}$, the supports of $\mathcal{P}_T(\mathbf{E})$ and $\mathcal{P}_{T'}(\mathbf{E}')$ are disjoint and

$$\operatorname{Mor}_{\mathbf{D}_c^b(V)}(\mathcal{P}_T(\mathbf{E}), \mathcal{P}_{T'}(\mathbf{E}')[k]) = 0, \qquad \text{for all } k.$$

For larger V, let S be a minimal stratum in $V \cap (\overline{T} \cap \overline{T'})$ and set $U = V \setminus S$. By Lemma 4.1 (which applies by the inductive hypothesis)

$$\operatorname{Mor}_{\mathbf{D}_c^b(V)}(\mathcal{P}_T(\mathbf{E}), \mathcal{P}_{T'}(\mathbf{E}')[k]) \cong \operatorname{Mor}_{\mathbf{D}_c^b(S)}(i_S^* \mathcal{P}_T(\mathbf{E}), i_S^! \mathcal{P}_{T'}(\mathbf{E}')[k]).$$

for $k \le 1$. Since S is dominated by a maximal stratum in $\overline{T} \cap \overline{T'}$ which is thus strictly smaller than T and T', the intersection cohomology vanishing and covanishing conditions apply to yield $i_S^* \mathcal{P}_T(\mathbf{E}) \in \mathbf{D}_c^{\le p(S)-1}(S)$ and $i_S^! \mathcal{P}_{T'}(\mathbf{E}')[k] \in \mathbf{D}_c^{\ge p(S)+1-k}(S)$. Now $p(S) - 1 < p(S) + 1 - k$ for $k \le 1$, so $\operatorname{Mor}_{\mathbf{D}_c^b(S)}(i_S^* \mathcal{P}_T(\mathbf{E}), i_S^! \mathcal{P}_{T'}(\mathbf{E}')[k]) = 0$ by (2.1). $\square$

We conclude that a non-trivial extension can only exist when $T = T'$, $T \prec T'$, or $T \succ T'$.

Lemma 4.3. *Let $V_T = (Y \setminus \overline{T}) \cup T$ and similarly for T'.*

(i) *If $T = T'$,* $\operatorname{Mor}_{\mathbf{D}_c^b(V_{T'})}(\mathcal{P}_T(\mathbf{E}), \mathcal{P}_T(\mathbf{E}')[1]) \cong \operatorname{Mor}_{\mathbf{D}_c^b(T)}(\mathbf{E}, \mathbf{E}'[1])$.

(ii) *If $T \prec T'$,* $\operatorname{Mor}_{\mathbf{D}_c^b(V_T)}(\mathcal{P}_T(\mathbf{E}), \mathcal{P}_{T'}(\mathbf{E}')[1]) \cong \operatorname{Mor}_{\pi_1(T)}(E, H^{p(T)}(\kappa_T \mathcal{P}_{T'}(\mathbf{E}')))$.

(iii) *If $T \succ T'$,* $\operatorname{Mor}_{\mathbf{D}_c^b(V_{T'})}(\mathcal{P}_T(\mathbf{E}), \mathcal{P}_{T'}(\mathbf{E}')[1]) \cong \operatorname{Mor}_{\pi_1(T')}(H^{p(T')-1}(\kappa_{T'} \mathcal{P}_T(\mathbf{E})), E')$.

Proof. The case $T = T'$ follows from Lemma 4.1. For $T \prec T'$ we calculate

$$\begin{aligned}
&\mathrm{Mor}_{\mathbf{D}^b_c(V_T)}(\mathcal{P}_T(\mathbf{E}), \mathcal{P}_{T'}(\mathbf{E}')[1]) \\
&\cong \mathrm{Mor}_{\mathbf{D}^b_c(T)}(\mathbf{E}[-p(T)], i_T^! \mathcal{P}_{T'}(\mathbf{E}')[1]) && \text{(by Lemma 4.1)} \\
&\cong \mathrm{Mor}_{\pi_1(T)}(E, H^{p(T)+1}(i_T^! \mathcal{P}_{T'}(\mathbf{E}'))) && \text{(by (3.6) and Lemma 2.1)} \\
&\cong \mathrm{Mor}_{\pi_1(T)}(E, H^{p(T)}(\kappa_T \mathcal{P}_{T'}(\mathbf{E}'))) && \text{by (3.8).}
\end{aligned}$$

The case where $T \succ T'$ is similar. □

Lemma 4.4. *Assume $S \prec T$ and $S \prec T'$. Let V be an open union of strata with $S \subset V$ a minimal stratum and set $U = V \backslash S$. A morphism $\phi_U \in \mathrm{Mor}_{\mathbf{D}^b_c(U)}(j_S^* \mathcal{P}_T(\mathbf{E}), j_S^* \mathcal{P}_{T'}(\mathbf{E}')[1])$ prolongs to a morphism ϕ_V over V if and only if*

$$H^{p(S)-1}(\kappa_S(\phi_U)) = 0 \text{ in } \mathrm{Mor}_{\pi_1(S)}(H^{p(S)-1}(\kappa_S \mathcal{P}_T(\mathbf{E})), H^{p(S)}(\kappa_S \mathcal{P}_{T'}(\mathbf{E}'))). \tag{4.1}$$

If this prolongation exists it is unique.

Proof. We apply $\mathrm{Mor}_{\mathbf{D}^b_c(V)}(\mathcal{P}_T(\mathbf{E}), \cdot)$ to the distinguished triangle $i_{S*} i_S^! \mathcal{P}_{T'}(\mathbf{E}') \to \mathcal{P}_{T'}(\mathbf{E}') \to j_{S*} j_S^* \mathcal{P}_{T'}(\mathbf{E}') \xrightarrow{[1]}$ which yields, after using adjointness, the long exact sequence

$$\begin{aligned}
&\mathrm{Mor}_{\mathbf{D}^b_c(S)}(i_S^* \mathcal{P}_T(\mathbf{E}), i_S^! \mathcal{P}_{T'}(\mathbf{E}')[1]) \longrightarrow \mathrm{Mor}_{\mathbf{D}^b_c(V)}(\mathcal{P}_T(\mathbf{E}), \mathcal{P}_{T'}(\mathbf{E}')[1]) \longrightarrow \\
&\qquad \mathrm{Mor}_{\mathbf{D}^b_c(U)}(j_S^* \mathcal{P}_T(\mathbf{E}), j_S^* \mathcal{P}_{T'}(\mathbf{E}')[1]) \longrightarrow \mathrm{Mor}_{\mathbf{D}^b_c(S)}(i_S^* \mathcal{P}_T(\mathbf{E}), i_S^! \mathcal{P}_{T'}(\mathbf{E}')[2]).
\end{aligned}$$

Since $i_S^* \mathcal{P}_T(\mathbf{E}) \in \mathbf{D}_c^{\leq p(S)-1}(S)$ by (3.5) and $i_S^! \mathcal{P}_{T'}(\mathbf{E}')[1] \in \mathbf{D}_c^{\geq p(S)}(S)$ by (3.6), the leftmost term is 0 by (2.1). This shows that ϕ_V is unique if it exists. It exists if and only if ϕ_U maps to zero in the last term, which by a similar calculation and using Lemma 2.1 is

$$\mathrm{Mor}_{\pi_1(S)}(H^{p(S)-1}(i_S^* \mathcal{P}_T(\mathbf{E})), H^{p(S)-1}(i_S^! \mathcal{P}_{T'}(\mathbf{E}')[2])).$$

By (3.7) and (3.8) this is the group of morphisms in (4.1). □

Note that the map on link cohomology at S which needs to be zero in order to prolong an extension is the map from the highest degree of link cohomology occurring in $i_S^* \mathcal{P}_T(\mathbf{E})$ to the lowest degree of link cohomology that does *not* occur in $i_S^* \mathcal{P}_{T'}(\mathbf{E}')$.

This section is summarized in the following

Proposition 4.5. $\mathrm{Ext}^1_{\mathbf{P}(Y)}(\mathcal{P}_T(\mathbf{E}), \mathcal{P}_{T'}(\mathbf{E}'))$ *is zero unless $T = T'$, $T \prec T'$ or $T \succ T'$.*

(i) *For $T = T'$, $\mathrm{Ext}^1_{\mathbf{P}(Y)}(\mathcal{P}_T(\mathbf{E}), \mathcal{P}_T(\mathbf{E}'))$ is isomorphic to a subgroup of*

$$\mathrm{Mor}_{\mathbf{D}^b_c(V_T)}(\mathcal{P}_T(\mathbf{E}), \mathcal{P}_T(\mathbf{E}')[1]) \cong \mathrm{Mor}_{\mathbf{D}^b_c(T)}(\mathbf{E}, \mathbf{E}'[1]).$$

(ii) *For* $T \prec T'$, $\mathrm{Ext}^1_{\mathbf{P}(Y)}(\mathcal{P}_T(\mathbf{E}), \mathcal{P}_{T'}(\mathbf{E}'))$ *is isomorphic to a subgroup of*

$$\mathrm{Mor}_{\mathbf{D}^b_c(V_T)}(\mathcal{P}_T(\mathbf{E}), \mathcal{P}_{T'}(\mathbf{E}')[1]) \cong \mathrm{Mor}_{\pi_1(T)}(E, H^{p(T)}(\kappa_T \mathcal{P}_{T'}(\mathbf{E}'))).$$

(iii) *For* $T \succ T'$, $\mathrm{Ext}^1_{\mathbf{P}(Y)}(\mathcal{P}_T(\mathbf{E}), \mathcal{P}_{T'}(\mathbf{E}'))$ *is isomorphic to a subgroup of*

$$\mathrm{Mor}_{\mathbf{D}^b_c(V_{T'})}(\mathcal{P}_T(\mathbf{E}), \mathcal{P}_{T'}(\mathbf{E}')[1]) \cong \mathrm{Mor}_{\pi_1(T')}(H^{p(T')-1}(\kappa_{T'} \mathcal{P}_T(\mathbf{E})), E').$$

In either case, the group of extensions consists of those morphisms ϕ *that recursively satisfy*

$$H^{p(S)-1}(\kappa_S(\phi|_{U_S})) = 0 \text{ in } \mathrm{Mor}_{\pi_1(S)}(H^{p(S)-1}(\kappa_S \mathcal{P}_T(\mathbf{E})), H^{p(S)}(\kappa_S \mathcal{P}_{T'}(\mathbf{E}'))).$$

for all S *satisfying* $S \prec T$ *and* $S \prec T'$. *Here* V_S *is an open union of strata containing* S *as a minimal stratum and* $U_S = V_S \setminus S$; *the condition above allows one to uniquely extend* ϕ *from* U_S *to* V_S.

Remarks. (i) The condition (4.1) needed to prolong an extension can be nontrivial even in the case $T = T'$; one example of an extension of coefficient systems failing this condition (and hence not prolonging to an extension of perverse sheaves) is [14, Examples 2.2.5 and 2.7.1]. The point is that the intermediate extension function $\mathbf{E} \mapsto \mathcal{P}_T(\mathbf{E})$ (3.4) is not exact even though it preserves injective and surjective maps [14, §2.7].

(ii) In §5.2 we will define a more refined notion of constructibility in which local systems on a stratum T are associated to algebraic representations of a reductive group L_T. All the results of this section hold in this context except one should replace $\mathrm{Mor}_{\pi_1(T)}$ by Mor_{L_T} as needed. In addition, since algebraic representations of a reductive group are semisimple, there are no extensions between simple objects when $T = T'$.

5 Reductive Borel-Serre compactification

The reductive Borel-Serre compactification $\widehat{X}$ first appeared in work of Zucker [46] and has grown in importance far beyond its original use. We recall it following [18], focusing mainly on its structure rather than its construction.

5.1 Stratification

For G a reductive algebraic group defined over $\mathbf{Q}$ and Γ an arithmetic subgroup, let $X = X_G = \Gamma\backslash G(\mathbf{R})/A_G(\mathbf{R})K$ be the corresponding locally symmetric space. Here A_G is the maximal $\mathbf{Q}$-split torus in the center of G, and K is a maximal compact subgroup of $G(\mathbf{R})$.

For a parabolic $\mathbf{Q}$-subgroup S of G, let $L_S = S/N_S$ denote its reductive Levi quotient, where N_S denotes the unipotent radical of S; it is also defined over $\mathbf{Q}$. We have an almost direct product $L_S = M_S A_S$ where $M_S = {}^0L_S = \bigcap_\chi \ker \chi^2$

in which χ runs over characters of L_S defined over $\mathbf{Q}$. Let $\Gamma_{N_S} = \Gamma \cap N_S$ and $\Gamma_{L_S} = (\Gamma \cap S)/(\Gamma \cap N_S)$ be the induced arithmetic subgroups. Starting from L_S and Γ_{L_S} we again obtain a locally symmetric space X_S. If S' is Γ-conjugate to S, then $X_{S'}$ and X_S may be canonically identified.

The reductive Borel-Serre compactification of X is the stratified pseudomanifold $\widehat{X} = \coprod_S X_S$ (where S ranges over the Γ-conjugacy classes of parabolic $\mathbf{Q}$-subgroups of G) endowed with an appropriate topology. The closure of a stratum X_S is denoted $\widehat{X}_S$; it is indeed the reductive Borel-Serre compactification of X_S and its strata X_T correspond to Γ-conjugacy classes of parabolic $\mathbf{Q}$-subgroups having a representative $T \subseteq S$.

From now on we will abuse notation by using the same letter S to denote a stratum X_S, the corresponding Γ-conjugacy class of parabolic $\mathbf{Q}$-subgroups, and a representative of that class. Thus the partial order $S \preccurlyeq T$ on strata corresponds on parabolic subgroups to $S \subseteq {}^{\gamma}T$ for some $\gamma \in \Gamma$.

The *depth* of a stratum S of $\widehat{X}$, $\operatorname{depth} S$, is the maximal length d of a chain $S = S_0 \prec S_1 \prec \cdots \prec S_d = X$. If $S \preccurlyeq T$, the *relative depth* of S viewed as a stratum of $\overline{T}$, $\operatorname{depth}_T S$, is the maximal length of a chain $S = S_0 \prec S_1 \prec \cdots \prec S_d = T$. We have $\operatorname{depth}_T S = \dim A_S - \dim A_T$.

An algebraic representation E of L_S induces a representation of Γ_{L_S} and hence a local system $\mathbf{E}$ on S.

5.2 Reductive constructibility

A constructible complex of sheaves $\mathcal{S}$ on a pseudomanifold has the property that for all strata S, the cohomology sheaves of $i_S^*\mathcal{S}$, $i_S^!\mathcal{S}$, and $\kappa_S\mathcal{S}$ are all finitely generated locally constant and hence are associated to finite-dimensional representations of Γ_{L_S}. Furthermore the maps on cohomology induced by the distinguished triangle $i_S^!\mathcal{S} \to i_S^*\mathcal{S} \to \kappa_S\mathcal{S} \xrightarrow{[1]}$ are morphisms of Γ_{L_S}-representations. A *reductively constructible* complex of sheaves $\mathcal{S}$ on $\widehat{X}$ is as above but has been enriched with extra structure so that all of these locally constant sheaves arise from algebraic representations of L_S and that the morphisms above are morphisms of L_S-modules. Morphisms between such sheaves must also induce morphisms of L_S-modules on the cohomology sheaves over a stratum.

Rather than construct the derived category of reductively constructible sheaves $\mathbf{D}^b_{rc}(\widehat{X})$ directly as suggested above, we note instead that if one starts with the category of $\mathscr{L}$-modules constructed in [41] and pass to the homotopy category, one obtains the desired category of reductively constructible sheaves. (In the homotopy category of $\mathscr{L}$-modules, every quasi-isomorphism is already an isomorphism and thus one does not need to localize further.) We will discuss the details elsewhere.

Exactly as in §3 one can define a perverse t-structure on $\mathbf{D}^b_{rc}(\widehat{X})$ and obtain a category of perverse sheaves which we again denote $\mathbf{P}(\widehat{X})$. The description of simple objects is the same except we start with $\mathbf{E}$ on T coming from an algebraic representation of L_T. The fact that the pushforward functors in the definition of the simple perverse sheaves do preserve reductive constructibility is a consequence of the theorems of Nomizu, van Est, and Kostant which we recall in §6.

The results in §4 all remain true with identical proofs however one must replace $\pi_1(S)$-morphisms with L_S-morphisms in Lemmas 4.3 and 4.4 and Proposition 4.5.

Enriching our objects $\mathcal{S}$ so that $H(i_S^*\mathcal{S})$ has an action of the central split torus $A_S \subset L_S$ is new data. However aside from this, passing from the constructible category to the reductively constructible category is not as major a change as it may appear at first glance due to the Borel density theorem [8] and Margulis superrigidity [31].

5.3 Links

We recall the link of a stratum S of $\widehat{X}$ following [18] (see also [42, §7]).

For any parabolic $\mathbf{Q}$-subgroup S, the Levi quotient L_S acts (via a lift to S) by conjugation on the Lie algebra $\mathfrak{n}_S$ of N_S. Though this action depends on the lift, the weights by which A_S acts on $\mathfrak{n}_S$ are well-defined and have a unique basis denoted Δ_S. If $S \subseteq S'$, then $N_{S'} \triangleleft N_S$ and the group $A_{S'}$ may be naturally viewed as a subgroup of A_S. We let $\Delta_S^{S'} \subseteq \Delta_S$ be those weights that restrict trivially to $A_{S'}$. The correspondence $S' \mapsto \Delta_S^{S'}$ is an order preserving bijection between parabolic $\mathbf{Q}$-subgroups containing S and subsets of Δ_S. Restriction to $A_{S'}$ yields a bijection between $\Delta_S \setminus \Delta_S^{S'}$ and $\Delta_{S'}$. Note that $\Delta_S^G = \Delta_S$.

Let $|\Delta_S|$ be a topological simplex with vertices indexed by the elements of Δ_S. Give $|\Delta_S|$ the stratification by its open faces $|\Delta_S^{S'}|^\circ = \operatorname{int} |\Delta_S^{S'}|$ indexed by $S' \supsetneq S$. Note that

$$\dim |\Delta_S| = \operatorname{depth} S - 1.$$

The topological link of the stratum $S \subset \widehat{X}$ is

$$\operatorname{Lk}_S = (\Gamma_{N_S} \backslash N_S(\mathbf{R}) \times |\Delta_S|)/\sim \tag{5.1}$$

where if $t \in |\Delta_S^{S'}|^\circ$, $(n,t) \sim (n'n,t)$ for all $n' \in N_{S'}(\mathbf{R})$. Thus the intersection of the link of S with higher strata $S' \supset S$ is

$$\operatorname{Lk}_S \cap S' = \Gamma_{N_S^{S'}} \backslash N_S^{S'}(\mathbf{R}) \times |\Delta_S^{S'}|^\circ \tag{5.2}$$

where $N_S^{S'} = N_S/N_{S'}$ and $\Gamma_{N_S^{S'}} = \Gamma_{N_S}/\Gamma_{N_{S'}}$.

More generally, the topological link of S viewed as a stratum of $\overline{T}$ (for $S \preccurlyeq T$) is

$$\operatorname{Lk}_S^T = \left(\Gamma_{N_S^T} \backslash N_S^T(\mathbf{R}) \times |\Delta_S^T|\right)\Big/\sim \tag{5.3}$$

and

$$\dim |\Delta_S^T| = \operatorname{depth}_T S - 1.$$

If $S \preccurlyeq S' \preccurlyeq T$, we have again

$$\operatorname{Lk}_S^T \cap S' = \Gamma_{N_S^{S'}} \backslash N_S^{S'}(\mathbf{R}) \times |\Delta_S^{S'}|^\circ. \tag{5.4}$$

6 Link cohomology in $\widehat{X}$

To actually calculate extensions we need to understand the link cohomology functor $\kappa_S = i_S^* j_{S*}$ on a stratum S. We calculate the link cohomology of $\mathcal{P}_T(\mathbf{E})$ for strata S of relative depth 1, 2, and 3, following the formula given in [41, §5.5] (see also [42, §18.4]).

6.1 Kostant's theorem

The calculation of $H(\kappa_S \mathcal{P}_T(\mathbf{E}))$ given later will involve the Lie algebra cohomology $H(\mathfrak{n}_S^T, E)$ where $\mathfrak{n}_S^T$ is the Lie algebra of N_S^T. The adjoint action of L_S on $\mathfrak{n}_S^T$ (via choice of a lift) induces an action of L_S on $H(\mathfrak{n}_S^T, E)$ which is independent of the lift. We begin by recalling Kostant's theorem [27] which gives a decomposition of $H(\mathfrak{n}_S^T, E)$ as an L_S-module.

We first consider $T = G$ and let E be an irreducible algebraic representation of G. Choose a Cartan subalgebra for the Lie algebra $\mathfrak{l}_S$ of L_S; it lifts to a Cartan subalgebra of the Lie algebra $\mathfrak{g}$ of G. Choose an order on the roots of $\mathfrak{g}$ so that the roots in $\mathfrak{n}_S$ are all positive. Let ρ be one-half the sum of the positive roots of $\mathfrak{g}$. Let $W = W^G$ be the Weyl group of the root system of $\mathfrak{g}$ with corresponding length function $\ell(w)$. Let $W^S \subseteq W$ be the subgroup generated by the simple reflections in roots of $\mathfrak{l}_S$. Let $W_S \subseteq W$ be the set of unique minimal length representatives of cosets in $W^S \backslash W$. Thus there is a product decomposition $W = W^S W_S$. For E an irreducible algebraic representation of G with highest weight λ, Kostant's theorem says

$$H(\mathfrak{n}_S, E) = \bigoplus_{w \in W_S} H^{\ell(w)}(\mathfrak{n}_S, E)_w[-\ell(w)]$$

where $H^{\ell(w)}(\mathfrak{n}_S, E)_w$ is an irreducible L_S-module and has highest weight $w(\lambda + \rho) - \rho$.

Consider now $H(\mathfrak{n}_S^T, E)$ where $S \prec T$ and E is an irreducible algebraic representation of L_T with highest weight λ. We replace G above by L_T. Thus let ρ^T be one-half the sum of the positive roots of $\mathfrak{l}_T$, let W^T be the Weyl group of the root system of $\mathfrak{l}_T$. We have a decomposition $W^T = W^S W_S^T$, where W_S^T consists of the minimal length representatives of the cosets in $W^S \backslash W^T$. Kostant's theorem here says

$$H(\mathfrak{n}_S^T, E) = \bigoplus_{w \in W_S^T} H^{\ell(w)}(\mathfrak{n}_S^T, E)_w[-\ell(w)]$$

where $H^{\ell(w)}(\mathfrak{n}_S^T, E)_w$ has highest weight $w(\lambda + \rho^T) - \rho^T$.

6.2 Relative depth 1

Let $\mathbf{E}$ be a local system on T corresponding to an algebraic representation E of L_T. By (5.3), since $\operatorname{depth}_T S = 1$, the link Lk_S^T is simply the compact nilmanifold $\Gamma_{N_S^T} \backslash N_S^T(\mathbf{R})$. Thus

$$H(\kappa_S \mathcal{P}_T(\mathbf{E})) = H(\Gamma_{N_S^T} \backslash N_S^T(\mathbf{R}); \mathbf{E})[-p(T)] \cong H(\mathfrak{n}_S^T, E)[-p(T)]. \qquad (6.1)$$

Here we use the theorem of Nomizu and van Est [36, 45] for the last isomorphism with Lie algebra cohomology. In fact the action of L_S on $\mathfrak{n}_S^T$ induces an action on $H(\mathfrak{n}_S^T, E)$ and the isomorphism in (6.1) is an isomorphism of L_S-modules.

If we combine this with Kostant's theorem we obtain

$$H(\kappa_S \mathcal{P}_T(\mathbf{E})) \cong \bigoplus_{w \in W_S^T} H^{\ell(w)}(\mathfrak{n}_S^T, E)_w[-\ell(w) - p(T)]. \tag{6.2}$$

Note that to go from here to $H(i_S^* \mathcal{P}_T(\mathbf{E}))$, according to (3.5) and (3.7), we need to truncate this cohomology, leaving just degrees $< p(S)$. Thus the sum will now only include w satisfying $\ell(w) + p(T) \leq p(S) - 1$, or

$$\ell(w) \leq p(S) - p(T) - 1 = \bar{p}_T(S).$$

Thus for relative depth 1

$$H(i_S^* \mathcal{P}_T(\mathbf{E})) \cong \bigoplus_{\substack{w \in W_S^T \\ \ell(w) \leq \bar{p}_T(S)}} H^{\ell(w)}(\mathfrak{n}_S^T, E)_w[-\ell(w) - p(T)] \tag{6.3}$$

(compare §3.3).

6.3 Relative depth 2

If $\operatorname{depth}_T S = 2$, the simplex $|\Delta_S^T|$ is a 1-simplex. The endpoints correspond to the two intermediate strata: $S \prec Q_1, Q_2 \prec T$. The cohomology of the interior of the link is the same as (6.2), however each term may be truncated at Q_1 or at Q_2 or at both — see (3.5) as well as (6.3). The terms that remain are those that are not truncated at either side or (with an additional shift by -1) those that are truncated at both sides. This corresponds to the fact that the cohomology of a 1-simplex relative to either one of its endpoints vanishes, while the cohomology relative to both endpoints is one dimensional in degree 1.

To precisely express these side truncations, write (for $i = 1$ or 2)

$$W^T = W^{Q_i} W_{Q_i}^T = W^S W_S^{Q_i} W_{Q_i}^T.$$

In fact $W_S^T = W_S^{Q_i} W_{Q_i}^T$ and for $w \in W_S^T$ we decompose $w = w^{Q_i} w_{Q_i}$ accordingly. Define

$$\ell_{Q_i}(w) = \ell(w_{Q_i}) \qquad (w = w^{Q_i} w_{Q_i}).$$

Then we have the formula

$$\begin{aligned} H(\kappa_S \mathcal{P}_T(\mathbf{E})) \cong & \bigoplus_{\substack{w \in W_S^T \\ \ell_{Q_1}(w) \leq \bar{p}_T(Q_1) \\ \ell_{Q_2}(w) \leq \bar{p}_T(Q_2)}} H^{\ell(w)}(\mathfrak{n}_S^T, E)_w[-\ell(w) - p(T)] \\ & \oplus \bigoplus_{\substack{w \in W_S^T \\ \ell_{Q_1}(w) > \bar{p}_T(Q_1) \\ \ell_{Q_2}(w) > \bar{p}_T(Q_2)}} H^{\ell(w)}(\mathfrak{n}_S^T, E)_w[-\ell(w) - p(T) - 1]. \end{aligned} \tag{6.4}$$

6.4 Relative depth 3 and higher

One can similarly write an explicit formula for $H(\kappa_S \mathcal{P}_T(\mathbf{E}))$ when $\operatorname{depth}_T S = 3$ however there are many more possible configurations of truncations on the 6 intermediate strata Q; these are illustrated in [42, §18.4, Figure 24] and result in additional degree shifts of 0, -1, and -2.

There is a general formula for arbitrary depth that involves the intersection cohomology of $|\Delta_S^T|$ associated to a special perversity:

$$\bar{p}_{T,w}(Q) = \bar{p}_T(Q) - \ell_Q(w) = p(Q) - p(T) - 1 - \ell_Q(w) \qquad (S \preccurlyeq Q \preccurlyeq T).$$

Let

$$I_{\bar{p}_{T,w}} H^{\cdot}(|\Delta_S^T|) = H^{\cdot}(\mathcal{I}_{\bar{p}_{T,w}}\mathcal{C}(|\Delta_S^T|, \mathbf{Z}))$$

be the hypercohomology of the corresponding Deligne sheaf (3.10). Then

$$H(\kappa_S \mathcal{P}_T(\mathbf{E})) \cong \bigoplus_{w \in W_S^T} H^{\ell(w)}(\mathfrak{n}_S^T, E)_w[-\ell(w) - p(T)] \otimes I_{\bar{p}_{T,w}} H(|\Delta_S^T|). \qquad (6.5)$$

7 Example Computations

To illustrate the results of §4, we calculate all extensions between simple perverse sheaves for the reductive Borel-Serre compactification $\widehat{X}$ of the locally symmetric spaces associated with $G = \mathrm{Sp}(4, \mathbf{R})$. Since $\operatorname{Ext}^1_{\mathbf{P}}(\mathcal{P}_T(\mathbf{E}), \mathcal{P}_{T'}(\mathbf{E}')) = 0$ unless $T \prec T'$ or $T \succ T'$, we can assume both T and T' are standard parabolic $\mathbf{Q}$-subgroups.

For brevity we only consider $p = p_-$ and we give complete details for $T \succ T'$; the results for $T \prec T'$ are summarized in §7.4.

7.1 Preliminaries

Here the root system is type C_2 with simple $\mathbf{Q}$-roots $\Delta = \{\alpha_1 = e_1 - e_2, \alpha_2 = 2e_2\}$. The standard parabolic $\mathbf{Q}$-subgroups P_I correspond to subsets $I \subseteq \{1,2\}$ $(\Delta^{P_I} = \{\alpha_i\}_{i \in I} \subseteq \Delta)$ and so form a lattice

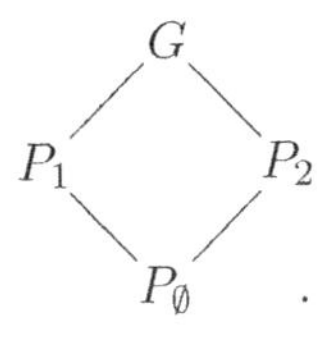

The real dimension of X is 6, while strata corresponding to P_1 and P_2 have dimension 2, and that corresponding to $P_\emptyset$ has dimension 0. Thus the perversity values $p_-(S)$ are

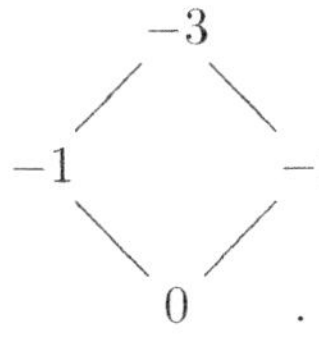

To calculate link cohomology we need the Weyl group which is generated by the reflections s_1 and s_2 in the simple roots modulo the relation $(s_1s_2)^3 = e$:

$$W = W_{P_\emptyset} = \{e, s_1, s_2, s_1s_2, s_2s_1, s_1s_2s_1, s_2s_1s_2\}.$$

For the intermediate strata we have

$$W^{P_1} = W^{P_1}_{P_\emptyset} = \{e, s_1\} \qquad \text{and} \qquad W^{P_2} = W^{P_2}_{P_\emptyset} = \{e, s_2\}$$

and

$$W_{P_1} = \{e, s_2, s_2s_1, s_2s_1s_2\} \qquad \text{and} \qquad W_{P_2} = \{e, s_1, s_1s_2, s_1s_2s_1\}.$$

7.2 The case $T = G$

We first set $T = G$ and E an irreducible algebraic representation of G with highest weight λ. The associated classical perversity $\bar{p}_G(S) = p(S) - p(G) - 1$ has values

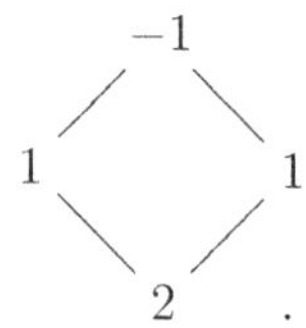

We look for nonzero extensions in $\operatorname{Ext}^1_{\mathbf{P}_-}(\mathcal{P}_G(\mathbf{E}), \mathcal{P}_{T'}(\mathbf{E}'))$ for $T' \prec G$, where E' is an irreducible algebraic representation of $L_{T'}$. To start constructing such an extension, according to Lemma 4.3(iii), we need an $L_{T'}$-module morphism

$$H^{p(T')-1}(\kappa_{T'}\mathcal{P}_G(\mathbf{E})) \longrightarrow E'.$$

We calculate the above link cohomology. We begin with $T' = P_\emptyset$ for which we use (6.4). In degree $p(T') - 1$, a contribution from the first direct sum of (6.4) requires $\ell(w) = \bar{p}_G(P_\emptyset) = 2$ and no truncation on either side. This means $w = s_1s_2$ or s_2s_1. Now $\ell_{P_1}(s_1s_2) = 1 \leq 1 = \bar{p}_G(P_1)$ (since $s_1s_2 = (s_1)(s_2) \in W^{P_1}W_{P_1}$) and $\ell_{P_2}(s_1s_2) = 2 \not\leq 1 = \bar{p}_G(P_2)$ (since $s_1s_2 = (e)(s_1s_2) \in W^{P_2}W_{P_2}$). Thus s_1s_2 cannot contribute to the link cohomology since the corresponding class is truncated on one side but not the other. Similarly s_2s_1 does not contribute since $\ell_{P_1}(s_2s_1) = 2 \not\leq 1$ and $\ell_{P_2}(s_2s_1) = 1 \leq 1$.

A contribution in degree $p(T') - 1$ from the second direct sum of (6.4) requires $\ell(w) = \bar{p}_G(P_\emptyset) - 1 = 1$ and truncation on both sides. Thus $w = s_1$ or s_2. For $w = s_1$, $\ell_{P_1}(s_1) = 0 \not> \bar{p}_G(P_1) = 1$ and $\ell_{P_2}(s_1) = 1 \not> \bar{p}_G(P_2) = 1$ and thus does not contribute, and similarly for $w = s_2$.

Thus $H^{p(P_\emptyset)-1}(\kappa_{P_\emptyset}\mathcal{P}_G(\mathbf{E})) = 0$ and

$$\operatorname{Ext}^1_{\mathbf{P}_-}(\mathcal{P}_G(\mathbf{E}), \mathcal{P}_{P_\emptyset}(\mathbf{E}')) = 0$$

for all E'.

We now consider the link cohomology at $T' = P_1$ or P_2. Since we are interested in degree $p(T') - 1$, a contribution from (6.2) requires $\ell(w) = \bar{p}_G(T') = 1$. Thus

$$H^{p(T')-1}(\kappa_{T'}\mathcal{P}_G(\mathbf{E})) = \begin{cases} H^1(\mathfrak{n}_{P_1}, E)_{s_2} & T' = P_1, \\ H^1(\mathfrak{n}_{P_2}, E)_{s_1} & T' = P_2. \end{cases} \tag{7.1}$$

This means that $\operatorname{Ext}^1_{\mathbf{P}\_}(\mathcal{P}_G(\mathbf{E}), \mathcal{P}_{P_1}(\mathbf{E}'))$ can only be nonzero (in which case it is $\mathbf{C}$) when E' has highest weight $s_2(\lambda + \rho) - \rho$. By Lemma 4.4, this extension will actually exist provided the map of L_{P_1}-modules

$$\phi\colon H^{-2}(\kappa_{P_1}\mathcal{P}_G(\mathbf{E})) = H^1(\mathfrak{n}_{P_1}, E)_{s_2} \longrightarrow E'$$

(which is an isomorphism), induces the zero map

$$H^{p(P_\emptyset)-1}(\kappa_{P_\emptyset}\mathcal{P}_G(\mathbf{E})) \longrightarrow H^{p(P_\emptyset)}(\kappa_{P_\emptyset}\mathcal{P}_{P_1}(\mathbf{E}')). \tag{7.2}$$

However we have already seen that $H^{p(P_\emptyset)-1}(\kappa_{P_\emptyset}\mathcal{P}_G(\mathbf{E})) = 0$, thus this condition is satisfied.

So we find that

$$\operatorname{Ext}^1_{\mathbf{P}\_}(\mathcal{P}_G(\mathbf{E}), \mathcal{P}_{P_1}(\mathbf{E}')) = \begin{cases} \mathbf{C} & \text{if } E' \text{ has highest weight } s_2(\lambda+\rho)-\rho, \\ 0 & \text{otherwise.} \end{cases} \tag{7.3}$$

A similar argument shows

$$\operatorname{Ext}^1_{\mathbf{P}\_}(\mathcal{P}_G(\mathbf{E}), \mathcal{P}_{P_2}(\mathbf{E}')) = \begin{cases} \mathbf{C} & \text{if } E' \text{ has highest weight } s_1(\lambda+\rho)-\rho, \\ 0 & \text{otherwise.} \end{cases} \tag{7.4}$$

7.3 The case $T = P_1$ and P_2

Fix $i = 1$ or 2, and say E is an irreducible algebraic representation of L_{P_i} with highest weight λ. By Lemma 4.3(iii)

$$\operatorname{Ext}^1_{\mathbf{P}\_}(\mathcal{P}_{P_i}(\mathbf{E}), \mathcal{P}_{P_\emptyset}(\mathbf{E}')) \cong \operatorname{Mor}_{L_{P_\emptyset}}(H^{p(P_\emptyset)-1}(\kappa_{P_\emptyset}\mathcal{P}_{P_i}(\mathbf{E})), E')$$

for any irreducible algebraic representation E' of $L_{P_\emptyset}$. Since $L_{P_\emptyset}$ is just the maximal torus, $E' = \mathbf{C}_\chi$ where χ is the character by which $L_{P_\emptyset}$ acts. We calculate by (6.2) that

$$H^{p(P_\emptyset)-1}(\kappa_{P_\emptyset}\mathcal{P}_{P_i}(\mathbf{E})) = \bigoplus_{\substack{w \in W^{P_i}_{P_\emptyset} \\ \ell(w) = \bar{p}_{P_i}(P_\emptyset)}} H^{\ell(w)}(\mathfrak{n}^{P_i}_{P_\emptyset}, E) = H^0(\mathfrak{n}^{P_i}_{P_\emptyset}, E)_e$$

since $\bar{p}_{P_i}(P_\emptyset) = 0$. Thus we find

$$\operatorname{Ext}^1_{\mathbf{P}\_}(\mathcal{P}_{P_i}(\mathbf{E}), \mathcal{P}_{P_\emptyset}(\mathbf{E}')) = \begin{cases} \mathbf{C} & \text{if } E' = \mathbf{C}_\lambda, \\ 0 & \text{otherwise.} \end{cases} \tag{7.5}$$

7.4 Results for $T \prec T'$

Let E' be an irreducible algebraic representation of $L_{T'}$ with highest weight λ. Similarly to the above, one may check that for $T \prec T'$, the only nonzero extensions are as follows:

$$\operatorname{Ext}^1_{\mathbf{P}_-}(\mathcal{P}_{P_\emptyset}(\mathbf{E}), \mathcal{P}_{P_1}(\mathbf{E}')) = \begin{cases} \mathbf{C} & \text{if } E = \mathbf{C}_{s_1(\lambda+\rho)-\rho}, \\ 0 & \text{otherwise.} \end{cases} \tag{7.6}$$

$$\operatorname{Ext}^1_{\mathbf{P}_-}(\mathcal{P}_{P_\emptyset}(\mathbf{E}), \mathcal{P}_{P_2}(\mathbf{E}')) = \begin{cases} \mathbf{C} & \text{if } E = \mathbf{C}_{s_2(\lambda+\rho)-\rho}, \\ 0 & \text{otherwise.} \end{cases} \tag{7.7}$$

$$\operatorname{Ext}^1_{\mathbf{P}_-}(\mathcal{P}_{P_1}(\mathbf{E}), \mathcal{P}_{G}(\mathbf{E}')) = \begin{cases} \mathbf{C} & \text{if } E \text{ has highest weight } s_2 s_1(\lambda+\rho)-\rho, \\ 0 & \text{otherwise.} \end{cases} \tag{7.8}$$

$$\operatorname{Ext}^1_{\mathbf{P}_-}(\mathcal{P}_{P_2}(\mathbf{E}), \mathcal{P}_{G}(\mathbf{E}')) = \begin{cases} \mathbf{C} & \text{if } E \text{ has highest weight } s_1 s_2(\lambda+\rho)-\rho, \\ 0 & \text{otherwise.} \end{cases} \tag{7.9}$$

References

[1] A. Ash, D. Mumford, M. Rapoport, and Y. Tai, *Smooth compactification of locally symmetric varieties*, Math Sci Press, Brookline, 1975.

[2] J. Ayoub and S. Zucker, *Relative Artin motives and the reductive Borel-Serre compactification of a locally symmetric variety*, Invent. Math. **188** (2012), no. 2, 277–427.

[3] W. Baily and A. Borel, *Compactification of arithmetic quotients of bounded symmetric domains*, Ann. of Math. **84** (1966), 442–528.

[4] A. Beĭlinson, *On the derived category of perverse sheaves*, K-theory, arithmetic and geometry (Moscow, 1984–1986), 1987, pp. 27–41.

[5] A. Beĭlinson, J. Bernstein, and P. Deligne, *Faisceaux pervers*, Analyse et topologie sur les espaces singuliers (I), CIRM, 6–10 juillet 1981, 1982.

[6] A. Beĭlinson and J. Bernstein, *Localisation de g-modules*, C. R. Acad. Sci. Paris Sér. I Math. **292** (1981), no. 1, 15–18.

[7] G. Böckle and C. Khare, *Mod l representations of arithmetic fundamental groups. II. A conjecture of A. J. de Jong*, Compos. Math. **142** (2006), no. 2, 271–294.

[8] A. Borel, *Density properties for certain subgroups of semi-simple groups without compact components*, Ann. of Math. **72** (1960), 179–188.

[9] A. Borel et al., *Intersection cohomology*, Birkhäuser, Boston, 1984.

[10] A. Borel and H. Garland, *Laplacian and the discrete spectrum of an arithmetic group*, Amer. J. Math. **105** (1983), no. 2, 309–335.

[11] J.-L. Brylinski and M. Kashiwara, *Kazhdan-Lusztig conjecture and holonomic systems*, Invent. Math. **64** (1981), no. 3, 387–410.

[12] D. C. Cohen, M. Goresky, and L. Ji, *On the Künneth formula for intersection cohomology*, Trans. Amer. Math. Soc. **333** (1992), no. 1, 63–69.

[13] M. A. A. de Cataldo and L. Migliorini, *The Hodge theory of algebraic maps*, Ann. Sci. École Norm. Sup. (4) **38** (2005), no. 5, 693–750.

[14] ______, *The decomposition theorem, perverse sheaves and the topology of algebraic maps*, Bull. Amer. Math. Soc. (N.S.) **46** (2009), no. 4, 535–633.

[15] A. J. de Jong, *A conjecture on arithmetic fundamental groups*, Israel J. Math. **121** (2001), 61–84.

[16] V. Drinfeld, *On a conjecture of Kashiwara*, Math. Res. Lett. **8** (2001), no. 5-6, 713–728.

[17] D. Gaitsgory, *On de Jong's conjecture*, Israel J. Math. **157** (2007), 155–191.

[18] M. Goresky, G. Harder, and R. MacPherson, *Weighted cohomology*, Invent. Math. **116** (1994), 139–213.

[19] M. Goresky and R. MacPherson, *Intersection homology theory*, Topology **19** (1980), no. 2, 135–162.

[20] ______, *Intersection homology II*, Invent. Math. **72** (1983), 77–129.

[21] ______, *Weighted cohomology of Satake compactifications*, 1988. Centre de recherches mathématiques, preprint #1593.

[22] M. Goresky and Y.-S. Tai, *Toroidal and reductive borel-serre compactifications of locally symmetric spaces*, Amer. J. Math. **121** (1999), 1095–1151.

[23] M. Kashiwara, *Semisimple holonomic $\mathcal{D}$-modules*, Topological field theory, primitive forms and related topics (Kyoto, 1996), 1998, pp. 267–271.

[24] D. Kazhdan and G. Lusztig, *Representations of Coxeter groups and Hecke algebras*, Invent. Math. **53** (1979), no. 2, 165–184.

[25] ______, *Schubert varieties and Poincaré duality*, Geometry of the Laplace operator (Proc. Sympos. Pure Math., Univ. Hawaii, Honolulu, Hawaii, 1979), 1980, pp. 185–203.

[26] S. L. Kleiman, *The development of intersection homology theory*, Pure Appl. Math. Q. **3** (2007), no. 1, Special Issue: In honor of Robert D. MacPherson. Part 3, 225–282.

[27] B. Kostant, *Lie algebra cohomology and the generalized Borel-Weil theorem*, Ann. of Math. **74** (1961), 329–387.

[28] J.-S. Li and J. Schwermer, *On the Eisenstein cohomology of arithmetic groups*, Duke Math. J. **123** (2004), no. 1, 141–169.

[29] E. Looijenga, *L^2-cohomology of locally symmetric varieties*, Compositio Math. **67** (1988), 3–20.

[30] E. Looijenga and M. Rapoport, *Weights in the local cohomology of a Baily-Borel compactification*, Complex geometry and Lie theory (Sundance, UT, 1989), 1991, pp. 223–260.

[31] G. A. Margulis, *Discrete subgroups of semisimple Lie groups*, Springer, Berlin, 1991.

[32] Y. Matsushima and S. Murakami, *On vector bundle valued harmonic forms and automorphic forms on symmetric Riemannian manifolds*, Ann. of Math. **78** (1963), 365–416.

[33] T. Mochizuki, *Asymptotic behaviour of tame harmonic bundles and an application to pure twistor D-modules. I*, Mem. Amer. Math. Soc. **185** (2007), no. 869, xii+324.

[34] ———, *Asymptotic behaviour of tame harmonic bundles and an application to pure twistor D-modules. II*, Mem. Amer. Math. Soc. **185** (2007), no. 870, xii+565.

[35] ———, *Wild harmonic bundles and wild pure twistor D-modules*, Astérisque **340** (2011), x+607.

[36] K. Nomizu, *On the cohomology of compact homogeneous spaces of nilpotent Lie groups*, Ann. of Math. **59** (1954), 531–538.

[37] M. Rapoport, *On the shape of the contribution of a fixed point on the boundary: The case of* $\mathbf{Q}$*-rank* 1 (*with an appendix by L. Saper and M. Stern*), The zeta functions of Picard modular surfaces, 1992, pp. 479–488.

[38] C. Sabbah, *Polarizable twistor $\mathcal{D}$-modules*, Astérisque **300** (2005), vi+208.

[39] M. Saito, *Modules de Hodge polarisables*, Publ. Res. Inst. Math. Sci. **24** (1988), no. 6, 849–995.

[40] ———, *Mixed Hodge modules*, Publ. Res. Inst. Math. Sci. **26** (1990), no. 2, 221–333.

[41] L. Saper, *$\mathscr{L}$-modules and micro-support*, 2001. to appear, arXiv:math.RT/0112251.

[42] ———, *On the cohomology of locally symmetric spaces and of their compactifications*, Current developments in mathematics, 2002, 2003, pp. 219–289.

[43] ———, *$\mathscr{L}$-modules and the conjecture of Rapoport and Goresky-MacPherson*, Formes automorphes, I — actes du semestre du Centre émile Borel, printemps 2000, 2005, pp. 319–334.

[44] L. Saper and M. Stern, *L_2-cohomology of arithmetic varieties*, Ann. of Math. **132** (1990), 1–69.

[45] W. T. van Est, *A generalization of the Cartan-Leray sequence, I, II*, Indag. Math. **20** (1958), 399–413.

[46] S. Zucker, *L_2 cohomology of warped products and arithmetic groups*, Invent. Math. **70** (1982), 169–218.

[47] ———, *Satake compactifications*, Comment. Math. Helv. **58** (1983), 312–343.

[48] ———, *Locally homogeneous variations of Hodge structure*, Enseign. Math. (2) **27** (1981), no. 3-4, 243–276.

[49] ———, *On the reductive Borel-Serre compactification. III. Mixed Hodge structures*, Asian J. Math. **8** (2004), no. 4, 881–911.

Hodge Theory and L^2-analysis
ALM 39, pp. 583–597

Nonlinear Harmonic Forms and Indefinite Bochner Formulas*

Mark Stern†

Abstract

Motivated by the Hopf conjecture, we introduce *nonlinear harmonic forms* — minimizers of the L_2 energy in a fixed de Rham cohomology class, subject to a nonlinear constraint. We prove Bochner theorems and basic existence and regularity results.

1 Introduction

The link between topology and geometry provided by harmonic forms and L_2-cohomology has played a fundamentally important role in differential and algebraic geometry. In this note, we introduce a new object to geometric analysis, which we call a *nonlinear harmonic form*. We define these forms as closed differential forms z which

- represent a fixed class in de Rham cohomology, and
- minimize $\|z\|_{L_2}^2$ subject to a natural nonlinear constraint $p(z) = 0$.

Because the energy is the usual L_2 energy, the nonlinear harmonic forms satisfy rich Euler Lagrange equations. The constraints we consider in this paper are polynomial. Our initial regularity results require that p is diffeomorphism invariant. Our first example, which arises from our investigations into the Hopf conjecture, imposes the constraint $p(z) = z \wedge z$. These forms are in some ways intermediate between surfaces and harmonic forms, and their analysis is similar to that of harmonic maps. It is not difficult to prove minimizers exist, but their regularity theory remains to be developed. For example, we have the following theorem.

---

*Dedicated to Steve Zucker on the occasion of his 65th birthday conference.

†Duke University, Department of Mathematics. E-mail: stern@math.duke.edu, partially supported by by NSF grant DMS 1005761.

Theorem 1.1. *Let M^n be a compact Riemannian manifold. Let f be a p-form with $f \wedge f = 0$. Let h be the harmonic representative of f. Set*

$$Q := \{y \in H_1 : (h + dy) \wedge (h + dy) = 0, \quad \text{and } d^*y = 0\}.$$

Let $E(y) = \|h + dy\|^2$. Let $\nu_0 := \inf\{E(y) : y \in Q\}$. Then there exists $y \in H_1$ such that $E(y) = \nu_0$. Moreover $z := h + dy$ lies in the Morrey space $L^{2,n-2p}$.

In this note, we establish only the most elementary properties of nonlinear harmonic forms. A careful analysis of their singularities requires further study.

Our interest in these nonlinear harmonic forms was first sparked by our application of a new indefinite Bochner formula to investigation of the Hopf conjecture. Although we have not fruitfully applied nonlinear harmonics to this conjecture, we include a discussion of the Hopf conjecture, which may be of independent interest.

Conjecture 1.2. *(Hopf) $S^2 \times S^2$ does not admit a metric of strictly positive sectional curvature.*

In order to apply harmonic form techniques to this question, we consider a stronger conjecture.

Conjecture 1.3. *A compact oriented 4-manifold with positive sectional curvature with nonvanishing second betti number has definite intersection form.*

As $H^2(S^2 \times S^2)$ has an indefinite intersection form, the Hopf conjecture follows immediately from Conjecture 1.3.

Let h be a harmonic 2- form on a compact oriented Riemannian 4-manifold of positive sectional curvature. We may decompose h as

$$h = h_+ + h_-,$$

where

$$*h_\pm = \pm h_\pm.$$

Conjecture 1.3 implies that

$$|h_+||h_-| \equiv 0. \tag{1.4}$$

In a local orthonormal frame, the Bochner formula gives

$$-\frac{1}{2}\Delta|h|^2 = |\nabla h|^2 - R_{ijkl}h_{kj}h_{il} + \mathrm{Ric}_{il}h_{lq}h_{iq}. \tag{1.5}$$

Positivity of the sectional curvature implies positivity of the Ricci term in (1.5) but does not imply positivity of the remaining curvature term without additional assumptions. Equation (1.4) suggests we replace $|h|^2$ in the the Bochner formula with $|h_+|^2|h_-|^2$. This leads to an indefinite Bochner formula.

In a neighborhood of a point where $|h_+|^2|h_-|^2 \neq 0$, we can choose an oriented orthonormal coframe $\{w^i\}_{i=1}^4$ and frame $\{e_i\}_{i=1}^4$ in which h takes the form $h = aw^1 \wedge w^2 + bw^3 \wedge w^4$, with $a > b$. In this frame, we have (see (2.11))

$$-\frac{1}{4}\Delta \ln(|h_+|^2|h_-|^2) = R_{1331} + R_{2442} + R_{2332} + R_{1441} + |II|^2 - \frac{1}{2}|T|^2, \tag{1.6}$$

where II is a generalized second fundamental form and T is a vectorfield which vanishes if and only if the distribution spanned by $\{e_1, e_2\}$ and the distribution spanned by $\{e_3, e_4\}$ are both integrable.

The only curvature terms which appear in (1.6) are sectional curvatures, making such exotic Bochner formulas appear well suited to studying metrics of positive sectional curvature, if we can control the torsion terms. In order to eliminate half the torsion terms, we study nonlinear harmonic forms subject to the constraint $z \wedge z = 0$. For such forms, we obtain the integrability of the distribution spanned by $\{e_3, e_4\}$, eliminating a summand of T. Moreover, the Euler Lagrange equations for the nonlinear harmonic still imply (1.6). In fact, these equations can be derived (6.15) using only the fact that 0 is a critical point of $\|\phi_t^* h\|^2$, for all smooth one parameter families of diffeomorphisms satisfying $\phi_0(x) = x$. This does not suffice to settle (1.3) but suggests that the nonlinear harmonics deserve careful study.

I would like to thank Daniel Stern for explaining the significance of Bettiol's work and for many useful discussions. I also thank Hugh Bray and Robert Bryant for helpful discussions.

2 An indefinite Bochner formula

Let h be a strongly harmonic 2 form on a Riemannian 4-manifold M. In a neighborhood of a point where h is neither self dual nor antiself dual, we may choose an oriented local orthonormal frame $\{e_i\}_{i=1}^4$ and dual frame $\{w^i\}_{i=1}^4$ so that

$$h = aw^1 \wedge w^2 + bw^3 \wedge w^4. \tag{2.1}$$

Let Σ_a and Σ_b denote the distributions spanned by $\{e_1, e_2\}$ and $\{e_3, e_4\}$ respectively. Write

$$\nabla_{e_i} e_j = \gamma_{ij}^k e_k, \text{ and } c_{ij}^k := \gamma_{ij}^k - \gamma_{ji}^k. \tag{2.2}$$

Then the torsion vectors

$$T_a := c_{12}^3 e_3 + c_{12}^4 e_4 \text{ and } T_b := c_{34}^1 e_1 + c_{34}^2 e_2 \tag{2.3}$$

vanish if and only if the respective distributions Σ_a and Σ_b are integrable. Define the generalized second fundamental form of Σ_a and Σ_b by

$$II_a := \sum_{i,j=1,2} \sum_{k=3,4} \frac{1}{2}(\gamma_{ij}^m + \gamma_{ji}^m) w^i \otimes w^j \otimes e_m, \tag{2.4}$$

and

$$II_b := \sum_{i,j=3,4} \sum_{k=1,2} \frac{1}{2}(\gamma_{ij}^m + \gamma_{ji}^m) w^i \otimes w^j \otimes e_m. \tag{2.5}$$

The generalized mean curvature vectors of these second fundamental forms are given by

$$H_a = (\gamma_{11}^3 + \gamma_{22}^3)e_3 + (\gamma_{11}^4 + \gamma_{22}^4)e_4, \tag{2.6}$$

and

$$H_b = (\gamma_{33}^1 + \gamma_{44}^1)e_1 + (\gamma_{33}^2 + \gamma_{44}^2)e_2. \tag{2.7}$$

In this notation, $dh = 0$ gives

$$0 = a_3 - aH_a^3 + bT_a^4,$$

$$0 = a_4 - aH_a^4 - bT_a^3,$$

$$0 = b_1 - bH_b^1 + aT_b^2,$$

$$0 = b_2 - bH_b^2 - aT_b^1,$$

and $d^*h = 0$ gives

$$0 = a_1 - aH_b^1 + bT_b^2,$$

$$0 = a_2 - aH_b^2 - bT_b^1,$$

$$0 = b_3 - bH_a^3 + aT_a^4,$$

$$0 = b_4 - bH_a^4 - aT_a^3.$$

Setting $f = \ln(\sqrt{a^2 - b^2})$, gives

$$\nabla f = H_a + H_b; \tag{2.8}$$

so f is a generalized area function. Taking second derivatives, we have

$$f_{11} = \gamma_{33,1}^1 + \gamma_{44,1}^1,$$

$$f_{22} = \gamma_{33,2}^2 + \gamma_{44,2}^2,$$

$$f_{33} = \gamma_{11,3}^3 + \gamma_{22,3}^3,$$

$$f_{44} = \gamma_{11,4}^4 + \gamma_{22,4}^4.$$

Hence

$$\sum_j f_{jj} = \gamma_{33,1}^1 + \gamma_{44,1}^1 + \gamma_{33,2}^2 + \gamma_{44,2}^2 + \gamma_{11,3}^3 + \gamma_{22,3}^3 + \gamma_{11,4}^4 + \gamma_{22,4}^4. \tag{2.9}$$

In this frame, the sectional curvatures are given by

$$R_{ijji} = \gamma_{jj,i}^i + \gamma_{im}^i \gamma_{jj}^m - \gamma_{ij,j}^i - \gamma_{jm}^i \gamma_{ij}^m - (\gamma_{ij}^m - \gamma_{ji}^m)\gamma_{mj}^i. \tag{2.10}$$

We may choose our frame so that at a given point $\gamma_{i3}^4 = 0 = \gamma_{i1}^2$. In such a frame adapted to the given point, we have

$$R_{1331} + (\gamma_{11}^3)^2 + (\frac{\gamma_{12}^3 + \gamma_{21}^3}{2})^2 - (\frac{c_{12}^3}{2})^2 + (\gamma_{33}^1)^2 + (\frac{\gamma_{34}^1 + \gamma_{43}^1}{2})^2 - (\frac{c_{34}^1}{2})^2 = \gamma_{33,1}^1 + \gamma_{11,3}^3.$$

Inserting this and related equations for other sectional curvatures into (2.9) yields our indefinite Bochner formula. At a critical point,

$$-\Delta f = R_{1331} + R_{2332} + R_{1441} + R_{2442} + |II_a|^2 + |II_b|^2 - \frac{1}{2}|T_a|^2 - \frac{1}{2}|T_b|^2. \tag{2.11}$$

If the torsion vectors in the above expressions vanished or were dominated by the second fundamental form terms, Conjecture 1.3 would follow immediately from (2.11). Wherever h is neither self dual nor anti-self-dual, we have

$$-\Delta f = \kappa + R_{1221} + R_{3443} - 2R^{\Sigma_a}_{1221} - 2R^{\Sigma_b}_{3443} - \frac{1}{2}|T_a|^2 - \frac{1}{2}|T_b|^2, \tag{2.12}$$

where κ denotes the scalar curvature of M, and R^{Σ_a} and R^{Σ_b} denote the Riemann curvature of the leaves of the respective foliation when the distributions are integrable. Otherwise they are defined by the Gauss equations in terms of the second fundamental form and the curvature of M by the same relations as the integrable case.

The likelihood of applying these formula to the Hopf conjecture is somewhat diminished by results of Renato Bettiol.

Definition 2.13. Let $\{v_j\}_{j=1}^4$ be any orthonormal basis of T_pM, M a 4−manifold. $\frac{1}{2}(R(v_1, v_3, v_3, v_1) + R(v_2, v_4, v_4, v_2))$ is called the *biorthogonal curvature* associated to the plane σ spanned by $\{v_1, v_3\}$ and its orthogonal complement $\sigma^\perp$.

Bettiol [1] has proved that $S^2 \times S^2$ admits metrics with positive biorthogonal curvature. The curvature terms appearing in (1.6) occur in biorthogonal pairs, and therefore more information is needed to apply this formula to Conjectures 1.2 or 1.3.

We now introduce a special case of nonlinear harmonic forms in order to remove some of the torsion terms in (2.11).

3 Heuristic properties of nonlinear harmonic forms

In this section, in order to anticipate the fruits of future regularity analysis, we formally investigate properties of minimizers of $\|z\|^2$ in a fixed degree 2 cohomology class on a compact 4-manifold, subject to the constraint that $z \wedge z = 0$. Again, we emphasize that the discussion in this section is heuristic only. In subsequent sections we prove several of these heuristic results.

Let f be a closed 2-form satisfying $f \wedge f = 0$. Suppose we consider closed 2-forms z cohomologous to f such that $\|z\|^2_{L_2}$ is minimal subject to the constraint $z \wedge z = 0$. Then formally $\langle d^*z, b\rangle_{L_2} = 0$ for all b satisfying $z \wedge db = 0$. Hence d^*z is perpendicular to the kernel of $e(z)d$, where $e(\phi)$ denotes exterior multiplication on the left by ϕ. Thus d^*z is in the image of $d^*e^*(z)$. That is

$$d(*z - \mu z) = 0, \tag{3.1}$$

for some function μ, which may be viewed formally as a Lagrange multiplier. Where $z \neq 0$, let $\{e_i\}_{i=1}^4$ and $\{w^j\}_{i=1}^4$ be an oriented orthonormal frame and dual frame so that $z = aw^1 \wedge w^2$, with $a > 0$. Then $dz = 0$ implies

$$0 = a_3 - a(\gamma^3_{11} + \gamma^3_{22}), \tag{3.2}$$

$$0 = a_4 - a(\gamma^4_{11} + \gamma^4_{22}), \tag{3.3}$$

$$0 = ac_{34}^1 = ac_{34}^2. \tag{3.4}$$

So, $\{e_3, e_4\}$ span an integrable distribution. The constraint on d^*z gives

$$0 = a(c_{12}^3 - \mu_4), \tag{3.5}$$

$$0 = a(c_{12}^4 + \mu_3), \tag{3.6}$$

$$0 = a_1 - a(\gamma_{33}^1 + \gamma_{44}^1), \tag{3.7}$$

$$0 = a_2 - a(\gamma_{33}^2 + \gamma_{44}^2). \tag{3.8}$$

Equations (3.2), (3.3), (3.7), and (3.8) can again be expressed more succinctly in the notation of (2.8) as

$$\nabla \ln(a) = H_a + H_b. \tag{3.9}$$

Equations (3.5) and (3.6) imply

$$aT\mu = 0, \tag{3.10}$$

where

$$T := c_{12}^3 e_3 + c_{12}^4 e_4.$$

Equations (3.5) and (3.6) also imply that we may view T as the Hamiltonian vector field for μ restricted to the leaves of the foliation associated to the integrable distribution spanned by $\{e_3, e_4\}$, equipped with the symplectic form $w^3 \wedge w^4$ pulled back to the leaves. Let ϕ_s denote the one parameter family of diffeomorphisms generated by aT on the space $a > 0$. Equations (3.10) implies that $\phi_s^* \mu = \mu$. We also have $\phi_s^* z = z$ since $i_T z = 0$ and z is closed, where i_X denotes interior product with X. The volume form is also invariant under ϕ_s since

$$di_{aT}dvol = d(c_{12}^3 aw^1 \wedge w^2 \wedge w^4 - c_{12}^4 aw^1 \wedge w^2 \wedge w^3) = d(d\mu \wedge z) = 0.$$

Hence, the nonlinear harmonic form comes equipped with a foliation with a symplectic structure, and a distinguished Hamiltonian whose Hamiltonian vector field preserves both the nonlinear harmonic form and the volume form.

Returning to Bochner formulas, let now $f = \ln(a)$. At a critical point p of f, the mean curvature of the leaf through p is zero. The Bochner formula at the critical point becomes

$$-\Delta f = R_{1331} + R_{2332} + R_{1441} + R_{2442} + |II_a|^2 + |II_b|^2 - \frac{1}{2}|d\mu|^2. \tag{3.11}$$

This is essentially identical to (2.11). Once again the curvatures appear in biorthogonal pairs. Let f have a maximum at $p \in M$. Considering the positivity of the full hessian rather than simply its trace does not appear to be useful.

4 Existence of quadratic nonlinear harmonic forms

In this section, using standard techniques, we prove the existence of nonlinear harmonic forms in L_2 subject to quadratic constraints. Let H_s denote the Sobolev space of differential forms ϕ with $(1+\Delta)^{s/2}\phi \in L_2$ and norm $\|\phi\|_{H_s} = \|(1+\Delta)^{s/2}\phi\|_{L_2}$.

Theorem 4.1. *Let M^n be a compact Riemannian manifold. Let $f := (f_1, \dots, f_r)$ be an r-tuple of closed differential forms with*

$$p^k(f) := \sum_{i,j} p^k_{ij} \wedge f_i \wedge f_j = 0,\ 1 \le k \le D,$$

for some smooth differential forms p^k_{ij}. Let h_i denote the harmonic representative of f_i. Set

$$Q := \{y = (y_1, \dots, y_r) \in H_1 : p^k(h+dy) = 0,\ 1 \le k \le D \text{ and } d^* y_i = 0\}.$$

Let $E(y) := \|h+dy\|^2 := \sum_i \|h_i + dy_i\|^2$. Let $\nu_0 := \inf\{E(y) : y \in Q\}$. Then there exists $y \in H_1$ such that $E(y) = \nu_0$.

Proof. Let $\{y^m\}_{m=1}^\infty \subset Q$ be a minimizing sequence for E. Then the sequence is bounded in H_1. By passing to a subsequence, we may assume that $y^m \overset{H_1}{\rightharpoonup} y$ and $y^m \overset{L_2}{\to} y$, as $m \to \infty$, for some $y \in H_1$. Then for all smooth forms ϕ and $\forall k$,

$$\begin{aligned} 0 &= \lim_{m\to\infty} \int \phi \wedge p^k(h + dy^m) \\ &= \int \phi \wedge p^k(h+dy) + \lim_{m\to\infty} \int \phi \wedge p^k(dy^m - dy) \\ &= \int \phi \wedge p^k(h+dy) - \lim_{m\to\infty} \int (-1)^{deg(\phi\wedge p^k_{ij})} d(\phi \wedge p^k_{ij}) \wedge (y_i^m - y_i) \wedge (dy_j^m - dy_j) \\ &= \int \phi \wedge p^k(h+dy). \end{aligned}$$

Hence $y \in Q$. Next we observe that

$$\nu_0 = \lim_{m\to\infty} \|h + dy^m\|^2 = \|h+dy\|^2 + \lim_{m\to\infty} \|dy^m - dy\|^2$$

implies $E(y) = \nu_0$, and $\lim_{m\to\infty} \|dy^m - dy\|^2 = 0$. In particular, $y^m \overset{H_1}{\to} y$. □

We call

$$z := h + dy \tag{4.2}$$

a nonlinear harmonic representative of f. Observe $d^*z \in H^1_{-1}$.

It is not clear that we may replace the minimizing sequence by a sequence of smooth y_n. Smooth minimizing sequences would likely significantly aid regularity theorems. See [3]. We cannot yet rule out, however, the possibility that the minimal energy for $y \in H_1$ is strictly smaller than the infimum of energies for smooth y. In particular, we have not established that smooth forms satisfying the constraint are dense in the set of H_1 forms satisfying the constraint. Similar issues arise in the study of harmonic maps.

4.1 Price monotonicity

Recall the Morrey norm defined by

$$\|u\|_{L^{p,\mu}} := (\sup_{x\in M, 0<r<\text{diam}(M)} r^{-\mu}\int_{B_r(x)}|u|^p dv)^{\frac{1}{p}}. \tag{4.3}$$

Then the Morrey space $L^{p,\mu}$ is defined to be the subset of L^p with finite $L^{p,\mu}$ norm.

Theorem 4.4. *If the nonlinear harmonic form z defined in (4.2) is homogeneous of degree p, and if the coefficients p^k_{ij} are constants, then $z \in L^{2,n-2p}$.*

Proof. Let Y be a smooth vectorfield. Let ϕ_t be the one parameter family of diffeomorphisms generated by Y. Then ϕ_t preserves H_1 and since the p^k_{ij} are constant, $p^k(f^*\zeta) = f^*p^k(\zeta) = 0$ if $p^k(\zeta) = 0$. Hence $\|\phi_t^* z\|^2_{L_2} \geq \|z\|^2_{L_2}$. In particular (see (6.10) or [4, p. 142])

$$0 = \int\langle z, (-\frac{1}{2}div(Y) + Y^i_{;j}e(w^j)i_{e_i})z\rangle. \tag{4.5}$$

For smooth z, this is equivalent to

$$\langle d^* z, i_Y z\rangle = 0. \tag{4.6}$$

Equation (4.5) allows us to prove monotonicity formulas in the usual fashion. We include the details for the convenience of the reader.

Fix geodesic coordinates $\{x^i\}$ centered at some $p \in M$, and for some function y choose

$$Y := y(\frac{r^2}{2s})x^j\frac{\partial}{\partial x^j}.$$

Then

$$\begin{aligned} div(Y) &= \frac{1}{\sqrt{g}}\frac{\partial}{\partial x^j}(\sqrt{g}y(\frac{r^2}{2s})x^j) \\ &= ny(\frac{r^2}{2s}) + \frac{r^2}{s}y'(\frac{r^2}{2s}) + \frac{r}{2}y(\frac{r^2}{2s})\frac{\partial\ln(g)}{\partial r} \\ &= ny(\frac{r^2}{2s}) - 2s\frac{d}{ds}y(\frac{r^2}{2s}) + \frac{r}{2}y(\frac{r^2}{2s})\frac{\partial\ln(g)}{\partial r}, \end{aligned}$$

and

$$\begin{aligned} Y^i_{;j}e(w^j)i_{e_i} &= y(\frac{r^2}{2s})e(dx^j)i_{\frac{\partial}{\partial x^j}} + y'(\frac{r^2}{2s})\frac{r^2}{s}e(dr)i_{\frac{\partial}{\partial r}} + y(\frac{r^2}{2s})x^m\Gamma^i_{jm}e(dx^j)i_{\frac{\partial}{\partial x^i}} \\ &= y(\frac{r^2}{2s})e(dx^j)i_{\frac{\partial}{\partial x^j}} - 2s\frac{d}{ds}y(\frac{r^2}{2s})e(dr)i_{\frac{\partial}{\partial r}} + y(\frac{r^2}{2s})x^m\Gamma^i_{jm}e(dx^j)i_{\frac{\partial}{\partial x^i}}. \end{aligned}$$

Hence

$$s^{p-1-\frac{n}{2}}(-\frac{1}{2}div(Y)+Y^i_{;j}e(w^j)i_{e_i})z = (\frac{d}{ds}(s^{p-\frac{n}{2}}y)-2s^{p-\frac{n}{2}}\frac{dy}{ds}e(dr)i_{\frac{\partial}{\partial r}}+s^{p-1-\frac{n}{2}}yw)z,$$

where $w := -\frac{r}{4}\frac{\partial \ln(g)}{\partial r} + x^m \Gamma^i_{jm} e(dx^j) i_{\frac{\partial}{\partial x^i}}$. Inserting this expression into (4.5) and multiplying by $e^{\lambda s}$ gives

$$\frac{d}{ds}(e^{\lambda s} s^{p-\frac{n}{2}} \int y|z|^2 dv) = e^{\lambda s} s^{p-\frac{n}{2}} \int (2\frac{dy}{ds}|i_{\frac{\partial}{\partial r}} z|^2 + y\langle z, (\lambda - \frac{w}{s})z\rangle) dv. \tag{4.7}$$

Integrate this equality in s from σ^2 to τ^2 to get

$$\begin{aligned} & e^{\lambda\tau^2}\tau^{2p-n} \int y(\frac{r^2}{2\tau^2})|z|^2 dv - e^{\lambda\sigma^2}\sigma^{2p-n} \int y(\frac{r^2}{2\sigma^2})|z|^2 dv \\ & = \int_{\sigma^2}^{\tau^2} e^{\lambda s} s^{p-\frac{n}{2}} \int (2\frac{dy}{ds}|i_{\frac{\partial}{\partial r}} z|^2 + y\langle z, (\lambda - \frac{w}{s})z\rangle) dv ds. \end{aligned} \tag{4.8}$$

Replace y with a sequence $\{y_n\}$ of C^1 monotone decreasing functions, supported in $[0, 1]$ and converging to the characteristic function of $[0, \frac{1}{2}]$. Using the support of y_n and the fact that $|w| = O(r^2)$, we see that we can pick $\lambda > 0$ depending on geometric data so that $\lambda > \frac{w}{s}$ on the support of y. Hence, the right hand side of (4.8) is nonnegative, yielding

$$e^{\lambda\tau^2}\tau^{2p-n} \int_{B_\tau} |z|^2 dv \geq e^{\lambda\sigma^2}\sigma^{2p-n} \int_{B_\sigma} |z|^2 dv. \tag{4.9}$$

This monotonicity relation immediately implies $z \in L^{2,n-2p}$. □

In the study of harmonic maps and Yang Mills connections, monotonicity relations such as (4.9) are one of the main ingredients in establishing regularity results.

5 Euler Lagrange equations

Write

$$P(z) = (p^1(z), \ldots, p^D(z)),$$

with $p^k(z)$ defined as in Theorem 4.1, still assumed to be quadratic. Let $V = P^{-1}(0)$. Let $z(t) = h + db(t)$ be a curve in V, with $b(t)$ a C^1 curve in H_1. In components, write $z_i(t) = h_i + db_i(t)$, with h_i harmonic and $d^* b_i = 0$. Then

$$0 = \frac{d}{dt}P(z(t)) = DP_z d\dot{b} = 0,$$

where DP_z denotes the derivative of P at z. Let K_z denote the kernel of $DP_z d$ in H_1.

Definition 5.1. We call K_z the *formal tangent space* to V at z. We say V is *unobstructed* at z if for each $v \in K_z$, there is a differentiable curve $c(t)$ in V so that $c(0) = z$ and $\dot{c}(0) = v$.

For simplicity, we now assume that $\dim M = 4$ and that the p_{ij}^k are closed. Then since $H_1 \subset L^4$ and $z \in L_2$, we may view DP_z as a bounded linear map from $H_1 \to H_{-1}$. In the quadratic case under consideration, this map is an algebraic operator given by the adjoint of left multiplication by the product of smooth forms and components of z. For example, when $p(z) = z \wedge z$, $DP_z = 2e^*(z)$. Hence dDP_z defines a bounded linear map from $H_1 \to H_{-2}$. Here we have used $dp_{ij}^k = 0$ to substitute dDP_z for $DP_z d$. We let d^* and DP_z^* denote the formal L_2 adjoint of these two operators, and $(dDP_z)^{*_H} = (1+\Delta)^{-1}DP_z^* d^*(1+\Delta)^{-2}$ denotes the Hilbert space adjoint of $dDP_z : H_1 \to H_{-2}$. Then

$$K_z = (\mathrm{Im}\ (dDP_z)^{*_H})^{\perp_{H_1}} = ((1+\Delta)^{-1}DP_z^* d^* H_2)^{\perp_{H_1}}. \tag{5.2}$$

Similarly, if $E(h+db) := \|h+db\|^2$, then

$$\frac{d}{dt}E(b(t)) = \sum_i \langle d^* z_i, \dot{b}_i \rangle_{L_2} = \sum_i \langle (1+\Delta)^{-1} d^* z_i, \dot{b}_i \rangle_{H_1} := DE_{b(t)}\dot{b}.$$

If $z = h + db$ is an E minimizer, then

$$DE_b v = 0, \ \forall v \in K_z,$$

such that there exists a differentiable curve $\beta(t)$ in V with $b = \beta(0)$ and $v = \dot{\beta}(0)$.

Thus, if $z = h + db$ minimizes E *and* K_z *is unobstructed*, then the Euler Lagrange equation becomes

$$d^* z = \lim_{j\to\infty} DP_z^* d^* \mu_j, \ \text{in } H_{-1}, \tag{5.3}$$

for some sequence μ_j in H_2. When the image is closed, we can, of course, replace the limit with $d^* z = DP_z^* d^* \mu$, some $\mu \in H_2$.

Determining when the tangent space is unobstructed is a basic aspect of the regularity theory of nonlinear harmonic forms. We first examine the information following from diffeomorphism invariance of the constraints.

6 Diffeomorphism variations

In this section we explore first and second variation formulas associated with variation by diffeomorphism. The resulting first variation formula is standard and underlies Price's inequality. The second variation formulas are perhaps less familiar.

If $z \wedge z = 0$, or more generally, if the constraint $p(z)$ satisfies $\phi_t^* p(z) = p(\phi_t^* z)$, for the smooth one parameter family of diffeomorphism ϕ_t with $\phi_0(x) = x$, then $\phi_t^* z$ defines a curve in the constraint space, which we can use to verify some of the formal Euler Lagrange equations. Observe that when z is smooth and nonvanishing, then locally there exist coordinates so that if $z \wedge z = 0$ then

$$z = dx^1 \wedge dx^2,$$

and smooth variations of z are given by variations of the functions x^1 and x^2. In particular, the variation is given by the Lie derivative of z by the vectorfield

$$X := \dot{x}^1 \frac{\partial}{\partial x^1} - \dot{x}^2 \frac{\partial}{\partial x^2}.$$

So, we see that in this instance not much information is lost in considering only variations by diffeomorphisms. Let z be a minimizer for $\|\phi_t^* z\|^2$, for all one parameter families of diffeomorphisms ϕ_t generated by a vectorfield. Then we have

$$\|z\|^2 \leq \|\phi_t^* z\|^2 = \int_M z \wedge \phi_{-t}^*(*\phi_t^* z). \tag{6.1}$$

Hence we deduce a second variation formula (and the first variation formula leading to Price's inequality) by expanding

$$*_t := \phi_{-t}^* * \phi_t^*$$

to second order in t. Express $*$ as ,

$$* = \pm c(dvol),$$

where, $c(dvol)$ denotes clifford multiplication by the volume form, and the sign depends on dimension and degree. In a local oriented orthonormal frame $\{v_i\}_{i=1}^n$ and dual coframe $\{\omega^i\}_{i=1}^n$, we write

$$c(dvol) = c(\omega^1)\cdots c(\omega^n) = (e(\omega^1) - i_{v_1})\cdots(e(\omega^n) - i_{v_n}),$$

where i_v denotes interior multiplication by v. It is also convenient to define

$$\hat{c}(\omega^i) = e(\omega^i) + i_{v_i} = e(\omega^i) + e^*(\omega^i).$$

Observe that

$$\phi_{-t}^* e(\omega^j)\phi_t^* = e(\phi_{-t}^*\omega^j),$$

and

$$\phi_{-t}^* i_{v_j}\phi_t^* = i_{d\phi_t v_j}.$$

Hence

$$\begin{aligned} \phi_{-t}^*(c(\omega^j)\phi_t^* &= (e(\phi_{-t}^*\omega^j) - i_{d\phi_t v_j}) \\ &= c(\omega^j) - t(e(L_X\omega^j) - i_{L_X v_j}) + \frac{t^2}{2}(e(L_X^2\omega^j) - i_{L_X^2 v_j}) \\ &=: c(\omega^j) - tL_X c(\omega^j) + \frac{t^2}{2}L_X^2 c(\omega^j). \end{aligned} \tag{6.2}$$

Here L_X denote the Lie derivative with respect to X. On differential forms,

$$L_X = \{e(\omega^m)\nabla_{v_m}, X^k e^*(\omega^k)\} = e(\omega^m)e^*(\omega^k)X^k_{;m} + \nabla_X. \tag{6.3}$$

On vector fields,

$$L_X = \nabla_X - X^m_{;k} v_m \otimes \omega^k. \tag{6.4}$$

For convenience, fix a point p and choose an orthonormal frame and coframe which is radially covariant constant (in geodesic coordinates) in a neighborhood of p. Then we have

$$\nabla_Y v_j = \frac{1}{2}R(r\frac{\partial}{\partial r}, Y, v_i, v_m)v_m + O(r^2).$$

In particular, $(\nabla_X v_i)(p) = (\nabla_X^2 v_i)(p) = 0$. In this frame, at p, we have

$$\begin{aligned}\phi_{-t}^* c(\omega^j)\phi_t^* &= c(\omega^j) - t(e(X^j_{;m}\omega^m) + i_{X^m_{;j}v_m})\\ &+\frac{t^2}{2}((X^k_{;m}X^j_{;k} + X^k X^j_{;mk})e(\omega^m) + (X^k X^m_{;jk} - X^m_{;k}X^k_{;j})i_{v_m}) + O(t^3)\\ &= c(\omega^j) - t(e(X^j_{;m}\omega^m) + i_{X^m_{;j}v_m})\\ &+\frac{t^2}{2}(((\nabla_X X)^j_{;m} - R(X, v_m, v_j, X))e(\omega^m)\\ &+((\nabla_X X)^m_{;j} - R(X, v_j, v_m, X) - 2X^m_{;k}X^k_{;j})i_{v_m}) + O(t^3). \end{aligned} \tag{6.5}$$

We expand this two different ways. Let A and B denote the skew and symmetric summands of ∇X respectively.

$$\begin{aligned}\phi_{-t}^* c(\omega^j)\phi_t^* &= c(\omega^j) - t(B^m_j\hat{c}(\omega^m) - A^m_j c(\omega^m))\\ &+\frac{t^2}{2}(((A^2+B^2)^j_m + (X^k A^j_{;m})_{;k} - X^k_{;k}A^j_{;m})c(\omega^m)\\ &+(\{A,B\}^j_m + X^k B^j_{;mk})\hat{c}(\omega^m)) + O(t^3),\end{aligned} \tag{6.6}$$

and

$$\begin{aligned}\phi_{-t}^* c(\omega^j)\phi_t^* &= c(\omega^j) - t(B^m_j\hat{c}(\omega^m) - A^m_j c(\omega^m))\\ &+\frac{t^2}{2}(((A^2+B^2+\{A,B\})^m_{;j} - \frac{1}{2}(\nabla_X X)^m_{;j} + \frac{1}{2}(\nabla_X X)^j_{;m})c(\omega^m)\\ &+(-X^m_{;k}X^k_{;j} + \frac{1}{2}(\nabla_X X)^j_{;m} - R(X, v_m, v_j, X)\\ &+\frac{1}{2}(\nabla_X X)^m_{;j})\hat{c}(\omega^m)) + O(t^3).\end{aligned} \tag{6.7}$$

These two expressions lead to two different expressions for $*^{-1}*_t$. From (6.6), we have

$$\begin{aligned}*^{-1}*_t &= I + tB^j_m c(\omega^j)\hat{c}(\omega^m) + \frac{t^2}{2}(-X^k B^j_{m;k}c(\omega^j)\hat{c}(\omega^m)\\ &+(\sum_{i,m} B^i_m c(\omega^i)\hat{c}(\omega^m))^2) + O(t^3).\end{aligned} \tag{6.8}$$

From (6.7) we have

$$\begin{aligned}
*^{-1}*_t = I + tB^j_m c(\omega^j)\hat{c}(\omega^m) + \frac{t^2}{2}([(A^2+B^2)^j_{;m}) + R(X, v_m, v_j, X) \\
-\frac{1}{2}(\nabla_X X)^j_{;m} - \frac{1}{2}(\nabla_X X)^m_{;j}]c(\omega^j)\hat{c}(\omega^m) \\
+(\sum_{i,m} B^i_m c(\omega^i)\hat{c}(\omega^m))^2) + O(t^3).
\end{aligned} \tag{6.9}$$

These computations are simplified by diagonalizing B and skew diagonalizing A (not simultaneously). It is also useful to note that the degree preserving summand of $c(\omega^j)\hat{c}(\omega^m)$ satisfies

$$c(\omega^j)\hat{c}(\omega^m) = c(\omega^m)\hat{c}(\omega^j).$$

The $O(t)$ term in either (6.8) or (6.9) yields the first variation formula.

Proposition 6.10. *First variation formula. Let $\|\phi_t^* z\|^2$ have a critical point at $t = 0$ for all smooth 1 parameter families of diffeomorphisms generated by vector fields X. Then*

$$0 = \langle z, (X^j_{;m} + X^m_{;j})c(\omega^j)\hat{c}(\omega^m)z\rangle_{L_2}. \tag{6.11}$$

Suppose now z is a differential k-form satisfying (6.11). Choose a frame $\{v_j\}_j$ at p such that the bilinear form $Q(v,w) := \langle i_v z, i_w z\rangle$ is diagonal. In this frame, we have

$$\langle i_{v_j} z, i_{v_m} z\rangle(p) = 0, \text{ for } j \neq m. \tag{6.12}$$

In such a frame, (6.11) becomes

$$0 = 2\int \sum_j X^j_{;j} |i_{v_j} z|^2 dv - \int div(X)|z|^2 dv. \tag{6.13}$$

Integration by parts yields

$$0 = \langle X, \sum_j 2v_j(|i_{v_j} z|^2)v_j - \nabla|z|^2\rangle - 2\langle X, \sum_{j,m} \gamma^j_{mm}(|i_{v_j} z|^2 - |i_{v_m} z|^2)v_j\rangle. \tag{6.14}$$

Hence (no sum over j)

$$v_j(|i_{v_j} z|^2 - \frac{1}{2}|z|^2) = \sum_m \gamma^j_{mm}(|i_{v_j} z|^2 - |i_{v_m} z|^2). \tag{6.15}$$

In dimension 4, Q is a scalar multiple of the metric at p if and only if $z(p)$ is self dual or antiself dual. Otherwise Q has 2 distinct eigenvalues at p. In higher dimensions, the strong assumption that Q has at most two distinct eigenvalues at each point leads to new Bochner type formulas analogous to (2.11), except they involve divergences of vectorfields which are not gradient vectorfields. This eigenvalue condition is not diffeomorphism invariant; so, we see no immediate application of this observation.

In dimension 4 for z a 2-form, (6.15) is equivalent to (2.8) or (3.9). Thus, in this case, the form satisfies half of the equations given by $dz = 0$ and the d^*z

equation. The assumption $dz = 0$ gives 2 more equations. We therefore verify half the Euler Lagrange equations formally satisfied by this type of nonlinear harmonic, including all the equations needed for the Bochner formula (3.11) to hold weakly. If the constraint equations are not diffeomorphism invariant, then the allowed vectorfields X generating diffeomorphisms in the first variation must be correspondingly constrained. For example, if the constraint equation coefficients are functions (only) of a symplectic form, then the first variation formula holds for locally hamiltonian vectorfields. If we restrict to gradient variations only, the first variation formula reduces to a single second order equation.

Proposition 6.16. *First variation formula for gradient flow minimizing 2-forms in a compact 4-manifold. Let z be a 2-form in a compact oriented 4-manifold such that $\|\phi_t^* z\|^2$ has a critical point at $t = 0$ for all smooth 1 parameter families of diffeomorphisms (with $\phi_0 =$ identity) generated by gradient vector fields X. Write $z = aw^1 \wedge w^2 + bw^3 \wedge w^4$ as in Section 3. Then*

$$0 = (\Delta - 2\Delta_D^*)(a^2 - b^2). \tag{6.17}$$

Similarly, one obtains a single second order equation in the case of locally Hamiltonian vector fields.

The second variation inequality now arises from evaluating the t^2 terms in the two expressions for $\langle z, *^{-1} *_t z\rangle$, when z satisfies the first variation equality, and using the first variation equality to simplify. We give 2 expressions for the resulting second variation inequality.

Proposition 6.18. *Second variation formula. Let $\|\phi_t^* z\|^2$ have a local minimum at $t = 0$ for all smooth 1 parameter families of diffeomorphisms, ϕ_t generated by gradient fields X. Then*

$$0 \leq \langle z, (-X^k B^j_{m;k} c(\omega^j)\hat{c}(\omega^m) + (\sum_{i,m} B^i_m c(\omega^i)\hat{c}(\omega^m))^2) z\rangle_{L_2}, \tag{6.19}$$

and

$$0 \leq \langle z, ([(A^2 + B^2)^j_m + R(X, v_m, v_j, X)]c(\omega^j)\hat{c}(\omega^m) + (\sum_{i,m} B^i_m c(\omega^i)\hat{c}(\omega^m))^2) z\rangle_{L_2}. \tag{6.20}$$

Observe that these inequalities also apply to unconstrained harmonic forms. In this case, (6.20) gives vanishing theorems when we have a sufficient supply of vector fields with controlled covariant derivative. For example, Grassmann manifolds viewed as spaces of unitary projections have the functions $F_C(p) = \operatorname{tr} pC$, for each hermitian matrix C. (See for example [6, Theorem 1.4.12].) Using $X = \nabla F_A$ in the second variation formula, with A running over an orthonormal basis of hermitian matrices, one can obtain numerous cohomology vanishing theorems, which, however, can be proved more easily by other techniques.

Orientation preserving isometries preserve the Hodge star operator. Hence for X a Killing field, the second variation should be trivial. For Killing fields, $B = 0$. More generally, one expects the A term to drop out of the second variation formula, as it does in (6.19) but not (6.20). This suggests (6.20) might contain information about Killing vectorfields.

Suppose that X is a Killing vectorfield. Then $B = 0$, and the second variation inequality becomes an equality, yielding

$$0 = \int \sum_j (|\nabla_j X|^2 - R(X, v_j, v_j, X))(2|i_{v_j} z|^2 - |z|^2) dv, \tag{6.21}$$

generalizing the integral form of the Bochner formula for Killing fields. Choosing $z = dvol$ returns the usual formula.

References

[1] R. Bettiol, *Positive Biorthogonal Curvature on* $S^2 \times S^2$, http:// arXiv:1210.0043.

[2] J. Eels and J. Sampson, *Harmonic mappings of Riemannian manifolds*, American Journal of Mathematics **86** (1964), no. 1, 109–160.

[3] Y. Meyer and T. Riviere, *A Partial regularity result for a class of Stationary Yang-Mills Fields in high dimension*, Rev. Mat. Iberoamericana **19** (2003), 195–219.

[4] P. Price, *A monotonicity formula for Yang Mills fields*, J. Manuscripta Math. **43** (1983), 131–166.

[5] R. Schoen and S.-T. Yau, *Lectures on Differential Geometry*, International Press, Somerville Mass, 1994.

[6] Y. Xin, *Geometry of Harmonic Maps*, Birkhauser, Boston, 1996.

Advanced Lectures in Mathematics (ALM)

Executive editors: Shing-Tung Yau, Kefeng Liu, and Lizhen Ji

www.intlpress.com/alm

ALM 39: Hodge Theory and L^2-analysis *

ALM 38: Complex Geometry from Riemann to Kähler–Einstein and Calabi–Yau *

ALM 37: Proceedings of the Sixth International Congress of Chinese Mathematicians, Volume II

ALM 36: Proceedings of the Sixth International Congress of Chinese Mathematicians, Volume I

ALM 35: The Legacy of Bernhard Riemann After One Hundred and Fifty Years, Volumes I and II

ALM 34: Some Topics in Harmonic Analysis and Applications

ALM 33: Introduction to Modern Mathematics

ALM 32: Handbook of Group Actions, Volume II

ALM 31: Handbook of Group Actions, Volume I

ALM 30: Automorphic Forms and L-functions

ALM 29: Selected Expository Works of Shing-Tung Yau with Commentary, Volume II

ALM 28: Selected Expository Works of Shing-Tung Yau with Commentary, Volume I

ALM 27: Number Theory and Related Areas

ALM 26: Handbook of Moduli, Volume III

ALM 25: Handbook of Moduli, Volume II

ALM 24: Handbook of Moduli, Volume I

ALM 23: Recent Developments in Geometry and Analysis

ALM 22: Differential Geometry: Under the Influence of S.-S. Chern

ALM 21: Advances in Geometric Analysis

ALM 20: Surveys in Geometric Analysis and Relativity

ALM 19: Arithmetic Geometry and Automorphic Forms

ALM 18: Geometry and Analysis, No. 2

ALM 17: Geometry and Analysis, No. 1

ALM 16: Transformation Groups and Moduli Spaces of Curves

ALM 15: An Introduction to Groups and Lattices: Finite Groups and Positive Definite Rational Lattices

ALM 14: Handbook of Geometric Analysis, No. 3

ALM 13: Handbook of Geometric Analysis, No. 2

ALM 12: Cohomology of Groups and Algebraic K-theory

ALM 11: Recent Advances in Geometric Analysis

ALM 10: Trends in Partial Differential Equations

ALM 9: Automorphic Forms and the Langlands Program

ALM 8: Recent Developments in Algebra and Related Areas

ALM 7: Handbook of Geometric Analysis, No. 1

ALM 6: Geometry, Analysis and Topology of Discrete Groups

ALM 4: Variational Principles for Discrete Surfaces

ALM 3: Computational Conformal Geometry

ALM 2: Asymptotic Theory in Probability and Statistics with Applications

ALM 1: Superstring Theory

* Published out of sequence.

Made in the USA
Monee, IL
08 May 2021

67944472R10341